GEOGRAPHY OF NEW YORK STATE

A New York State Study

GEOGRAPHY OF

NEW YORK STATE

JOHN H. THOMPSON, *Editor*

SYRACUSE UNIVERSITY PRESS

Copyright © 1966, 1977 by Syracuse University Press, Syracuse, New York

ALL RIGHTS RESERVED. THIS BOOK, INCLUDING ILLUSTRATIVE MATERIALS, MAY NOT BE REPRODUCED IN ANY FORM WITHOUT PERMISSION IN WRITING FROM THE PUBLISHER.

FIRST EDITION 1966
Paperback Edition 1977
Second Paperback Printing 1980

Title-spread photographs courtesy of N.Y.S. Dept. of Commerce (left) and Buffalo Area Chamber of Commerce

Library of Congress Cataloging in Publication Data
Thompson, John Henry, 1919-   ed.
    Geography of New York State.
    (A New York State study)
    Includes bibliographies and index.
    1. New York (State)—Description and travel—
1951-   I. Title.
[F125.T5   1977]   330.9'747'04   77-4337
    ISBN 0-8156-2182-5

MANUFACTURED IN THE UNITED STATES OF AMERICA

Preface

NEW YORK STATE is extremely challenging from a geographic point of view and worthy of considerable attention by geographers. Few geographers, though, have studied it and even fewer have written about it. Until now there has been no general geography of the state available (with the exception of public school textbooks written for the most part by nongeographers). *Geography of New York State* is intended to fill—at least partially—this longstanding void. It should be useful to anyone interested in the state's development and potential.

Each of the volume's five parts could almost be a book in itself, yet the parts are all related and are better understood in the context of the whole volume than as separate entities.

Part One presents the geography of five important elements of the resource base: land forms, climate, water, vegetation, and soils. It sets the scene upon which the human geography of the state evolved.

Tracing New York's development from Indian times to the present, Part Two places matters of geographical interest in historical perspective. It provides an interpretative background for the discussions that follow.

Part Three examines economic activities today, dividing them into three sectors: the primary sector—including agriculture, mining, lumbering, and fishing; the secondary sector—manufacturing and construction; and the tertiary sector—sales and services. A chapter on the state's economic strength concludes this part.

Part Four points up the fact that the differences between urban and rural areas in New York are so great that it might be said to have two landscapes, one urban and one rural. It also presents recommendations for establishment of planning and development regions.

Part Five explores the historical development, industrial status, and economic potential of the state's seven great urban systems—Albany-Schenectady, Binghamton, Buffalo, New York, Rochester, Syracuse, and Utica.

Because of its broad approach, it would have been almost impossible for one person to write the entire book. Thus, early in the planning stage I decided that, although I could prepare some portions, the book would have to be a cooperative venture. Fourteen experts around the state and two from beyond its borders enthusiastically accepted my invitation to write on topics with which they were particularly familiar. Without their cooperation this volume could not have been produced. On the other hand, the employment of so many contributors means that a variety of writing styles and substantial range in level of treatment are bound to result. Thus, at the outset all contributors assured me a free hand in editing, which I have employed in an effort to achieve balance and cohesiveness. For the benefit of the volume as a whole, I expanded some chapters and trimmed others, and I must ask the contributors to exhibit forebearance if they think their original version superior to the final one.

Variation in documentation from chapter to chapter was so great that I decided generally to minimize it and to include a short list of selected references at the end of each chapter. The only exceptions to this procedure are Chapters 7, 8, and 9, where extensive documentation seemed essential. (The selected references for these chapters, to which notes in the text refer, appear at the end of Chapter 9.)

An extensive and varied collection of over one hundred maps will be a unique attraction to readers. They serve as an indispensable representation of the state's geography and a vital addition to the written text. Syracuse University and the Syracuse University Press lent an important helping hand by setting up a cartographic fund to defray the cost of drafting the maps. They were prepared over a period of several years by graduate students of geography working largely within the financial framework of the cartographic fund. Among graduate students who contributed most to the cartography are Shin-yi Hsu, Harley E. Scott, Larry R. Martin, and Robert J. Tata. John Fonda made improvements in some of the maps and redrafted several others just prior to publication.

Among the many people who contributed to this book, Donald W. Meinig and David J. de Laubenfels must be especially praised for taking on writing responsibilities at late dates so that the project could be completed.

Vice President Eric H. Faigle and a former colleague, Eleanor E. Hanlon, both long-time teachers of New York State geography at Syracuse University and very knowledgeable about the state, gave encouragement and guidance at crucial stages in the project.

The cooperation of Frank E. Richards of Phoenix, New York, the publisher of the *Richards Atlas of New York State*, to which Robert J. Rayback (the author of Chapter 6) and I contributed, is gratefully acknowledged. Chapters 1, 6, 10, and 11 have drawn on textual and illustrative materials in the atlas, and Mr. Richards has kindly granted permission to reproduce two maps, Figures 69 and 73.

I want to express the deepest appreciation to Richard T. Lewis, who, as my research assistant for a year and a half, assembled many of the statistical data for the economic sections, actually developed some of the measurement techniques used in Chapter 12, and assembled a substantial number of the tables. Mary Kishman deserves special credit for her dedication to typing responsibilities on both preliminary and final forms of the manuscript.

The editing and writing of *Geography of New York State* have been enjoyable and professionally rewarding. The effort will prove doubly justified if readers find the book enlightening.

JOHN H. THOMPSON

Syracuse, New York
February 1966

Contents

List of Contributors xi

List of Figures xiii

List of Tables xv

INTRODUCTION—John H. Thompson 1

PART ONE: THE RESOURCE BASE 17

1. Land Forms—George B. Cressey 19
2. Climate—Douglas B. Carter 54
3. Water—Arthur R. Eschner 79
4. Vegetation—David J. de Laubenfels 90
5. Soil—David J. de Laubenfels 104

PART TWO: THREE AND A HALF CENTURIES OF CHANGE 111

6. The Indian—Robert J. Rayback 113
7. The Colonial Period—D. W. Meinig 121
8. Geography of Expansion—D. W. Meinig 140
9. Elaboration and Change—D. W. Meinig 172

PART THREE: ECONOMIC ACTIVITIES TODAY 197

10. The Primary Sector—John H. Thompson 201
11. The Secondary Sector 232
    John H. Thompson
12. The Tertiary Sector—John H. Thompson 255
    *with a section on Recreation and Conservation
    by* Henry G. Williams, Jr., *and*
    Roger Thompson
13. Strength of the State's Economy 311
    Sidney C. Sufrin *and* John H. Thompson

PART FOUR: LANDSCAPES AND REGIONS 331

14. The Urban Landscape 333
    David J. de Laubenfels,
    John H. Thompson *and* John E. Brush
15. The Rural Landscape 358
    John H. Thompson
16. Planning and Development Regions 370
    John H. Thompson

PART FIVE: THE GREAT URBAN SYSTEMS 383

17. Albany-Schenectady—HOWARD J. FLIERL 385
18. Binghamton—JOSEPH E. VAN RIPER 396
19. Buffalo 407
    KATHERYNE THOMAS WHITTEMORE
20. New York—ROBERT MCNEE 423
21. Rochester—ROBERT B. HALL, JR. 458
22. Syracuse—DAVID J. DE LAUBENFELS 469
23. Utica—SIDLEY K. MACFARLANE 480

A LOOK FORWARD—JOHN H. THOMPSON 489

Appendix A. Selected Summit Elevations 501
Appendix B. Principal Water Bodies 503
Appendix C. Principal Rivers 506
Appendix D. County Data 510
Appendix E. Urban Place Data 512

Index 525

Contributors

JOHN E. BRUSH, coauthor of the chapter on the urban landscape, is Professor of Geography at Rutgers University. He holds the B.A. from the University of Chicago and the M.S. and Ph.D. from the University of Wisconsin. The author of *The Population of New Jersey*, he has contributed articles to many books and journals. He was a Guggenheim Fellow in 1957/58.

DOUGLAS B. CARTER, Professor of Geography at Southern Illinois University, wrote the chapter on climate. Formerly in the geography department at Syracuse University, Mr. Carter holds the B.A. from Eastern Washington College, a certificate from the University of Chicago Institute of Meteorology, and the M.A. and Ph.D. from the University of Washington. He has published *Three Water Balance Maps of Eastern North America*, edited *Fresh Water Resources*, and contributed to *Investigating the Earth*. He is a contributing editor of *Publications in Climatology*.

GEORGE B. CRESSEY was Maxwell Professor of Geography at Syracuse University until his death in 1963. He held the B.S. from Denison University, the Ph.D. in geology from the University of Chicago and the Ph.D. in geography from Clark University. Of major value in the chapter on land forms, which Mr. Cressey contributed to this volume, is the map of land form regions originated by him and widely used by other scholars. Among his publications are *China's Geographic Foundations*, *Asia's Lands and Peoples*, and *Soviet Potentials*.

ARTHUR R. ESCHNER, the author of the chapter on water, is Associate Professor of Forest Influences and Acting Director of the Water Research Institute for Forest Lands and Industries at the State University College of Forestry at Syracuse University. Before coming to Syracuse, Mr. Eschner was with the U.S. Forest Service doing research in watershed management. He holds the B.S. from the New York State College of Forestry, the M.S. from Iowa State College, and the Ph.D. from the State University College of Forestry at Syracuse. He has contributed to various journals and U.S. Forest Service publications.

HOWARD J. FLIERL, who wrote the chapter on the Albany–Schenectady Urban System, is Professor of Geography at the State University of New York at Albany. Coauthor of *Living in New York*, he holds the B.S. from the State University of New York College at Buffalo and the M.S. and Ph.D. from Syracuse University.

ROBERT B. HALL, JR., Professor of Geology and Geography at the University of Rochester, contributed the chapter on the Rochester Urban System. The author of *Japan: Industrial Power of Asia*, he was Fulbright Scholar in Japan in 1954/55 and in 1961/62. He holds the B.A., M.A., and Ph.D. from the University of Michigan.

DAVID J. DE LAUBENFELS, author of the chapters on vegetation, soil, and the Syracuse Urban System and coauthor of the chapter on the urban landscape, is Associate Professor of Geography at Syracuse University. After graduating from Colgate University, he earned the M.A. and Ph.D. at the University of Illinois and held a postdoctoral fellowship at Johns Hopkins University. Among the journals, atlases, and books to which he has contributed are *One World Divided*, *The North American Midwest*, and *Reader's Digest Atlas*.

SIDLEY K. MACFARLANE, Dean of Utica College of Syracuse University, wrote the chapter on the Utica Urban System. Mr. Macfarlane, who holds the B.A. and Ph.D. from Syracuse University and the M.A. from Clark University, contributed to *Industrial Renewal: Determining the Potential and Accelerating the Economy of the Utica Urban Area*.

ROBERT MCNEE, who taught at the City College of New York for a decade, is now Head of the geography department at the University of Cincinnati. He contributed the chapter on the New York Urban System to this volume. Among his publications are *Focus on Economic Activity* and the forthcoming *Proceedings of Urban Sprawl Conference*. Mr. McNee graduated from Wayne University and earned the M.A. and Ph.D. at Syracuse University.

D. W. MEINIG wrote the chapters on the colonial period, expansion, and elaboration and change. Professor of Geography at Syracuse University, Mr. Meinig holds the B.S. from Georgetown University and the M.A. and Ph.D. from the University of Washington. He is the author of *On the Margins of the Good Earth*, as well as numerous articles on historical and cultural geography.

ROBERT J. RAYBACK, an adopted member of the Wolf Clan of the Oneida Indians, is the author of the chapter on the Indian. Professor of History at Syracuse University, he holds the B.A. from Western Reserve University and the M.A. and Ph.D. from the University of Wisconsin. Among his publications are *Millard Fillmore: A Biography of the President*, *History of Our United States*, and *Richards Atlas of New York State*.

SIDNEY C. SUFRIN, coauthor of the chapter on the strength of the state's economy, is Professor of Economics at Syracuse University. He has written extensively on labor and economic development. Among his books are *What Price Progress?*, *Issues in Federal Aid to Education*, *Administering the National Defense Education Act*, and *Labor Policy and the Business Cycle*. He is also coauthor of several books, including *The New St. Lawrence Frontier*. He holds the B.A. from the University of Pennsylvania and the Ph.D. from Ohio State University.

JOHN H. THOMPSON, who made the greatest single contribution to this volume as both author and editor, is Professor of Geography at Syracuse University. He is the author of *New England Excursion Guidebook* and coauthor of *Manufacturing in the St. Lawrence Area of New York State* and of *The Economic Status of Upstate New York at Mid-Century*. His articles have appeared in many books and journals. Mr. Thompson holds the B.A. from Clark University, the M.A. from the University of Colorado, and the Ph.D. from the University of Washington. He was Visiting Fulbright Professor in Economic Geography at Kobe University in Japan in 1954/55.

ROGER THOMPSON, coauthor of the section on recreation and conservation in the chapter on the tertiary sector, was until recently Executive Secretary of the New York State Recreation Council. Now Research Analyst with the New York State Senate Finance Committee, Mr. Thompson holds the B.S. and Ph.D. from the State University College of Forestry at Syracuse University, and the M.S. from Syracuse University. He has published numerous articles in professional journals.

JOSEPH E. VAN RIPER, Professor of Geography at the State University of New York at Binghamton, wrote the chapter on the Binghamton Urban System. He earned the B.A. and Ph.D. at the University of Michigan and the M.A. at Syracuse University. From 1960 to 1962 he was Smith-Mundt Visiting Professor at the American University of Beirut. He is the author of *Man's Physical World*.

KATHERYNE THOMAS WHITTEMORE, the author of the chapter on the Buffalo Urban System, is Professor Emeritus of Geography and Director of the Arts and Sciences Division of the State University College at Buffalo. She has written articles for various journals and newspapers, as well as two books, *Asia, the Great Continent* and *The United States and Canada*. Mrs. Whittemore holds the B.A. from Vassar and the M.A. and Ph.D. from Clark.

HENRY G. WILLIAMS, JR., coauthor of the section on recreation and conservation in the chapter on the tertiary sector, is Professional Research Associate for Land Use Planning at the State University College of Forestry at Syracuse University. He is the author of *Chenango Area Development Study*, and has done planning studies for more than thirty municipalities in New York State. Mr. Williams holds the B.A. from Utica College of Syracuse University, and the M.A. from Syracuse University.

Figures

1. Economic Ties with the Nation 4–5
   (a) Origin of Commodities Consumed in New York State 4
   (b) Destination of Commodities Produced in New York State 4
   (c) New York State in Relation to the National Core Area 5
   (d) New York State Inbound & Outbound Air Passengers 5
2. Population, 1960 8–9
3. Per Cent Change of Population, 1950–60 10
4. An Example of Overlay Analysis 11
5. Area Analysis Tripod 12
6. Development Graph, New York State 14
7. Stages of Glacial Development 22
8. Underlying Rock Formations 24
9. Land Form Regions 26
10. Niagara Falls 40
11. Cross Section of Lake George 41
12. North–South Profile of Long Island 43
13. Land Form Categories 48
14. Excessive Slope 49
15. Surface Configuration 50–51
16. Average Solar Radiation for Central Park & Sayville 55
17. Reference Map for Climatic Stations 58
18. Mean Monthly Temperature: January 60
19. Mean Monthly Temperature: July 61
20. Hypothetical Diagram of a Mid-Latitude Cyclonic Storm 65
21. Average Annual Precipitation, 1931–55 68
22. Mean Seasonal Snowfall (in back-cover pocket)
23. Mean Annual Water Surplus 73
24. Climatic Regions 75
25. Average Annual Runoff 80
26. Average Annual Discharge of Principal Streams 82–83
27. Mean Monthly Flow 84
28. Forests in the Land Use Picture of New York 90
29. Natural Vegetation Zones 92
30. Extent of Present Forest 99
31. Forest Land Ownership 100
32. Sawmills, 1952 and 1953 101
33. Soil Regions 107
34. Agricultural Potential 109
35. Indian Occupance in the Early 17th Century 114
36. Known Indian Settlement Sites 116
37. New Netherland, 1656 122
38. Hudson Estuary and Various Colonial Settlements 123
39. New York Counties, Colonial Era 125
40. Province of New York, 1771 132
41. New York, 1775 133
42. Major Land Tracts 142
43. Spread of Settlement 146–47
44. Turnpikes and Roads 157
45. N.Y.C.–Albany–Pittsburgh Routes, 1818 158
46. Canals, c. 1855 160
47. Railroads, c. 1855 164
48. Population, 1850 170
49. Foreign-born Population, 1855 171
50. Railroads, 1893 173
51. Railroad Trunk Lines and Two Local Systems, c. 1895 174
52. New England Connections, 1890's 175
53. Railroad Connections to the Coal Fields, 1890's 176
54. Population, 1880 180
55. Electric Railroads, c. 1920 182
56. Paved Roads, c. 1926 183
57. Foreign-born Population, 1920 185
58. Foreign-born "Profiles," 1920 186
59. Superhighways, 1963 187
60. Airline Routes, 1933 and 1963 188
61. Percentage of the Total State Population in the New York City Region 191
62. Transportation: Geographic Characteristics (Diagrammatic) 192–93
63. Agricultural Trends in New York State 202
64. Dairying 205
65. All Fruit 206
66. Crop Combinations 210
67. Agricultural Regions 212–13
68. Trend in Land Use on Farms, 1850–1959 220
69. Mineral Production (in back-cover pocket)
70. Forest Condition and Use Regions 228
71. The Adirondack Park 229
72. Relative Significance of the Major Industry Groups, 1961 233

73. Manufacturing in New York State (in back-cover pocket)
74. Manufacturing Regions   236
75. Regional Growth Trends, 1947–61   237
76. Manufacturing over the Years, 1869–1954   244–45
77. Curve of Industrial Growth   247
78. The Industrial Aging Process   248
    (a) National or Regional Scale   248
    (b) Metropolitan Scale   248
79. Employment Change by Major Industry Groups, 1947–61   252
80. Selected "Growth" and "Loss" Industries in New York State, 1947–61   253
81. Tertiary Surpluses and Deficits—New York Counties, 1960   258
82. (a) Tertiary Potential for Urban Centers, 1960   259
    (b) Tertiary Surpluses and Deficits for Urban Centers, 1960   259
83. Tertiary Relationships—Syracuse SMSA, 1960   262
84. Hypothetical Diagram of an Evolving Transportation Network   266
85. Railroads, c. 1960   268
86. Comparative Highway Traffic Flow, c. 1960   269
87. Passenger Originations from New York State Airports, 1963   274
88. (a) Transportation Potential for Urban Centers, 1960   275
    (b) Transportation Surpluses and Deficits for Urban Centers, 1960   275
89. Wholesale Trade, 1954 and 1958   278
90. Retail Trade, 1954 and 1958   279
91. (a) Service Potential for Urban Centers, 1960   280
    (b) Service Surpluses and Deficits for Urban Centers, 1960   281
92. Trade Centers and Trade Areas, 1st and 2nd Order   282
93. Trade Centers and Trade Areas, 3rd and 4th Order   283
94. Electric Service Areas and Power Facilities   286–87
95. Major Recreation Attractions   293
96. Recreation Regions   306
97. Economic Development, Levels and Trends   325
98. General Economic Health (by counties)   326
99. General Economic Health for the 1950's (using isopleths)   327
100. (a) Hypothetical Concentric Zone Diagram of a Typical Urban System   335
     (b) Cross Section from A to B Showing Hypothetical Land Values   335
101. Major Urban Systems of New York State   337
102. Hypothetical Sector Diagram of a Typical Urban System   341
103. Hypothetical Multiple Nuclei Diagram of a Typical Urban System   343
104. Rank Size Diagram of New York State Cities   349
105. Theoretical Diagram of the Six Functional Classes of Urban Places in New York State   350
106. Larger Urban Centers, Levels in Functional Hierarchy   351
107. Growth Tendency of Urban Systems in New York State   353
108. An Example of Areal Political Complexity, Syracuse, New York   355
109. Rural Landscapes, c. 1960   360–61
110. In the Beginning   365
111. The First Summer   365
112. A Decade Later   366
113. The Land Is Tamed   367
114. The Productive Rural Nonfarmer Takes Over   367
115. Composite Uniform Regions   372
116. A Hypothetical Diagram Combining Nodal & Uniform Regions   377
117. Composite Nodal Regions Superimposed on Composite Uniform Regions   378
118. Planning and Development Regions (As Recommended by New York State Office for Regional Development)   379
119. Planning and Development Regions (Resulting from Fitting Geographic Reality to County and Town Boundaries)   380
120. Albany-Schenectady-Troy Metropolitan Area   386
121. Critical Site Elements of Albany-Schenectady   388
122. Binghamton Metropolitan Area   396
123. Critical Site Elements of Binghamton   397
124. Buffalo Metropolitan Area   408
125. Critical Site Elements of Buffalo   411
126. Orientation Map of New York Urban System   423
127. New York Metropolitan Area   426–27
128. Critical Situation Elements of New York   428
129. Concentric Growth Rings of the New York Urban System   445
130. Generalized Land Use: New York Urban System   447
131. Rochester Metropolitan Area   459
132. Critical Site Elements of Rochester   460
133. Syracuse Metropolitan Area   470
134. Critical Site Elements of Syracuse   472
135. Utica-Rome Metropolitan Area   481
136. Critical Site Elements of Utica-Rome   483
137. Manufacturing Plants: Utica Area   486
138. Projected Land Use to c. 2000 (in back-cover pocket)

Tables

1. New York State as a Part of the United States 6
2. Geologic Time Table 25
3. Mean Cloud Cover and Number of Clear Days 59
4. Mean Monthly Temperature at Representative Stations, 1931–55 59
5. Normal Heating Degree-Days 63
6. Prevailing Wind Directions 64
7. Average Number of Thunderstorms 65
8. Average Wind Speed 66
9. Mean Monthly Precipitation, 1931–55 69
10. Average Number of Days with Measurable Precipitation 69
11. Average Water Balances, Ogdensburg and Brookhaven 72
12. Data for the State's Climatic Regions 76
13. Populations by County, Race, Region—Selected Colonial Censuses 130
14. Population, 1780–1960 189
15. Foreign-born Population Classified by "Mother Tongue," 1960 190
16. Estimated New York Gross State Product, 1960 199
17. The Primary Sector, 1954 and 1958 201
18. Value of Agricultural Production and Sales, 1954 and 1959 209
19. County Output of Minerals 222
20. Growth of Mineral Industries 223
21. Major Items in New York State Mineral Production 223
22. Manufacturing Production 234–35
23. Manufacturing in Metropolitan Areas 236
24. The Manufacturing Regions 237
25. Trends in Value Added by Manufacture for New York State and Selected Areas, 1947–61 251
26. The Tertiary Sector, New York State 255
27. Tertiary Activities in New York State Counties 257
28. Tertiary Activities in Urban Centers in New York State 260–61
29. Tertiary Structure for Selected Central Places, 1960 263
30. The Tertiary Sector Dissected 276–77
31. Electric Power Generating Stations 288
32. Variation in Electric Power Costs 290
33. Recreational Data 295–99
34. Average Total Tax Bills for Three Hypothetical Corporations in Selected States, 1957 312
35. Tax Efforts 314
36. Strength of New York's Economy 316–17
37. Rank of United States and Nine States in Strength of Economy as Measured by Selected Indicators 320–21
38. Rank of New York Counties for Nine Indicators of Economic Health 322–23
39. Status of Selected New York Counties According to Nine Indicators of Economic Health 324
40. Land Use Percentages for the Three Rural Landscapes and New York State, c. 1960 358
41. Planning and Development Regions 381
42. Major Manufacturing Concerns of the Albany-Schenectady Urban System 393
43. Population of Binghamton for Selected Years 399
44. Population Trends within the Binghamton Urban System 403
45. Major Manufacturing Concerns of the Binghamton SMSA 404
46. Major Manufacturing Concerns of the Buffalo SMSA 420
47. Population of Counties Included in the New York Urban System 424
48. Population Projections for Counties Included in the New York Urban System 456
49. Major Manufacturing Concerns of the Rochester SMSA 466
50. Relative Growth Rates of the Large Urban Systems 471
51. Estimated Population in Each of the Suburban Sectors of Syracuse, 1960 474
52. Major Manufacturing Concerns of the Syracuse SMSA 478
53. Population Trends of Selected Mohawk Valley Places 482
54. Population Trends in Utica and Rome 482
55. Major Manufacturing Concerns of the Utica-Rome SMSA 485

56. Population of New York State, 1930–1960; Projections, 1970–2000   492
57. Population of the Great Urban Systems, 1930–60; Projections 1970–2000   493
58. Population of New York Counties, 1930–60; Projections, 1970–2000   494
59. Population of Larger Urban Places Outside the Great Urban Systems, 1960; Projections, 1980 and 2000   495
60. Number of New Yorkers Working in the State and in the Great Urban Systems, 1960; Projections 1970–2000   497
61. Number of New Yorkers Working in the State and in the Great Urban Systems (by Sector), 1960; Projections, 1970–2000   498

INTRODUCTION

JOHN H. THOMPSON

Introduction

New York is an important and interesting state. It stands pre-eminent in business and manufacturing, and only in recent years has California become a competitor to its number-one rank in population. Within its borders are found not only the nation's largest city but inaccessible wilderness areas as well. In some places farms flourish; in others they have been largely abandoned. Attractive mountain scenery is within a few hours' drive of lowland plains. Snowfall, the heaviest east of the Rocky Mountains, clogs roads and provides excellent skiing upstate on the same days that winter rains are soaking Long Island. Some parts of the state have experienced a long and continuing history of vigorous economic success; other parts have not done so well. This book deals with the physical, historical, and economic geography of the state. It discusses the many differences from place to place and attempts to explain these differences and point up their significance. It should help both the casual reader and the serious student understand the state, and by improving understanding make it a more enjoyable and profitable place in which to live, do business, or just travel.

New York as a Part of the United States

New York is only a medium-sized state; however, in terms of economic importance to the nation, it is of much higher rank (see Table 1). Its location, astride heavily used routes of commerce in the older Northeast, gave it an early importance which has been well maintained. Furthermore, the location within its borders of New York City enhances the state's image as a vital part of the greater nation.

With but slightly over 1 per cent of the nation's area, New York has 9 per cent of the population, 10 per cent of the manufacturing, and even larger portions of the trade and service activities. With nearly 12 per cent of the nation's personal income, the state can probably be estimated to contribute at least 10 per cent of the Gross National Product (GNP). Nearly 10 per cent of all college students are being educated here, too. Within the state only such activities as agriculture, mining, and lumbering are nationally insignificant. It is interesting to note that those sectors of the economy—manufacturing, trade, and services—that are standouts, are city-builders; while agriculture, mining, and lumbering, activities of less significance in this state, keep the countryside filled with people. The impact of this situation will become more apparent as this geographic interpretation unfolds.

Although New York is involved most heavily in trade with the northeastern part of the country, it does buy materials from all states and sell to all states. Figure 1 (*a* and *b*) provides some indication of these spatial relationships. As a national supplier of goods from clothing to war material, New York's role is vital; as a purchaser of goods from coal to food, it must depend on other parts of the country. Being a part of the economic core area, or principal economic axis, of the United States (see Fig. 1*c*) means that relatively favorable transport costs exist from most New York State points to about 70 per cent of the national market. It also means that goods are moving in and through the state in great quantities, thus stimulating business. In 1825 the Erie Canal linked New York harbor with the Great Lakes. Recently the St. Lawrence Seaway has provided extended access to the Great Lakes. New York has been fortunate to have two "coasts," the Atlantic and the Great Lakes–St. Lawrence. Certainly market and transport advantages have always been two of the state's outstanding attributes.

New York City, by itself, gives New York superlative qualities unmatched by any other state. It is not only the country's largest city but also the world's leading port and greatest manufacturing, retail, wholesale, and financial center. So great is the city's impact on the state that, although it occupies but a tiny part of the area, it tends to dominate many functions ranging from manufacturing to politics. It contributes heavily, too, to the state's extensive air traffic generation (Fig. 1*d*).

A Population Map as a Starting Point for Geographic Analysis

If the uneven distribution of population over any area is thoroughly understood, much becomes

ECONOMIC TIES WITH THE NATION

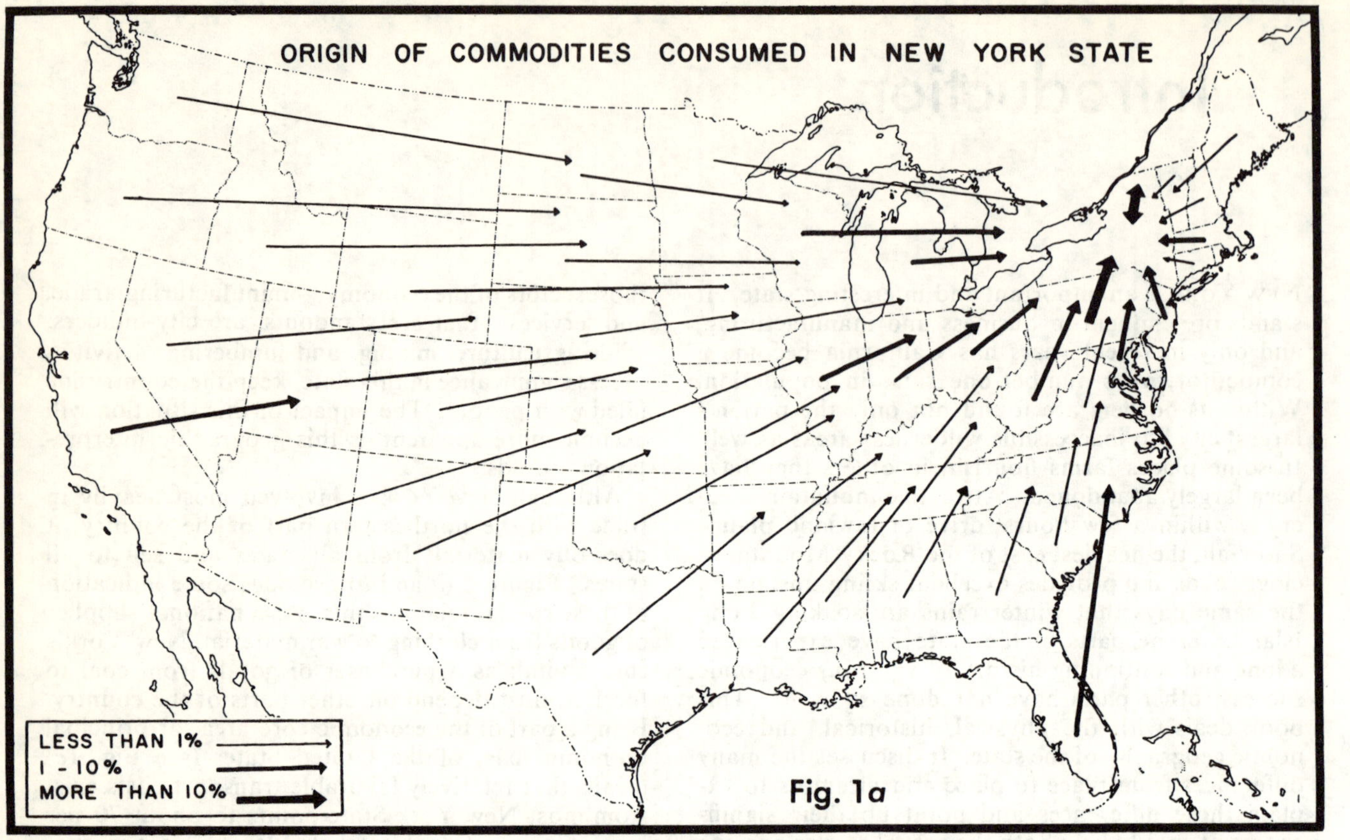

ORIGIN OF COMMODITIES CONSUMED IN NEW YORK STATE

LESS THAN 1%
1 – 10%
MORE THAN 10%

Fig. 1a

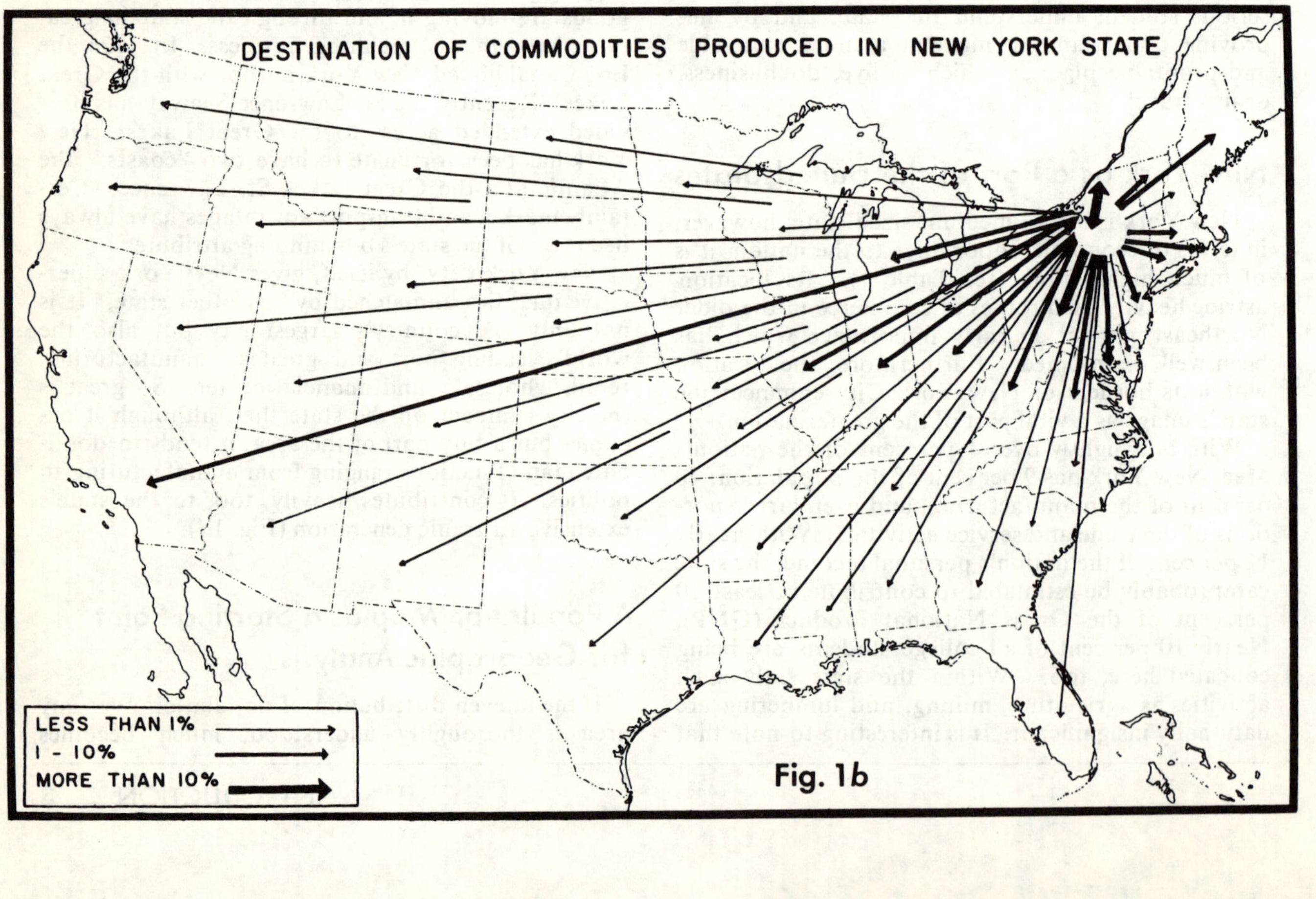

DESTINATION OF COMMODITIES PRODUCED IN NEW YORK STATE

LESS THAN 1%
1 – 10%
MORE THAN 10%

Fig. 1b

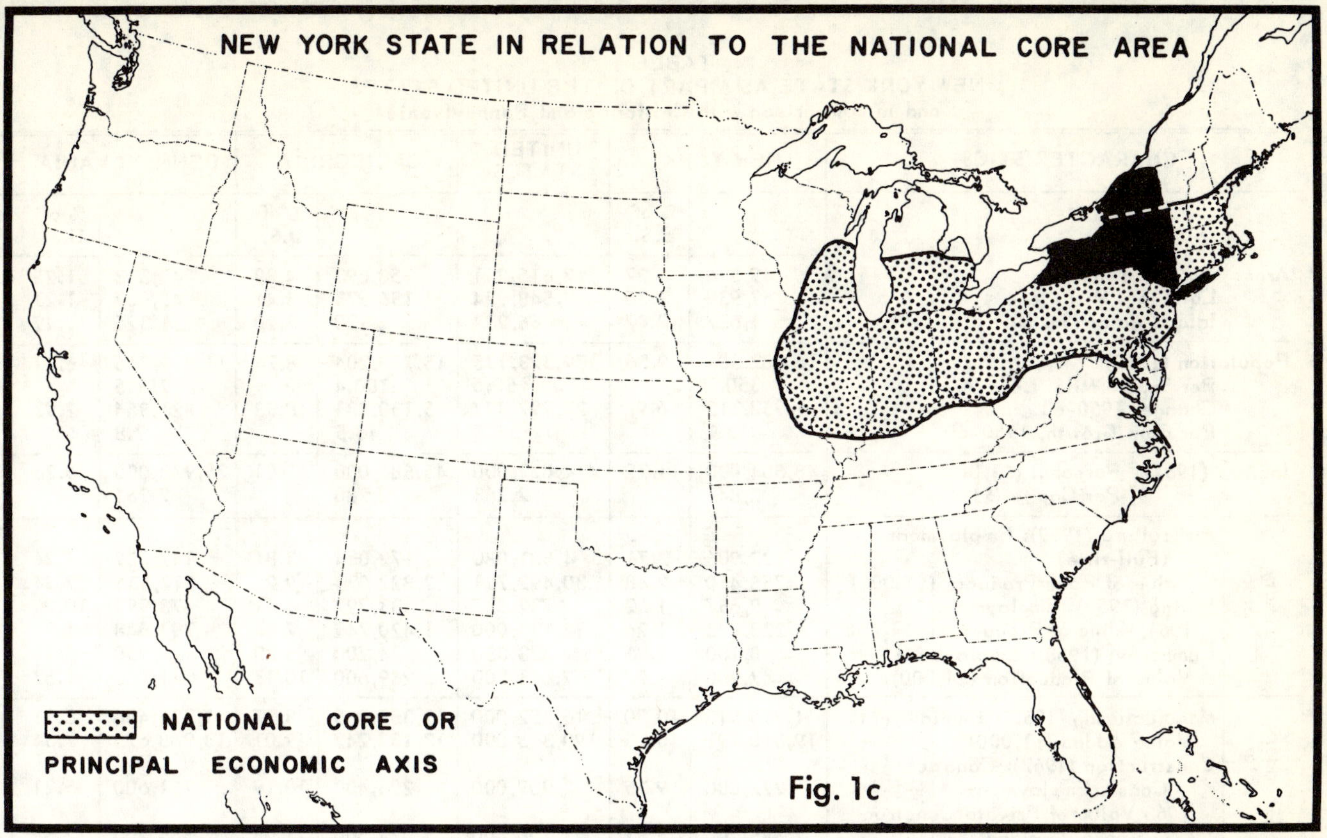

NEW YORK STATE IN RELATION TO THE NATIONAL CORE AREA

:::::: NATIONAL CORE OR
PRINCIPAL ECONOMIC AXIS

Fig. 1c

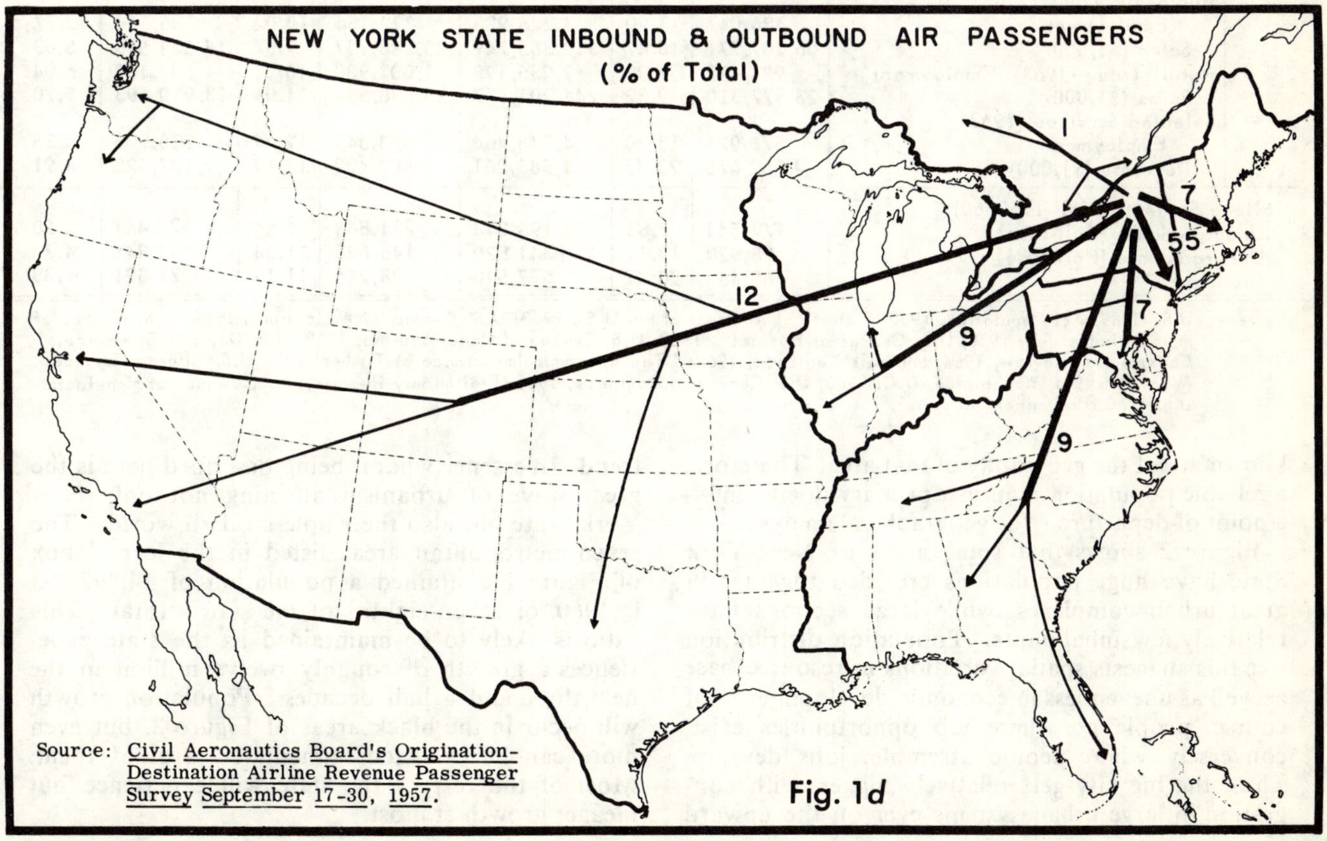

NEW YORK STATE INBOUND & OUTBOUND AIR PASSENGERS
(% of Total)

1
7
55
12
9
7
9

Source: Civil Aeronautics Board's Origination-Destination Airline Revenue Passenger Survey September 17-30, 1957.

Fig. 1d

INTRODUCTION 5

TABLE 1
NEW YORK STATE AS A PART OF THE UNITED STATES
and in comparison with California and Pennsylvania

CHARACTERISTICS	NEW YORK	% of U.S.	UNITED STATES	CALIFORNIA	% of U.S.	PENNSYLVANIA	% of U.S.
Area: Total (Sq. Mi.)	49,576	1.37	3,615,211	158,693	4.39	45,333	1.25
Land (Sq. Mi.)	47,939	1.35	3,548,974	156,573	4.41	45,007	1.27
Inland Water (Sq. Mi.)	1,637	2.47	66,237	2,120	3.20	326	.49
Population (1960):	16,782,304	9.36	179,323,175	15,717,204	8.76	11,319,366	6.31
Per Square Mile (Land Area)	350.1		50.5	100.4		251.5	
Change, 1950–60	1,952,112	6.97	27,997,377	5,130,981	18.33	821,354	2.93
Per Cent Growth, 1950–60	13.2		18.5	48.5		7.8	
Income (1961): Personal ($1,000)	48,504,000	11.72	414,022,000	45,586,000	11.01	25,933,000	6.26
Per Capita ($)	2,848		2,263	2,780		2,261	
Primary Sector							
Agriculture (1959): Employment (Full-time)	122,906	2.71	4,521,080	172,034	3.81	147,609	3.26
Value of Farm Products ($1,000)	755,410	2.48	30,492,721	2,822,071	9.25	712,535	2.34
Mining (1958): Employment	9,657	1.32	732,637	33,792	4.61	73,993	10.09
1961 Value of Production ($1,000)	228,983	1.26	18,131,000	1,420,749	7.84	791,648	4.37
Lumbering (1958): Employment	8,100	1.91	423,350	24,700	5.83	8,950	2.11
Value of Production ($1,000)	47,200	1.77	2,661,200	269,600	10.13	41,850	1.57
Secondary Sector							
Manufacturing (1963): Employment	1,716,513	10.50	16,352,000	1,352,065	8.27	1,326,456	8.11
Value Added ($1,000)	19,510,191	10.25	190,395,000	17,157,242	9.01	13,968,675	7.34
Construction (1962): Contract Const. Employment	272,000	9.35	2,909,000	296,400	10.19	151,600	5.21
1963 Value of Private Construction Authorized ($1,000)	1,485,200	6.81	21,796,400	5,074,500	23.28	565,200	2.59
Tertiary Sector							
Wholesale Trade (1963): Employment	426,047	13.20	3,226,928	323,184	10.02	186,114	5.77
Sales ($1,000)	66,207,771	18.47	358,385,749	35,386,137	9.87	18,043,645	5.03
Retail Trade (1963): Employment	994,172	9.98	9,956,198	1,007,980	10.12	601,167	6.04
Sales ($1,000)	23,977,310	9.82	244,201,777	26,888,554	11.01	13,910,693	5.70
Selected Services (1963): Employment	576,024	13.46	4,278,464	523,541	12.24	228,905	5.35
Receipts ($1,000)	9,062,675	20.33	44,586,261	5,882,420	13.19	2,187,523	4.91
College Students (Fall 1959–60):							
Undergraduate (Full-time)	172,543	7.86	2,196,214	211,868	9.65	122,481	5.58
Undergraduate (Part-time)	88,920	13.12	341,820	146,625	21.64	32,345	4.77
Graduate	69,463	20.32	677,510	38,246	11.19	21,821	6.38

Sources: U.S. Census of Population, 1960; Statistical Abstract of the U.S., 1963; U.S. Census of Agriculture, 1959; U.S. Census of Mineral Industries, 1958; U.S. Census of Business, 1958; U.S. Census of Manufacturing, 1958; U.S. Dept. of Commerce, *Construction Review*, 1964; Minerals Yearbook, 1963; "The Economic Importance of Timber in the U.S." Forest Service, Misc. Pub. 941, Washington, D.C. 1963; U.S. Census of Business, 1963 (Preliminary Reports); U.S. Census of Manufacturing, 1963 (Preliminary Reports).

known about the geography of that area. Therefore, a reliable population map is a primary document—a point of departure—for geographical analysis.

Figure 2 shows that some areas of New York State have huge populations crowded together in great urban complexes, while large sections have relatively few inhabitants. Population distribution like this suggests spatial variations in resource base as well as unevenness in economic development. Of course, people go where job opportunities exist; conversely, where people assemble, jobs develop. Thus, the big city gets relatively bigger, with congestion in large urban systems ever on the upward trend. In essence what is being described here is the great wave of urbanism affecting not only New York State but also the whole modern world. The seven metropolitan areas listed in the legend box of Figure 2 contained a population of 14,392,793 in 1960, or seven-eighths of the state's total. This ratio is likely to be maintained as the state experiences a growth of roughly twelve million in the next three and a half decades. Population growth will occur in the black areas of Figure 2, but even more can be expected immediately around them. Most of the rest of the state will experience but meager growth at most.

Historically, population trends of urban centers have generally been of three types. All villages and cities, of course, began as small places. In the case of the first type the community grew to a few hundred in size and then declined and in certain instances disappeared. The second type grew larger but reached a plateau and became static. Some cities in this second category attained populations of several tens of thousands. The third type kept right on growing to a hundred thousand and more. The seven great urban systems are products of this last type. Why a few areas experienced so much growth and others did not is a question of considerable importance, and one for which answers will be sought in this volume. At this point each reader can probably identify his own home town as belonging to one of the three types and perhaps offer some explanation for its status.

Population trends in rural areas have been quite different. Fifty or a hundred years ago there were actually more people scattered through rural New York than now. The urban systems were relatively, and actually, much less significant then.

New York filled up with farms before the turn of the century. Since then, the rural areas have been emptying as poorer and smaller farms are abandoned and the early farmer's descendants go to cities to seek work. By the time the present generation of farmers retires, at least another third of the state's farms will have disappeared and rural farm populations will have dwindled still further.

Depopulation of rural areas has been continual and often rapid, but this trend now seems to be slowing down. Areas not too far from urban centers are receiving ever-growing numbers of what the U.S. Census Bureau calls rural nonfarmers. The number of these people in any given locality depends on the distance from, and economic vitality of, nearby cities as well as the quality of the roads which must take most of them to the city jobs. Ordinarily they are simply individuals who work in the city but prefer living in the country.

There is another group, too, that will stem the tide of rural depopulation, particularly in the areas beyond easy driving distance to city jobs. Collectively, these people might be referred to as the unproductive rural nonfarmer group, to differentiate them from those rural nonfarmers who work and are thus in the productive segment of society. Often unemployed or on welfare, they occupy houses in the countryside wherever rental or occupance costs are low enough.

Figure 3 shows what has been happening in recent years. The doughnut rings immediately around cities are growing impressively, while most rural areas are not doing nearly so well. A forecast for spatial unevenness in economic growth as well as in population growth is in order. The large urban systems are where they are partly because of favorable locational attributes, such as terrain and transportation; but now that they are large their very largeness, diversity, and dynamism do more to sustain their growth than many of the advantages that initiated growth. The Hudson–Mohawk–Lake Plain belt from New York City to Buffalo via Albany, Syracuse, and Rochester contains most of the state's population and most of its productive capacity and dynamism. Future growth will focus largely on the big centers of that belt, not the smaller ones between them. Even though these smaller centers are along the major transport arteries so important to the belt, and otherwise have most of the locational attributes of their larger neighbors, they lack the growth potential that seems to be the asset of the big city today.

Geographic Themes and Their Application

Some of the questions this book will attempt to answer are becoming apparent. Where are the people? Why are they so unevenly distributed? What causes some parts of the state to be economically healthy while others are suffering setbacks? What is likely to be the development pattern of the future? Do differences in the resource base from place to place in the state provide answers, or are variations in man's heritage, abilities, energies, and institutions more important?

The following brief examination of the nature of modern geography and some of its more important themes will set the course for this book. Careful scrutiny of the ideas in the paragraphs immediately following should aid the reader as he proceeds through the book from thought to thought and from chapter to chapter.

PLACE CONSCIOUSNESS, A PREREQUISITE TO GEOGRAPHIC UNDERSTANDING

Place consciousness might be defined as that mental state which exhibits curiosity about the nature and causes of differences from place to place; or, more simply stated, a curiosity about where places are, why they are there, and the significance of their location.

The extent of place *un*consciousness of some Americans is phenomenal. Take, for example, the case of a Syracusan who decided to spend his retirement years in Florida. After purchasing a home there he was unable to answer a query as to whether it was on the Atlantic Coast or the Gulf Coast. Further questioning about his trip to Florida to

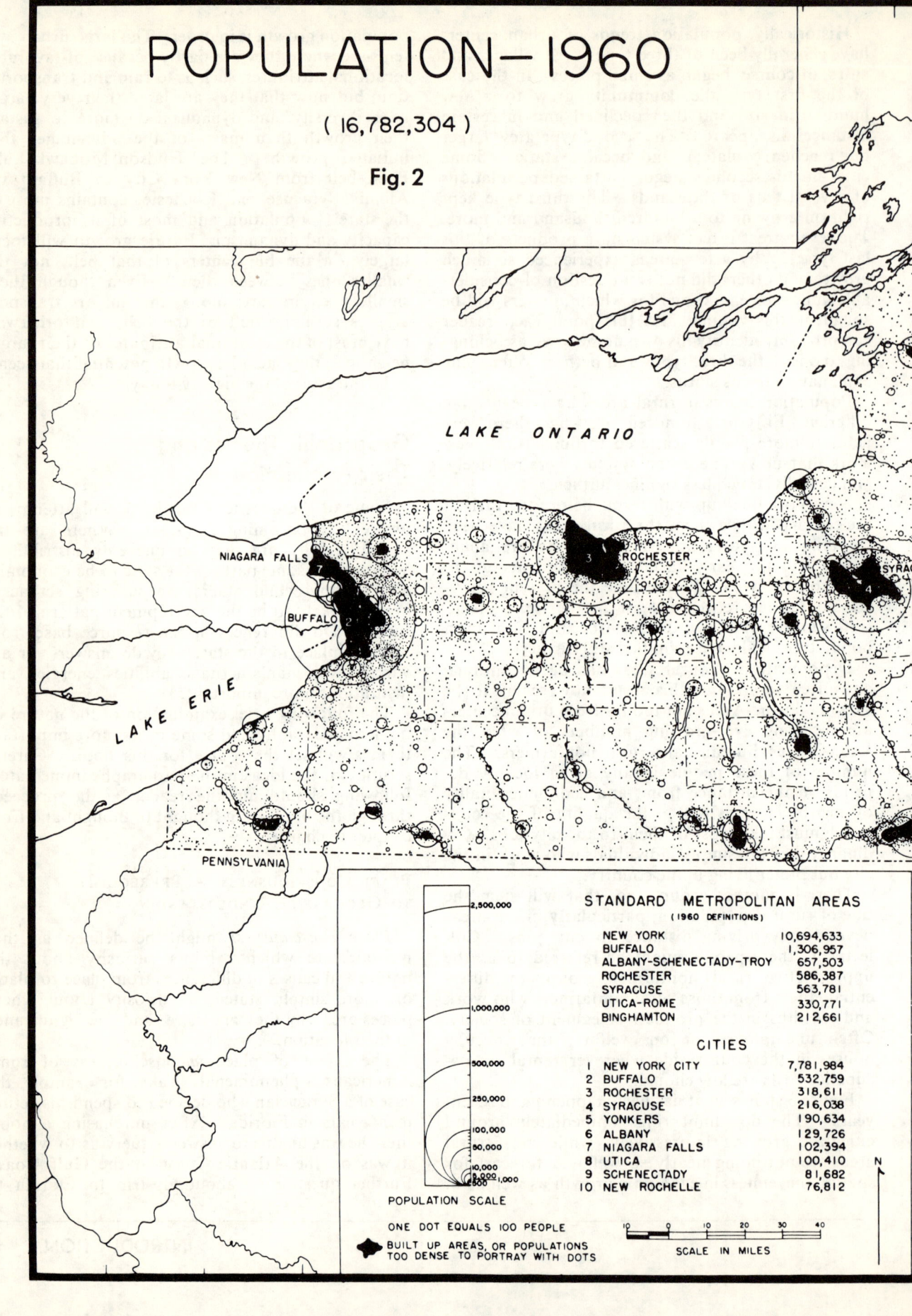

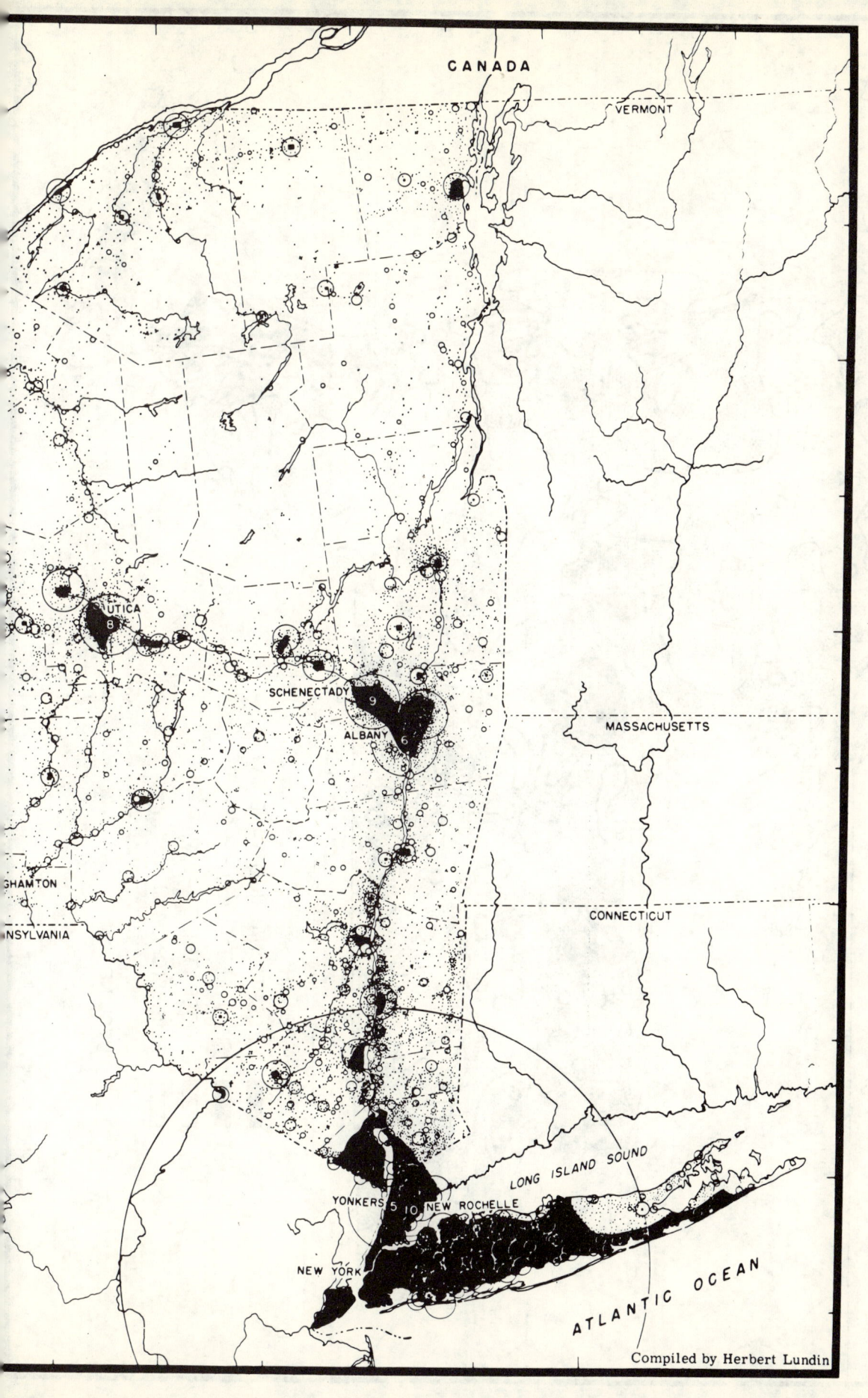

Compiled by Herbert Lundin

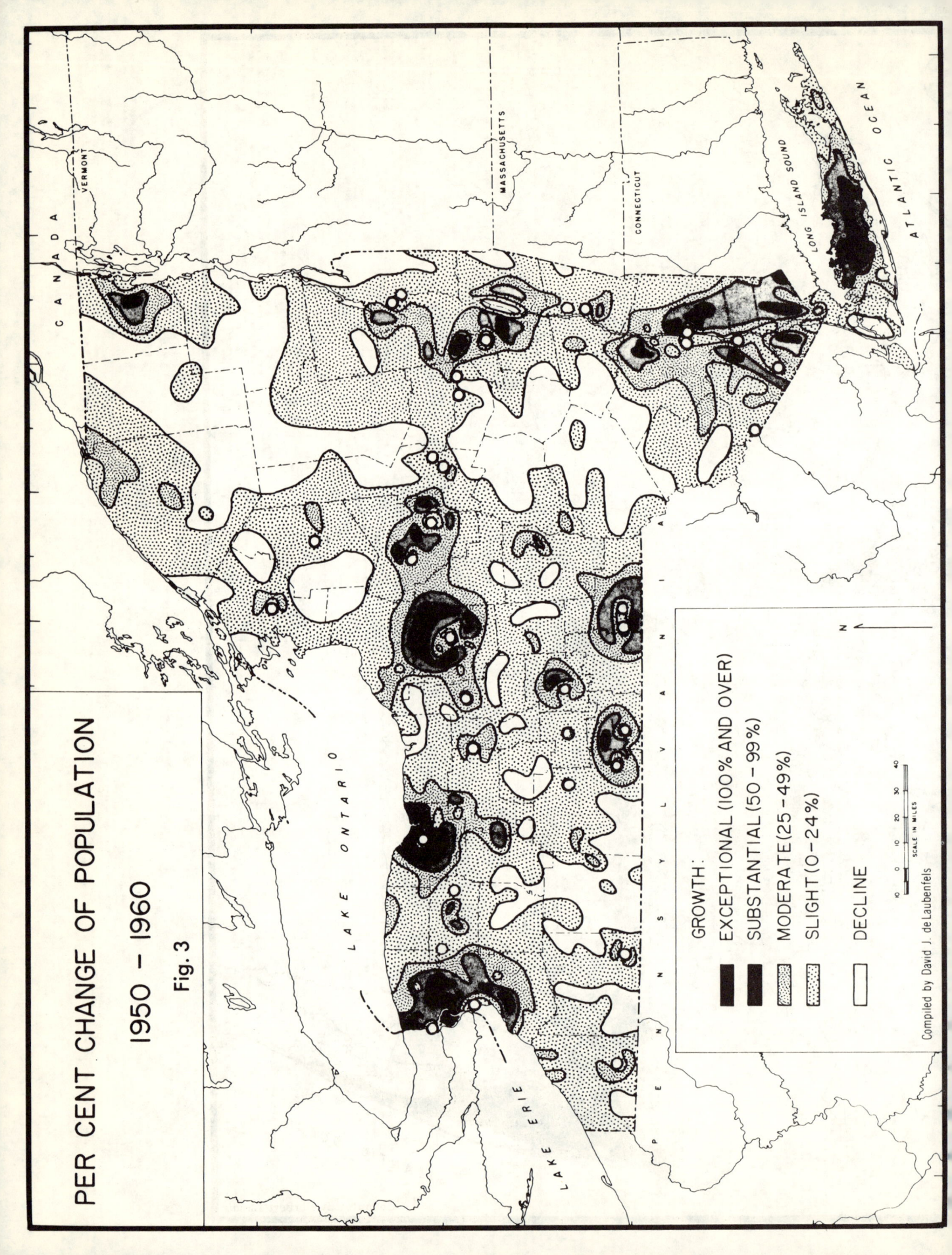

purchase the retirement property revealed that he had made the several-thousand-mile journey almost without being interested in or seeing a thing en route. Only the names of two or three cities through which he passed remained as remembrances of what might have been a fascinating auto tour through some of America's most varied and interesting sections. The lack of place consciousness when traveling, reading the news, or just thinking, is a serious deficiency and one that causes people to miss a great deal in life.

This volume uses maps to an extent unequaled by any other book on New York State. It is natural that maps should be widely used, for they are the finest tool known for focusing on the importance of place. The maps are designed as an integral part of the book and will make the reading of every chapter more meaningful. They should be visually compared with one another, for in this way similarities of patterns of different phenomena will become evident and relationships between the things portrayed on the maps can be implied. The many maps and related textual material can help develop in the careful reader a place consciousness that will serve him well even years after he has laid this book aside.

A simple exercise in using maps, in fact overlaying maps, to locate a site for building a new house will illustrate the practical importance of place or spatial information. Suppose, for example, you the reader were planning to build a new home. And suppose you insisted that your home: (1) have a good view,

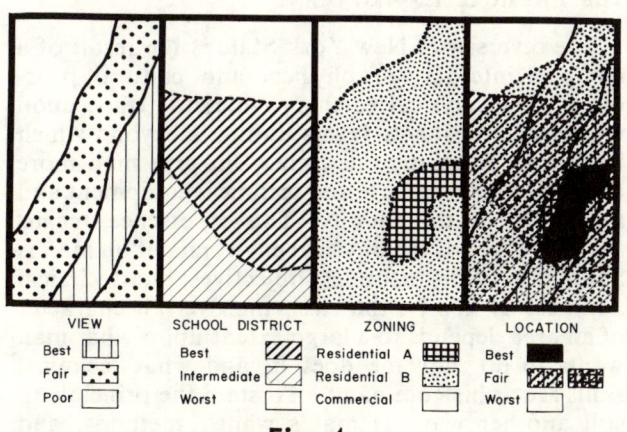

AN EXAMPLE OF OVERLAY ANALYSIS

Fig. 4

(2) be in the best school district in the locality, and (3) be located in an area zoned by your community for Class A residences. The first three simple maps in Figure 4 hypothetically portray the spatial characteristics of the criteria designated as being important. Now overlay these three maps and find the best area, in terms of the three criteria, for your house. It would be where Best View, Best School District, and Residential A Zoning spatially coincide, or only the darkly shaded area on the fourth map.

Maps containing other spatial information such as the location of roads or city water might have been added to the analysis. In any case, consciousness of the significance of this kind of information and desire to know about it are attributes everyone will find useful and interesting, whether he is building a house or attempting to understand his state, his country, or the world.

THE INTERESTS AND AIMS OF MODERN GEOGRAPHY

Place consciousness is vital to geographic understanding. *Geography* may be defined as the field of learning dealing with the arrangement and association of phenomena on the surface of the earth. To put it another way, geography deals with man's occupation and use of the earth by interpreting the spatial characteristics of things that affect them. No two places in New York State, or on the entire surface of the earth, are exactly alike. Differences are due to a wide variety of causes such as variation in the physical resource base, economic trends, political conditions, or man's capabilities and interests. Geography describes the differences and similarities and forecasts their consequences.

The reader might stop for a moment to think about two places he knows well, attempting to identify and explain some of the differences between the two places. Because of its spatial focus, geography has been called the science of place. Just as history deals with the sequence of human affairs, so geography deals with spatial characteristics and associations on the face of the earth. Together geography and history form a base upon which a better understanding of the other social sciences may be built.

Geography does not devote itself exclusively to the study of the physical environment, or even only to the consideration of relationships between man and his physical surroundings, as is so often thought by nongeographers. Investigations of these kinds are a part of geography, as will be seen in this volume, but they are not all of it. A given geographic investigation may deal with the spatial characteristics of physical, cultural, material, or nonmaterial phenomena. There can be a geography of manufacturing or voting habits, just as well as a geography of land forms or climate.

Geography's focus is always spatial; the things it interprets may vary considerably. The expression "geography of an area" correctly refers, then, to the study of the arrangement and association (spatial characteristics) of phenomena found in the area. In

order to understand thoroughly or interpret an area such as New York State, a geographer traditionally investigates relevant spatial characteristics of the physical environment and the cultural environment, realizing that both give character to the area. He also realizes that the physical environment is relatively stable but that the cultural environment is ordinarily highly dynamic. This results in an ever-changing geography and makes a historical perspective important to current understanding or forecasting.

Figure 5 will clarify this line of thought. The diagram uses the analogy of a tripod, the top of which, area understanding or analysis, is supported by three legs: the cultural leg, the physical leg, and the time-factor leg. The implication is clear: remove one of these legs and the tripod (area analysis) collapses. In a general way this volume places New York State on a tripod and fashions three legs to support its interpretation.

Man's Interrelationship with the Physical Environment

The diversity of New York State is the result of a complex interplay of physical and cultural processes. The physical earth provides the stage upon which man must act. It sets up limits beyond which he cannot go at any one time. But give man more energy, understanding, and technical equipment and he can extend these limits. This, in essence, means that the effect of the physical environment on man is a function of his attitudes, objectives, and technical abilities. If this be the case, the over-all character of an area depends to a large extent upon what man wants to do, how he does it, and what technical skills are at his command. To state the principle in still another way, as man's wants, methods, and skills change, he makes different use of the physical earth upon which he lives and brings about major changes in the character, or general geography, of places.

Imagine, for example, the differences between the skyscraper-dominated Manhattan of today and the sparsely settled, rocky woodland of 1626 purchased from the Indians for twenty-four dollars; or the contrasts between the modern, thriving Erie-Ontario

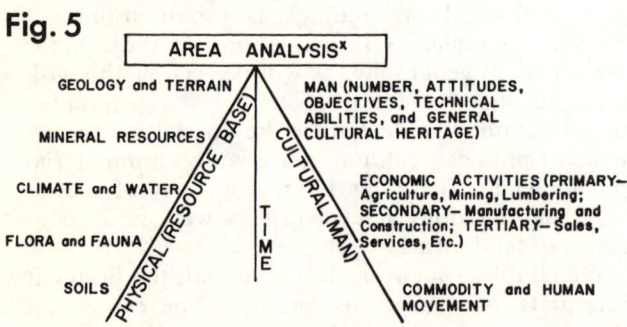

AREA ANALYSIS TRIPOD

Fig. 5

*A rational procedure for sequence of investigation might begin at the top of the physical leg and proceed downward followed by a similar treatment of the cultural leg. The specific handling of these kinds of material of course would depend on the purpose.

12   GEOGRAPHY OF NEW YORK STATE

CONTRASTS IN A STATE OF GREAT DIVERSITY. No state in the Union exhibits more extreme ranges in degree of economic development and population concentration than New York. On the left is the wilderness country near Mount McIntyre in the heart of the Adirondacks. Permanent residents are few and opportunities for economic development are limited. *U.S. Forest Service.* On the right is the world's most congested locality, New York City. As the "Gateway" and "Primate" city of America, it continues to grow and become even more congested. *Port of New York Authority.*

Lowland upon which Syracuse, Rochester, and surrounding farms are now situated and the primeval forest land described so aptly in the works of James Fenimore Cooper. Man has wrought great changes; he has pushed ever farther apart the limits set by the physical environment. What will happen in the future to change the geography of these areas may be even more striking than what has happened in the past. Man will be the active agent. The more advanced the society, the more dominant is the cultural leg in supporting the area analysis tripod.

LOCATIONAL ADVANTAGE AND
SPATIAL INTERACTION

Every farm, town, city, factory, railroad, or resort has certain characteristics and experiences certain successes or failures because of its location in relation to other factors. A farm's location relative to terrain, soil quality, roads to market, or amount of rainfall may strongly determine its success. A city's position relative to major transport facilities, water supplies, harbors, electrical power supplies, and other cities is bound to be important. A factory's location relative to raw materials, markets, transportation, and power strongly affects its production costs and profit potential. Some areas in New York State have evolved as favorable places for many kinds of economic activity, others for few kinds. As a result, some areas are growing rapidly while others are not. Advantages in location at one time may not persist as the geography (spatial characteristics) of raw material sources and markets changes. This means that parts of the state that once may have had relative locational advantages, and as a result prospered, may no longer have these advantages and be languishing.

As implied by the concept of locational advantage, in a commercially active society like that of New York State, a given place does not exist in isolation but is related functionally to other places. It is constantly involved in spatial interaction with these other places in some way. Geography, therefore, in its concern with where places are, why they are there, and the significance of their location, must examine as well how their location and development relate to the influences exerted by other places. To illustrate: part of the success in dairying in northern New York is due to the market for fluid milk in metropolitan New York; the factories in Rochester sell most of their products outside the city; Syracuse University draws a large part of its student body from distant places.

Enterprises like those cited, which function in such a way as to bring money into a local area from the outside, obviously generate growth in the local area and may therefore be referred to as *basic*. By contrast, those enterprises which simply result in redistributing money locally may be termed *nonbasic*. Ordinarily, it may be assumed that nonbasic enterprises do not result in economic growth. Some examples would be barbershops, dry-cleaning establishments, most restaurants, and most repair and service establishments. This dichotomy of activities, based on two different kinds of geographical markets, is useful in analyzing how, and to what extent, spatial interaction exists between places. It is useful particularly in showing the spatial interaction between cities or between cities and their surroundings.

As a result of functional ties between places, *geographical axes* of outstanding lines of movement for goods and people connect main parts of a functional system. Most nations have what might be called a *principal economic axis* (or *core area*, if the main lines of movement become complex and laterally diffused) connecting the two leading economic areas. Of significance to our concern in this volume has been the development of a wide core area or principal economic axis in the United States between the Middle Atlantic Seaboard around New York City on the east and Chicago on the west. New York State is favorably situated in reference to this axis. Geographical axes of smaller dimensions and local nature develop, too, giving a more detailed geographical orientation to parts of our functional system. For example, in New York State axes up the Hudson Valley, out the Mohawk Valley, and across the Ontario Lake Plain, down the St. Lawrence Valley, and across the Southern Tier have given diverse orientations to movement and diverse growth potentials to the many parts of the state.

DEVELOPMENT THEORY

Every part of the world, New York State being no exception, seems to go through an economic metamorphosis from an undeveloped to a developed status. The livelihood structure, or the way men make their living, changes as this development takes place. First there is a situation where nearly everyone is engaged in the *primary sector* of the economy (agriculture, mining, forestry, and fishing). Later the economy becomes rather evenly oriented between the *secondary sector* (manufacturing and construction) and the *tertiary sector* (sales, services, and related activities), with the primary sector falling well behind. Finally, the tertiary sector strongly dominates the secondary sector, and the primary sector becomes of surprisingly little importance. The extent of dominance of the various sectors may be measured by the per cent of the total labor force engaged in each. In New York State the primary sector employed almost 80 per cent of the labor

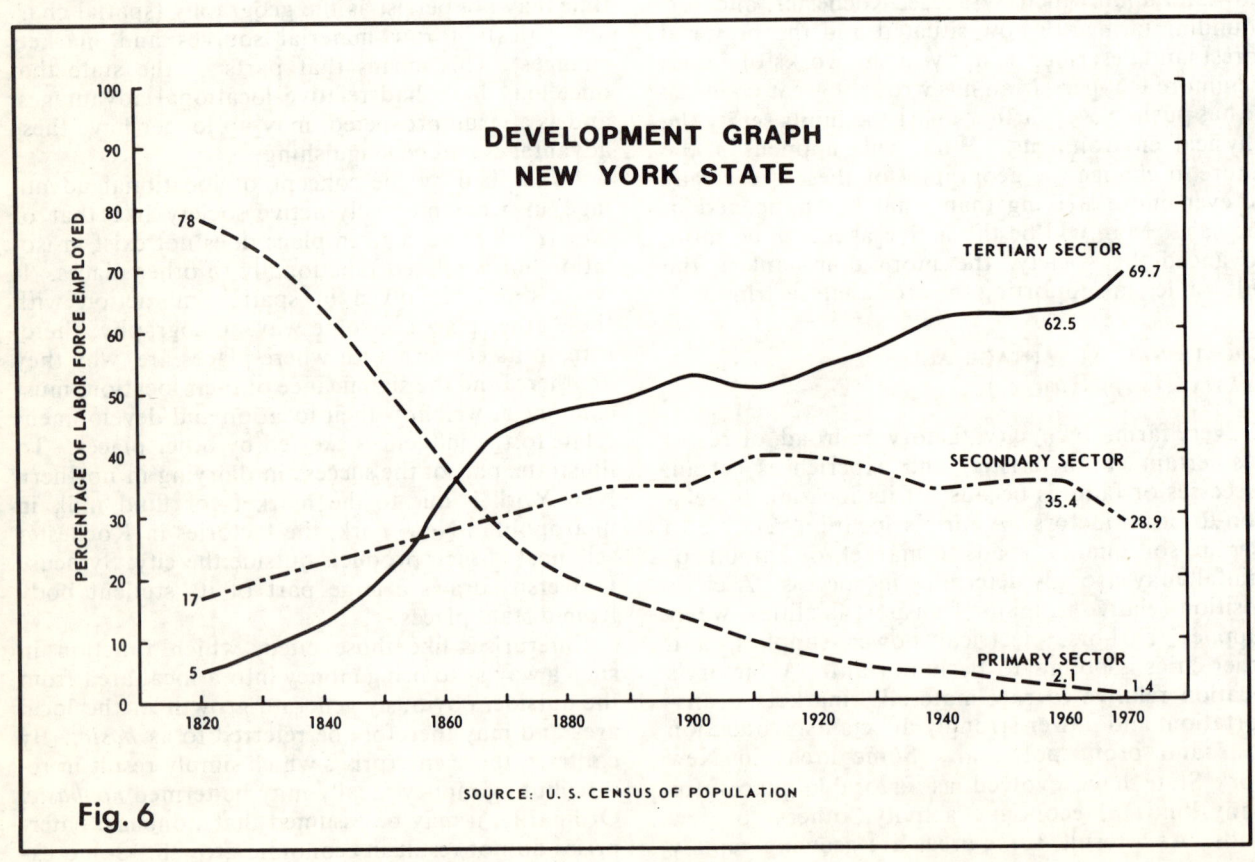

Fig. 6

force in 1820 but only about 2 per cent in 1960. On the other hand, the tertiary sector increased from 5 per cent of the labor force in 1820 to over 60 per cent in 1960. The secondary sector approximately doubled from 17 per cent in 1820 to 35 per cent in 1960. The development graph (Fig. 6) portrays what has happened in the state.

New York State's present stage of development suggests that its over-all future will be tied almost exclusively to the tertiary and secondary sectors of the economy, which are made up largely of city-oriented activities. This supports the above forecast of increasing urban dominance in the state.

As the population and labor force double during the next forty years or so, many more new jobs will be found in the tertiary sector than in the secondary. In fact, it is likely that this ratio will be three or four to one. The already small part of the labor force in the primary sector is likely to become even smaller.

The Regional Concept

In geography two kinds of regions are identified: uniform and nodal. A *uniform region* is an area of any size which approaches homogeneity in terms of the criteria by which it is defined, i.e., land forms, agriculture, or soils. A *nodal*, or *functional*, *region* is composed of a focus surrounded by an area that is functionally related to it. For example, a city, a school, or even a supermarket may be the focus of a nodal region. Whether regionalizing is done on a uniform or nodal basis, it is a device for spatial classifying and in this volume will be extensively employed. In geography, the term *area* refers only to a portion of earth space with no implication of homogeneity or focus.

Any effort at regionalizing should be predicated on two premises: (1) that phenomena are organized on the surface of the earth in such a way as to produce a mosaic of regions, each exhibiting certain homogeneous qualities and/or internal cohesion that will differentiate it from surrounding areas; and (2) that identifying and mapping these regions is a useful procedure for discovering order in earth space.

That nearly everything on the surface of the earth varies from place to place is well understood. Because there is a limit to the diverse detail the human mind can grasp if it is to arrive at a broad understanding of earth space, it is desirable, if not necessary, to group similar small areal units into larger units. It would be difficult, for example, to arrive at an over-all understanding of New York State based on such details as the difference in soil from one part of a farmer's field to another or slope variation in an area the size of a city block. Soil and slope conditions, as well as other phenomena, must be generalized into larger areal units that have some degree of similarity if over-all understanding of the state is to be attained. Such units can be identified for most phenomena that make up the geography of the state. Their size and the level of generalization depend on purpose.

Since this volume uses middle-level ranges of generalization, it divides the state into anywhere from three to a dozen or more units, depending on phenomena. Each unit exhibits homogeneous or nodal qualities in terms of specified criteria and may correctly be referred to as a region. Such regionalizing helps considerably in examining and understanding areal differentiation in the state. Land form regions, agricultural regions, manufacturing regions, and trade areas or regions, to mention a few, are identified with the above objective in mind. Later on these are further grouped into regions with over-all physical and cultural similarities. These composite regions exhibit similarities in history of use, current problems, and economic potential, and thus may provide a framework for planning and development.

Should anyone want to divide any of the regions used in this book into smaller regions based on a finer mosaic of detail, it would be perfectly possible to do so. The divisions would again be based on specified criteria, and the smaller the regions identified, the greater would be the detail dealt with and theoretically the closer one would come to complete homogeneity.

In the case of uniform regions no matter what level of generalization is applied or what size the regions, complete homogeneity probably is seldom achieved. Instead, uniform regions generally have cores that are clearly distinct from the cores of adjacent regions. Marginal portions of regions may be less distinct and in fact may merge almost imperceptibly with margins of adjoining regions. That this happens in no way destroys the validity of the regional concept or its utility in geographic analysis.

Geographical Persistence

Present spatial patterns exhibit relic features that have persisted through time in spite of changes in attitudes, objectives, and technical abilities of people, forms of land use, and the regional economy. Despite radical changes in such things as crops, vehicles, and types of houses, this geographical persistence expresses itself in the shape and size of properties and field boundaries, and in the location of roads, houses, and other structures. The reader may want to make observations of his own in the field to see to what extent geographical persistence is present. This volume deals primarily with the broader aspects of geographical persistence such as geographical axes and general interregional relationships.

Organization of the Book

This book is divided into five parts. Part One, "The Resource Base," Part Two, "Three and a Half Centuries of Change," and Part Three, "Economic Activities Today," represent rather closely the three legs of the area development tripod. They present facts and interpret the complex interplay of the physical and cultural environment through time and evaluate the spatial variation or development potential.

Part Four, "Landscapes and Regions," focuses on the great contrasts between urban and rural areas in trends and problems and presents a comprehensive nodal and uniform regionalization of the state. The regionalization is used as a basis for recommended planning and development units.

Part Five, "The Great Urban Systems," analyzes the seven largest urban systems of the state in terms of origin and development, function, and prospects.

Selected References

Ackerman, Edward A. *Geography as a Fundamental Research Discipline*. ("Department of Geography Research Paper," No. 53.) Chicago: Univ. of Chicago, 1958.

——— et al. *The Science of Geography*. Report of the Ad Hoc Committee on Geography of the National Academy of Sciences, Earth Science Division. Washington, D.C.: National Academy of Sciences, 1965.

Bogue, Donald J. *The Population of the United States*. Glencoe, Ill.: Free Press, 1959.

Broek, Jan O. M. *Geography: Its Scope and Spirit*. ("Social Science Seminar Series.") Columbus, Ohio: Charles E. Merrill Books, 1965.

Hartshorne, Richard. *The Nature of Geography: A Critical Survey of Current Thought in the Light of the Past*. Lancaster, Pa.: The Association of American Geographers, 1939. First published in the *Annals of the AAG*, XXIX, Nos. 3-4 (1939), 171–645.

———, *Perspective on the Nature of Geography*. ("Association of American Geographers Monograph Series," No. 1.) Chicago: Rand McNally, 1959.

Hoosan, David J. M. "The Distribution of Population as the Essential Geographical Expression," *Canadian Geographer*, No. 17 (Nov., 1960), pp. 10–20.

James, Preston E., and Clarence F. Jones (eds.). *American Geography: Inventory and Prospect*. Syracuse: Syracuse Univ. Press for the Association of American Geographers, 1954.

———. *New Viewpoints in Geography*. Twenty-ninth Yearbook of the National Council for the Social Studies. Washington, D. C.: National Council for the Social Studies, 1959.

McCarty, Harold H., and James B. Lindberg. *A Preface to Economic Geography*. Englewood Cliffs, N.J.: Prentice-Hall, 1966.

Thomas, William L. (ed.). *Man's Role in Changing the Face of the Earth*. Chicago: Univ. of Chicago Press, 1956.

Zelinsky, Wilbur. *A Prologue to Population Geography*. ("Foundations of Economic Geography Series.") Englewood Cliffs, N.J.: Prentice-Hall, 1966.

———. "Rural Population Dynamics as an Index to Social and Economic Development: A Geographic Overview," *The Sociological Quarterly*, IV, No. 2 (Spring, 1963), 99–121.

PART ONE

THE RESOURCE BASE

THE resource base, or physical environment, of New York State varies considerably from place to place, and this variation is reflected in the use New Yorkers have made of their state. The Adirondacks, for example, have sufficiently rough terrain, vigorous climate, poor soils, and an off-the-beaten-track location to make agriculture and the development of cities extremely difficult; on the other hand, they have supplied forest and mineral products, and are attractive for many types of recreational activities. The large lowland stretching from Utica to Buffalo and lying between Lake Ontario and the Appalachian Upland has fostered growth of great urban populations, and the suitable soils and climate there sustain perhaps the state's best agricultural area.

The Hudson-Mohawk route from the interior of the continent focuses on an excellent natural harbor at the mouth of the Hudson River. Because of the route's location it has contributed substantially to the growth of the greatest urban system in the nation, New York City. Steep slopes, unresponsive soils, short growing seasons, and poor drainage make it difficult to farm in the Appalachian Upland or in the Tug Hill plateau.

Without in any way overstating the importance of the resource base or understating the significance of man, it is apparent that human use of the land and the geography of economic development respond to the resource base. This part of the volume presents the geography of five important elements of the resource base: land forms, climate, water, vegetation, and soils. (Minerals, a sixth element of the resource base, are treated under the subject of mining in Chapter 10.)

CHAPTER 1   GEORGE B. CRESSEY

Land Forms

THE SURFACE of any area is the stage upon which the human drama is played. To know how land forms vary from place to place thus is an aid to understanding many aspects of the geography; to know why they vary increases understanding. In this chapter land form regions are identified and described, and a brief review of the fascinating geologic past is given. Also, certain interesting and well-known features are singled out for special attention.

The Glacial Story

The latest episodes in land form development are often the most interesting and the easiest to identify. Thus, in trying to understand New York State's surface features it may be appropriate to begin with events of the last hundred thousand years and postpone until later in the chapter the story of the earlier millions of years. This most recent geologic time has been the age of glaciers.

ICE FROM THE MOUNTAINS AND ICE FROM THE NORTH

New York now has heavy snow every winter, but it all melts during the following summer. On the top of Mount Marcy pockets of snow may last into July; with an abnormally severe winter and an exceptionally cool summer, some of the snow from one winter might carry over into the next winter, thus creating a permanent snow field. The compaction and recrystallization of this snow over a period of many years would produce glacial ice.

There is only limited evidence as to the date when the glacial stage began in New York State, but probably it was between 50,000 and 100,000 years ago. Its disappearance, more easily identified, occurred not more than 8,000 or 10,000 years ago. During that interim climatic conditions suitable to glacial development prevailed. The higher peaks of the Adirondacks, and subsequently the Catskills, developed mountain glaciers, with long tongues of ice projecting down the valleys toward the lowlands. Much larger quantities of ice formed on either side of Hudson Bay in Canada. As these great Canadian ice masses became thicker and thicker, eventually developing broad domes with depths of a mile or two, the margins moved outward, burying the land to the south.

The glaciers of New York thus had two original sources: highlands within the state and the much more important sources northward in Canada. As glaciation developed, these two sources of ice merged and one continuous continental glacier covered the state, overtopping the highest mountain, Mount Marcy, and producing a nearly featureless ice surface, sloping slightly to the south. All of this, if it could have been viewed, certainly would have looked like present-day interior Greenland.

EVIDENCE OF GLACIATION

Scattered about the surface of the state are stones that are "out of place" or do not appear to "belong" where they are found. Most of the state is underlain by sedimentary rocks such as limestone, shale, or sandstone, yet on the surface one finds pebbles and even boulders of igneous and metamorphic rocks quite unlike any nearby bedrock. The closest location of origin for these pebbles and boulders may be identified as the Adirondacks or even Canada. If this be the case, it is clear that these are glacial importations, correctly called *erratics*.

Elsewhere there are unique hills and other topographic features which are clearly not the result of running water or wind. Many of the present stream patterns, too, appear abnormal and can be explained only by glaciation.

Continental glaciers do most of their work—especially depositional work—within ten miles of their margins. Thus, it may be assumed that the major glacial features of New York were formed at times when an ice front was nearby. Since the glacier slowly advanced across the state at least twice, and as slowly retreated each time, there were four instances when any given place was in the active zone.

The limit of a glacier is determined by the balance between supply of ice and melting. When more ice moves forward than melts, the front advances; where melting exceeds supply, the front retreats. In both cases, and at all times except with final stagnation, the marginal ice is in continual forward

GLACIALLY SCOURED PEAKS in the Adirondacks. Here is the high peak country of the Adirondacks. Mountain glaciers helped shape the peaks. More significantly, continental glaciers reached such thickness over New York that they completely buried and overrode even these highest promontories. Lack of soil and severe climatic conditions prohibit forest growth on the peaks. Photo shows the forests of Colden and Algonquin Mountains. *N.Y.S. Dept. of Conservation.*

movement. *Ice retreat* refers only to the northward shift of the terminus, not to a reversal in the ice movement. Probably most of the depositional forms we now see date from the last waning stage or the last retreat, for in its vigorous activity at this time it erased many souvenirs of earlier movements. Although in the Midwest four separate glacial advances and retreats are recognized, there is certain evidence of only two such cycles in New York. However, the two invasions and retreats were so complex, with periods when melting exceeded advance and other times when the front remained stationary for many years, that there was ample time and opportunity for a rich variety of glacial features to be developed. Only a small area in the southwestern part of the state and parts of Long Island and Staten Island remained free of ice at the time of maximum development.

The most widespread souvenir of glacial deposition is *ground moraine*, or *till*, a veneer of unassorted clay, pebbles, and stones which covers most of the state. In places it may be 100 feet thick or more; elsewhere it is thin or absent. Nowhere is it stratified or does it show evidence of being laid down by running water. Ground moraine is material carried along by the advancing glacier and dropped wherever the ice melted. Some of the debris has been moved only a few miles, bulldozed by the ice front; some has been transported hundreds of miles as is shown by crystalline rocks carried southward from Canada.

Most ground moraine surfaces are chaotic and irregular and without organized slopes. Where overloaded ice overrode its own till sheets, the debris may have been shaped into elongated half-egg-shaped forms, the most symmetrical of which are called *drumlins*. Nowhere in the world are these features better developed than in the area between Rochester and Syracuse. Drumlins are like an inverted boat, with a ridge line parallel to the direction of ice movement. Usually the north side, against which the ice pushed, is steep, while the crest line slopes gradually to the south. Drumlins range in height from 20 to 100 feet or more, and in length may reach several hundred yards. Their composition is similar to that of ground moraine.

Even Lake Ontario and the drumlins to the south are related genetically. As the glacier moved southward from Canada, it dug into the soft shales and sandstones which underlie Lake Ontario, converting what was probably only the westward extension of an ancient St. Lawrence Valley into a shallow lake basin. From this excavated depression the ice moved uphill across a gently rising country to the south. The upper surface of the ice had a southward downhill slope, for otherwise the ice would be stagnant; but the bottom of the ice was moving uphill, away from the deep center of accumulation. Due to the scouring action of the ice, its lower portion had incorporated excessive amounts of ground-up rock. It may be that the bottom 500 feet were half-rock and half-ice; at any rate, the basal part of the glacier was overloaded, and as it advanced across the drumlin area mentioned above some of its load adhered to earlier ground moraines. Drumlins, thus representing a plastering-on process, formed as the ice overrode and shaped its own debris.

SAND AND GRAVEL DEPOSITS are part of the glacial heritage. In many places all over the state melt waters from glaciers laid down valuable sand and gravel beds. In some instances the deposition took the form of deltas in temporary lakes; elsewhere it is associated with kames, valley trains, or ancient beaches. A farm such as this one in Tioga County has an added source of income as road-building or construction results in a local market for the sand and gravel. *N.Y.S. Dept. of Commerce.*

The volume or weight of glacial debris moved within New York State is incalculable, but the ton-miles certainly reach impressive figures. Some of this load was deposited beneath the ice, as with drumlins or ground moraine; but some was carried within the mass of the ice-rock mixture and accumulated at the edge or front of the glacier where the ice melts. Every ice margin which marked the more or less stationary edge of the glacier for a number of years is bordered by a line of debris known as a *terminal moraine*. Strictly speaking, there is but one true terminal moraine, that which represents the maximum limit of glaciation; all others are recessional moraines, though identical in their formation. These linear moraines are dumps made up of whatever the ice carried. Some of the material is water-laid and thus sorted, but most is like ground moraine without stratification.

The true terminal moraine in eastern United States lies in Pennsylvania and Long Island, but several recessional moraines cross central New York State. One of the best developed winds across the southern hills of the Finger Lake area. Another, farther north, extends from Lake Erie to the St. Lawrence Valley roughly along the edge of the Ontario Lowland. Shorter recessional moraines have been traced in various parts of the state.

Great amounts of water poured out from the ice in summer, for temperatures were well above freezing even tens of miles behind the ice margin. These glacial rivers carried a heavy load of debris, laying down outwash plains, as across Long Island, or narrow valley trains, as in the hills of central New York. Blocks of ice were sometimes covered by these materials, and after melting depressions called *kettles* resulted. Hills composed of stratified outwash materials between kettles are known as *kames*.

Where the ice front faced a southward slope, as toward the Hudson, Susquehanna, or Ohio rivers, glacial melt waters could easily drain away. Once the ice front receded north of a drainage divide, however, and while the St. Lawrence Valley was still blocked by ice, lakes south of the ice margin developed, as in enlarged Lake Ontario, or in ancient Lake Albany, or the Champlain Sea (Fig. 7). *Lacustrine* (deposited in lakes) deltas and sand plains as well as beaches here replace the *fluviatile* (deposited on land) outwash features of southern New York. Tremendous quantities of lacustrine sands almost completely surround the Adirondacks, being particularly prevalent in the area west of Albany. Both lacustrine and fluviatile features contain important quantities of sand and gravel so necessary to road-building, concrete work, and the like.

After the ice front retreated north of the Susquehanna divide, but before it reached the Ontario Plain, a series of temporary channels cut across the previous north-south drainage to carry melt waters eastward to the Mohawk-Hudson river system. These "cross channels" carried water from glacial Lake Erie to the Hudson; several are well developed in the hills near Syracuse.

Where glacial ice moves rapidly, it may be a powerful agent of erosion. As indicated above, this is especially true near the margin of continental ice sheets and may be especially important where moving ice is concentrated in valleys. Valley glaciers thus cut the spectacular feature known as Yosemite Valley in California. As the ice sheet advanced across the Ontario Plain and encountered the northward-facing escarpments of the Appalachian Upland, it found a number of preglacial stream-cut valleys roughly aligned with the direction of ice movement. While the ice front was still near the escarpment, long lobes crowded tens of miles down these valleys, thus concentrating the flow and providing the necessary thickness for a deep scouring

STAGES OF GLACIAL DEVELOPMENT

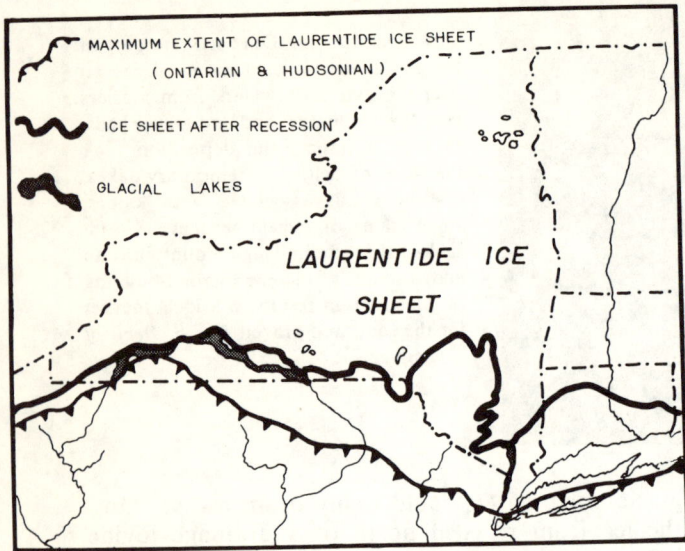

It is apparent here that only southern Long Island and south-westernmost New York escaped glaciation. Ponded melt water just beyond the receded ice front flowed south into the Allegheny and Susquehanna rivers, as well as down the Hudson Valley. The highest peaks in the Adirondacks, which were completely covered during the maximum ice advance, are now shown as <u>nunataks</u>, or small protuberances, through the ice.

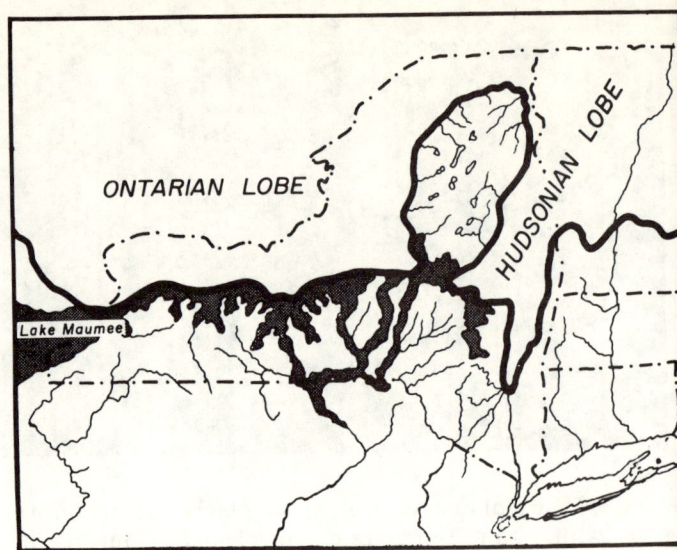

Glacial Lake Maumee in the Erie basin is now draining to the Wabash and Mississippi rivers. The ice has receded to the northern edge of the present-day Finger Lakes, but drainage is still south through the many valley outlets. Numerous glacial lakes have developed adjacent to the ice front and along the northern edge of the Appalachian Upland. A large expanse of the Adirondacks is now exposed, although the higher elevations support small mountain glaciers.

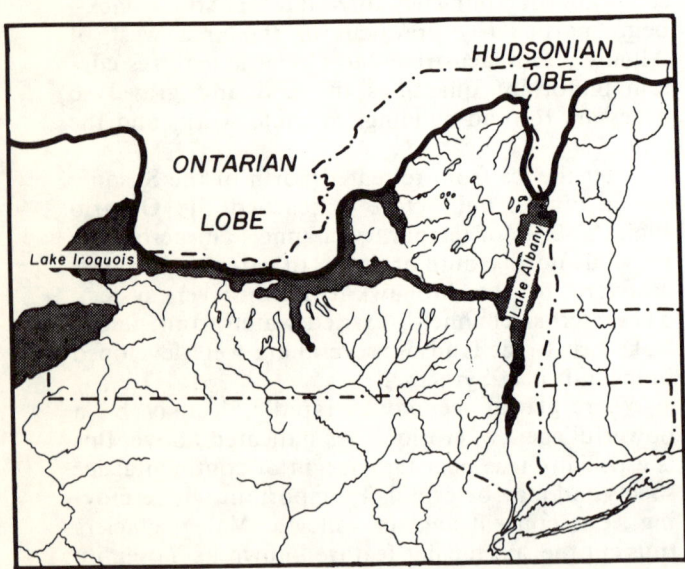

Glacial Lake Iroquois now occupies the Ontario basin and melt waters are flowing east through the Mohawk Valley to Lake Albany and from there down the Hudson River. Water from the other Great Lakes is now flowing north over a young Niagara Falls and east through the Mohawk outlet. A readvance of the ice that occurred after this stage temporarily covered Lake Iroquois again.

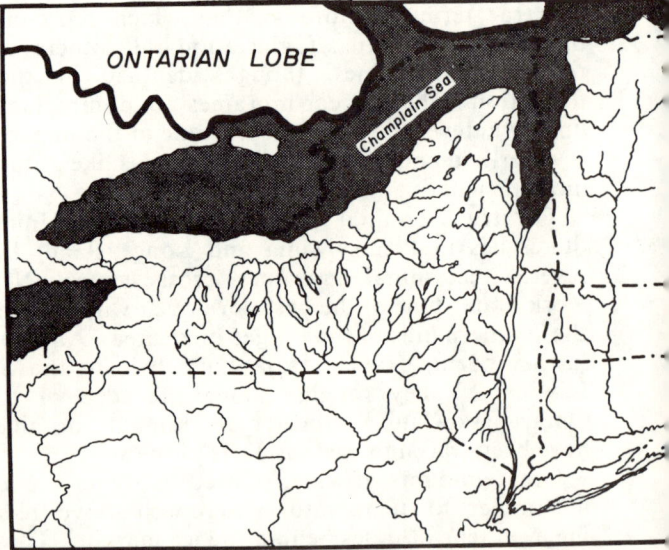

The Ontario ice lobe is retreating northward and the Hudsonian ice lobe has almost disappeared, freeing a passageway to the Gulf of St. Lawrence. The Champlain Valley and Ontario basin are depressed areas filled by the Champlain Sea. In viewing the position of the many glacial lakes and water courses during these stages, it becomes clear why such wide distribution of lake sediments, beach and delta deposits, and widespread occurrences of water-deposited sands and gravel exist in the state.

Adapted from: H.L. Fairchild, "Stages in the Waning of the Ice Sheet in New York State," <u>New York State Museum Bulletin</u> 160 (1909); and R.F. Flint, <u>Glacial and Pleistocene Geology</u>, New York: Wiley, 1957.

Fig. 7

action. In several instances ice deepened the previous valleys by 1,000 feet or more. Deep valleys like that along Onondaga Creek south of Syracuse or those crossed in numerous places by U.S. Highway 20 in central and western New York were formed. A number of these valleys were either sufficiently scoured out or dammed on one end so that lakes resulted. These are the Finger Lakes. Lake Cayuga, 381 feet above the sea, has its bottom 54 feet below sea level. The glacial erosion that formed Lake Seneca is even more impressive. This lake, 444 feet above the sea, reaches a depth of 174 feet below sea level, and a drilling 600 feet below its bottom did not reach the rock floor of the old valley. In all there are eleven Finger Lakes in the state, identified by their finger-like shape and origin in glacial scour.

Before the Glaciers Came

The history of the glacial epoch has been told in terms of a few tens of thousands of years, but some rocks in New York State are over a billion years old. The oldest rocks occur in the Adirondacks, near the New England border, and in the lower Hudson Valley.

About 550 million years ago much of eastern North America was a nearly featureless plain underlain by these ancient igneous and metamorphic (crystalline) rocks. With the exception of the Adirondacks much of New York State seems to have been a low-lying, gently rolling expanse of land only slightly above sea level. But at about this time the area sank and the sea spread across it. The Adirondacks and much of New England escaped this submergence and, as a matter of fact, these areas subsequently remained above the sea.

A sea of varying depth lay over the area for most of the following 325 million years, during which time great quantities of sediment were washed into it from the Adirondacks, New England, and parts of Canada. The result was a vast accumulation of bed after bed of sand, mud, lime, and salt many thousands of feet in total thickness. Salt deposition undoubtedly occurred under shallow, lagoon-like conditions and in periods of regional aridity. The bottom of the sea sank progressively to make room for this vast accumulation.

These sediments hardened into sandstones, shales, limestones, and even rock salt. In the rocks are numerous remains of marine life which permit easy identification of geologic age and quick comparison with sedimentary rocks elsewhere around the world. These are the rocks of the Paleozoic Era (see Table 2).

Most of the sedimentary rocks of New York State belong to the middle periods of the Paleozoic Era, especially the Ordovician, Silurian, and Devonian. During these periods broad interior seas extended from the crystalline land mass of the Adirondacks and New England westward into the Mississippi Valley. Many thick formations of sandstone, deposited near shore, merged westward into shales and then limestone in the clearer water far from the ancient shore line.

All of New York except the higher Adirondacks was submerged during the Ordovician Period, with limestones being the principal rocks formed in the widespread seas. The period was brought to a close by the Taconic mountain-building episode in the eastern edge of the state.

The Silurian Period was also a time of submergence, with the accumulation of several thousand feet of sandstones and limestones. Among the unique beds is the salt formation which outcrops in Syracuse and continues buried beneath much of southwestern New York. Devonian beds, the most widespread in the state, occur on the surface in the Catskills and all over southwestern New York. Their maximum thickness is 4,000 feet.

The Paleozoic Era ended with the Appalachian Revolution (220 million years ago), which formed the Appalachian Upland to the south and permanently raised most of the state above sea level. Triassic sandstones and lava flows exposed in an area next to New Jersey represent continental deposits rather than marine formations. Cretaceous sediments are present only in Long Island and Staten Island. Elsewhere the state remained above the sea and was the site of erosion rather than deposition. During this time a widespread peneplain developed across much of New York. (A *peneplain* forms when the forces of erosion continue uninterrupted for a long enough period to reduce a land area to almost a plain surface.) This was then upwarped at the end of the Mesozoic Era, and our present scenery is the result of the dissection by water and ice of this ancient erosion surface.

Figure 8 indicates the distribution of underlying rock types over the state today. These, of course, are commonly covered with glacial till but are exposed to view in enough places to facilitate their mapping and study. Roughly a quarter of New York—namely the Adirondacks and smaller areas in the southeast—is underlain by crystalline (igneous and metamorphic) rocks. Sandstones occupy nearly another quarter, as in the Catskills and the Tug Hill area. These erosion-resistant sandstones have been responsible for "holding up" the highland elevation of the Catskills and Tug Hill. Limestones are of lesser extent but outcrop in long strips, especially around the margins of the Appalachian Upland where they form escarpments. These limestones are widely used in building, highway construction, and the cement industry. The remainder of the state, roughly half of the total area, is covered by shales or by shaly sandstones.

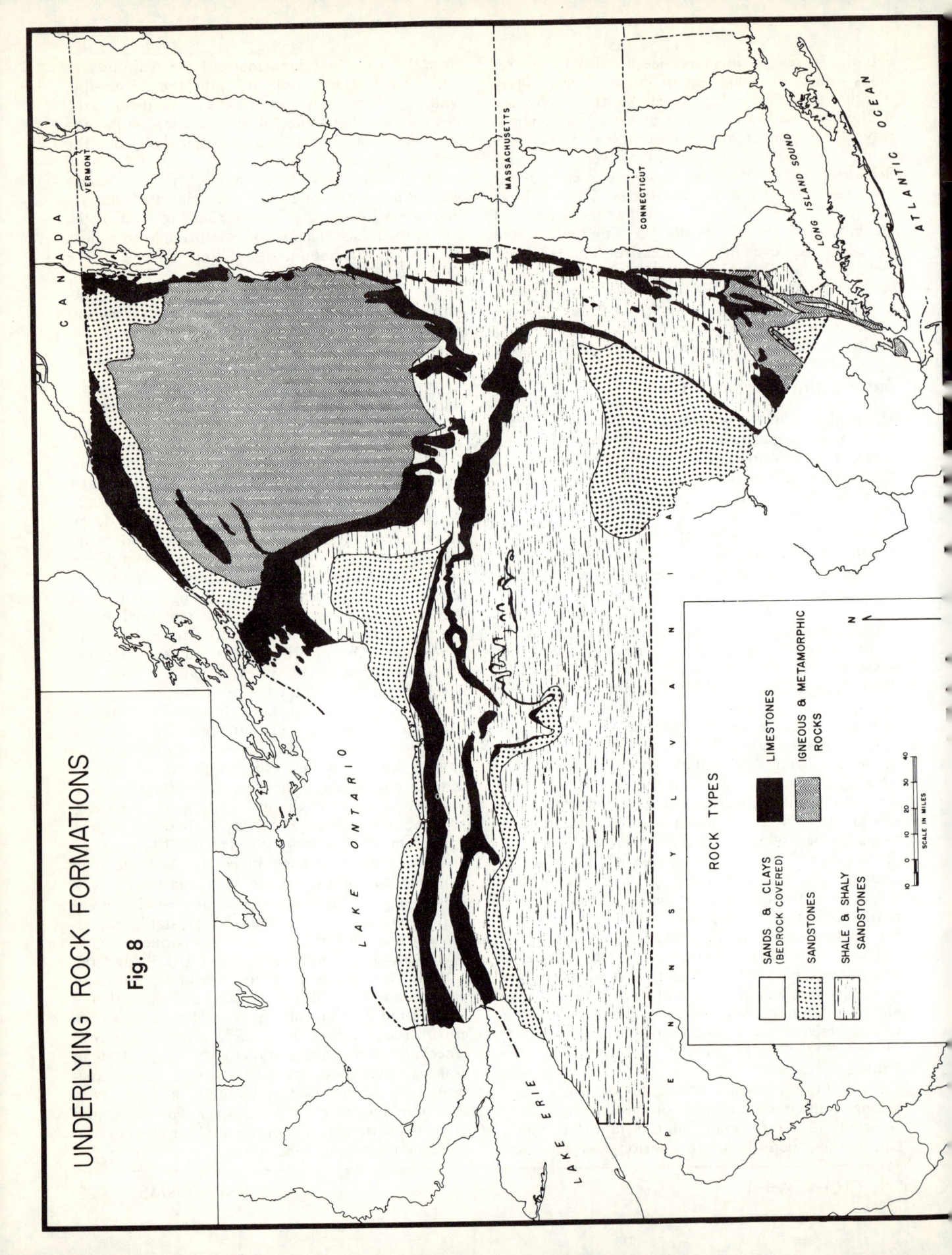

Land Form Regions

As has been pointed out, during hundreds of millions of years of geologic history there occurred great upward and downward movements of land, periodic encroachment of the seas, deposition of thousands of feet of sediments in the seas, and the sculpturing of the land by glaciers and running water. The result is a diversity in land forms equaled by few of our states. Within the boundaries are portions of many of the major land form regions of eastern North America. Nine regions with twenty-eight different subdivisions are shown in Figure 9. In each case the classification represents a distinct type of land form, resulting from distinct geologic structures and erosional development.

ADIRONDACK UPLAND

The Adirondack Upland consists of an ancient domed Pre-Cambrian erosion surface, perhaps even a peneplain, with *monadnocks* (erosional remnants) forming the higher, more rugged features such as Mount Marcy. Ancient crystalline rocks similar to those of the Canadian Shield in Canada prevail. Intense glacial scour has removed most of the original soil and, in general, smoothed the land surface. Some of the eroded material now chokes the pre-glacial valleys, deranging the stream patterns and producing numerous lakes. Lacustrine deposits occur around the margins of the upland. Three subdivisions are identified.

The Adirondack Mountain Peaks subdivision includes the highest and most rugged section. Summit elevations are above 3,000 feet and *local relief* (difference in elevation within 25 miles) exceeds 2,000 feet. Mount Marcy (5,344 feet) and Algonquin Peak (5,114 feet) are the highest points.

The Adirondack Low Mountains subdivision surrounds the Adirondack Mountain Peaks. Some summits here are over 2,000 feet, but local relief is generally below 1,000 feet. Hundreds of glacial lakes such as Saranac, Tupper, Raquette, and the Fulton Chain make up in beauty for what is lacking in local relief.

The Western Adirondack Hills subdivision forms a broad zone of foothills, in part covered by sandy lacustrine deposits laid down in glacial lakes. Some of the state's richest mineral deposits—such as iron, lead, zinc, and talc—are mined from the ancient crystalline rocks in the southern part of St. Lawrence County.

ST. LAWRENCE–CHAMPLAIN LOWLANDS

This region is part of the lowland corridor extending northeastward from the Great Lakes to the sea. Limestone and sandstone make up most of the bedrock, but over the bedrock are a variety of marine and lake deposits. In places these deposits form good soils for agriculture; in other places they are so sandy as to be worthless. There are three subdivisions.

The St. Lawrence Marine Plain is a flat to gently rolling strip along the St. Lawrence River. Marine clays predominate. The St. Lawrence Hills sub-

TABLE 2
GEOLOGIC TIME TABLE

Era	Period	Epoch	Duration (Million Years)	Began (Million Years Ago)	Mountain-Building Disturbance
Cenozoic	Neogene	Pleistocene Pliocene Miocene	1 11 13.7	1 12 25.7	
	Paleogene	Oligocene Eocene Paleocene	8.3 21 15	34 55 70	Laramian Revolution
Mesozoic	Cretaceous Jurassic Triassic		65 45 40	135 180 220	Nevadan Orogeny Palisades Disturbance
Paleozoic	Permian Pennsylvanian Mississippian Devonian Silurian Ordovician Cambrian		55 55 25 55 20 60 80	275 330 355 410 430 490 570	Appalachian Revolution Hercynian Chain (Europe) Culmide Disturbance (Europe) Acadian Mtns., and Bretonian (Europe) Caledonian Disturbance (Europe) Taconic Disturbance Vermont Disturbance
Pre-Cambrian			2,730	3,300*	Killarney Revolution (Canada)

*Oldest known rock is 3.3 billion years old; molten body of the earth probably dates to 4 billion years ago.

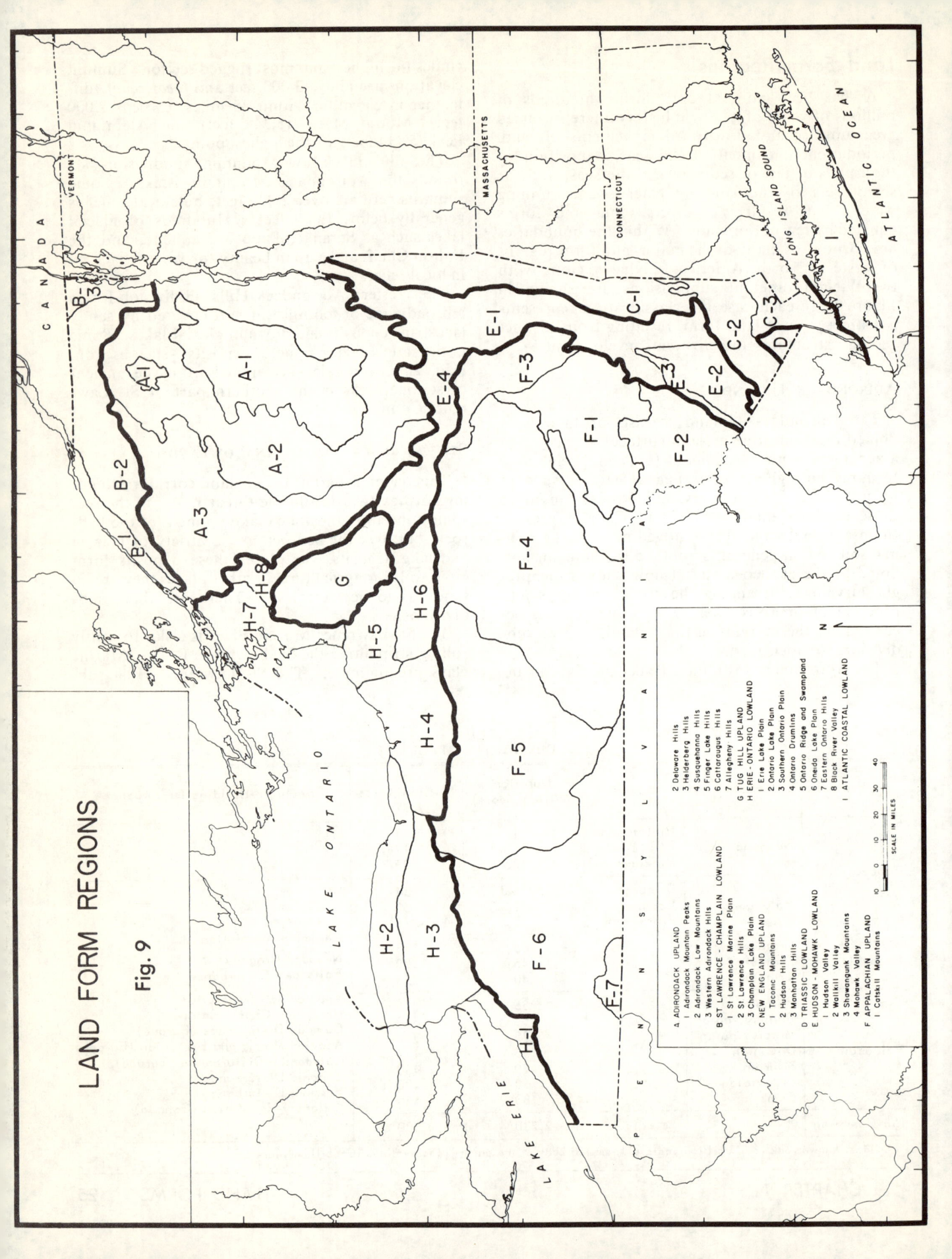

Fig. 9 — LAND FORM REGIONS

ADIRONDACK OVERVIEW. Not all of the Adirondacks are rugged. Much of the region is characterized by low mountains or hills. Forests predominate except around settlements. Here the village of Saranac Lake gives evidence of being supported by tourism and wood-using industries. *N.Y.S. Dept. of Commerce.*

WHITEFACE HIGHWAY. The Adirondacks are not excessively penetrated by roads but this road up Whiteface Mountain (4,867 feet) provides easy access to some of the best Adirondack scenery in the state. View is in a southeasterly direction. *N.Y.S. Dept. of Commerce.*

division contains a bit rougher country than the Marine Plain and is situated between it and the Adirondacks. It is underlain largely with sandstones covered with glacial drift. The Champlain Lake Plain, low and relatively flat, has limestone as its principal bedrock. Marine clays cover the limestone.

New England Upland

The New England Upland is an area of ancient crystalline rocks which barely reaches the New York border in the form of the Taconic Mountains and then extends south across the lower Hudson Valley into Pennsylvania. Manhattan Island, too, is in the region. This is part of the upland area that supplied so much sedimentary material to the Paleozoic seas of New York State. It has three subdivisions.

The Taconic Mountains extend in a north–south direction along the eastern border of the state. Their picturesque ruggedness reaches a maximum across the border in Vermont. Local relief exceeds 2,000 feet, valleys are narrow, and slopes are steep. Rocks are of many varieties, but all are very old, faulted, and altered by stress and strain in the earth's crust.

The Hudson Hills subdivision, commonly called the Highlands of the Hudson, is composed of crystalline rocks. The surface has been eroded by running water into rather rugged terrain. Summit levels are about 1,000 feet above the Hudson River, which flows through the area in a steep-sided gorge extending from Storm King, past West Point, to a point just north of Peekskill.

The Manhattan Hills, low in elevation, developed on complex ancient rocks. They include Manhattan Island and most of Westchester County. From the latter comes a substantial part of New York City's water supply.

Triassic Lowland

The Triassic Lowland is a small, low, rolling area underlain with shales, soft sandstones, and hard lavas. It is one of a few areas in the state that have

THE AUSABLE RIVER AND WHITEFACE MOUNTAIN. The Ausable River, one of the most beautiful streams in the Adirondack Region, is famous for trout fishing. Whiteface Mountain wears a cap of snow in this fall season. *N.Y.S. Dept. of Commerce.*

THE LAKE PLACID AREA. This panorama from the top of Whiteface Mountain shows Lake Placid and hundreds of square miles of Adirondack Country. The lakes, mountains, and forests have attracted tourists and summer vacationists for generations and should become progressively more important for this kind of use as the population in the Northeast rises. *N.Y.S. Dept. of Commerce.*

experienced post-Paleozoic sedimentation, the Triassic Period being the oldest of the Mesozoic Era. It is bounded on the northwest by the Ramapo Mountains, which are part of the Hudson Hills, and its eastern edge is composed of the lava trap rocks that form the Palisades along the Hudson River.

HUDSON–MOHAWK LOWLAND

This is a region of particular significance because it has provided the easy access route between New York City and the interior of the country. Its soft sedimentary rocks and overlying glacial deposits have been eroded so as to result in a variety of terrain. Four subdivisions are discernible.

The Hudson Valley forms a 10- to 20-mile-wide lowland lying between the Helderberg Escarpment and Catskills on the west and the Taconics on the east. North of Albany it is wide and flat, covered with glacial lake deposits. Farther south the valley narrows and contains a complex of hills and terraces underlain with highly folded sedimentary rocks. From the standpoint of rock structure this subdivision is a northward extension of the Ridge and Valley Province occurring in Pennsylvania and southward.

The Wallkill Valley is a broad, open valley covered with glacial drift and drained by the Wallkill River, which flows northeast to the Hudson.

CHAPTER 1   LAND FORMS

St. Lawrence Valley. View from near Fort Covington looking northwest over the St. Lawrence Marine Plain. The island-studded St. Lawrence is entirely in Canada at this location, the boundary being several miles this side of the river. Large areas of trees usually indicate either poor drainage or sandy soil. *N.Y.S. Dept. of Commerce.*

The Champlain Lake Plain narrows to a thin belt of relatively level farm land west of Lake Champlain and east of the Adirondacks. Here, near Crown Point, that thin belt ends as the mountains in the distance reach the shore of the lake. *N.Y.S. Dept. of Commerce.*

THE HUDSON HILLS or Highlands near Bear Mountain. Where the Hudson River breaks through the crystalline Hudson Hills is one of the most beautiful stretches of river valley in America. Ocean-going vessels navigate the river here on their way to Albany many miles upstream. *N.Y.S. Dept. of Commerce.*

Structurally it is a southern extension of the Hudson Valley.

The Shawangunk Mountains, although not a lowland, are structurally a subdivision of the Hudson–Mohawk region. They form a steep-sided ridge rising 1,000 feet on the west side of the Wallkill Valley. Like many of the ridges in the Ridge and Valley Province to the south, the Shawangunks are composed of sharply folded sediments. A cap of hard sandstone conglomerate has provided protection against erosion.

The Mohawk Valley subdivision is an east–west lowland drained by the Mohawk River and lying between the Adirondacks on the north and the Appalachian Upland on the south. Although the Mohawk flows in a fairly narrow inner valley, the subdivision is generally 10 to 30 miles wide. Underlain by soft shales, it has been eroded to a depth of perhaps 1,000 feet below the higher country to the north and south. At Little Falls it narrows to a deep gorge where the Mohawk River has cut its way through a preglacial drainage divide. Since earliest settlement began, this has been a major artery of commerce, containing first the Erie Canal, then the railroad, and later the New York State Barge Canal and major highways, culminating in the construction of the New York State Thruway in the 1950's.

APPALACHIAN UPLAND

This is the largest of the land form regions in New York, occupying nearly half the state. The entire area is underlain with Paleozoic sedimentary rocks dipping slightly toward the south and west. All but the extreme southwest has been glaciated and, as in the case of the Finger Lakes, deeply scoured.

Around the northern and eastern margins a series of resistant beds form rather pronounced escarpments, the most prominent of which is the Helderberg Escarpment southwest of Albany.

Severe erosion by streams as well as glaciers has produced a highly variable surface. Conditions range from the rolling hills near the Finger Lakes to the low mountains of the Catskills. Local relief in some areas is less than 500 feet; in others, considerably more than 3,000 feet. Much of the upland carries traces of a Cretaceous peneplain, and the Catskills may well be monadnocks rising above that peneplain. Their existence is due to resistant underlying sandstones. Seven subdivisions exist.

The beautiful Catskill Mountains have summit elevations from 2,000 feet to about 4,000 feet above the sea. The resistant sandstones underlying the area may be delta deposits from a Paleozoic sea. Both stream and glacial erosion have resulted in deep dissection. A striking characteristic of Catskill topography is its coarse texture. Valleys are relatively few, and the intervening masses correspondingly bulky. The paucity of small tributaries is thought to be due to the permeability of the sand-

THE APPALACHIAN UPLAND. With its tree-clothed slopes and cultivated valleys, the Appalachian Upland is attractive to look at but is often a difficult place in which to make a living. Some farms succeed, but many have failed in recent years. The result is an increasing acreage of brush and forest land and a decreasing of cultivated fields. *N.Y.S. Dept. of Commerce.*

CHEMUNG VALLEY near Waverly. Flat-bottom lands along the Chemung River respond to modern agricultural needs, but the surrounding uplands have largely grown back to woodland. Note the generally level appearance of the Appalachian Upland skyline. *N.Y.S. Dept. of Commerce.*

stones, which causes water to soak in rather than run off.

The Delaware Hills subdivision occurs in the upper reaches of the Delaware River drainage basin. In contrast to the Catskills it has a local relief of only a few hundred feet.

The Helderberg Hills form the northeastern corner of the Appalachian Upland region. Rising nearly 2,000 feet above the Hudson and Mohawk valleys, the Helderberg Hills terminate in sharp northeastward-facing scarps. These scarps are the result of very resistant capping limestone.

The Susquehanna Hills subdivision is drained largely by the upper Susquehanna River and its tributaries. Shales are the most common bedrock, but sandstones and limestones are present too. Divides are at heights of 1,700 to 2,100 feet, and, although the surface of the plateau appears rather even when viewed from a distance, close examination shows but limited level land. Valleys are numerous and narrow enough to make farming difficult.

The Finger Lake Hills subdivision occupies the central and lowest portion of the Appalachian Upland and, of course, is characterized by deeply glaciated valleys, some of which contain the famous Finger Lakes. The uplands between the lakes are relatively level and afford considerable good land for farming.

The Cattaraugus Hills constitute a relatively flat-topped upland with deep intervening valleys. Drainage is in three directions: to the Ohio River system and the Gulf, to the Susquehanna system and the Atlantic, and to Lake Ontario via the Genesee River.

The Allegheny Hills subdivision, now largely in Allegany State Park, was not glaciated. Its terrain is thus a bit more angular, its valleys less irregular in direction, and its bedrock more exposed than in other parts of the state.

TUG HILL UPLAND

The plateau-like Tug Hill Upland is an outlier of the Appalachian Upland Region. It drops off 1,800 to 2,000 feet to lowlands in all directions and is underlain by Paleozoic sandstones, limestones, and shales which dip gently westward. Bad drainage, poor soils, and excessive snow have limited agriculture and caused most of those farms which were established to be abandoned. It is one of the least settled parts of the state today.

ERIE–ONTARIO LOWLAND

The Erie–Ontario Lowland is a region of low relief lying just south of the two great lakes. In part it is on featureless old lake bottoms formed beneath glacial lakes; elsewhere there are rolling hills representing ground moraine deposits. A unique aspect is the drumlin belt between Rochester and Syracuse. Eight subdivisions are cited.

The Erie Lake Plain forms a narrow strip of land rising but a hundred feet or so above Lake Erie and lying between it and the Cattaraugus Hills. It exhibits beaches and other features of higher postglacial levels of the lake. The area is particularly well known for its grape production.

The Ontario Lake Plain, like the Erie Lake Plain, is dominated by lacustrine features. It is bounded on the south by the Niagara limestone escarpment. It, too, is known for specialty crops, particularly fruit.

The Southern Ontario Plain is a rolling area covered with glacial drift and lying between the Ontario Lake Plain on the north and the Cattaraugus Hills on the south. Much of its western portion is now occupied by the rapidly expanding Buffalo Urban System.

The Ontario Drumlins subdivision is almost everywhere dominated by the half-egg-shaped glacial features called drumlins. There are literally thousands of drumlins, in places so close together that they give the area a distinctly hilly appearance. The Syracuse Urban System occupies its southeastern portion.

The Ontario Ridge and Swampland, lying eastward from the city of Oswego, is a poorly drained area characterized by many swamps and ridged ground moraine. It has proved generally resistant to economic development.

The Oneida Lake Plain, a nearly featureless plain south of Oneida Lake, contains broad swamps and mucklands.

The Eastern Ontario Hills subdivision includes an area of low hills composed of glacial drift. Notable sand dunes occur along the eastern shore of Lake Ontario.

The Black River Valley forms a lowland lying between the Tug Hill and Adirondacks and, as such, is important as an access route between the Mohawk Valley and the North Country. Once occupied by a glacial lake, the valley has many lacustrine sand deposits.

ATLANTIC COASTAL LOWLAND

The Atlantic Coastal Lowland occurs in the state only on Long Island and Staten Island. Here is found the terminal moraine of the great ice sheet, which in places rises over 300 feet above sea level. In fact, without this glacial "dump" most of Long Island simply would not be what it is today. Southward from the crest of the terminal moraine on Long

TYPICAL CATSKILL SKYLINE. As seen from the lowlands near Woodstock, this is reminiscent of Rip Van Winkle Country. Set aside as a state park, the Catskills will remain a public heritage to use and enjoy. *N.Y.S. Dept. of Commerce.*

THE DELAWARE HILLS near Deposit. Here, where the west branch of the Delaware emerges from the Catskills, local relief is substantial and the upland surface is generally in slope much too steep for agriculture. Farms in the valleys must compete with an ever growing demand for highway right of ways. Sometimes these right of ways will cut farm units in half, leaving both halves as uneconomic units. The highway here is newly completed Route 17. *N.Y.S. Dept. of Commerce.*

Island and sloping toward the sea is a broad outwash plain.

Selected Physiographic Features

The origins of fifteen well-known and interesting physiographic features in the state will now be summarized. If this exposure to land form interpretation whets the appetite, the reader may want to seek further information on these and other features by browsing through the list of selected references at the end of the chapter.

THE GREAT LAKES

Few topics are more interesting or have received more attention than the origin of the Great Lakes. Although Lakes Superior, Huron, and Michigan do not bound New York, they nevertheless affect Lakes Erie and Ontario as well as the Niagara and St. Lawrence rivers. In order to gain proper perspective in the study of the Great Lakes, one must regard conditions prior to, during, and after glacial advance.

Prior to the advance of ice lobes across this section of the continent, the topography had been much altered by erosion. In association with an upward movement of the land, stream action deepened and widened valleys in the area now occupied by the Great Lakes, but there appears to be no evidence that any major lakes were formed in these channels prior to being filled by glacial ice.

The lobes of ice that moved southward in the Pleistocene Epoch were definitely influenced by the preglacial erosion channels, and erosion of the underlying rock was carried to an unusual depth. There is little doubt of the tremendous force of this ice mass whose thickness was at least 1,000 to 2,000 feet thicker in the valleys than on the adjacent land ridges. With increased ice thickness in the valleys,

THE CATTARAUGUS HILLS near Hornell. Less rugged than near Waverly, the upland surface supports more successful farms. The best farm land, though, is still in the valleys. Settlements and transportation facilities tend to locate in the valleys, too. Hornell has traditionally been a major railroad center, as evidenced by the freight-marshaling yard in the foreground. *N.Y.S. Dept. of Commerce.*

THE DELAWARE VALLEY near Hancock. Although in the Delaware Hills, this area takes on many of the characteristics of the Catskill foothills. Little flat land is available in the valley, and the massive, rounded hills contain no farms. The future of the area, like that of the Catskills, would seem to be in tourism. *N.Y.S. Dept. of Commerce.*

THE ERIE-ONTARIO LOWLAND north of Syracuse. The lowland is generally rather good for agriculture, but residential, commercial, and industrial facilities find it attractive, too. Here Syracuse suburbs, including various warehousing and manufacturing firms, are superseding agriculture as principal users of the land. *N.Y.S. Dept. of Commerce.*

the rate of power for downcutting increased substantially (see Fig. 7).

In the final stage of glaciation, the ice was gradually reduced by melting and the lands became exposed once again. Many valleys were filled by glacial moraine and the great volume of glacial melt water was forced, in many cases, to develop new channels of egress, for their normal drainage pattern was blocked by the still present but now thinner ice mass to the north. Waters that had previously flowed to the St. Lawrence were forced to the southeast and east through the Susquehanna and Mohawk River valleys. Although the St. Lawrence drainage area was not a major drainage channel in preglacial times, it assumed major importance in postglacial times because of the depression of the land by the ice. When the St. Lawrence area became free of ice, the streams, whenever possible, returned to their former stream channels, but some developed new channels.

The tremendous weight of ice which extended from Long Island almost to the junction of the Mississippi and Ohio rivers compressed the underlying rocks. An elastic rebound of this rock compression has caused a minor tilting of the land, which is not everywhere uniform. There is evidence that the rebound became effective soon after ice removal. At the present time, the northern side of the Great Lakes, especially Lake Superior, is rising more rapidly than the south shore, causing a spilling effect of the water body on the south. The tilting of this land could conceivably eventually force a restriction of water flowing north from Lake Erie toward Lake Ontario. Thus, the future of Lake Ontario and Niagara Falls may depend upon structural changes in the earth as well as the rate of stream erosion. All of this will not happen overnight, however.

The process of headward erosion by the Niagara River continues to seek sea level in its ultimate aim to drain a portion of the Great Lakes, but large portions of the beds of all the lakes, except Erie, are below sea level, so there would still remain large lakes in the basins of Ontario, Michigan, and Superior, and a smaller remnant of Lake Huron would exist.

The Great Lakes and St. Lawrence Seaway, stretching a distance of 1,340 miles from Montreal to Duluth, constitute the world's largest fresh-water, navigable inland waterway. The Great Lakes region is now undergoing a major developmental transformation in order to make many of the "lake cities" major ports of entry for international transportation. Although the recent development of the St. Lawrence Seaway has failed to reap the expected economic revenue, the completion of five locks on the Welland Canal and an improved tax structure are expected to increase trade.

CHAPTER 1   LAND FORMS   37

The lakes have long been important in the transfer of iron ore, wheat, and wood products. It is expected that many new products formerly prohibited by the high cost of transportation will now arrive at midwestern ports from overseas. The major obstacle, a frozen waterway four or five months of the year, will continue to plague what would be a lucrative business proposition in a warmer climate.

The fishing industry, which for many years was of great importance, has now diminished to a minor occupation. The sea lamprey (eel) which destroyed great numbers of trout and whitefish, is now being scientifically exterminated by marine biologists. If this effort is successful and if catch allowances are held at a low enough level to allow for proper propagation, increased fish populations will develop.

The trend of boating and water recreation to which the Great Lakes contribute heavily does not seem to have reached a peak.

Niagara Falls

Niagara Falls is one of the most spectacular and thrilling highlights in sightseeing and for years has been one of the major natural attractions in this state. The thunderous roar and sight of millions of gallons of water rushing over a precipice some 167 feet high on the American Falls and 158 feet high on the Canadian Falls is long remembered. The American Falls are 1,000 feet wide and are separated from the 2,500-foot-wide Canadian Falls by Goat Island.

The American Falls are nearly at right angles to the river flow, and only 6 per cent of the water

The Erie Lake Plain south of Dunkirk. Dunkirk on Lake Erie (in the left background) is a small industrial city which has not spread extensively into the surrounding farm land. Vineyards may be seen occupying many of the fields. The New York State Thruway cuts across the photo from right to left. *N.Y.S. Dept. of Commerce.*

NIAGARA FALLS, a major tourist attraction, is unquestionably the most famous physical feature in the state. The American Falls are at the left and the Horseshoe or Canadian Falls are at the right. *N.Y.S. Dept. of Commerce.*

passes over them. The Canadian Falls have eroded in a more irregular form and have developed in an arc shape, hence the name Horseshoe Falls. Their greater width and larger water flow 'make them more spectacular.

The gorge of the Niagara River extends about 6½ miles downstream from the present position of the falls. This gorge was formed because of erosional differences in the underlying rocks and also because of the tilting of these sedimentary beds. Between the north end of Niagara Gorge and Fort Niagara, the river flows quite smoothly over a wide valley with no resemblance to the gorge, falls, and rapids found upstream.

The steady upstream migration of Niagara Falls has produced the 6½-mile-long Niagara Gorge. The structure of the rock formations being attacked by the running and falling waters may be seen in Fig. 10. Since the strata which determine the falls are dipping toward the south, that is, upstream, the height of the falls will diminish as they retreat farther upstream. In fact, by the time they have retreated 2 miles farther, they will be nothing but a series of rapids.

The migration rate at the Canadian Falls at this time is perhaps 4 or 5 feet per year, while at the American Falls it is but 1 foot per year. Scientists have calculated that a maximum of 8,800 years have elapsed since the river first started to carve its channel and that it will be another 75,000 years, discounting uplift, before the Niagara River will cut back into Lake Erie and cause it to be drained.

Skyscraper observation towers, helicopter flights, and elevator shafts to the base of the falls now accommodate some of the five million people who visit the area each year. The development of huge hydroelectric power plants has been the basis for considerable industrial development nearby.

NIAGARA FALLS

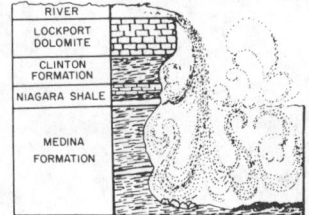

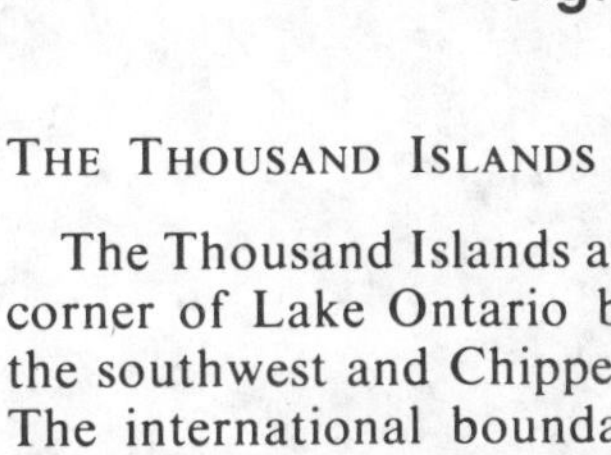

Cross section of Niagara Falls showing the undercutting action of the water in the less resistant lower beds. The overlying Lockport dolomite, being most durable, serves as a capping layer above the weaker formations. The rate of upstream migration of the falls depends on the rate of undercutting and subsequent falling of the capping layer.

Fig. 10

THE THOUSAND ISLANDS

The Thousand Islands are located at the northeast corner of Lake Ontario between Wolfe Island on the southwest and Chippewa Bay on the northeast. The international boundary line between Canada and the United States follows an irregular line through them. The tremendous volume of ice once covering this area caused a depression of the land. The combination of retreating ice and greatly depressed land allowed the melt waters to flood through and erode a previously highly irregular land mass. This land mass became the Thousand Islands passageway of the St. Lawrence River. A resistant rock belt, which extends north–south across the Thousand Islands area was irregularly dissected, and the higher elevations of this rock belt became the islands as we see them today. Contrary to popular opinion, it is unlikely that the present stream valley of the St. Lawrence ever served as a major outlet for the Great Lakes prior to the Ice Age. Interestingly enough, this lowland area, so depressed by the glacial ice, is now slowly rising to regain its former position. If this rising continues, it is possible that the waters in the far distant geologic future may be drained away so that the islands are islands no more.

The Thousand Islands area is one of the most colorful and attractive natural locales in New York. It has long been recognized as ideal for summer homes, boating, and fishing. The State of New York now has a number of parks and camp sites so summer visitors can enjoy the cool but clear summer weather. The islands are of great variety in size and shape; some are mere pinnacles, others are large. The winter months cloak the area in deep snow and ice, and only a small number of inhabitants occupy island homes throughout the year.

Since this area is in the path of the St. Lawrence Seaway, the shipping lanes are now used more than ever before. The sight of a passing freighter is no longer a highlight of the day.

THE THOUSAND ISLANDS AREA. As the broad St. Lawrence River emerges from Lake Ontario, it is dotted with some 1,700 islands. The area is a veritable boating wonderland. Most of the islands have summer houses and a few support year-round residences. The St. Lawrence Seaway channel follows the buoys in the right foreground. *N.Y.S. Dept. of Commerce.*

Mount Marcy

Mount Marcy, the highest point in New York State at an elevation of 5,344 feet, is located in the east-central part of the Adirondacks. The area surrounding it for an airline distance of 22 miles in any direction contains an abundance of mountains above 3,000 feet in elevation.

Mount Marcy is one of many monadnocks formed in this area as a result of the planation during the Paleozoic and Mesozoic eras. Most of these mountains are composed of resistant, dark, intrusive igneous rocks and younger, subsequently intruded, lighter igneous rocks.

At the beginning of the Cenozoic Era a great uplift occurred which slowly raised the monadnock-studded peneplain thousands of feet above its former elevation. The erosion initiated during that time has continued to the present, carving additional mountains, hills, and valleys. The huge amount of loose earth materials picked up by the ice worked like tools to leave scratch marks and gouges on the underlying rock. These scratches, or striations, now show the general direction of ice movement on the slopes of Mount Marcy and other exposed places. The many beautiful lakes surrounding Mount Marcy may also be attributed to the work of glaciers.

With its well-marked foot trails and camping facilities, Mount Marcy is a haven for vacationists. In the fall the weather turns brisk and the mountains are colored by unsurpassably beautiful autumn foliage. In winter the ski development at nearby Whiteface, with its double chair-lift system, the longest in the East, draws thousands of ski enthusiasts. The internationally famous Olympic Bobsled Run on Mount Van Hoevenberg at Lake Placid is in this area, too.

Lake George

Lake George, which trends in a north-northeast direction for a distance of about 32 miles, has a maximum width of 3 miles. The 187-foot depth of this lake is unusual in the Adirondacks.

The lake waters are confined by irregular escarpment walls that reach heights of 2,000 feet on the east and west, and by glacial moraine dams at the north and south ends of the lake. The present overflow of crystal-clear water passes into Lake Champlain, but evidence points to a southern preglacial drainage channel to the Hudson River. The glacial dams have impounded the waters which surround the well-vegetated rock islands, and the lake shore line has an appearance of submergence. Although glacial sands and boulders are abundant adjacent to the north-south shore line, the great bulk of rock has fallen from the steep escarpments.

Studies of the geologic history and structure demonstrate the importance of faulting in this area. As may be seen in Figure 11, the great number of nearly vertical faults have allowed some earth segments to drop while others remained stationary or rose. This faulting did not occur all at one time or even during a short period of earth history. It is difficult to date because no rocks younger than early Paleozoic age are present, but it is likely that faulting began in the Pre-Cambrian Era and has continued intermittently to the present day. The clean and only slightly weathered rock walls found in certain parts of the escarpment seem to indicate rather recent fault action, although exceedingly little rock movement has taken place since the ice left this valley some 8,000 years ago. Some minor movements have disrupted loose glacial material to demonstrate that this area has not reached a point of equilibrium.

Long prior to the white man's settlement, the Indians used the waters of Lake George for passage between the Hudson River and Lake Champlain. The value of this passage was illustrated in the Indian tribal wars as well as the Revolutionary War.

The town of Lake George is now developing into a major resort area. During the past decade or so, with the popularity of boating and fishing, the area has become transformed. Much of the Lake George shore line and many islands are owned by New York State. The Lake George Beach and Battlefield state parks attract thousands of sightseers, swimmers, and campers each year, and there are fifty-seven state-owned islands of sufficient size to permit active public use, forty-eight of which are available for camping.

Palisades and the Lower Hudson

The Palisades are situated along the west side of the Hudson River northward from Hoboken. Bordering the river in the Palisades area are rocks of two different types. The eastern or Manhattan side is composed of an igneous and metamorphic complex of Pre-Cambrian age, while the western

CROSS SECTION OF LAKE GEORGE

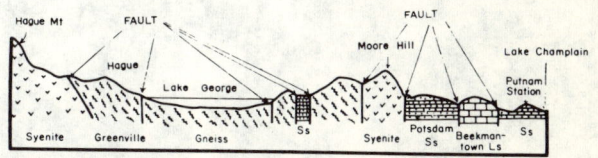

Section across the middle part of Lake George from Lake Champlain to Hague Mountain, a distance of ten miles. This section is a classic example of the fault block system on the eastern periphery of the Adirondacks. Lake George occupies a <u>graben</u>, a down-thrown block between two faults. The vertical scale is exaggerated.

Fig. 11

side of the gorge is composed of basalt of Triassic age.

Triassic basalt forced its way between previously formed sedimentary beds in the form of a sill. The Hudson River has carved through this sill of basalt to form the sheer cliffs that we now call the Palisades. Jointing systems, caused by original cooling and solidification of the sill, contributed to the columnar appearance of the Palisades.

During the Pleistocene Epoch, a great amount of glacial melt water loaded with sediment was carried down the Hudson River. In the narrow stretches, confined by rock walls near the Palisades, the erosive power of the river was great, and it deepened the gorge with little sediment being deposited. Much of the glacial material was carried on to the continental shelf in a preglacial canyon which extends 50 miles out to sea and reaches a depth of 8,000 feet below sea level. Although this canyon was not caused by glacial action, turbidity currents from glacial melt water may have enlarged its size in places. The Hudson River actually is an estuary of the Atlantic Ocean, being in a drowned condition as far north as Troy. The river channel reaches a depth of 700 feet near West Point.

The highlands adjacent to and overlooking the Hudson gorge are used today more than ever before. In the past decade this area has been the scene of a tremendous building and development program, both for commercial establishments and private housing. The state has established several parks, such as Tallman Mountain and Rockland Lake, to preserve the scenic beauty of the west shore of the Hudson River. This chain of parks cover almost

LAKE GEORGE (from the west). Bordered by steep mountains which rise 2,000 feet above its 45 square miles of crystal-clear waters, Lake George must indeed be classed as one of New York's most spectacular water bodies. Camping, boating, and other recreational activities are all important. *N.Y.S. Dept. of Commerce.*

60,000 acres and attract over six million people annually.

Long Island

Long Island, the largest island adjoining the United States proper, extends 118 miles east-northeast from the mouth of the Hudson River. It is composed of low plateaus on the north side of the island, longitudinal ridges of glacial moraine through the central parts of the island, and gently sloping plains to the south of the moraines. The shore lines are composed of steep bluffs that are constantly eroded by sea-water action on the weathered and poorly consolidated materials.

Previous to the advance of the glacial ice sheets, the island was a nearly level area composed of poorly cemented Cretaceous sediments. As the ice advanced across it from the north during the Wisconsin stage of the Pleistocene Epoch, a new feature, the ridge known as Ronkonkoma, was formed. This ridge is the terminal moraine, as is a second ridge, Harbor Hill, which was formed by a later readvance of ice. The third ice advance did not leave a ridge of any magnitude, as it was concerned with the modification of the Harbor Hill Moraine. There is no decisive evidence to prove any ice advance beyond the Ronkonkoma morainal ridge. Although much erosion has occurred since the last ice advance, these ridges stand 200 feet above the plateau and 391 feet above the sea at this time. Figure 12 depicts the plateau and ridges of Long Island.

Numerous parts of the island are covered by ground moraine as well as kettle ponds and eskers. Kellis Pond near Bridgehampton and Lake Ronkonkoma near St. James are kettle ponds, and an esker may be noted near Smithtown Branch. Some of the lower areas on the island have been slowly filled in with vegetation and have become potential bogs.

Perhaps the most outstanding feature of Long Island is the number of beautiful beaches, especially those parallel to the south shore. The present beaches—such as Jones, Rockaway, and Long—are the result of wave action and longshore currents which have developed barrier beaches, lagoons, and spits, with sand bars, out of materials deposited earlier by glacial melt waters. The constant migration of the bars is a cause for much concern, and costly corrective measures are needed to limit real estate destruction.

The Jones Beach State Park, a former sandy area with sparse sea grass growth, was purchased by New York State in 1927 and was ready to be used in 1929. New York has spent millions of dollars in making this a seaside recreational and cultural center. Millions of people visit this and nearby beaches annually.

One of the most important truck garden spots in the United States is on Long Island. The hills of glacial moraine and some of the glacial outwash plains of the southern half of the island are covered with rich soil. The climate and nearby market are instrumental in the present development.

Perhaps the most infrequently mentioned but economically significant products made available as a result of glaciation are sand and gravel. The relative importance of these deposits is pointed up by the fact that in Nassau and Suffolk counties over fourteen million dollars' worth of these materials was processed in 1962.

NORTH-SOUTH PROFILE OF LONG ISLAND

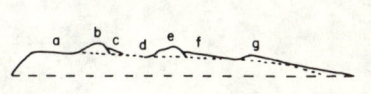

a) - Manhasset Plateau
b) - Harbor Hill moraine
c) - Outwash from ice producing Harbor Hill moraine
d) - Manhasset surface
e) - Ronkonkoma moraine
f) - Outwash from ice producing Ronkonkoma moraine
g) - Manhasset Ridge projecting above outwash

Fig. 12

Finger Lakes Region

The Finger Lakes are in the west-central section of the state. Of the eleven lakes in the area only seven are recognized by popular terminology to be Finger Lakes. In order from west to east, they are: Canandaigua (area 16.57 square miles), Keuka (17.43 square miles), Seneca (66.7 square miles), Cayuga (66.4 square miles), Owasco (10.3 square miles), Skaneateles (13.8 square miles), and Otisco (3.5 square miles). The other four, lying to the west, are Conesus, Hemlock, Honeoye, and Canadice.

There is strong evidence based on stream patterns that these basins were all preglacial stream valleys. Two of the major lakes, Seneca and Cayuga, each 40 miles long and about 2½ miles wide, drained to the north, while the drainage of the remaining five is questionable but may have been southward.

One of the most interesting aspects is the great depth of the basins that contain Cayuga and Seneca lakes. It is likely that some time prior to the glacial advance the land was elevated and dissected by streams into deep and broad valleys. The advancing ice was confined in the valleys to a greater depth than on the surrounding uplands, and as a result depressed and scoured out the valleys to their present depth.

The valley sides are now straight and tributaries entering the valleys hang above the valley floor,

SENECA LAKE (looking north from Watkins Glen). Stretching north-south through the Appalachian Upland for 35 miles, Seneca Lake covers nearly 67 square miles and reaches a depth of 618 feet. Its bottom is well below sea level. Much of its shore line still remains undeveloped for recreation. *N.Y.S. Dept. of Commerce.*

sometimes by several hundred feet. Nothing but a glacier could have produced land features of this nature.

Before the retreat of the glacier, the ice remained at the south end of the Finger Lakes for a considerable length of time and deposited a large recessional moraine, which functions as an obstacle to southward draining of the lakes. This moraine, called the Valley Head Moraine, is the drainage divide between rivers draining to Lake Ontario and those belonging to the Susquehanna system. The Genesee River is an exception and rises well to the south of this divide.

Taughannock Falls is located on the west side of Cayuga Lake about 8 miles northwest of Ithaca. This falls, with a vertical drop of 215 feet, has the distinction of being the highest waterfall east of the Rocky Mountains. The stream forming it flows over a resistant formation, the rock joint planes of which favor a steep fall.

This beautiful waterfall owes its existence to a former glacial lake, which once stood at a very much higher level in the Cayuga basin. The older preglacial tributary valley was filled by deltaic sediments. As the land dropped to a lower level, the former tributary stream found it easier to seek a new channel into the Cayuga Valley. It is most likely that this new course did originate during an interglacial stage. In any case, it is a true hanging valley waterfall.

The Finger Lakes region is without doubt one of the most picturesque locations in the country. Many central New York residents living in the larger cities have their summer home or camp on one of these beautiful lakes. The lakes, with their many sporting events, are popular with people of all ages, whether for water skiing or sitting quietly in a rowboat fishing. Waters of these lakes are also the source of water supply for cities such as Syracuse, which uses Skaneateles Lake water.

Landscapes of the Genesee

The Genesee River, with headwaters in Pennsylvania, flows northward to Lake Ontario and has four distinct landscapes along its course. In the first, from the headwaters to Portageville, the river flows in a preglacial drift-filled valley where the gradient is low and the valley well smoothed. At Portageville the second landscape takes shape as the river enters a 25-mile-long postglacial rocky gorge, the sides of which are almost perpendicular and rise to a height of 800 feet above the river. This gorge has been referred to as the "Grand Canyon of the East" and has been made into Letchworth State Park. At "Portage Falls" the river drops over three steps, having heights of 66 feet, 110 feet, and 96 feet. The third landscape occurs as the stream passes out of its gorge at Mount Morris and flows upon a wide preglacial moraine-filled flood plain. In the fourth landscape the river passes through a 7-mile-long postglacial gorge, starting at Rochester and extending into Lake Ontario at Windsor Beach. This gorge contains three waterfalls which drop in steps of 98 feet, 20 feet, and 105 feet. Water power sites here contributed to the founding and growth of the city of Rochester.

Allegany State Park

Allegany State Park, on the border of New York and Pennsylvania, covers an area of approximately 100 square miles. It is the only part of the state never glaciated or covered with glacial outwash.

In contrast to the Adirondacks, the area is underlain with sedimentary rocks. When the eastern part of the United States was undergoing a period of erosion during the Cretaceous Period, mountains in this area were peneplained. The land was subsequently slowly uplifted almost 2,000 feet, and stream action began again the downcutting movement and slowly widened the valleys. This erosion has continued to the present time.

The ice sheet made its farthest advance and stopped just to the north of the park area. It blocked the Allegheny River and its tributaries in their preglacial flow to the north. Later glacial melt waters deposited huge amounts of material which further blocked the tributary valleys and filled the main valley. Temporary ponds formed in the tributary valleys and deltaic deposits were laid down in those ponds. The melt water in the valleys was forced to the south through a minor valley and it was only at this time that the Allegheny River developed its present course to the Ohio River. Since then, it has eroded most of the glacial material out of its valley, leaving several terraces which show the height of the original valley filling.

This area is a splendid contrast in glaciated-nonglaciated conditions, and as such is of great interest to those interested in glacial or physiographic studies. The ordinary vacationist, too, will find the many trails, scenic drives, and improved roads throughout Allegany State Park much to his liking. The park is open all year, with facilities for winter sports enthusiasts.

Howe Caverns

The exploration of underground caverns has long been a fascination to the curious and the adventure-seeker. There are few caverns in the northeastern section of our country, but one that has been highly publicized and commercialized for a long time is Howe Caverns. Howe Caverns is located about 38 miles southwest of Albany. A second cavern, Secret Caverns, is within 1 mile of the same locality.

The entrance to Howe Caverns is 1,100 feet above sea level, and two elevators are used to carry tourists some 156 feet below the surface to the cavern floor. A winding passageway 3 to 6 feet wide, 10 to 100 feet high, and about 1½ miles long is the main attraction. The maximum depth reached in the cavern is about 200 feet.

The origin of this and most other caverns is from the work of ground water slowly percolating through the joint system of host rocks. These joints, or cracks, are gradually widened at the surface by rain water, which contains carbon dioxide. The carbon dioxide dissolves a minute amount of the host rock, almost always limestone, and the material percolates through the joint blocks to a lower level. Over an extended period of time the surface blocks may be worn away or dissolved, with a depression resulting. The depth of cavern development depends upon the elevation of the outlet through which the water escapes after entering the cavern. This level may fluctuate with time because of the changing ground water level and possible variation in land movement. For this reason caverns are better developed at some depths than at others. The irregularity in size and shape of underground stream channels inside the cave may be caused by slight variations in rock texture, chemical composition, jointing, and the amount of water available.

An outstanding feature of many limestone caves is stalactites, which hang from the cavern roofs. Their varied shapes and forms are developed by slight amounts of calcium carbonate waters dripping from the roof of the cavern. If an excessive amount of water drips from the roof, some will fall to the cavern floor and build a deposit upward, called a stalagmite. Sometimes stalactites and stalagmites join to form a column.

In 1929 an electrical system for lighting and communication was installed at Howe Caverns. The

THE SECOND LANDSCAPE OF THE GENESEE. The photograph on the left shows the "Grand Canyon of the East," where the Genesee River cuts nearly 800 feet into the sedimentary rocks of the Appalachian Upland. The Middle Falls in the photo on the right cascades 110 feet over rock ledges. The area is a state park. *N.Y.S. Dept. of Commerce.*

colored lighting in certain areas of the cavern has added much beauty to the unusual cave deposits. This beauty can be seen to best advantage during a gondola-style boat ride along a 1_4-mile-long lake known as Lake of Venus. The reflections at this spot are mirror perfect.

The Helderberg Escarpment

John Boyd Thacher Park is located 14 miles west of Albany and 22 miles south of Schenectady in Albany County. The outstanding feature at this locality is the Helderberg cliff or eastern escarpment of the Helderberg Plateau. A splendid view of the glacial lake plain (ancient Lake Albany), may be seen 900 feet below and to the north.

The escarpment is composed of sedimentary rocks which once extended northward to flank the crystalline rocks of the Adirondacks. After ancient Devonian seas spread across this area, materials were deposited which later formed limestone, shale, and sandstone. These Devonian beds were uplifted and dipped gently toward the south. In the time since this uplift, stream erosion has been cutting into these beds and wearing them away to form an inner lowland. This inner lowland is now between the

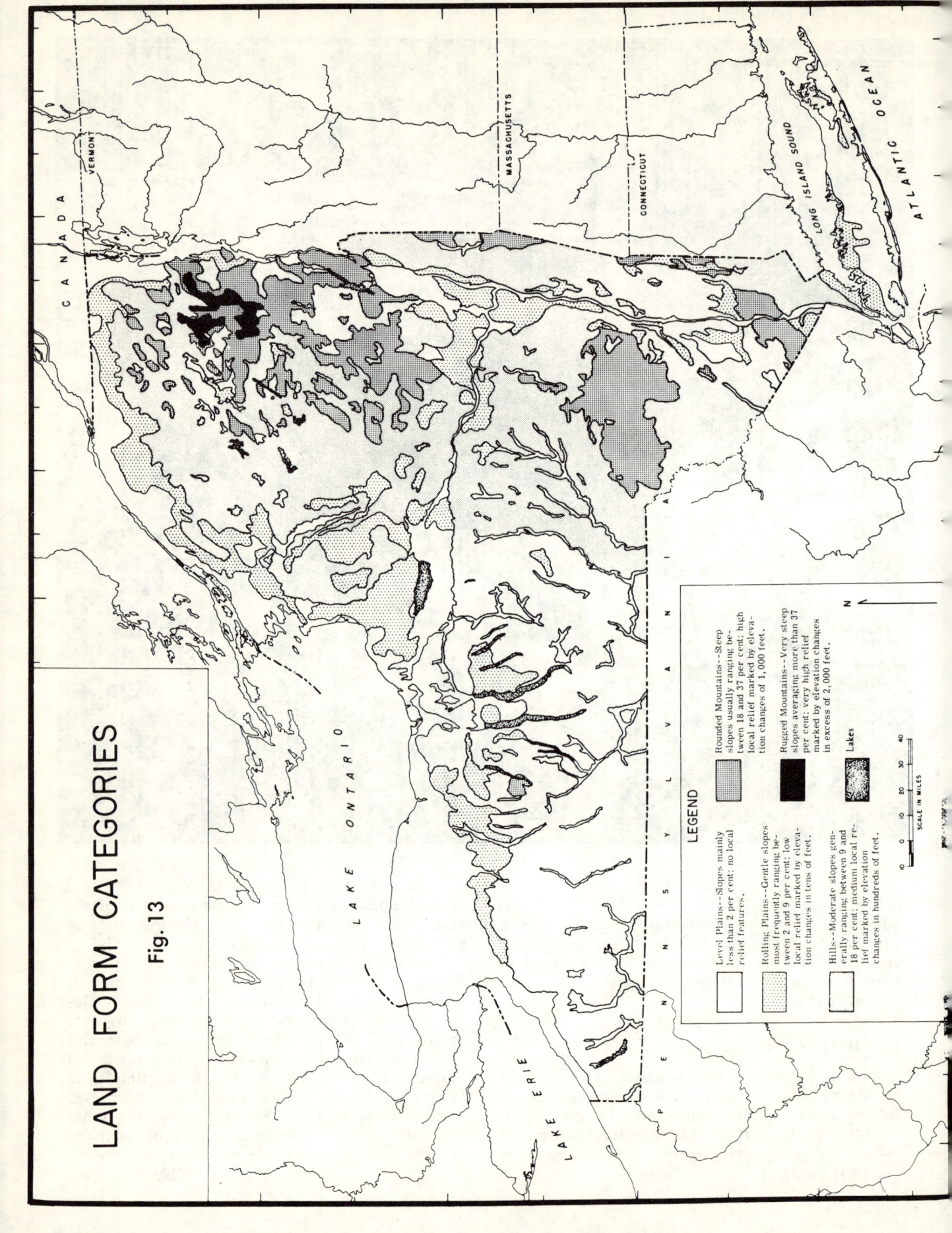

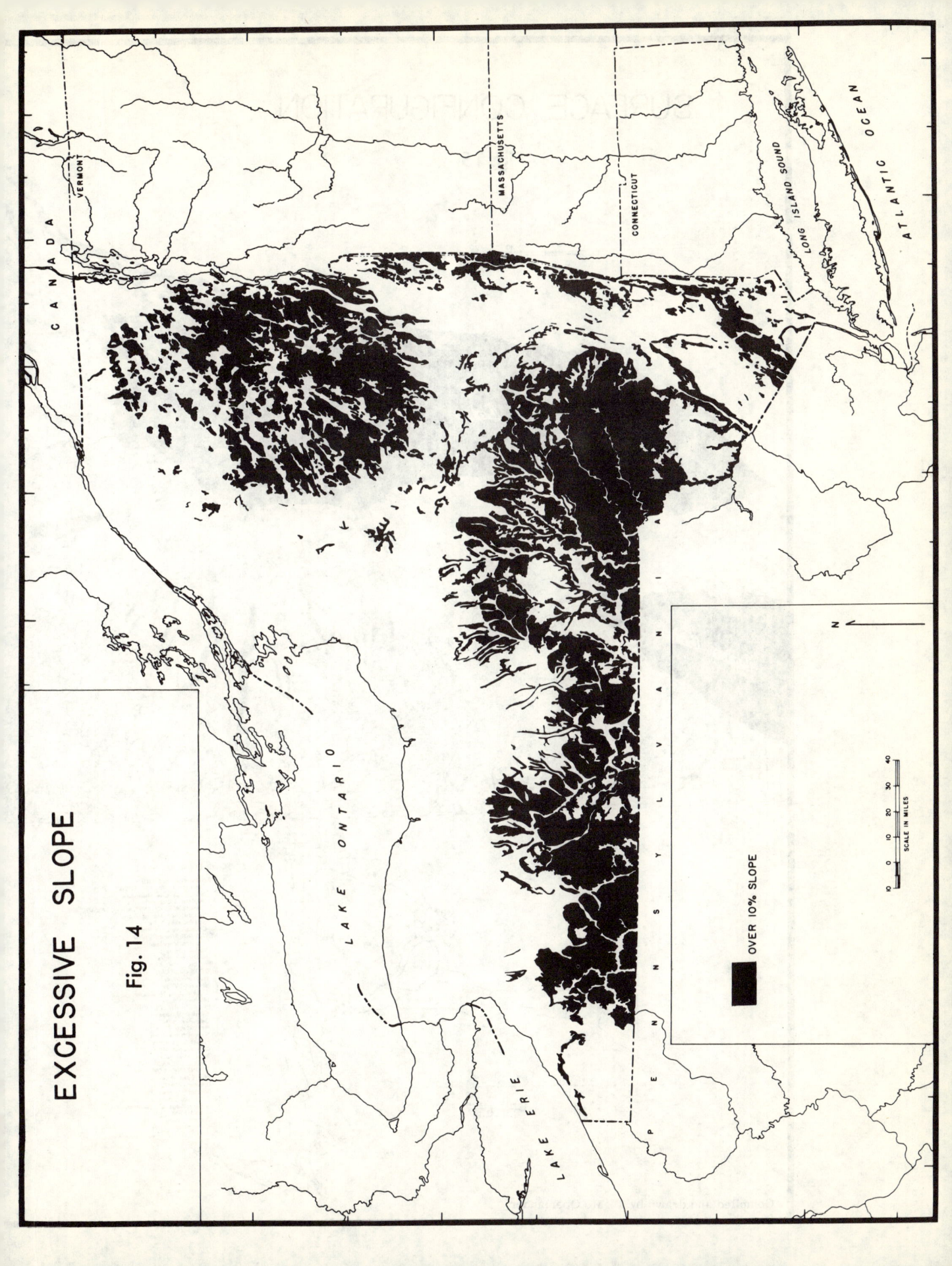

SURFACE CONFIGURATION

Fig. 15

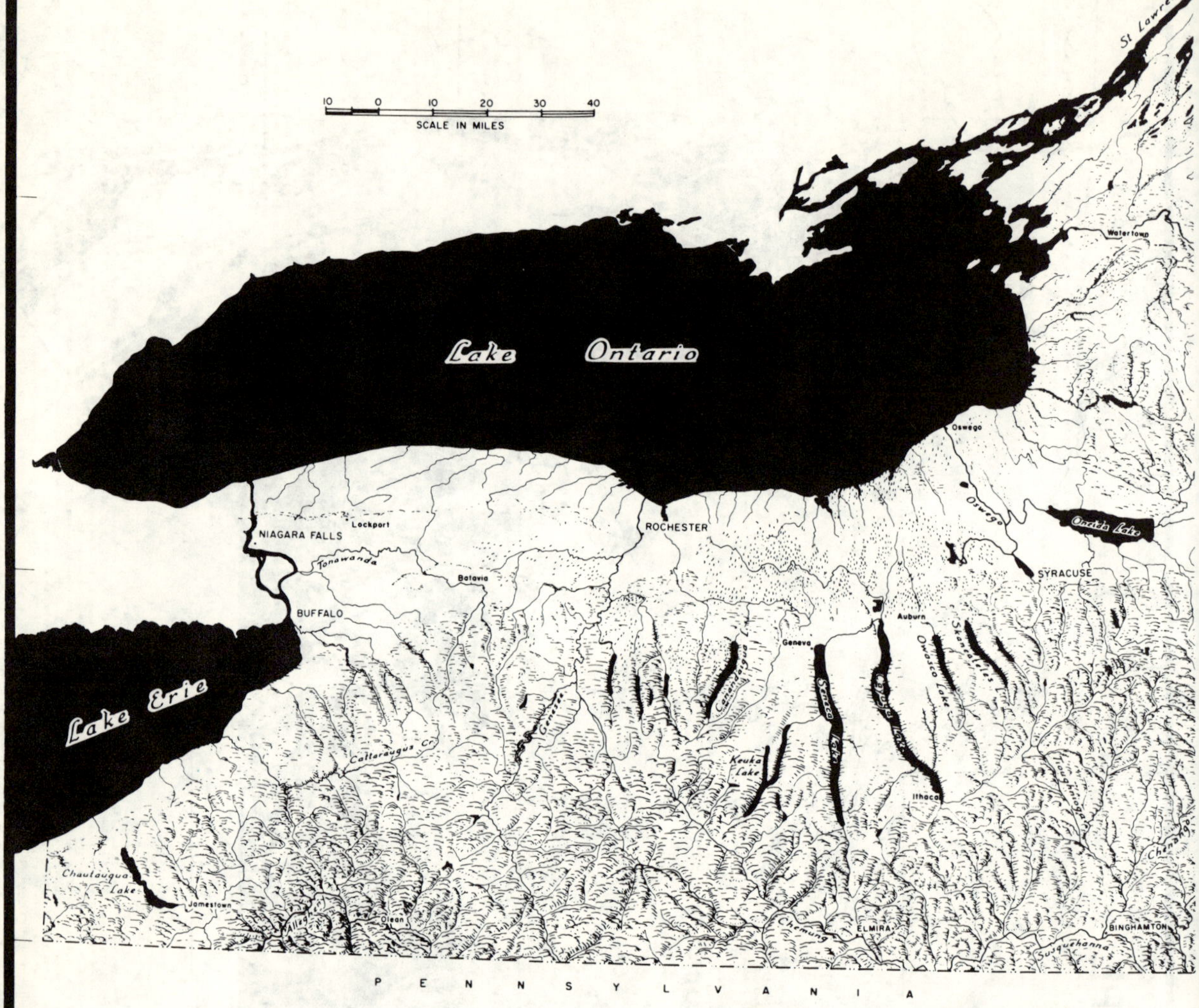
10 0 10 20 30 40
SCALE IN MILES
St Lawrence
Watertown
Lake Ontario
Oswego
Oswego
Oneida Lake
SYRACUSE
NIAGARA FALLS
Lockport
ROCHESTER
Tonawanda
Batavia
BUFFALO
Auburn
Geneva
Owasco Lake
Skaneateles
Lake Erie
Genesee
Canandaigua
Seneca Lake
Cayuga Lake
Keuka Lake
Cattaraugus Cr.
Ithaca
Chenango
Chautauqua Lake
Jamestown
Olean
Chemung
ELMIRA
BINGHAMTON
Susquehanna
P E N N S Y L V A N I A

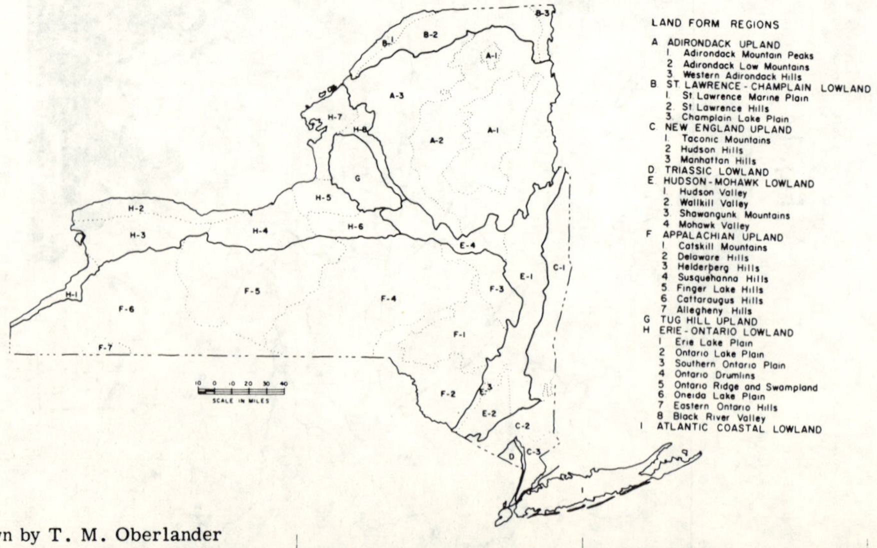
B-3
B-1
B-2
A-1
A-3
H-7
H-8
A-2
A-1
G
H-5
H-2
H-3
H-4
H-6
E-4
C-1
E-1
F-3
H-1
F-5
F-4
F-6
F-1
F-7
E-3
F-2
E-2
C-2
D
C-3
I
10 0 10 20 30 40
SCALE IN MILES
LAND FORM REGIONS
A ADIRONDACK UPLAND
1 Adirondack Mountain Peaks
2 Adirondack Low Mountains
3 Western Adirondack Hills
B ST LAWRENCE - CHAMPLAIN LOWLAND
1 St Lawrence Marine Plain
2 St Lawrence Hills
3 Champlain Lake Plain
C NEW ENGLAND UPLAND
1 Taconic Mountains
2 Hudson Hills
3 Manhattan Hills
D TRIASSIC LOWLAND
E HUDSON - MOHAWK LOWLAND
1 Hudson Valley
2 Wallkill Valley
3 Shawangunk Mountains
4 Mohawk Valley
F APPALACHIAN UPLAND
1 Catskill Mountains
2 Delaware Hills
3 Helderberg Hills
4 Susquehanna Hills
5 Finger Lake Hills
6 Cattaraugus Hills
7 Allegheny Hills
G TUG HILL UPLAND
H ERIE - ONTARIO LOWLAND
1 Erie Lake Plain
2 Ontario Lake Plain
3 Southern Ontario Plain
4 Ontario Drumlins
5 Ontario Ridge and Swampland
6 Oneida Lake Plain
7 Eastern Ontario Hills
8 Black River Valley
I ATLANTIC COASTAL LOWLAND

Compiled and drawn by T. M. Oberlander

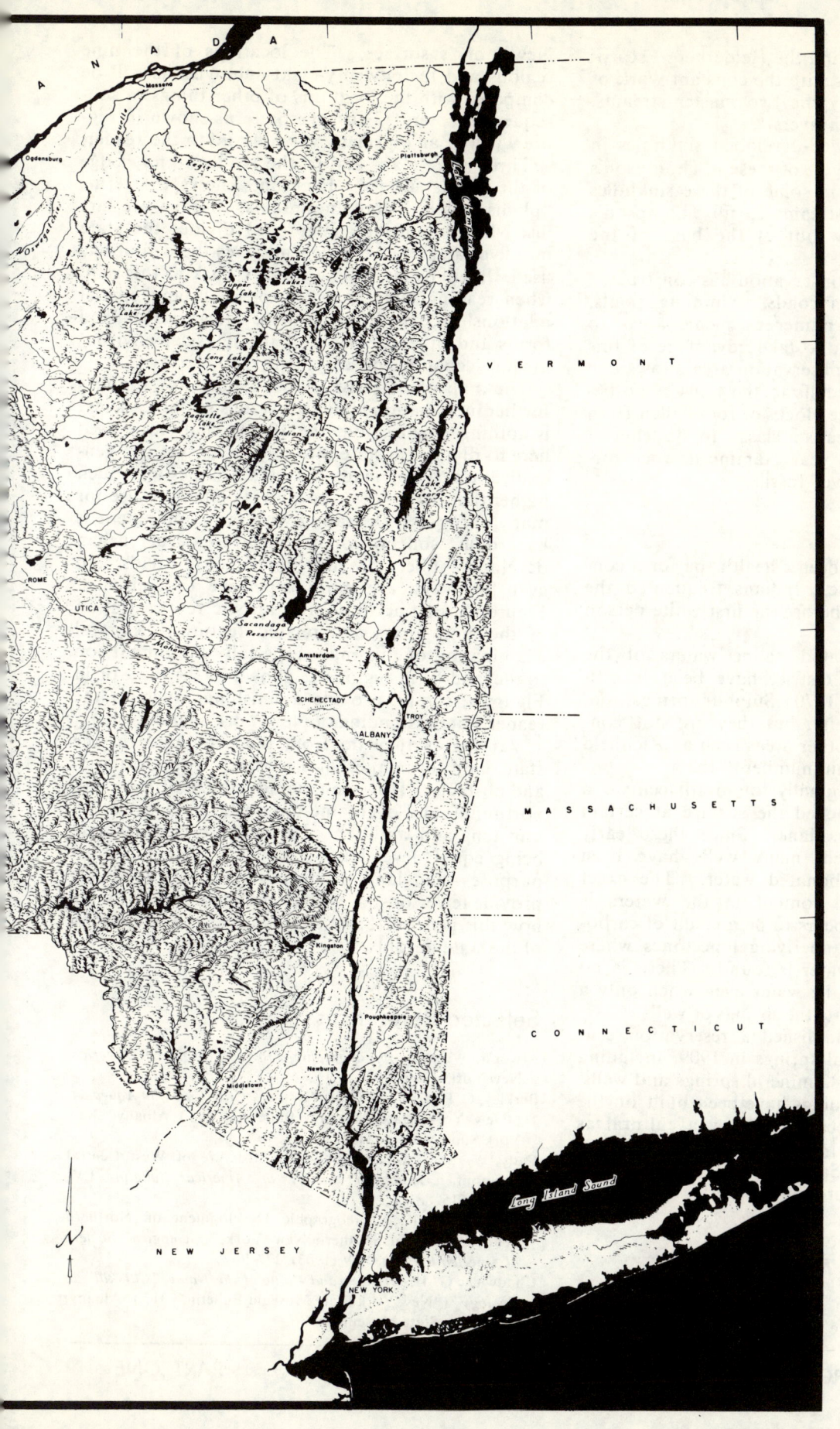

Adirondack Mountains and the Helderberg Escarpment. Erosion continues with the constant work of degrading, controlled by the two master streams, the Mohawk and Hudson rivers.

This area has many well-developed sinkholes in limestone beds. The largest of these is Thompson's Lake. The drainage from some of these sinkholes flows by underground streams to fill Thompson's Lake, while others flow out of the base of the escarpment.

The Department of Conservation has constructed a network of macadam roads, swimming pools, camping grounds, and numerous picnic areas to enable the general public to take advantage of this unusual state park. An observation area allows one to look down the sheer cliffs to the talus slope below, which contains large blocks of rock fallen from the scarp face. This area is a classic to students of geology; hundreds each year examine its rock formations and well-preserved fossils.

SARATOGA SPRINGS

Saratoga Springs has been a health spa for a considerable length of time. Indians frequented the springs for many years before the first white person arrived on the scene.

The unusual carbonated, saline waters of the Saratoga-Ballston Spa district have been investigated by scientists since 1770. Sulphur springs, too, are found in this vicinity, but they are not considered unusual, since other areas such as Richfield Springs have an abundant number of them.

The springs were originally found adjacent to a fault where waters reached the surface at certain places along the fault plane. Since these early springs were discovered, many wells have been drilled to tap this carbonated water. The exact derivation of chemical content in the waters is questionable, but it appears to be a result of carbonation related to the underlying limestones where most of this type of water is found. There is no uniformity in taste of the water even when only a few hundred feet separate the springs or wells.

New York State established a reservation centered about the mineral springs in 1909, including 1300 acres of land and 83 mineral springs and wells. A great variety of buildings have been built on the reservation and this location is now a cultural as well as health center. Mineral waters are still bottled commercially at Saratoga in large quantities and shipped to points around the world.

Land Forms and Man

Using the land form region as a spatially differentiating tool, we have presented the variations in New York's surface. The locations of the nine regions and their twenty-eight subdivisions will be compared with the locations of other things in subsequent chapters. Thus, the patterns shown in Figure 9 will be an especially valuable aid to the reader.

Three more maps add to the reader's knowledge of the state's surface. Everyone knows the terms "plains," "hills," and "mountains," but few are sure just what they mean. Figure 13 divides the state into five categories. The legend gives the definitions used. It will be interesting to look back at this map when reading later chapters to see if any spatial relationships exist between these categories of land forms and other elements of the physical environment or economic development.

The map of excessive slope, Figure 14, supplies further information about the state's surface. There is nothing magic about the 10 per cent slope used here to divide the state into two parts, yet the generalization can be made that when slopes are much higher than 10 per cent the land is difficult to use for many purposes. Operating with farm equipment is hazardous; soil erosion is likely to occur; urban development becomes more involved; railroads, and even highways, are more expensive to construct. Figure 15 is an accurate and artistic representation of the land forms and related features. Careful examination of this map, especially in comparison to the land form regions as shown in the inset or in Figure 9, should provide a finishing touch to the reader's understanding of New York's land forms.

Varying as they do from place to place in the state, land forms present a mosaic of opportunities and obstacles. How we take advantage of these opportunities and overcome the obstacles depends on our aims and abilities. Certainly, everything else being equal, level land is easier to use for most purposes than land in steep slopes, and lowlands provide fewer obstacles than highlands. Let us see how this generalization stands up as the geography of the state unfolds.

Selected References

Atwood, W. W. *Physiographic Provinces of North America.* New York: Ginn, 1940.

Berkley, C. P. *Geology of the New York City (Catskill) Aqueduct.* ("New York State Museum Bulletin," 146.) Albany: New York State Museum, 1911.

Cady, W. M. "Stratigraphy and Structure of West Central Vermont," *Geological Society of America Bulletin*, LVI (1945), 515–88.

Campbell, M. R. "Geographic Development of Northern Pennsylvania and Southern New York," *Geological Society of America Bulletin*, XIV (1903), 277–96.

Chadwick, G. H. *History and Value of the Name "Catskill" in Geology.* ("New York State Museum Bulletin," 317.) Albany: New York State Museum, 1936.

Clark, J. M., and D. D. Luther. *Geology of the Watkins and Elmira Quadrangles.* ("New York State Museum Bulletin," 81.) Albany: New York State Museum, 1905.

Cushing, H. P., and R. Ruedemann. *Geology of Saratoga Springs and Vicinity.* ("New York State Museum Bulletin," 169.) Albany: New York State Museum, 1914.

Eardley, A. J. *Structural Geology of North America.* New York: Harper, 1951.

Emmons, W. H., *et al. Geology: Principles and Processes.* 5th ed. New York: McGraw-Hill, 1960.

Fairchild, H. L. *Glacial Waters in Central New York.* "New York State Museum Bulletin," 127.) Albany: New York State Museum, 1909.

———. *Pleistocene Marine Submergence of the Hudson, Champlain and St. Lawrence Valleys.* ("New York State Museum Bulletin," 209-10.) Albany: New York State Museum, 1918.

Fenneman, N. M. *Physiography of Eastern United States.* New York: McGraw-Hill, 1938.

Finch, V. C., *et al. Fundamentals of Physical Geography.* New York: McGraw-Hill, 1961.

Flint, R. F. *Glacial and Pleistocene Geology.* New York: Wiley, 1957.

———. "Glacial Geology of Long Island," *Geological Society of America Bulletin*, LXIV (1953), 897-919.

Fuller, M. L. The Geology of Long Island. ("U. S. Geological Survey Professional Paper," 82.) Washington, D. C.: Govt. Printing Office, 1914.

Geologic Map of New York, 1961. ("Map and Chart Series," No. 5.) Albany: State Museum & Science Service, Geological Survey, 1962.

Goldring, Winifred. *Guide to the Geology of John Boyd Thacher Park and Vicinity.* ("New York State Museum Handbook," 14.) Albany: New York State Museum, 1933.

Grabau, A. W. *Niagara Falls and Vicinity.* ("New York State Museum Bulletin," 45.) Albany: New York State Museum, 1901.

Hall, J. *Second Report, Geological Survey of New York.* Albany, 1838.

Johnson, D. W. "Date of Local Glaciation in the White, Adirondack and Catskill Mountains," *Geological Society of America Bulletin*, XXVIII (1917), 543-52.

———. *Stream Sculpture on the Atlantic Slope.* New York: Columbia Univ. Press, 1931.

Kemp, J. F. *Mineral Springs of Saratoga, New York.* ("New York State Museum Bulletin," 159.) Albany: New York State Museum, 1912.

——— and H. L. Alling. *Geology of the Ausable Quadrangle.* ("New York State Museum Bulletin," 261.) Albany: New York State Museum, 1925.

——— and ———. *Geology of the Mount Marcy Quadrangle.* ("New York State Museum Bulletin," 229-30.) Albany: New York State Museum, 1920.

Leverett, F. *Outline of the History of the Great Lakes.* 12th Report of the Michigan Academy of Science. Lansing: Michigan Academy of Science, 1910.

——— and F. B. Taylor. *The Pleistocene of Indiana and Michigan and the History of the Great Lakes.* ("U. S. Geological Survey Monograph," 53.) Washington, D.C.: Govt. Printing Office, 1915.

Lobeck, A. K. *A Popular Guide to the Geology and Physiography of Allegany State Park.* ("New York State Museum Handbook," 1.) Albany: New York State Museum, 1927.

MacClintock, P., and H. G. Richards. "Geology of Long Island," *Geological Society of America Bulletin*, XLVII (1936), 289-338.

Miller, W. J. *The Adirondack Mountains.* ("New York State Museum Bulletin," 193.) Albany: New York State Museum, 1917.

———. *Geologic History of New York State.* ("New York State Museum Bulletin," 255.) Albany: New York State Museum, 1924.

Newland, D. H., and H. Vaughan. *Guide to the Geology of the Lake George Region.* ("New York State Museum Handbook," 19.) Albany: New York State Museum, 1942.

Rich, J. L. *Glacial Geology of the Catskills.* ("New York State Museum Bulletin," 299.) Albany: New York State Museum, 1934.

Rogers, J. "Stratigraphy and Structure in the Upper Champlain Valley," *Geological Society of America Bulletin*, XLVIII (1937), 1573-88.

Spencer, J. W. "Origin of the Basins of the Great Lakes of America," *Amer. Geol.*, VII (1891), 86-97.

Tarr, R. S. *The Physical Geography of New York State.* New York: Macmillan, 1902,

Tolman, C. F. *Ground Water.* New York: McGraw-Hill, 1937.

Veatch, A. C., and A. H. Smith. *Atlantic Submarine Valleys of the United States and the Congo Submarine Valley.* ("Geological Society of America Special Paper," 7.) New York: Geological Society of America, 1939.

Wallace, E. R. *Descriptive Guide to the Adirondacks.* Syracuse: Forest Pub. House, 1899.

Work Projects Administration, New York Writers' Program. *New York: A Guide to the Empire State.* New York: Oxford Univ. Press, 1940.

CHAPTER 2   DOUGLAS B. CARTER

Climate

The physical state of the atmosphere, particularly its day-by-day disposition, is ordinarily referred to as *weather*. *Climate* is the long-range composite of weather, but it is more, too. Since climate deals mainly with conditions over a long period while weather is concerned with more momentary events, a slowly acting but grossly important process such as evaporation assumes drastically greater significance for climate than for weather, thus giving climate a more complex dimension than just a sum-total of weather. Furthermore, climate might be correctly thought of as including moisture and temperature conditions in the soil as well as in the atmosphere. If this broader concept of climate is carried to its logical conclusion, climate should be defined as the long-range physical state of the atmosphere and its relation to the earth's surface.

Climate's effects on the physical earth and on man are widely evident. Some aspects of the physical environment such as amounts and rates of plant growth, muddiness of ground, magnitude and timing of stream flow, development of soils, and virtually any other condition of the earth's surface which depends upon heat and moisture are extensively governed by climate. It follows that climate's involvement in man's activities is bound to be extensive. Seasonal activities, particularly such things as summer and winter recreation and agriculture, feel its impact. It is involved in the availability and storage of raw materials, the supply and demand features of power, problems of transportation, construction of buildings, nature of clothing sold, and in some aspects of labor and market vitality. Even the physical and psychological condition of individuals may be affected. Frequently climate is an outstanding asset. When it functions as a handicap, man is challenged to ameliorate its handicapping influences through use of heating, refrigeration, air-conditioning, irrigation, and other measures.

This chapter is concerned with the main climatic processes that govern the behavior of the atmosphere over New York, the measurements of weather and climatic conditions that reveal the operation of these processes, and the aspects of climate that appear to be controlled by other features of the environment. These considerations help in understanding the geography of the state by explaining the differences in climate from place to place.

Climatic Processes, Elements, and Controls

Three processes are largely responsible for the nature of climate at any place: (1) gains and losses of energy, (2) gains and losses of moisture, and (3) movement of air with its burdens of heat, moisture, and storminess into and away from the place. Each of these processes operates as an economy of incoming and outgoing amounts, and all three are interdependent. They produce the contrasts in such climatic elements as temperature, precipitation and humidity, winds, and soil moisture adequacy.

Values of these climatic elements vary over the state because the processes of air movements and energy and moisture exchanges are different at various locations. Latitude, land and water distribution, mountain barriers, elevation, and storms control the values of climatic elements by affecting the continuing process of air movements and energy and moisture exchanges. Thus, they are ordinarily referred to as climatic controls.

New York's Climate in Brief

New York is a moderately large state with considerable variation in terrain, elevation, and exposure to water bodies. Thus it may be assumed that areal variations in energy, moisture, and air movements result in variation in climate. Actually variation does exist in the state, but it is not as great as sometimes thought. Generally, mean annual temperatures are in the 40's, and mean January temperatures are below freezing, in the 20's or teens. A good deal of cloudiness prevails and the state is under frequent impact from frontal action producing an ever changing day-to-day weather situation and generally humid conditions. Surface winds blow from every direction, but over the entire state there is a persistence of air from the southwest, west, and northwest.

Climatic changes from season to season are substantial. In summer the state is dominated by warm, moist air generally originating over the Atlantic. Although the greatest totals of precipitation occur then, it is the season of least cloudiness. Energy received in summer is significantly greater than in other seasons and it is used so abundantly in nature to vaporize moisture through evaporation and transpiration processes that the precipitation is insufficient to replenish soil moisture as rapidly as it is depleted. Droughts of short duration are therefore characteristic in most parts of the state. Temperatures are surprisingly even from place to place, and there is a minimum of difference between upland and lowland locations.

Both warm, moist air and cool, dry air are common in autumn, the former resulting in considerable precipitation but the latter giving rise to many days of beautiful, clear weather. Calms and fogs occur most frequently in this season. As in summer, there is relatively little variation in temperature over the state.

Winter, with associated cold air masses from the interior of the continent, brings the greatest temperature contrast to the state. The coastal area experiences temperatures that dart now above, now below, freezing. Upstate, temperatures above freezing are less frequent, so snow falls and accumulates to significant depths. Very low temperatures, comparable to conditions in Maine and Minnesota are occasionally experienced in upstate localities, particularly in the Adirondacks and their western fringe. Storms with well-developed fronts are most frequent in winter, and the number of days with precipitation and overcast skies is exceedingly large.

Spring retains a large measure of the temperature contrasts of winter because southern coastal areas accumulate more energy than more northerly parts of the state. Cloudiness is common. It is interesting the way Lakes Erie and Ontario regulate the temperature of air passing over them in the spring. They may not prevent frosts when very cold air flows in from the interior of the continent, but at other times the removal of energy from the air by the lakes keeps it cool and delays spring plant growth. The especially vulnerable stages of plant development, such as fruit-tree blossoming, are thereby delayed until the hazard of frost is appreciably reduced. The result is a concentration of fruit production along the lake shores.

Energy

In a sense the atmosphere acts as an energy tank, absorbing excesses of energy flowing into it during the daytime and releasing energy at night. The atmosphere gains very little energy directly from sunshine, allowing most incoming solar radiation to pass through to the ground. The earth thus warmed radiates energy day and night, and its radiation, of longer wave length than solar radiation, is absorbed almost entirely by the atmosphere. Radiation of energy from the earth is the principal means of energy loss from the ground and energy gain for the atmosphere. Cloudiness is an important factor in reflecting away solar radiation, but it also regulates another facet of the energy economy—the capture of energy radiated upward from the earth's surface. The atmosphere radiates its heat downward to earth as well as outward to space. This exchange between atmosphere and ground is continual, so more energy is received from atmospheric radiation than from direct solar radiation, which is interrupted at night and by clouds. At the earth's surface some energy is utilized for vaporizing moisture and a rather tiny portion is consumed in the convective heating of the atmosphere. Temperature records indicate the nature of the balance between incoming and outgoing energy, though they give no hint of the magnitude of these energies.

Solar Radiation

An atmosphere containing many impurities transmits fewer calories of solar radiation per unit of time than a clear atmosphere. This principle is illustrated in Figure 16, where average amounts of solar radi-

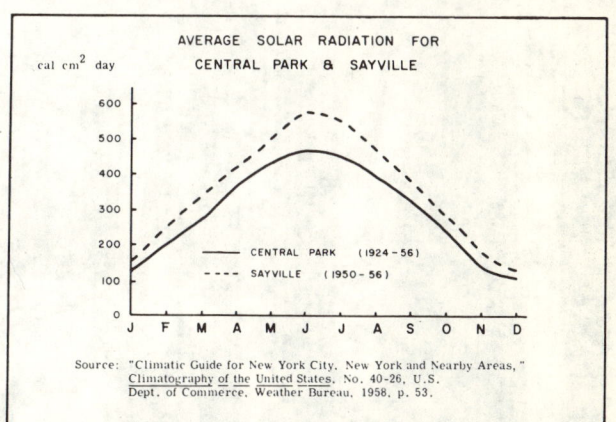

Fig. 16

ation at the Central Park Observatory in New York City and at Sayville, Long Island, are compared. In common with other large urban centers, the city receives less radiation—in this case about 10 per cent less—than suburban and rural areas nearby because of larger amounts of impurities in the air.

The data shown in Figure 16 are roughly representative of solar radiation receipts throughout the whole state. At Ithaca numerical values are similar to the Central Park record. Northern parts of the

STRIKING SEASONAL CONTRASTS in climate give the New York environment pleasing diversity. Heavy snows in winter, and warm but not hot summers, can especially be appreciated in rural settings such as these. *N.Y.S. Dept. of Commerce*.

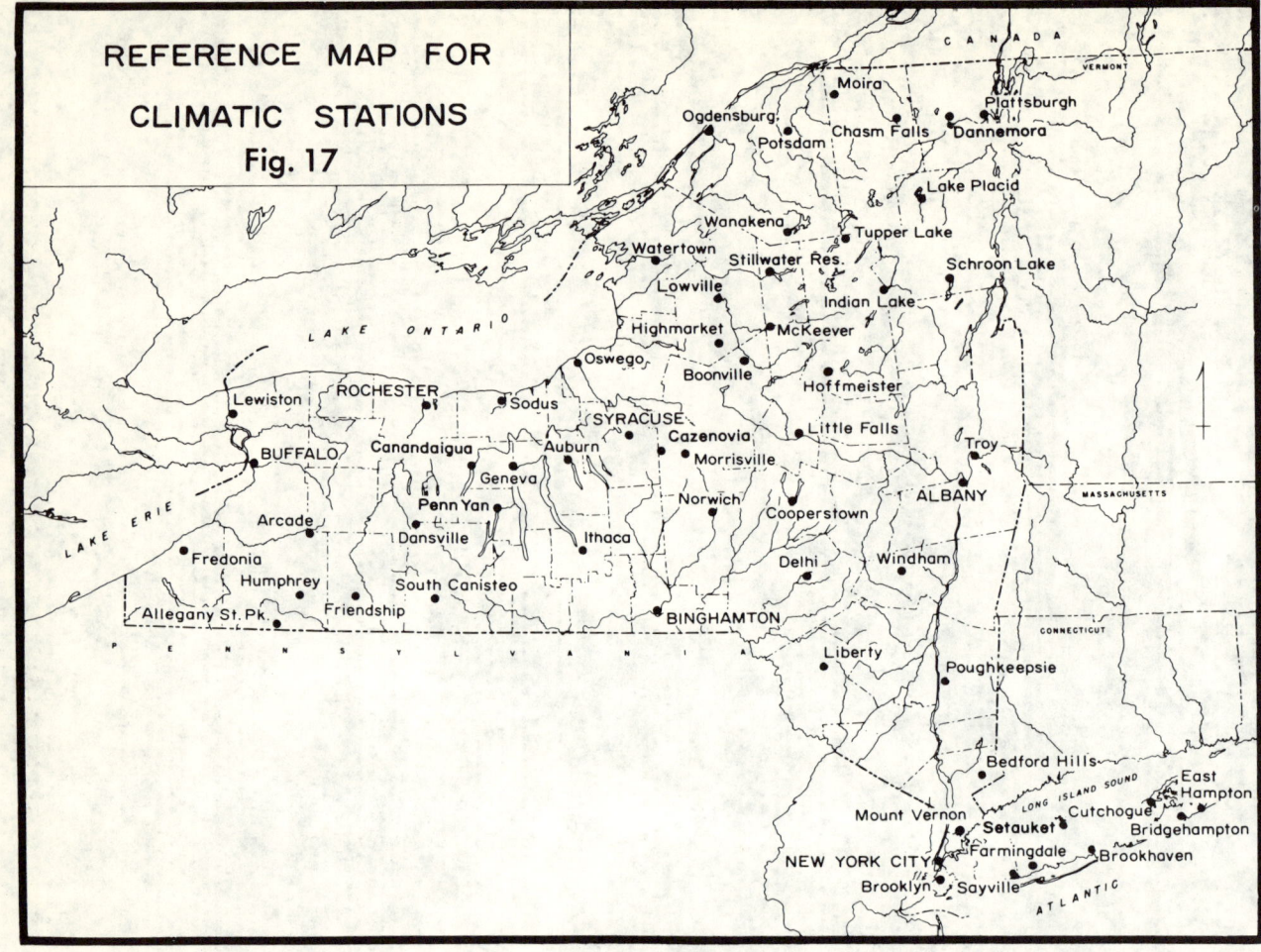

state, though adequate records are lacking, probably receive nearly as much solar radiation, the difference in latitude being insufficient to produce appreciable variation.

Solar radiation intensity and duration change markedly with the seasons, and there are considerable contrasts in the amount of radiation absorbed by slopes which face in various directions. Higher elevations experience more intense radiation than lowlands, and wherever there is greater frequency of cloudiness solar radiation is depleted.

The annual total of solar radiation received on every square centimeter of horizontal surface and summed up from the curves in Figure 16, comes to a fabulous total of 109,000 calories for Central Park and 130,000 calories for Sayville. While this is less than half the solar radiation received in Iraq and Arabia, the area with the world maximum, it is twice the amount that reaches the ground in Fairbanks, Alaska. The importance of 130,000 calories on every square centimeter can be better appreciated if expressed in familiar terms: if all that solar energy could be used exclusively to melt snow, it would liquefy all the freshly fallen snow, uncompacted, that might cover the state to a depth of 500 feet.

The Effect of Cloudiness

Clouds are extremely effective reflectors of solar radiation. They also intercept and delay the loss of heat by radiation from the earth. Occurrences of cloudy and clear days are summarized in Table 3. There is a very high incidence of cloudiness in upstate cities during winter; fewer than 10 per cent of the days have clear weather. Cloudiness at Buffalo, Rochester, and Syracuse during winter attains near-record proportions, with an average of more than eight-tenths of the sky covered. As a result, less than one-third of the possible sunshine is received in upstate cities during November, December, and January. During December hardly a quarter of the weak energy supply manages to penetrate the clouds.

Autumn in New York City has the maximum frequency of clear skies. Elsewhere in the state

TABLE 3
MEAN CLOUD COVER* AND NUMBER OF CLEAR† DAYS
(Sunrise to Sunset)

	Cloud Cover													Clear Days												
	J	F	M	A	M	J	J	A	S	O	N	D	Yr	J	F	M	A	M	J	J	A	S	O	N	D	Yr.
Albany	7.0	6.8	6.7	6.9	6.9	6.4	6.0	5.8	5.6	5.8	7.1	6.8	6.5	5	5	7	5	5	5	6	8	9	9	4	5	73
Binghamton	7.6	7.2	7.1	6.9	6.6	6.3	6.2	6.1	6.0	6.3	7.5	7.9	6.8	4	4	5	6	7	6	6	7	8	8	4	3	68
Buffalo	8.1	7.5	7.0	6.5	6.1	5.6	5.3	5.4	5.6	6.2	7.7	8.2	6.6	2	3	5	7	8	8	10	9	9	7	3	1	72
N.Y.C.: Central Pk.	6.2	5.8	5.8	6.0	5.8	5.7	5.6	5.5	5.2	5.0	5.9	6.0	5.7	8	8	9	7	8	7	8	9	10	11	9	9	103
Rochester	8.1	7.8	7.2	6.7	6.7	5.9	5.5	5.6	5.6	6.1	8.0	8.1	6.8	2	3	6	8	9	10	11	11	10	8	3	2	83
Syracuse	7.9	7.5	7.1	6.6	6.0	5.7	5.4	5.6	5.8	6.3	7.7	8.1	6.6	3	4	5	6	8	8	9	9	8	7	3	2	72

*Cloud cover as expressed out of a possible total of 10.
†A clear day is defined as one with less than .2 average cloud cover.

TABLE 4
MEAN MONTHLY TEMPERATURE AT REPRESENTATIVE STATIONS, 1931–55
(In Degrees Fahrenheit)

Location	Station	J	F	M	A	M	J	J	A	S	O	N	D	Yr.
St. Lawrence Valley	Ogdensburg	18.0°	19.7°	29.4°	43.8°	55.8°	65.5°	70.9°	68.9°	60.8°	50.2°	37.4°	23.0°	45.3°
Champlain Lowland	Dannemora	18.1	19.0	28.1	41.7	55.1	64.5	69.3	67.2	59.4	48.6	35.1	21.6	44.0
Adirondacks	Indian Lake	17.0	17.0	25.8	38.1	51.1	60.0	64.4	62.3	55.0	44.7	32.5	19.8	40.6
Adirondacks	Stillwater Res.	14.8	14.3	24.2	38.4	51.9	61.2	65.9	64.1	56.3	45.6	32.5	18.2	40.6
Adirondacks	Lake Placid	15.2	15.6	25.1	38.4	51.5	60.8	65.1	62.8	55.1	45.0	31.7	18.3	40.4
Black River Valley	Lowville	18.6	19.0	28.6	42.6	55.1	64.5	69.1	66.9	59.1	48.5	35.6	22.1	44.1
Lake Ontario Lowland	Oswego	25.4	25.2	32.9	43.8	54.7	64.6	70.8	69.4	62.1	52.1	40.5	28.9	47.5
Lake Erie Lowland	Fredonia	28.6	27.5	34.8	46.2	57.4	68.1	72.5	71.0	64.7	53.9	41.9	31.6	49.9
Western App. Upland	Allegany St. Pk.	26.1	25.3	33.1	44.5	55.3	64.0	67.7	66.1	59.8	49.9	38.1	27.7	46.5
Finger Lakes Area	Geneva Exp. Sta.	27.0	26.4	34.6	46.5	58.2	68.1	72.9	70.8	63.4	52.8	40.9	29.6	49.3
Eastern App. Upland	Morrisville	21.5	20.8	29.6	42.1	53.8	63.1	67.7	65.7	58.0	47.8	36.3	24.0	44.2
Eastern App. Upland	Delhi	24.1	23.8	32.1	43.9	55.5	54.3	68.7	66.7	59.7	49.5	37.8	26.0	46.0
Mohawk Valley	Little Falls Res.	21.5	21.7	30.7	44.0	56.6	65.6	70.6	68.5	60.8	50.1	37.4	24.4	46.0
Hudson Valley	Poughkeepsie	27.6	28.3	37.6	49.3	60.7	69.5	74.7	72.3	64.4	54.0	42.2	30.4	50.9
Hudson Hills	Bedford Hills	30.1	30.5	38.7	49.7	61.0	69.5	74.7	72.6	65.4	55.0	42.9	32.3	51.9
Long Island	Bridgehampton	32.3	31.9	37.9	46.5	56.2	65.2	71.4	70.7	64.4	55.2	45.1	34.9	51.0
Long Island	Setauket	33.4	32.7	39.5	50.2	59.8	67.9	73.8	72.4	66.7	57.1	46.5	36.0	53.0
New York City	Central Park	32.7	32.5	40.9	50.5	61.9	71.1	76.1	74.3	68.3	57.6	46.6	35.9	54.0

Source: U.S. Dept. of Commerce, Weather Bureau, "Climates of the States—New York," *Climatography of the United States,* No. 60-30 (February, 1960).

summer is the season with most clear weather, with autumn following close behind.

Air Temperature

Measured in the shade in standardized shelters with comparable instruments, air temperature is readily available from an abundance of locations in the state. As stated previously, it is not a direct measure of the disposition of the energy received from the sun, yet the averages of air temperature are a useful index of the energy available for growth of plants and of the heat load on structures and mammals. On the average, air temperature, like solar radiation, varies seasonally, although the extremes of the average temperature march are delayed thirty to fifty days after the summer and winter extremes of radiation. Table 4 shows the mean annual and monthly temperatures for representative climatic stations.

It is apparent from the table that temperatures behave rather similarly across the state. Every climatic station, except those on Long Island or Manhattan Island, has a mean January or February temperature below freezing and all have a mean temperature of 64° F to 72° F in July. From the coldest mean January record in the Adirondacks to the warmest on Long Island Sound, is 18° F. In July the difference is 12° F. In the variations from one station to another are found the consistent effects of environmental factors which contribute to local determinations of the heat, moisture, and air movement exchanges between the earth and the atmosphere.

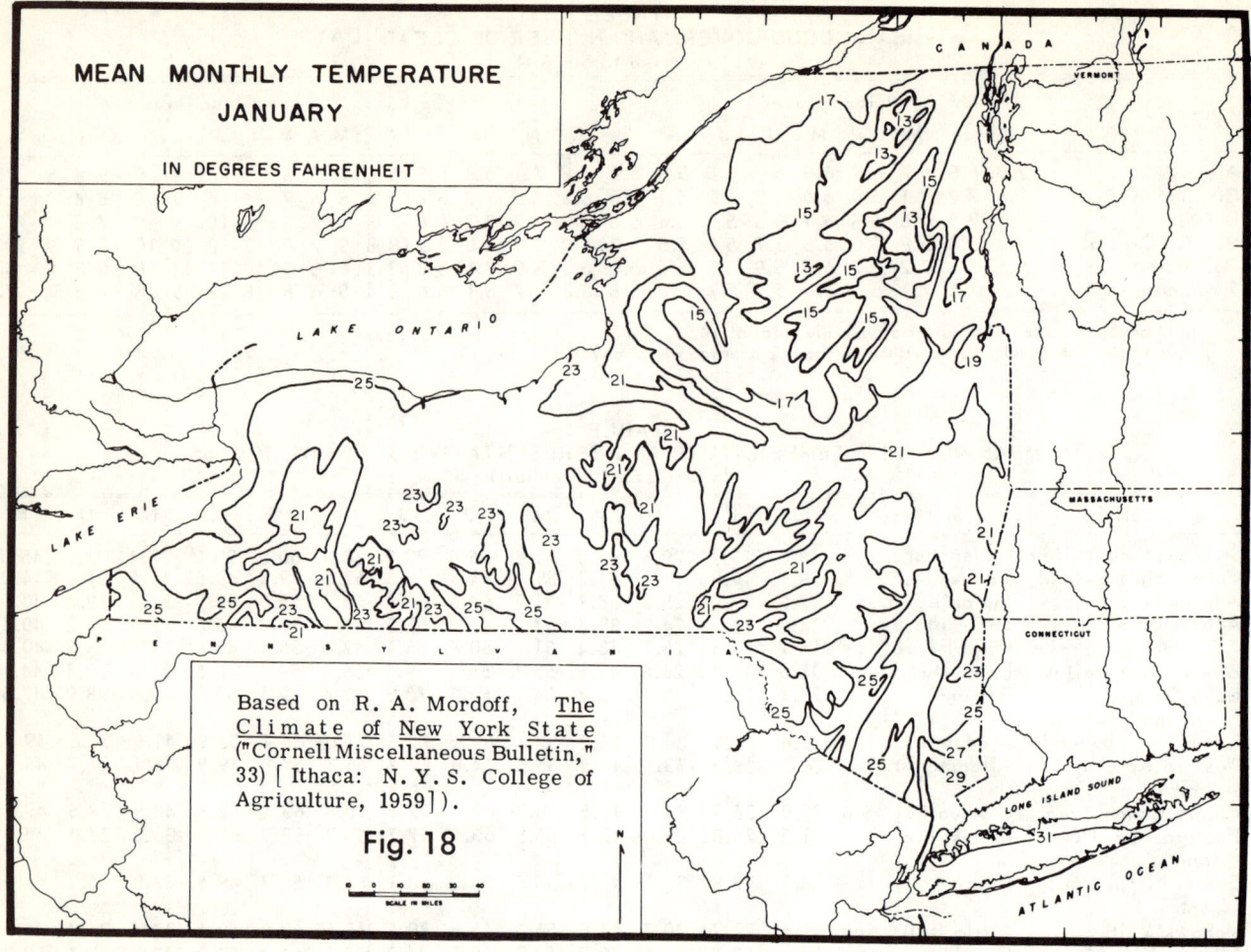

Fig. 18. Based on R. A. Mordoff, *The Climate of New York State* ("Cornell Miscellaneous Bulletin," 33) [Ithaca: N.Y.S. College of Agriculture, 1959]).

The apparent influence of latitude is demonstrated in Table 4 by the regular increase in mean annual temperature southward among the lowland stations. The relatively big change in temperature from month to month should be ascribed largely to the effect of latitude or position of New York on the globe and on the continent. There is a large difference in the effectiveness of the sun's rays for heating in summer and winter here because of the angles at which the energy is received. A given quantity of energy is spread over large areas in winter or small areas in summer and persists for short or long durations. The East Coast position of New York assures that the usual flow of air to it from over the continent will result in more heated air in summer and more cooled air in winter than is characteristic of West Coast locations at similar latitudes.

Elevation is unmistakably significant in temperature comparisons. The stations with the lowest average temperatures in the state are Stillwater in winter and Indian Lake in summer. Lake Placid is intermediate at both seasons, but its annual average is lower than either Stillwater's or Indian Lake's. These generally low temperatures for Adirondack stations are clearly due to their elevation. Higher locations, for which there are no records, undoubtedly have even lower averages.

The presence of large water bodies to serve as special storages for energy is responsible for the retardation of warming in spring and of cooling in autumn in some areas, e.g., Oswego, Fredonia, and Setauket. The magnitude of the effect is conveniently demonstrated in Table 4 by comparing temperatures of Oswego and Geneva. These stations' mean monthly temperatures for February are only 1.2° F. apart. The difference increases to 3.5° F in May and June, when Geneva warms more rapidly than Oswego. From August to January Geneva's temperature average is never more than 1.4° F. warmer than Oswego's, and in November Oswego almost equals the mean at Geneva. The climatic significance of Oswego's proximity to large Lake Ontario is confirmed by the more pronounced lag of temperature there than at Geneva, where the effect of water bodies is negligible.

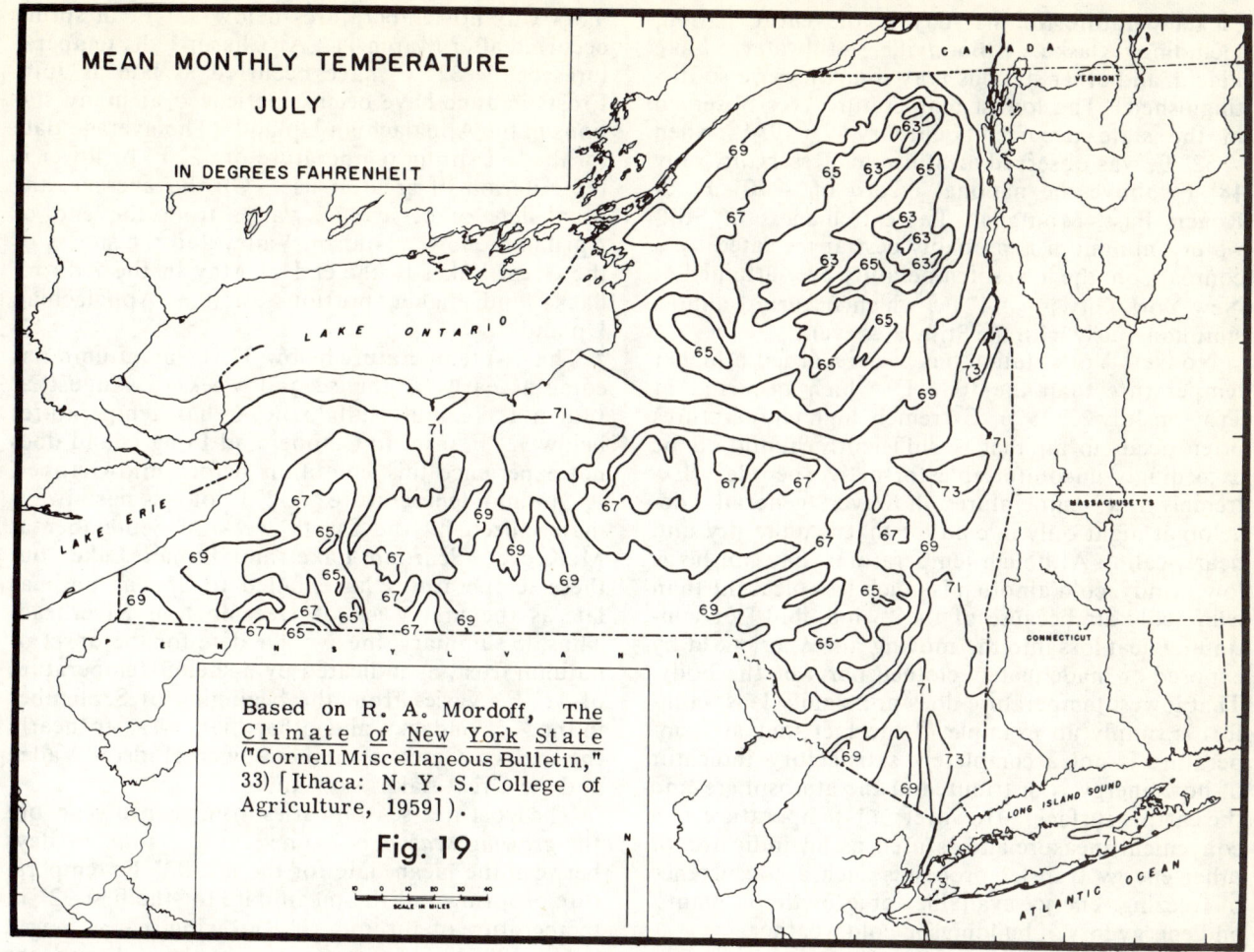

Fig. 19. Mean Monthly Temperature July in Degrees Fahrenheit. Based on R. A. Mordoff, The Climate of New York State ("Cornell Miscellaneous Bulletin," 33) [Ithaca: N. Y. S. College of Agriculture, 1959]).

The barrier to air movement which the Catskills and Adirondacks constitute is seemingly unimportant to the average distribution of temperature in the lowlands around them. While barriers are the least effective factor in the distribution of average temperature, the most effective are elevation and latitude. The effect of large water bodies is only moderately, and locally, important.

Mapping Mean Monthly Temperatures

So many factors bear on the distribution of temperature that maps of monthly temperatures are never simple. January mean monthly temperatures in New York, which are mapped in Figure 18, exhibit a range of nearly 20°F. This is due mainly to elevation's effect and constitutes an impressive difference—equivalent to the latitudinal difference in temperature between two sea level stations as far removed as the cities of New York and Quebec. Since elevation affects the mean temperature so greatly, the pattern of monthly temperature necessarily resembles somewhat the contours of elevation. A layering of temperature is therefore apparent. In parts of the state where the relief is abrupt, as in the valleys of the Hudson and Mohawk rivers, the temperature pattern for January exhibits abruptness in its gradients. Where the presence of water bodies that store much heat is added to the relief effect, as on the western slope of Tug Hill or in the lower Hudson Valley, the most dramatic gradients of monthly temperature for the entire year are found.

The July mean monthly temperature is presented in Figure 19. The range of temperature across the state in July is only about 10° F. Abrupt changes in elevation are matched by steep gradients of temperature, crowding the isotherms along their extent. In the lee of Lake Ontario on the western slope of the Tug Hill Upland, the elevation effect and the lake effect on temperature tend to counteract each other, diminishing the thermal gradient in comparison with January.

Extreme Temperatures

Occasionally, when a cold air mass invades the state, some New York station may record the coldest temperature for that day in the entire nation, including Alaska. Boonville, Stillwater, Lake Placid, and other stations may sometimes be so distinguished. The lowest temperature ever observed in the state occurred February 9, 1934, when −52° F, was observed at Stillwater Reservoir, only 18° F. above the national record of −70° F. at Rogers Pass, Montana. The extremeness of Stillwater's minimum is probably best appreciated by a comparison; the lowest temperature ever attained at New York City is −14° F., which occurred almost simultaneously with the Stillwater event.

No New York station has ever recorded a higher temperature than the 108° F. which occurred at Troy on July 22, 1926. Extremely high temperatures often occur in air that is sufficiently humid to be exceedingly uncomfortable for active people. Extremely low temperatures, however, generally develop at night only in cold air that is quite dry and nearly calm. Although temperatures may not be as low, windy, cold air often is much less pleasant than calm, cold air because of the "wind chill" or continuing heat loss into the moving air experienced by exposed or inadequately clothed parts of the body. That lowest temperature does not signify least comfort is simply an example of the fact that air temperature is not a completely satisfactory indicator of how energy is distributed in the atmosphere and the earth's surface. However, air temperature is a convenient measure and is used as an indicator of other energy transfer processes such as occurrence of freezing, energy available for growth of plants, and energy loss of buildings in cold weather.

Frost Occurrence

Freezing is said to have occurred: (1) when vegetation freezes or frost crystals form on the ground though the temperature in the weather shelter is warmer than freezing and (2) when shelter temperatures are below freezing regardless of whether there is frost damage among local types of vegetation. The first of these is not an objective observation supported by an instrument record, so it is not incorporated in the summary of climatological data. The climatological record which represents freezing conditions is 32° F. or lower, measured in the shelter, but other temperatures may be indicative of frost damage to certain plants. Because some hardy plants are not as easily affected as others, climatological summaries report the mean date of the occurrence of temperatures of 16° F., 20° F., 24° F., 28° F., and 32° F. Usually the ground is even colder than the shelter when nocturnal freezing occurs. The shelter temperature is usually colder than the levels above which fruit-tree blossoms repose.

Freezing conditions are certain to occur every winter throughout the state. In one year at New York City no temperatures below 32° F. in spring occurred after March 11. At Chasm Falls temperatures below 32° F. have occurred as late as July. Frosts in June have been experienced at many stations in the Appalachian Upland. The average date for the last spring temperature of 32° F. or lower is the criterion of greatest use. On the average, the usual date of occurrence varies from the end of April in the lower Hudson Valley and the shores of the Great Lakes to the end of May in the Adirondacks and higher portions of the Appalachian Upland.

The first temperature below 32° F. in autumn has come as early as the second week of August at Indian Lake, but Buffalo never has temperatures below 32° F. prior to October and Long Island does not experience this condition before mid-October. An autumn temperature of 32° F. or less has always materialized by the fourth week of September at McKeever, Schroon Lake, and Indian Lake, but these temperatures have failed to appear until as late as the first week of December in Riverhead. Thus, in summary, the average date for the onset of autumn frost, as indicated by a shelter temperature of 32° F., varies from the beginning of September in the Adirondacks and other high areas to nearly the end of October in the Lower Hudson Valley and the Great Lakes Lowland.

The frost-free season varies from year to year, but the *growing season* is defined as the time in days between the mean date for the last 32° F. temperature of spring and the mean date for the first 32° F. temperature of autumn. The growing season is well below 100 days in some parts of the Adirondacks and between 100 and 120 days on the higher portions of the Appalachian Upland. On the Erie-Ontario Lowland and along the Hudson–Mohawk Lowland it approximates 180 days while northeastern Long Island experiences as much as 220 days free of frosts.

Degree-Days

The concept of degree-days is concerned with the difference between the mean temperature and a reference value determined over a period of time. There are two types of degree-days: heating degree-days and growing degree-days. *Heating degree-days* are utilized to represent the heat loss of buildings where the mean daily temperature is below the temperature selected as a standard of comfort. *Growing degree-days* are reckoned as the sum of the departures of mean daily temperature from a low threshold temperature chosen to represent germination conditions.

Heating degree-days are obtained by subtracting the daily mean temperature from 65 whenever the daily mean is below 65° F. and accumulating the

TABLE 5
NORMAL HEATING DEGREE-DAYS

	J	F	M	A	M	J	J	A	S	O	N	D	Yr.
Albany	1,318	1,179	989	597	246	50	0	24	139	443	780	1,197	6,962
Binghamton	1,218	1,100	927	570	240	48	0	36	141	428	735	1,113	6,556
Buffalo	1,225	1,128	992	636	315	72	16	30	122	433	753	1,116	6,838
N.Y.C.: Central Park	1,001	910	747	435	130	7	0	0	31	250	552	902	4,965
Rochester	1,249	1,148	992	615	289	54	9	34	133	440	759	1,141	6,863
Syracuse	1,225	1,117	955	570	247	37	0	29	117	396	714	1,113	6,520
L. Placid (1962)	1,575	1,472	1,161	777	384	155	157	144	394	664	1,113	1,436	9,431

result. For example, a mean daily temperature of 39° F. indicates that 26 heating degree-days occurred in one day. Summing the heating degree-days over the year results in the totals for New York State locations given in Table 5.

Except for Lake Placid, Albany has the greatest total of heating degree-days in the year, with Buffalo and Rochester close behind. Syracuse and Binghamton require nearly as much heating, while New York City is appreciably warmer. Fragmentary records for other stations indicate that over 8,000 heating degree-days accumulate at the extreme northern border of the state and the 9,431 heating degree-days for Lake Placid indicate the extremely cold nature of Adirondack locations. In conterminous United States outside New York State, only Maine, the extreme northern parts of Vermont and New Hampshire, northern portions of Minnesota, Wisconsin, Michigan, Montana, and all of North Dakota exceed the 8,000 level. Heating degree-days are useful to fuel suppliers who study a building's fuel consumption in past years and thus are able to plan for deliveries each year by watching the rate of accumulation of heating degree-days. The relation of fuel consumption to heating degree-days is altered somewhat by the behavior of the wind. High winds ventilate the surface of a building, increasing its heat loss. Thus again there is evidence that the use of temperature records to represent energy transfers is subject to certain errors, though the practical value of convenient temperature records is outstanding.

Growing degree-days are obtained by subtracting a "zero" value, representing the minimum temperature for growth, from the mean daily temperature. Totals for each day of the year are then summed. Since the minimum temperature for growth of crops is not the same for all crops, there are many "zero" values and thus many kinds of growing degree-days. Their accumulation is not published in the regular publications of the Weather Bureau. Yet growing degree-days are widely used in modern agriculture. Seed suppliers for cash crops sometimes provide a service to their customers which uses the degree-day idea, but they call it a "heat unit." A certain number of heat units, or growing degree-days, is required to mature a given crop. Since some years accumulate growing degree-days more rapidly than others, the number of days from planting to harvest differs from year to year. To insure that crops will be harvested in orderly procession, freezing and packing plants use the growing degree-day that applies to its crops in order to plan for plantings. If 100 acres of harvest is desired per day in August, plantings of 100 acres must be spaced in May by the equivalent of an August day's heat. Thus growing degree-days occurring on the average between days at harvest time become the basis for reporting planting dates in the spring. Table 12 provides data on growing degree months for New York stations. The concept of the *growing degree month* is exactly the same as for growing degree-days except that the "zero" value, in this case 40° F., is subtracted from the mean monthly temperatures instead of from the mean daily temperatures. Both are summed.

EVAPOTRANSPIRATION, AN ENERGY-USER

It might appear from what has been said that energy is involved mainly with processes that affect air temperature. In reality, only a meager part of the energy reaching the earth directly affects air temperature. The preoccupation of climatologists with temperature phenomena is a consequence of their ready availability and the utter lack of other measures of energy. As a matter of fact, the disposition of the main amount of the energy gained by the ground is for the vaporization of water in evapotranspiration. Evapotranspiration, a key factor in the moisture supply, is the largest consumer of energy. Unfortunately, it is not widely measured, but, since it can be estimated rather well, it will be dealt with later when moisture is discussed.

Air Movement

Air movement, or wind, with its associated transfer of heat, moisture, and storminess, strongly influences the weather and climate of any locality. Interestingly enough, wind directions and storm conditions in New York are little affected by the

TABLE 6
PREVAILING WIND DIRECTIONS

	J	F	M	A	M	J	J	A	S	O	N	D
Albany	WNW	WNW	WNW	WNW	S	S	S	S	S	S	S	S
Binghamton	NW	NW	NW	NW	NW	NW	W	E	NE	NW	W	W
Buffalo	WSW	SW	SW	SW	SW	SW	SW	SW	S	S	S	WSW
New York City	NW	NW	NW	NW	NW	SW	SW	SW	SW	SW	NW	NW
Rochester	WSW	WSW	WSW	WSW	SW	SW	SW	SW	SW	SW	WSW	WSW
Syracuse	NW	WNW	WNW	WNW	WNW	WNW	W	NW	S	E	WSW	WSW

condition of the ground. On the contrary, winds and storms are largely the result of the state of the general circulation of the atmosphere, particularly the layer of air well above the effect of friction with the ground.

DIRECTION OF MOVEMENT AND
STORM MECHANISMS

In the latitude of New York the general circulation is represented by a rapidly moving eastward stream of air. The surface component of this circulation is popularly referred to as the westerlies. Unlike the generally uniform flow of air aloft, the surface air over New York is distributed by eddy flows hundreds of miles in diameter. These eddy flows, which have low-pressure centers, are called *mid-latitude cyclonic storms* and receive surface winds spiraling inward in counterclockwise fashion. Because upper air conditions over central and eastern North America generally are conducive to intensification of the mid-latitude cyclonic storms as they move eastward or northeastward over New York, these storms are important weather producers. Between succeeding cyclonic storms high-pressure eddies, or *anticyclones*, produce outward-moving, clockwise, spiral patterns of air movement.

Both the high-pressure and low-pressure eddies, with their entire compass range of wind direction, drift along toward the east, guided in their course essentially by the winds of the general circulation above. As a consequence, surface wind directions in New York State are capricious; every direction is represented part of the time at all stations and, unlike tropical conditions, no one direction completely dominates all others. In Table 6 prevailing directions at selected stations are presented. Directions associated with the prevailing eastward drift of disturbances in the general circulation tend to occur most often. Thus west, southwest, and northwest directions together are the majority. When the circulation aloft is least active, local topographic features may direct the gentler winds. Then local breezes produced by heating and cooling of slopes may dominate much of the time.

The mechanism of the mid-latitude cyclonic storm brings together air masses from strikingly different source regions and with strikingly different characteristics. The contact line between two such contrasting air masses is called a *front*. It is along these fronts where warmer, lighter air masses rise over the colder, heavier air that abrupt weather changes and precipitation are most likely to occur. Characteristically, southern and eastern portions of these storms will have northward-flowing air and be showery, cloudy, and relatively warm, while western and northern portions, dominated by southward-flowing air, are inclined to be clear and cold (see Fig. 20). Cyclonic storms, whose centers move across the central Atlantic states south of New York, may contribute a flow of stormy, moist air from the Atlantic to New York State in the spiraling counterclockwise winds north of the storm centers. When these storms pass north of the state over Ontario and Quebec, there usually is a flow of warm, moist air of tropical or subtropical origin which crosses the state on its way to the center of the storm and produces showers as it goes.

Small-scale storms with only local effects very often are instigated by larger frontal features. For example, a cold front is often associated with a line of local thunderstorms which it sweeps along its path. The individual cells of the thunderstorms wax and wane, but almost the whole area over which the cold front passes experiences thunderstorms and rain. According to Table 7, thunderstorms are plentiful in New York during June, July, and August. New York has more thunderstorms than the West Coast of the United States or any part of Canada but fewer than central or southeastern United States. Thunderstorms move according to the direction of the winds aloft. Usually they progress from southwest toward northeast. In some parts of the state, where their most frequent direction of movement coincides with the axis of river valleys such as the St. Lawrence, it is erroneously believed that the valleys guide thunderstorm travel.

Other storms of small extent include tornadoes and squalls. Tornadoes, very small in size but extremely intense and destructive, tend to occur in late spring and early summer, though they are rare in New York. Between 1880 and 1960 only sixty-six tornadoes were reported in the state. Squalls, winds of considerable intensity and usually of short dur-

Hypothetical Diagram of a Mid-Latitude Cyclonic Storm

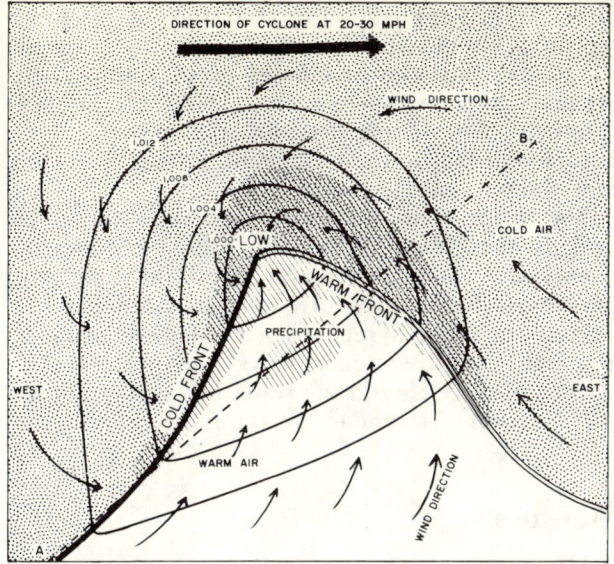

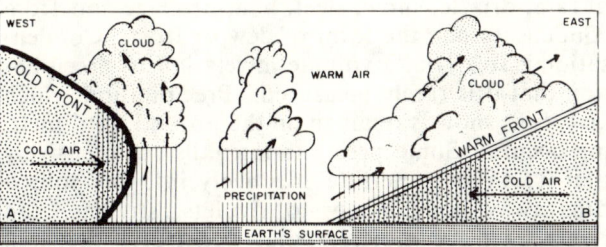

Fig. 20

ation, are not reported in climatological summaries, but they occur frequently in various parts of the state when air is unstable and turbulent. The effect of the Great Lakes on cold winter air produces noteworthy snow squalls. Although a few such squalls cover large areas, most are of local extent only. Rochester, Buffalo, and Syracuse have about 25 per cent more days with precipitation during December, January, and February than stations distant from the Great Lakes. The snow squalls at these cities may last for several hours per day without resulting in a measurable amount of precipitation; on other occasions heavy snow may occur.

TABLE 7
AVERAGE NUMBER OF THUNDERSTORMS

	J	F	M	A	M	J	J	A	S	O	N	D	Yr.
Albany	*	*	1	2	4	7	8	5	3	1	*	*	31
Binghamton	*	*	1	2	4	6	8	6	3	1	*	*	31
Buffalo	*	*	1	2	4	5	6	5	3	2	1	*	30
N.Y.C.: Battery	*	*	1	2	4	6	7	6	3	1	1	*	31
Rochester	*	*	1	2	4	5	7	5	3	1	*	*	28
Syracuse	*	*	1	2	4	6	7	6	3	1	*	*	30

*Occasional storm recorded.

Hurricanes, larger in size than tornadoes, frequently pass close enough to the state to affect its weather. A direct hit by a hurricane may be disastrous because its effects are widespread and its total damage in the state probably is greater than in any other kind of violent storm. Between 1901 and 1955 more than fifteen hurricanes caused damage to property on eastern Long Island.

VELOCITY OF AIR MOVEMENT

Even though violent storms such as hurricanes, tornadoes, thunderstorms, and squalls do occur, average wind velocities in the state are not high. The momentum of the upper atmosphere is transferred downward through the lower air and is dissipated in friction against the surface of the earth. Consequently, in the layer of air near the ground, the speed of the wind is much less than speeds aloft.

At the ground the average wind speed may vary more between nearby sites with different exposures than between sites with comparable exposures at widely separated localities. In Table 8, where average wind speeds for selected stations in the state are summarized, wind speed at Buffalo is shown to be relatively great. While it is difficult to assure comparability of sites for wind measurements, the speeds at Buffalo are very nearly the same as speeds measured at the southernmost tip of Manhattan Island. The effect of adjacent water bodies on wind speeds at Buffalo and The Battery is apparent, while the decrease between The Battery and protected Central Park is indeed abrupt. Central Park wind speeds are like those at Syracuse, indicating that air movement is slowed as much in passing through southern Manhattan as it might be by surface friction in reaching miles farther inland.

Maximum average wind speed comes in March at all cities except Buffalo and Rochester. For Buffalo the maximum speed occurs in December and January, which coincides with the time of maximum contrast between water and air temperatures, before the ice on shallow Lake Erie has collected in quantity. At Rochester the turbulence induced by contrast in lake and air temperature lasts somewhat longer, but is less severe since a significant part of deep Lake Ontario only rarely freezes over. The average wind speed at Buffalo is so high that it must be regarded as a special factor in the interpretation of temperature there.

A much greater volume of air will be heated or cooled when wind speed is high than when it is low. Thus, changes of temperature in a large volume of rapidly moving air will be less than those that would occur in the smaller volumes stirred by lighter winds. Comparison of mean temperatures, freezing conditions, and degree-days at Buffalo with those of other places is made difficult by this effect of high wind. Large heat losses from buildings and clothing when it is cold, and little likelihood of early autumn frost or extremely low winter temperatures, is characteristic of Buffalo. Also, few late spring frosts or high summer temperatures occur at Buffalo. Extreme wind speeds near the ground of 91 m.p.h. have been observed at Buffalo and of 113 m.p.h. at New York City. Such high speeds, however, are so rare in the state that special precautions relative to construction and outdoor storage of lumber, containers, or any loose items of low density ordinarily are not taken. In the upper air, at 20,000 feet and above, wind speeds regularly exceed 150 m.p.h.

The occurrence of very low wind speeds is disadvantageous in many communities because of the inability of calm air to remove a continuing discharge of automobile fumes, trash burner smoke, industrial smokes and fumes, chimney emissions, and radioactive materials. When low wind speeds persist, they are nearly always associated with anticyclonic weather in which the normal reduction of temperature with height is reversed. In such instances, warmer air is found aloft, having descended from the very high atmosphere and having been warmed as it came under greater pressure. This constitutes a *thermal inversion*—the inversion of the usual temperature structure—which limits rising smoke, etc., since it signifies that air of very low density lies over the surface air. Smoke and hot gases then rise only to the level of the low-density air where their buoyancy is matched or exceeded. Often the inversion layer confines emissions to a depth of one or two thousand feet. Over a large city the volume of air beneath the warm layer which is replaced by low velocities is too small to accommodate normal emissions. Atmospheric pollution is the result. Industrial control of emissions during periods of low velocity and inadequate volumes of air transport is generally quite effective. The release of radioactive materials at Brookhaven and Tuxedo Park in the southeastern part of the state ceases entirely during these periods. The output of industrial effluents at major centers such as Buffalo is constantly surveyed by a state agency and recommended reductions are made when necessary. The introduction of new industrial processes may increase the burden for a time, as the adoption of direct oxygen application in steel-making has done recently. Yet in virtually every community the main supplies of particles and gases to the atmosphere are made by individual citizens. Their automobiles, chimneys, and refuse fires are the major contributors, and their output varies little from one kind of weather to another.

The occurrence of heavy fogs is confined to periods of little air movement. Binghamton with twenty-seven and Albany with twenty-three per year have more frequent occurrences than other major cities in the state. Syracuse has the fewest. August, September, and October mark the greatest frequency of heavy fog in the state, but spring constitues a period of secondary maximum.

Moisture

Moisture is derived from precipitation in the form of rain, drizzle, snow, sleet, hail, or glaze and from condensation in the form of dew or frost. Condensation is so trivial a volume in relation to precipitation that it is rarely measured. Precipitation which arrives as snow is retained on the ground until melting occurs. Some precipitation falls so intensively that it cannot penetrate effectively into the soil; it runs away immediately to rivulets, streams, and rivers. Of the moisture that penetrates the soil, about half is held by the capillary spaces in the soil and the remainder temporarily fills the larger spaces but drains out within a day or two as gravity water to the subsoil and ground water table. The moisture held in the capillary spaces is the effective precipitation since it alone contributes to the growth of plants. All of this moisture is lost from the soil by evaporation and transpiration.

Measures of moisture are rather limited. Only rainfall and snowfall are regularly recorded throughout the state. Although evaporation from pans is measured at special sites, results are not applicable to losses from the ground, so to get at these losses evaporation and transpiration (*evapotranspiration*) losses must be estimated. Procedures for estimating potential evapotranspiration have been developed

TABLE 8
AVERAGE WIND SPEED (Miles Per Hour)

	J	F	M	A	M	J	J	A	S	O	N	D	Yr.
Albany	9.9	10.6	10.7	10.4	9.1	8.1	7.3	6.7	7.4	8.0	8.9	9.3	8.9
Binghamton	6.9	7.1	7.2	6.9	5.9	5.2	4.7	4.6	4.8	5.5	6.4	6.6	6.0
Buffalo	17.4	16.4	15.9	14.8	13.2	12.5	12.1	11.7	12.8	14.1	16.4	17.0	14.5
N.Y.C.: Central Park	11.1	11.1	11.3	10.9	9.1	8.4	7.9	7.9	8.3	9.3	10.3	10.6	9.7
N.Y.C.: Battery	16.4	16.7	17.1	15.2	13.5	12.8	12.1	11.7	12.4	13.9	15.7	16.2	14.5
Rochester	11.0	11.1	10.9	10.1	8.9	8.0	7.5	7.0	7.6	8.4	9.9	10.4	9.2
Syracuse	11.1	11.3	11.4	11.0	9.3	8.4	8.1	7.8	8.4	9.3	10.8	10.9	9.8

and the water balance (defined on page 70) at any station can be computed. The water balance computations provide information relative to water deficiency and water surplus in the soil as dictated by average amounts of energy and moisture received.

MEAN ANNUAL PRECIPITATION

Two parts of the state have more than 50 inches of precipitation annually: the Tug Hill and the south-central Adirondacks, and the southern Catskills. The southwestern portion of the Appalachian Upland is almost as wet. The isohyetal lines in Figure 21 tend to conform to the pattern of elevation wherever relief is prominent because land barriers greatly affect the distribution of precipitation. Yet no single elevation corresponds to the same amount of precipitation at very many places. Maximum precipitation zones are found on the southwest sides of the upland areas in the state but below the summit levels. Figure 21 shows that in the lee of these areas are lowlands with rather small amounts of precipitation. While the difference in elevation between uplands and lowlands is not great, the leeward precipitation minimum resembles the rainshadow of more mountainous areas in western North America. The western shore of Lake Champlain receives only slightly more than half the annual precipitation found on the windward, southwest side of the Adirondacks. The middle and northern parts of the Hudson Valley have significantly less precipitation than the highlands on either side of the valley, and the Finger Lakes region is noticeably drier than the western part of the Appalachian Upland or the Catskills. Even the Tug Hill Upland provides considerable contrast in precipitation with the lower Black River Valley just to the northeast of it. The correspondence between the land forms maps (Fig. 9 or 13) and the precipitation map (Fig. 21) is here underscored.

Average annual precipitation in both Rochester and Syracuse is distinctive nationally. Rochester shares with Detroit the distinction of having less precipitation than any other large city in eastern United States; and Syracuse has more days with measurable precipitation than any other large city in all the conterminous forty-eight states.[1] This combination of many days with precipitation but a low annual total indicates a relatively low average intensity of precipitation.

MEAN MONTHLY PRECIPITATION

In the twenty-five-year period, 1931–55, average monthly precipitation was reasonably uniform at most stations over the state, with 2.5 to 3.5 inches falling in every month in most places. Furthermore, at most stations the amounts per day are so nearly uniform that the differing number of days in the months accounts for a significant part of the variations in monthly precipitation totals. February, therefore, tends to be the month with least precipitation at many stations, as Table 9 illustrates. Buffalo is a noteworthy exception to this. Its minimum of 2.45 inches occurs in July, while its maximum of 3.09 inches falls in November. At the extreme southeastern end of the state on Long Island, Bridgehampton has a regime similar to Buffalo's, with 2.59 in July and 4.79 in November. That the minimum occurs in July at both stations is a testimony to the lack of convective heating over the water surfaces, which elsewhere is effective in aggrandizing or initiating precipitation conditions. Many stations affected by air trajectories from over the large water bodies adjacent to the state have small totals of precipitation during June. Yet other stations, too, which may be little affected by the altered climatic processes over water bodies, have reduced precipitation during June. At stations on the west side of the Adirondacks, at those on the north side of the Appalachian Upland, and at places along the Erie–Ontario Lowland, June is a month of less precipitation than May. At some of these stations the decrease could be attributed to the difference in the duration of the months, but at Highmarket, Ogdensburg, Lowville, and Oswego the decrease from May to June is more than half an inch. On the east side of the Adirondacks, the upper and central Hudson Valley, and the Catskills, the month of June has large amounts of precipitation, nearly the same as July totals, which are maximum for the year. Here the lake effect is not apparent.

The effect of the Great Lakes and the Atlantic Ocean on precipitation totals is not always similar, although simultaneous maximums occur in the vicinity of each body of water. Buffalo, Highmarket, Lowville, Oswego, and Fredonia have maximums in autumn when the waters of the lakes are capable of transferring to transient, cooler air quite large amounts of water vapor and heat. At stations near these but not much affected by the lakes, October is one of the drier months. On Long Island November stands out after a not very rainy October as the month with maximum precipitation. While the effect of the ocean on air passing over it is important, the intensity of nascent storms, especially in November, which move northeastward essentially parallel to the coast of the middle Atlantic states, is important, too. November storms have been responsible for exceptional rains, high winds, and highest tides along the coastal area.

Although there is relatively little contrast in

---

[1] C. K. Vestal, "The Precipitation Day Statistic," *Monthly Weather Review*, LXXXIX, No. 2 (February, 1961), 31–37.

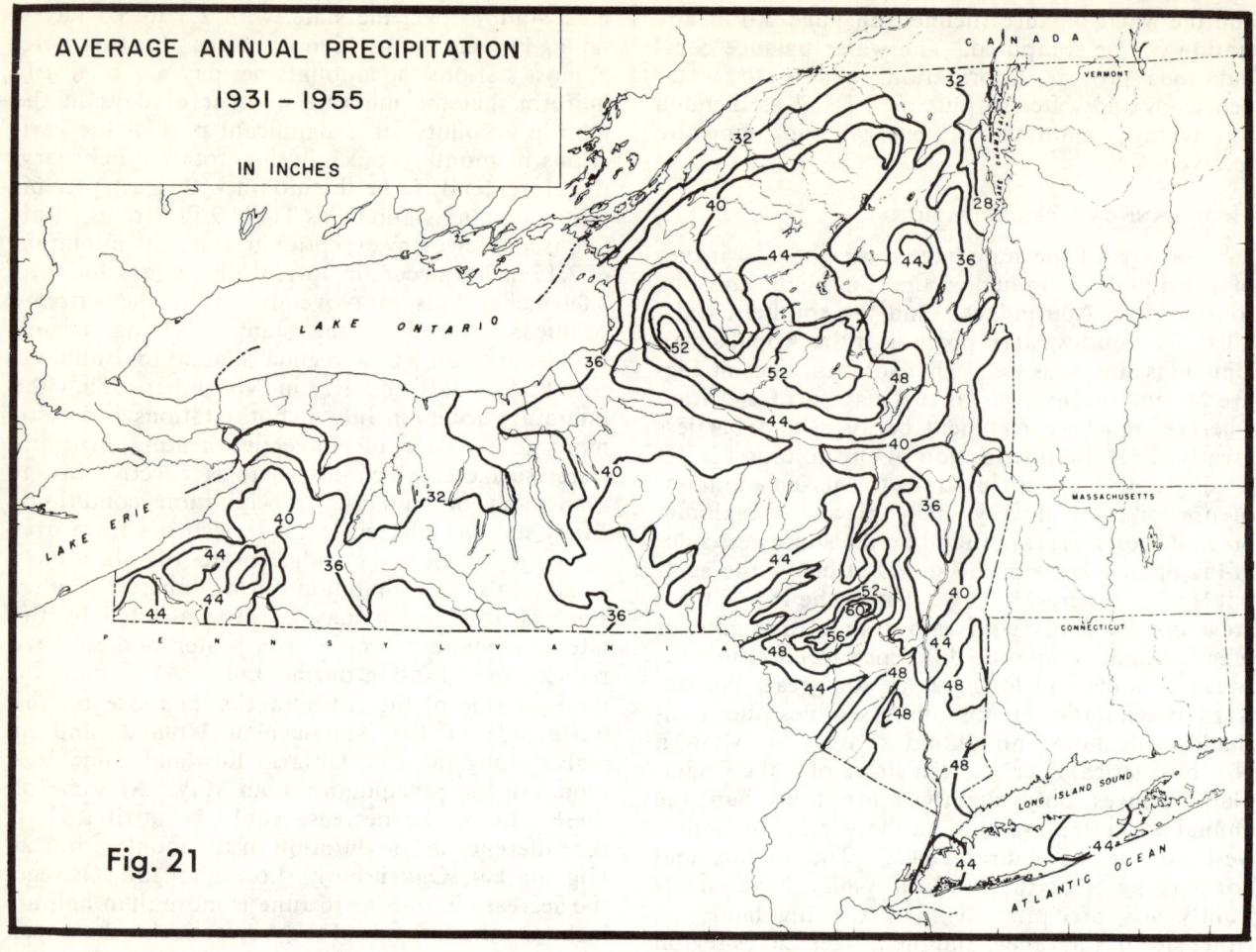

Fig. 21

amount of monthly precipitation throughout the state, the months differ considerably in the number of days with recorded precipitation. Tabulations of days with precipitation are available only for the first-order stations where hourly observations are made. In Table 10 the Syracuse, Rochester, and Buffalo records indicate that measurable precipitation occurs three days in five during winter. In addition, there are some other days with precipitation when snow flurries accumulate only a trace of snow. Although precipitation is frequent, it is not abundant; hardly any places other than Long Island and the upland locations record more than three inches per month during winter. For Syracuse and Rochester winter months have the least precipitation of the year, though the frequency of precipitation days is greatest. During winter the number of days with measurable precipitation in New York City is a third less than in the upstate cities.

Throughout the state summer has fewer days with precipitation than either spring or winter. Autumn generally has still fewer days with precipitation except for places affected by squalls and showers off the lakes.

Intense Precipitation

When precipitation falls so intensively that it cannot all penetrate the soil, a portion immediately runs off in rivulets and streams. Intense precipitation is therefore of considerable interest in designing bridges, culverts, and drainage ways and in considering available supplies of water in the soil. In the period 1941–50 all but a few stations recorded more than 3 inches in a single day, while a substantial number of places have experienced over 6 inches. Ithaca has recorded nearly 8 inches and New York City 9.55 inches in a twenty-four-hour period. Yet Buffalo, Rochester, Syracuse, Binghamton, and Albany have never experienced as much as 5 inches per day in the many years for which there are records.

Intense precipitation records at all stations have been set in the months between May and December, with July the most common. Thus, precipitation is the most intense in the period when soil moisture is low, and the contribution of intense precipitation to runoff is less than it would be for the same rainfall reaching the ground in spring and early summer when soil moisture is at or near capacity.

Snowfall

Boonville measures 204 inches of average annual snowfall. Though its record is short, this is more than any other station in the state. Certainly larger amounts must fall in remote uplands. The Tug Hill Upland west of Boonville probably receives at least 225 inches, which would make it the snowiest place in the United States east of the Rocky Mountains. Mean seasonal snowfall, shown in Figure 22, is greatest in the same elevated areas where average annual precipitation is maximum, but the snowfall pattern is considerably different in other respects from the mean annual precipitation pattern (Fig. 21). For example, Figure 22 shows that almost all valleys receive considerably less snowfall than adjacent highlands. This feature is due partly to the augmented precipitation at higher elevations and partly to the greater frequency of snow at places with colder temperatures. Every declivity is characterized by a *snow shadow*, a sharp decrease in snowfall from upland to valley. Especially marked are decreases in the valleys of the Delaware, Schoharie, Genesee, and Black rivers.

The amounts of snowfall are only slightly affected by latitude, elevation being the principal factor. While elevation is important in the snowfall distribution, the barriers posed by the Catskills and other portions of the Appalachian Upland, the Tug Hill, and the southwestern Adirondacks are even more clearly differentiated in Figure 22. The conspicuous effects of Lakes Erie and Ontario stand out, too. Snow squalls from the lakes are largely responsible for the outstanding maxima on the Tug Hill Upland and on the western portion of the Appalachian Upland south of Buffalo. The instability imparted to the air over the lakes is due to moisture enrichment and to heating, both factors which tend to make surface air less dense than overlying air. Some squalls begin before reaching shore, while others start inland. Relief barriers further intensify snow squalls. Sometimes the intensity of snowfall from these squalls rivals the intensity of summer showers. In December, 1958, 72 inches of snow, equivalent to about 7 inches of precipitation, fell at Oswego in twenty-four hours. Often snow squalls contribute only a dusting of flakes intermittently throughout

TABLE 9
MEAN MONTHLY PRECIPITATION, 1931–55 (In Inches)

Station	J	F	M	A	M	J	J	A	S	O	N	D	Yr.
Ogdensburg	1.91	1.80	2.38	2.45	2.86	2.47	3.05	2.66	2.99	2.40	2.52	2.29	29.78
Dannemora	2.29	2.13	2.43	2.94	3.28	3.52	3.55	3.25	3.29	2.69	2.55	2.60	34.52
Stillwater Res.	3.94	3.68	4.29	4.01	4.07	3.70	4.86	3.90	4.55	4.57	4.43	4.67	50.67
Lake Placid	3.04	2.87	3.36	2.74	3.24	3.59	3.95	3.55	3.79	2.87	2.98	3.21	39.19
Lowville	2.79	2.41	3.18	3.11	3.12	2.68	3.27	3.16	3.17	3.40	3.44	3.32	37.05
Highmarket	4.08	3.77	4.38	4.08	4.25	3.29	4.43	4.04	4.89	4.98	4.65	4.59	51.43
Oswego	2.86	2.63	3.09	2.74	2.93	2.15	2.67	2.53	2.71	3.28	3.07	3.10	33.76
Fredonia	2.54	2.14	3.23	3.15	3.07	3.05	2.95	3.35	3.78	3.51	3.39	2.70	36.86
Allegany St. Pk.	2.79	2.68	3.58	3.43	4.36	4.18	4.25	3.59	3.93	3.46	3.70	3.16	43.11
Geneva Exp. Sta.	2.20	2.28	2.98	2.74	3.13	3.09	3.01	2.82	2.67	3.00	2.58	2.37	32.87
Morrisville	2.38	2.40	2.96	3.03	3.71	3.61	3.84	3.80	3.54	3.66	2.94	2.83	38.70
Delhi	2.74	2.48	3.03	3.19	4.15	3.88	4.69	4.36	3.52	3.45	3.55	2.85	41.89
Little Falls Res.	2.93	2.34	3.05	3.38	3.61	4.00	4.56	4.00	4.34	3.57	3.28	2.89	41.95
Poughkeepsie	2.89	2.45	3.21	3.57	3.79	3.69	4.13	4.01	3.59	2.92	3.49	3.05	40.79
Bedford Hills	3.43	2.81	4.08	4.08	4.24	4.15	4.91	4.88	4.11	3.55	4.00	3.90	48.14
Bridgehampton	4.20	3.48	4.41	3.60	3.53	2.96	2.59	4.65	3.58	3.41	4.79	3.95	45.15
Setauket	3.86	2.99	4.29	3.76	3.64	3.59	3.23	4.33	3.76	3.19	4.24	3.60	44.48

TABLE 10
AVERAGE NUMBER OF DAYS WITH MEASURABLE PRECIPITATION

	J	F	M	A	M	J	J	A	S	O	N	D	Yr.
Albany	12	11	12	13	13	11	11	9	9	9	10	11	131
Binghamton	15	13	14	14	13	12	12	11	10	11	13	14	152
Buffalo	19	17	16	14	13	11	10	10	11	12	15	18	166
N.Y.C.: Central Park	12	10	12	11	11	10	11	10	8	8	9	10	122
Rochester	19	17	17	14	13	11	11	10	10	12	15	18	167
Syracuse	18	17	17	15	13	12	11	10	11	12	15	17	168

HEAVY SNOW IN THE APPALACHIAN UPLAND. From 60 to 200 inches of snow blanket the uplands, providing the basis for dozens of winter sports centers. *N.Y.S. Dept. of Commerce.*

the day, which is measurable as a depth of snow but is too poor in water content to be measured as a contribution to precipitation totals. One result of the persistent snow squalls is an amazing total of snow at stations with low elevations. Syracuse, Rochester, and Buffalo are the snowiest large cities in the United States. Of these, Syracuse ranks first by a small margin.

The parts of New York with more than 100 inches of snowfall per year are popularly called *snow belts*. They lie in the lee of Lakes Erie and Ontario and extend onto the adjacent uplands. In the snow belts at low elevation there are many roads to be cleared of snow, a job requiring elaborate equipment and organization. The New York State snow belts associated with Great Lakes snow squalls are the farthest equatorward of all lowland snow belts in the world.

EVAPOTRANSPIRATION AND POTENTIAL EVAPOTRANSPIRATION

Evapotranspiration, the loss of water to the atmosphere through the processes of evaporation and transpiration, is the reverse of precipitation. It is the only means by which moisture gets into the atmosphere, and it is the main consumer of energy gained by the ground through radiation. Evapotranspiration increases as the net energy increases, provided there is no shortage of moisture. Accordingly, evapotranspiration is regulated by both available energy and available moisture. The concepts of *potential evapotranspiration* and *actual evapotranspiration* are introduced here to distinguish between the moisture loss when soil moisture supplies are adequate and the moisture loss when soil moisture supplies are less than adequate. In other words potential evapotranspiration is the amount of water needed to expend the available energy. Actual evapotranspiration cannot exceed potential evapotranspiration, but neither can it exceed the available moisture supply. By comparing *available moisture* (precipitation plus soil moisture storage) and potential evapotranspiration, it is possible to arrive at satisfactory estimates of actual evapotranspiration. Potential evapotranspiration is thus a practical item as well as the key factor in understanding gains and losses of energy and moisture.

THE WATER BALANCE

Comparison of potential evapotranspiration and available moisture determines actual evapotranspiration. This comparison is called the water balance, and it is convenient to view it as a simplified bookkeeping procedure. Potential evapotranspiration (PE) is regarded as expenditure, demand, or need; precipitation (P) constitutes the monthly supply. Whenever the need (PE) is greater than supply (P), the period is dominated by soil-drying conditions, but if need is less than supply, a soil-wetting condition exists. The soil serves as a storage (or bank) for moisture, but its capacity to store moisture is limited by available capillary spaces in the root zone. The most representative soil moisture storage limit, we assume, is a depth sufficient to store 4 inches of rainfall, though specific soils have both less and more. The water balance for soils with 4 inches of storage at Ogdensburg and Brookhaven, given in Table 11, illustrates the procedure for computing actual evapotranspiration (AE), water deficit (D), and water surplus (S).

For Ogdensburg, Table 11 shows that the period through April is a time of wetting conditions since P exceeds PE. We begin the computation of the water balance in the first month of the drying season, May, with the knowledge that soil moisture in storage must be at its capacity of 4.0 inches from the long wetting season. Thus, the comparison of water supplies (P) in May with water need (PE) indicates that 0.4 inches must be obtained from soil moisture storage if the actual evapotranspiration is to attain the potential rate. Soil moisture is depleted by 0.4 inches from the 4.0 inches of the previous month, leaving a storage of 3.6 inches. Actual evapotranspiration proceeds at the potential rate, and there is no shortage of moisture (D) and no surplus (S). During June and July continued high demands for moisture relative to supply further tax the soil moisture reserve until it is exhausted. In July 2.2 inches of soil moisture carried over from June are expended along with 3.1 inches of current precipitation for a total of 5.3 inches of actual evapotranspiration. This is 0.3 inches short of the potential expenditure and constitutes a measure of the water deficit. August and September suffer a partial supply of moisture with consequent lack of storage. October is the beginning of the wetting season when there is enough to satisfy the maximum need for moisture of 1.6 inches, leaving 1.2 inches excess to be contributed to soil moisture storage. November and December continue the wetting of the soil, which culminates in December with the attainment of its capacity of 4.0 inches. However, only 0.6 inches of storage can be accommodated in December, so the remaining 1.6 inches are surplus. Surplus dominates the remaining months of the wetting season.

The computation for Brookhaven is quite similar except that there is both more moisture need and more moisture supply. Yet the distribution of precipitation during summer months is slightly more in harmony with needs at Brookhaven than at Ogdensburg and there is less water deficit. Brookhaven and Ogdensburg represent quite different kinds of climate, especially in winter, but their water balances show that Brookhaven's richer precipitation goes mainly for streamflow, while its larger demand for moisture is largely satisfied. The water balance bookkeeping procedure for each station in the state would show conditions falling largely between the Ogdensburg and Brookhaven examples, though there are some with greater energy in Long Island and the New York City area and several with less energy in the elevated portions of the state. The progress of the water balances is almost everywhere of the same general character: a wetting period predominates, but there is a drying season when soil moisture is depleted. At most stations the average amount of precipitation in summer is ordinarily sufficient to obscure the amount of occasional water deficits.

Water Deficit

The droughtiest portion of the state is the northwest corner. From Buffalo to Rochester and northward to the Lake Ontario shore, droughts representing 10 to 20 per cent of the annual moisture requirement are common. Other zones with prevailing shortages during summer are the Central Finger Lakes, the St. Lawrence Valley, and the Champlain Lowland. Water deficit is not large in the lower Hudson Valley or on Long Island.

Soils that retain less moisture than the 4 inches assumed in the water balance computation experience deficit more often. On sandy soils the soil moisture capacity may be so small for certain crops that the precipitation of summer is not sufficiently well spaced through the month to satisfy needs and there is both deficit and surplus within an actual month. Irrigation has been found to be practical, even essential, for valuable crops grown on soils with low moisture-holding capacities, and eventually probably will prove rational for overcoming the late summer deficit periods in many parts of the state.

Water Surplus

Annual totals of average water surplus (totals for all months showing surpluses, as on Table 11) are substantial for all of New York, which means runoff will be substantial. More than 30 inches of annual water surplus emanates from parts of its Tug Hill Upland and the southwestern Adirondacks (see Fig. 23). Most of the Catskills, and the Adirondacks generally, as well as western parts of the

Appalachian Upland have more than 20 inches of water surplus. The above areas constitute essentially those parts of the state which have over 45 inches of precipitation per year. Toward the higher elevations, where precipitation increases a few inches, water surplus increases several inches because there is both more moisture to provide a surplus and less consumed for evapotranspiration.

Areas with the least water surplus are found in the northern part of the state, in the Finger Lakes area, and in the Lake Champlain lowland. All have less than 10 inches of average water surplus to contribute annually to stream flow. While 10 inches is not much for New York, it is an impressive quantity of water, more than Michigan, "the water wonderland," produces in any part of its Lower Peninsula.

Much of the water surplus occurs as snow and is released sporadically when melting occurs. The retention of surplus on the uplands in the form of snow could be an important deterrent to flooding if melt is due to local solar radiation receipts. However, melt often is almost cataclysmically rapid, even at moderately high elevations, when humid, warm air arrives over the state from a tropical origin.

Water surplus must leave the place where it occurs since it is surplus to what can be stored and evaporated. Some of the surplus penetrates the soil, already saturated with soil moisture in the capillary spaces. The surplus first floods the non-capillary pore spaces, then empties downward to the ground water table later to seek its way to streams and rivers. Another part of the water surplus runs overland to join directly the surface rivulets and streams.

Climatic Regions*

All of New York State has what is usually referred to as a Humid Continental Climate. Nevertheless, significant variations occur within the state. The Adirondack area indeed has a different climate from Long Island. The climate of the Finger Lakes area is different from that of the higher Appalachian Upland nearby. Terrain, and particularly elevation, has great impact on New York's climate. Noticeable changes in the climate in short distances can almost always be correlated with a sharp change in elevation. Frequently variations from one side of a substantial topographic feature such as the Adirondacks or Catskills to the other side or even from a valley to the top of an adjacent high ridge may be as great as those between distant corners of the state. Because of this impact of terrain, the land form regions identified in Chapter 1 are spatially similar to any set of climatic regions. These similarities are clear if Figures 13 and 24 are compared. Six climatic regions in all are shown in Figure 24 and described below. Data for representative stations are shown in Table 12.

Central to the regionalization is the consideration of human use potential. Climate, we will see, affects

*This section on climatic regions added by the editor.

TABLE 11
AVERAGE WATER BALANCES, OGDENSBURG AND BROOKHAVEN (In Inches)

Ogdensburg

	J	F	M	A	M	J	J	A	S	O	N	D	Yr.
PE*	0	0	0	1.2	3.2	4.6	5.6	4.7	3.2	1.6	.3	0	24.4
P	2.1	2.1	2.3	2.2	2.8	3.2	3.1	2.8	2.7	2.8	2.5	2.2	30.8
ST	4.0	4.0	4.0	4.0	3.6	2.2	0	0	0	1.2	3.4	4.0	
AE	0	0	0	1.2	3.2	4.6	5.3	2.8	2.7	1.6	.3	0	21.7
D	0	0	0	0	0	0	.3	1.9	.5	0	0	0	2.7
S	2.1	2.1	2.3	1.0	0	0	0	0	0	0	0	1.6	9.1

Brookhaven

	J	F	M	A	M	J	J	A	S	O	N	D	Yr.
PE	0	0	.3	1.4	3.1	4.8	5.7	5.2	3.5	2.0	0.7	0	26.7
P	3.8	4.0	5.0	3.8	4.0	2.9	3.6	3.4	3.4	3.6	4.2	4.2	45.9
ST	4.0	4.0	4.0	4.0	4.0	2.1	0	0	0	1.6	4.0	4.0	
AE	0	0	.3	1.4	3.1	4.8	5.7	3.4	3.4	2.0	.7	0	24.8
D	0	0	0	0	0	0	0	1.8	.1	0	0	0	1.9
S	3.8	4.0	4.7	2.4	.9	0	0	0	0	0	1.1	4.2	21.1

*Abbreviations: PE, potential evapotranspiration; P, precipitation; ST, soil moisture storage at the end of the month; AE, actual evapotranspiration; D, water deficit; and S, water surplus.

For calculation of potential evapotranspiration see C. W. Thornthwaite, "An Approach Toward a Rational Classification of Climate," *Geographical Review*, XXXVIII (1948), 55-94.

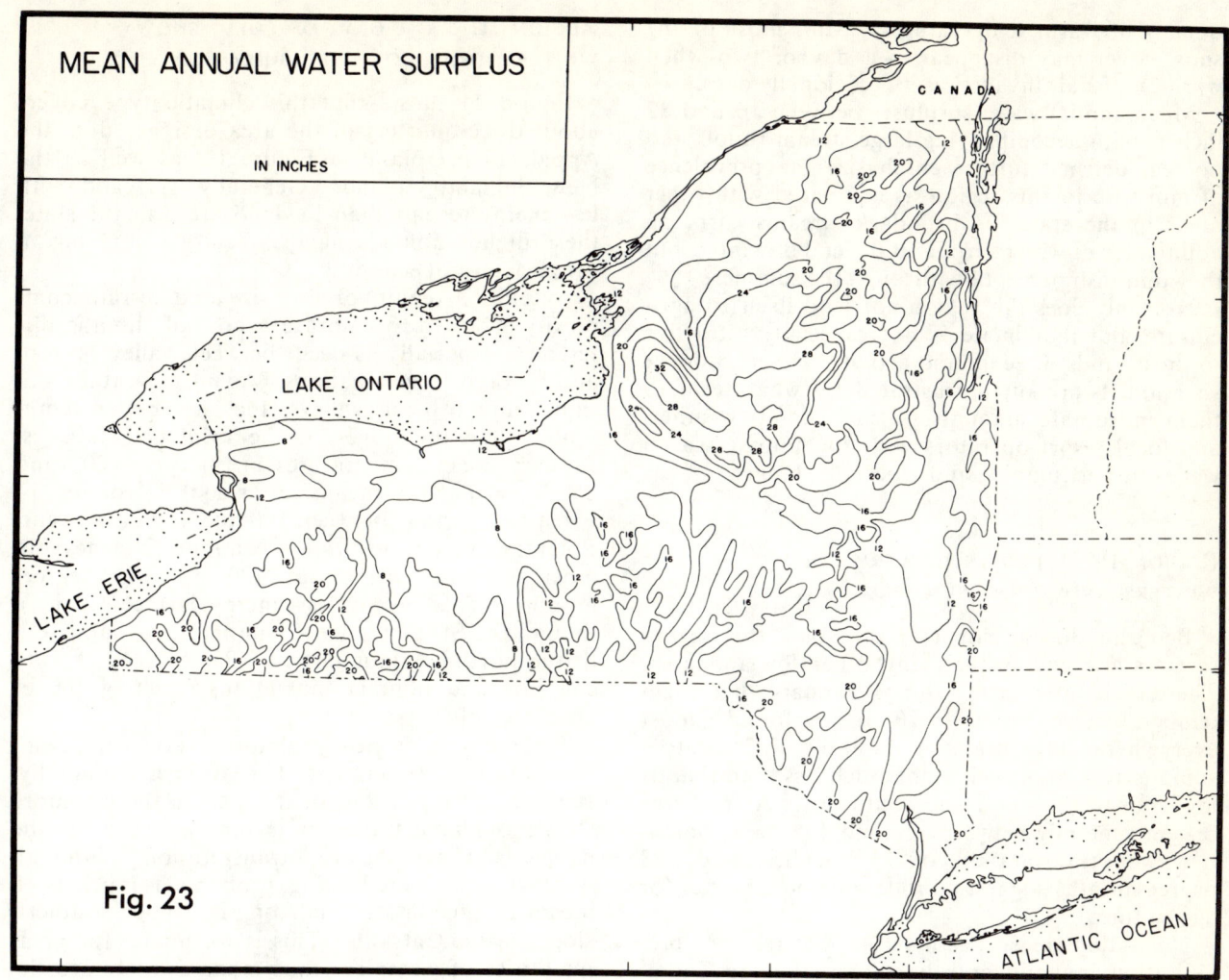

Fig. 23

natural vegetation and soil. The latter in turn has a strong impact on agriculture. In setting up the climatic regions, characteristics likely to have bearing on agricultural success are given special attention.

REGION I—A REGION OF EXTREMELY COLD, SNOWY WINTERS AND VERY COOL, WET SUMMERS

This area is high in both elevation and latitude. Thus it receives the least available energy of all parts of the state. The distribution of energy within it varies locally as the relief varies, but essentially no part of the region has adequate energy in the short season of reliable growth for successful agriculture. The average duration of the frost-free season ranges from 85 to 140 days, yet energy accumulates slowly in this time and amounts to a smaller total than during the same period elsewhere in the state. The potential evapotranspiration is only 21 inches annually and there are approximately 100 growing degree months. Winters are cold, with mean January temperatures of about 15° F. Mean July temperatures of 65° F. are the coolest in the state. Minimum temperatures below −50° F. have been recorded. The region is also the wettest in the state. All parts average over 35 inches of precipitation annually and the southwestern portions exceed 50 inches. Hoffmeister is the wettest weather station in the state, receiving 53.84 inches per year. Snowfall is high (90–165 inches) throughout the Adirondacks, and even more (130–225 inches) falls on the Tug Hill, which lies in the lee of Lake Ontario and receives the brunt of the snow squalls attributed to the heat and moisture contributed to the air by Lake Ontario. To the advantage of winter sports, snow remains on the ground in this region longer than in other parts of the state because of the lesser frequency of warm temperatures. Still, most of the snow cover may disappear in a day or two when warm, moist air invades, as it occasionally does.

Mean annual water surpluses between 16 and 32 inches, with accompanying large stream runoff and no real deficits, further emphasize the prevalence of moisture in this region as compared with other areas in the state. Adirondack streams carry an abundance of water surplus at the end of winter, but they diminish perceptibly through the summer.

Not only does this region not lend itself to agriculture, but its climate is not especially attractive to most kinds of economic activity. Even summer vacationists are often frustrated by what seem to them inordinate amounts of cool, damp weather, and local resort operators wish the tourist season were warmer, sunnier, and especially, longer.

REGION II—A REGION OF VERY COLD WINTERS AND SUNNY SUMMERS

Being in the far north of the state, this region receives less energy than any other lowland area. There is so little energy during January that mean temperatures of less than 20° F. are found almost everywhere. July means of about 69° F. indicate a summer receipt of energy not much less than that of lowlands to the south and southwest. A frost-free season generally between 135 and 155 days, potential evapotranspiration of 23.5 inches, and 128 degree months suggest considerable possibilities for agriculture.

No stations receive as much as 40 inches of precipitation annually, and the shores of Lake Champlain are one of the driest sections of the state. Since moisture which converges into storms arrives almost always from the southwest, south, or southeast, the Adirondacks cast a rain shadow over the lowlands. The result is less cloudiness, less precipitation in both amount and frequency, and more sunshine, especially in summer. Substantial amounts of the precipitation fall in the form of snow (60 to 100 inches) which stays on the ground for a long season.

Mud is the climate's legacy, especially in spring. Although mud is prevalent in other parts of the state, it is most distinctive here where slowly melting snows over frozen ground produce quantities of it. Lack of sufficient energy to dry out the ground effectively prolongs the period of muddiness. This handicaps early spring work in the fields but does not prohibit a widespread dairying economy. Most of the farms are dairy farms, although some vegetable and fruit production is carried on in the vicinity of Lake Champlain, where drier spring soil conditions persist.

REGION III—A REGION OF COLD, SNOWY WINTERS AND COOL, WET SUMMERS

Figure 24 shows that this climatic type covers about three-quarters of the area designated as the Appalachian Upland in Figure 13, as well as the New England Upland. Generally wet, and with less energy receipt than lowland areas in the state, these higher uplands have difficulty competing in the agricultural economy.

Actually a region of this size and terrain complexity is bound to embrace a host of climatic differences, especially as occur between valley bottom and exposed ridge locations. As most weather stations tend to be in valleys, they are better represented than the ridges in the available statistics. January mean temperatures are between 20° and 25° F., while July means are a cool 67° or 68° F. Frost-free periods between 100 and 150 days occur. Potential evapotranspiration is about 22 inches and most stations have about 120 growing degree-months. This means less energy is received here than in the St. Lawrence–Champlain Lowland well to the north. Minimums of −25° to −40° F. are common and summer maximums reaching 100°F. are extremely rare.

Precipitation is heavy almost everywhere, with few stations receiving less than 40 inches annually. In the western portions of the plateau there is more than enough moisture even in summer to satisfy the demands for average evapotranspiration. Although weather records are lacking, probably as much as 60 inches of precipitation fall on the upper southern slopes of the Catskills. This is fortunate, for great quantities of water are thus made available to the reservoirs and streams serving the urban Atlantic Seaboard. All of this climatic region, particularly the eastern portion, contributes significantly to stream flow, with perhaps half of the precipitation being gathered by streams. More water surplus is available per square mile in this region than anywhere else in the state except for the southwestern Adirondacks and the Tug Hill Upland.

Snowfall is substantial but is of course much greater and stays on the ground longer at high elevations than in the valleys. The western part of the upland, which lies athwart the path of storms crossing Lake Erie, receives more than 100 inches of snow a year; much of it comes in the form of snow squalls similar to those striking the Tug Hill east of Lake Ontario.

No weather stations are found in the New York portion of the New England Upland, but records in Massachusetts suggest a climate quite similar to that of the eastern part of the Appalachian Upland.

This climatic region has experienced perhaps more farm abandonment than anywhere else in the

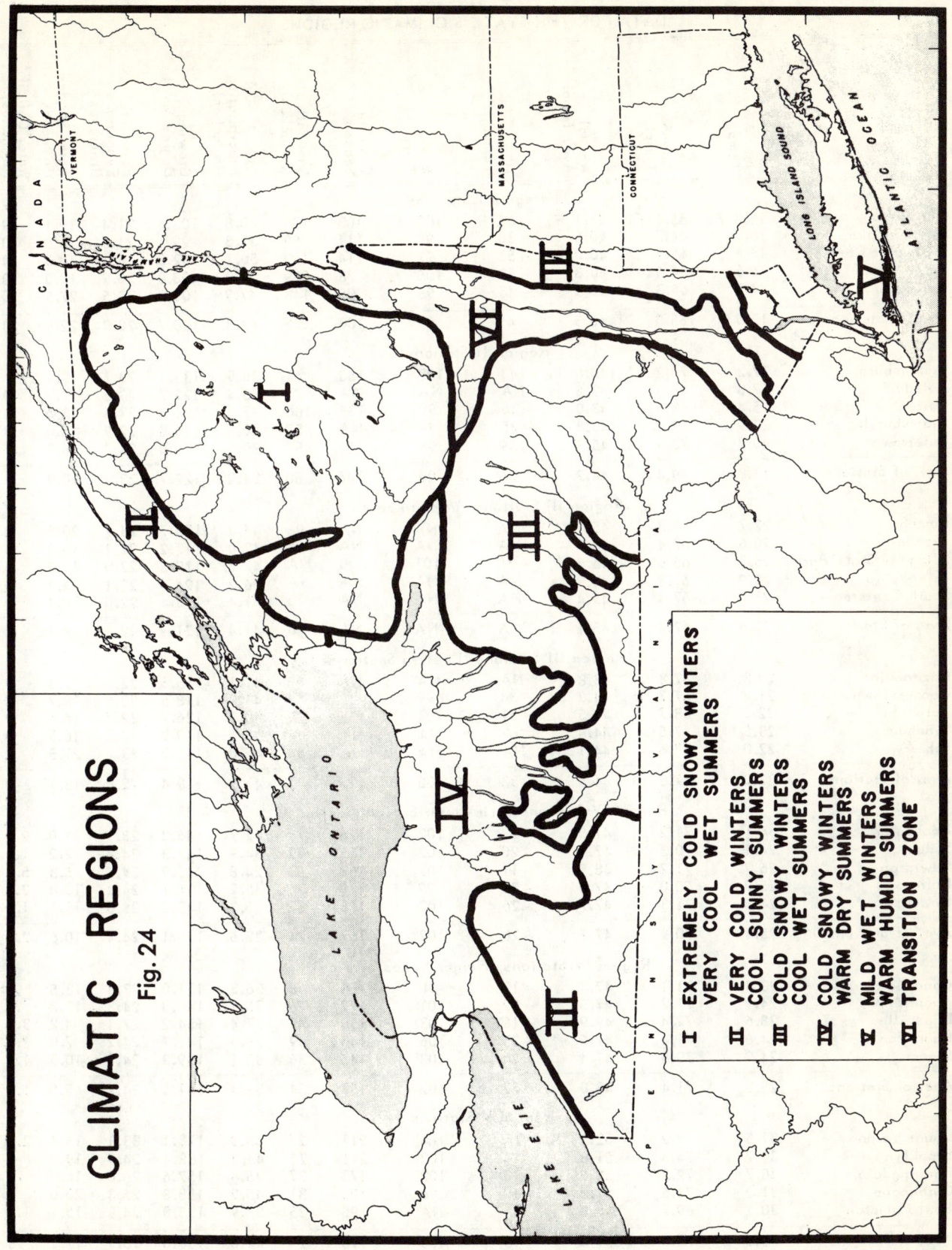

CHAPTER 2 CLIMATE 75

TABLE 12
DATA FOR THE STATE'S CLIMATIC REGIONS

Place	Jan. Ave. Temp.	July Ave. Temp.	Annual Ave. Temp.	Min. Temp.	Max. Temp.	Growing Season (In Days)	Snowfall (In Inches)	Ave. Annual Precipitation (In Inches)	Growing Degree Months*	Potential Evapotransp. (In Inches)	Water Surplus (In Inches)	Water Deficit (In Inches)
Region I Stations												
Wanakena	15.8°F.	65.1°F.	41.1°F.	−45°F.	100°F.	103	133	40.8	102.6	21.4	19.1	0
Tupper Lake	16.0	64.8	40.5	−38	98	111	88	37.3	99.1	21.3	16.0	0
Stillwater Res.	14.4	65.4	40.0	−52	97	114	150	50.7	102.0	20.3	30.0	0
Indian Lake	16.6	64.6	40.8	−42	105	78	89	39.9	98.8	20.9	19.0	0
McKeever	14.9	65.3	40.3	−48	95	96	148	47.9	101.0	20.5	27.5	0
Ave. of Stations	15.5	65.0	40.5	−45	99	100	122	43.3	100.7	20.9	22.8	0
Region II Stations												
Ogdensburg	17.2	70.2	44.6	−43	100	153	61	30.9	133.1	23.7	9.4	2.1
Potsdam	17.8	68.6	43.5	NA	NA	NA	77	33.4	122.7	23.2	9.9	0
Moira	15.4	68.5	43.0	−36	99	135	106	37.0	121.4	23.6	13.4	0
Plattsburgh	17.9	69.3	43.8	−25	99	NA	57	30.4	124.8	22.9	5.0	.8
Watertown	20.7	70.4	45.9	−39	99	151	99	39.5	136.8	23.8	16.3	.7
Ave. of Stations	17.8	69.4	44.2	−36	99	146	80	34.2	127.7	23.4	10.9	.7
Region III Stations (Western Section)												
Humphrey	22.7	68.3	45.4	NA	NA	NA	96	44.4	125.7	23.7	20.8	0
Arcade	20.6	67.4	44.0	NA	NA	NA	86	42.0	117.2	22.1	19.8	0
Allegany State Park	25.2	66.6	45.6	−35	101	90	NA	42.8	119.2	22.0	19.2	0
Friendship	20.9	67.3	44.9	NA	NA	NA	NA	36.2	124.0	22.1	14.3	0
South Canisteo	23.5	67.4	45.4	NA	NA	NA	75	41.4	123.4	22.0	19.4	0
Ave. of Stations	22.6	67.4	45.1	NA	NA	NA	86	41.4	121.9	22.4	18.7	0
Region III Stations (Eastern Section)												
Cazenovia	20.8	67.3	43.8	NA	NA	NA	95	41.1	115.5	22.9	17.3	0
Cooperstown	21.4	67.7	44.3	−34	99	123	69	41.0	118.5	22.5	18.7	0
Norwich	22.3	68.7	45.5	−31	101	127	72	40.2	126.7	22.7	16.5	0
Windham	23.5	67.5	44.8	NA	NA	NA	57	38.6	119.2	22.2	16.5	0
Liberty	22.0	67.6	44.1	NA	NA	NA	64	45.0	117.2	23.7	21.5	0
Ave. of Stations	22.0	67.8	44.5	−33	100	125	71	41.2	119.4	22.8	18.1	0
Region IV Stations (Erie-Ontario Plain Section)												
Rochester	25.1	71.2	47.5	−22	102	176	78	32.7	136.3	23.9	11.0	2.2
Sodus	24.1	70.3	47.9	−20	102	155	82	34.9	143.3	24.4	12.2	1.8
Lewiston	26.7	71.7	48.5	−11	107	157	51	24.8	143.9	24.7	5.3	5.1
Buffalo	25.2	70.0	47.0	−21	99	179	76	35.2	134.1	24.1	13.4	2.2
Syracuse	24.3	71.3	47.7	−26	102	168	83	35.3	143.0	25.0	11.1	1.9
Ave. of Stations	25.1	70.9	47.7	−20	102	167	74	32.6	140.1	24.4	10.6	2.6
Region IV Stations (Finger Lakes Section)												
Canandaigua	23.4	71.1	47.4	−13	98	NA	48	36.3	131.0	23.3	13.5	.5
Auburn	24.9	71.2	47.5	−32	101	172	76	35.7	140.0	24.3	12.3	.4
Dansville	28.6	72.4	49.9	−15	103	150	42	27.4	154.2	27.1	4.2	3.7
Penn Yan	24.6	71.4	47.6	−25	106	146	49	29.5	141.1	24.2	7.0	1.6
Ithaca	24.9	70.8	47.4	−24	103	145	56	33.1	139.2	24.7	10.3	1.7
Ave. of Stations	25.3	71.4	48.0	−22	102	153	54	32.4	141.1	24.7	9.5	1.6
Region V Stations												
Mount Vernon	31.5	74.2	52.5	NA	NA	211	27	43.3	173.1	33.1	15.4	2.0
Brooklyn	30.7	74.3	51.8	−14	102	211	25	43.2	168.4	24.9	19.4	0
Farmingdale	30.7	72.2	50.8	−14	102	173	37	45.4	157.6	26.6	18.7	0
Cutchogue	31.2	71.6	51.0	−10	96	196	31	45.2	158.8	25.3	20.0	0
East Hampton	30.1	69.7	48.8	NA	NA	190	25	37.9	135.0	24.8	15.1	1.9
Ave. of Stations	30.8	72.4	51.0	−13	100	196	29	42.6	158.6	26.9	17.7	.8

NA: Not Available.
*Cumulative average monthly degrees over 40°F.
Source: U.S. Weather Bureau publications.

state. Farmers tried to use these lands, but the low energy supply, deep snows, poor soils, and steep terrain generally have proved difficult to contend with.

Region IV—a region of cold, snowy winters and warm, dry summers

This climatic region is warmer both in summer and winter than the higher plateaus to the south but is sufficiently dry in the summer to exhibit a water deficit. Its greater receipt of energy enhances the range of agricultural possibilities, even though droughty conditions in some summers have an adverse effect.

Mean January temperatures of 25° F. and July averages of 70° or 71° F. are characteristic. Frost-free seasons of from 150 to 180 days are longest in upstate. Potential evapotranspiration of between 23 and 27 inches and about 140 growing degree months further illustrate the relatively large amount of energy available for crop production. Substantial absolute ranges in temperatures from minimums of −20° to −30° F. to maximums of 100° to 106° F. occur.

As a whole this is one of the driest areas in the state, with annual precipitation varying from 25 to 40 inches. Although summer is the season of maximum moisture supply, it is also the season of even greater moisture need, so soil moisture is depleted and deficits of small proportion occur in most years. There are many occasions when agriculture could be greatly aided by irrigation, and as a matter of fact some irrigation is carried on at many farms. Water surpluses, as would be expected, are small, so that only about one-fourth of the annual precipitation escapes as stream flow. Snowfall varies from 40 to 70 inches near the Finger Lakes to over 80 inches on the Erie–Ontario Plain, where snow squalls from the lakes are active. Especially sunny and dry, the Finger Lakes district is reasonably well suited to horticulture, and plants with deep root systems, such as the grape, are widely grown. The immediate Great Lake shores, because of the likelihood of less frost damage along them, support the state's outstanding fruit belt. Wheat and other small grains do well in the Genesee Country south of Rochester; elsewhere dairying predominates.

Region V—a region of mild, wet winters and warm, humid summers

The largest energy receipt in the state occurs in this region, with the mainland area experiencing even higher temperatures than Long Island, which is more frequently ventilated by cool winds from the ocean. January mean temperatures of 30° F. are 15° F. higher than in the Adirondacks and 5° or 6° F. higher than in the Finger Lake–Lake Plain region. Less difference between this area and upstate is expressed by July mean temperatures. Frost-free seasons approximate 200 days; potential evapotranspiration, 27 inches; and growing degree months, 158. For all three of these indicators of energy receipt, much lower figures characterize outer Long Island as compared to a place like Mt. Vernon on the mainland. Actually, although the frost-free season on Long Island is one-sixth longer than that at Troy, 150 miles up the Hudson Valley, potential evapotranspiration is but 5 per cent greater.

This region receives about 43 inches of precipitation annually and has a water surplus of approximately 18 inches. Both of these amounts are greater than for any other lowland in the state except the lower Hudson Valley. Because of the relatively warm temperatures in winter no more than 40 inches of snow falls on this region per year, and much of Long Island receives less than 25 inches. This means there is little snow on the ground in winter.

Much of the area has become urbanized, so a discussion of the relation of climate to agriculture is more or less academic. On the other hand, the high energy receipt, generous precipitation, and long growing season favor Long Island as a vegetable-producing area. Because of the almost insatiable demand for potatoes and other vegetables in metropolitan New York City, Long Island farms find it profitable to produce these items intensively even in the face of ever encroaching urbanism. The mild climate, too, is an asset to the great urban development. Snow removal costs are at a minimum as far as New York State locations are concerned, as are heating costs.

Region VI—a transition region

With an abundance of energy relative to surrounding uplands, the Hudson Transition Region comprises a gradually changing extension into the interior of the warmer conditions of the New York–Long Island Region. Although drier than the uplands to the east and west, the Hudson Valley has generous amounts of precipitation in the south. To the north dryness increases, reaching major problem proportions on the sandy soil areas near the confluence of the Mohawk and Hudson rivers.

Both energy and moisture variations make this a region of appreciable transition rather than one of similarity. Mean January temperatures ranging from 20° to 30° F., July means of from the high 60°'s to 74° F., and frost-free periods of from less than 150 days to more than 200 days illustrate this transitional nature. So does a variation in growing degree months from 130 to 165. The record maximum temperature in the state of 108° F. was recorded at Troy. Annual precipitation, too, varies considerably, with stations exhibiting a range of from 35 to more than 45 inches. Snowfall is no more than 35 or 40 inches in the south but double that figure as this region approaches the Adirondacks in the north.

A transition like this would be expected to reflect itself in a considerable range of agricultural possibilities. This it does, with agricultural economies varying from dairying to fruit.

In Retrospect

Substantial energy receipt and a generally adequate moisture supply accented by elevation and the Great Lakes characterize New York's climate.

Compared to the remainder of the state, the coastal area has less cloudiness, less frequent storms, more energy, and higher temperatures with greater freedom from frost. These are the generally milder conditions that characterize seaboard locations southward from Cape Cod.

Cool and cloudy, the mountains and the higher Appalachian Upland receive relatively meager energy supplies. Upstate lowlands are much affected by the harsh climates of the interior of the continent and in fact are climatically like these interior regions. Especially during winter, great extremes in temperature are experienced. In spring and autumn the Great Lakes modify thermal conditions significantly in their immediate vicinity.

The moisture situation is not drastically different throughout much of the state. Average water deficits are small and surpluses are substantial, the epitome of humid climates. Somber conditions dominate during winter in upstate, where record-high frequencies of days with precipitation, extremely low totals of sunshine receipts, and record snowfall occur. Data in Table 12 highlight climatic conditions in the state.

The geography of agriculture as presented in Chapter 10 will further make clear the role climate plays in the state's economy. Chapters on water, vegetation, and soils, and even discussions of the historical geography, make clear that climate is no insignificant element in the resource base of New York State.

Selected References

Carter, Douglas B. "The Water Balance of the Delaware Valley," *Publications in Climatology*, XI, No. 3 (1958), 249-69.

Critchfield, Howard G. *General Climatology*. Englewood Cliffs, N. J.: Prentice-Hall, 1960.

Frederick, Ralph H., et al. *Freezing Temperatures in New York State*. ("Cornell Miscellaneous Bulletin," 33.) Ithaca: New York State College of Agriculture, 1959.

Hare, F. K. *The Restless Atmosphere*. London: Hutchinson Univ. Library, 1961.

Landsberg, Helmut. *Physical Climatology*. 2nd ed. DuBois, Pa.: Gray Printing Co., 1958.

Meinzen, Oscar E. (ed.) *Hydrology*. New York: Dover, 1942.

Mordoff, R. A. *The Climate of New York State*. ("Cornell Extension Bulletin," 764.) Ithaca: New York State College of Agriculture, 1949.

Muller, Robert A. "Some Economic Consequences of Deep Snowfalls in Northern New York." Unpublished Master's thesis, Department of Geography, Syracuse University, 1960.

Sutton, O. G. *The Challenge of the Atmosphere*. New York: Harper, 1961.

Thornthwaite, C. W. "An Approach Toward a Rational Classification of Climate," *The Geographical Review*, XXXVIII, No. 1 (1948), 55-94.

—— and G. R. Mather. "The Water Balance," *Publications in Climatology*, VIII, No. 1 (1955), 1-104.

U. S. Department of Agriculture. *Climate and Man. 1941 Yearbook of Agriculture*. Washington, D.C.: Govt. Printing Office, 1941.

U. S. Weather Bureau. *Climatic Summaries for New York State*. Washington, D. C.: Govt. Printing Office, various dates.

CHAPTER 3   ARTHUR R. ESCHNER

Water

No resource is more important than water. It is used for drinking, recreation, industry and power, navigation, and waste disposal. Fortunately, the over-all quantity of water is no real problem in New York because of the humid climate. Having enough water of proper quality at the right place and at the right time, however, often is a problem, and one that will concern more and more people as time goes on.

Throughout the state's history the population has used water for all conceivable purposes; watercourses have even had considerable influence on the geography of transportation and cities. Only fairly recently though have there been such great concentrations of people, such high per capita consumption of water, and such great per capita demand for waste disposal. The great wave of urbanism is multiplying the complexity of water resource management at an alarming rate. Water, an almost ubiquitous and essentially free resource in the past, will become much more precious in the future. Supplying cities with good water, keeping rivers and lakes clean and beautiful, and protecting ground water from depletion will be problems familiar to nearly everyone. This chapter discusses the water supply and demand situation in New York State and elaborates some of the problems associated with it.

Although the chemical formula for water, H_2O, is common knowledge, and is used even in casual conversation, the apparent simplicity of the resource is deceptive. This common, colorless, odorless, tasteless substance is actually complex. It may contain as many as twenty different chemical compounds without ordinarily being considered impure. Even in its purest natural form, as falling rain or snow, it contains dissolved gases and a variety of substances and organisms washed out of the atmosphere. The most useful and generally most abundant physical form in New York is as a liquid, but it does of course occur commonly as a gas and a solid; usually two, and often all three, forms are present simultaneously in our environment. It is the only widely occuring natural compound that exhibits these properties.

In addition, water in nature is indestructible, cycling endlessly from sea, to sky, to land, and back to the sea. This fundamental fact makes a simple adding up of the volume of water in lakes, rivers, and underground formations, or the amount which falls as precipitation, incomplete as an inventory of the resource. In New York State water is nearly inexhaustible, and in essence the amount available is in many respects more nearly a function of the price people are willing to pay than anything else. At the present nation-wide going rate of 5 cents a ton, delivered, water is "cheaper than dirt." If the cost were to be increased sufficiently, almost any quantity and quality of water could be made available at a given place in time.

Supply

New York's 40-inch average precipitation feeds 70,000 miles of streams and 3.5 million acres of lakes and ponds. Its western border lies in the world's largest system of fresh water, the Great Lakes, which normally flow at an average rate of over two hundred thousand cubic feet per second.

The water supply for New York's residents and industries is derived at present from two general sources, surface water and ground water. These two types of water indicate useful general distinctions, but they are not strictly independent sources.

Surface Water

Surface water is the water in streams, lakes, and reservoirs. Although some of it may have fallen directly on the water bodies or run over the surface of the land into a stream or lake, by far the larger portion has commonly entered the ground after falling from the clouds and later percolated into the water bodies. The water that enters the soil and is not removed by evaporation or extracted by the roots of plants returns to the surface downslope from its point of entry according to a pattern dictated by the subsurface soil and geological conditions.

Ground water, water that occurs in a saturated zone within the soil and rock mantle, acts as the

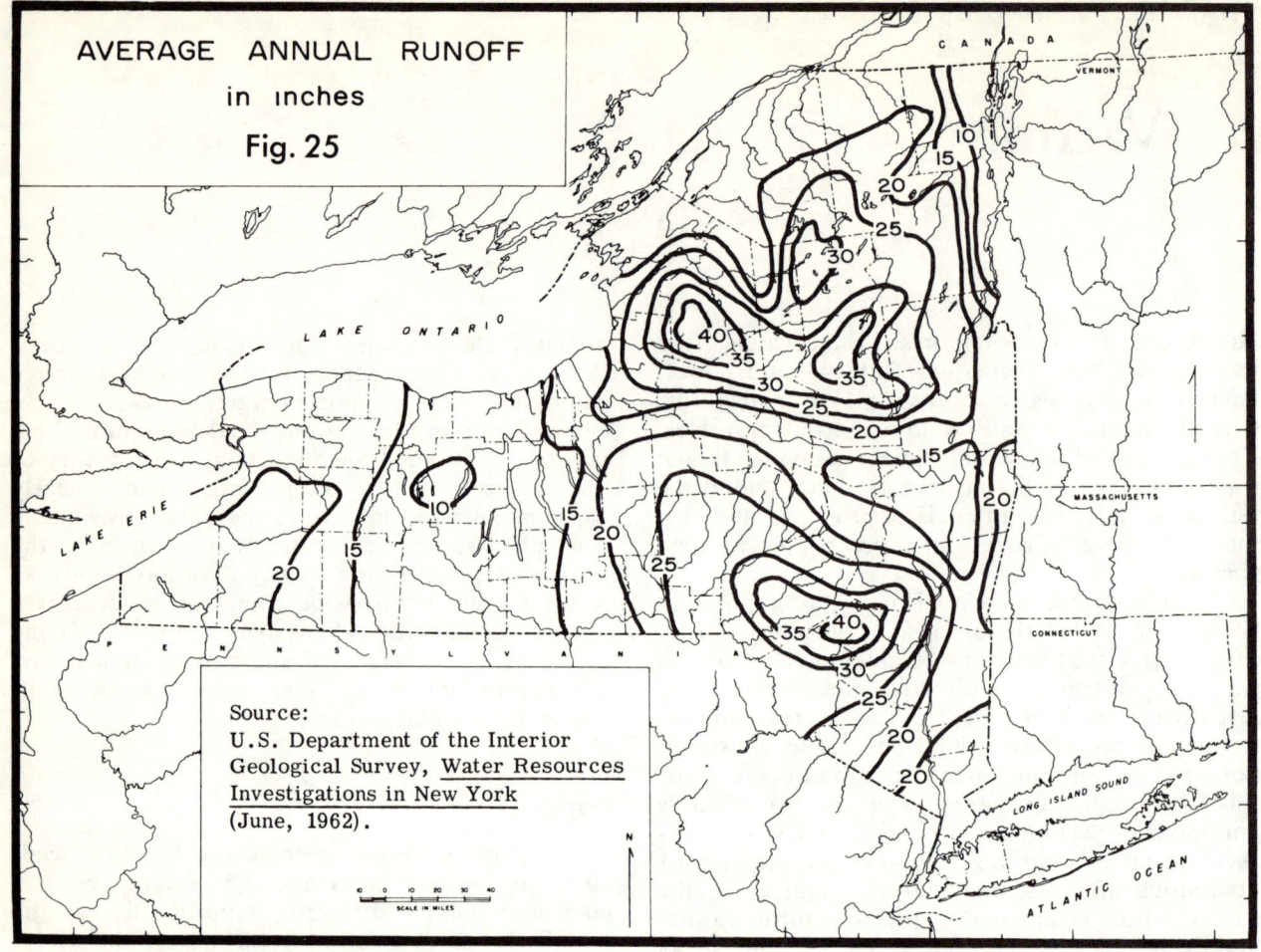

AVERAGE ANNUAL RUNOFF in inches
Fig. 25

Source:
U.S. Department of the Interior Geological Survey, Water Resources Investigations in New York (June, 1962).

most dependable, steady source of water for surface streams. It is the primary source of streamflow during long rainless periods.

The general pattern of average annual runoff in streams is shown in Figure 25. This pattern is similar to that in Figures 21 and 22, which show mean annual precipitation and mean seasonal snowfall, and illustrates the primary dependence of streamflow on precipitation. The average discharge of the principal streams of New York (Fig. 26) indicates the large volumes of surface water naturally available on the Great Lakes and along major rivers. New York City has diverted a considerable portion of the flow from the Delaware and Hudson river tributaries in the Catskills to modify the natural flow patterns in that area.

Obviously annual totals and average rates of flow are only part of the information important in determining the availability of water. Seasonal variation of runoff is significant too because it indicates when streamflow may be a problem because of excess or deficit. Figure 27 shows the mean monthly flow in per cent of annual flow for five representative watersheds in New York. All the streams show a common tendency toward high flow in spring and low flow in late summer. Some differences in the regularity exist between streams in the various land form regions. The massive contribution of melt from a persistent snow pack in the Tug Hill and Adirondack areas results in over 25 per cent of the annual runoff there occurring in the month of April. By way of contrast, the Beaverkill and Wappinger Creek in the Catskills and Hudson Lowlands have a maximum flow of less than 20 per cent in any one month. The most widely fluctuating as well as the most stable streamflow are exhibited by the two stations in the Appalachian Upland. Owego Creek has a March runoff of 23 per cent of the annual total, while its August flow is but 1 per cent. The Beaverkill's March discharge is 20 per cent and its August discharge is over 3 per cent of the annual total.

Surface water, in whole or in part, supplies the requirements of over eleven million New Yorkers in about seven hundred communities. Four hundred water utilities, public and private, are involved.

Ground Water

Ground water in usable quantities is available from wells in essentially all sections of the state. It is found both in bedrock formations and in the unconsolidated deposits that cover them. The most productive wells are usually those that draw from sand and gravel. These deposits are highly variable; their location, composition, and arrangement depend upon the manner in which they were formed. Productive sand and gravel aquifers commonly result from deposition of glacial melt waters in valleys or as deltaic and beach deposits in glacial lakes. Many of the valleys of the Appalachian Upland, for example, have excellent water-bearing sands and gravels, and lake deposits occur widely on the periphery of the Adirondacks. Long Island, too, has sand and gravel aquifers. Heavy precipitation and streams continually recharge these kinds of aquifers, making immense water withdrawal from them possible.

Carbonate rocks, limestones, dolomite, and marble form the most productive water-bearing bedrock in the state. Figure 8 indicates the geographic occurrence of this kind of rock. In some areas, hydrogen sulfide, sulfates, and salt occurring in the water make it somewhat less than high quality. Other types of bedrock, as well as unconsolidated clay and silt deposits, generally yield less water than carbonate rocks and sands and gravels.

Ground water is used by more than four million New York State residents. About 8 per cent of the fresh water used, 630 million gallons per day, is from wells tapping subterranean supplies. It is the sole source for 647 public water systems and contributes in part to 106 others in the state. Most farms and other users not in villages or cities depend on wells.

Other Potential Supplies

In addition to the two conventional sources of water mentioned above, modern technology is rapidly reducing the obstacles to the utilization of two other sources of water that have been recognized as potentially valuable for many years: desalted sea water and reclaimed waste water.

Desalinization of sea water has long been an attractive prospect for water-users situated near the ocean or on tidal estuaries. This process, in theory, should make available an almost limitless supply. A number of different techniques are currently being explored for separating the fresh water from the salty brines. They include distillation, electrodialysis, ion exchange, reverse osmosis, and freezing. The most important common characteristic of all these techniques is a high energy requirement. Power sources proposed for the desalinization have included most of the conventional ones as well as solar energy, burning garbage, and atomic power. An atomic pilot plant that will supply power to produce one million gallons of water per day is being constructed in the town of Riverhead on Long Island. The location of desalinization plants at sea level, however, requires that the entire fresh water production from them be pumped, and adds an additional cost of perhaps 20 cents per thousand gallons to the still relatively high production cost of 30 to 50 cents per thousand gallons. This is considerably higher than most water costs today, but is not necessarily out of reason for certain users.

Sewage renovation or waste water reclamation is not as attractive a prospect as desalinization of sea water to most people but may become a more practical financial investment. To some degree it is already being done; many cities withdraw water from streams and ground water aquifers into which others have introduced wastes. With treatment this water can be made generally usable. The lower energy requirement for removing the usually less than 2 per cent total solids from sewage, as compared to the 3.5 per cent total solids in sea water, is the primary reason why this is being considered for wide application. It seems reasonable to expect that there will be a shift of emphasis in the near future from simple waste disposal to the deliberate processing of sewage for reuse. It is becoming more and more apparent that using clean water once, and then discarding it, is shortsighted and wasteful.

Weather modification, or cloud-seeding to augment natural rainfall also has been considered by some as a potential method of increasing the supply of water. Results to date have been inconclusive and interest has waned. Serious legal questions, in addition to the physical problems, at present limit the use of this method.

Demand

The water needs of New York State are practically impossible to define precisely. All the uses are not measured or even estimated with any regularity or accuracy. Even if they were measured, it would still be difficult to determine the amount of water required, for many of the uses are nonconsumptive. That is, the water is used, then subsequently returned to the stream or underground system from which it was taken, and so may be used again. A simple total of the amount of water needed by a series of individuals, communities, or industries is thus usually an overstatement of the actual amount of water required. However, it has been estimated that 20 billion gallons of water are required daily to meet the needs of New York State's people; approximately 1,100 gallons per person. The ordinary individual himself, of course, does not use 1,100

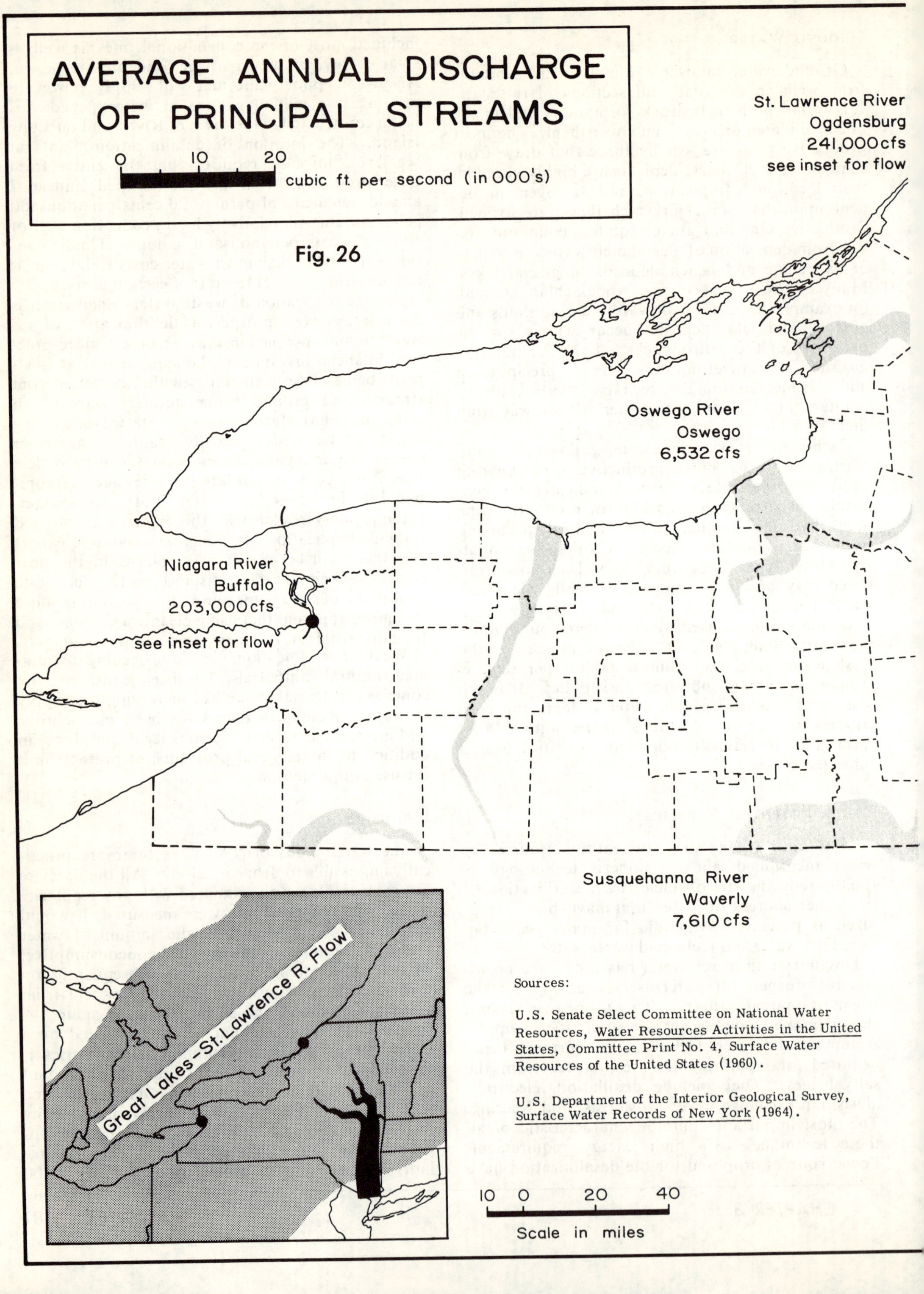
AVERAGE ANNUAL DISCHARGE OF PRINCIPAL STREAMS
0 10 20
cubic ft. per second (in 000's)
St. Lawrence River
Ogdensburg
241,000 cfs
see inset for flow
Oswego River
Oswego
6,532 cfs
Niagara River
Buffalo
203,000 cfs
see inset for flow
Susquehanna River
Waverly
7,610 cfs
Great Lakes-St. Lawrence R. Flow
10 0 20 40
Scale in miles

Fig. 26

Sources:

U.S. Senate Select Committee on National Water Resources, Water Resources Activities in the United States, Committee Print No. 4, Surface Water Resources of the United States (1960).

U.S. Department of the Interior Geological Survey, Surface Water Records of New York (1964).

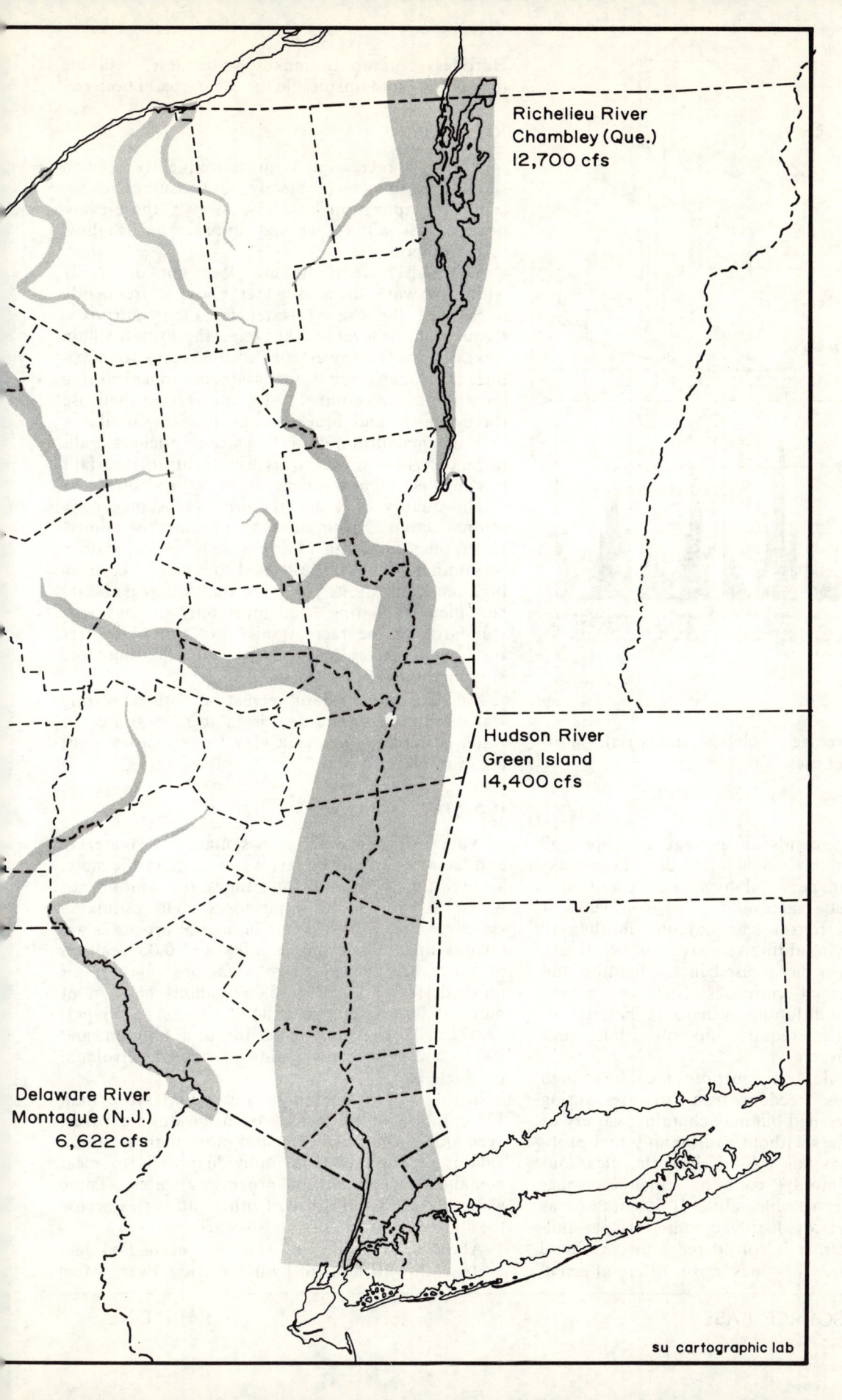

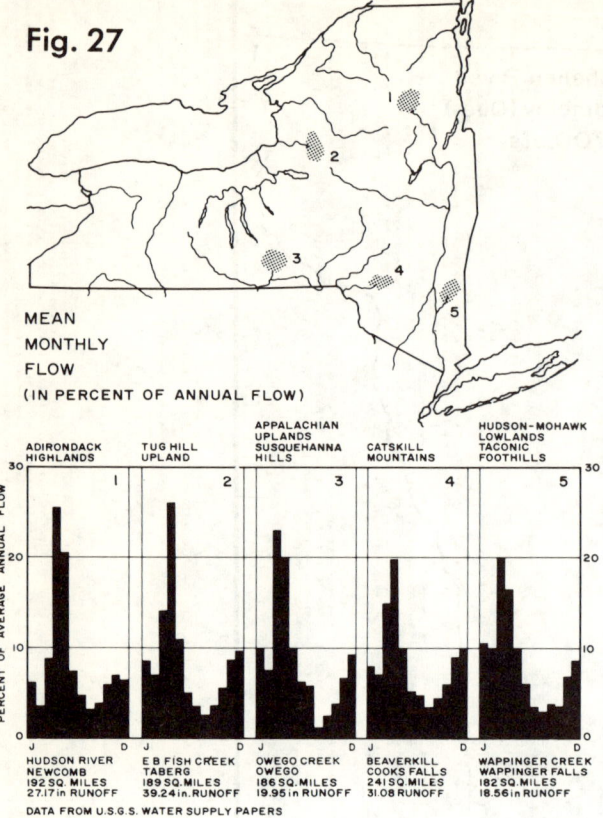

Fig. 27

gallons; this average includes industrial, agricultural, and other uses.

DOMESTIC

Each person demands an average of about 50 gallons of water a day. This is for drinking, cooking, cleaning, and personal hygiene. About one-fifth of the domestic water used is lost from leaks of various types in transmission, adding another 10 gallons per day. In addition an average of 10 gallons per person per day is used in fire-fighting and for other municipal purposes, such as street-cleaning. Air-conditioning systems in homes, offices, and factories require unknown but huge amounts of cool water.

Water used for domestic purposes must be of high quality as far as freedom from disease-causing organisms is concerned but may contain a variety of dissolved substances without causing any real problems. Water that has passed through calcareous formations commonly contains calcium, magnesium, and other soluble salts. If it contains as much as 240 parts of dissolved materials per million parts of water, it is considered *hard*. Ground water supplies are the ones most often affected. Hardness commonly makes water less desirable for washing and unsuitable for use in steam boilers.

RECREATION

Water for recreation is most frequently used *in situ* and is almost completely nonconsumptive, except for evaporation losses. However, these evaporation losses may be substantial from shallow lakes and ponds.

Although recreational use does not ordinarily withdraw water from a water body, it frequently does limit the use of water for other purposes. Regulating the level of lakes (e.g., the Fulton Chain Lakes in the southwestern Adirondacks) to maximize their recreational and aesthetic appeal during the summer may compete with and restrict their use for domestic and industrial purposes. Similarly, maintenance of minimum flows over Niagara Falls during daylight hours to assure beauty of the falls has an effect on power generation in the vicinity.

The quality of water required for some recreational activities approaches that needed for domestic supplies, although a fairly wide range is possible. Swimming should be restricted to waters free from high concentrations of bacteria, whereas water suitable for boating need meet only the aesthetic standards of the recreationist and should be free of obnoxious colors, odors, and floating and suspended material.

Much of our outdoor recreation with its strong water-centered orientation is in itself a source of water pollution. Oil and sewage from pleasure craft are examples.

INDUSTRY AND POWER

Water is perhaps industry's main raw material, and factories are the largest water users in the state. They withdraw about four times the amount of water used for domestic purposes. The estimated water requirements of some industrial processes are outstanding. For example, 1,000 to 10,000 gallons of water are needed to manufacture one ton of finished steel; 10,000 to 75,000 gallons per ton of pulp; 25,000 to 80,000 gallons per ton of paper; 32,000 to 320,000 gallons per ton of aluminum; and 200 to 500 gallons per gallon of refined petroleum products.

In the Buffalo area industries pump more than 125 million gallons per day from the Buffalo River, even more from Lake Erie, and more than a million gallons per minute from individual wells to meet demands for cooling and processing water. These water systems, and those of other industries across the state, are largely privately owned.

About 6,000 gallons of water are needed to generate a kilowatt hour of hydroelectric power. Most of the hydroelectric power generated in New York is developed along the Niagara and St. Lawrence rivers. There are also numerous smaller generating stations ringing the Adirondack Mountains. More water is used in the state for generating electrical power than for all other uses combined. Practically all of the water withdrawn for power-generating, however, is returned in usable form to the watercourse from which it was taken. As was true of recreation though—another nonconsumptive use— the needs for a firm power supply restrict prior withdrawals for consumptive uses.

NAVIGATION

The use of New York's waters for water-borne commerce has always been of major importance. Explorers, traders, and settlers used the waterways as the earliest and easiest natural routes for travel. The Hudson, Mohawk, and St. Lawrence rivers and the Erie Canal were among early routes to the interior of the state. In spite of a decline in use of older canals, water transportation is still significant on two routes. The Hudson River has a 27-foot channel, adequate to carry seagoing vessels, as far as Albany. The St. Lawrence Seaway and Great Lakes form a second route over which increasing numbers of ocean-going vessels move. The strategic location of the Great Lakes between industrial regions and areas producing raw materials, the depth and stability of the channels, the size of the lake vessels, and efficient loading and unloading equipment are all important in maintaining the position of the Great Lakes–St. Lawrence route as a waterway of the first rank.

IRRIGATION

Although not generally recognized, agriculture is one of the major consumptive water-users in the state. In large measure this use goes almost unnoticed, for the water is used where it falls as rain or snow. However, irrigation of farm lands in New York has been increasing rapidly despite our generally humid climate. At least 90,000 acres were irrigated on an estimated four thousand farms in the early 1960's, and expansion upward is occuring at a rapid rate.

A variety of sources are being used for irrigation water. Most of the farmers who irrigate in Upstate New York have built ponds to store water. On Long Island ground water is the principal source. Streams are used by a large number of irrigators, but many of the streams are inadequate sources during dry periods when water is most needed.

WASTE DISPOSAL AND TRANSPORT

Many people feel that waste disposal in water in any form is a disgrace. Others feel that the use of water as a carrier of waste far outweighs a number of other potential values, such as commercial or sport fishing. These are extreme positions. Municipal and industrial waste disposal systems have been developed on the basis of using water bodies as the recipients of unwanted materials of many forms. Most people agree that some reduction and care in releasing wastes is desirable, rather than using the full assimilating capacity of the water body; no user has the right to pollute water so that it cannot be reused. The distinction between releasing wastes and polluting is important. Waste disposal is a necessary function that can be tolerated if regulated; pollution interferes with reasonable reuse of the water.

All of New York's streams and lakes carry waste materials. Where the amount of waste introduced into a stream is of a type and amount that the stream can break down or harmlessly assimilate in a reasonable distance, the problem is minimal. Well-aerated, turbulent streams of large volume are able to absorb more wastes with less damage than sluggish low-volume streams. However, even water bodies with extremely large capacities for dilution and rejuvenation, such as some of the Great Lakes and the Hudson River, are becoming, or have already become, essentially "open sewers" choked with massive amounts of domestic and industrial wastes. Herein lies one of the major water problems not only of New York State but of the world.

Problems

The water resource problems of New York are not unique. They are the almost universal ones of quantity available at the time and place of need, and quality suitable for the uses to be served. As the state's great population and general economy rise though, and as that population and economic activity become increasingly concentrated in the great urban systems, the problems of water supply and water quality intensify. Increases in per capita demand for water and waste disposal also contribute to the intensification.

QUANTITY

Although there is probably no likelihood of an over-all shortage of water for New York State, increased demands in some areas at certain times do, or will, create problems. Prolonged droughts such as the one in the early 1960's deepens concern over the adequacy of the supply for an increasing population. New types of everyday equipment such as air-conditioning systems, garbage disposal units, and automatic dishwashers, as well as increases in all types of uses, continually increase demands. Withdrawals in the year 2000 are projected to be at three or four times the present rate.

The traditional solution to problems of deficient water quantity has been the development of additional storage and withdrawal capacity from good upland streams. New York City is the prime example of this approach. Its upstate system of reservoirs provides the bulk of the 1.25 billion gallons a day used by its residents. There are indications that this solution may not be extended indefinitely—or even much further. Resistance by New Jersey to the "export" of water from the Delaware River basin to New York City in the 1930's has resulted in the formation of the Delaware River Basin Commission and an interstate agreement setting limits to withdrawals by New York from the Delaware. New York City may now divert 490 million gallons per day from tributaries of the Delaware which feed its Neversink and Pepacton reservoirs. When the Cannonsville Reservoir is completed the city will be able to divert 800 million gallons per day. At present it is supposed to release water from its reservoirs in quantities adequate to maintain a minimum rate of flow in the river near Montague, New Jersey, of 985.6 million gallons per day. With the completion of the Cannonsville project this volume will be increased to 1.13 billion gallons per day.

The recent drought in the Delaware River basin, which started in late 1961 and has continued for four years—to the time of this writing—has been the most intense in the history of the basin. In the middle of June, 1965, New York City stopped making releases into the Delaware River when it became apparent that its reservoirs would empty in late summer if it did. In the course of the hearings and investigations that followed, the city was allowed more flexibility in its reservoir releases.

Some changes in water management, long suggested as alternatives to the extension of its reservoir system, such as use of the Hudson River, universal metering, and a vigorous leak detection and control program, are in the offing. It is estimated that the city's people waste approximately 300 million gallons a day in one way or another. It might be concluded from this that water costs in the city are so low that conservation does not pay.

Where water is considered a commodity subject to economic laws and managed by investor-owned water companies, costs are more closely related to the expenses involved in storage, treatment, and transmission, and effects of the recent drought have not been so sharply felt. In 1964 and 1965, of fifty instances of water-use restriction in New York State, only five involved investor-owned companies. This suggests that, if the pricing structure were more realistic, it is possible that gross adjustments would come about in allocation of water between uses and among users and that many so-called shortages would disappear. The popular conception of water as essentially inexhaustible and essentially free needs scrutiny.

Sometimes reduction in recharge resulting from use patterns has adverse effects on quantity. For example, extensive suburban development with attendant impermeable roofs and hard-surfaced roads has cut recharge to the ground water aquifer of the East Meadow Brook basin on Long Island by an estimated sixty-three thousand gallons daily. This is particularly important because of the extensive local dependency on ground water. In some areas on Long Island where withdrawal has been excessive, salt water has invaded the aquifer. Short sewer lines and seepage basins injecting waste water into the aquifer to act as a barrier to impede the entry of salt water offers some limited hope as a partial remedy.

The recent decline in water levels on the Great Lakes has generally been attributed to the deficient precipitation since 1960. Dredging, diversion of water, and the regulation of outflows from Lakes Superior and Ontario have also had some effect. Between 1952 and 1964 Lake Erie's average level dropped 4.3 feet. As a result, hydroelectric power production near Niagara Falls was reduced two billion kilowatt hours from the expected thirteen billion mark. Low levels have also had an effect on shipping, requiring vessels to carry less than normal cargo capacity to make passage between Lake Huron and Lake Erie.

In contrast to quantity deficits there are occasions when surpluses produce problems in the state. Floods are most frequently caused by the annually recurring spring snow melt or spring rain on snow or saturated and frozen soil. Extreme meteorological events, such as the combination of hurricanes Connie and Diane in August, 1955, also cause floods but occur much less frequently. The areas most often affected by floods are the narrow valleys of the Appalachian Upland and the Erie–Ontario Lowland.

No satisfactory data are available which give a comprehensive picture of flood damage in New York, but an indication of the approximate magnitude of the problem might be gained from the amount of money spent by the U.S. Army Corps of Engineers in flood protection from 1936 to 1960: 85 million dollars. Such an outlay should at least equal reductions in flood damages as a result of the protection program.

Some areas, such as the Adirondacks and the Tug

Hill, do not generally suffer much flood damage, not because floods do not occur, but because flood plains of the rivers are not heavily populated or used. These areas do contribute to snow-melt floods in the Black River Valley and elsewhere, however.

Floods caused by ice jams are not infrequent in western New York on such streams as the Buffalo River, Cattaraugus Creek, and lower reaches of tributaries of the Niagara River. This type of flood is not necessarily dependent upon an unusually high rate of streamflow but instead on ice conditions. A disastrous flood of this nature was caused by an ice jam at the mouth of Cattaraugus Creek in March, 1963, when the discharge was not extraordinarily high.

QUALITY

Water may be of poor quality in its natural state, or it may have been made that way by human action. More and more it is deterioration caused by human action which is plaguing the state water supplies.

Objectional tastes, odors, and color-producing substances may be found naturally in waters derived from swamps, marshes, and peat bogs. Aquatic life, particularly algae, may contribute offensive characteristics, too, under largely natural or undisturbed conditions. And ground water from areas with soluble salts, or carbonate or sulfate rocks, such as are found across New York from Buffalo to Albany, frequently have high chloride content, hardness, or unpleasant odors. The encroachment of saline water into some of Long Island's aquifers mentioned above is a natural form of ground water quality deterioration, aggravated by man's activities in withdrawing excessive amounts from the aquifer.

Many of the most objectionable and widespread aspects of water quality in New York are a direct result of man's introduction of wastes into water bodies often in excess of the water's natural purifying capability.

Many wastes, such as sewage, add nutrients to water which enhance biotic growth. This process is often spoken of as enrichment, or, if it is carried too far, as overenrichment. Of course, organic wastes can also, and usually do, contain coliform bacteria, which indicate a potential health hazard. *Pollution* is a general term meaning different things to different people, but, as pointed out above, in essence it denotes lowering the quality of water so as to interfere with reasonable reuse.

Ever since the beginning of European colonization in New York, there has probably been a certain amount of water quality deterioration related to agriculture. Clean cultivation without soil-conserving practices on steeply sloping land continues to contribute quantities of silt to streams, especially during periods of high surface runoff. Streams draining the finer-textured soils of the Appalachian Upland are more commonly affected than those draining more level land or the coarser-textured soils of the Adirondacks.

In recent years there has been a growing awareness of the potential damage to water quality from insecticides and herbicides used in agriculture and forestry. Although there have been no dramatic cases of damage to water quality or aquatic life in New York State which compare with the spectacular fish kills on the Mississippi, an uneasy feeling is growing about the potential chronic effects of continued exposure to these and other chemicals being introduced into our water supplies. This concern has been expressed with increasing frequency by people in public health, and with good reason. We are almost completely ignorant of the chronic physiological effects caused by ingesting minute amounts of many of these materials. There is mounting evidence that the treatment given most of our water supplies to make them bacteriologically safe is grossly inadequate to remove substantial amounts of these chemicals.

Municipal and industrial wastes which tend to deteriorate water quality are usually distinctly different in character but commonly occur in close geographic association. Primary treatment of municipal sewage, if properly carried out, usually may remove 40 to 70 per cent of the suspended solids and 20 to 40 per cent of the biochemical oxygen demand (BOD). Oxygen is needed for the decomposition of solids in the natural purification process so under some conditions excess demands from sewage can exceed the supply in the water. Secondary treatment may remove an additional 25 to 45 per cent of the solids and 55 to 65 per cent of the BOD. The effluent is then commonly chlorinated to kill almost all of the bacteria.

Foaming synthetic detergent wastes have been an increasingly apparent water pollution problem in many of New York's streams and ground water supplies. An estimated 400 million pounds of detergent are used annually in New York State. These compounds have been resistant to natural decomposition processes in the soil and water, and their foaming qualities make them easily observable. Recognition of the fact that detergents are only one of many compounds in waste water gives these detergents some indicator value of the total problem. Recently, however, the industry has been developing foamless detergents, which tend to obscure the problem but not eliminate it. Efforts are also being made to develop synthetic detergents that break down more readily in soil, water, and waste treatment plants, and tests of new detergents are being carried out on Long Island.

Wastes produced by industry are extremely variable and may be quite complex. They may contain a variety of chemicals which remain essentially unchanged in the water: acids or alkalis, organic matter, heat, and oils among other things. The amount of these substances discharged to streams, lakes, and ground water is as variable as the materials themselves and is often related to the cost of the materials and the feasibility of reuse or sale as a by-product of the industrial process.

Hudson River Case

A number of large cities discharge untreated or improperly treated sewage into the Hudson River system. Just to cite principal offenders: Utica dumps about 15 million gallons of sewage daily into the Mohawk; 60 million gallons are added in the vicinity of Albany; and New York City dumps about 400 million gallons into its harbor. The massive pollution from these and almost 120 other prime sources has made the river unfit in some places even for boating, not to mention drinking and swimming. The movement upstream of about 20 billion gallons of water by every incoming flood tide further complicates the problem. The possible change in the position of the saline waterfront with projected plans for withdrawing water from the river at Chelsea to supply New York City is of concern to places such as Poughkeepsie which are currently using the Hudson after a complicated, expensive process of purification, filtration, coagulation, precipitation, sedimentation, treatment with activated charcoal, and disinfection with chlorine gas.

Lake Erie Case

Overshadowing the problem of diminution of supply caused by nature and man in the Great Lakes basin is the serious pollution problem developing in the lakes, especially in Lake Erie. Pollution is now approaching a point at which correction may be practically impossible. The main problems in Lake Erie, the shallowest of the five lakes, are overenrichment—especially with phosphates and nitrates, bacterial pollution, and industrial wastes. Overenrichment has promoted algae growth; as the algae die and sink to the bottom, decomposition uses up the oxygen, with the result that there is severe depletion of oxygen in about one-quarter of the lake. Limited dispersal of various wastes results in concentrated "slugs" which retain their identity, drifting about the lake for considerable distances and periods of time. Pollution from ships of all nations has brought with it the possibility of spreading exotic diseases in the area. The striking increase in pleasure boating on the Great Lakes and across New York State has led to a suggestion by the state health commissioner for legislation to control the discharge of wastes from these and commercial vessels into inland waters. The pollution of Lake Erie has damaged recreation, commercial and sport fishing, and water supply. Only if the wastes introduced into it can be reduced to a level that does not continue the process of excessive deterioration can the utility of the lake as a water supply be assured for the future. This will require years and billions of dollars, but it would seem that action must begin immediately. The alternative would be to let Lake Erie deteriorate into a great big seething cesspool whose qualities would eventually be passed on to the waters of Niagara Falls, Lake Ontario, and the St. Lawrence River. New Yorkers cannot afford to let this happen.

Oneida Lake Case

Oneida Lake naturally contains large quantities of nutrients conducive to biotic growth, such as nitrogen and phosphorus, and its shallow waters and warm summer temperature result in widespread plant growth and a high fish population. As a result it has long been recognized as one the finest fishing lakes in this part of the country.

Direct pollution from lakeshore homes and businesses, and large additions of waste materials from five main influent streams have caused a superabundance of nutrients, or overenrichment, and excessive biotic growth. When algae remains decompose on the shores or in shallows, they create a nuisance, particularly to recreation areas. Also, as in Lake Erie, as biotic remains die and sink to the bottom, the process of decomposition uses up oxygen in the water, limiting the supply available to most forms of life at depth. Certainly aesthetic qualities of the lake are lessened. Thus, although waste additions cause Oneida Lake's water to be unsuitable for drinking because of contamination from coliform bacteria, they create even greater problems due to overenrichment and excessive biotic growth. Similar problems occur in Chautauqua Lake, certain bays on Lake Ontario, and elsewhere. Excessive biotic growth, usually together with silting, can in time literally fill up bodies of water, leading to their eventual disappearance from the landscape.

Case of the Long Island Duck Industry

The leaching of soluble elements from fertilized farm lands or the dumping of farm wastes into water bodies may add to overenrichment of receiving waters with attendant problems similar to those in the Oneida Lake case. For example, duck-raising on a large scale developed around Great South Bay on Long Island after the oyster industry had become well established in the bay. Although pollution of the bay may have several sources, it is recognized that wastes from the duck-raising areas fertilized the nearby waters, producing a heavy algae bloom which could not be used as food by the oysters or may even have produced materials harmful to them. The several-million-dollar shellfishing industry disappeared from the vicinity.

Cleaning Up

Approximately two-thirds of New York State's residents live in areas affected by pollution and their increasing awareness of the problem has led them to support a recent state-sponsored 1.7 billion dollar clean water program, one billion dollars of which is a bond issue to help finance construction of sewage treatment works.

It is apparent from the complexity of the water resource and the many-faceted problems associated with it that more effective use of water hopefully will come as a result of a coordinated approach on a broad front. Certainly the idea that water ordinarily must be used more than once is a good one, and we are going to have to learn to live with it. When a shirt is soiled, one does not throw it away, one washes it. Water must be looked at similarly. In cases when it must be used for waste disposal, its quality should not be reduced below the level at which it can be economically reclaimed. It would probably also be a good idea to establish something like water users' associations, or other systems of water management, whose responsibility would be to safeguard the quantity and quality of water within their area of jurisdiction. This kind of thing is done in the Ruhr district of Germany with excellent results. There, all water-users must become members of an association, and both their voting weight and their contribution to the association's budget are based upon their water use and the degree to which they cause pollution. Both consumption and pollution are calculated in population equivalents. Keeping the Ruhr clean costs 10 million dollars a year; private firms contribute 2.8 million dollars and municipalities and other government bodies pay a like amount. The balance comes from about 100 waterworks that use the river. No doubt this kind of management pushes the cost upward, but it is a necessary price of a proper sustained yield plan for water. New York State will spend billions on water management in the years ahead. Water will not forever be "cheaper than dirt."

Selected References

Blake, N. M. *Water for the Cities*. Syracuse: Syracuse Univ. Press, 1956.

Bordne, Erich F. *Water Resources of a Western New York Region: A Case Study of Water Resources and Use in the Genesee Valley and Western Lake Ontario Basin*. Syracuse: Syracuse Univ. Press, 1960.

"The Crisis in Water," *Saturday Review* (October 23, 1965), pp. 23–46.

"Filth in the Great Lakes: What Can Be Done About It," *U. S. News and World Report* (December 13, 1965), pp. 58–61.

Heath, Ralph C. *Ground Water in New York*. ("U.S. Geological Survey Bulletin," G. W.-51, in cooperation with the New York Water Resources Commission.) Albany: n. p., 1964.

La Sala, A. M., Jr., W. E. Harding, and R. J. Archer. *Water Resources of the Lake Erie–Niagara Area, New York*. ("U. S. Geological Survey Bulletin," G.W.-52, in cooperation with the New York Division of Water Resources.) Albany: The Division, 1964.

MacKiehan, K. A., and J. C. Kammerer. *Estimated Use of Water in the United States, 1960*. ("U.S. Geological Survey Circular," 456.) Washington, D.C.: Govt. Printing Office, 1961.

Martin, Roscoe C. *Water for New York: A Study in State Administration of Water Resources*. Syracuse: Syracuse Univ. Press, 1960.

Mt. Pleasant, R. C., M. C. Rand, and N. L. Nemerow. *Chemical and Microbiological Aspects of Oneida Lake, New York*. Report No. 6, Department of Civil Engineering, Syracuse University, 1962.

Robinson, F. L. *Floods in New York, Magnitude and Frequency*. ("U.S. Geological Survey Circular," 454.) Washington, D.C.: Govt. Printing Office, 1961.

Salvato, Joseph A., and Arthur Handley. *Water Resources in Rensselaer County*. Rensselaer County Health Department. Troy, N.Y.: New York State Department of Health, 1961.

Temporary State Commission on Water Resources Planning. *Water Resources Management. Six-Year Progress Report*. State of New York Legislative Document No. 27, 1965.

———. *Water Resources Planning and Development*. The Two-Year Report. State of New York Legislative Document No. 42, 1961.

U.S. Department of Health, Education, and Welfare. *The Hudson River and Its Tributaries*. ("Public Health Service Publication.") Washington, D.C.: Govt. Printing Office, 1965.

———. *Municipal Water Facilities—Communities of 25,000 Population and Over*. ("Public Health Service Publication," No. 661) Washington, D.C.: Govt. Printing Office, 1964.

U.S. Department of Interior, Geological Survey. "Water Resources Investigations in New York." Washington, D.C.: Govt. Printing Office, 1962.

"Water Problems in the United States," *U. S. News and World Report* (October 25, 1965), pp. 66–75.

CHAPTER 4   DAVID J. DE LAUBENFELS

Vegetation

THE original forests in New York covered nearly the entire land area of the state. Today the vegetation is still dominated by forests. Forests take a long time to grow, longer here than in many other areas, because of the relatively cool climate. It is well to remember this time factor when we look around us, because many areas not being actively farmed in the state do not yet have a forest cover. Given time, the unused rural areas of today will all develop into forest again by passing through various stages of recovery. The extensive forests that do exist include a few small areas of virgin trees left over from the early days. Most of the forest we have now, however, is growing either because farm land has been abandoned and the forest has grown back, or because lands that were cut over for lumber, but not used for farming, have regrown. A relatively small area has been replanted by man.

In 1880, the year when acreage in farms reached its maximum in the state, there were 24 million acres in farms, or almost five-sixths of the total area. Much of this land was actually cleared. By 1950 less than 13 million acres remained in farms. Much of these 11 million acres dropped from farming in seventy years may be assumed to have reverted to forest. Currently about 250,000 acres of farm land is being abandoned annually in the state, and a substantial proportion of this acreage will become forest land, thus expanding our forests rapidly. Figure 28 portrays the robust role forests play in the land use picture. Only Maine and Pennsylvania, among the northeastern states, rank ahead of New York in forested area. Hardwood species predominate on nearly 85 per cent of the commercial forest land in New York, with sugar maple being the major species.

The state's forests are not uniform and, in order to gain an appreciation of the variations that occur, this chapter treats first the factors in forest distribution. Then it describes each forest type or zone and identifies the distributional patterns. Finally, economic aspects will be noted through a discussion of the use of the forests.

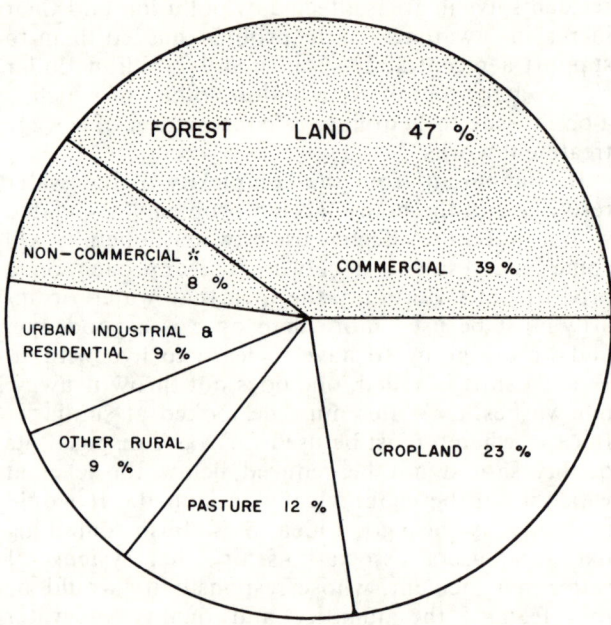

FORESTS IN THE LAND USE PICTURE OF NEW YORK

* Forest land withdrawn from timber utilization by law and forest land too poor to support growth of commercial value or too inaccessible for economical management.

Source: New York State Soil and Water Conservation Needs Inventory, Temporary State Commission on Water Resources Planning, 1962.

Fig. 28

Factors in Forest Distribution

Forest is a general term which includes the possibility of considerable variation from place to place in an area the size of New York. None of New York's many different tree species grows throughout the state, but each flourishes in those areas where it is well suited to compete with the other trees. Usually several tree species grow together to form a characteristic *association;* the more hardy kinds of tree turn up as members of several associations. Each association develops in response to three major variables: climate, soil, and the degree and kind of disturbances that have occurred in the past. Because of this last factor, patterns of forest must not be thought of as static but in a state of flux as the vegetation slowly recovers from one kind of disturbance or another. Early stages of this recovery may well be represented by vegetation associations with few or no trees.

CLIMATE

Differences in climate account for the broader patterns of vegetation in the state, the plant cover responding primarily to the number of days and the temperatures during the growing season. These two items can be essentially combined through use of growing degree months (introduced in Chapter 2). A growing degree month is a temperature efficiency index consisting of the sum of all average monthly temperatures over 40° F. The value of 40° F. is taken because plants do not ordinarily grow at lower temperatures. Either higher temperatures or longer growing seasons (frost-free period) would make this figure larger. Empirically, plant distribution in the state correlates better with this index than with other commonly available climatic data. The correlation is also better than with potential evapotranspiration, another indicator of growth potential. It is difficult to experiment with long-lived trees to determine the true critical values of survival or of effective competition with other trees, and there are many tree species that would have to be studied in this way. As a result, forest zone boundaries can not often be explained rigorously; it is necessary to explain them empirically. The fact that the amounts of moisture are not considered critical stems from recognition of a general adequacy of precipitation over the state. Figure 29 displays the growing degree month data for more than 150 places superimposed on a map of natural vegetation zones. When the effects of terrain are overlooked, zones of energy receipt or temperature efficiency run roughly east–west across the state. That is, in the south there are long, warm summers, while with an increase in latitude the summers tend to be shorter and cooler until, in some more northerly areas, a pleasantly cool but disappointingly short summer is experienced. Differences in elevation, too, are important, the higher places being the cooler. Coolest summers occur in the higher Adirondacks. The actual position of each temperature zone is further modified by water bodies, sheltered position, and local terrain. Temperatures are moderated and the frost-free season lengthened by proximity to the Atlantic Ocean or to the larger inland lakes. Areas south and east of highland masses seem to be sheltered and so escape the more severe effects of cold continental air masses. In hilly areas cold air drains into low places, shortening the frost-free season, while south-facing slopes get more direct sunlight and, as a result, warmer average conditions. All parts of the state are reasonably moist but suffer periodic droughts when rainfall fails to supply the needs of the plants for several weeks. A greater drought hazard occurs when temperatures, and therefore water needs, are higher. The vegetation zones to be described below show a close spatial correspondence to the growing degree month data in Figure 29.

SOIL

Differences in soil, especially the water-retaining capacities but also the chemical content, produce striking local differences in vegetation. A series of plant formations based on the degree of drainage can be thought of as a varying response to the different wetnesses of the soil. Thus the many low, poorly drained places, usually resulting from glaciation, have special kinds of vegetation. Where the standing water is shallow there are marshes; many are small, but some are like the Montezuma marsh in mileage. Seasonally wet areas display a marsh meadow kind of vegetation made up of grasses and sedges. With better drainage, shrubby plants invade the meadows and different kinds of bushes and bushy trees eventually take over the area, the species depending partly on the climatic zone. Where the water is deep and stagnant, as for example in the myriad kettles caused by the glaciers, there develops a mat of vegetation called a bog. Sphagnum moss and sedges together with various specialized shrubs live in, and make, the bog environment. As the bog fills in, bushy plants become established much the same as happens in the swamp meadow. In either case the final result is a swamp forest of trees like larch, spruce, and fir that are tolerant of or require a high ground water table.

Special soil conditions also may be the result of too little water, shallow, rocky soil, or high concentrations of certain chemicals like lime. The great masses of glacial lacustrine sand that ring the Adirondacks and extend east–west along various beach ridges in other parts of the state account for the bulk of the soils deficient in water. Bare sands tend to be invaded by tough grasses and a heath type of vegetation. Soon various *xerophytic* (able to grow with a limited water supply) trees such as pitch pine, gray birch, and chokecherry fill in, eventually forming an open scrub forest. Where deeper soils now

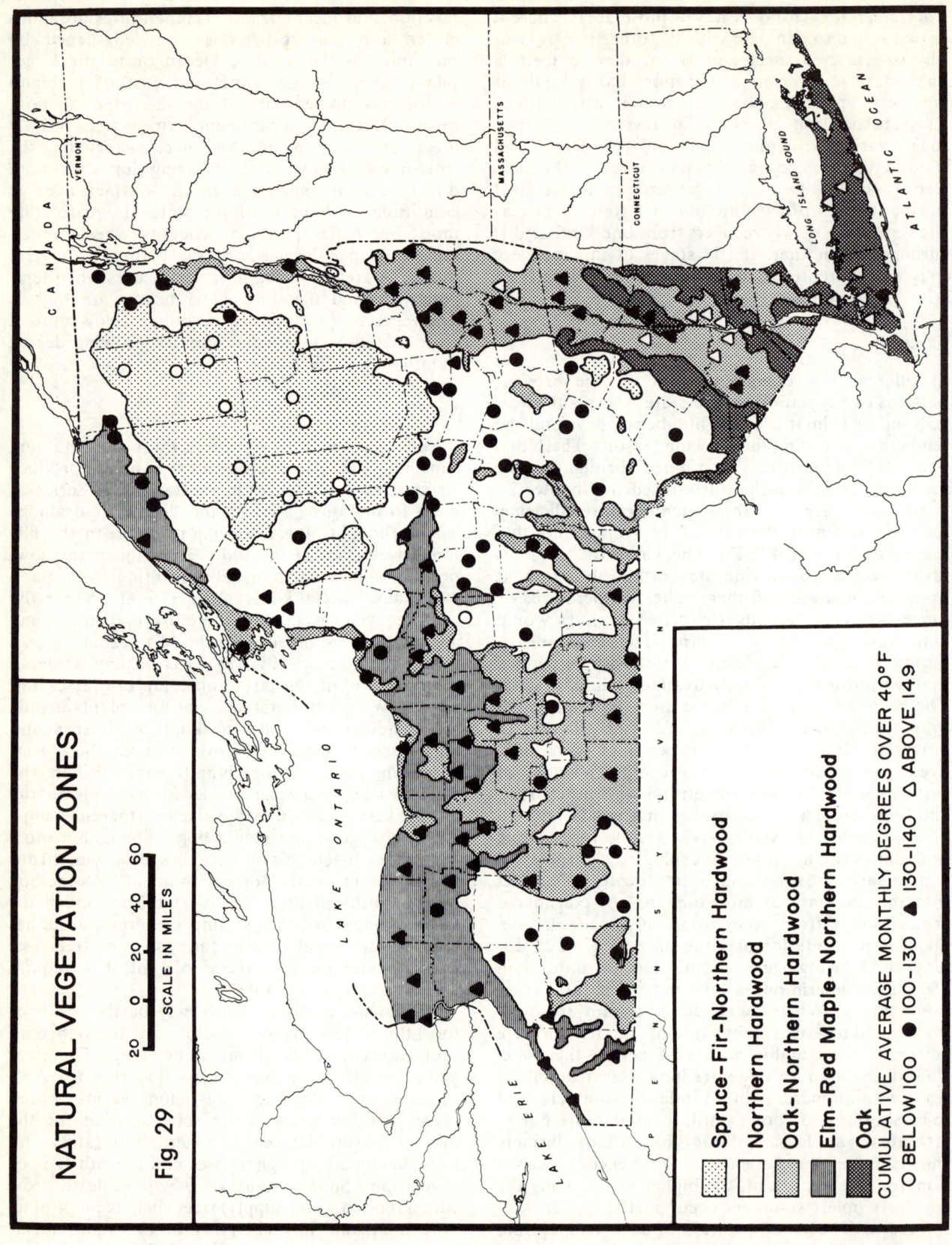

Fig. 29 NATURAL VEGETATION ZONES

exist over sand, pure stands of white pine often were originally found. Even more deeply developed soil seems to eliminate the drought effects of the underlying sand and allows a general *mesophytic* (grow under medium moisture conditions) forest cover to become established. Rocky places are pioneered by lichens and mosses, followed in some areas by vegetation similar to that over sand. With deeply developed soil, rocky areas lose any distinctiveness. The widespread occurrence of high-lime soils in the state is characterized by still other secondary vegetative associations. Cedar, for example, often vigorously invades open fields over limestone.

DISTURBANCE

Most of the forests of New York are in some stage of recovery from the impact of man. Where the area is attractive to economic use, fields, pastures, yards, cities, and other things use it, occupying from some to all of the ground surface. All but a small bit of the rest has been used and re-used over the past two centuries. If not farmed, the land was at least cut over for lumber. Again, from time to time, valuable trees have been cut out until the area may be sadly depleted as far as these valuable species are concerned. Perhaps, through carelessness, a forest fire occurred and further degenerated the vegetative cover. Finally, large areas are partly used by sporadic pasturing. Obviously, human use (or misuse) can favor the growth of tough and commercially useless plants. Recovery from disturbance follows the general patterns of plant types and successions on unfavorable soils. Trees may quickly establish themselves in abandoned cultivated fields, especially white pine, hawthorn, white ash, and other sun-tolerant species. Abandoned pastures and cutover areas have a mat of grass, weeds, and brush in them. Large abandoned pastures may preserve prairie-like conditions for a considerable time near the center, but moving in from the edges bushes and scrubby trees like hawthorn or locust tend to develop a scrubby forest cover. Red maple and American elm take over the more moist sites. This sort of succession may take many years, but always after a time forest trees begin to close in on the land.

What a waste of time and land it is to wait for vegetation slowly to evolve from disturbed or extreme conditions to a full and healthy forest cover! More and more, man is coming to realize that proper forest management can short-cut the development of valuable forest growth. By discouraging useless waste in land use and discouraging the growth of low-value plants, while encouraging or even planting valuable tree species, man can reduce somewhat the long time it takes the forest to grow. The critical question often is: Can he do this economically?

Forest Zones of New York

Five forest zones are identified in Figure 29. Any more detailed map of vegetation would become a mosaic of many different plant associations. Even leaving out the associations related to edaphic conditions and succession, there is still local variation in the relative amounts of the common and less common species. Looking at the state from a broader view, one can say that there is a group of trees that can be called the northern hardwoods which occupy all of the state except the extreme southeast and the higher Adirondacks. This group of northern hardwoods is dominated by beech and sugar maple, with many associated trees such as ash, basswood, cherry, hemlock, birch, white pine, red maple, and occasional oaks. The forest zones in the state are best described by their relationships to the northern hardwood complex. They are: (1) the oak zone, an area in the southeastern part of the state dominated by oaks and largely without other species of the northern hardwood complex, (2) oak–northern hardwood zone, where northern hardwoods are associated with considerable amounts of oak, (3) northern hardwood zone, where this complex is strongly dominant, and (4) spruce-fir zone, found in highland areas where spruce and fir are mixed with the northern hardwoods and in places even outnumber them. On the Erie-Ontario Lowlands and in the St. Lawrence Valley, a great deal of elm and red maple and some oak are mixed with the northern hardwoods. This really is a subtype of the oak–northern hardwood zone, but it is referred to separately as the elm–red maple–northern hardwood zone.

Because of the topography, the zones do not form continuous regions but are separated into several parts; each one, big or small, forms a distinct vegetation region.

OAK ZONE

Several species of oak occupy the warmer regions and thinner soils southeast of the Catskills and out onto Long Island where there are at least 150 growing degree months. All factors—low altitude, low latitude, nearness to the Atlantic, and shelter by the Catskill Mountains from cold winds—combine to make this part of the state much warmer than all the rest. July temperatures average about 72° or 73° F. and there are over 170 days without frost. It was here that farming first took hold in the state. Today the better lowland soils still are intensively farmed, but the widespread rocky and sandy areas have been

INTERIOR OF A NORTHERN HARDWOOD FOREST. A mixture of large trees and small ones and evidence of previous logging is typical. Selective logging of maple, birch, ash, or cherry may be economical. A heavy ground cover of ferns, small trees, and low brush, as shown here, is commonly found. *N.Y.S. Dept. of Conservation.*

abandoned. The high density of urban population in the same area, however, has meant that recreational uses and general pressure on the forest has been widespread. As a result, most of the forest land is scrubby and there is much open land being invaded by bushes. Add to all this the fact that here the best forest tree was the chestnut, wiped out a generation ago by blight.

Within the oak zone there are several common secondary associations. On the sands and sterile rocky soils of Long Island and the ridges that cross the lower Hudson, scrubby oaks grow with pitch pine. Sometimes pure stands of pitch pine develop. These dry areas are especially subject to fire, a good reason why they remain scrubby and deeper soils are no longer present. Red cedar springs up in pastures, abandoned fields, and on thin, rocky soils over limestone. Red maple, aspen, fire cherry, hickory, and black locust also pioneer on old fields here and there. In poorly drained areas some of the denser stands of oaks are found today but with a jungle of tangled vines and weeds. On parts of Long Island near the growing metropolis are some forest remnants that have more southerly elements mixed in, like the tulip poplar and sweet gum.

Oak-Northern Hardwood Zone

A transition zone where oak and northern hardwood intermingle or alternate with each other extends into many parts of the state. This combination flourishes on the moister and deeper soils in the Hudson Valley, at low and intermediate elevations among the hills across the Appalachian Upland, especially in the Finger Lakes area, and at lower elevations northward, until in the northernmost parts of the state it strings along only the very lowest plains. The somewhat cooler summers or shorter growing seasons in these areas limit the growing degree months to a range of 130–150. Northern hardwoods can not only survive, but also compete successfully, with the more moderate amounts of heating, whereas oaks, requiring more heat, seem still to find plenty of energy for their needs.

Direction of slope is responsible for sharp local variations within the zone, significant because a great deal of steep terrain across the center of the state coincides with this region. Instead of mixed stands of oak and northern hardwoods, there may be alternating stands of each. Slopes facing south and southwest receive more sun and support nearly pure stands of oak or oak mixed with hickory and other trees. The north and northeast slopes receive the least sun and tend to resemble more northerly vegetation zones.

New and different secondary formations appear along with the cooler temperatures. White pine thrives on sandy or droughty soils, which occur frequently south and east of the Adirondacks, although pitch pine and scrub oaks, too, may be found. White pine also springs up quickly in abandoned fields and its more common occurrence in the more eastern parts of New York may well point to earlier abandonment in those areas. Unwanted cropland used to revert directly to forest rather than be converted to pasture before final abandonment, as is the current practice, and white pine does not pioneer well into abandoned pasture. The oak-northern hardwood zone is ideally suited for the American elm, provided the vegetative cover is open. These elms crowd along fences and into pastures, while they follow the scrubby red maples into poorly drained areas. Their wine-glass shape is so characteristic a companion of man and his fields, it is a shame that many are being killed by the Dutch elm disease. Red cedar, white ash, hawthorn, and locust may be seen invading abandoned pastures in this zone.

Elm-Red Maple-Northern Hardwood Zone

The widespread poorly drained areas together with nearly complete removal of the natural forest on the Ontario and St. Lawrence plains have produced regional forest characteristics sufficiently different to give these areas another zone designation although their forests are very similar to those in the oak-northern hardwood zone. This zone is distinguished by the relative frequency of American elm and red maple, although both oak and northern hardwoods are still present in less abundance. The amount of oak is especially reduced because the well-drained soils, which it prefers, have all been cleared and are generally still in use for growing crops or for pasture. The level to rolling topography of this part of the state, its good soils, and its accessibility to major transport routes made the area attractive to farm settlement. On the other hand, the extensive poorly drained areas provide a special habitat, along with abandoned fields in some sections, for elm and red maple. The prevalence of these trees, then, is due not to climate but to the accident of disturbance and edaphic conditions, which are further related to the surface configuration. The fact that forests here are so similar to the oak-northern hardwood zone is not surprising if the reader will turn back to the preceding chapter and observe the similarity in climate of the two areas.

Northern Hardwood Zone

A broad zone, with between 110 and 130 growing degree months, is dominated by northern hardwoods. Average July temperatures are 67° or 68° F. and the growing season is seldom more than 150 days. On the other hand, mean July temperatures do reach at least 66° F. and there are four months without frost. Occasional oaks are found, but there does not seem to be enough energy received over the summer to allow them to compete successfully. On the other hand, the climate appears to be too warm and dry for spruce and fir. The stands of northern hardwoods extend south across the Pennsylvania border and the higher parts of the Appalachian Upland. The lower areas around the Finger Lakes and the Ontario Plain are not in this zone because they are warm enough to allow oaks and other trees to mix in. Extending to the north, the northern hardwoods ring the Adirondacks at lower elevations.

The northern hardwood zone is far from uniform, being made up of a large group of trees but clearly dominated by beech and sugar maple. In warmer areas basswood, white ash, and black cherry are regular associates; to the north they are less common. In cooler areas yellow birch becomes a third dominant with beech and maple. Hemlock, white pine, and white cedar—all evergreens—are abundant among the northern hardwoods but are not at all evenly distributed. Hemlock is most characteristic of moist, shady slopes and ravines, at times forming nearly pure groves. There is considerable evidence that hemlock was once more common, but the attractiveness of this species to lumbermen and the use of the bark as a source of tannin made it among the first trees to be cut in early days. White cedar may grow along with hemlock, or may crowd into the abandoned fields and poorly drained areas, especially on limy soils. Alder and larch appear in wet areas, too. White pine occupies the same sorts of sites it prefers in the oak–northern hardwood zone, as do occasional elm and red maple.

SPRUCE–FIR ZONE

The coolest parts of New York State favor the growth of spruce and fir. Although the northern hardwoods continue to be present, there is a greater abundance and variety of softwoods, represented by a group of species not only tolerant of, but in some cases requiring, cooler conditions. July mean temperatures in the higher and northerly parts of the state are below 66° F. and growing seasons are less than 125 days. Growing degree month totals of about 100 also illustrate the coolness of the climate. The highest parts of the Catskills, the Tug Hill area, and the Adirondacks in general fall within this zone. Spruce, fir, and larch, as well as white pine, hemlock, and white cedar all flourish in the ubiquitous poorly drained areas, with red spruce dominating the assemblage. Occasional bogs in other parts of the state contain related vegetation. All of the softwoods have, however, been greatly reduced in number because of the demand for them as lumber. At the same time the less desirable northern hardwoods have become more abundant. As a matter of fact, repeated cutting and burning of the forest and, more recently, the mushrooming of forest plantations make it difficult to find any truly representative stand of naturally growing trees. At higher elevations and in more remote locations, the softwood element becomes more common, balsam fir occupying large areas of well-drained soils. The few peaks rising above about 4,500 feet are beyond the tree line and display patches of tundra and bare rock.

The Adirondacks contain little good soil and thus are generally not farmed. This favors large-scale lumber activities and, unfortunately, large-scale fires. Forests recover but grow especially slowly in the cool conditions here. It is now a general practice to plant relatively fast-growing softwoods on the open lands; the species being used are red pine, Scotch pine, white pine, and white spruce. The exotic Scotch pine, interestingly, is admirably suited to the local conditions and is everywhere escaping into the wild vegetation. Red pine and spruce are so commonly planted that many of the trees seen along roads are not wild at all. Extensive areas of abandoned or burned land, particularly on poorer soils in the lower Adirondacks and beyond, toward Canada, are occupied by scrubby second- or third-growth hardwoods. To the north aspen, gray birch, and bushy forms of cherry predominate, while paper birch is more common to the east.

Forest Lot, Tract Size, and Ownership

The size of individual areas in natural vegetation and the proportion of total area covered are as important in describing and explaining vegetation as is the species composition. Where most of the surface area is in some improved land use, the kind of natural vegetation has little or no meaning. Not only may the relative area be unimportant, but small stands of plants tend to have a diminished composition. The habitats of various species may be largely eliminated and, furthermore, small wood lots have different environmental characteristics than large forests. On the other hand, a continuous cover over a large area is more subject to forest fire and clear-cutting by lumbermen. Intensity of natural vegetation distribution varies in New York State from practically nothing to complete forest cover. Figure 30 clarifies this point.

Heavy forests grow mostly in upland and steep areas such as the Catskills, the Adirondacks, the Tug Hill, and the Hudson Hills. Almost all the spruce-fir zone is included, as well as many low,

A Good Stand of White Pines. Relatively pure stands of white pine are rare in the state. The white pine was originally widespread and in great demand by the lumber and shipbuilding industries. Reforestation is somewhat hazardous because of diseases to which the tree is subject. *N.Y.S. Dept. of Conservation.*

sandy areas such as Long Island. A heavy forest is a piece of land of which at least two-thirds supports forest. In such areas are to be found forest tracts in excess of a thousand acres and farm wood lots exceeding fifty acres. Some counties are almost entirely forest. The strongest statements about misuse and degeneration of the forest are applied where forest covers most of the land. The statements are true and all the more noticeable because forest dominates the landscape. Nearly all of the large areas occupied by forest in the state are in stages of regrowth. They have been cut over for lumber, burned, or abandoned from farming recently enough so that the forest has not had time to recover to a stage of mature stability. But fortunately sufficient luxuriance and natural beauty have been achieved for the needs of general recreation.

Areas where forest and wood lot cover less than a third of the surface occupy low, relatively level land with good soil. By far the largest such area in New York State is the Erie-Ontario Lowlands. Other areas extend eastward along the Mohawk depression, northeastward along the St. Lawrence, and southward around the Finger Lakes. What this suggests is that forests in the recent past were confined largely to the most undesirable places such as swamps, river banks, steep slopes, or areas of generally poor soils. The formerly dominant trees of the better sites are largely gone, and the remaining forest species represent extreme edaphic types. Thus it is not surprising that the elm–red maple zone coincides rather closely with mostly farmed land (compare Figs. 29 and 30). It is interesting to note that very little sparsely forested land occurs in the warmest or coolest parts of the state.

Alternating forest and farm land are characteristic of hilly terrain of intermediate steepness. Some more level terrain of poorer soil is included, too. A great deal of New York State has forest and farm land in roughly equal proportions, in which case the most characteristic land use is usually dairy farming. Some of the best stands of trees occur here, where wood lots and forest tracts are of respectable size but where continuous forest does not exist. The stands are too small for sustained lumbering activity and the forest fire risk is not too great. Signs of disturbance are far from lacking, however, and essentially all of the forest is second or third growth. Even the more established tracts are subject to selective cutting for more valuable species and may be grazed by dairy cattle. There is much abandoned pasture land, full of hawthorn, cherry, and various bushes and weeds, that is being surrendered to nature. More and more, rather than surrender the land, it is being planted to trees. It may make good sense to use fields not really worth pasturing for long-range crops of Christmas trees and pulpwood, or by planting enhance their value for recreation. It makes little sense to wait for the forest ever so slowly to become re-established.

Figure 31 shows the proportion of forest land in different ownership categories. The state Forest Preserve, which accounts for the bulk of noncommercial forest land, is situated in the Adirondacks and the Catskills. The small percentage (6 per cent) of government-owned commercial forest land is made up of state and county forests outside the preserve. It is especially interesting to note that nearly half the forests in the state are not owned by farmers and forest industries but by rural nonfarmers, city people and others. These owners as well as farmers tend to have small parcels. This is illustrated by the fact that 55 per cent of the privately owned forest land is in parcels under 100 acres in size.

Use of the Forest

There are several ways that the forests of New York State contribute to the local economy, either through the farmers, through companies whose purpose is to use the forest products, or through private and public recreation. Different trees yield lumber, pulpwood, sugar, Christmas trees, or fence posts, while the forest itself provides an environment conducive to relaxation.

Lumber

In the year 1850 New York ranked first among states in lumber production, accounting for 30 per cent of the lumber cut in the United States. For about forty years the annual cut was in excess of one billion board feet and the state remained among the first ten in production. In these early years pine, spruce, and hemlock, cut from virgin timber stands, were most important. The old forests were largely harvested prior to World War I; since 1915 New York has produced only about 1 per cent of the nation's lumber.

Only in the Adirondacks do wood-using industries employ a substantial part of the labor force today, and there only because little other industry exists. Slightly more than half of the timber cut in the Adirondacks is still softwood, elsewhere the softwoods are almost insignificant. After the good-sized softwoods were gone, attention was directed toward use of the hardwoods, especially maples, oaks, black cherry, and yellow birch. Whereas softwoods are used for lumber and paper pulp, hardwoods have more specialized uses, being converted into flooring, furniture, bowling pins, and other products. A moderate amount of hardwood timber is cut in all parts of the state, except in the big urban

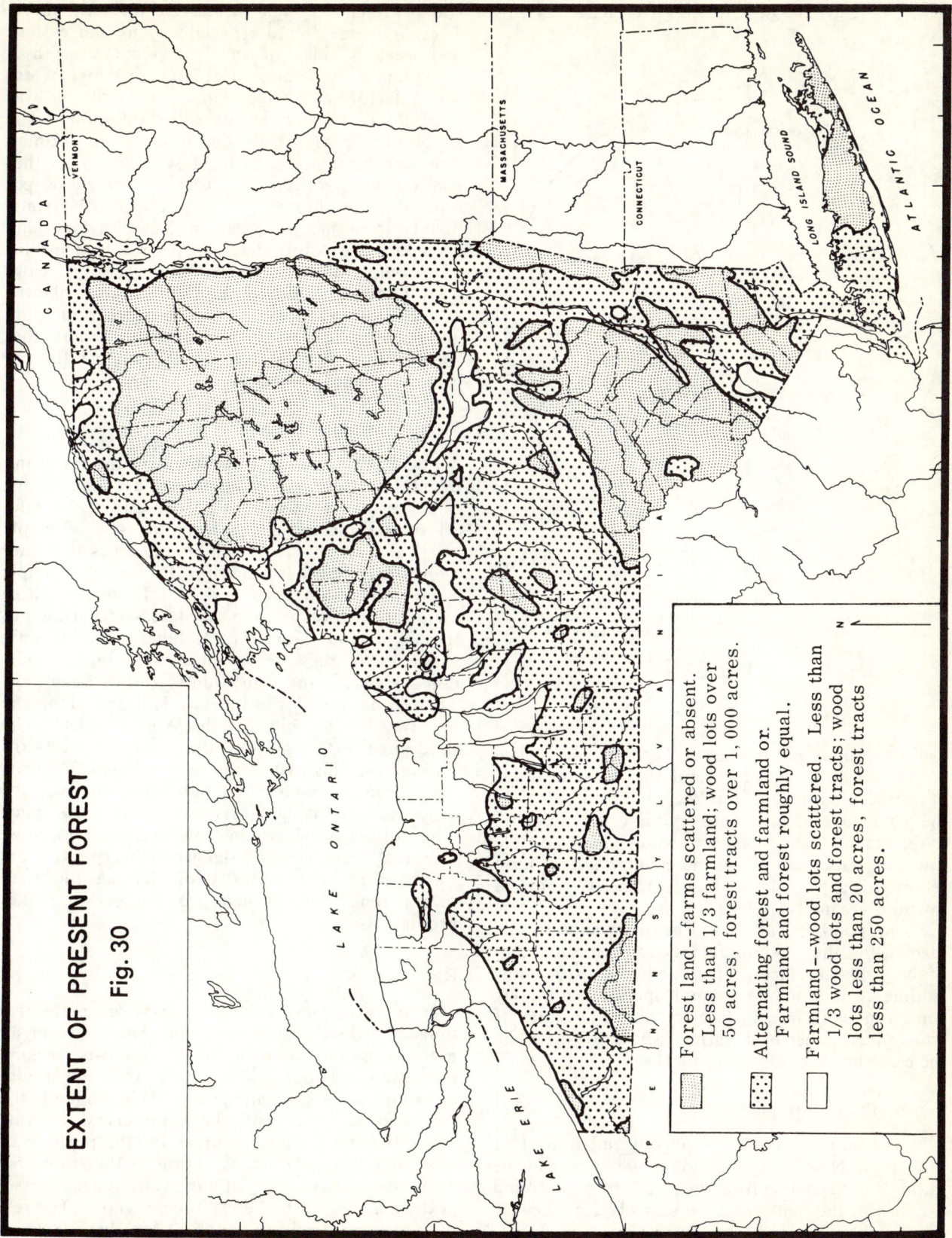

CHAPTER 4 VEGETATION 99

FOREST LAND OWNERSHIP IN NEW YORK

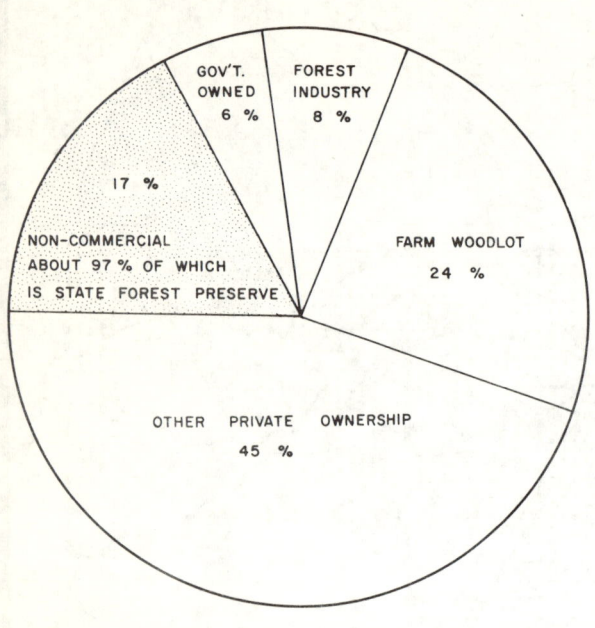

Source: *New York Facts*, Publication of the New York Forest Industries Committee and the American Forest Products Industries, Inc., Washington, D.C., 1962.

Fig. 31

agglomerations and the heavily farmed zones. The largest sawmills cluster around the densest stands of northern hardwoods but become more scarce as the percentage of forest preserve and certain types of recreational land increases. Many groups of idle sawmills mark areas where merchantable trees have recently disappeared. In Figure 32 active and idle sawmills are indicated by symbols.

Farm wood lots are a source of revenue to the farmer, but they seldom amount to even 5 per cent of the gross farm income in any county. From time to time a small mobile sawmill operation may buy some timber from a farm wood lot. The sale of white pine lumber from farms is noteworthy along the eastern side of the Adirondacks.

Wood Pulp and Paper

Wood pulp production is an old and important industry in New York, with about 30 per cent of the nation's pulp coming from the state's forests around the turn of the century. Now scarcely 2 per cent is produced here.

A fairly large number of pulp and paper mills still operate, although there were formerly more. These mills are found especially along the eastern and western sides of the Adirondacks, with a secondary concentration near Niagara Falls. Water was a factor in the location of the mills, as was accessibility to supplies of softwood pulp. The supply of pulp has greatly diminished in recent times, thus causing many mills to close. Pulp mills that continue in operation obtain approximately 60 per cent of their supplies of pulp outside the state, mainly from nearby areas in New England and Canada. The Adirondacks provide only one-third of the pulp now consumed within the state. Greater uses of hardwood pulp in the future could make the state's forests a more important resource to the pulp and paper industries. It is encouraging to note that some large mills have converted to almost 100 per cent use of hardwoods.

Miscellaneous Uses

Fence posts are cut to be used on the farm and may also be sold. Only small trees are required for this type of use. Cedar is particularly in demand. Substantial income comes from the sale of maple sugar and syrup. In the favorable areas there are many "sugar bushes" developed for regular maple sugar production each spring. There are three major production areas in the state, concentrated in the higher and snowier parts but necessarily only where maple trees are common. The most important area lies in the high country east of Lake Erie. The second area lies in the Tug Hill and along the northwestern foothills of the Adirondacks. The third area extends from the Catskills westward through the hills east of the Finger Lakes. Increasingly farmers and other landholders all over the state are producing Christmas trees. So much interest has been shown in growing these trees, however, that there is now a danger of overproduction. Furthermore, the necessity of shaping the trees several times during their growing period makes profits illusive.

Recreation

In and near New York State are large concentrations of urban population from which swarm an ever increasing number of vacationists seeking relaxation and sport. They want lakes and mountains and forests. Anticipating their demand, the state very early set aside large preserves. Private interests have established areas for the purpose of recreation, too. The constitution of the state provides that a large part of the Adirondacks be set aside in a park to be kept "forever wild." The rest of the area is in private hands. The Adirondacks as a whole have enormous recreation potential. Lands

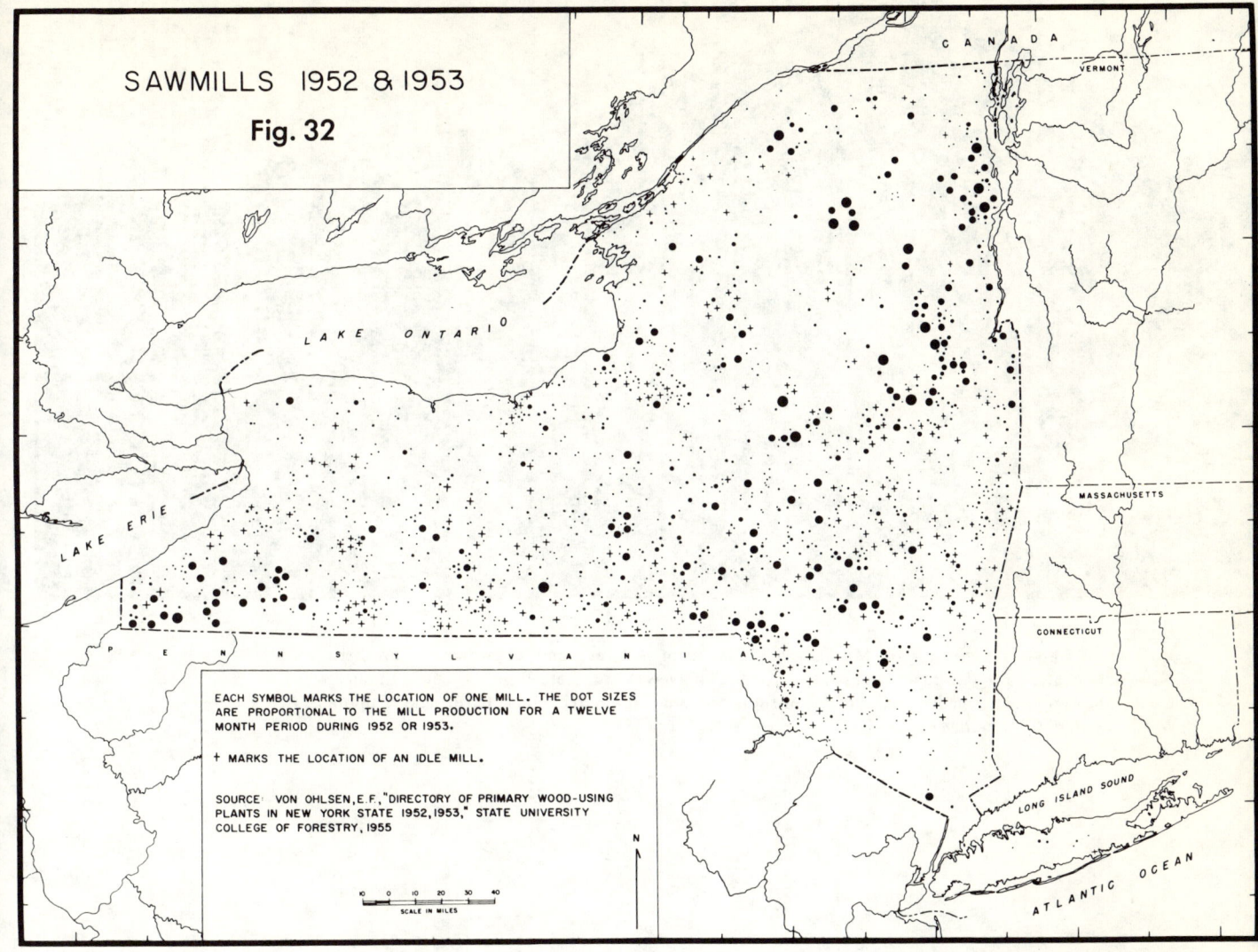

generally were denuded of any valuable or large trees before being acquired by the state but are now covered with second-growth forests. Mile after mile of the private land belongs to forest industries, and lumber and pulpwood are regularly harvested. There also is a large forested state park in the Catskill Mountains with extensive recreation opportunities. This area is used even more intensively than the Adirondacks because of its proximity to New York City. Allegany State Park along the Pennsylvania border in western New York is famous for its limited stands of old-growth trees.

Some Concluding Thoughts

Because of continuing abandonment of marginal farm land slowly followed by invasion of trees on that land, the forest area in New York is expanding. The forests could be used increasingly for wood products and recreation. With good forest management, both of these uses can be facilitated. It will be hypothesized here, however, that the latter, although difficult to evaluate in dollars and cents, will greatly outstrip the former in the years ahead. Tremendous acreages of abandoned farm land and forest land are being bought each year, largely by nonfarmers, for private recreation land. In most cases, the purchasers are simply individuals who want to own a piece of woodland for their private enjoyment. With most of the people living in urban areas and an expected doubling of the population in the next thirty-five or forty years, it seems logical to presume that this desire to own woodland for purposes other than wood production will increase. This new type of owner will pay local taxes and spend money at local markets. He may in this way make forests pay without cutting a single tree.

PULPWOOD BEING ASSEMBLED at Mechanicville. North of Albany, and elsewhere in northern New York State, pulp and paper industries make use of New York State forests. Often, though, local wood supplies are inadequate and out-of-state or Canadian sources must be depended upon. *N.Y.S. Dept. of Conservation.*

SCOTCH PINE PLANTATION in the Adirondacks. Scotch pine will grow in all parts of the state and is widely planted, especially for Christmas trees. It is a fast grower to Christmas tree size, but must be pruned or shaped. It is suitable for pulp, too. *N.Y.S. Dept. of Conservation.*

Selected References

Armstrong, G. R., and J. C. Bjorkbom. *The Timber Resources of New York*. Upper Darby, Pa.: Northeast Forest Experiment Station, U.S. Department of Agriculture, 1956.

Braun, Lucy E. *Deciduous Forests of Eastern North America*. Philadelphia: Blakiston, 1950.

Bray, William L. *The Development of the Vegetation of New York State*. ("N. Y. State College of Forestry Technical Publication," 3.) Syracuse: State University College of Forestry at Syracuse University, 1915.

Harshberger, John W. *Phytogeographic Survey of North America*. Leipzig: W. Engelmann, and New York: G. E. Stechert, 1911. Reprinted Weinheim: H. R. Engelmann, and New York: Hafner Pub., 1958.

Küchler, A. W. *Potential Natural Vegetation of the Conterminous United States*. ("American Geographical Society Special Publication," 36). New York: American Geographical Society, 1964.

Larson, Charles C. *Forest Economy of the Adirondack Region*. ("State University College of Forestry Bulletin," 39.) Syracuse: State University College of Forestry at Syracuse University, 1956.

Ricknagel, A. B. *The Forests of New York State*. New York: Macmillan, 1923.

Shelford, Victor E. *The Ecology of North America*. Urbana: Univ. of Illinois Press, 1963.

Stout, Neil J. *Atlas of Forestry in New York*. ("State University College of Forestry Bulletin," 41). Syracuse: State University College of Forestry at Syracuse University, 1958.

CHAPTER 5   DAVID J. DE LAUBENFELS

Soil

Soil is an indispensable resource. From it comes most of the food we eat, either directly as plant products or indirectly by way of animals. This means that a substantial portion of the minerals and other critical substances which maintain human life is derived from soil. Not only are soils important to the consumer of food, but their productive potential is critical to the producer of food, the farmer. The success of agricultural ventures and the general economic health of rural areas are bound to depend on soil quality.

Variation from Place to Place

Like other aspects of New York's physical environment, soil is not everywhere the same. Of major importance is its variation in quality. In some places it is productive; in others it is not. In some places it responds effectively to fertilization; in others it does not. Reasons for such areal diversity are many. Climate, natural vegetation, terrain, parent material, and drainage conditions—all these are influential in soil development. Terrain, climate, and vegetation vary considerably from one part of the state to another, so soils are bound to vary, too.

Parent material and drainage conditions are likely to exhibit an even more highly detailed mosaic of variation than climate and vegetation, with significant differences sometimes existing in distances of but a few yards. Because glacial ice sheets moved across the state only a few thousand years ago, New York soils are young as soils go. Relatively fresh parent material lies close to the surface, usually not seriously depleted of soluble minerals as is often the case of older soils in unglaciated humid areas. Young soils strongly reflect the nature of the material from which they were formed and thus vary as glacial deposits, and the even more recent deposits of rivers and wind, vary. Swampy, poorly drained areas are common, but nearby excessively drained and dry knobs or hillsides may occur. Such diversity in water conditions produces soil differences, even though climate, parent material, and other factors are the same.

The better soils in New York are likely to have developed on adequately drained and relatively level sites which originally supported a predominance of deciduous broad-leaf forests. Parent material derived from limestone or glacial drift with much limestone in it is an asset, too, for it contains lime, an important soil ingredient. Often a favorable combination of the above conditions exists on alluvial valley bottoms, but frequently, too, it is found on the broader, glaciated plains or the more level uplands in the state.

Modern farming techniques can enable man successfully to overcome one or more soil deficiencies. For example, artificial drainage of poorly drained but otherwise good land is common, as is the liming of low-lime soils. Thus, in analyzing the geography of soils and seeking to understand the future they hold for the state, it is necessary to remember that use depends not only on inherent qualities but also on the response of soils to man's management.

Soil Properties

Soil is a complex of organic and inorganic materials suitable to the growth of plants. It forms slowly on the surface of the earth as a result of climate, plants, and chemical action working on whatever material is exposed. Color, texture, structure, degree of acidity, drainage, profile and depth are all important properties which differentiate soil from place to place and give it varying productive capability levels.

Color

To the casual observer, one of the most obvious soil properties is color. Soil colors in New York range through various shades of brown and gray; some are even black. Degrees of gray, black, and dark brown are likely due to the amount of organic material present, while reddish-brown colors may be due to concentrations of iron oxides. Where whitish or light-gray soils are found, both organic materials and iron oxides are scarce. Although darker soils are usually thought of as being best for agriculture, there are instances when this is not the case.

TEXTURE

Texture refers to the prevailing size of soil particles. Some of the commonly recognized textural classes, from the largest particle size to the smallest, are sand, silt, and clay. Sandy soils do not retain water or organic material well, while fine clays tend to be impermeable to the passage of water and difficult for plant roots to penetrate. The better soils have medium textures or a combination of textures not dominated by extremes. These are often referred to as loams, silt loams, or sandy loams.

STRUCTURE

Structure refers to the arrangement of soil particles. A good soil structure is attained by the association of soil particles into groups called *floccules*. Space between floccules permits easy passage of water and air and penetration of plant roots. A soil with a good structure appears granular or friable and is not ordinarily subject to serious compaction when walked on or run over with heavy machinery.

DEGREE OF ACIDITY

Soil acidity is one of the main limiting factors in crop production in the state. The degree of acidity depends in part on the amount of organic materials which break down into organic acids in the presence of bacteria and water, and in part on the supply of the most common acid neutralizer, lime. In a humid climate soluble lime is readily leached away as it slowly forms from the breakdown of parent material, and therefore humid climate soils are normally of acid character. Some parent materials, such as sand, may supply no lime at all. On the other hand, certain sections of New York have soils derived from glacial till containing large amounts of limestone. Such soils have lime and therefore nonacid conditions in the subsoil and are less strongly acid or even neutral in the topsoil. The acid soils have acid conditions in both topsoil and subsoil.[1] Because most crops, especially grains and hay, do better on soils that are neutral or nearly so, it is important to differentiate between these kinds of areas and those of greater acidity.

DRAINAGE

Proper soil drainage is very important to soil development and usage. Drainage is related to texture, structure, and slope. Excessively drained soils retain little water for plant growth and soil development. Poorly drained soils retain too much moisture for most cultivated plants. Excess water interferes with the breakdown of organic material and may result in the accumulation of a black mass of partially decayed and undecayed plant material called *muck*. Local variations in terrain and drainage produce sharp differences in soil, even though the parent material is the same. Such changes of soil character in short distances—perhaps in a single field—make careful and detailed soil analysis essential to proper use of the land.

PROFILE

If the soil-forming processes continue long enough, contrasting layers, or *horizons*, are formed. The succession of these layers, from the surface down to the underlying parent material, is called the *soil profile*. Three horizons, often with subdivisions, are commonly recognized. The profile may vary in depth from a few inches to many tens of feet. On top is a layer enriched by humus derived from forest litter, grass roots, and the like. In the humid climate of New York there is an excess of water moving through the soil which carries soluble materials downward. This process is called leaching. Therefore, the upper layer (the A horizon, or topsoil, as it is popularly called) tends to be high in dark-colored organic material but often relatively low in soluble salts. Beneath the topsoil is a layer, or zone, which has been enriched by the leaching process. This is the B horizon, or subsoil. It is likely to be dense, with accumulations of aluminum and iron oxides and fine clay particles; almost certainly it is lighter in color because of the paucity of organic material. In some of the state's acid soils derived from glacial till a portion of the B horizon, called a *fragipan*, is commonly so tightly packed and impermeable and pore space is so limited that little water and air are available to roots. This zone can limit plant growth and cause serious drainage problems. Below the B horizon is the little-changed upper portion of the parent material, the C horizon. Soils with well-developed profiles may be thought of as *mature*, those lacking profiles as *immature*.

DEPTH

The A and B horizons of soils in New York are usually not more than two feet deep, and frequently much more shallow than this. Where water erosion has carried away upper soil layers as they form on steep slopes or where glacial scouring exposed bedrock at the surface, perhaps nothing that could be called A and B horizons exists at all. Similarly, recently deposited alluvium on river flood plains, or dunes produced by wind, because of the short time

---

[1] Degree of acidity is measured by what is known as the pH scale, which ranges from about 4 under extremely acid conditions to about 10 under extremely alkaline conditions. Soils with a pH range of 6.6 to 7.3 are referred to as neutral; those from 5.6 to 6.5 as slightly to medium acid; and those from 4.5 to 5.5 very strongly to strongly acid.

lapse since formation, may show no horizon development. The important thing is that sufficient depth of plant-supporting material exists for the kind of plants growing.

Of the soil properties discussed, those which are probably most significant to New York State's geography and the long-range productive potential of the land are acidity, texture, drainage, and depth. The nature of parent material and terrain, because they affect a number of the above properties, are basic to the differentiation of the state's soil regions.

Soil Regions

Soil scientists have identified many hundreds of kinds of soil, or soil *types*, in New York State. Groups of soil types that regularly occur in close proximity may be combined into soil *associations*. Normally an association is dominated by soils with similar parent material but with a variety of drainage and site characteristics. There are eighty associations in the state.[2] Similar associations can be further grouped together into six broad categories which closely resemble in concept and level of generalization what could be considered soil regions. Although local variation within any one of these regions is appreciable, each exhibits considerable similarity throughout, particularly in terms of parent material, terrain location, acidity, texture, depth, and over-all usability. Figure 33 shows that some of the regions form broad zones across the state, as regional divisions of this scale usually do; others are highly fragmented or occur only in scattered locations.

REGION I—LIMY SOILS ON GLACIAL TILL
OVER UNDULATING TO ROLLING TERRAIN

Limestone outcrops in the state supplied materials for the glaciers to grind up and spread as they moved southward. (See Fig. 8 for limestone locations.) Massive beds of limestone form the northern and eastern escarpment of the Appalachian Upland. A lower series outcrops around the southern margins of the Adirondack Highlands and westward across the Ontario Lowland. In a few places lime-poor glacial drift has completely covered the limestone bedrock. But of more importance is the fact that the glaciers have spread broken and ground-up limestone, mixed into the till, in significant quantities twenty miles or more south of the outcrops. As a result there was produced a broad belt of deep parent material high in lime and extending from Lake Erie across the state into the Hudson Valley.

Smaller areas with similar conditions occur in parts of Jefferson, St. Lawrence, and Clinton counties in northern New York. On these limy tills have developed some of the state's most productive soils.

Best conditions for agriculture occur where the terrain is not too steep, where the soils are deep, and where drainage is good. When these three conditions are met, high percentages of the land are still cleared of forest and actively farmed. Early transportation and the growth of large nearby city markets have fostered a relatively successful farming economy. There is considerable diversity in agricultural production. Although dairying occupies the bulk of the area, market-gardening, fruit-growing, and poultry-farming are all important locally.

REGION II—LIMY SOILS ON GLACIAL LAKE
SEDIMENTS OVER LEVEL TO UNDULATING TERRAIN

These soils are somewhat like those of the first region described except that the parent material is largely glacial lake sediments rather than glacial till. The result is finer texture, fewer rocks, and generally a somewhat lower lime content.

During the waning stages of glaciation, waters were impounded in portions of the state when ice blocked normal drainage channels down the Mohawk-Hudson valleys or the St. Lawrence (see Fig. 7). Deposits of medium- and fine-textured sediments were laid down in many places. Seldom do they occupy a single large area; usually they form strips or small areas along what was the margin of the glacial lakes into which they were deposited. By nature of their formation they exhibit a level to slightly undulating surface. Largest areas of occurrence are on the Erie–Ontario Lowland and in the Hudson and St. Lawrence valleys. Some valley bottom filling occurred in the Appalachian Upland too. The soils that evolved on this glaciolacustrine parent material are of good texture, have liberal supplies of lime in their subsoil, but vary greatly from place to place in drainage characteristics, depending on whether the surrounding land generally slopes away from them or rises above them.

Use potential, where natural drainage is good, or where artificial drainage is feasible, compares favorably with the soils of Region I. In fact, because of generally level terrain and locations on low-lying sites near the lakes or otherwise suitable places for fruit and vegetables, some of the soils of this region may even be more intensively used than those of Region I.

REGION III—ALLUVIAL SOILS IN
VALLEY BOTTOMS

When the glaciers retreated from New York their melt waters deposited sediments in many of the

---

[2] See Marlin G. Cline, *Soils and Soil Associations of New York*, Cornell Extension Bulletin 930, New York State College of Agriculture, State University of New York, Cornell University, 1957.

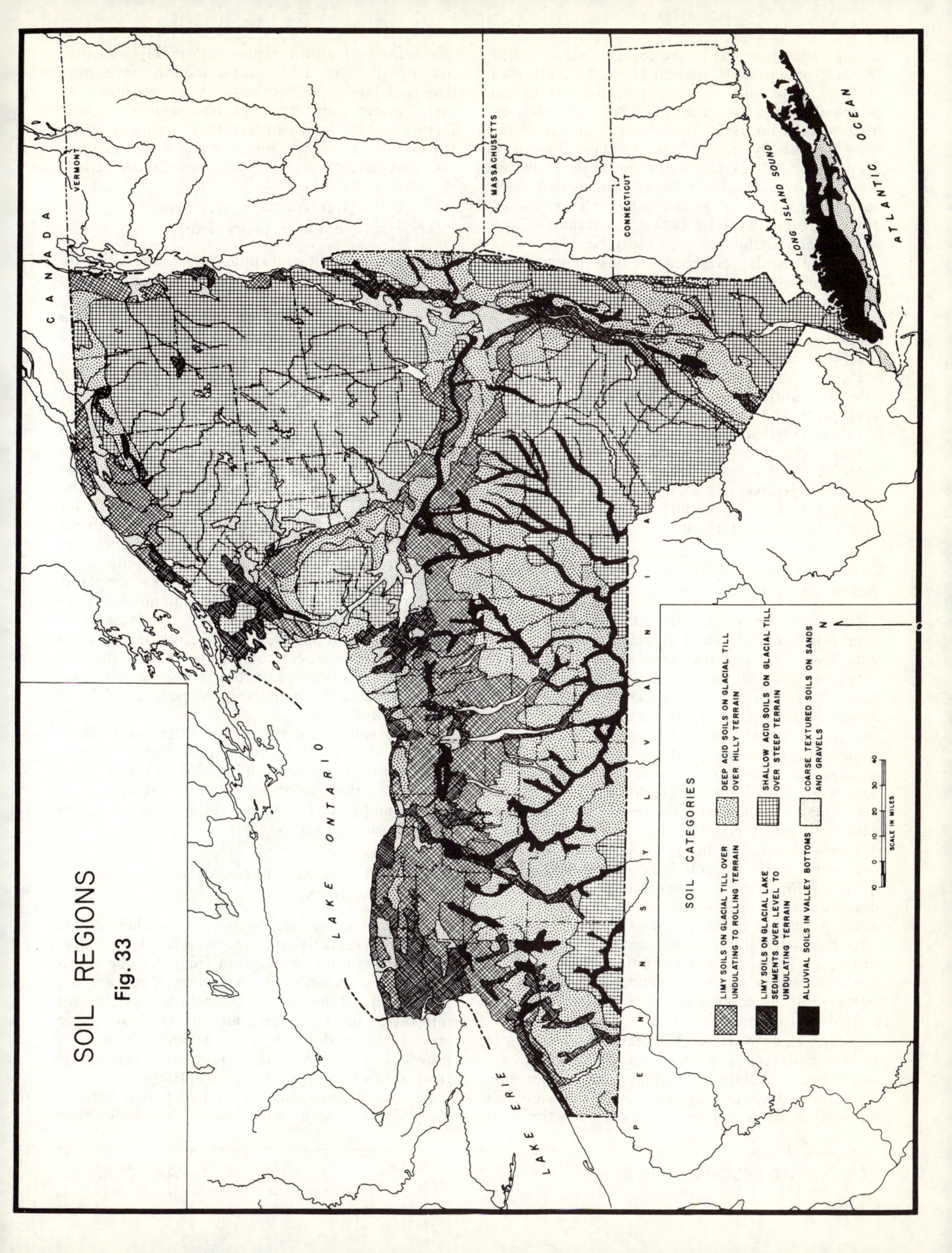

state's valley bottoms in the form of valley trains. In postglacial periods, streams have added alluvium at flood time until a substantial portion of the principal valleys developed narrow, but relatively level, alluvial bottoms. Important examples of this show up clearly in the Appalachian Upland in Figure 33. Similar conditions occur in the Mohawk-Hudson Lowland. Long Island, too, is essentially covered by a skirt of glacial outwash deposits laid down in front of the ice between moraines. Little postglacial alluvium has been added here.

Although soils developed on this parent material commonly are immature and lack well-developed profiles, they frequently have excellent productive potentials. They are of good texture, responsive to fertilization, occupy relatively level sites, and are free of stones. Drainage problems, both poor and excessive, do occur depending on subsoil conditions and water table levels. Thus, in some places soils may have to be artificially drained, in others irrigated.

Hay and grains are raised, and in many places specialty crops do well. These include numerous kinds of vegetables and some fruit. The soils of this region, for example, support the bulk of the potato production of Long Island.

Region IV—Deep Acid Soils on Glacial Till over Hilly Terrain

Glacial till, derived less from limestone and more from nonlime rocks such as shale, sandstone, and slate, covers most of the Appalachian Upland, the area east of the Hudson, and other sections of the state. Particularly in the western two-thirds of the Appalachian Upland in substantial portions of the Mohawk-Hudson Lowland, and north of the limestone outcrops on the Ontario Lowland are large areas of deep acid soils derived from this kind of parent material. In this soil region glaciers left steep places here and there which have thin soils and elsewhere produced pockets of soil that are poorly drained. Characteristically, though, the soils are deep and exhibit moderately good drainage. Fragipan development is widespread and does cause drainage difficulties where terrain suggests that drainage should be no problem. Acidity necessitates heavy fertilization and generally is a serious problem. The worst acid conditions occur in areas farthest from limestone till sources. A substantial portion of the terrain is steep, which does not contribute to agricultural success.

Nearly all of the region has been cultivated at one time or another, but a great deal of land has been abandoned. Acidity requiring excessive fertilizer costs, and steep terrain resulting in erosion and cultivation problems, seem to have been the main causes for agricultural failure. Probably well over half of all farm fields in the region have been damaged moderately to severely by erosion. In many places poor drainage has been a decisive factor too. On land still farmed, dairying is by far the dominant activity, but in some localities potatoes and cabbage are popular and successful crops.

Region V—Shallow Acid Soils on Glacial Till over Steep Terrain

Igneous and metamorphic rocks and resistant sandstones underlie the higher and steeper portions of the state. In fact, these rocks are largely responsible for the greater elevation and ruggedness. This is the case of the Adirondacks, the Catskills, highlands near the Hudson, the Tug Hill, and parts of Allegany State Park near Olean in western New York. In general the till sheet was not too thick over these areas (it was actually absent in some sections near Olean). In places glacial scouring left essentially bare rock surfaces as in higher sections of the Adirondacks and Catskills; in other locales preglacial drainage was interrupted and poor drainage resulted. This is particularly common in the Tug Hill and portions of the Adirondacks. In the case of these two areas and the highlands near the Hudson, glaciation produced stony conditions almost equal to those of much of New England. Cooler climates, heavier snowfall, and more coniferous trees in the natural vegetation all adversely affected soil development, too. The over-all result is soil with extreme acid conditions, shallow profiles, rockiness, and drainage problems, situated in portions of the state with steep slopes and difficult climatic conditions.

Such a soil region clearly has little potential for farming. It has never been attractive to farmers; those who attempted farming did not stay long. Therefore, the region is almost exclusively in forests, and forestry and recreation seem to be the most logical uses today.

Region VI—Coarse-Textured Soils on Sands and Gravels

Sand and gravel, originating as beaches, deltas, and marginal lake-bottom deposits in glacial lakes, and later as dunes downwind from the above, are surprisingly common in the state. They are dispersed around the base of the Adirondacks, extend into the Ontario Lowland, and are especially widespread west and north from Albany. Sands and gravels, too, were piled up in moraines, seen now as rows of low hills in southern Long Island.

Soils developing on this kind of parent material are characteristically very poor. They suffer from

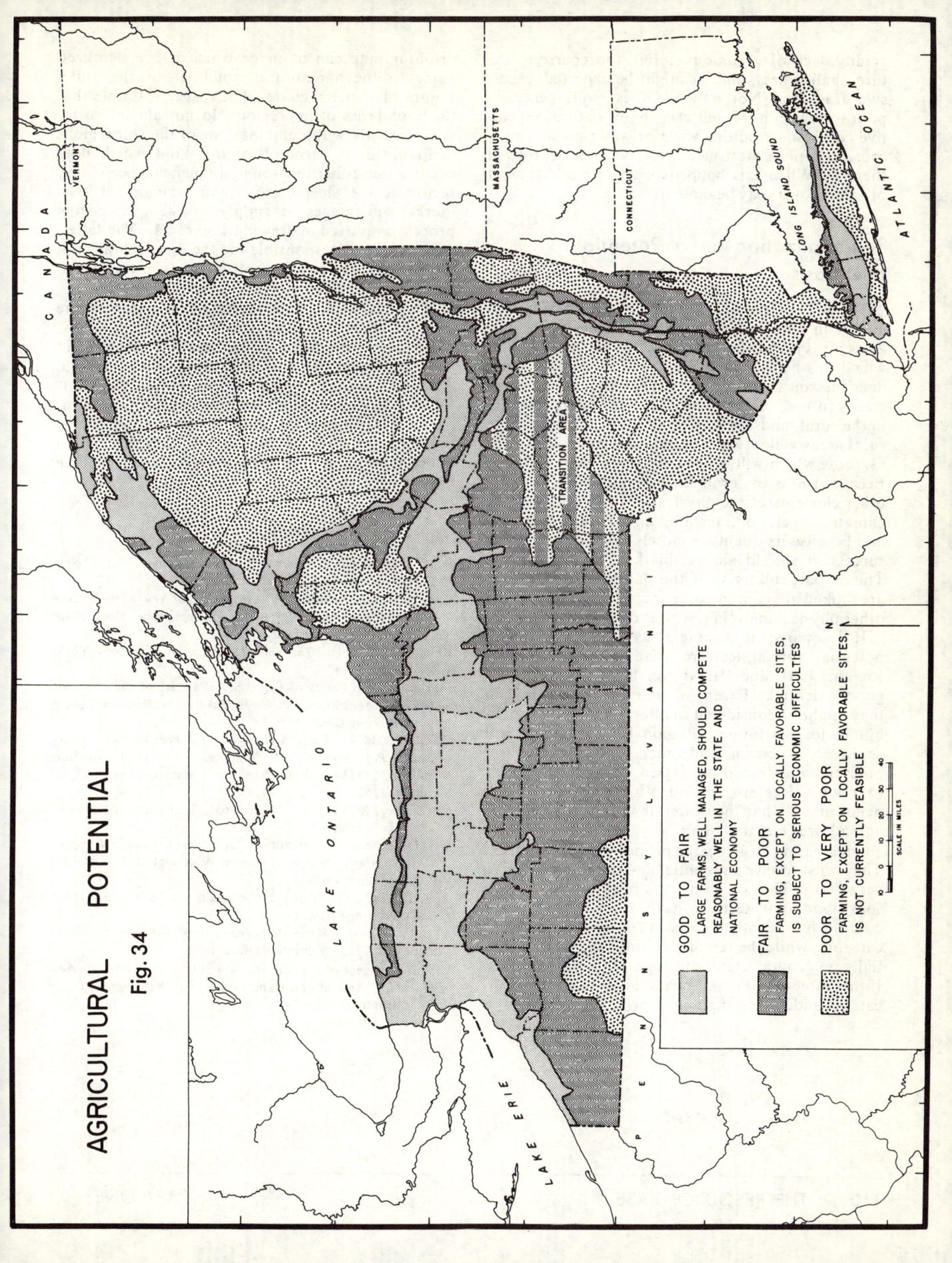

acidity, lack of organic material, too coarse a texture, shallowness, and, as might be expected, excessive drainage. Not only are these soils generally poor for agriculture, but clearing of natural vegetative cover and resulting wind or water erosion may cause complete destruction of what little topsoil there is. When this happens even re-establishment of forest cover may be difficult.

Regionalization on a Potential Use Basis

The soils of six soil regions have been discussed largely in terms of origin and inherent characteristics. It is perhaps potential use of the regions with which this book should be most concerned. What does the soil map (Fig. 33) suggest concerning the state's future? Does it help in forecasting changes in the rural landscape? Can it aid in spotting those rural areas which will experience economic growth vs. those which will exhibit economic deterioration? Because soil is the product, in a sense, of so many other elements of the physical environment such as climate, vegetation, terrain, and parent material, and because its quality is closely related to farming success, it should aid in this kind of forecasting. This is particularly so if the patterns in Figure 33 are looked at again in comparison with those in the other physical maps in previous chapters.

It is apparent that some areas have much greater potential for agriculture than others. Clearly Regions I, II, and III, above, have comparatively good potential. Except in unusual circumstances, they might be considered to offer good to fair possibilities for the future. Region IV, because of soil acidity, hilly terrain, and other handicaps, has less potential and should be classed no better than fair to poor. Regions V and VI clearly have such deficient soil that they offer few opportunities for agriculture, but they may be valuable for forestry and recreation or, if located properly, for urban uses. They must receive a potential designation of poor to very poor. Thus, certain parts of the state, from a soils standpoint, seem to have a fair to good agricultural future, other parts fall in the poor to fair category, while the remainder will experience essentially no economic gain through ordinary soil uses. If this three-way breakdown is accepted and implications added from figures in preceding chapters, a resulting map can be made which is a generalized guide to the agricultural potential of the state. Figure 34 is such a guide. The reader will note that the boundaries of the regions do not always correspond exactly with combinations of the soil regions in Figure 33. Variation of this kind which does occur is due either to strongly influencing secondary factors such as steep terrain, poor drainage, or local market advantages or simply to the generalizing process employed in drawing Figure 34. The latter, for example, is responsible for the disappearance of the narrow, good valley-bottom areas in the Appalachian Upland.

It will be useful and interesting to compare the agricultural potential map with others that follow, particularly the maps of agricultural regions in chapter 10 and of rural landscapes in chapter 15. Also, now at the end of Part One, it would be fruitful for the reader to retrace his steps a bit, to compare the various maps in these early chapters, to use them as building blocks for constructing an understanding of the geography of the resource base.

Selected References

Bennett, Hugh H. *Elements of Soil Conservation.* 2nd ed. New York: McGraw-Hill, 1955.

Cline, Marlin G. *Soils and Soil Associations of New York.* ("Cornell Extension Bulletin," 930.) Ithaca: New York State College of Agriculture, May, 1955.

Kellogg, Charles H. *The Soils That Support Us.* New York: Macmillan, 1949.

Marbut, C. F. "Soils of the U.S." Part III of the *Atlas of American Agriculture*, ed. O. E. Baker. Washington, D.C.: Govt. Printing Office, 1936.

Olson, Gerald W. *Using Soil Surveys for Problems of the Expanding Population in New York State.* ("Cornell Extension Bulletin," 1123.) Ithaca: New York State College of Agriculture, 1964.

Soil Science, XCVI, No. 1 (July, 1963). Entire issue devoted to soil classification.

U.S. Department of Agriculture. *Soil Classification, a Comprehensive System, 7th Approximation.* Washington, D.C.: Govt. Printing Office, 1960.

———. *Soil. 1957 Yearbook of Agriculture.* Washington, D.C.: Govt. Printing Office, 1957.

———. *Soils and Men. 1938 Yearbook of Agriculture.* Washington, D.C.: Govt. Printing Office, 1938.

U.S. Department of Agriculture, Soil Survey Staff. *Soil Survey Manual.* ("Agricultural Handbook," 18.) Washington, D.C.: Govt. Printing Office, 1951.

PART TWO

THREE AND A HALF CENTURIES OF CHANGE

PART ONE, the survey of the resource base, has suggested that some parts of the state are much easier for man to use than other parts. The type, extent, and success of use of course depends on man's interests and capabilities. As these human qualities change through time, it may be presumed that man's interrelationships with the resource base of the state change, too. The next four chapters consider human occupance of the state for the last three and a half centuries. In them changing patterns of locational advantage and spatial interaction become apparent, and dominant economic activities will be seen to pass from the hunting and primitive agricultural stages of the Indian and earliest European settlers to the urban-building manufactural and commercial forms of later years. The concept of geographical persistence becomes evident through numerous examples.

Special attention is given to the spread of population over the state and to the differential growth in population from place to place. The development and function of the transportation networks are dealt with at length because transportation has played such a critical role not only in accounting for differential population growth but also in determining the patterns of agricultural and manufactural activities. Part Two will demonstrate that all the way from the time of Indian occupance to the present the state as a whole has progressed not so much because of a favorable resource base but rather because of an excellent location on the best access route between two marketing areas, the early and successfully settled East Coast and the rich and diversely productive interior of the continent.

These chapters provide historical perspective on matters of geographical interest and in so doing set an interpretative background for the discussion of current economic activities dealt with later in the volume.

CHAPTER 6   ROBERT J. RAYBACK

The Indian

THE AMERICAN INDIAN was the first to use the land that is now New York State. His aims and his technical abilities were different from those of the white men who came later, so the relationships he developed with the land were different. There is at least one similarity though; even in his early period of occupance the Indian gained wealth from the trade originating outside the borders of the state.

In the Beginning

Indian occupation of New York has had a long history, possibly dating back to 7000 B.C. or earlier. Following retreat of Pleistocene glaciers the first human inhabitants of the state, Paleo-Indian Hunters, lived a nomadic life, hunting game animals—perhaps including the now extinct mammoth and mastodon. Evidence of this very old occupance within New York's borders rests upon discovery of fluted projectile points or weapon heads rather than upon carbon dating of organic remains, although the latter technique has been used accurately to date similar artifacts in other parts of the United States.

The Paleo-Indian Hunters were followed by people of the Archaic Stage. Carbon dating methods have determined that this stage extended from about 3500 to 1000 B.C. The Archaic Indians were less nomadic, made use of a much wider variety of plants and animals, and developed new types of projectile points and other tools. They learned how to grind and polish stone and make tools good enough to fell and shape trees. The Archaic Stage came to a close with the introduction of stone cooking vessels and the appearance of the first pottery.

The final period of native Indian culture in New York State is called the Woodland Stage. This stage, lasting from 1000 B.C. to European Conquest, was characterized by complex styles of pottery forms and decoration and by the introduction of agriculture and village life. The Woodland Stage reached its culmination with the Iroquois, who lived in villages, carried on trade, and evolved the concept of political organization and union.

Europeans who discovered the American natives identified them as Indians, that is, inhabitants of India. After recognizing their mistake, Europeans next identified the natives as a separate race of mankind—red men. Again they were mistaken. Actually, the American Indian has a red skin only when he applies a pigment for ceremonial and decorative purposes. Really he is brown-hued and closely resembles modern Mongolians and Koreans not only in color but also in facial and hirsute features. It appears that the American Indian was originally an Asian, stemming from the same parental heritage as the modern Mongolian people.

Some time around 20,000 or 30,000 years ago emigrants from Asia crossed the Bering Strait (or over a land bridge that may have existed at that time) from Siberia to Alaska. This migration probably came in waves over hundreds of years and brought to the New World a number of different groups. All, however, were culturally primitive. None had progressed beyond the hunting and fishing stage. These people pursued the large animals of the late Pleistocene Epoch: hairy mammoths, giant sloths, and the antlered family of caribou, elk, and deer who lived in the cold, wet region that the glacier had released to nature. Verdant foliage provided ample nourishment for herds of these herbivorous beasts. Man, it appears, first arrived in New York as he kept pace behind these moving animals who sought foodstuffs in the revegetated area south of the receding glacier.

In the early nineteenth century, farmers who broke the wilderness to cultivation disturbed the former Indian village sites. To be sure, these pioneers assiduously collected the easily available artifacts in cigar boxes, but so amateurish were their efforts that their collections have not proved very useful in reconstructing the past. Massive movements of earth for the twentieth century's roadway and building projects, moreover, have eliminated many other sites. No antiquities law protected the artifacts of prehistoric times. Enough archeological evidence exists, particularly of the Archaic and Woodland stages, however, to demonstrate that the Indian culture of New York matured, improved, and grew more sophisticated through the centuries. Bows and arrows were added to the artillery of stone tomahawks, stone knives, and javelins. Ce-

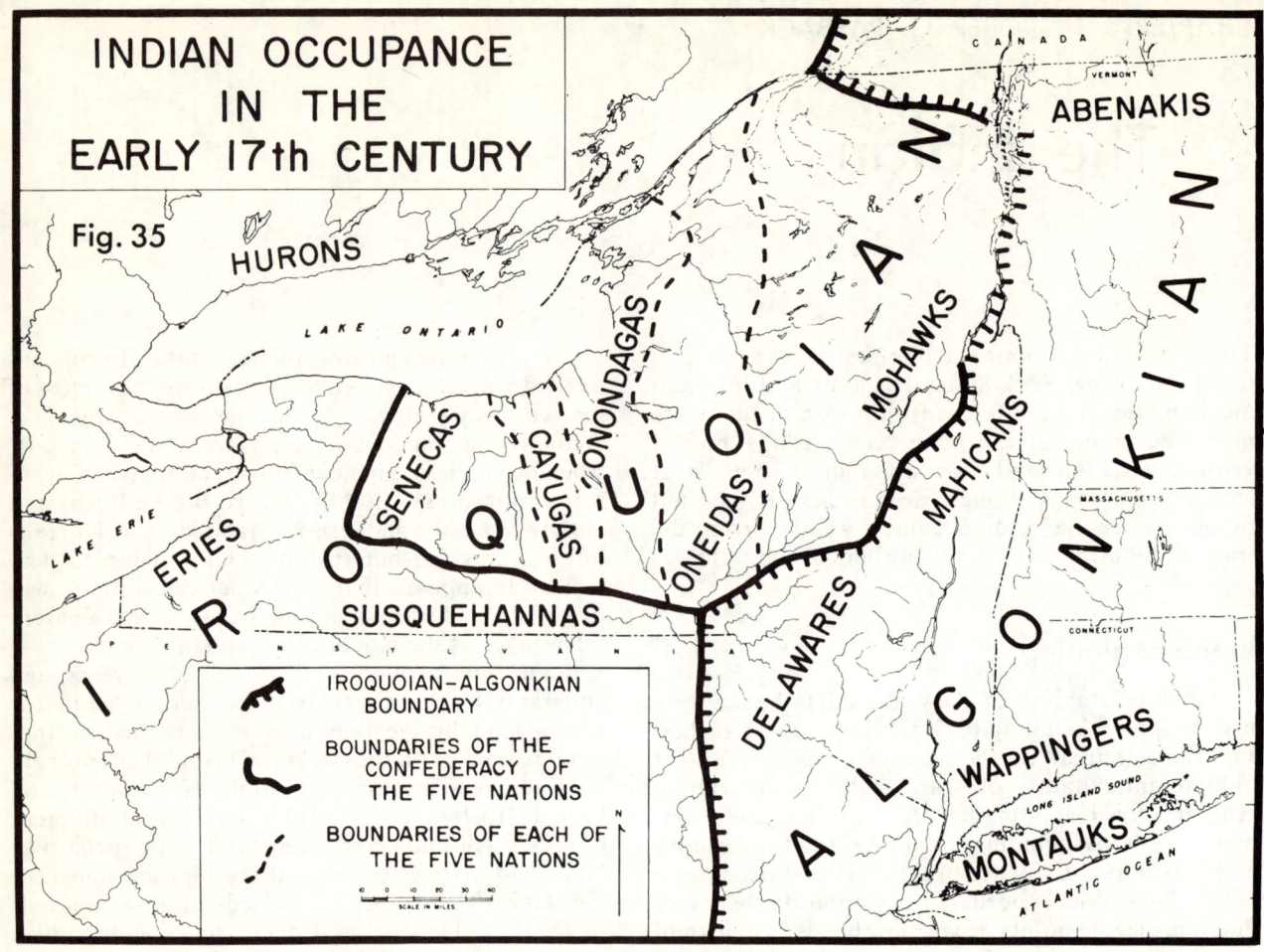

ramic pots made cooking and carrying water easier, hunting and fishing as the primary food providers gave way to the more efficient and reliable techniques of raising crops, and fortified villages of hand-constructed dwellings displaced the cave and natural ledge as a place to live. Unfortunately, the Indians had not learned to write and so left no written record to simplify the task of reconstructing their past.

Two Groups: The Iroquoian and Algonkian

More, of course, is known about the geography of the Indian culture after the Europeans arrived on the scene. At that point in time, New York housed two different groups of Indians: one spoke a number of dialects classified as Algonkian; the other group spoke Iroquoian (Fig. 35). Many of their customs differed, but their clothing and crops were similar. Each group occupied distinct areas. The Algonquins made the Hudson Valley and the Mohawk Valley westward to Schoharie Creek their homeland in New York. Elsewhere, their linguistic relatives dominated most of Canada, New England, New Jersey, Delaware, and Virginia. The New York Iroquois possessed, but did not occupy, all that the Algonquins could not retain. The Iroquois also had close linguistic relatives outside of New York, in Pennsylvania, Ohio, Canada, and far to the south in North Carolina and eastern Tennessee.

The Algonquins, it is believed, had lived along the North Atlantic Coast for centuries, the last of them having come from the West before the year 1000. The Iroquoian groups probably lived at this time somewhere in the Midwest and migrated to the New York area in the thirteenth or fourteenth century, at that time overcoming resistance of prior occupants. The five Iroquoian tribes of central and western New York, allegedly with the encouragement of Hiawatha, evolved the Confederacy of the Five Nations about 1570. The Onondaga became the major element in the federation and their chief was the main leader. The Mohawk Nation was known as the Keeper of the Eastern Door, and the

Seneca Nation as the Keeper of the Western Door. The Oneida and Cayuga tribes were also members. The confederacy probably reached its peak of power and effectiveness after the Dutch settled the lower Hudson Valley. No other Indian political organization ever developed was as efficient in both peaceful trade ventures and war endeavors. One after another, the Indians in the vicinity were overrun by the well-organized and powerful forces of the Five Nations. The Mohicans were conquered in 1626, the Hurons dispersed in the period 1647-49, the Eries in 1654, and the Susquehannas in 1673-75. To guarantee better trade advantages, the Five Nations successfully extended their control to the vicinity of Lake Michigan and even beyond to the Mississippi. In 1714 the Five Nations added a sixth member, the Tuscaroras, who were forced to flee from their home grounds in North Carolina.

How Many Indians?

No one will ever know the exact Indian population of New York for any era previous to the regular census of the United States government. Intelligent guesses about this figure, however, range from 15,000 to 75,000 as a maximum for any given date. When Samuel de Champlain first reported the strength of the Iroquois, who were his enemies, he claimed that they could put 8,000 warriors in the field. Of course this was an exaggeration based upon the imagination of the victims of the Iroquois.

Subsequent responses of the New York Iroquois to calls of the colonial government to join in the fight against the French never exceeded 1,000. Since no manpower levy ever included all the able-bodied, it would seem generous to believe that the Iroquois had more than 4,000 warriors, or probably an average of one to a family. The average historic Iroquois household seems to have contained about five members. This would make the sixteenth-century Iroquois population about 20,000. Since a balance of physical strength between the Iroquois and their Algonquin neighbors existed when the Europeans arrived on the scene in 1609, this would indicate that the New York Algonquins also may have numbered about 20,000 for a total New York Indian population of 40,000.

The figure of 20,000 Iroquois is partially substantiated by information from the Clinton-Sullivan expedition during the Revolutionary War. This was a punitive campaign, which razed forty Iroquois villages—nearly every one. A census of the long-houses destroyed—drawn from soldiers' diaries—sets forth the distinct possibility that they could have housed 20,000 people. Reason exists, furthermore, to believe that the western nations of the Iroquois Confederacy were at least as populous in 1779 as in 1609. The Mohawks, on the other hand, had probably suffered considerable depletion by the time of the Revolution.

Where Did Indians Live?

The map of Indian settlements (Fig. 36) shows known places which the Indians of all languages and cultural stages occupied in the thousands of years before New York State assigned them reservations. It is a reference map of all known villages for all times and locates over 1,000 former sites. At any given date, however, no authority believes that there were ever over 70 or 80 in use on the New York scene. Except for small peripheral settlements and hunting stations, populations of the Five Nations tended to congregate in about a dozen localities. The largest of these concentrations, which probably included several sites, had perhaps 1,000 inhabitants. Thus an image one gets is of a sparsely populated area in which man had not yet begun to use effectively most of the land at his disposal. This is as would be expected, as the Indian's way of life simply could not support dense populations. Interestingly enough, the Indian and European had things in common when it came to choosing parts of the state in which to live. Both rejected the Adirondacks, the Catskills, the Shawangunk, and the Tug Hill. Both settled populations only thinly in the North Country and the Appalachian Upland. Both sought fresh-water streams for their settlements and, as a matter of fact, found difficulty in using clay soils or steep slopes for agriculture. Here, however, the similarity ends.

Though Indians of all cultural stages rejected the harsh mountains and plateau country, they commonly chose specific sites for settlement which Europeans have deemed of only modest value. Rarely did the Indians locate a village at the mouth of a major stream or at the confluence of two significant rivers. Rather, they timidly rejected focal positions of this kind. For the most part, they placed their villages on lesser streams. The Mohawk villages, clustered on the Mohawk River near modern Fonda, and settlements at the site of Binghamton were exceptions. Locations like Buffalo, Niagara Falls, Rochester, Syracuse, Oswego, Watertown, Utica, Schenectady, and Albany, while not rejected entirely by the Indians, were of minor importance. The Indians even found the Long Island Sound shores of Westchester County and Long Island more desirable than Manhattan or the east and west banks of the Hudson River. All this simply points out that sites of especial value to white man's modern trade held less importance for the Indian. Protection from potential enemies was more important.

This does not mean that the Indians were not

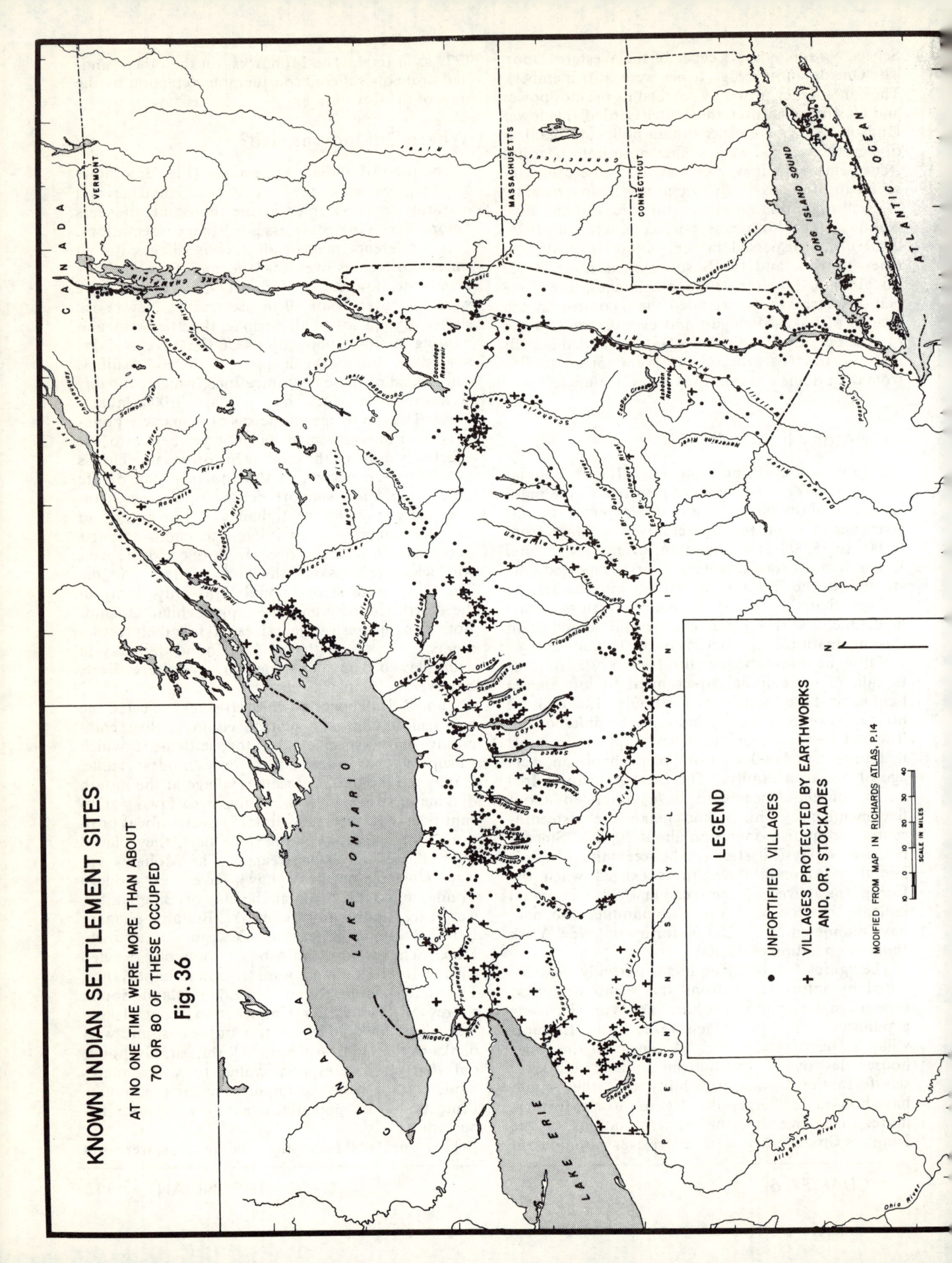

travelers. The Iroquois had made a fortuitous choice in selecting New York for their homeland. They settled in a land in which navigable water could be used for travel in nearly every direction. They had water access to the Great Lakes, the St. Lawrence Gulf, New York's harbor, the Delaware and Chesapeake bays, and the Mississippi Valley. The Iroquois used the mobility this water network provided to full advantage in war and peace. They located their villages within easy walking distance of it. Usually they cached their canoes close to navigable water, while selecting sites off the "beaten track" for permanent living areas.

What Were Indian Villages Like?

Archeological remains indicate that the Indians had three types of villages. One was an open, unfortified hamlet, and the other two were strongly fortified with either earthworks or stockades, or both.

Possibly there was only one form of barrier, the stockade, and occasionally this might have been supported with earthworks. There are indications that, after constructing a palisade around their village, the Indians dug a moat along the outside of the wall and threw up the excavated soil into a steep embankment against the stockade. The trench added height to the wall from the outside, and the steep slope of the embankment discouraged ascent. In the event of an assault, the attacking force was thrown off-stride as it hit the moat. These were as much as six feet deep, and some appeared to be capable of receiving water.

All earthworks that protected villages, however, may not have supported stockades. Some could have been independent breastworks. In a majority of remains of this type, no evidence of stockading exists. Protection without a wooden stockade could have been obtained by erecting an earthen wall six or seven feet high with the diggings from a five- or six-foot trench. The piling would have created a ten- to twelve-foot wall of earth, possibly surmounted by brambles and thicket. Such a barrier probably would have proved adequate as a protection against the storming techniques and fighting philosophy of the attacking Indians, who relied more on surprise and ambush than on cunning and perseverance.

The open, unfortified village was a product of peace and security. Not fearing an attack, the natives placed their homes where convenience, including companionship, beckoned. In troubled times and areas, however, dispersion invited danger. Settlement then took the form of closely built, irregularly arranged groups of from 10 to 140 dwellings, all nestled behind stockades. As far as possible, houses and fortifications were located in the center of the cultivated lands, preferably on a hillock or a rising slope, which gave the sentries advantages.

In the village's center, the residents maintained a plaza large enough for public assemblies. Elsewhere, they packed their cabins close together in spite of the persistent hazard of fire. The streets resulted from passageways that the fickle builders left between the dwellings, though they were careful to leave a broad, empty area between the cabins and the stockade wall. The homes of New York's Indians, therefore, can be said to be an adjustment to economic and political activities of their particular group. Those who advanced economically built more elaborate establishments; those who lived on the verge of war adjusted to a garrison state.

They Were Agriculturists

When Europeans arrived in New York the Algonquins and Iroquois made their living in approximately the same fashion. Both had adjusted remarkably to the settled ways of agriculture. Their nomadic life was in the past, and forest resources—flora and fauna—supplemented, albeit in an important fashion, the crops of the field.

Maize, or corn, production underwrote the economy, and, except for slashing and burning the forest to clear the field, the women were the farmers. In the spring they went out under a female overseer to plant, in turn, the fields of each family, and in the fall they bore the harvest home for a husking bee. They grew corn with two-foot ears and kernels in five colors. The stalks stood as high as a man. Between the rows they planted many varieties of beans and squash. The Iroquois called these three plants (corn, beans, squash) "Our Supporters" and believed they were guarded by three spirit-sisters. As a result, the "sisters" were among the most revered beings in the polytheistic world of Iroquoian mythology. The religion of the Indians was animistic and based upon the phenomena of nature.

The advantages of farming had encouraged the Indians to create tools—a measurement of civilization's growth. After fire and axe had done their part in clearing the land, big wooden hooks—rakes—were used to drag the brushwood to heaps where it was burned. The women then hilled up earth for corn-planting with the aid of tools—wooden spades and hoes. Sometimes both were made of the shoulder blade of a deer fastened to a wooden handle. Then, with a digging stick, a planter would make eight or nine holes in the hill, drop a seed in each, and tap in some loose dirt.

Hunting and Fishing

To supplement the women's farming activity, the men went fishing and hunting. The most useful tools they had for fishing were spears, or harpoons, and nets. Although a keen archer could shoot a large fish with an arrow, most fishermen favored the spear as their weapon. By this means a skilled fisherman might snare three hundred eels in a single night. Fish hooks, though known, were not used to any great extent before Europeans introduced the Indians to the curved, barbed metal hook.

When fishing in groups, the Indians usually abandoned both arrow and spear and used nets. To augment their catch, they constructed elaborate weirs behind which they drove the fish in great numbers. The women of the tribe preserved the fish and eels when caught in quantity by smoking them slowly over hickory fires.

Some of the hunting tools and weapons the Indians used were inherited from the distant past. Their most necessary implement was the knife, and they made the simplest of these in a few minutes by picking up a thin stone or shell and giving it an edge. On the other hand, the highly prized axe, or celt, with its carefully chipped and polished cutting edge and fitted handle, sometimes took a lifetime to manufacture. Sons honored them as legacies and warriors took them as spoils of war. Of the articles that improved the Indians' ability to gather meat from the forest, the most romantic were bows and arrows. These the hunters made in different sizes and with varying power. They favored red cedar for the bow, which, after careful selection, was hardened with fire, shaped by scraping, and strung with a hempen cord or deer sinew.

The arrows were made from straight-grained wood and the arrowheads could be prepared from a number of materials. Stone, either flint or chert, was the most popular material, even though it required extreme patience to chip a blank into the necessary flat, triangular shape. Because suitable stone was not available everywhere, bone or horn was fitted to the shafts either as hollow points or solid pieces inserted into the wood. Occasionally, a tribe used wooden arrowheads.

Yet, for all the craftsmanship that went into the manufacturing of bows and arrows, it was the traps, snares, and nets for wild fowl that accounted for the success of New York's forest hunter. Of these devices he had a wide assortment, upon which he depended heavily for his annual meat supply.

In sum, the Indian economy at the opening of historic times in New York was primitive but effective. In most ways the natives, like the other denizens of the countryside, struggled against the relentless cyclical demands of nature. They had, however, made more than an ecological adjustment to their environment. They had begun to shape nature's bountifulness to their own ends.

Areal Specialization and Trade

Also, they had made a sophisticated advance: a certain amount of areal specialization of production had begun to appear. Long Island Indians, for example, adjusted easily to clam-digging. This not only provided a food supply but left as residue the basic ingredient of wampum, seashells, especially the purple-colored lip of the quahogs. The Oneida Indians possessed flint quarries and were soon engaged in turning out "blanks" for arrowheads, far in excess of their needs. The Hurons and northern and eastern Algonquins stripped great sheets of bark from white birch and from it manufactured canoes, light in weight and balanced. These canoes were much superior in many ways to the heavy elm-bark or dug-out contrivances used by the Indians of the non-birch-bearing country. The Senecas, meanwhile, exploited the flood plains of the Genesee Valley and produced more corn than they needed, trading it to their neighbors. The Neutrals and the Petuns in Canada to the west grew tobacco and hemp beyond their requirements and set up trade with the Iroquois.

The over-all result of this areal specialization was a surprisingly wide exchange of surpluses. When this exchange began, no one knows. Yet by the opening of the historic era the Indians had learned to transport heavy burdens long distances. Sometimes their journeys exceeded 1,000 miles. Water travel they solved easily with canoes. Land movement was less readily eased, since they did not have the wheel. It continued to require brawn. Sometimes they lightened the load with a "bretelle" or travois—a wheelbarrow without a wheel, which could be pulled on its point without placing the entire weight of its burden on the carrier. Another device was the burden strap, which made it possible for an individual to carry the weight of packaged burdens on his shoulders. In winter the problems of land transportation were more easily resolved. The traveler used snowshoes and could move through the forest without hindrance. He also dragged a heavily laden sled, a toboggan-like affair, about a foot wide and six or seven feet long.

Europeans found that the natives had well-traveled trade routes, and items of trade which each group desired or exchanged were fixed and allocated for specific tribes. In other words, before the coming of white men, the Indians had already sampled the fruits and joys of stepping beyond the bounds of self-sufficiency. Their trade, to be sure, was modest by modern standards, but the items of trade were highly prized. As a result, the intertribal trails which had started as animal runs or warpaths deepened with the steady tread of moccasins and became fixed lines of communication. As a good example of geographical persistence, Europeans later used these trails, improved them, and, in the twentieth century, paved them. The original road west across the state, for example, started at Utica, the jumping-off place at the height of navigation on the Mohawk River. It was the Genesee Road and replaced an Indian trail. It became New York's Route 5. This was only symbolic of what had taken place in the Hudson Valley and could take place throughout the state. Consequently, in modern New York, most major roads follow Indian trails. Exceptions, of course, are the modern superhighways and limited-access routes of post-World War II days.

That white men used what red men had placed on New York's landscape is an important concept for students of society to grasp. More important, however, is to recognize that the Indians had reached the stage of development where the interchange of goods was a means of improving their standard of living.

The Indians knew that their comforts depended on outside resources and held trading in great esteem. Every tribe had its trade routes and its special interest areas, each willing to protect its advantages with its whole being, including war.

The Europeans and Change

When the Europeans, bent on trading, appeared in the area with their fabulous trade items, the impact on the existing commercialism of native Indians was tremendous. Quickly tribes vied with each other to become the middlemen in the transfer of goods from Europeans to the rest of the continent. Soon (1608) at Montreal, the Hurons established their right to peddle all French goods westward. At Albany a bit later (1626) the Iroquois, through the Mohawks, won a monopoly to merchandise all Dutch goods westward. The resulting competition between Hurons and Iroquois for the Great Lakes area customers who wanted European artifacts shaped the destiny of a continent and gave New York its outline for emergence.

The character of the Indian use of New York and the nature of their culture also changed quickly after the French, Dutch, and English arrived. By the mid-seventeenth century, the Algonquin tribes of the Hudson Valley and Long Island had faded from the scene. Either Kieft's War (1643–45) had eliminated them or they moved westward and lost their tribal identity. Those that stayed in the area hovered on the edge of the European community and fell victim to its cultured attractions.

In contrast to the Algonquins, the Iroquois entered into the "golden era" of their history. As middlemen in a trade between Dutch and English, on one hand, and the Indians of the interior, on the other, the Iroquois prospered, and their mode of living changed. The commercial exchange in which they became involved was called "the fur trade." But the Indians were not interested in pelts. They sought white man's manufactured products and offered furs for them. Quickly the trade produced enough profits so that the Iroquois women stopped using clay pots and bought iron skillets and kettles. Fur robes were discarded for woolen blankets. Bone needles disappeared as steel ones came into use. Steel drills produced wampum at an alarming rate, and inflation set in. Wrought-iron tomahawks and knives replaced the old stone weapons. The bow and arrow fell victim to the blunderbuss and musket. Soon the Indians lost their knowledge of making arrows and only a few old-timers practiced the craft of flaking flints.

In other words, the self-sufficiency of the Indian economy completely disappeared, and the Iroquois lived in the vortex of European commercialism. They did not resist this change in their way of life. Rather, they avidly welcomed it. So eagerly did they embrace their prosperity that in order to preserve it they fought a series of wars with their Indian neighbors. Between 1626 and 1653 they attacked and dispersed the Mohicans, Hurons, Petuns, Neutrals, and Eries. They disciplined the Ottawas in 1656 and eliminated the Susquehannas in 1675. All of these neighbors had challenged the Iroquois monopoly of trading Dutch and English goods westward or had siphoned off the fur pelts of the interior to French markets, leaving the Iroquois without adequate exchange commodities.

The "golden age" of the Iroquois lasted into the 1680's. At that point the French entered into serious competition for the furs of the interior. The Iroquois were not equipped to overcome this European rival, and, no longer able to return to their primitive self-sufficiency, they began to decline in strength and prosperity. The descent was painful and lasted a hundred years. The end came when they chose the wrong side in the Revolutionary War.

Prophecy for New York

In retrospect the Iroquois experience seems to have prophesied the future of New York. The Iroquois rose to pre-eminence, not because of their fur resources or because of any other natural gifts. Rather, their success was founded on their ideal location between two marketing areas: the European-occupied East Coast and the interior of the continent. They squatted on the only natural access to the interior between the St. Lawrence River and the southern end of the Appalachian Highlands, and they took advantage of the commercial opportunity this situation created. Later, New Yorkers have in a sense relived the Iroquois success, making the interior of the continent their economic reserve.

Selected References

Beauchamp, W. M. *Aboriginal Occupation of New York*. ("New York State Museum Bulletin," 32.) Albany: New York State Museum, 1900.

———. *Aboriginal Place Names of New York*. ("New York State Museum Bulletin," 108.) Albany: New York State Museum, 1907.

———. *A History of the New York Iroquois*. ("New York State Museum Bulletin," 78.) Albany: New York State Museum, 1905.

———. "The Origin and Early Life of New York Iroquois," *Oneida Historical Society, Transactions for 1887–89*. Oneida: Oneida Historical Society, 1889. Pp. 119–42.

Brennan, Louis A. "A Short Evaluation of the Current State of Knowledge of New York Prehistory Stated in Terms of the Problems Raised by It," *The Bulletin, New York State Archaeological Association*, No. 26 (1963), pp. 5–15.

Eiseley, Loren C. "The Paleo-Indians: Their Survival and Diffusion," in *New Interpretations of Aboriginal American Culture History, 75th Anniversary Volume of the Anthropological Society of Washington, D.C.*, ed. Betty J. Meggers and Clifford Evans. Washington, D.C.: Anthropological Society of Washington, 1955. Pp. 1–11.

Kroeber, A. L. *Cultural and Natural Areas of Native North America*. Berkeley: Univ. of California Press, 1947.

McNeish, R. S. "Iroquois Pottery Types, a Technique for the Study of Iroquois Prehistory," *Bulletin of the Natural History of Canada*, No. 124 (Ottawa, 1957).

Parker, A. C. "Outline of the Algonkian Occupancy of New York," *New York State Archaeological Association Researches and Transactions*, IV, No. 2 (1923), 48–80.

Rayback, Robert J. (ed.) *Richards Atlas of New York State*. 1st ed. Phoenix, N.Y.: Frank E. Richards, 1957–59.

Ritchie, William A. *The Archaeology of New York State*. Garden City, N.Y.: Natural History Press, 1965.

———. *The Chance Horizon, an Early Stage of Mohawk Iroquois Cultural Development*. ("New York State Museum Circular," 29.) Albany: New York State Museum, 1952.

———. *Indian History of New York State*. 3 vols. Albany: New York State Museum, 1950.

———. *An Introduction to Hudson Valley Prehistory*. ("New York State Museum and Science Service Bulletin," No. 367.) Albany: New York State Museum, 1958.

———. *The Pre-Iroquoian Occupations of New York State*. ("Rochester Museum of Arts and Sciences Memoir," No. 1.) Rochester: The Museum, 1944.

———. "Radiocarbon Dates on Samples from New York State," assembled by Frederick Johnson, *Memoirs of the Society for American Archaeology*, No. 8 (1951), pp. 30–31.

Roecker, Robert. "Waneta and Lamoka Lakes," *New York State Conservationist* (December–January, 1953–54), p. 25.

Speck, F. G. "The Iroquois," *Bulletin of the Cranbrook Institute of Science*, XXIII (1945), 1–94.

White, Marian E. *Iroquois Culture History in the Niagara Frontier Area of New York State*. ("Anthropological Papers, Museum of Anthropology, University of Michigan," No. 16.) Ann Arbor: University of Michigan, 1961.

Yawger, R. N. The Indian and the Pioneer. 2 vols. Syracuse, 1893.

CHAPTER 7   D. W. MEINIG

The Colonial Period, 1609–1775

This chapter and the two following provide historical perspective on certain features of geographical interest over the whole span of the European era in New York State. Special emphasis has been given to: (1) the initial spread of population over the state and subsequent patterns of growth and decline; (2) the distributions of various groups of people whose cultures were significantly different during some period of time; (3) the development of, and functional patterns within, various transportation networks; (4) certain general patterns of agricultural and industrial activities. It should be noted that further coverage of these and other topics, often in the perspective of a considerable span of time, will be found in later chapters.

Although these chapters attempt to view certain developments over a period of three and a half centuries, they remain no less geographical for being so apparently historical. For the primary focus is always upon phenomena of basic geographical interest viewed in geographical perspective—upon areal integrations and differentiations, upon spatial arrangements and relationships. The lengthened historical perspective merely allows an examination at a different scale of the continually changing results of those same processes which are responsible for those patterns of man and the earth studied by geographers of the present.

These chapters provide an informative overview, but if they do no more they will have accomplished but half their purpose, for it is an equal hope that they will prompt a deeper interest in the four topics emphasized and a lively curiosity about a host of questions left unanswered.

The colonial period, a span of more than 150 years in New York (1609–1775), is of fundamental geographical significance. It is important, however, not because of its duration, but simply because a knowledge of beginnings is essential to the study of regional geography. More than three centuries ago the first European colonists established certain patterns and implanted certain characteristics. Some of these have obviously continued to the present, others have persisted somewhat obscurely, while still others have been altered or effaced. In any case, knowledge of such colonial features is basic to analyses of the present, for an understanding of the origin of those which persist is a necessary element in explanation of the present. Also, those features which are known to have existed but are no longer evident provide a basis for the measure of geographical change, illuminating the processes that underlie areal patterns and characteristics both past and present.

Implicit in such statements is the suggestion that even seemingly simple and obvious geographic patterns are likely to be complex in origin and development. Presumably it is unnecessary to re-emphasize that "nature" does not determine the patterns that man superimposes. Indeed, colonial America was an excellent example of how different groups of people established quite different patterns in often basically similar physical environments. Even within New York, for example, a Manhattan focus and a Hudson axis were by no means predetermined, however simple and likely they might seem. The emphasis in this chapter is upon features that bear upon such matters: the selection of sites, the establishment of areal patterns, the testing of resources, and the varied groups of people involved. In each case, while the scale of developments is small, the implications are large.

New Netherland

The year 1609, when Henry Hudson sailed as far as he could up the river that would later bear his name, may be taken as the beginning of the Dutch period. Other explorers and traders soon followed, but it was another fifteen years before any substantial colonization was undertaken. Between 1624 and 1626 forts were established both upriver and on Manhattan, and a sizable group of colonists was sent over from Europe. It should be noted that New Netherland would encompass the lower Delaware Valley as well as the Hudson (encompass, in their terminology, the "South" River as well as the "North"), and thus there were Dutch settlements and activities beyond the limits of New York.

Fig. 37. New Netherland, 1656

BEGINNINGS

On November 7, 1626, the Dutch government received in the following words its first news of a colonization undertaken by the West India Company on the coast of North America:

Yesterday, arrived here the Ship, the Arms of Amsterdam, which sailed from New Netherland, out of the River Mauritius, on the 23rd September. They report that our people are in good heart and live in peace there; the women also have borne some children there. They have purchased the Island of Manhattes from the Indians for the value of 60 guilders; 'tis 11,000 morgens in size. They had all their grain sowed by the middle of May, and reaped by the middle of August. They send samples of summer grain; such as wheat, rye, barley, oats, buckwheat, canary seed, beans and flax.

The cargo of the aforesaid ships is:—7246 Beaver skins, 178½ Otter skins, 675 Otter skins, 48 Minck skins, 36 Wildcat skins, 33 Mincks, 34 Rat skins, considerable Oak timber and Hickory. [O'Callaghan, *Documents*, I, 37][1]

It is unlikely that this rather terse account was as unexciting and mundane as it may seem to a modern reader. Compared, for example, with reports emanating from a nearby English venture it was altogether auspicious, for Plymouth, though five years older, was still struggling for subsistence and had not as yet reaped enough of some of these crops to send even a sample home. And, although Virginia was at the moment in the frenzy of a tobacco boom, the long agony of her starvation period was only just past. Certainly all previous experience had suggested that colonial ventures in this part of America were likely to be matters of high risk and low return. To be told, therefore, that a convenient base had been peacefully obtained and that both man and land were yielding well and to receive a valuable cargo ought to have given more than routine satisfaction.

These three topics of first report—locale, production, and export—not only were central to the practical concerns of the moment but also proved indicative of important characteristics of subsequent development. The choice of Manhattan as the main base was by no means inevitable, although it was certainly appropriately within the context of activities already established at that time. Henry Hudson had opened a major axis for development seventeen years earlier when he sailed the *Half Moon* up the "North River" (for some years also called the "Mauritius," after Prince Maurice), and soon after explorations of the East River and Long Island Sound had provided another axis. Figure 37 portrays the level of knowledge about this area in 1656. Somewhere near the apex of the North and East rivers and within the magnificent estuarine harbor was obviously the most advantageous location. By 1626 these were established trafficways yielding a brisk Indian trade and had been reconnoitered for sites and resources. Far up the North River, near where Hudson had found "it to bee at an end for shipping to goe in" [Jameson, 39], Fort Orange (now Albany) was the latest in a sequence of posts which for a dozen years had been drawing from rich fur country well to the west. Around the estuary a few settlers and traders had made at least tentative beginnings on Staten, Nut (Governors), and Long Islands (see Fig. 38). The selection of Manhattan was made, therefore, within an area well known and in response to a pattern of activities well under way.

A successful harvest from the first trials was news of fundamental importance. To be sure, Hudson had stated that the land was "the finest for cultivation that I ever in my life set foot upon" [Jameson, 49], but worldly-wise Hollanders likely understood that explorers and discoverers were prone to speak in superlatives. The descriptions of Indian fields, "great quantity" of maize and beans, and offerings of pumpkins, grapes, and tobacco were more certain indications of fertility but not necessarily for European crops. Virginia, too, had been

Fig. 38

---

[1] Citations in this chapter refer to Selected References on pp. 195–96.

CHAPTER 7 THE COLONIAL PERIOD, 1609–1775

praised by her explorers, and her Indians were obviously successful agriculturists, but after twenty years of experiment the government there was still offering a prize for a successful crop of wheat. Probably it was not fully appreciated at the time just how fortunate this Dutch venture was to have taken hold in just that portion of America—sandwiched between harsh New England and subtropical Virginia—where most of the familiar crops and practices of the European homeland were very well suited. The success of that first harvest was not simply fortuitous, however, for the expedition had been designed to give agriculture a fair chance:

> As the country is well adapted for agriculture and the raising of everything that is produced here, the... gentlemen resolved to take advantage of the circumstance, and to provide the place with many necessaries, [and]... to ship thither... one hundred and three head of livestock—stallions, mares, bulls and cows —for breeding and multiplying, besides all the hogs and sheep that they thought expedient.... Country people have also joined the expedition, who take with them all furniture proper for the dairy; all sorts of seed, ploughs and agricultural implements are also present, so that nothing is wanting. [Jameson, 79]

Such a policy was as sound as it was uncommon in colonial schemes of the time, which generally displayed a remarkable nonchalance about basic subsistence matters and rarely recruited experienced farmers or gave proper priority to the shipment of animals and tools. The initial policy of the West India Company was described a few years later as follows:

> The farmer, being conveyed with his family over seas to New Netherland, was granted by the company for the term of six years a Bouwerie, which was partly cleared, and a good part of which was fit for the plough.
> The Company furnished the farmer a house, barn, farming implements and tools, together with four horses, four cows, sheep and pigs in proportion, the usufruct and enjoyment of which the husbandman should have during the six years, and on the expiration thereof, return the number of cattle he received. The entire increase remained with the farmer. The farmer was bound to pay yearly one hundred guilders and eighty pounds of butter rent for the cleared land and bouwerie. [O'Callaghan, *Documents*, I, 371]

This writer of 1650 then noted (and other sources essentially agree) that "the country people who obtained the above-mentioned conditions all prospered." He might well have added, "and no other colony in North America can make that claim" (with the possible exception of Maryland).

The list of furs in the return cargo was important because it indicated that a profitable staple in much demand in Europe was available from the first, without frantic experiments and fruitless investments. Officials did occasionally express regret at the paucity of valuable resources in the area, and furs alone did not make the company wealthy (though even a generally pessimistic report admitted that "the trade carried on there in peltries, is right advantageous"). But the company had other areas of richness—Brazil, Caribbean islands, and especially a tropical sea full of Spanish ships. New Netherland was a minor part in the over-all operations, and while this may have condemned it to relative neglect it saved it from attempts to make it something it could not be, such as a supplier of valuable exotic crops or precious minerals. It thereby escaped, for example, such insufferable pressures as the London Company put upon Virginia colonizers in the years before they "discovered" tobacco. In the long run New Netherland may have been a net drain on the company coffers, but its furs at least provided some income from the very first, and this saved the rest of its development from distortion.

In retrospect, then, that first report can now be seen to have indicated some very basic features of colonial development: a Manhattan focus, an agriculture successful from the start, and a source of the important commercial item, fur.

Learning and Settling

As more than one contemporary account pointed out, it was not only because the Dutch had laid first claim to it but "on account of the similarity of climate, situation and fertility, that this place is rightly called New Netherland." [Jameson, 293] The degree of similarity in the first item was something of a surprise, and, the company implied, a disappointment:

> Although it ought to be, in point of climate, as warm and as well adapted for the cultivation of fruits at least, as the furthest frontiers of France towards Spain; yet it has been found much colder, and as much subject to frost and other inconveniences as these; nay, as more northern countries. [O'Callaghan, *Documents*, I, 40]

It was the singular fact that "it is a good deal colder there than it ought to be according to the latitude" which made "the climate and seasons of the years ...nearly agree with ours [in Holland]." [Jameson, 50] This "latitudinal" reasoning was, of course, a commonplace of the time and was the root of many inappropriate expectations from colonizations on the American coast. After all, Manhattan was "Mediterranean" in latitude, south of southernmost France, and it was rather bewildering to find that "it freezes and snows severely in winter." It was through such trans-Atlantic experiences that the basic differences between east coast and west coast locations within the world climatic pattern were first revealed. It would be a long time before the reasons for such important variations would be well understood, although the writer who noted that the northwest wind of winter in New Netherland "corresponds with our N.E. because it blows across the country from the cold point" [Jameson, 269] was making a start.

The analogy between homeland and colony could not, of course, be pushed too far, and experience soon modified uncritical enthusiasm. Within a few years prospective emigrants were being cautioned that "the lands...are not all level and flat, and adapted to raising of grain, inasmuch as they are, with the exception of some few flats, generally covered with timber, in divers places also with large and small stones." [O'Callaghan, *Documents*, I, 365] Extravagant expectations from the fertility of virgin ground were curbed by warnings that the settler "must clear it well, and till it, just as our lands require" [Jameson, 132]; and, although there was plenty of grass in the meadows and forests, "when made into hay [it] is not so nutritious for the cattle as here [Holland]" [Jameson, 104], a discovery which would in time lead to heavy reliance upon imported plants rather than native grasses in the development of pastures in eastern North America. But despite any minor differences and deficiencies the really important fact was that there was no "starvation time" in the annals of New Netherland, indeed apparently no crop failures at all. Wheat, rye, barley, oats, peas, and beans, apples, pears, cherries, and peaches all thrived; within a few years beer, butter, and cheese were in ample supply; and it was not long before food surpluses were being exported to Dutch stations in the West Indies.

David De Vries, who knew the area well, observed in 1642 that this was "a pleasant and charming country, if only it were well peopled by our nation" [Jameson, 219], and thereby he gave emphasis to another important fact about New Netherland. It was not then well peopled and never would be while under Dutch rule. The reasons for this are several, but in general they may be said to be less a matter of the difficulties and dangers abroad than the comforts and securities of home. Holland was simply too prosperous, tolerant, and tranquil, too "pleasant and charming" itself to produce any considerable body of discontented, and New Netherland was too lacking in sources of quick wealth, such as gold or tobacco, to lure the greedy and speculative. The restrictions of company rule may have been an added hindrance in the early years, for although they recruited and subsidized emigrating farmers, they also maintained some very rigid controls. The most notable colonization scheme was the patroon system [Jameson, 90 ff.], though it became famous more for what it attempted than for what it accomplished. Initiated in 1629 by the West India Company, "the better to people their lands,

and to bring the country to produce more abundantly," the program offered large grants in the Hudson Valley to those who would undertake the establishment within them of at least fifty adult settlers within four years. The dimensions were precise in terms of river frontage (four leagues [approximately sixteen miles] on one side or two leagues on each side) and remarkably vague in terms of interior extent ("so far into the country as the situation of the occupiers shall permit"). All this was a good reflection of how completely river-bound were the perspectives and practicalities of the time. Several patroons were initiated, but only one was successful, that of Van Rensselaer around Fort Orange (see location in Fig. 39). Rensselaerswyck originally lay back from the west bank and was later extended inland from the east as well. It had a good location, but its success was more a measure of the abilities, resources, and interests of its proprietor than any special qualities of the area. The inherent deficiency of the patroon system was its reliance upon tenant farmers who were bound within a modified manorial structure. Later other schemes were offered, including small grants to individual farmers, and some increase in immigration resulted but still hardly commensurate with either the hopes of the company or the land available.

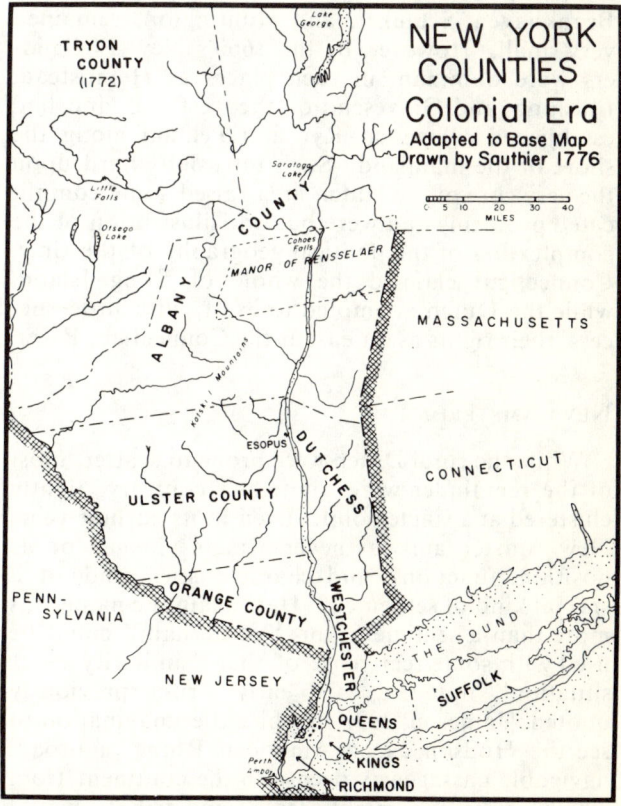

Fig. 39

CHAPTER 7   THE COLONIAL PERIOD, 1609–1775

The great majority of the settlers lived within a few miles of Manhattan: on adjacent Long Island, Staten Island, just north beyond the Harlem, and on the west bank of the Hudson. Prominent villages included Breuckelen, Flatlands, Flatbush, New Utrecht, New Harlem, and Bergen (see Fig. 38). Until late in the Dutch period there was not a single village between Manhattan and Rensselaerswyck. There were a number of farmers and traders scattered here and there, but the nearest thing to a cluster was the sixty or seventy people around Esopus (now Kingston). That group was also a good example of the disinclination of the rural Dutch to form villages. They finally did at Esopus, but with great reluctance even though under direct orders of the governor to do so and faced with an ominous Indian situation.

Difficulties with the Indians had in fact led in 1656 to a proclamation ordering all rural settlers to collect into villages "after the fashion of our New England neighbors." It was not necessary to look far to find these examples, for New Englanders had intruded directly into New Netherland very early. The first real villages were English settlements on Long Island. Southold and Southampton, established about 1640, were so far to the east as to be viewed as more a nuisance than a threat, and a scattering of settlements farther west, such as Brookhaven, Setauket, and Huntington, remained very small. However, by the 1660's New Englanders were dominant in such places as Hempstead, Flushing, and Gravesend on the west end, and had established villages at Rye and Pelham along the shore of the mainland. Such intrusions hard upon the very core of the Dutch area raised some complicated problems and were but one illustration of the complexities of the political geography of the time. Connecticut claimed the whole of Long Island, while the Dutch attempted to exert, with little success, their rights as far east as the Connecticut River.

New Amsterdam

While the rural Dutch were prone to scatter, most of the remainder were, quite the contrary, tightly clustered at a single point. Even in its earliest years New Amsterdam was never a mere village, for its position, functions, and characteristics made it a special kind of settlement. Here again the name was more than a sentiment, for "Amsterdam" could be used with some real sense of that "similarity of... situation" noted by the early writer previously quoted. [Jameson, 293] It took little imagination to see the Hudson as an American Rhine, a broad, navigable passageway deep into the continent from which, in time, a mighty traffic and tribute would naturally accrue to those who commanded the portal. Further, the whole situation around that entry, the intricate complex of land and sea, salt water and fresh, with its islands, bays, and coves, its narrow channels and radial river estuaries (Fig. 38) must have greatly accentuated the sense of familiarity in those used to plying the canal and river-laced lands about the Scheldt and Zuider Zee. Here were a new Amsterdam and new Netherland in a very real geographic sense.

The very site of this first town would seem to make it a likely focus of trade, yet it was not necessarily a commanding one. The unrestricted navigability of the Hudson and other waterways made direct export and import from scattered landings quite possible. Certainly until there arose a real need for some centralized marketing functions and facilities it is conceivable that this whole region could have developed a very dispersed pattern of trade more like that of Tidewater, Virginia, which was commercial and prosperous but essentially townless for many years. It is important, therefore, to note that the focus upon this point was not left to chance but was consciously designed by man from the very first. The West India Company planned New Amsterdam to be the military, political, and economic center. With regard to the latter, for example, Article XII of the patroon charter states, in part: "And inasmuch as it is the intention of the Company to people the island of the Manhattes first, all fruits and wares that are produced on the North River and lands lying thereabout shall, for the present, be brought there before being sent elsewhere." [Jameson, 92]

The following article granted to patroons the privilege of trading along the coast from Florida to Newfoundland, "provided that they do again return with all such goods as they shall get in trade to the island of the Manhattes, and pay five per cent duty to the Company." Rigid control of colonial trade was, of course, a commonplace of mercantilist practice, but its particular application in this area is of important and peculiar geographic significance. In spatial terms, the important features of the West India Company's policy were the deliberate allocation of locational priorities and functions and the conscious establishment of a focus and tributary area—in short, the implicit design of a nodal region (see Introduction). One of the notable features of life in New York State today is the chronic complaint by upstate interests of their involuntary "subsidy" of the metropolis. Those who are concerned about such matters might well ponder that the roots go back nearly three and a half centuries.

As for the actual commodities of this trade, a visitor of 1661 described the receipt of "beaver, otter, musk and other skins from the Indians and from the other towns in the River and Country inhabitants thereabouts" [Jameson, 423–4]; of beef, pork, wheat, butter, tobacco, and wampum from Long Island; of beef, sheep, wheat, flour, biscuit, malt, fish, butter, cider-apples, iron, tar, and wampum from New England; of tobacco, oxhides, beef, pork, and fruit from Virginia. In return New Amsterdam, drawing primarily from the cargoes of the "7. or 8. big ships" which each year arrived from Europe, sent out "Holland and other linnen, canvage [canvas], tape thrid, cordage, brasse, Hading cloth, stuffs, stockings, spices, fruit, all sorts of iron work, wine, Brandy, Annis, salt, and all useful manufactures." As this account shows, there was heavy traffic by Dutch traders with neighboring English colonies and New Amsterdam became the principal stopover for English vessels en route between New England and the Chesapeake and Carolina.

Broadly similar in situation and alike in entrepôt function, New Amsterdam was also in appearance a miniature of its namesake. The handsome detailed sketches and plans which have been preserved (see Kouwenhoven) are a vivid affirmation of Thomas Wertenbaker's description of its likeness to the trading towns of Holland:

There were the same curving streets lined with quaint houses, the same use of every open space for gardens or orchards, the same canals running through the heart of the town, the same sky-line with its tiled roofs, church tower and picturesque windmill, the same waterfront with its wharves and slips, and protectoring batteries. [Wertenbaker, 40]

The wall (now the site of Wall Street) separating town from country left only the very tip of the island for urban development, but by the end of the Dutch era it was far from full of buildings. Yet one of the striking features of New Amsterdam was how its design and architecture reflected that of land-scarce Holland with multistory narrow frontage buildings in solid rows along some streets. It does not seem unreasonable to suggest a cultural continuity between the narrow step-gabled three- or four-story brick house and shop of the 1660's and the jagged, towering skyline of the 1960's—to see each as an expression of tradition, habit, and aesthetics as well as the price of land.

Being a shipping crossroads and having an easy tolerance (with a few notable lapses) for religious and national differences, New Amsterdam was always a polyglot place. French-speaking Walloons were among its first inhabitants, Flemish and English groups came soon after, and in time a remarkably varied population accumulated. Father Jogues, a visitor of 1646, was told that there were men of eighteen different languages on Manhattan, which number must have included most of the dialects of maritime Western Europe and a few from Angola and the Guinea coast as well, for there were a number of Negro slaves. He also reported Calvinists, Catholics, English Puritans, Lutherans, and Anabaptists, and he had not covered the full list, for Anglicans, Quakers, and Huguenots were well represented in the colony, and were later augmented by a group of Jews who fled Pernambuco in Brazil when the Portuguese retook it from the Dutch.

The Legacy of the Dutch

Such a great variety is all the more remarkable in view of the total populations of the town and colony. In 1664, at the end of Dutch rule, New Amsterdam evidently contained no more than 1,500 residents, and the whole of New Netherland but about 8,000. Comparison with 25,000 in a New England of nearly the same age reveals how marginal the whole effort had been to the interests of the company and the nation.

Yet, though the numbers were few and the directly visible imprint would be almost wholly effaced over the years, a legacy remained. Certainly one of the most obvious was the mantle of place names spread over the whole region of direct control. Bronx (Bronck's) and Rensselaer are among the best-known examples of the numerous surnames imprinted on the land, Brooklyn (Breuckelen) and Harlem represent the ties to homeland localities, while "kill" attached to streams and channels remains as a generic term peculiar to the area. More important was the broad geographic framework, laid out if never very fully developed, stretching from the Mohawk to the Atlantic, from Schaenheckstede (Schenectady) on the Groote Vlachte (Big Flat) to Staten Island overlooking Sandy Hook. The key settlements established near either end of that axis would function ever after as key settlements along that same route. Further, something of the economic and cultural character of the centuries to follow was clearly discernible: an agriculture successful and productive but secondary to commerce, a hinterland inextricably bound to the port, the dominant role of interregional and intercontinental traffic, the cosmopolitanism and easy receptivity and free intermingling of nations and sects, the strong pressures from the overflowing tide of New England colonization.

New Netherland was a minor but characteristic product of the Dutch during their most creative era. Dutch domination of the seas was brief in time but world-wide in extent, touching every continent. Most of her numerous footholds were lost long ago, and almost none remain under her flag today, yet within a span of fifty years in the seventeenth century Dutch companies laid the foundations of three of the world's great cities and ports: Capetown, Jakarta (formerly Batavia), and New York City. It is a colonial legacy in which the Dutch may well take pride, and not least in the basic impetus and character which, despite their very brief tenure, they gave to the last.

New York

In the summer of 1664 an English fleet arrived in the harbor and received the reluctant surrender of the Dutch governor at New Amsterdam who had no forces with which to resist. The northern portion of New Netherland was organized into the colony of New York. Nine years later it was recaptured by a Dutch fleet but within a few months was restored to English rule by the treaty ending the last in the series of Anglo-Dutch wars. In New York international contention soon shifted to the interior where for seventy years the frontier was the focus of the sporadic savage outbursts of the so-called French and Indian wars in which England and France vied for continental supremacy. A dozen years after the final settlement of that rivalry in 1763 every district of New York was rent by the civil strife of the Revolution. No attempt will be made in this chapter to analyze the critical framework of international strategies or to examine the details of military campaigns (such matters are well covered in numerous history texts and well depicted in *Richards Atlas of New York State*). Political and military events have been considered here solely in terms of their impact upon other basic internal geographic patterns of the colony.

ENGLISH CONTINUITY: 1664–1700

Though obviously politically momentous, locally the events of 1664 had little immediate geographic significance. The abrupt change in flags and officials was neither accompanied by, nor did it clearly foreshadow, any great change in the extent, functional patterns, or characteristics of settlement in the colony. That New York, Kingston, and Albany became English in name only was an indication that the political change itself was a fortuitous result of a complicated European struggle and that the new rulers came with no special intentions or programs for development. The Dutch residents were confirmed in most of their rights and properties, the trade pattern was not drastically altered, and there was no great influx of English settlers. Thus, the regional pattern of New Netherland was not warped into or overlain with a specifically English pattern but remained intact to be elaborated and developed in full continuity with the past. Such a situation was far from inevitable. Political changes often do introduce basic geographic changes, and New Netherland had a population so small, a settlement so scattered, and resources so slightly developed that it was, in a sense, a fragile creation whose structure and features could easily have been abruptly altered. Had, for example, this new English acquisition been amalgamated with Connecticut or Massachusetts, opened to a flood of settlers from the east, and imprinted with their polity and patterns, the cultural geography of the Hudson Valley would have been quite altered in character.

Such pressures from New England actually did exist of course, as the Dutch had already found out, and it was only the fact that the Hudson Valley was continued as a separate colony, that its officials zealously retained control of the disposal of its lands, and that such lands were chiefly allocated in large grants to favored individuals that those Yankee pressures were contained, for the most part, in the hills back some distance from the river. Though patroon-like in scale these English grants were not in the main a simple continuation of Dutch policies, for the requirements of settlement and improvement faded away and they became more often a means of speculation than of actual colonization. Few grants were made in the early years of English rule, but near the end of the century the demand increased and the governors became more lavish. After half a century, by 1714, practically the entire Hudson Valley from Saratoga to the sea and most of the Catskills were in private hands. Although some governors took great care and conferred title only after careful surveys, others were indifferent and extravagant. Thus many grants contained no stipulation as to size and were defined by very uncertain boundaries. Some were so vague that even the general location was doubtful, and they became what Cadwallader Colden, a later surveyor general, called "ambulatory grants": "the patentees claiming, by virtue of the same grant, sometimes in one part of the Country, and sometimes in another, as they are driven from one place to another by others claiming the same lands with more certainty." As a result of such practices, not only was New York parceled out among relatively few landholders, but much of her acreage was for many years clothed with doubt and confusion as to ownership. Colden, in his report on "The State of the Lands in the Province of New York, in 1732," well summarized an important consequence:

Tho this Country was settled many years before Pennsylvania, and some of the Neighboring Collonies, and has many advantages over them, as to the situation and conveniences of Trade, it is not near so well cultivated, nor are there near such a number of Inhabitants as in the others, in proportion to the quantity of Land; and it is chiefly if not only where these large Grants are made where the Country remains uncultivated—tho they contain some of the best of the Lands, and the most conveniently situated. And every year the Young people go from this Province, and Purchase Land in the Neighbouring Colonies, while much better and every way more convenient Lands lie useless to the King and Country. The reason of this is that the Grantees themselves are not, nor never were in a Capacity to improve such large Tracts and other People will not become their Vassals or Tenants. [O'Callaghan, *Documentary History*, I, 381]

Although under such circumstances New York could have little attraction for English farmer-colonists, who now had such a wide choice of possible locations along the American coast, some English shippers and merchants were quick to see the advantages of New York City. A resident of 1669 noted that several houses had been bought by New Englanders as "there are several people in and about Boston which have inclination to come hither and live" and also that "many from Barmoodas and Barbadoes intend to remove hither; some are come as agents and have already bought some houses and plantations." [O'Callaghan, Documentary History, III, 183–84] Such movements were not large but gathered strength through the years and together with the necessary officials and their associates did begin to give a greater English ingredient to the continuing cosmopolitan character of the port and capital. Elsewhere there was little change. The governor of 1678 summarized the population as being "except in & neere New York [City] of Dutch Extraction & some few of all nations, but few Servants, much wanted & but very few slaves." [O'Callaghan, *Documentary History*, I, 91] In 1689 another governor stated that during his seven years' tenure "there has not come over into this province twenty English Scotch or Irish family's.... But of French there since my coming here several familys come both from St. Christophers & England & a great many more are expected as alsoe from Holland are come several Dutch familys" [O'Callaghan, *Documentary History*, I, 161] New Paltz was at that time the main center of the French Protestants who had arrived, and New Rochelle would soon be established by some of those anticipated. The main new Dutch settlement was at Tappan, begun the year before, and there was sporadic augmentation in older areas.

The governor who reported this failure to attract the English and the continued influx of others saw it as "another great argument of the necessity of adding to this government the neighbouring English Colonys, that a more equal ballance may bee kept here between his Majesty's naturall born subjects and foreigners which latter are the most prevailing part of this Government." In this way he related it to a problem of political geography which was vexing in other ways, especially in the control of trade. That problem was rooted in the only immediate geographic consequence of the English takeover. English recognition of East Jersey as a separate colony broke the political unity of the Hudson at the most critical point, for it opened the estuary to competition for trade, circumvention of customs, and contradictions in claims to such ambiguous areas as Staten Island (the ambiguity arising from the problem of defining exactly where the mouth of the Hudson was). If such a situation were allowed to continue, lamented the governor of New York, "wee are like to bee deserted by a great many of our merchants whoe intend to settle there." Because Jersey had no customs, he feared that they would capture much of the Indian trade. Ships already had begun to use Perth Amboy as a provision center and even to "break bulk there and run their goods into that Colony with intent afterwards to import the same privately and at more leisure into this Province." He unsuccessfully sought permission to erect "a small Fort with twelve guns upon Sandy-Hook" as an immediate means of rectifying the situation. His fears of competition would, in time, prove greatly exaggerated, but the splitting of the lower Hudson and its estuary between two political units would give rise to nagging problems of many sorts ever after.

Trouble with her eastern neighbor was more chronic and complicated, although control of trade, too, was an important part of the problem. The New Englanders on Long Island continued to live as if they were part of New England and tended to ignore regulations from New York City. Southampton was commonly described, along with New York City, as one of the "principall places of Trade," even if much of it were technically illegal, for the produce of eastern Long Island "was frequently carried to Boston and notwithstanding of the many strict rules and laws made to confine them to this place." [O'Callaghan, *Documents*, III, 797] Despite this chronic *de facto* situation, the King refused to grant a petition from the towns of that area asking that they be shifted to the jurisdiction of Connecticut, and he thereby confirmed New York's *de jure* control over the whole island. The mainland boundary issue with Connecticut, complicated by the penetration of New Englanders clear to Mamaroneck and beyond, was nominally settled in 1683, although the exact line remained in some dispute for years.

Internally the pattern of political geography was given more formal expression by the creation in 1683 of ten counties, whose names remain prominent on the map of today and whose boundaries

TABLE 13
POPULATIONS BY COUNTY, RACE, REGION—SELECTED COLONIAL CENSUSES

County	1698	1723	1749	1771
New York*	4,937	7,248	13,294	21,862
Richmond*	727	1,506	2,154	2,847
Kings*	2,017	2,218	2,283	3,623
Queens*	3,565	7,191	7,940	10,980
Westchester*	1,063	4,409	10,703	21,745
Suffolk	2,679	6,241	9,384	13,128
Orange	219	1,244	4,234	10,092
Ulster	1,384	2,923	4,810	13,950
Dutchess		1,083	7,912	22,404
Albany	1,476	6,501	10,634	42,706
TOTAL	18,067	40,564	73,348	163,337
White	15,897	34,393	62,756	143,474
Black	2,170 (12%)	6,171 (15%)	10,592 (15%)	19,863 (12%)
*New York City region as a percentage of the total	12,307 (68%)	22,572 (55%)	36,474 (50%)	61,057 (37%)

Source: *Census of the State of New York for 1855* (Albany, 1857), pp. iv–vii.

would remain unaltered until near the end of the colonial period (Fig. 39).

But, aside from these political matters, the basic patterns were little changed through the remainder of the seventeenth century. "New York is the Metropolis, and hath nothing to support it but trade," summarized the governor in 1691. [O'Callaghan, *Documentary History*, I, 407] As in Dutch days it was still the only authorized foreign port. Its traffic was chiefly in "Flower [flour], bread, pease, pork and sometimes horses" to the West Indies in return for rum and molasses, and in furs, linseed oil, and bone to England in exchange for manufactured goods. The commercial fleet regularly based here was listed as nine or ten three-mast vessels (c. 80–100 tons each), two or three ketches (c. 40 tons), and about twenty sloops (c. 20–25 tons); six or seven of the last were employed in the river trade. The only other towns were Kingston and Albany, all other settlements being referred to as "country villages." In eastern Long Island, which was dotted with such communities, the land was largely in pastures, and fishing and whaling were prominent activities; much of the central portion was described as "altogether barren," while the western end was the principal agricultural district of the colony. Settlement in the Hudson Valley was still intermittent, although many of the best sites directly along the river had been taken; the biggest expansion had been up the Wallkill, near Kingston, an enlargement of the old Esopus district, now as before the leading farming area upriver. Albany, incorporated as a city in 1686, was still wholly Dutch and still lived off its monopoly of the fur trade. Colonization in these northern reaches suffered a severe setback in 1690 when Schenectady, by then a solid community of some eighty "well built and well furnished homes," was sacked and burned by a raiding party of French and Indians. The village was reoccupied, but a good many in the area withdrew to safer locales. As a result the census returns showing 1,476 for Albany County near the end of the century were considered to reflect a reduction of perhaps 500.

The enumerations of 1698 give a useful summary of these patterns (Table 13), a total of only about 3,000 north of Westchester suggests how very sparsely the Hudson Valley was occupied, the more so when perhaps a third of these could be allocated to the immediate vicinity of Albany. On the other hand about a quarter of the population lived on Manhattan and two-thirds lived in the five counties adjacent to that focus, a grouping functionally significant even then, for all were within a day's journey of the capital by sloop. The total of 18,067, little more than a doubling in the thirty-four years of English rule, reveals the relatively slow pace of development.

NEW PEOPLES AND PATTERNS: 1700–50

Some acceleration of growth is indicated by the population of 40,564 in 1723 (Table 13), a more than doubling in twenty-five years. Further, there had been a rather important change in the proportionate distribution. New York County now had only 18 per cent of the total, and the five counties most directly tributary to the "metropolis" contained only 55 per cent. The Hudson Valley was now clearly a prominent area of growth, its expansion more attributable to the huge families of the rural Dutch than to immigration. There was, however, a sporadic influx from abroad and one particularly important addition.

The most notable of the newcomers of the time were the Palatine Germans, destined to stamp an indelible mark upon several localities in New York, as indeed they did in Pennsylvania and North Carolina. Fleeing the complex religious and economic difficulties of their Rhenish homeland, several thousand Palatines took refuge in England. In order to provide some support for them and at the same time ease England's dependence upon the Baltic for tar, turpentine, and resin, the British government sponsored a naval stores scheme in New York which would make use of Palatine labor. The first party was dispatched to Quassaick Creek in 1709, and there formed the nucleus of Newburgh (see Figs. 40 and 41 for place names). The main body arrived in the following year and was sent to camps on either side of the river above Kingston. The over-all program quickly foundered and was soon abandoned, leaving the Germans to fend for themselves. Some families remained in the vicinity of the original camps, eventually forming such communities as Rhinebeck, Germantown, and Katsbaan. Most, however, moved elsewhere, the largest group migrating to the middle Schoharie Valley, where they believed they had been granted the right to settle. They soon had land in crops and several villages established, from among which the communities of Schoharie and Middleburgh have evolved. Conflicts over land rights soon brought a further dispersal from that valley, this time through the aid of the governor to land along the Mohawk in the Canajoharie district (e.g., Palatine Bridge) and farther up at what became known as German Flats (later Herkimer).

About two thousand Palatines came to New York at this time, and it is important that this migration be seen not simply as a minute fore-shadowing of later mass movements but within the context of its times. These Germans, constituting some 7 or 8 per cent of the total of the receiving population, were the largest single European group to migrate into New York during the colonial period. Furthermore, the local colonization in the Middle Hudson, Schoharie, and Mohawk were made on a scale far greater than the usual beginnings. Arriving as large families rather than individuals, these new immigrants multiplied even in the early years of difficulty and they immediately became an important ingredient in the growth of the colony.

The frontier colonizations by these Germans west of Albany represented an important alteration in the geography of settlement. Prior to this movement Schenectady had been rebuilt, Dutch farmers had gradually spread westward to the vicinity of Amsterdam, and a number of patents had been issued for lands in the Mohawk Valley, but the memory of the 1690 massacre lingered and it was a slow expansion. German settlement beyond was therefore an important new thrust westward and one fostered by the governor in the hope that there would be "a Barrier against the sudden incursions of the French." The Schoharie settlement, too, was a significant departure from the earlier pattern. Behind the Helderbergs, remote from the Hudson, and even well back from the Mohawk, it was not only another westward salient but the first sizable penetration into the hill country of Upstate New York. Still another important initiative in this general direction was the establishment of Fort Oswego, at the mouth of that river, on Lake Ontario in 1727. It was designed to be a trading post as well as a garrison, and leading Albany merchants, who had lost their legal monopoly over the fur trade, soon had agents there. Such developments gave a wholly new sense of security in this direction, for the first time made the Mohawk an important trafficway for the colonists rather than just the Indians, and gave the first firm westward extension to the long-established Hudson Valley axis.

Another twenty-five years produced further changes, which, if less radical to the over-all pattern, were at least equally portentous. The mid-century (1749) total of 73,448 for the whole colony represented only a slight slackening in rate of growth and no obvious alterations of the broad regional patterns (Table 13). The continued vigor of expansion in the upriver counties (and of Suffolk on eastern Long Island) had reduced the portion of population of the five counties around New York City to almost exactly half of that total. The greatest proportionate growth was in Dutchess County, and that was a reflection of the flood tide of New Englanders into the Colony. New England encroachment was of course a century old, but now it became greatly increased in magnitude and broadened in pattern. Where the Dutch had followed fertile valleys reaching back east from the Hudson, as along the Fishkill, the two expansions met about midway between the border and the river, but in the hill country neglected by the Dutch, as in southern Dutchess (now Putnam) and northern Westchester counties, the New Englanders pushed clear to the Hudson itself. They, together with some newer British immigrants, even gained a few footholds across the river. The ethnic character of Newburgh, for example, was almost completely altered within a decade, when most of the Palatines,

Fig. 40. Province of New York, 1771

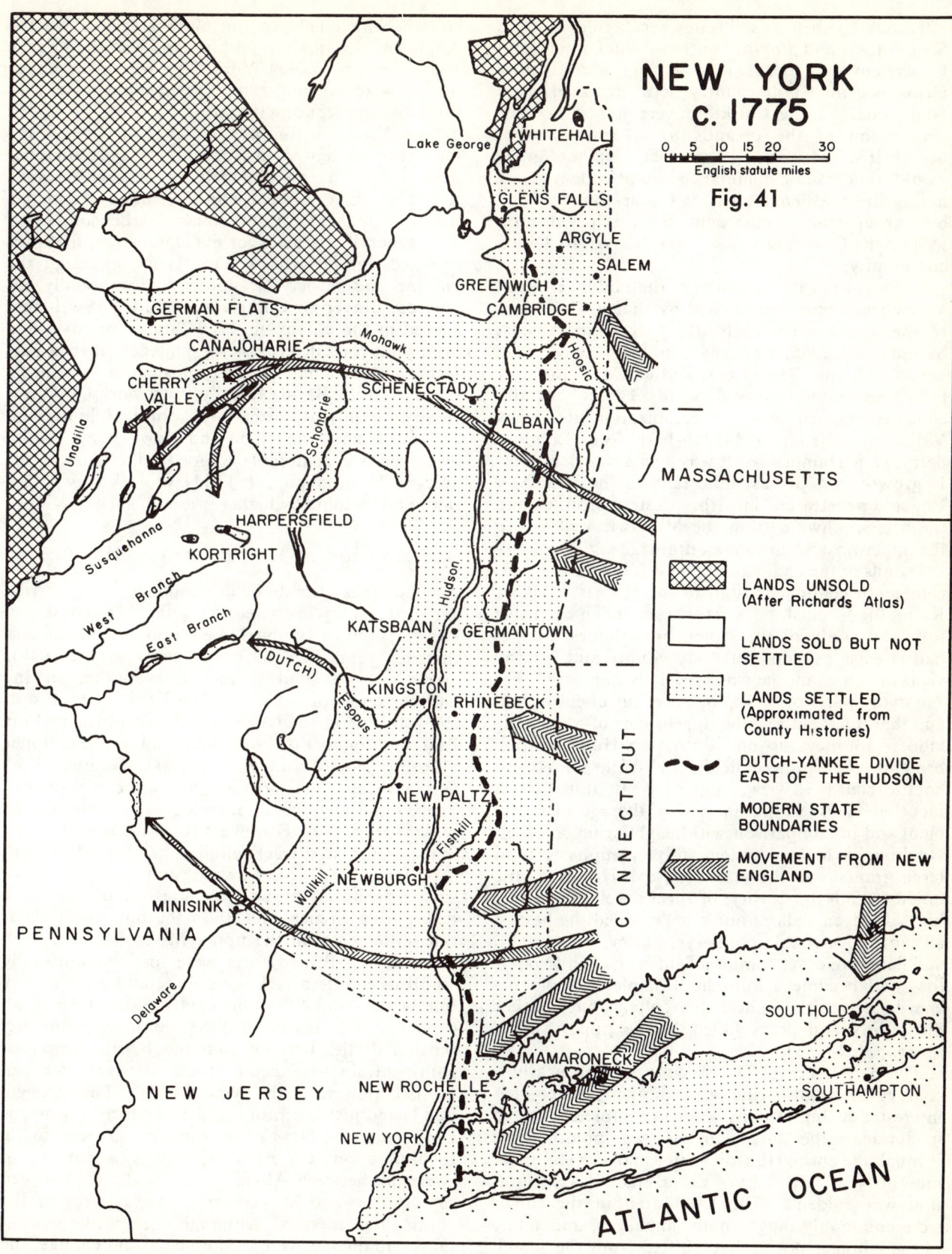

CHAPTER 7 THE COLONIAL PERIOD, 1609–1775

attracted by better soil elsewhere, sold out to Scotch-Irish and English settlers. In Orange and Ulster counties the Dutch, having spread their farms over the whole country between the Hudson Highlands and the Catskills, were joined in their penetration of the uplands by a sprinkling of Scotch-Irish pioneers, who, as their numbers grew, would soon assume dominance in that endeavor. A notable new settlement was at Cochecton, founded by a group from Connecticut, the first penetration well up the Delaware Valley from the nearest Dutch community.

The vigor of this expansion in the middle Hudson Valley was not quite matched by that in the upper reaches. Another French attack, this time against Saratoga in 1745, had again blunted the spread beyond Albany. The Mohawk salient was gradually broadened but not thrust forward. The only significant new element was the colonization at Cherry Valley by a group of Scotch-Irish from Londonderry, New Hampshire. It was a small nucleus slow in growth, but it was important as the first New England penetration into this western frontier and the first southward from the Mohawk Valley into the uppermost Susquehanna drainage.

Details of the character of New York in the mid-eighteenth century abound in the reports of Peter Kalm, the Swedish botanist who made an extensive tour and study of the American colonies. Kalm had at least a glimpse of every county and lengthy visits in some, and he was especially impressed that "so many parts of New York are still uncultivated, and that it has entirely the appearance of a frontier-land." Journeys up and down the Hudson often brought "fine plowed fields, well-built farms and good orchards in view," but only intermittently, a fact that surprised him in view of the age of settlement and in comparison with neighboring colonies. He thought the land system with its emphasis upon large grants was the main cause. However, the quality, if not the density, of rural development was high. Hudson Valley flour was "reconed the best in all North America"; corn, rye, barley, oats, hemp and flax were other common field crops; while potatoes had become a household staple. Apple orchards thrived throughout the colony, but peaches did poorly and pears failed in the upper valley. [Benson, I, 143, 331, 335]

The persistence of Dutch culture after nearly a century of English rule was another matter that interested Kalm. Albany was still completely Dutch in character; the Dutch language was predominant through the entire Hudson Valley and was still common in New York City. Yet a gradual acculturation was evident. Dutch children in the towns were commonly taught both languages, and many of the younger people had shifted from the Dutch church, which held to the mother tongue, to the Anglican. Among the visible signs was the change in architecture in New York City, where the newer houses were no longer built in the old style with the gable end fronting the street.

New York City impressed Kalm as one of the most thriving centers in America: "In size it comes next to Boston and Philadelphia, but with regard to fine buildings, opulence and extensive commerce, it vies with them for supremacy." [Benson, I, 31] The general character of its commerce had not changed significantly over the years, although the volume and variety of goods had markedly increased. Governor Clinton reported 157 vessels registered in the colony in 1749 (and a number had been built for sale elsewhere), and described a pattern of traffic, that touched on all four continents of the Atlantic Basin. Locally, New York City drew trade not only from its own political hinterland but also from Connecticut and New Jersey. Kalm described a considerable wagon traffic from as far away as Trenton, attracted to New York City rather than Philadelphia by higher prices.

Acceleration of Developments: 1750–70

All of these trends of development were carried forward rapidly in the early 1750's. Dutchess and Ulster counties doubled their populations in less than ten years. In the north, however, no major advances were made before still another in the chronic series of wars with the French broke out. English forts at the Oneida-Mohawk portage (Fort Bull and later Fort Stanwix) and on the upper Hudson (Fort Edward) and Lake George (Fort William Henry or Fort George) were established to protect the two great entryways into the settled regions (Fig. 40). However, the government failed in its attempt to attract colonists northward to form a buffer behind Fort Edward, and the hazards of such a position were soon demonstrated again by a devastating French raid upon the outermost Palatine settlements on the upper Mohawk. Although the outlying British forts were lost, an unusually vigorous counterattack not only repulsed the enemy but by autumn of 1760 brought the surrender of all Canada. The treaty of 1763, confirming the expulsion of the French from the North American continent, finally ended a chronic struggle between the two powers in the New World. The struggle had lasted just a century and though by no means confined to the New York frontier had persistently focused upon it. Broadly, it was a contest of strategies between Albany and Montreal—between the Hudson and St. Lawrence—for control of the continental interior. Although the specific contestants and the means and motives would change, the general pattern of that geographic rivalry has continued in one guise or another for another two centuries.

The next fifteen years, from 1760 to 1775, mark a great pulsation in the developing pattern of settlement. At long last freed of the dangers of French and Indian attack, the frontiers were pushed forward with unprecedented vigor until abruptly halted and then repelled by the agonies of the American Revolution, which became a civil war of major proportions in New York. There was growth in every district and expansion along every wilderness margin, but two major thrusts produced the most significant alterations in the geography of the colony.

One was into the upper Hudson Country, beyond Albany. Dutch farmers had long gradually expanded up the river and even well up some of the tributaries, such as the Hoosic, but it was a slow, sporadic process. Now New Englanders, having already spilled over their borders into all the back country east of the river below Albany, rapidly planted their farms and villages—e.g., Cambridge, Greenwich, Salem—northward into the upper basin (Fig. 41). Viewed more broadly, it was but a small part of a more general northward expansion which, before the Revolution, would spread Yankee settlers over nearly the whole southern half of Vermont (at the time claimed as part of New York) and up the Connecticut Valley nearly to Quebec. There were also a number of special colonization schemes in this region, most of them attempts by patentees to get some development started on their lands. Among the more notable were the colony of Highland Scots at Argyle, and the beginnings at Whitehall (then Skenesborough) near the southernmost reaches of Lake Champlain. The latter lay just beyond the margin of the more generally occupied country. Farther north a cluster of settlers had gathered near the fort at Ticonderoga, and still farther north numerous patents had been issued and a few families were dispersed among various points along or near Lake Champlain. About the only real indication of a community as yet, however, was that fostered by a grantee at Willsboro by the falls of the Bouquet.

The second notable thrust was southwesterly into the upper Susquehanna basin. It was a continuation in people, pattern, and direction of the tiny beginnings at Cherry Valley. Almost entirely a New England movement, though perhaps more by Scotch-Irish than older Yankee stock, the pattern was one of tiny, isolated communities sprinkled over the broad region of hilly country lying between the Unadilla and the Catskills, with even some penetration well into the latter, as at Harpersfield and Kortright. Although some of these settlers had been actively recruited by land patentees, most came on their own initiative, usually as small voluntary groupings based on kinship, former neighborhood, or church affiliation.

The Mohawk settlement axis was broadened by the same kind of movement and people. Such town names as Galway (Saratoga County), Perth and Broadalbin (Fulton County), Knox and New Scotland (Albany County) mark where these British newcomers filled in toward the higher hill country north and south of the Dutch-German settlements in the main valley. These older residents continued to spread their farms into new ground but more slowly, steadily, and contiguously. Despite the removal of the French, the western point of colonization was extended but a very short distance beyond Herkimer.

Along and within the southern Catskills much the same kind of expansion (a slow Dutch movement with an increasing number of Scotch-Irish in the vanguard) was taking place but without much alteration of the general regional pattern of settlement. One notable exception was the migration of a small group of Dutch farmers up Esopus Creek and across the center of the Catskills into the valley of the East Branch of the Delaware.

On the Eve of the Revolution

The census of 1771, the last before the Revolution, clearly reflected the growth in numbers and the general changes in pattern (Table 13). An acceleration in the rate of growth was evident and the considerably more than doubling in but twenty-two years was obviously made possible only by a marked increase in the number of immigrants. The regional pattern of that growth clearly reflected the colonizations in the Hudson Valley and northern frontiers. Orange, Ulster, and Dutchess counties had more than doubled in population, while Albany had a fourfold increase. The latter, in fact, had spurted forward so rapidly that it now had almost twice the total of New York County, the next most populous. Such growth and, more importantly, its spread over so wide an area led in the next year to the creation of Tryon County to include much of the Mohawk Valley and a broad, vaguely delimited wilderness beyond (Fig. 39). Also, in the north Charlotte County was soon created to cover the Champlain area and lands to the east, into what later became Vermont. The New York City region was now reduced from 50 to 37 per cent of the total population, a significant shift in the proportional pattern.

Although the metropolitan region as a whole lagged behind percentage growth elsewhere, New York City lost none of its commanding urban and commercial position. This was not only because growth upriver was also an expansion of its own hinterland but also because virtually all of that growth had been spread over the countryside. Albany was still the only other incorporated community, and its size, perhaps three thousand, was hardly commensurate with the increases in regional population. Schenectady, somewhat enlivened by the greater Mohawk traffic, and Kingston, the river focus of the Wallkill district, were still about the only others that might be classified as anything beyond country villages.

The existence of so few centers of population was not simply the inevitable accompaniment of a subsistence economy. Actually there was a thriving commercial life, but it, too, was largely rural in character, and these products of farm and forest could usually be hauled but a short distance to a river landing for shipment. Thus, most of the ordinary service functions common to regional commercial centers such as storing, handling, sorting, grading, and forwarding could be concentrated at New York City. Although no longer the only authorized foreign port, it retained in fact as much dominance as ever. In 1770 an Albany entrepreneur pioneered the first direct voyage from his city to the West Indies, but this did not become a common practice.

Governor Tryon's "Report on the State of the Province of New York" in 1774 [O'Callaghan, *Documentary History*, I, 737] provides an illuminating summary of the items and pattern of commerce at the end of the colonial period. The staple commodities of trade were of three sorts: (1) "Natural produce" of: (*a*) Farms—Wheat, Indian Corn, Oats, Rye, Pease, Barley and Buck Wheat, Flaxseed, Live Stock, Pork, Beef, Beeswax; (*b*) Forests— "Masts & Spars, Timber & lumber of all sorts," "Pot & Pearl Ashes"; (*c*) the Hunt—Furs and Skins; (*d*) Mines—Iron Ore (nearly all from Ancram in present Columbia County). (2) Manufactures of Local Produce: Flour, Pig and Bar Iron, Soap and Candles, Hats, Shoes, Cordage, Cabinet Ware, Tanning, Malting, Brewing, Ship Building. (3) Manufactures of Imported Produce: "Distilling of Rum and Spirits, Refining of Sugar, making of chocolate."

The principal areas traded with were Britain, other British settlements on the North American continent, and the British and Dutch West Indies. New York merchants dealt extensively in the re-export of West Indian goods ("Log Wood and other Dye woods and Stuffs, Sarsparilla, Mahogany, Cotton, Ginger & Pimento, some Raw Hides") and of Carolina naval stores ("Tar, Pitch & Turpentine"). More occasionally vessels traded directly with Africa, bringing slaves back to the West Indies; with Madeira and the Canary Islands, returning with wine; with Europe, especially when there was a market for grain; and with Gibraltar and Minorca, the two British Mediterranean outposts of the time. Outbound cargoes were chiefly grain and lumber products to all of these, except Africa, where rum and English goods were traded. As for imports, other than of goods for processing or re-export, the governor summarized by stating that practically all articles available on the British market were "in some proportion imported here, except such as are among our staple Commodities."

Tryon's report of 709 vessels employed in the sea trade of the colony in 1772, as compared with 477 ten years before, and 157 in 1749, was a good measure of the remarkable growth of commerce. The governor gave no account of the river traffic, but other sources indicate a marked increase there, too, with much larger sloops and schooners having become common after about 1760. Similarly, on the Mohawk the canoes of the fur traders had been superseded by larger batteaux of several tons capacity. The Albany-Schenectady road had become a well-worn portage, avoiding the more devious river route and transshipment at Cohoes Falls; upriver, Little Falls was another obstacle. As yet batteaux did not often ascend much farther than Little Falls, except to serve military outposts, but the governor called attention to the long-apparent fact that at Fort Stanwix (now Rome) "a short cut across the carrying place there might be made into Wood Creek which runs into the Oneida Lake, and thence thro' the Onondaga [Oswego] River into Lake Ontario." There was as yet much less traffic on the upper Hudson, but here, too, he pointed out, beyond Fort Edward "it seems practicable to open a passage by Locks &c. to the Waters of Lake Champlain which communicate with the River St. Lawrence." He went on to state that "to surmount these obstructions an Expense will be required, too heavy for the Province at present to support," but added that "when effected would open a most effective inland navigation, equal perhaps to any as yet known." It was, of course, a prescient observation, but a century and a half of experience with these waters and portages had made it by now a reasonable one.

GEOGRAPHICAL IMPACT OF THE REVOLUTION

Governor Tryon was of course not aware that his report of 1774 was a description at a critical point in the geographical development of New York. The Revolution would markedly reshape the regional patterns of settlement, distorting and reducing those extant at the outbreak and laying the groundwork for radical extensions upon its close. The war was peculiarly hard on New York. The old Hudson-St. Lawrence pattern of rivalry was now revived with the same old harsh consequences for the frontier.

Nearly all of the country north and west beyond Schenectady suffered in some degree. The Mohawk Valley was heavily and repeatedly ravaged by British or Indian forces, as was the Schoharie, and virtually all of the Susquehanna settlements were abandoned. Far to the south the Minisink settlement (Port Jervis) along the Delaware in western Orange County was raided by a party from the interior. Kingston was burned by a British army. The New York City region persisted as a focal point of conflict throughout the war, many peripheral localities suffered, and a considerable portion of the city itself was burned. Furthermore, because of the nature of its colonial society New York was probably more disrupted internally than any other colony. Just how many thousands of Loyalists were killed and/or left is not certain, and just how their departure altered the patterns of population cannot be reconstructed in detail, but the change in some areas was certainly considerable. New York City's population was reportedly reduced by half, many of the manor houses in the Hudson Valley estates were abandoned, and every district must have been affected in some degree.

But the impact of the war was not wholly negative. The long series of campaigns served to advertise the province. In total a good many thousands of soldiers tramped over a good many districts of interior New York, and not a few of these men, British as well as American, had a good eye for country and liked what they saw. This was especially true of the Finger Lakes Country and Susquehanna Valley. Here American campaigns provided the first mass contact with a large area whose fertility was attested to by the very purpose of those operations—the destruction of Indian food supplies. No doubt more than one soldier who set fire to a cornfield decided that he was burning a better crop than he had ever seen at home and resolved to come back after the war and claim a piece of such land for his own. Further, the war was a tedious struggle and one so recurrently focused upon New York that colonization remained almost completely in abeyance. Such a prolonged disruption of the process of settlement expansion, which had been gathering ever greater momentum, could not help but build up pressures which, once relieved, were likely to burst forth in unprecedented strength, radically extending the old bounds of settlement. Finally, independence gave new local political powers and prompted or accelerated numerous forces which, together, would initiate a breaking up of the big landed estates in the older areas, and in the new areas would facilitate dealing with the Indians, negotiating the conflicting territorial claims of neighboring colonies, and pave the way for a rapid disposal of public lands. Thus, the stage was being set for a whole new era of settlement and development of New York.

Geographical Significance of the Colonial Period

Looking back over the whole century and a half of that Dutch-English colonial period, one can assess something of its over-all geographical importance. Taken as a whole, these years marked a stage of development whose accomplishments can usefully be measured against later times and against those of other colonies at the same time. Something of the former will appear in subsequent chapters; as for the latter, considering its relative size, age, and later pre-eminence, New York occupied a surprisingly modest place among the thirteen colonies. Its estimated total population of 185,000 in 1775 made it rank seventh, with fewer people than its small neighbor, Connecticut. Nor was the geographical extent of its settlement especially impressive in comparison. No other colony had anything like a Hudson River to allow such easy penetration into the interior, yet settlers of North Carolina, Virginia, and Pennsylvania had made much farther westward advances, while Yankees of New England had spread up the Connecticut Valley, far outdistancing the Yorkers in the parallel Hudson–Champlain trough. In this broader view, therefore, New York's growth appears to be relatively slow in pace and constricted in area. In contrast, however, is the position of New York City, which with a population of about 23,000 had forged well ahead of Boston (c. 15,000) and was second only to Philadelphia (c. 33,000).

Internally the most conspicuous general features have to do with the selection of sites and elaboration of an areal framework, with the tapping of resources, and with the introduction of particular groups of colonists. The establishment of the main axis from Albany to the estuary must not be accepted as simply an inevitable response to the obvious lineaments of nature. Certainly any group settling in that valley would make some use of that waterway, but political unity was an important factor in allowing it to persist from the first as a thoroughfare. Had Connecticut, for example, been able to maintain her claim for an extension of boundary northwest from her present extreme southwest corner, the river valley would have been bisected above Peekskill. It is quite conceivable that concern for revenues and privileges would have fostered a port within that jurisdiction which could have materially usurped some of the functions of New York City and changed the patterns of development at least in this early period. The maintenance of the political integrity of the Hudson Valley (except for the Jersey shore) throughout the colonial period was therefore of basic geographical significance. Manhattan and Albany were early and firmly established as the terminals of the Hudson axis, and the tentacles of trade fanning out from each in the colonial era were more than hints of future patterns.

New York City by this time had clearly demonstrated its power to draw trade both from its neighbors and from overseas; as an international entrepôt it was unrivaled in America. The fur trade had brought areas distant to the north and to the west into focus on Albany (and established the pattern of Hudson–St. Lawrence rivalry as well), settlers and batteaux had given a more substantial, if local, emphasis to such salients, while their future extension as navigable thoroughfares was well within practicable vision. In addition, Yankee trails across the Berkshires were beginning to emphasize the crossroads function of the locale. Beyond these primary sites, however, surprisingly little differentiation in function and importance had taken place. With the exception of Schenectady, and perhaps Kingston, there was little to suggest which villages or river landings would rise above their neighbors.

Certainly within the area occupied a considerable testing of resources of sea, soil, and forest had been made. The New Englanders on Long Island, sharing a frontier with their mainland kin, were, like them, extensively engaged in whaling, offshore fishing, and oystering. The Hudson fishery was known but much less exploited. In agriculture, although but a small fraction even of the area within farms was actually in cultivation, considerable assessment of soils and local climates had been made. Except for the failure of some of the less hardy fruits in the upper valley, however, there had been very little regional differentiation in crops or livestock. Dutch-German and British agricultures were somewhat different in tools and techniques but not sharply contrasting in basic crops and animals. The most commonly recognized broad regional difference was that along the Hudson margins, between the lowlands of the upper river and the highlands south of Newburgh. Tryon's *Report* of 1774 suggested that during colonial times man had already accentuated nature's differentiation: "The soil in general is much thinner and lighter in the Southern, than in the Northern Parts and having been longer under Culture and subject to bad Husbandry, is much more exhausted." [O'Callaghan, *Documentary History*, I, 739] Despite, however, a rather lengthy testing of the land in some districts, the results from any particular field, farm, or locality would commonly be measured only against those obtained from lands nearby. Only in postcolonial times, when settlement had spread over whole new regions, would it be possible to evaluate the Hudson area—valley and upland—in broader perspective. Figure 34 supports the colonists' opinions of the better nature of the middle reaches of the Hudson Valley.

The earliest farms tended to seek open ground—Indian old fields or meadows—but almost from the first in every locale expansion necessitated a clearing of the woodlands. In this way the farmer became a forester as well, for it was rarely a case of simple destruction; at least some of the wood was used for fuel, fencing, and building. In time some became commercial farmer-foresters, engaging in farming in the summer and forestry in the winter. With the establishment of water-powered sawmills, lumbering also became a specialty in itself. Mills were located at numerous water-power sites over the settled country by the time of the Revolution, and the volume of forest products produced had steadily increased. Yet contact with the two areas which would in time become the great lumber specialty regions had only just been made. Not until the 1760's was any penetration of the Catskills made and then more in search of farm land than forests. More significant was the settlement at Glens Falls, where the dense and valuable forests of the Adirondacks are accessible from streams which would prove suitable for log-driving. By 1775 only a very slight beginning in exploitation had been made, but it represented the attainment of a position strategic to one of the great resource areas of the colony.

Certainly one of the most important features of the colonial era was the creation of culture areas. The primary differentiation was of course that between European and Indian areas, a geographical pattern which was in a state of continual change from the beginnings of colonization until the end. The New York situation, like that in most of the other colonies, was a demonstration that this particular differentiation would be a sharp one, reflecting little intermingling of peoples or extensive enclaves of one group within the other. On the eve of the Revolution Tryon estimated that all of the shattered tribes of Long Island and the Hudson Valley could muster no more than 300 "Fighting Men" and that they were "in general so scattered and dispersed, and so addicted to wandering that no certain account can be obtained of them." [O'Callaghan, *Documentary History*, I, 746] The once mighty Mohawks had been reduced to "Two Villages on the Mohawk River and a Few Families at Schoharie," totaling 406 persons. All the rest of the Indians lived beyond the European settlement area, the westernmost salient of which lay on the margins of the lands of the Oneidas.

The pattern of culture areas within the European

region was of course much more significant, for it was more nearly permanent. Broadly the colony could be divided into three regions: the Dutch areas, including the closely akin German and Huguenot settlements; the Yankee (New England) areas; and the highly cosmopolitan center of New York City. The first two areas were, of course, not entirely uniform within nor everywhere sharply separated one from another, yet the areas of dominance were clear enough to allow such a geographic generalization to be made with some accuracy at the scale appropriate to this coverage (Fig. 41). These two areas were unlike each other in actual pattern. The Dutch area was a unit, whereas the Yankee area was in two parts, one to the east—the Long Island and New England borderlands, and the other to the west—the Susquehanna and Delaware settlements. To say that these were more nearly permanent does not of course mean that they would remain unchanged, only that they would persist in some degree. In fact all three types of areas would be significantly modified in later decades, and such changes are one of the most basic features of the historical geography of the state.

The full import of these culture areas cannot, however, be appreciated at this scale, for it is in the details of life and landscape that they take on full meaning. They had an obvious geographic significance in the colonial era simply because each contained a distinct kind of people with its own ways of thinking and doing. That some of those habits, methods, and attitudes have been transmitted through the generations to some of the people in the area today seems rather certain, even if rather difficult to prove absolutely. Such continuity in human characteristics within a particular area is in itself an important kind of geographical persistence.

However, even where that kind of heritage has been broken, where there has been a replacement of one culture group by another, there remains another kind of geographical persistence. For each of these colonial cultures put its distinctive imprint upon the face of the earth; in the names on the land, in houses and barns, homesites and farmsteads, in churches and inns, hamlets and towns, in the layout of fields, fences, and property boundaries, in the patterns of roads and streets, millsites and wharves, and in a good many other features as well. Not all of that imprint remains, certainly, but some of it does and much of the rest has only been obscured rather than erased. For those who know how to read the evidence, not only in the countryside but often in the heart of the city as well, it is clear that this colonial past remains intermingled with—indeed, in many ways still gives shape to—the geography of the present.

CHAPTER 8   D. W. MEINIG

Geography of Expansion, 1785–1855

IN THE YEARS after the Revolution New York underwent remarkable expansion and growth. It was a period of great geographical creativity in which a whole array of new areal patterns were established. It was during this time that settlers not only spread into nearly every nook and corner suitable for habitation but also began to concentrate into distinct regional patterns that would tend to persist strongly through another century of development. On a more detailed level it was a time during which thousands of villages were established, out of which differentiations in growth very soon revealed certain sites as focal points of greater development, setting the initial framework for the elaboration of urban hierarchies. It was also the period during which the resources of forest and mine were comprehensively tapped and patterns of exploitation established, the aboriginal landscape was domesticated, the variations in soil and climate were tested with crops, and an agricultural regionalism first appeared.

The seven decades allotted to this period may be conveniently subdivided into two parts. The first, 1785 to 1825, covers the main era of actual expansion of colonists into every suitable district. By 1825 the rural regions were by no means completely filled, nor was the process of colonization ended, but every district had been entered, no large blocks of farmland remained unsold, and in general the basic patterns of farms and villages had been set. The second part, involving the last decades, is characterized not only by rapid growth but also by the rapid differentiation of the rather uniform distributions of initial colonization into markedly uneven patterns of growth and specialization. It was an era very closely associated with radical improvements in transportation. The turnpike movement brought the rapid evolution of a road network and, more importantly, allowed the establishment of stage and mail service into every major locale, thus creating an actual functional network. The canal era opened the way for long-distance bulk traffic and spurred commercialization and regional specialization in agriculture and concentration and elaboration in industry. By 1855 the canal-building period was nearly over and the newest transportation innovation, the railroad, after twenty-five years of experimentation and only local development, had been linked into its first real system spanning the state.

The year 1855 is of no special significance in itself as a terminal date except that an elaborate state census of that year provides a useful statistical measure of many features. But it does fall within an important transition in the nature and character of developments. Behind lies the era of colonization and the setting of a geographic framework of development; ahead lies the era of burgeoning industrial, commercial, and urban growth, elaborated very largely within that framework.

Preparation for Expansion

To prepare the way for a general expansion of settlers into new regions of the state, three major, and in some ways interrelated, problems had to be resolved: (1) territorial disputes with neighboring states, (2) removal of the Indians and extinction of their rights to their former lands, and (3) establishment of policies regarding the sale of public lands to private owners. For the most part, these matters were settled in the 1780's. Because there was considerable local variation and the details were often very complicated, only a general picture can be given.

New York was involved in several territorial disputes, the two most important ones, directly or indirectly, being with Massachusetts. Each was rooted in vague and contradictory provisions of old colonial charters, and, while that sort of problem was hardly new in type, these particular issues were of major scope and significance. One, involving the whole of what became Vermont, was a quarrel with New Hampshire, which got started well before the Revolution. New Hampshire, having been organized out of Massachusetts territory, regarded its proper western boundary as being an extension northward of that of its parent, that is, a line approximately parallel with and twenty miles east of the Hudson. New York refused to recognize such a claim, but, though supported by British court interpretations, its attempts to exert jurisdiction were ignored or repulsed. Both governments granted titles to lands in the area, but, in keeping with past practices, those of New York were chiefly large grants to speculators, those of New Hampshire more to individuals or groups of actual colonists. In the end those in actual possession triumphed. In 1777 the New Englanders west of the Connecticut River proclaimed the new state of Vermont, and the refusal of New York to recognize this creation did not retard its development. Those who attempted to colonize under the auspices of New York were threatened, harassed, and in some cases had their farms destroyed and were driven from the area. By 1790 Yankee settlers had spread over so much of the land, the government was functioning so effectively, and New York was so obviously helpless in opposition that it finally granted recognition and the boundary with Vermont was formalized.

The other conflict was over lands west of the Mohawk and Catskills, where Massachusetts held to the legality of the unlimited inland extension of her early charter. In face of the reality of New York's development along the Hudson axis, Massachusetts made no serious pretense of insisting upon sovereign control over such a detached area, but having suffered in the Revolution and desperately needing money and room for her expanding population, she steadfastly held out for some share in disposal of such lands. In 1786 a compromise was reached. New York was to have sovereign rights to the whole region, but a "pre-emption line" was to be drawn due south from Sodus Bay on Lake Ontario; Massachusetts was given property rights to all west of that line, New York to all east except for a special area in which Massachusetts had already initiated surveys and negotiations for sale (the area which became known as the "Boston Ten Towns"). These agreements, together with the solution during these years of several lesser disputes with its southern neighbors, finally gave precision to the political limits of New York on the map.

The treaty ending the Revolutionary War contained no provisions dealing with the Indian situation. Yet, because the war had drastically altered the situation, it was inevitable that major changes would follow. As most of the Indians had supported the British they were regarded as enemies, as they had suffered heavy losses they were obviously vulnerable, and as they actually occupied but a tiny fraction of the land to which they claimed tribal rights there was now little hope of blocking settler expansion. By various means and for very minor material compensation the tribes were induced, in some cases directly by the state and in others by the holders of large land grants, to relinquish all their rights to the land and to settle within small reservations. A number of these were established west or north of the Mohawk, seven of which (Oneida, Onondaga, Tonawanda, Tuscarora, Cattaraugus, Allegheny, St. Regis) remain today, though mostly considerably reduced from their original size (See Fig. 42).

Actually, the government had committed itself to foster expansion into some of these lands well before negotiations with the Indians had begun. As a stimulus to recruiting during the Revolution, the state, having plenty of land but precious little money, had pledged a handsome bonus in land to each soldier, the amount ranging from 600 to 6,000 acres depending upon rank. A large block of more than 1,500,000 acres was designated for this purpose in the Oneida-Onondaga districts. This became known as the "New" Military Tract because an "Old" Military Tract had been hastily laid out on the northern slope of the Adirondacks in an effort to accomodate those claimants who grew impatient with the delays in getting the intended area cleared of Indian claims and ready for allocation (Fig. 42). As the rights of such lands were transferable, speculation became rampant. That not a single war veteran selected an allotment in the Old Military Tract was testimony both to the remoteness and relative undesirability of those lands and to the degree to which such men had peddled their rights. Some did receive lots in the new tract when it was finally opened in 1791, but there, too, most of it fell into the control of speculators anxious to profit by resale.

The state also sponsored several other schemes to aid those with special claims for indemnity. In the far northeast corner a tract was set aside for refugees from Canada and Nova Scotia who had sympathized with the Revolution. On the Susquehanna the Clinton Township was reserved for the "Vermont sufferers," New Yorkers who had been repulsed from their farms or claims in that interstate struggle. In these, as in such special surveys as the Chenango Twenty Towns and the St. Lawrence Ten Towns, the avowed purpose was to transfer land directly to intending settlers. In few places, however, was such a result predominant and in some it was almost completely absent. And in fact the state catered directly to speculative pressures by auctioning much of its land to the highest bidder. These were in blocks varying from a few thousand acres to a few million. By far the greatest purchase was that of Macomb, embracing all of the unsold land between the central Adirondacks and the St. Lawrence.

Massachusetts also elected to dispose of her acreage wholesale. The entire region west of the Pre-emption Line was purchased by a syndicate headed by Phelps and Gorham, though financial stress soon forced them to forfeit part and to resell

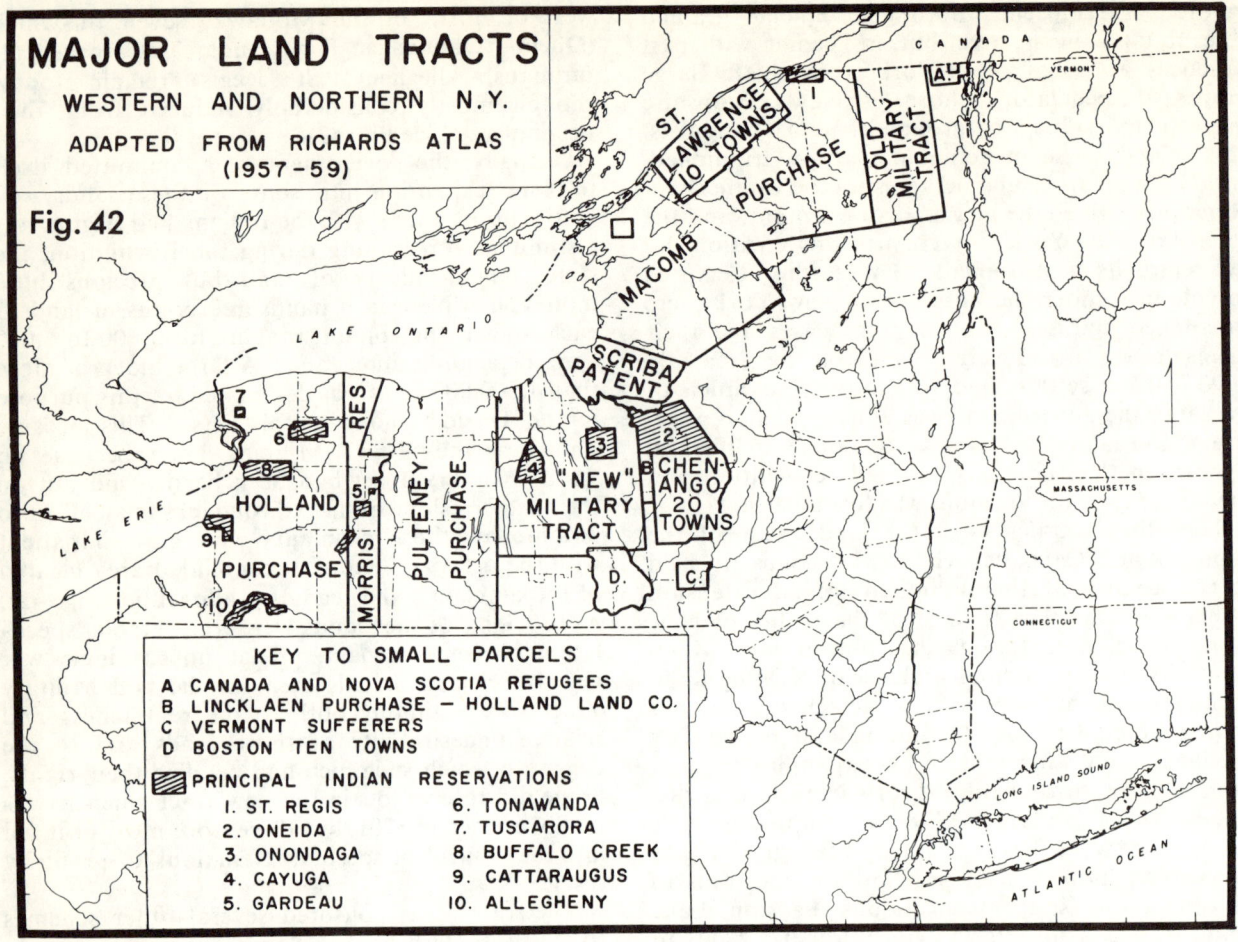

Fig. 42. MAJOR LAND TRACTS, WESTERN AND NORTHERN N.Y. ADAPTED FROM RICHARDS ATLAS (1957-59)

KEY TO SMALL PARCELS
A CANADA AND NOVA SCOTIA REFUGEES
B LINCKLAEN PURCHASE — HOLLAND LAND CO.
C VERMONT SUFFERERS
D BOSTON TEN TOWNS

PRINCIPAL INDIAN RESERVATIONS
1. ST. REGIS
2. ONEIDA
3. ONONDAGA
4. CAYUGA
5. GARDEAU
6. TONAWANDA
7. TUSCARORA
8. BUFFALO CREEK
9. CATTARAUGUS
10. ALLEGHENY

the rest to other speculators. The area became divided into three main blocks, each a broad strip from Lake Ontario to Pennsylvania. These were, from east to west: the Pulteney Purchase, controlled by an English syndicate; the Morris Reserve, which was soon further subdivided; and the Holland Land Company area, controlled by a group of Dutch bankers who also had some smaller holdings in central New York, most notably the Lincklaen Purchase around Cazenovia (Fig. 42).

The frenzy of speculation that characterized the marketing of New York lands was by no means wholly detrimental to the actual settlers. To be sure, in many cases the title to land changed hands many times before actual colonization and if each took his profit the intending farmer could be faced with a considerable fee. But, in fact, much money was lost as well as made in these speculations. The original price in most areas was but a few cents per acre, and, even so, many land-jobbers were precariously in debt and had to sell at least part of their holdings quickly in order to cover their contracts. Furthermore, although there were soon plenty of settlers seeking the better lands, rare was the one who had enough money to buy a farm outright. Hence, necessarily much of the land was ultimately marketed in farm-size lots at a reasonable price on extended credit.

Most of the more financially stable of the big landowners hoped to derive their profits not from a quick resale to other land-jobbers but from the steady rise in land values as their districts became increasingly settled and developed. Some spent a great deal of money on building roads and bridges, clearing forests, erecting sawmills, gristmills, and initiating villages, a practice that became known as the "hotbed" method from its purpose of accelerating greatly the rate of growth and thereby the rise in the value of lands. Even in such areas the early settlers were lured by cheap prices, with the hope of being able to charge much more once the area began to attract large numbers. But these schemes rarely succeeded as planned. Land costs could not become really high simply because there was a constantly expanding supply and area of choice. Even when the first sales were being made in New York, farms were being marketed in Ohio, and long before New York was filled excellent lands were available as far west as Illinois.

In fact, ultimately the general results were less different than might be expected from the federal land-marketing system, which was being initiated contemporaneously in Ohio. Much of New York was surveyed in more or less rectangular patterns not too different from the newly adopted federal township and range scheme farther west. And it is well to remember that technically there was no "free land" in America at the time and there would be none for more than half a century, not until the Homestead Act of 1862. Probably the most important distinguishing features of the New York situation were the very uneven and unstable prices and the fact that some of the best land was purposely held off the market for some years. On balance, New York settlers were not necessarily at a disadvantage. Even if few of those in control of land sales wanted to put in practice Judge William Cooper's axiom of "a moderate price, long credit, a deed in fee, and a friendly landlord," they found that they were unlikely to sell much land very fast unless they did follow it in some degree.

Geographically, the most important feature of the New York experience in land disposal was the significant variations from place to place in the methods and results. Many patterns of today can be traced directly back to these original surveys and to the earliest selection of sites and routes established in preparation for the settlement of the land. Some of the more significant features in the colonization of the several major districts will be discussed later in this chapter.

The Colonization Movement: General Features

The movement of population into the New York *frontier* (those areas colonized after 1776; see Fig. 43) was the greatest "land rush" of its time, and would continue to rank among the greatest within the whole famous "westward movement" in American history.

Magnitude

The first years of peace were largely ones of recovery and preparation. The ravaged margins of settlement were reoccupied and soon restored; within a few years a traveler through the Mohawk, well aware of the recent sufferings of that district, was amazed to find "every house and barn rebuilt, the pastures crowded with cattle, sheep, etc. and the lap of Ceres full." [O'Callaghan, *Documentary History*, II, 1105][1] But for a while there was no great surge westward. Colonists steadily infiltrated the Catskills region and edged out beyond the margins of the prewar frontier, but penetration into the farther districts was at first small and scattered. In 1790 there were only about 7,500 residents in the whole of central and western New York, and these were very largely in Oneida and Otsego counties. Another 1,500 had settled in the northern portion, nearly all within a short distance of Lake Champlain. These few thousands proved to be but the small vanguard of a mighty following. A report from Albany in the winter of 1795 that in three days 1,200 immigrant sleighs had passed through town bound for the west became a famous indicator, often cited by historians, of the magnitude of the movement. Three years later it was claimed that an average of more than a hundred sleighs a week passed through Geneva headed for the Genesee Country. Such local samplings were but a hint of what the census of 1800 would reveal: more than 100,000 inhabitants west of the old colonial settlement region. In addition there were more than 10,000 in the far north, still, however, largely in the Champlain area. In another decade these totals were increased to about 300,000 and 50,000. By 1810 settlers had spread, if only very thinly in many districts, over most of the state; only in the Adirondacks and Tug Hill and in the higher western portions of the Appalachian Upland were there large blocks within which settlement had not been at least initiated. By 1820 there were over 700,000 persons recorded in the areas settled after the Revolution, and the vacant lands had been much further reduced. Immigration and growth would continue in all of these districts for several more decades, but after 1820 it was very largely a matter of filling in the gross areal pattern already established.

Sources

"Settlers are continually pouring in from the Connecticut hive," reported Elkanah Watson while visiting the Utica area in 1788. Returning two years later, he found the influx "wonderfully increased," and wrote of the emigrants "swarming into these fertile regions in shoals, like the ancient Israelites, seeking the land of promise." [Durant, 366 ff.] The metaphors were apt and the reference to source indicative. This was the great era of the "Yankee invasion," an enormous outpouring from the East which prompted people then and for decades after to speak of Upstate New York as a "colony" of New England. Connecticut was a hive indeed, with surplus from single townships spreading over whole

---

[1] Citations in this chapter refer to Selected References on pp. 195–96.

districts of the New York frontier; Massachusetts sent out thousands not only to her own pre-emption lands but into every section; Rhode Islanders came by the shipload up the Hudson. Vermonters spread so rapidly and dominantly over the country beyond Lake Champlain that it was colloquially referred to as "New Vermont."

The reasons for this great movement were not complicated. In 1790 the New England states contained just under a million people. It was a population not only growing rapidly, as all American populations were, but rapidly outgrowing the available land. Only the most remote uplands and Maine remained as local Yankee frontiers; New York was closer for many and for all it offered better land. Pressures had been built up during the Revolution, and even before the war was over the attractions of this western country were becoming known. The burst of land speculation which immediately followed, much of it fostered by New England promoters, found an immediate response, and the Yankee social system, with its tight community bonds and close kinship networks, prompted an unusual degree of group migration and sustained such movement by means of the greatest lure of all: the enthusiastic reports from family and friends of their actual success on the frontiers.

New England was clearly the predominant source but certainly not the only one. Such areas as the New Military Tract, insofar as it was settled by former soldiers, drew of course from the older New York regions. Dutch and German settlers from the Hudson and Mohawk could be found in the frontier districts, but they were the dominant ingredient only here and there along the margins of their older areas—in the Catskills, adjacent to the Schoharie, broadening out of the Mohawk Valley—and even in these areas they had to compete with the swarming Yankees.

There were also other American source regions, principally Pennsylvania and New Jersey, from which colonists moved up the Susquehanna and Delaware into the southern border country and, here and there, penetrated well north into the Finger Lakes district. But even part of this movement was Yankee in character, with new settlers from Connecticut and eastern Long Island, or perhaps New Jerseyites but a generation removed from their New England roots.

Direct immigration from Europe was very small in total, but in a few localities it was dominant because of group migrations. Probably the most notable during these early years was the Welsh colony in the Remsen district north of Utica, which had enough cohesion to retain its native language in churches and newspapers for decades. A small group of Highland Scots at Caledonia in the Genesee was one of the very few in the more distant frontier. Several schemes to settle refugees from the French Revolution were very largely failures.

The Rev. John Taylor, traveling in 1802, found Utica "filled with a great quantity of people of all nations and religions," and viewed the Mohawk region such "a perfect Babel" that the speech "even of New England people is injured by their being intermingled with the Dutch, Irish and Scotch." [O'Callaghan, *Documentary History*, III, 685–7] But he was reporting from the only real corridor of mixture, and in general, in this state of colonization, frontier New York was considerably less cosmopolitan than the Hudson Valley; and Utica, the momentary "metropolis" of the West, had nothing like the variety of New York City. That kind of variety awaited the introduction of major influxes from abroad—awaited, in other words, the canal era and its quickened economic pace, which would bring in the Irish and the Germans as vanguards of the great migrations that would take place over the next century. Meanwhile, frontier New York was largely an extension of Yankeeland, and even the metropolis, the Hudson Valley, and Dutch Albany of older New York had become permanently altered by the same inundation. Furthermore, it was a Yankee "invasion" or "capture" or "conquest" in a sense difficult to appreciate today. Such military terms had a sharp meaning—triumphant to the one, bitter to the other—to peoples with fresh memories of the many and often violent boundary disputes and with strong feelings for the cultural differences between Yankees and Yorkers. Not until they had lived among one another for many years and had come to share a common distrust of new strangers would these antagonisms fade away. [See Fox]

Patterns

The influx of so many people in so short a time and their rapid spread into every agriculturally suitable district of the frontier may suggest an inundation, a sudden broad flood rolling uniformly out across the land. Such a metaphor is not wholly inappropriate to the general event, but when this swirling tide is examined in any detail patterns do appear and it is revealed to be not a single wave but a number of flows greatly differing in volume, in some places well channeled, in others branching and spreading, and at times converging and intermingling.

The main stream flowed through the Mohawk Valley, a funneling which in this instance was more the result of its accessibility to overlanders from New England than its connection with the navigable Hudson, although a good many emigrants from the coastal regions did sail upriver. The greatest volume passed through Albany, but there were also important lesser streams from southern Vermont via Schenectady, and from northern Connecticut via Catskill on the Hudson. Pouring through the narrow constriction between the Appalachian Upland on the south and the Adirondacks on the north, this main flow fanned out immediately beyond into numerous distributaries: northwesterly into the Black River Valley or down the Oswego River, southwesterly into the upper Susquehanna and Chenango districts. The main westward stream continued along the margins of the Appalachian Upland, touching the northern ends of the Finger Lakes, but with a steady, widespread trickling off into localities of special attraction. For a time this west-flowing tide stopped abruptly just beyond the Genesee at the border of the still closed Holland Land Company's holdings and turned to spread out north and south within that valley. With the opening of the Holland area in 1801 this whole pattern of flows was extended over the remaining western portion of the state.

The surge through the Mohawk was so great that it spread in some degree over the whole of western New York, even to the southern-most boundary. There it encountered another inflow, up from Pennsylvania channeled along the Susquehanna, or more directly north from the Wyoming Valley. As this northward movement reached New York, it turned and spread strongly east and west along the valleys from Binghamton to the Canisteo. From this lateral flow numerous but much feebler trickles reached well into the Finger Lakes district, making the latter a broad zone of intermingling of colonists with widely varying points of origin.

In the far north Vermonters migrated along another passageway around the northern edge of the Adirondacks, and as they moved on beyond the Raquette River they intermingled with their fellow Yankees from southern New England who were spreading northward from the Black River corridor on the southwest side of the Adirondacks.

The spread-of-settlement map (Fig. 43), though generalized, reveals further complexities in the actual geography of colonization. It gives some hint of the movement along certain corridors, but a more striking feature is the discontinuities—the appearance during each of the time periods of numerous clusters beyond the margins of the main block of settled country. A more intimate picture would only serve further to emphasize that, contrary to its usual general depiction, in detail the "westward movement" in America was never a "wave" rolling along a broad front but was rather a highly selective, uneven, fragmented pattern of advance. There were many factors involved. Certainly terrain was an important consideration, and most of the valleys were selected early and the highest uplands left until last. Yet Yankee migrants were by no means contemptuous of ground that would later be regarded as thin-soiled, stony, and of low farming quality. Further, much of the richest level ground was swampy and malarial in its pristine state and therefore avoided at first. The great variety of tracts, with their different terms of sale, facilities of improvement, degree of advertising, and dates of opening introduced further variations. Financial convenience, legal availability, and notoriety often outweighed actual physical accessibility and inherent quality in the sequence of settlement.

Certainly the most famous of the many districts being advertised during the early years of expansion was the Genesee Country. It is revealing of the interests and attitudes of the time to study the following description of the "peculiar advantages" emphasized in an account of the 1790's:

1. "The uncommon excellence and fertility of the soil.
2. "The superior quality of the timber, and the advantages of easy cultivation, in consequence of being generally free from underwood.
3. "The abundance of grass for cattle.
4. "The vast quantities of sugar-maple tree.
5. "The great variety of other fine timber, such as oak, hickory, black walnut, chestnuts, ash of different kinds, elm, butternut, basswood, poplar, pines, and also thorn trees of a prodigious size.
6. "The variety of fruit-trees, and also smaller fruits.
7. "The vast variety of wild animals and game.
8. "The great variety of birds for game.
9. "The uncommon abundance of very fine fish.
10. "The excellence of the climate in that region ... which is less severe in winter, and not so warm in summer, as the same latitudes nearer the sea.—The total exemption from all periodical disorders, particularly the fever and ague.
11. "The vast advantages derived from the navigable lakes, rivers, and creeks ... affording communication from the northern parts ... to Quebec, or ... to Albany; ... from ... the southern part ... to Northumberland, ... Maryland and Virginia ... and (with a portage of 12 miles) even to Philadelphia.
12. "But above all, the uncommon benefits these lands derive from the vicinity to the thick settled countries in New York and New England ... on the one hand, and Northumberland country in Pennsylvania on the other, from all which ... there must be an overflow of emigrants every year, until these lands are fully settled." [O'Callaghan, *Documentary History*, II, 1111–13]

As may be expected, there was a glow of optimism, some degree of exaggeration, and even a bit of untruth (perhaps not conscious) in this account. For example, the early settlers in the Genesee probably suffered more severely from "fever and ague" than those in any neighboring district, and the hope of great profits from tapping wild sugar maple would prove a chronic frustration. Yet there was much truth in the account, and, moreover, the whole list of advantages was actually applicable in

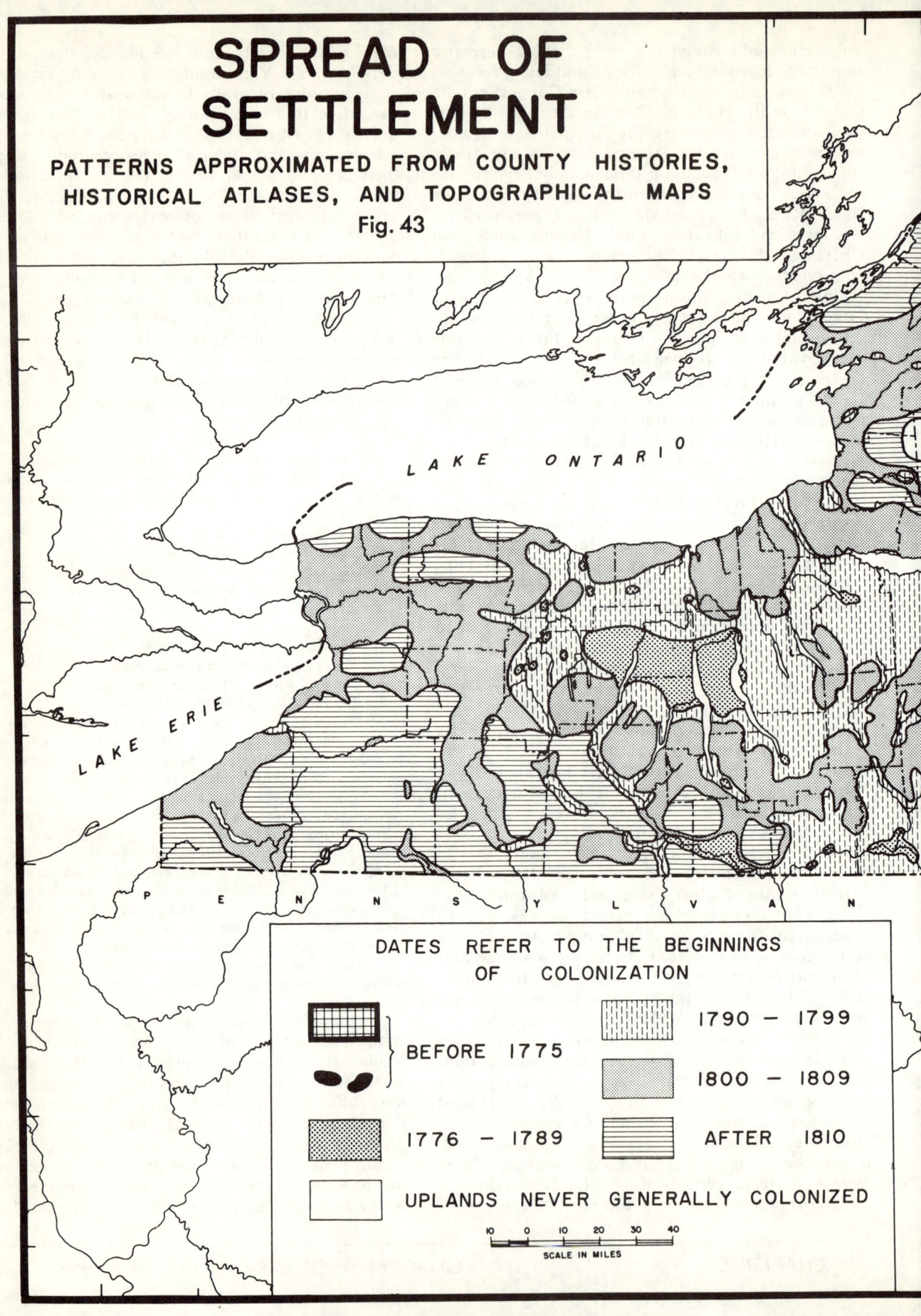

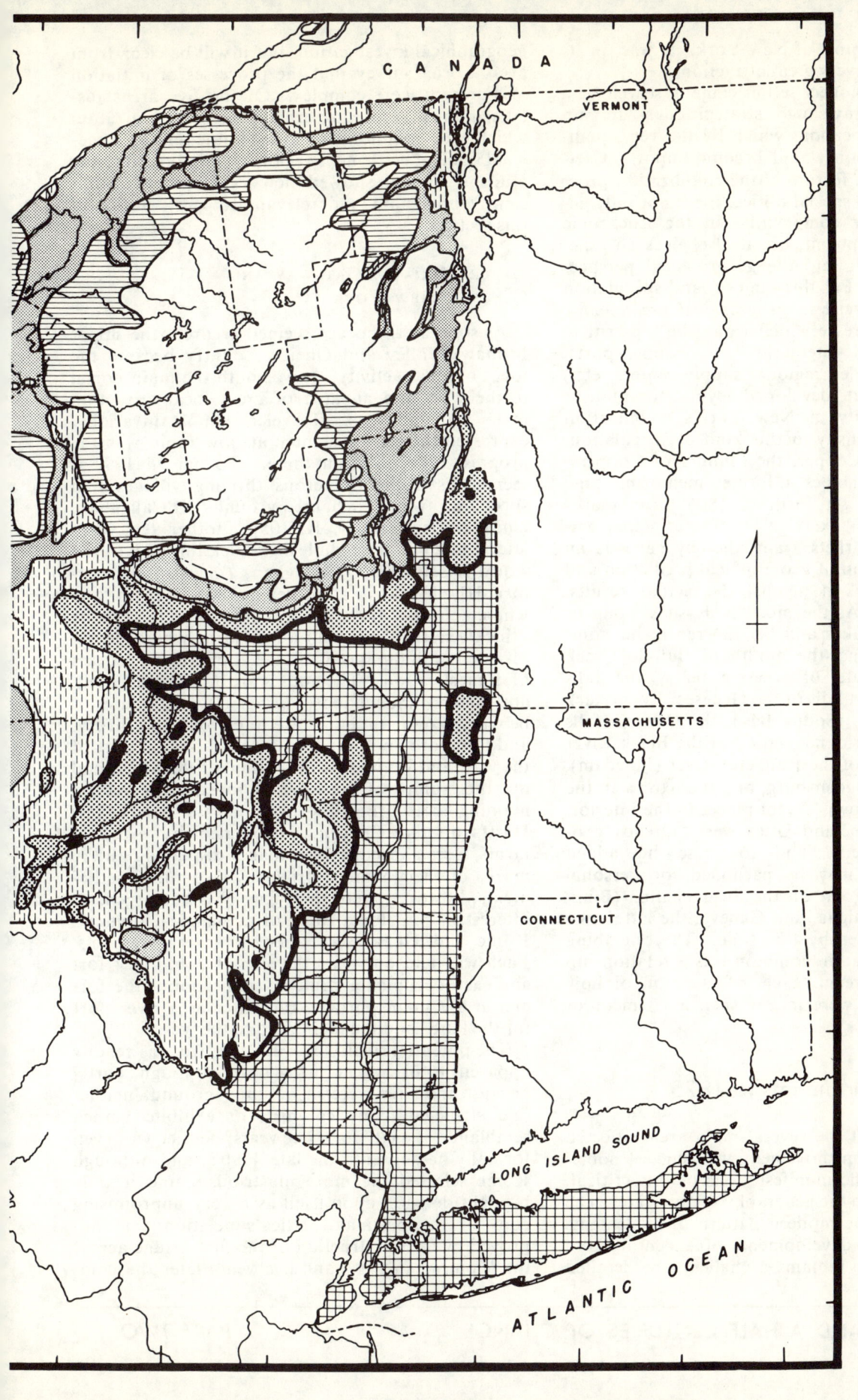

a general way to most of New York. It was, in its time, a frontier of very great attractions.

But not every pioneer settler came in search of a farm. A good many sought strategic sites—to gain control of those locations which by the very qualities of their position would become important settlements. This is a feature of the colonization process which deserves special notice, for it not only accounts for further complexities in the isochronic patterns of development, but it also gives an especially good insight into the geographical perspectives of the time. For this kind of land speculation was really an exercise in *geographical prediction*—the attempt to foresee which would be the critical points of "spatial interaction" (junctions, ports, manufacturing nodes, regional supply centers, etc.) once the whole state developed beyond the pioneer stage. The rapidity of New York's colonization gave a special intensity to this kind of speculation.

Judge William Cooper, the founder of Cooperstown and one of the best-informed men of his time on such matters, set forth in 1806 those places which he thought "likely to become principal emporia or great markets," and thereby gave us an example of this kind of geographical prediction and a chance to check it against the actual results. [Cooper, 14–18] As the most likely sites along or near the Great Lakes and St. Lawrence, he nominated the following: the mouth of Buffalo Creek (Buffalo), the straits of Niagara below the falls (Lewiston), the first falls of the Genesee (Rochester), the Greater Sodus (Sodus Bay), the Oswego falls (Fulton), the heads of navigation of the Black River (Brownville), and of the Raquette River (Potsdam), Sackets Harbor, Ogdensburg, and the oxbow in the Oswegatchie (Oxbow). As for places in the interior, he noted that Rome and Utica were "already considerably advanced," and to these he added Cooperstown (he may be pardoned for personal bias), Chenango Point on the Susquehanna (Binghamton), Canandaigua, and Geneva, the latter two also well established by that time. The one thing all of these sites had in common was a relationship with a lake or stream, a clear indication of how critical waterways were in any such assessment of that time.

Regional Developments to 1825

A closer look at the several main areas affected by this colonization movement will suggest something of the specific manifestations of this critical, formative period in the geography of the state. The degree to which the modern pattern of settlements still mirrors these developments of a century and a half or more ago remains a challenge to detailed geographical investigation, but it will be clear from even a brief survey that the processes of initiation and growth were complex. Of the five areas discussed, the first four are broad zones of colonization which were commonly regarded at the time as somewhat distinct settlement regions, while the fifth, the Hudson Valley, showed new developments sufficiently important and relevant to merit notice in this section.

Upper Mohawk Gateway and Central New York

As soon as the peace seemed secured, the upper Mohawk Valley and Oneida Country became the focus of great activity. It was both the main portal to the west and an area of great local attraction in its soil and sites. Travelers almost invariably expressed their astonishment at how rapid was the progress of civilization there. To one in 1792 it seemed as if the numerous thriving villages had sprung up from "enchanted ground" [O'Callaghan, *Documentary History*, II, 105]; another ten years later thought it "incredible how thick this part of the world is settled" [O'Callaghan, *Documentary History*, III, 689]; while a journey in 1805 revealed "the whole country" from Utica to Westmoreland to be "thickly settled" but with a landscape which clearly revealed both the speed and the youth of the process of domestication: "The houses are mostly well built, and many of them handsome; very few log houses to be seen. Young orchards are numerous and thrifty, and Lombardy poplars line the road a great part of the way; and yet we saw not a single field which had not the stumps of the original forest trees yet remaining in it" [Bigelow, 21]. Whitesboro, New Hartford, and Clinton, surrounded by excellent farms, were soon thriving villages displaying all the marks of their Yankee origins. Two other settlements, however, were the focus of greater general interest, if not, for a time, of actual development. Rome and Utica were alike in several respects. Each was laid out near the site of a military fort and each occupied a strategic traffic point, the former at the main portage, the latter as a river port for the lands and the main road to the west.

The potential rivalry between the two was readily apparent to observers of the time, although, partly because the sites were in swampy ground, neither one attracted many colonists or exhibited much semblance of a city for some years; neither was even formally named until the late 1790's, and although Rome selected the more illustrious name it only brought derision upon itself as a very unpromising copy of the original. Utica's position was improved by the construction of the first bridge across the Mohawk in 1792, and five years later the completion of the shallow canal across the portage westward from the Mohawk enhanced the traffic position of Rome. It appears that general opinion favored Rome as having "the most natural advantages" [Cooper, 18], but after 1800 it was increasingly clear that Utica was forging ahead in actual development. Rome was handicapped by the fact that the proprietor sought to lease rather than sell his lands outright. But there were other factors. The main trafficway, the Genesee Road, led west from the river landing at Utica, and in 1800 it was chartered as the Seneca Turnpike and improved as the main trunk line to the frontier. Furthermore, the immediate vicinity of Utica was better suited to farming and the industry of the times. Sauquoit and Oriskany creeks were soon lined with mills, and flour, lumber, and textiles became prominent products. In 1809 the first bank west of Albany was established at Utica, and in the next year the Black River Turnpike began to channel more firmly the traffic from the north. In 1813 Spafford described Utica as "the commercial capital of the great Western District, and the central point of all the great avenues of communications." With a population of 1,700, Utica was not much of a metropolis, but insofar as the "Western District" had a capital Utica was it and loomed large compared with the but fifty houses in Rome at that time.

Directly west of these gateways the large Oneida Indian reservation precluded colonization until after 1795, when it was greatly reduced in size. Thus, the hill land to the south—much of which was encompassed within the "Chenango Twenty Towns" surveyed in 1789 and marketed soon after—was settled earlier. These towns were settled almost entirely by New Englanders, and as there was little planning or special promotion and the country itself was relatively uniform in character, dozens of little villages grew up, none of which established any early pre-eminence over its neighbors. Sherburne, at the intersection of two turnpikes, and Norwich, with its handsome courthouse, became the largest within a decade or so, but neither was more than a modest village.

Just west of this block of surveys, however, was a prominent settlement which was energetically promoted. The Holland Land Company purchased a small tract in that area, and in 1793 Lincklaen, the resident agent, advertised that he had laid out a new settlement named Cazenovia (after the chief agent) at the outlet of a small, beautiful lake, and was engaged in "building mills, erecting a well assorted store, potash works, and opening different roads to Whitestown, the Salt-Springs, the Chenango river and Catskill." [Evans, 38] As with most such "hotbed" promotions, the results did not equal the hopes, yet Cazenovia quickly became a thriving village, soon added several more mills, and for thirty years was one of the most prominent settlements of central New York.

The New Military Tract, the next block of lands farther west, was one of the largest and most famous of the several districts. It was not long before some of its notoriety was due as much to its nomenclature as to its qualities, and many persons would no doubt have agreed with Timothy Dwight that there was "something singular, and I think ludicrous" in the dense grid of classical names imposed upon the townships. [Mau, 142] The most famous locale, however, was soon known by a name which was classical but thoroughly appropriate: Salina. The salt springs along Onondaga Lake were a focus of activity from the earliest days of colonization. In 1797 the state, in conjunction with Indian treaties, acquired title to the production area, fixed a duty on each bushel of salt produced, and assisted in organizing the industry. By 1804 about 100,000 bushels were being produced annually and marketed throughout central and western New York and adjacent Canada. It was swampy country notorious for summer fevers, and for some years settlement developed only enough to serve the barest needs of the workers. In time, as the industry rapidly expanded, several villages grew up near the various salt works: Salina, Liverpool, and Geddes. Another village, known by various names, such as South Salina, Milan, and Corinth, grew up a short distance from the salt works at a point near where a north branch of the Genesee Road crossed Onondaga Creek. As nearly all of the salt was shipped by water, and as the whole locale was unhealthful, this settlement grew but slowly. Manlius, a few miles east, at the junction of the Seneca and Cherry Valley turnpikes, was for a time a more flourishing trade center. But as overland traffic and the prospects for a trans-state canal increased, South Salina was resurveyed into more ample proportions, rechristened "Syracuse," and made ready for a more promising future. It was also assured of a more healthful one by the lowering of the outlet and level of Onondaga Lake to drain the swamps in 1822.

Colonists spread over the eastern Finger Lakes Country in the 1790's, and villages at or near the ends of the larger lakes, such as Skaneateles, Auburn, and Ithaca, soon became small trade and milling centers. As settlers filled in southward they crossed into Susquehanna drainage, and river landings where arks of farm produce could be launched for Pennsylvania or Baltimore became focal points of trade. Homer and Port Watson (near Cortland) on the Tioughnioga made up the principal settlement cluster on the upper tributaries, while Newtown (Elmira), Owego, and Chenango Point (Binghamton), at important valley junctions, were the main shipping centers along the southern border of the state. A narrow strip of that Southern Tier had been colonized somewhat earlier than lands to the north, partly by the strong surge of Yankees who came down the upper valley of the main Susquehanna, and partly by migrants who came upriver from Pennsylvania and New Jersey. However, except for exports, the area was so isolated for so long that even the leading villages lagged behind the growth of those along the roads, rivers, and lakes farther north. By 1820 it was certain that the Erie Canal project would greatly enhance the progress of the trunk-line strip across the north of this central portion of the state, and, by means of improved waterway connections, quicken the development of the Finger Lakes area. But, unless further links could be made on south to the Susquehanna, the retardation of the Southern Tier would simply become magnified.

THE GENESEE COUNTRY

West of the New Military Tract lay the large blocks of the Pulteney Purchase and the Morris Reserve. The area encompassed by these two tracts was commonly called the Genesee Country, though only a portion of it was actually part of the Genesee River basin. It quickly became the most famous sector of the New York frontier, in part because of lavish advertisement but also because of its actual qualities and the successes of its early farmers. The Genesee Flats, a level, nearly treeless, richly grassed strip of alluvium two to four miles wide and extending north some distance from around present Geneseo, was sufficiently unusual to attract much attention, although there was at first some disagreement among woodland pioneers as to the desirability of such land. But there was little doubt about the high quality of the broad zone of rolling hills between these flats and Seneca Lake. The rich Indian fields of that country had excited the admiration of soldiers, and shortly after the Revolution, and well before the big land-marketing campaigns were launched, Yankee settlers were moving in. Farther south the greater local relief—higher crests, steeper slopes, often thinner soils, and narrower valleys—proved much less attractive, and despite strenuous efforts to induce sales and settlement, developments long lagged well behind those to the north. The lands just south of Lake Ontario, often swampy or gravelly, were generally avoided at first, too.

Two of the most flourishing communities in the colonization era, Geneva and Canandaigua, were remarkably alike in origin, site, situation, and scale of development. Each was next to what had been a large Indian village in the midst of excellent farmlands, each was near the foot of one of the Finger Lakes and on the Genesee Road. Each also received the benefit of much publicity and subsidized facilities as part of land promotion efforts.

It is perhaps surprising that no comparable settlement developed early in the Genesee Valley proper. Williamsburg, founded in 1792 by Williamson, the chief Pulteney land agent, seemed to have an advantageous site at the mouth of Canaseraga Creek. But Williamson soon shifted his headquarters and efforts elsewhere, and instead of fulfilling an early traveler's prediction that it would "in all probability be a place of much trade" [O'Callaghan, *Documentary History*, II, 1107], Williamsburg was virtually abandoned within ten years. Geneseo (long known as Big Tree), a base of the Wadsworths, important Genesee landowners and agents, and Avon (formerly Hartford), where the main road crossed the flats, were other early villages with promising prospects, but neither enjoyed much growth.

In the south, however, Bath quite rivaled the two Finger Lake communities of Geneva and Canandaigua in fame and attention, if not in actual growth. It was wholly a creation of Williamson and was for ten years his headquarters. He selected the site, laid out a handsome design, named it after the English home of his employer, and began it with a set of elegant buildings. In Williamson's eyes, Bath was more than just an attractive locality; it was a site strategic within the framework of his development plans for the whole region. He had first entered the Genesee Country from Pennsylvania, and he was ever after oriented in that direction. His first project was to open a road from Williamsport, Pennsylvania, northward via the Tioga and Cohocton to Williamsburg, and he made strenuous efforts to attract emigrants through this portal. Soon he had Bath connected via a road along Keuka Lake with Geneva and the main Genesee Road. Williamson thus saw his favorite village as the junction of two great routeways; its development never equaled his hopes because these entryways were never of equal importance, and the Steuben County area proved far less attractive to colonists than did that of Ontario and Livingston counties.

In part Williamson's emphasis upon the southern district rested upon his belief that the Susquehanna provided the best outlet and Baltimore the best market for Genesee produce. Soon there was considerable traffic in that direction, but it was limited to the brief high-water season and was almost wholly an export movement. Both features inhibited the rise of major trade centers at strategic river sites. In the north the agent viewed Sodus Bay as the logical port and subsidized there a wharf, sawmill, flour mill, distillery, and hotel, but there, too, commerce was too small and intermittent to sustain any major growth. A traveler of 1800 found Genesee Valley settlers using Irondequoit and Braddock bays (now Rochester) as lake ports, neither of which was well suited, though he found the latter "a better situation and a more flourishing settlement." [Maude, 108–9] At the nearby falls of the Genesee he found nothing but one very poor gristmill. Three years later Williamson sold the site to Colonel Rochester, from Maryland, but it was too far north of the best farming district to be put to much use for some years.

In 1800 this entire tract had a population of about 15,000, of which probably more than two-thirds were settled within fifteen miles of the Genesee Road west from Geneva. Ten years later the total was about 50,000 and the proportionate distribution not much changed. In another decade the total passed 100,000 and much of the area was well beyond the raw pioneer stage. Although colonization had spread thickly and well beyond the early corridors, the early centers maintained their prominence. Spafford, in 1813, noted the "great amount of business" at Canandaigua and was sure it would "be the Metropolis of the Western Counties." Darby, a widely traveled Englishman, declared it to be "by far the most richly built town of its extent" he had ever seen and doubted "whether a more desirable village exists in the United States." [Darby, 132–33, 213] A local census at that time returned a total of 1,788 persons. Geneva had slightly fewer, and, although Bath was considerably smaller, it was still pre-eminent in the more sparsely settled south. The one significant new settlement at this time was at the Genesee Falls near Lake Ontario. Rochester had begun to resemble a real village only a few years before but had enjoyed a spurt of growth. In 1820 it numbered 1,502 persons, had flour, paper, and cotton mills, and was served by an official port at Charlotte on the lower Genesee. This whole development was a mark not only of natural advantages and local enterprise but also of the fact that the lower Genesee Country was becoming well settled, had a significant surplus product, and could begin to turn its attention to a new level of commerce and manufacture.

THE HOLLAND COMPANY LANDS IN WESTERN NEW YORK

Beyond the Genesee lay the main tract of the Holland Land Company, a huge block of more than two million acres, most of which was gradually sold in small units to intending settlers. Joseph Ellicott was resident agent from before the opening of sales in 1801 until 1821, and, despite the fact that his basic purpose was simply to sell land and his employers adopted a very conservative policy toward the subsidy of improvements, he exerted considerable influence upon certain basic patterns and upon the direction and rate of colonization. It would be interesting to determine how much of the basic geographic design of western New York still bears the mark of his efforts.

Acting upon the principle that "without a Survey no rational nor methodical plan of settlement for such an extended and valuable body of land can be devised," the company charged Ellicott with the task of having all of their land "accurately surveyed and laid into towns of six miles square, with a map of each town accompanied with field books descriptive of the land, waters, mill seats, Plains, Valleys, mines, minerals, etc." [Ellicott, I, 22] Such instructions were very much like those being given at this time by the official geographer of the United States to federal surveyors in the Ohio Country. Unlike his governmental counterpart, however, Ellicott had authority to modify the scheme to fit local conditions. Thus, towns were sometimes subdivided so as "to make each Lot...equally convenient to water," and river-fronting lots were often made narrower and longer so as to distribute the rich alluvial land among more settlers and to avoid leaving the company with the backlands, "which if disconnected with this Intervale [valley bottom] Land and lying destitute of Water might remain a long Time unsold." [Ellicott, I, 177]

Ellicott was constantly faced with making what were essentially geographic decisions. In order to set the sale prices and to carry out his orders to reserve some choice areas which were to be held for later sale in hopes of higher profits, he had to have some notion of the natural qualities of the several kinds of land in each of the rather widely varied parts of his domain. He also had to make judgments as to the desirability of various locations, which necessitated some idea of the probable pattern of routeways, ports, mill sites, and trade centers. Although a great conserver of his employers' money, he insisted that "to insure the speedy Sale and Settlement of A Tract of Country" nothing was more important than the opening of roads "in various Directions; regard being had to their leading in such a way as to promise to become leading Roads." [Ellicott, I, 330] Over the years he laid out and subsidized the improvement of a considerable portion of the present network. At first emphasis was given to emigrant entryways and export outlets—extensions of the Genesee Road (N.Y. Route 5) to Lake Erie, roads to Lewiston, to Lake Ontario at Oak Orchard Creek (Point Breeze), and to Olean on the Allegheny. As the land along the Genesee Road began to fill up, he opened another (Route Alternate 20), parallel but some distance south, in order to lead emigrants through another tier of townships. In time numerous others were added.

The Holland Company did not attempt to found and promote any great number of villages, but two were given special attention. Ellicott selected the location of his own headquarters and cut down the first tree to clear the ground for Batavia. He did much to help develop it into a community, subsidizing a sawmill, a gristmill, the first store, an inn, and, later, the county courthouse—all in the expectation that such facilities and aura of promise would be "seen by every Traveller" and would give "the Country a Reputation, and thereby induce Emigrants to turn their Attention" to this area as their best prospect. [Ellicott, I, 199] By 1810 he was taking pride in reporting to his superiors that "the village of Batavia is probably as well or better accommodated with Roads...than any other Village in [the] Region." [Ellicott, II, 42] Each spoke in the road pattern radiating from Batavia, which is such a prominent feature of today's map, was first opened by him.

Ellicott was also responsible for the reappearance of "New Amsterdam," this time as the symbol of Dutch initiative in the other end of New York. But, though he selected the site, designed its pattern, gave it its initial promotion, and built its courthouse, he never got the name to stick: in common parlance it was always called Buffalo, after its creek.

Although there were much good land and some excellent townsites in the Holland Company's tract, distance, the "intervening opportunity" of other well-advertised lands, and the slow development of roads and facilities were handicaps. At the end of the first decade, in 1810, there were about 16,000 inhabitants. Another ten years added 60,000 to the total, an acceleration of growth all the more impressive considering the dampening effect of the war with Britain, which resulted in the burning of all of the settlements from Buffalo north along the Niagara. By 1820 roads had been opened through all but the roughest areas, and there were at least a few colonists in nearly every district. The census of that year showed that in terms of modern county areas Genesee County (Batavia) was the most populous, with Erie County (Buffalo) ranking fourth, lagging behind both Wyoming and Chautauqua.

In view of their apparent prospects and active promotion, the growth of communities was slow. In 1810 Batavia and Buffalo were estimated to have about 400 people each. Three years later Spafford described Buffalo as growing rapidly, but noted that the nearby village of Black Rock was thriving, too, and "is deemed a better site for a great trading town than that of Buffalo." Darby found Buffalo in 1818 "composed in great part by one street following the ...road eastwards," and thought its situation neither "grand nor striking" but "extremely advantageous as a commercial depot." [Darby, 155–56] In 1820 the town of Buffalo totaled but 2,095, fewer than Batavia and about the same number as Canandaigua. It was of course ten years younger than the Finger Lakes "metropolis," but still it was a meager development in view of the potential which was widely recognized. For, as Ellicott had noted, it was

> the general Opinion of those who have the most correct Idea of Geography of the Western and Northwestern Part of the United States and Territory belonging to Great Britain, that New-Amsterdam [having thusly christened his child, "Buffalo" must have seemed to him like an unwanted nickname] from its commanding position near the Outlet of Lake Erie will become before many years elapse, in Magnitude among the most populous Villages in this State. [Ellicott, II, 50]

Some improvement of the bar at the mouth of Buffalo Creek was needed, but once accomplished Ellicott was certain that this village would become "the grand Imperium of the Western World."

But Black Rock was still thriving and there were efforts elsewhere seeking to capture at least part of the Erie trade. A group from Albany had recently laid out the village of Dunkirk, built a wharf, and extended a road inland. Farther on, the old landing at Portland still served the portage road to Mayville on Chautauqua Lake, from which the waters flowed to the Allegheny. Mayville, founded by Ellicott as a location for a district land office, was the leading village of this southwestern corner. Olean, too, was one of the oldest and as the head of navigation in the Allegheny was the focus of much traffic. Yet it had very little local trade, for most of its activity was concentrated in two or three months of late spring when water was high enough for navigation, there was virtually no upriver movement, and the local district was too rugged to attract many colonists.

Northern New York

The northern part of the state, the St. Lawrence–Champlain Lowlands extending from the eastern shores of Lake Ontario to Lake Champlain, was colonized by two different streams of migrants. One came directly west from Vermont along the Canadian border, the other came from the south by way of the Mohawk and Black rivers. As a result the area developed from the first within a framework of two districts, different in orientation, each bound to its nearest waterway.

Most of the lands in the Champlain district were patented before the Revolution, but few settlements had been commenced and all had to begin anew after the war. Colonization was fairly rapid as the area was relatively well known and accessible to Vermonters and other Yankees. Agricultural land was limited and farming remained an adjunct to forestry in most localities. The timber resources were rich, the many streams falling steeply from the Adirondacks provided plenty of mill sites, and the lake and the Richelieu River provided a means of shipment to the Canadian market. After 1800 a number of small ironworks were established to work local ores. The war of 1812-14 briefly stimulated these industries, and the initiation of the Champlain Canal soon after promised a reorientation and increase in trade. Plattsburgh, described by Spafford in his gazetteer as "handsomely laid out" and "a place of very considerable business," early emerged as the principal settlement, a position enhanced by the establishment in 1812 of an important Army barracks nearby.

Developments in the broad lowland lying along the northwest shoulder of the Adirondack Upland were complex and retarded in comparison with those in the Champlain district. In the late 1780's the state marked off the "St. Lawrence Ten Towns," a double-tiered block fronting on the St. Lawrence. These and the whole unsurveyed area between the two Salmon rivers (between Pulaski and Malone) were purchased as a speculation by Alexander Macomb, a wealthy fur-trader. Financial pressures forced Macomb quickly to peddle large subdivisions of this huge tract to other speculators, leading to great complexity in the surveys and patents. A considerable number of wealthy New York City merchants and financiers became involved in land development in this sector, many of them took a close personal interest in their holdings, and some of their names, such as Low, Pierrepont, Gouverneur, Clarkson, Ogden, and Waddington, remain prominent on the map. There were also others, such as Constable and Scriba, whose names on the land are today much less bold than were their attempts at colonization.

William Constable, in control of much of the Black River Valley, peddled a large tract, named Castorland after its presumed wealth in beaver, to a group of wealthy Frenchmen fleeing the Revolution in the early 1790's. However, the scheme was a failure; only a few French actually attempted to settle, and the rough realities of life in this remote wilderness were so unlike their idealized visions that almost none stayed. In 1802 the Rev. John Taylor reported little clusters of Yankee settlers from Massachusetts and Connecticut sprinkled along the entire valley.

George Scriba purchased a large tract lying between Oneida Lake and the Tug Hill Upland and bounded by Lake Ontario and the Oswego River on the west. He established Rotterdam (now Constantia) as his main village, Vera Cruz (now Texas) as a port on Lake Ontario, and built mills, stores, inns, a shipyard, and roads. He succeeded in spending a great deal of money but attracted very few settlers. Much of the land was swampy and relatively infertile, and the superior advantages of the Oswego River and port kept the main stream of traffic from passing through Scriba's lands.

In contrast, the lower Black River district had a set of natural advantages that made it an early focus of interest without expensive promotion efforts. It was easily accessible by water, was fronted by an excellent sheltered anchorage within Sackets Harbor, seemed destined to become the focus of routes from the interior, and had a whole series of magnificent water-power sites. Brownville, at the head of navigation, Watertown at the main lower falls, and Carthage at the head of the Long Falls, were early settlements of promise. Sackets Harbor achieved sudden fame in the War of 1812 and remained a naval depot and shipbuilding center afterward. The first American steamship on Lake Ontario was launched here in 1816 and put into service between Lewiston, Oswego, and Ogdensburg.

Just as the campaigns of the Revolution had served to advertise central New York, so the war with Britain gave much-needed publicity to the north. Immigration there, which had been very slow in comparison with that of other major districts, was greatly accelerated. Between 1815 and 1820 several major turnpikes were constructed to give access to the lands lying well back from the St. Lawrence northeast from Watertown. Where these roads crossed important streams at good mill sites, a number of small villages, such as Gouverneur, Canton, Potsdam, and Malone, began to grow. Ogdensburg owed its somewhat earlier development to the efforts of David Parish, who bought the whole village in its infancy and undertook major promotion. Parish built stores, warehouses, and schooners and soon had it established as the leading American port on the St. Lawrence.

The great majority of the settlers over all this northern country were from New England, the two streams—via the Mohawk and via Champlain—meeting and interpenetrating in the general area between Canton and Malone. Throughout this pioneer period the meager commercial life was strongly focused upon Montreal. The dangers of the political frontier retarded immigration during the years of deteriorating relations between the United States and Great Britain, but the outbreak of war soon accelerated developments. By 1820 the military danger was gone, but the commercial dependence upon a foreign outlet remained as an unstable factor. However, canals under construction and proposed were designed to remove that difficulty.

The Hudson Valley

The entire state was growing during these post-Revolution years, the older areas as well as the frontier. The population of the Hudson Valley increased by more than 150,000 between 1790 and 1820 and that of the metropolis itself by over 100,000. By 1820 New York City had achieved much of the kind of pre-eminence it would thereafter maintain: it was now the nation's largest city (123,706 to Philadelphia's 63,802) and most important port, it was certainly the chief financial center and probably the leader in total value of manufactures. Physically it was no longer clustered at the very tip of Manhattan, but was spread across the whole breadth of the southern end, with its wharves along the East River now beginning to be matched by similar developments on the opposite shore in Brooklyn. None of these features, however greater in magnitude, was a change in basic character from earlier times. Nor was the bustle of its harbor different, though it did display a new element that would play an important role in the evolution of the city and port: steam ferries plying the estuary and steamboats offering scheduled service to Albany. Soon scheduled packet service across the Atlantic would do much to clinch New York City's great surge ahead of her rivals.

Upriver all of the lands worth taking had been colonized, the good soils cleared, and the rural population was spread thickly over the countryside. Most of the actual increase in population had been rural, yet the more striking geographic change was the development of the river communities. The scheduled stops of the new steamboat service were a good indication of which were the important centers: Poughkeepsie, Newburgh, Kingston, Hudson, and Albany, with a steamboat connecting to Troy. Of these six cities, three were old, three new; the old were on the west bank, the new on the east. The three on the east were so clearly creations of New England enterprise that it is tempting to visualize them on the cultural map as thrusts of the westward-spreading Yankee movement—a breaching of the Dutch frontier in the hills east of the river and the seizure of strategic footholds on the Hudson itself. And despite the fact that many of these New Englanders actually arrived by sea and the river rather than overland, this view is apt in concept and meaning, for these new river communities were spectacular exhibits of Yankee enterprise and interests.

The city of Hudson was a remarkable creation. It was founded by a group of businessmen, traders, and whalers from Rhode Island and Nantucket who, having suffered heavily from the depredations of British cruisers during the Revolution, sought a more sheltered, secure base. Purchasing land at what had been Claverack Landing, a mere loading point for Dutch farmers, they laid out their new city and initiated its astonishingly rapid development. In three years it had a population of 1,500, numerous shops, and shipbuilding yards, was the home base of twenty-five seagoing vessels and one of the busiest ports on the Hudson. Its prosperity was further enhanced by its overland traffic, as it was the focus of rural roads and on an emigrant thoroughfare from New England with branches leading to the Schoharie and the Susquehanna.

The establishment and growth of Troy was little less extraordinary. Founded by New Englanders four years after Hudson, it was designed to be a Yankee competitor to Dutch Albany. Soon it had become the focus of a network of routes, the center of an impressive array of manufactures, and a thriving port. William Darby in 1818 praised the "elegant and spacious appearance" of its brick buildings and declared that it was "in point of wealth, business, population and extent, ... the third town in the state of New York." [Darby, 37]

Poughkeepsie was not wholly a new creation or quite so entirely a Yankee development. But though its roots go well back in time it was hardly even a village before the Revolution and it was not until about 1800 that it underwent rapid growth, based upon a Yankee influx, the rural prosperity of its hinterland, its growing number of mills, and its flourishing river trade.

Of those on the west side, Kingston was least changed. Darby was impressed by its rather staid "air of domestic comfort, plenty and ease" [Darby, 25], and though it had grown somewhat in size and commerce it was clearly essentially still the Dutch settlement of old. On the other hand, both Newburgh and Albany were bustling centers seeking to make the most out of their positions as ports for the traffic to and from the frontier. Newburgh, incorporated in 1800, took a keen interest in turnpike developments and sought to promote routes leading across to the upper Delaware and into the Finger Lakes as a short cut to the west. Albany, despite the competition of other river towns and of new outposts such as Utica, grew with every year. Not only was it still the great junction of water and land, but virtually every development in the frontiers of western New York as well as much of Vermont in some way enhanced its wealth. Spafford, in 1813, commented on its great trade with places far and near and expressed his doubt "if there be a place on this continent which is daily visited by so many teams." Its population of 12,630 in 1820 made it but a tenth as large as New York City, yet in the nation it was outranked only by the five largest Atlantic ports. It was itself of course an important Atlantic port; Spafford reported that 206 vessels—of which 70 were from states other than New York—paid yearly wharfage fees.

Transportation Development

No feature was more important to the course of development and to the emergent geography of New York during the seven decades following the Revolution than the extension and elaboration of transportation facilities. For the early colonists the most critical need had been for a means of getting bulk produce to a market; because for many years few had much money, the volume of merchandise imported to the interior districts had been small. Later, for villages which were specifically established at what were judged to be sites strategic for trade, it was critical to establish and keep improving access routes to an ever-widening tributary area. For those older and larger cities which had grown as terminals or transshipment centers, it was imperative to keep traffic flowing along established channels; for their rivals it was essential to disrupt or divert such flows. Transportation, therefore, was of critical importance to every settlement, old or new, large or small. All of these concerns were peculiarly geographic because all focused on connections between places, patterns of circulation, and spatial interaction.

During this era a great variety of mediums of transportation were tried and used with some success: river boats, arks, and rafts, the turnpike with its stage and mail coaches and freighting wagons, canals, river and lake steamboats, railroads, plank roads. Each of these had an impact which varied greatly among the districts of the state, for each was unique in its feasibility and efficiency under the different conditions of terrain, seasonality, and type of goods. As a result, each had its own special geographic pattern. The following coverage can be no more than a summary of a very complex topic. An attempt is made to describe the sequence of development of these various mediums, where appropriate to give some idea of the various functional parts of each system, to note something of the effects upon trade centers and route patterns as this sequence was elaborated, and to sketch the over-all geographic patterns of connections and traffic flow for the resultant complex system as it was near the mid-nineteenth century.

NATURAL WATERWAYS AND TURNPIKES

Throughout the colonial period settlement had been so closely associated with the Hudson and Mohawk that transportation was not a topic of great concern. The expansion of settlers in the post-Revolution decades inevitably brought it to prominence, despite the fact that compared with most interior regions of the time the New York frontier was remarkably well situated with respect to natural waterways. Promoters never failed to emphasize this fact as a major attraction, but nonetheless both the settlers and the merchants who sought to serve them soon found sufficient difficulties to make transportation a matter of persistent concern.

The natural drainage of western New York, containing four divergent systems, gave some sort of link to four quite separate markets: (1) east via the Mohawk to Albany and the Hudson, (2) north via Lake Ontario and the St. Lawrence to Montreal, (3) southeast via the Susquehanna to Harrisburg and Baltimore, and (4) southwest via the Allegheny to Pittsburgh (waterways westward via Lake Erie did not at this time lead to any market). Certain special features of this pattern of natural water routes allowed a considerable degree of competition to develop among the first three for the capture of the traffic from a broad zone between the Unadilla and the Genesee. Although the Finger Lakes drained northward, the very short portage (and later canal) at Rome allowed a relatively easy divergence of trade east to the Mohawk, while to the south it was but a short haul to the tributaries of the Susquehanna system which extended over such a broad zone. The fact that the upper waters of the latter were in many cases navigable from within a few miles of their source during the spring high water season served to intensify the competitive export situation. Such navigability was in part the result of the kinds of craft used, of which the "ark" was the most distinctive. In the words of Judge Cooper, the ark was

> neither strictly a boat nor a raft, but partakes of the nature of each; it is of the form of a lozenge.... It is hastily and cheaply constructed, and it is not even water tight, but the bottom is fitted up with light timber, so that it is buoyant enough to keep the grain and other perishable produce from being wet. [Cooper, 12–13]

Such vessels were a creative expedient, quite adequate to convey the bulk products of the frontier—potash, salt, cooperage, flour, beef, pork, hemp—to market and were themselves broken up and sold for lumber at downstream locations. A great amount of logs and rough lumber, too, was rafted down the Susquehanna system.

Although New York City, Montreal, and Baltimore thus competed for the trade of a large area of frontier New York, the relative position of each was dependent upon many factors. Because Ontario's agricultural development was much slower than that of New York, the produce of the latter did not face as much competition in the Montreal market as elsewhere and there was the added advantage of sufficient water for shipping throughout the ice-free season. On the other hand, the local market was rather limited, the winter blockade enforced a long embargo on exports, and all imports from Montreal had to pay a heavy American duty. Thus the traffic was largely one way and other political matters often caused further difficulties. The chief problems with the Baltimore market were the competition from closer agricultural and lumbering districts, the restriction of frontier exports to a few weeks of high water, and the conditions that made upriver movements almost wholly impracticable. As a result of these several factors, Albany remained the principal port for much of the produce from all except the extreme southern sectors of interior New York. It was, of course, a familiar orientation, for most of the settlers had come that way, but more important was the fact that there was river service both up and down the Mohawk through most of the year, and that Albany merchants were aggressive in cultivating the frontier trade. Furthermore, although these natural waterways were of basic importance in the export of bulk goods, the emigrant traffic, imports of merchandise, and (increasingly), exports as well moved by road. For, despite the general physical advantages of water carriage over land, a good road could often serve better than a sluggish or shoaly stream, and Albany interests were quick to see this as a means of enlarging their competitive advantage.

The development of improved roads in western New York was very largely the result of the turnpike movement which flared into sudden prominence at the start of the nineteenth century. The first turnpike was chartered in 1797; by 1811, 135 charters had been issued; and, although the failure of many companies had gradually led to more caution, ten years later 278 companies had been chartered for the construction or improvement and operation of a total of more than 6,000 miles of road, of which about two-thirds had actually been built. Such speculation could only thrive with some evidence of large profits, and the veritable mania of this movement was indicative of the heavy traffic on at least some of these chartered routes. New turnpikes were established well into the 1830's, and some of the old, well-maintained toll roads continued in business well into the railroad era.

It is difficult to give an adequate summation of the turnpike movement because the topic awaits detailed geographical research. Present secondary materials do not provide the basis for a map of all turnpikes or of those in operation at any particular date after the mania got under way. In some cases it is difficult to discover exact routes; in others it is hard to be sure just how much of a proposed route was actually constructed; in most cases it is impossible to get much information on the actual condition of the roadway (an important and highly variable factor), and in nearly all there is no really satisfactory measure of the volume of traffic. Nevertheless, the longer routeways, some of which were very famous, display certain patterns of general geographical significance which can be briefly noted. (See Fig. 44 for locations.)

The first turnpike charter was for a road between Albany and Schenectady, the oldest heavily traveled overland route in the state. Before that pike was completed in 1805 a great many others were under way, with chief attention quite naturally being directed to those routes of greatest existing traffic. A scheduled stage and mail service had been established across the state before 1800, the emigrant traffic was greater every year, and thus the main thoroughfare from Massachusetts through Albany, Utica, and beyond was quickly chartered and improved. Various companies held franchises to different sections. The old Genesee Road from Utica to Geneva and Canandaigua became the Seneca Turnpike and was soon extended through Batavia to Black Rock. This first great western turnpike crossed the Mohawk at Schenectady and followed along the river to Utica (N.Y. Route 5); soon a more direct but hillier road, the Cherry Valley Turnpike, west from Albany to Manlius (U.S. Route 20) was competing for the long-distance traffic.

Perhaps the most striking geographic aspect of the turnpike era was the intense competition among the several Hudson ports for the western trade. Hudson, Kingston, and Newburgh each had an interest in one or more thoroughfares passing around or through the Catskills and was anxious to advertise itself as being on "the shortest route" to the western country. Each also had its connections eastward with the New England turnpike network, the source regions for so much of the overland traffic. To some extent the more southerly routes were not directly competitive with the main roads west from Albany, as the two groups led into somewhat separated sectors, yet there was also some very conscious competition. Newburgh, for instance, helped subsidize the construction of a steamer on Cayuga Lake to link with the fast stage service from Newburgh to Ithaca in order to extend its reach into the Finger Lakes and Genesee

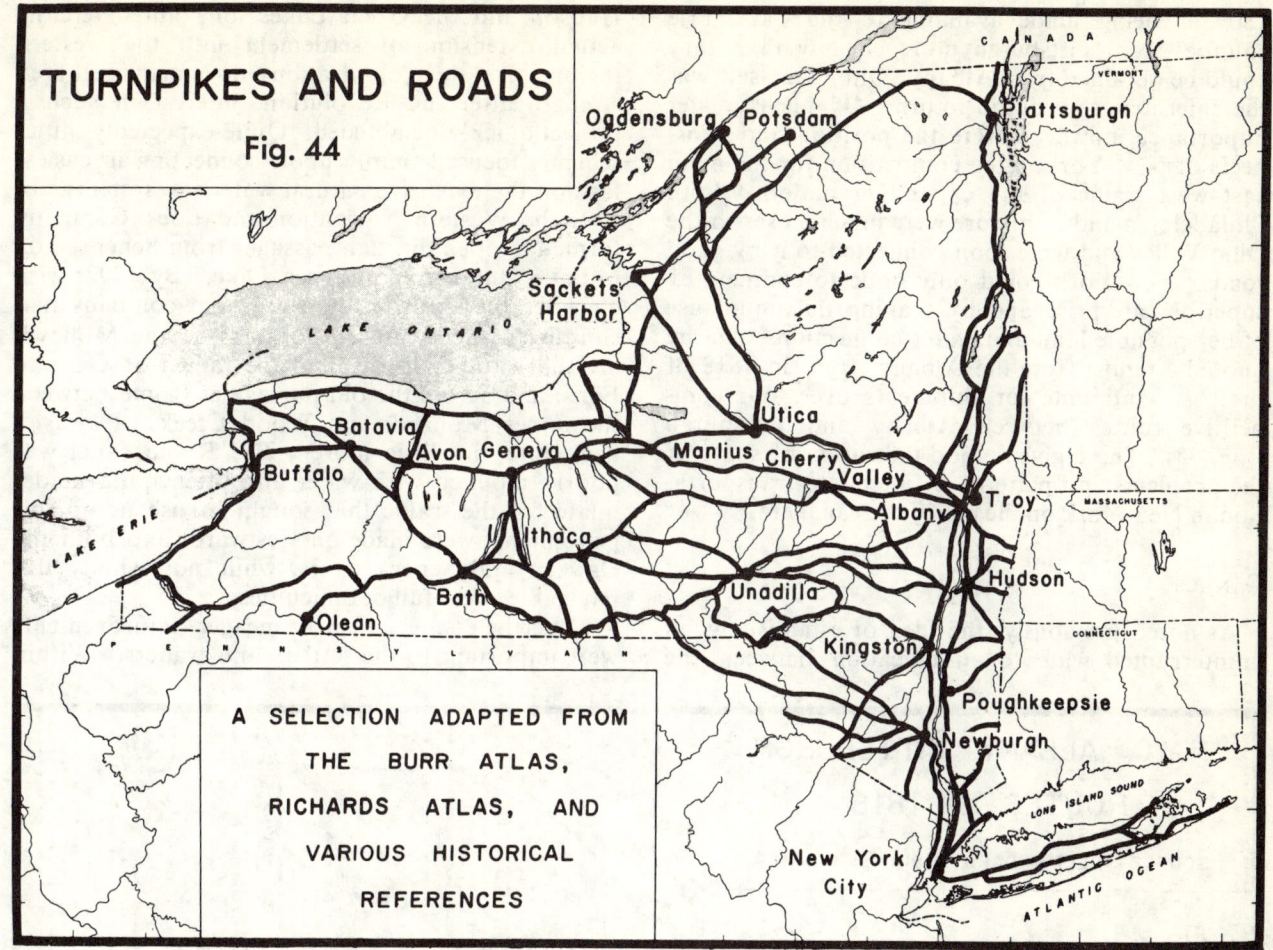

TURNPIKES AND ROADS
Fig. 44

A SELECTION ADAPTED FROM THE BURR ATLAS, RICHARDS ATLAS, AND VARIOUS HISTORICAL REFERENCES

Country. Nor did Albany rest content with her dominance of the Mohawk corridor, for an Albany-Susquehanna turnpike (N.Y. Route 7) reached directly into the southern region, cutting across the routes from rival ports. In general, the pattern was one of keen competition among several aspiring gateway cities.

The inherent geographic quality of the turnpike era, therefore, was its relatively open, competitive character. Not being severely limited as to grade and curvature, roads could be built through hilly and even mountainous country in such a fashion as to serve much of the movement of the times. For the rapidly increasing passenger and mail traffic, distance and speed became more important than level terrain. Furthermore, because the general condition of roads was very poor, a well-built and maintained turnpike might be able to compete very well even if its route were more devious and through rugged country. In this way such natural corridors as the Mohawk Valley lost some of their competitive advantage, and thus Albany was forced to share an increasing proportion of the western traffic with her downriver rivals.

Nevertheless, Albany retained her claim as the principal gateway to the west and the traffic both to and from those regions steadily increased. Although the great bulk of that trade was intrastate, there was also a steadily growing interest in the larger question of interregional trade between the Atlantic Seaboard and the Ohio Country. The development of the Ohio Valley was so rapid and its orientation downriver to New Orleans so divergent that the problem of strengthening the bonds between the Atlantic Seaboard and this burgeoning interior region became a matter of imperative national concern. New York was affected by Ohio developments in two ways. The lesser was the fact that the southwestern corner of New York was part of the Ohio drainage system, and with the rise of Pittsburgh that sector increasingly turned in that direction not only as an export outlet but as a source of imports as well. Such things as glass, castings, bar iron, nails, linens, and liquors were listed in

1809 as being made available to the Lake Erie counties from Pittsburgh more cheaply than they could be obtained from Albany. Onondaga salt was the chief item sent out in return. Of much greater importance, however, were the position and prospects of New York in the competition for the main east–west traffic of the expanding nation. Both Philadelphia and Baltimore were much closer to the Ohio Valley and were soon connected to it by good roads. New York could only hope to compete by superior enterprise and by making maximum use of her possible land and water connections. Darby quoted a report from the Albany *Argus* in 1818 of the costs and time for shipments over four competitive routes between Albany and Pittsburgh (Fig. 45). These give a good indication of some of the problems and patterns of interstate transportation in New York on the eve of the canal era.

CANALS

As noted previously, the idea of establishing an uninterrupted water communication between the Hudson and the Great Lakes long antedated the actual extension of settlement into the western region. As the colonization movement gathered strength after the Revolution, interest in such a project quickly heightened. Quite expectedly initial schemes focused simply upon connecting in easiest fashion the extensive natural waterway systems. In 1792 the Western Navigation and Lock Company formed to open through passages from Schenectady to Lake Ontario and Seneca Lake. By 1797, with considerable aid from the state, that company had completed short canals and locks on the Mohawk around Little Falls, around the rapids at German Flats, and across the old portage at Rome between the upper Mohawk and Wood Creek. It was a beginning, but little more. The Rome canal was poorly supplied with water and often quite inadequate for the traffic that sought to use it; no improvements were made on westward, and both the Oswego and Seneca rivers, while not wholly unnavigable, were full of difficulties.

Nevertheless, even such meager improvements were important to the settlers and traders. Within

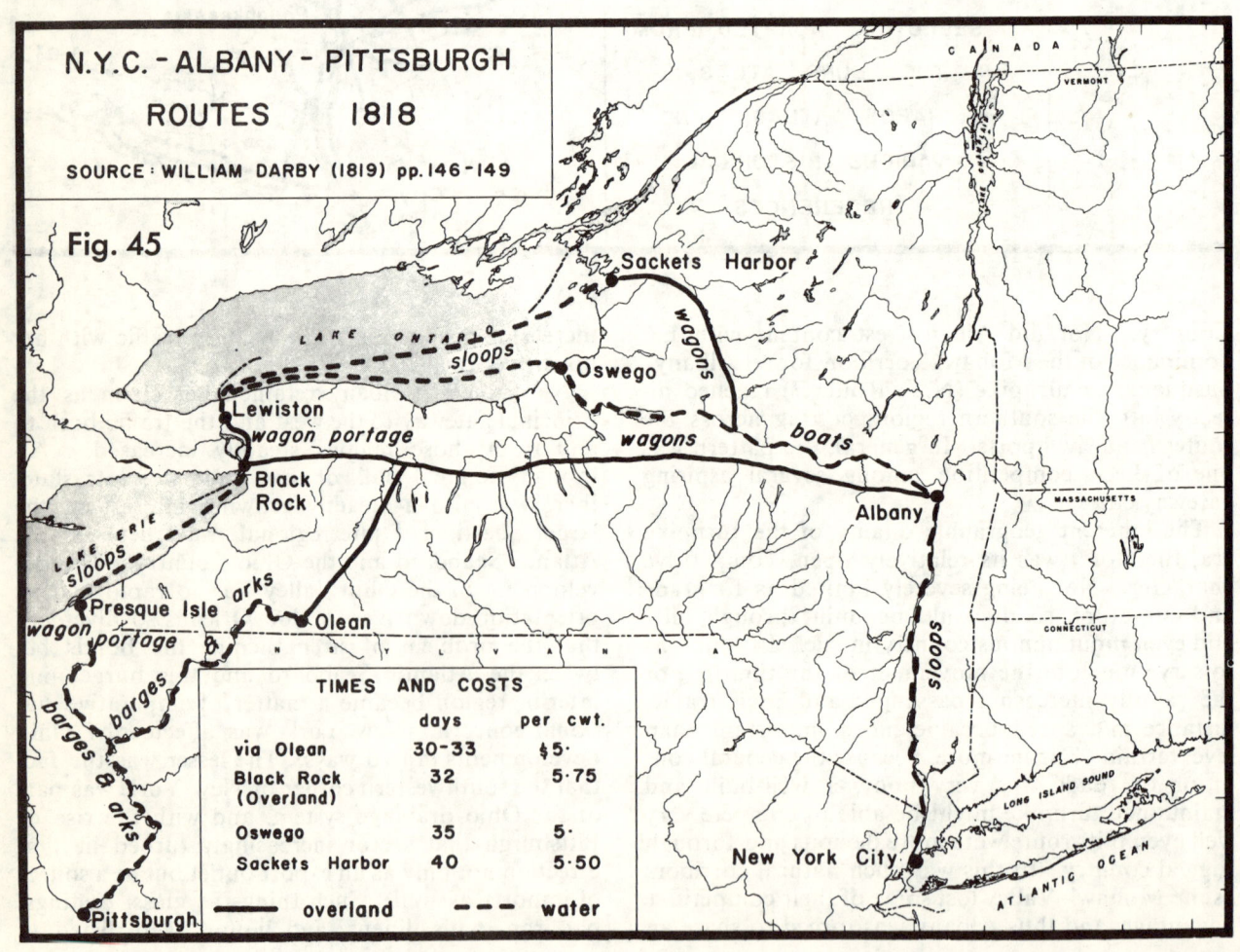

ten years turnpikes had been developed to such an extent as to offer intense competition to this waterway: The result was greatly reduced freight costs and a substantial increase in the volume of freight moved. Still the very nature of these early turnpikes and canals kept the charges so high as to make long-distance shipment of bulk products generally unprofitable. The expense of shipping wheat from the Genesee Country to New York City was often greater than the sale price. Thus the rapid growth of settlement in western New York revived the old Hudson–St. Lawrence rivalry. As Judge Cooper put it at the time (c. 1806): "The trade of this vast country must be divided between Montreal and New York, and the half of it lost to the United States, unless an inland communication can be formed from Lake Erie to the Hudson." [Cooper, 18]

The obvious solution was a radically improved waterway system, and the idea of a canal across the state quickly gathered interest. Within a few years preliminary surveys were run which proved the feasibility of several possible routes. The war with Britain delayed action, but at the same time the high costs and delays encountered in the movement of military supplies to the Ontario frontiers accentuated the need for a greatly improved facility. The interest of the state government was also enhanced by the prospect of greatly enlarging the market for Onondaga salt, on every bushel of which the state collected a fee.

By 1815 the project was being ardently debated, and it was clear that the issue involved a number of geographical questions. One was a clear case of political geography arising from divergent or conflicting regional interests. Farmers on Long Island and in the Hudson Valley not only would receive no benefit but feared they would be harmed by the competition from western grain cheaply produced from virgin soils. Surprising, perhaps, was the opposition of many New York City interests, rooted presumably in the assumptions that the western trade would accrue to them anyway and that the costs were disproportionate to any reasonable expectation of benefits. Final approval had to overcome the unanimous opposition of metropolitan legislators.

A more sharply focused geographical issue was whether the canal ought to be dug clear across the state to Lake Erie, or whether it ought to extend to Oswego on Lake Ontario with another short canal around Niagara Falls to give access to Lake Erie. The latter had much general public favor, for it was shorter and therefore was assumed to be cheaper. However, actual surveys suggested that the Oswego-Niagara scheme would be more expensive because of the terrain of those particular segments. There were other disadvantages to the Lake Ontario route: it would provide only indirect service to the most productive areas of western New York, it would necessitate two extra transshipments on goods moving between Lake Erie and the Hudson, and it still left Montreal in a position to compete with New York City for the export trade. The proponents of the direct route to Erie won the issue and in 1817 construction was authorized. The entire 364 miles of this Erie Canal, all but ten of which were an artificial waterway, were completed and triumphantly celebrated in 1825 (Fig. 46).

While every citizen of the state could appropriately celebrate and enjoy the admiration of the nation for the successful completion of what was for the time an unusually bold and gigantic undertaking, inevitably the actual impact was highly selective, enlarging the prospects of some centers and contracting those of others. It was not merely a case of being on the canal or not. Schenectady was on it, but, having long thrived on the transshipment business, seemed likely to suffer from the fact that the canal now provided a passage around Cohoes Falls and led to huge mooring basins at West Troy and Albany. In the western region such early leading centers as Geneva, Canandaigua, and Batavia were for the first time off the main trafficway and could not help but foresee some dampening of their prospects. On the other hand, Rochester, on the main artery, now found the value of its fine milling sites greatly enhanced by the radical improvement in the facilities for marketing its grain. Basil Hall, visiting in 1827, found its streets "crowded with people, carts, stages, cattle and pigs," and was impressed that

everything in this bustling place appeared to be in motion. The very streets seemed to be starting up of their own accord, ready-made, and looking as fresh as new, as if they had been turned out of the workmen's hands but an hour before, or that a great boxful of new houses had been sent by steam from New York, and tumbled out on the half-cleared land. [Mau, 270–71]

There were warehouses half-completed but half-full of goods and mills full of people at work "while at the top the carpenters were busy nailing on the planks of the roof." It was the greatest "boom town" of its day. There were others of lesser extent, such as Syracuse, Lockport, and Buffalo. The advantage of taking water from Lake Erie at a higher level at Buffalo than from the Niagara River at Black Rock had determined the location of the western terminus and thereby finally decided the chronic rivalry between those two settlements.

The Erie Canal prompted a frantic era of canal proposals and constructions not only in New York but throughout the northern states. The very physical accomplishment silenced a chorus of doubters

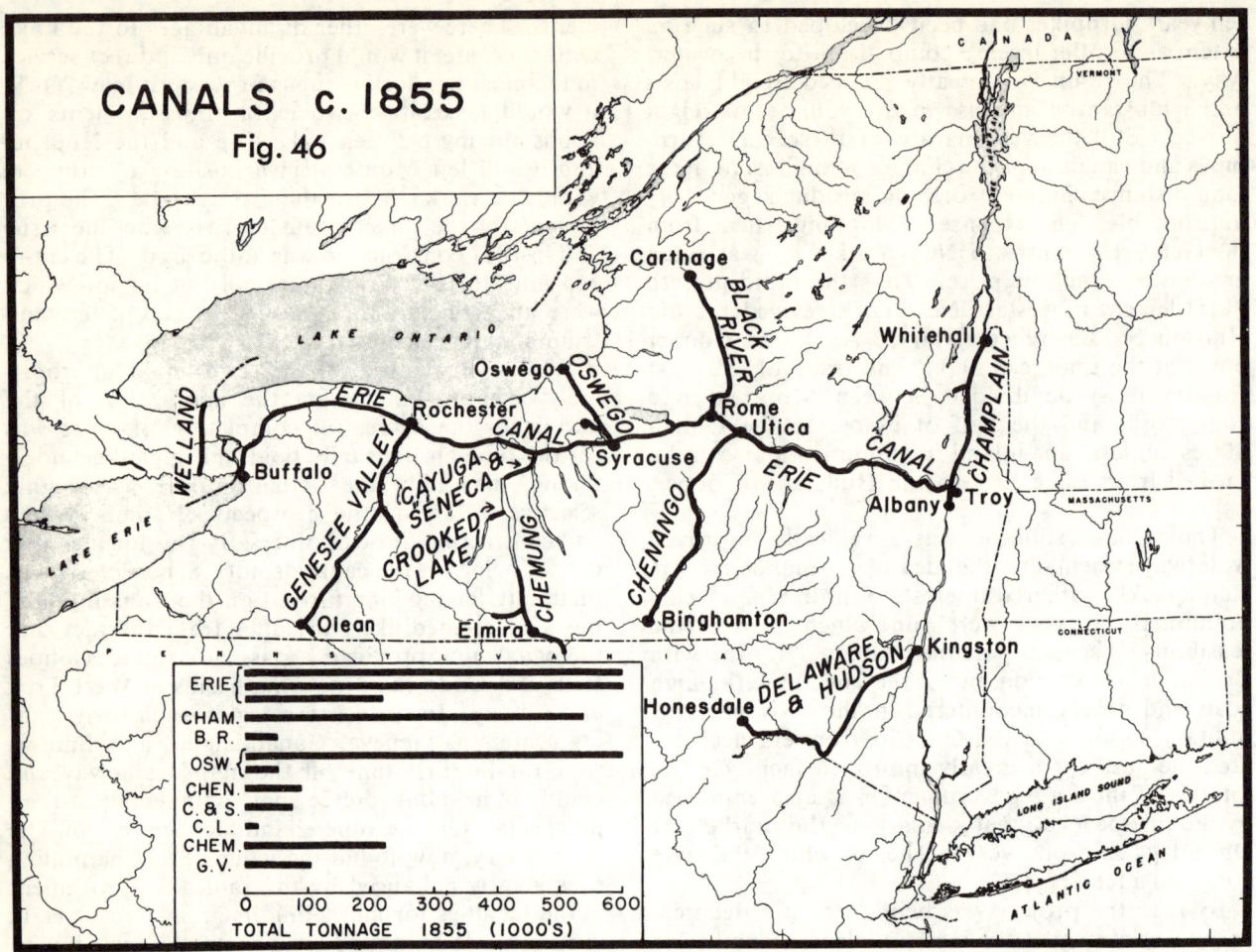

and enlarged the whole public vision of what was possible. Furthermore, it was an immediate and outstanding financial success for many years; the annual tolls exceeded the estimated revenues. Nationally, it was a major triumph of New York City over Philadelphia and Baltimore in the contest for the Ohio Country trade, and it spurred these cities into immediate imitation. Locally its success initiated a flurry of demands for the extension of tributary canals into districts bypassed, and over the next twenty years several were built.

Three of these feeder canals were actually authorized before the Erie was completed. Each also followed along routes already being used in part as natural waterways and thus were likely expansions involving no radical alteration of trade routes. The first of these, the Champlain Canal, was begun concurrently with the Erie as a long-contemplated improvement of the other great natural corridor leading from the Hudson Valley. It was viewed as a means of cheapening transport to the northeast corner of the state and also as a means of diverting the export traffic of the whole Champlain region from Quebec to the Hudson. The Oswego Canal, once removed from the dispute as to the best route for the main trunk line, was also soon authorized, an expression of the now greatly heightened confidence of New York in the competition for trade. Fear of Montreal had now faded, and a through route to Lake Ontario was seen as a means of capturing trade from Canadian shores as well as serving those of New York. A third feeder was the connection from the Erie Canal to Cayuga and Seneca lakes, completed in 1828.

As the traffic and revenues of the Erie and its branches continued to surpass all expectations, in the 1830's attention was directed to more audacious tributary projects, designed to provide waterway links to districts more difficult of access. To the north, as there was much interest in a connection with the Black River Country, numerous surveys were made. The most far-reaching proposal was for a canal from the Erie near Herkimer or Rome to Ogdensburg. Lesser schemes looked for a link with the lower Black River, but in either case there was some rivalry as to whether the canal ought to be built around the east or the west side of Tug Hill. Ultimately, an even shorter scheme was authorized: from Rome to Lyons Falls, with river channel improvements on to Carthage. It was a costly project (e.g., 109 locks in 35 miles) about which there was much doubt, and final approval was not only an expression of the optimism prevailing during the height of the "canal fever" and the genuine desire to aid the development of local districts but of the multipurpose nature of the project, for the prospect of providing a major addition to the water supply of the Erie was a critical factor in its favor. Construction began in 1839 but was not fully completed until 1855.

To the south of the Erie, there was not only the possibility of opening up other local districts but the attraction of making connections with other waterways to form a much-enlarged system. The most obvious possibility was where the Finger Lakes approached navigable waters of the Susquehanna and a Seneca Lake–Chemung Canal was proposed very early. Indeed it was a sufficiently prominent idea for the Western Inland Lock Navigation Company in 1792 to ask the legislature to prohibit any such southward connection for fear of diversion of traffic from the Mohawk route. That attempt failed, however, and as central New York developed and the Erie Canal was assured, any fear of diversion was far overbalanced by the advantages. The latter included widening the market for Onondaga salt and plaster, opening an import route for Pennsylvania coal (which, it was feared, would soon become critical because the fuel demands of the salt works were rapidly depleting the timber), and assisting the general development of this southern district. Rather than threatening a diversion from New York, it was now believed that it would be a means of capturing all the traffic presently exported from the state down the Susquehanna. Construction of this canal, plus a lateral feeder from the Chemung to Horseheads, was authorized in 1829 and completed in 1834.

The extension of a Branch from the Erie Canal into the Chenango Valley was early sought but long delayed, principally by doubts as to its potential earnings and as to an adequate water supply at the summit. The issue was finally decided in 1833, partly at least by the emphasis placed upon the potential coal traffic from Pennsylvania. Binghamton was firmly established as the southern terminus, but there was much competition among a half-dozen communities in the north. Utica, by an energetic campaign, narrowly won over Whitesboro. Navigation was opened in 1837.

In the far western section of the state there was a great variety of proposals focusing upon some sort of link with the Allegheny system. These varied from a canal from that river to Lake Erie, via the old Mayville–Chautauqua portage, to a direct route from the Erie Canal, such as along the old road from the Genesee Valley to Olean. Such widely divergent proposals reflected different interests: those urging the former were primarily concerned with establishing a waterway to Pittsburgh in order to facilitate the importation of coal; those favoring the Genesee were primarily concerned with serving a rich agricultural district and hoping to divert the Allegheny lumber trade toward Albany. A Rochester–Olean route with a branch to Dansville was selected in the following year, but work in the southern section was intermittent for years. When it finally did reach Olean in 1857, no locks were constructed to allow passage into the Allegheny, for Pennsylvania had not improved river navigation, the Erie Railroad had been completed six years before, and the need of a waterway to Pittsburgh had faded in importance.

All of these canals, as well as several short laterals, such as the Crooked Lake Canal between Keuka and Seneca lakes, and the Lake Oneida and Baldwinsville canals, which allowed alternate routes for Erie–Oswego traffic, were built and maintained by the state. The one important private canal was the Delaware and Hudson, constructed by an anthracite coal company to facilitate the marketing of its product in the Hudson Valley. This canal, extending from Honesdale, Pennsylvania, to Kingston, was completed in 1829. This over-all canal network was a relatively simple geographical system. It was primarily a single great trunk line with a series of feeders. In function, it was even simpler than the map suggested, for the traffic on the canals leading to the Delaware, the Susquehanna, and the Allegheny was very largely northbound out of the coal regions of Pennsylvania, making these essentially feeder lines rather than lateral linkages with other waterway systems. The case of the Champlain canal was similar. The completion of the Welland Canal in 1831 across the Niagara isthmus made the Welland–Lake Ontario–Oswego route a possible trunk line competitive with the western half of the Erie Canal. In later years, with the great increases in the movement of wheat from the western states, the development of flour-milling at Oswego, and a growing congestion at locks and terminals on the Erie Canal, this route functioned in this manner.

Thus the Hudson–Mohawk axis of traffic and development was enormously strengthened during the canal era, effectively overcoming the divergent export orientation of the natural waterway systems and to a considerable extent suppressing the competitive patterns initiated by the turnpike developments. Although stages still competed with fair success for the passenger trade, freight was so rigorously focused upon Albany and Troy as to blight the hopes of other aspiring gateway cities along the Hudson. And, although the canal was in general similar in route to the main older east-west turnpike, it was sufficiently different in detail in certain localities to have a major effect upon numerous young villages and cities.

The volume of canal traffic grew steadily through all of these years, but the general pattern of movement remained essentially stable. Of the total tonnage carried on all of the canals in 1853, the Erie alone accounted for 52 per cent, the Oswego and the Champlain were the most important feeders with 18 and 14 per cent respectively, and none of the others surpassed 6 per cent of the total. Eastbound tonnage received at the Hudson was more than four times as great as that westbound. Forest products accounted for 43 per cent of the total, and agricultural materials, 27 per cent; principal specific items were lumber, flour and wheat, wood, coal, stone-lime-clay, railroad iron, and salt. The Erie Canal was of course not only the trunk line of the New York system but also the eastern trunk of a waterway network that interconnected with the whole system of interior North America. The traffic drawn from that interior rapidly increased after the mid-1830's and by 1847 surpassed that drawn from New York. By the mid-1850's the tonnage gathered through Buffalo and Oswego from the West was more than three times that originating in New York, and it was claimed that the "traffic shed" reached very nearly to the Mississippi and Ohio rivers, encompassing the whole of the Great Lakes system. That divide was of course not a sharp one, and there was much interpenetration between the New York and the New Orleans trade areas, but the important role of the Erie Canal in greatly strengthening the east-west axis of the developing nation and projecting that axis across New York State was clearly evident.

RAILROADS

The railroad era began in New York while the "canal fever" was at its peak, and for twenty years thereafter both canals and railroads were being built. Because the trunk line had been completed first, such canal construction was merely the elaboration of an established system, while for the railroads it was a piecemeal development toward such a system. By the early 1850's the relationship between canals and railroads had evolved to a point where it was clear that the newer mode was no longer a mere adjunct to the waterways. By then a true railroad network, with both intrastate and numerous interstate connections, was apparent and its impact upon established trade patterns was being sharply felt.

The Erie Canal had hardly been completed when the Mohawk & Hudson Railway was incorporated to run between Schenectady and Albany. Construction was delayed until 1830 and completed in the following year. The fact that it terminated at each end in an inclined plane to reach the waterside was an expression of its purpose: to be an integral part of the main water system, allowing an efficient short cut around the tedious lockage and devious canal route via Cohoes. It was simply a more modern version of the long-established stage and freighting lines connecting Albany with the Mohawk boats, and its very appearance—the tiny *DeWitt Clinton* pulling a train of three stagecoaches fitted with flanged wheels—was a vivid expression of its continuity with previous modes.

By 1836, however, rails had been laid on west to Utica, paralleling the canal all the way, and with this the railroad began to be something more than an integral part of a canal system. It began to share the traffic, taking over much of the passenger and mail business, though by the law of the time forced to leave all of the freight to the state-owned waterways. Three years later a link was made into Syracuse, and by 1842 one could travel from Albany to Buffalo, though using the tracks of seven different companies. Through trains were soon established for the convenience of passengers.

Such fragmented development was an expression of the dominance of local interests at this time. Railroads quickly became the newest form of speculation, but they were also from the very start instruments of geographic rivalry; that is, they were used by cities and merchant groups to try to capture or control the trade of certain districts—to control spatial interactions. Indeed, the very first line was a device of Albany to reduce Troy's share of the terminal traffic from the Erie Canal, and a second line, the Schenectady & Saratoga, was an attempt by the same interests to monopolize the passenger traffic of that resort. Troy thereupon began a long and frustrating campaign to retaliate and compete. Her line to Schenectady was stunted by the dominance of Albany interests over the critical Utica connection, and her subsidized connections to the north and east proved little more effective. Troy, never completely shut out, continued to grow, but it became increasingly clear that Albany had effectively used the new instrument of the railroad to regain some of its old dominance over the pivotal trade position of the general locale.

As railroads were extended, inevitably competition on a much wider scale developed. As far as

New York was concerned, the linking of the several local railroad segments west from Albany merely created a long feeder to the Hudson River, for there was no railroad leading on to New York City. There was, however, a connection eastward to the rapidly expanding New England system and there were soon fears that this series of lines deep into New York would become a feeder of Boston instead of New York City. It seems clear that having enjoyed such a tremendous surge of development during the early canal era and having outdistanced her rival Atlantic ports in so many ways, New York City was rather complacent about early railroad developments. By the mid-1840's, however, the threat of Boston began to seem quite serious. A speaker in 1846 called attention to the trains departing from East Albany "loaded down and groaning under the burden of our own products and produce of the west; carrying our merchants and the merchants from distant States, that formerly thronged to New York, rapidly and *en masse* to the city of Boston." [Ellis, 133] The picture may have been exaggerated, but the threat seemed ominous enough, and in addition to the Albany gateway, New England interests were building another link with the west through Vermont, connecting with a railroad across northern New York to Ogdensburg. The idea, long held by many, that the railroad could never compete with Hudson River service was increasingly challenged. The railroad had brought a new sense of speed, timing, and regularity. The riverboats were no longer sufficient, and the winter ice blockade no longer tolerable for the many demands of the new age. Thus belatedly, but quickly once begun, two rail lines between New York City and Albany-Troy, one along the river, the other inland to the east, were completed by 1852.

In contrast to this complacency about the Hudson route, other New York interests had been promoting a more direct connection across the state to the Great Lakes for some years. The New York & Erie was chartered in 1832, and its potential benefit to so many districts isolated from the Erie Canal seemed so certain that the state supported the original survey and advanced money for its construction. Originally the main line was planned between Piermont on the Hudson and Dunkirk on Lake Erie, though after a few years of operation the main eastern terminal was shifted to Jersey City. Actual construction was long delayed but by 1848 was being rapidly carried forward, and Dunkirk was reached in 1851. In the meantime several short lines had been built in central and western New York and other segments were soon added. In 1853 the several companies whose lines linked Albany with Buffalo were consolidated into the New York Central and prepared to take full advantage of the recent regulations which allowed full freedom of competition with the canals.

The map of railroads as of 1855, therefore, reveals a considerable number of lines—a total of 2,300 miles—which gave at least some rail service to every major settlement region except the Catskills (Fig. 47). It is important to note, however, that in function it was not one network but two main ones, plus several other somewhat separate lines. The two principal networks were that of the New York & Erie and connecting lines of 6-foot gauge totaling about 950 miles, and the New York Central and connecting lines of 4 feet 8½ inches (the later "standard gauge"), a total of about 1,300 miles. Although each dominated a different section of the state, they were competitive not only for the through traffic from the western states to New York City, but for much of the trade of the richest parts of central and western New York, for the broad-gauge lines reached north to Syracuse, Canandaigua, Rochester, Batavia, and Buffalo.

This broad-gauge network was the most radical geographic feature of the railroad developments up to that date. Elsewhere, the railroad was to a large extent fairly directly complementary to or competitive with canals, greatly enlarging service but not drastically altering the general route patterns of traffic. But the New York & Erie provided a more direct link between the metropolis and Lake Erie, which had never before existed as an important freight route. It also cut across the Delaware and Susquehanna and severed most of the traffic which was formerly floated downstream and reoriented some of the Finger Lakes and Genesee trade directly southeastward to New York City. In this way, the whole Southern Tier of counties was opened to a great quickening of development. Sites that had long been only villages but now found themselves at important junctions between this new trunk line and other rail, canal, or river routes, such as Binghamton, Owego, Elmira, and Corning, suddenly sprouted into smaller cities. Such cities as Syracuse, Rochester, and Buffalo were affected, too, for they enjoyed a new alternative and highly competitive connection to the Atlantic Seaboard. In its first years of operation the New York & Erie carried considerably greater total tonnage than the New York Central (though of course the latter was competing with the Hudson and Erie Canal waterway). The fact that it had twice as much westbound traffic as the New York Central was indicative of its importance to New York City as a more direct trunk line to interior markets.

Among the lesser lines of significance was the Buffalo & State Line along the south shore of Lake Erie, a segment of a much larger network, principally in Ohio, with its own peculiar gauge of 4 feet

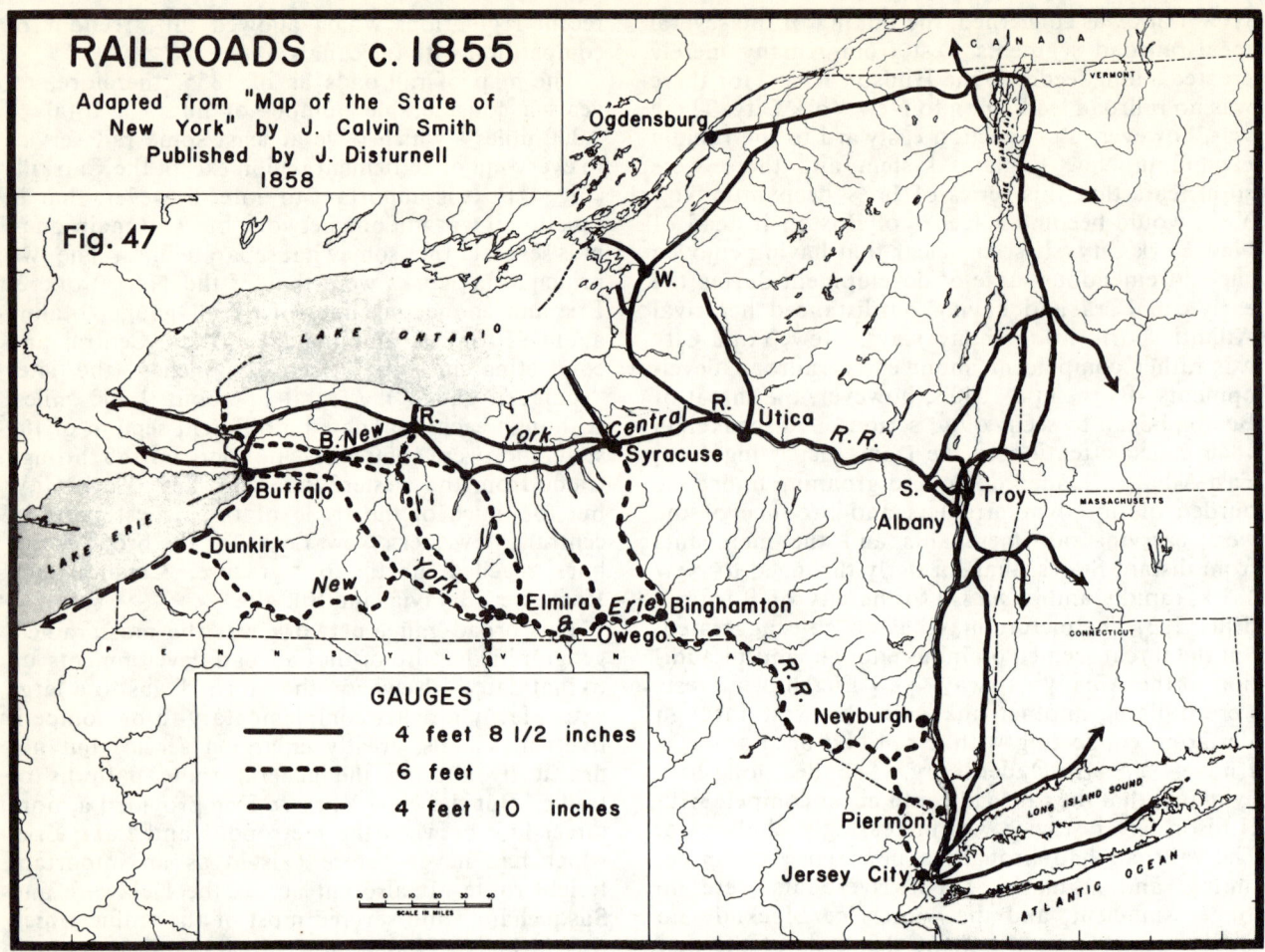

10 inches. Thus all traffic through Dunkirk and Buffalo had to change cars. Another noteworthy railroad extended across northern New York to Ogdensburg, a segment of a northerly trunk line from New England ports. Although isolated, the Long Island railroad, too, was part of a larger network of service. It was extended to the extreme tip at Greenport, where ferries across the Long Island Sound connected with a mainland railroad to give fast service between New York City and Providence and Boston; the north shore line via New Haven was at this time still incomplete.

Thus the railroad further altered the patterns of interconnection and circulation. Each medium of transportation had its special geographic pattern, reflecting not only the particular terrain features of the New York area but the spatial characteristics inherent in the very nature of the medium. The railroad was less limited by terrain than canals but more so than turnpikes, and thus very early in the railroad era two major trunk lines were developed as against the one canal trunk, but it proved impossible to equal the multiple incipient trunk lines of the turnpike era, despite some early attempts to do just that. Hudson, for example, early built a rail link with New England and earnestly promoted a railroad from Catskill on the Hudson directly northwest to Canajoharie in order to intercept some of the Albany traffic, but to no avail. Newburgh sought to become the terminus of the New York & Erie but only got a branch belatedly and could not hope to compete with Jersey City as a terminal. Thus the railroad era was little more favorable to the gateway aspirations of these Hudson River cities than was the canal era.

The Transportation Network c. 1855

While the above treatment has covered a sequence of transportation development, the complete transportation network of 1855 included many elements of all the mediums. Natural waterways were still of basic importance. The Hudson was crowded with vessels of many types and sizes, the river steamers had lost much of their passenger business to the railroads but were beginning to replace sailing vessels for freight. The Great Lakes, too, carried a thriving commerce. More than nine thousand vessels called at Buffalo in 1855. Steam had made its appearance early on Lake Erie, in 1818, but the prominence of heavy bulk cargoes in a predominantly one-way movement gave sail a strong advantage for much of the traffic. Lesser lakes, too, were important links in the system. Lake Champlain supported considerable traffic in lumber and iron. Cayuga, Seneca, Keuka, and Canandaigua lakes each had daily steamer service between ports at either end, and the first two had considerable northbound coal trade. Turnpikes still thrived where not yet paralleled by a railroad or canal; the Catskills were still dependent on them and, while elsewhere much of the long-distance traffic had been lost, many segments remained as important local feeders. At this very time still another transportation innovation was the object of great interest: the plank road. Though the basic idea had been long used in some form here and there in swamps (though usually with logs rather than planks), the completion of the first plank turnpike in 1846, from Syracuse to Central Square, touched off a new mania. By 1855 more than three hundred companies had been formed, several hundred miles of road built, and most cities had at least a few radiating into the countryside. Such roads were merely short feeders, almost wholly complementary to the railroad, and were themselves an expression of the new standards for speed and ease of travel fostered by the iron roads.

Despite the widespread and heavy impact of the railroad, the Erie Canal remained the great trunk route for bulk freight. In 1852 its tonnage was nearly four times that of the combined total of the trunk rail lines. Railroad competition would help make certain that the income from some of the later expensive canal feeders, such as the Black River and the Genesee, would never even cover their maintenance charges, but so far the competition between canal and railroad trunk lines had developed into a complementary service, each dominant in the movement of different kinds of products: heavy bulk goods of forests, farms, and mines by canal; passengers and lighter, higher-valued freight by rail.

The over-all transportation complex of the 1850's was therefore one of considerable diversity in its mediums, their interrelationships, patterns of service, and areas of competition. But there was also ample evidence at that time that it was by no means a stable complex for the many new companies, and the active extension of construction made it clear that the railroad pattern was far from complete.

Agriculture

There was very little regional variation in the pioneer agriculture of New York. In the recesses of the Catskills, on the St. Lawrence Lowland, in the Finger Lakes Country, or the Genesee Flats, the crops, animals, tools, and techniques of the early settlers were very nearly the same. A plot of wheat, another of corn, a patch of flax, a meager garden, perhaps a few fruit trees, a yoke of oxen, a few cattle and hogs were the common elements. In most areas the ax was the indispensable farm tool—to open the forest and, year after year, to hack its margins ever farther back. It was a process of domestication of the landscape which took a good many years.

Travelers were always struck by the disheveled appearance of the pioneer countryside, for American colonists had long ago learned the futility of trying to create clean-cleared fields immediately. And so stumps and roots and great logs were left encumbering the ground until eliminated by rotting and repeated burnings. Under such conditions the simplest tools and methods were often best. John Maude, journeying through the New Military Tract in 1800, noted that he rarely saw a plough; as soon as brush and timber had been cut and burned the land was harrowed and sowed with wheat, and other crops. Clearing was slow and enormously laborious, but it produced the raw material for one of the few products for which the pioneer found a ready cash market. With care in conserving the ashes a fair income could be obtained from the pearlash crudely extracted. Other forest products—logs, furs, maple sugar—were also common sources of income. Even when his fields were cleared and good harvests obtained, the settler remote from a natural waterway might find it difficult to enlarge his income, for few farm products could stand the expense of long-distance overland shipment. Thus the pressure for canals developed precisely at the time when central New York had reached the stage of having a rapid annual increase in the land ready for crops but no feasible means of marketing the harvest. Without improved transportation there could be no large-scale commercialization and thus no real regional specialization in agriculture.

The Hudson Valley and lower Mohawk of course had no such problem and the post-Revolution spurt of growth in the cities and trade along the Atlantic Seaboard soon prompted an increasing emphasis upon commercial production. Wheat was the principal crop and after 1800 these areas began to be known as one of the "granaries" of the nation. Cattle, sheep, and poultry became regular items of exchange, and shipments of butter and cheese from the Hudson counties increased. There was no intensive specialization, other crops continued to be raised, and most of the produce was still consumed on the farm, yet it was an important shift in emphasis and it brought a fresh prosperity to favored locales, in Dutchess, Orange, and Ulster counties especially.

The completion of the Erie Canal had an important impact upon the Hudson Valley as well as upon central and western New York. It rescued the latter region from its isolation and allowed it to shift out of the pioneer subsistence economy. Here, too, wheat became the major cash crop and before long a new "wheat belt" emerged, extending from Onondaga west to Lake Erie and lying between the rougher hill country of the upper Finger Lakes and the swampy or gravelly areas of the Ontario plain. It was good wheat country compared with any other of the time. The yields and the quality of the grain were high, and nowhere more so than in the Genesee Valley. "Genesee" thus became a mark of excellence in wheat and flour and the name was applied loosely to the produce of the whole region. To the pioneer gristmills were soon added whole clusters of larger mills at strategic sites and especially at Rochester, which may be viewed as the principal urban product of this agricultural regionalization.

The competition from such high-quality grain so cheaply produced from fresh land was in itself severe in the older region, but it was made more so by special difficulties in the Hudson Valley. Here farmers had long struggled against the ravages of the Hessian fly, and in the late 1820's, just as the volume of Genesee exports began to increase, another insect, the tiny "midge," began to infest the area and drastically reduce yields. Although various means, such as a change in the time of sowing, better cultivation, and the shift to more resistant varieties gradually proved fairly effective in reducing the dangers from these pests, in time most Hudson Valley farmers simply stopped raising wheat. There were not many alternatives. Rye was grown as a local bread grain; barley, primarily for brewing, became the leading cash grain in the Mohawk; elsewhere corn, oats, and hay became the principal crops. Although animals continued to be driven to the metropolitan market, here, too, there was keen competition from the western regions, for large cattle drives to Albany and New York City became important outlets for the growing surplus stock of the interior. Clearly, survival in this rivalry required some kind of specialization and intensification in the older region and the farmers turned increasingly to dairying. The early Dutch had provided comparatively good quality stock for that purpose and gradually Orange County butter and Herkimer cheese became the standards of excellence in dairy produce. A beginning was made in orcharding, too, in some locales; apples, peaches, and pears were commonplace in all agricultural districts but were now for the first time given careful attention as specialty crops. In Kings and Queens counties market-gardening was steadily expanded to meet mounting urban demands. Thus, in the second quarter of the nineteenth century, regional agricultural variations of some importance began to appear.

At mid-century some fairly definite regional agricultural patterns were discernible but by no means stable. A western wheat region was rather clearly evident, but its eastern border was gradually being shifted westward as farmers turned more to barley. Historically, an emphasis upon barley had been steadily spreading westward from the Mohawk. At this time New York grew nearly 70 per cent of the nation's barley, with the greatest concentration in Onondaga County. That barley district was a link in area and in type between the wheat district to the west and the dairying and specialty crops region to the east. Barley was not simply another grain used in place of wheat but was more of a specialty crop, raised for the exacting needs of the malting and brewing market. Dairying was the predominant source of income and fodder crops the principal farm acreage in the Hudson and Mohawk, but there were numerous specialties in various locales, such as orcharding along the Hudson, flax in Washington and Rensselaer counties, broom corn in Schoharie and Montgomery, and hops in the Otsego–Oneida area. Elsewhere, regional distinctiveness was less well defined. Throughout the southern region livestock was important, chiefly as meat animals, but the building of the Erie Railroad was expected to cause a shift in emphasis to dairying in the counties of the Delaware drainage basin. In the north, wheat, sheep, and dairying were combined in the Watertown–Potsdam district, but the remainder of the state was at a relatively low level of commercial development.

By the middle of the century, also, New York enjoyed remarkable pre-eminence in American agriculture. Though by no means the largest state, it had the greatest acreage of improved land, and as an even more significant sign of its leadership, the total value of farming equipment and machinery in use was nearly twice that of Pennsylvania, its nearest rival. New York was the leading producer of oats, barley, and buckwheat and was only barely surpassed in wheat and rye. It was the leader by a very large margin in butter and cheese, orchard and garden produce, potatoes and poultry, and also, by a smaller degree, in cattle, total value of livestock, and livestock slaughtered. In sheep and wool it had recently fallen slightly behind Ohio. It was an impressive accomplishment and accompanying it was a leadership in the development of improved machinery, experimentation with new methods, and dissemination of knowledge through many active agricultural societies.

Yet, despite this seemingly commanding position, the signs of change were already discernible. Just as the Genesee Country had forced the Hudson to shift its agricultural emphasis, so the rapid development of Ohio, Indiana, and Illinois was bringing heavy pressure upon western New York. Before 1840 New York flour was moving west out of Buffalo; after that date western grain came pouring in. In 1840 Buffalo received a million bushels of wheat from the West, ten years later it received over three and a half million, and in 1855, more than eight million. In a single generation the Genesee had become an "old" wheat region, its soils no longer fresh, its crops no longer free of pests and diseases. As a noted English agricultural scientist, who came to Syracuse in 1849 to deliver an address at the annual State Agricultural Society fair, observed:

This celebrated wheat region, *as a whole*, is gradually approaching the exhausted condition to which the more easterly wheat-growing, naturally poorer districts, had earlier arrived.... [It is] becoming unable to compete with the cheap wheat-growing virgin soils of the West. [Johnson, I, 172]

Nevertheless, he concluded that the condition and prospects of New York agriculture were very favorable, not simply because of natural conditions but also because of the intense progressive interest of so many agriculturists. By mid-century it was becoming apparent that fertile minds might be as important as fertile soils in the emerging competition of a national agricultural economy.

Manufacturing

Manufacturing, like agriculture, in the years following the Revolution was carried on chiefly as small-scale activities widely scattered through the settled districts. The gristmill, sawmill, and country forge were the main elements, and New York was especially fortunate in its abundance of small streams and forests to supply the necessary water power and charcoal. Cities and larger villages had a much greater array of workshops and craftsmen, but there was little semblance of factory organization, division of labor, and automatic machinery before 1820. By then the beginnings of a new technical era were only barely apparent and in only a few localities. The census of that year listed a mere 9,700 industrial laborers in the state, of whom a quarter were children. The best representatives of the most modern industry of the time were the several small cotton mills, such as those on the Oriskany and Sauquoit in Oneida County, in the hills of Columbia and Dutchess counties, and in New York City, a pattern reflecting the New England source of the inventions, organization, and leadership of that era.

The canals, making possible the long-distance movement of heavy bulk commodities, gave the first real stimulus to industrial concentration. Larger sawmills and flour mills began to cluster at the most advantageous sites, and larger ironworks developed at convenient assembly points of ore and fuel. The latter were also an expression of technological advances. After considerable experiment and long promotion, anthracite had been accepted as superior to charcoal for large forges and furnaces. The completion of the Champlain and the Delaware and Hudson canals allowed Adirondack ore and Pennsylvania coal to be brought together in the Hudson Valley and prompted the establishment of important ironworks, especially at Troy.

The more complex machines being introduced by the mid-nineteenth century led to greater concentration in many activities. Whereas wagons, being largely of wood with a few iron parts, could be built in a hundred workshops, the new cast-iron plows came from perhaps a score of special plants, and the more complicated reapers and threshers were produced only by a few well-equipped factories.

The second quarter of the nineteenth century, with its radical improvements in transportation, power, and automatic machinery and its rapid broadening of a fully national market marked the real emergence of the "industrial revolution" in America, and the patterns of industries in New York in the 1850's were an excellent exhibit of the general character of that phase. Manufacturing specialization at favorable localities, often the larger urban units, became increasingly important.

The lumber industry was a good example of this transition from the old to the new. There were still hundreds of small sawmills scattered over the state, little different in scale or equipment from their recent pioneer predecessors, yet there were also concentrations of large mills and a whole system of lumbering had been evolved to suit the special characteristics of New York. The principal focus of developments was in the upper Hudson; Albany–Glens Falls was the main axis, the "trunk line" of the lumber industry tapping the southern Adirondacks, the main logging region. The technique of log-driving down the numerous streams, many of which had been especially improved for the purpose, was a New York innovation. Glens Falls was the great assembly and sorting point as well as a major milling center. From here logs and lumber were rafted down the Hudson to Troy and Albany, which together constituted the greatest lumber-marketing center in America. Lumbermen were moving in on the whole perimeter of the Adirondacks, and the Black River Valley, the Raquette, and the Ausable were important secondary clusters of sawmilling. The Southern Tier of counties, too, continued to be of some importance, with marketing prospects somewhat improved by the new railroad lines. Thus, despite the continued existence of the small mills, the industry as a whole had developed into an elaborate specialized activity which concentrated upon exploiting the richest resources accessible by the most advanced techniques available at the time for distribution to the widest possible market.

Flour-milling reflected the same kind of development and even a phase beyond that of lumbering. Evolution beyond the country gristmill came quickly as western New York developed into a specialty wheat region. Rochester was the outstanding center within the producing area, and the canals allowed Albany and New York City mills to draw easily from the same region. By 1855, however, Rochester's output had been surpassed by Oswego and Fulton, and Buffalo, Lockport, and Ogdensburg had become important milling points, all chiefly based upon wheat imported via the Great Lakes from the new western wheat states. Thus flour-milling was already shifting from an industry located convenient to local raw materials to one strategically located within the whole national pattern of production and consumption areas.

The over-all manufacturing geography of the state at mid-century displayed two important diversified districts, several clusters each important in a particular industry, and a wide scattering of many smaller centers in which the volume and employment were small but which in total represented a considerable share of the state's industrial economy. Clustering was the new trend; wide scattering a thing of the past. New York City and Brooklyn made up the largest and most diversified district. Here were textile mills, garment factories, ironworks, specialized metal shops, woodworking plants of all kinds, the biggest shipyard turning out everything from river craft to ocean clippers, steamships and sailing vessels, breweries and distilleries, flour mills and meat-packing plants. Its plants probably produced the greatest variety of consumer goods in America. The second major district was centered in Albany and Troy, reaching out to Schenectady and Cohoes. This was the greatest American market center for lumber and barley, important in brewing and flour-milling, and one of the nation's major iron-manufacturing centers. Schenectady, long in the doldrums, was the home of a large locomotive works; Cohoes and Troy had important cotton mills. Elsewhere there were important industrial centers, but largely concentrated upon a single activity: Utica and its neighboring villages on cotton and woolen textiles, Syracuse on salt, Rochester with only two small cotton mills alongside its twenty-three flour mills. Some of the Hudson Valley counties, especially Columbia, Dutchess, and Orange, had a considerable variety of factories, but they failed to develop large clusters or large cities.

Thus New York's manufacturing patterns, like her agricultural patterns, were quite unstable at the time. Clearly they were in transition toward greater scale and variety of production and toward greater concentration in major urban places. Nationally her industrial position, however, was not comparable to her agricultural pre-eminence. Rich local resources were the basis of leadership in lumber and flour-milling, but it was already clear that those local resources would decline and future dominance of their processing was by no means assured. On the national map of manufacturing New York textile districts appeared to be but the western fringe and outliers of the larger northeastern region centered in New England. Such a geographic perspective is an accurate reflection of the developmental processes, for New England was the dynamic center of innovation and progress and New York was largely a borrower. This was true in the iron industry, too. Despite the fact that New York plants were based primarily upon New York ores, Pennsylvania was clearly the principal center of growth and technological development. New York's total value of manufactures was greater than that of any other state, but the New England states together well surpassed her, Pennsylvania was not far behind, and the industry of both of these neighbors was growing more rapidly. New York was certainly a major part of the rapidly developing national manufacturing belt, but the trends and patterns of the time did not indicate that it would become the very core of it.

Population Patterns

All of these material developments were in a sense encompassed within, and exhibited by, New York's population growth during these years. From a total of 340,120 in the first national census in 1790, ranking fifth among the states, New York reached nearly a million in 1810, not quite enough to end Virginia's long leadership as the most populous. A further doubling in twenty years to nearly two million secured her pre-eminence, and a total of 3,466,212 in 1855 was not only far ahead of the next state but a remarkable measure of sustained growth. Considering the magnitude of increase and the spread of people over the whole state the proportionate distribution of population had not been altered as greatly as might be expected. The metropolitan region (the five counties) contained nearly 29 per cent of the total, a reduction from 38 per cent in 1771, but not a marked one considering the very great expansion in the populated area. The "colonial area," the Hudson-Mohawk from Herkimer County to the sea, still contained about half of the total. The L-shaped Hudson—Mohawk—Lake Erie axis is discernible in Figure 48 and in fact contained nearly three-fourths of the population of the state, yet a comparison with Figure 2 shows that axis to be relatively faint and, aside from the obvious exception of the metropolis, the 1850 distribution seems characterized more by dispersal than concentration.

Expectedly in 1850 the key points on that axis, the terminal and the pivot, were the chief centers of population. This corridor of relatively dense settlement and traffic was of course only one of several focused upon New York City. The newly established trunk line across the southern part of the state, the New York & Erie Railroad route, was probably too new to have as yet had a major impact, but the importance of older traffic-ways had been magnified. For even though New York State was not itself the center of the Northeast industrial region, New York City was, with southern New England and eastern Pennsylvania in some degree tributary. Ironically for Philadelphia, every improvement of transportation across Pennsylvania was also an improvement of New York City's links with the West, and by 1855 New York City was firmly established as the great meeting point of the main trunk lines of America with the great trunk line of the high seas, the North Atlantic route.

Albany's growth, while steady, was not commensurate with the strategic position of its locale because that position was shared with several rivals. Troy was the principal contender and had enjoyed a greater rate of growth, though, having started much later, remained smaller; Schenectady, a much older rival, had grown very slowly and was little larger than Cohoes, a new city. This cluster of cities had a total population of over 100,000, making it the second largest urban region in the state.

The western terminal had grown slowly for some years and then enjoyed a rapid spurt toward mid-century. The whole town of Buffalo had a population of only 5,000 at the opening of the Erie Canal. Not incorporated until 1832, the city grew to about 20,000 by 1835. However, twenty years later it had nearly 75,000, and that was in large part a measure of its shift from being merely a local focus of traffic to a great intermediary between East and West within the main industrial and commercial belt of America.

Between these terminals a major differentiation in city growths was well established. Newburgh, Poughkeepsie, Kingston, and Hudson, failing to develop their own trunk lines to the West, remained local trade centers with further growth dependent upon manufacturing. Along the western leg of the axis, Utica, Syracuse, and Rochester were clearly established as principal cities, each an important node of trade and each with a particular manufacturing emphasis. Elsewhere, Oswego, Watertown, and Ogdensburg in the north and Binghamton and Elmira in the Southern Tier had emerged as the leading centers of their respective districts. Cazenovia, Geneva, Canandaigua, Batavia, and Bath, though each locally important, were proof that an early start and vigorous promotion were much less important than having sites persistently strategic within the developing patterns of commerce and industry.

The geographic pattern of culture groups developed through two stages during these seventy years. The first, associated with the great colonization movement following the Revolution resulted, broadly, in a vast extension of Yankee dominance over the north, center, and west, with a persistence of the old Yorker population (Dutch-German-English) in the Hudson-Mohawk and with New York City continuing as a great cosmopolitan center. A second stage began to appear with the increase in immigration from Europe, starting in the 1820's, expanding through the next decade or so, and rapidly rising by the 1850's. The Irish and the Germans made up the great majority of the total, with considerable numbers of English and Scotch also. Such a heavy influx of new foreigners tended strongly to bind the older elements together and, along with other factors, eased the lingering Yankee-Yorker antagonisms. In this way the deep-rooted cultural distinctiveness of the Hudson-Mohawk region tended to fade in significance, overshadowed by the new patterns of more keenly felt cultural differences.

This new pattern was very largely associated with the major urban centers along the Hudson-Mohawk axis. The map of the proportion of foreign-born in the total population of 1855 vividly illustrates this fact (Fig. 49). More than half of the population of the three main urban clusters was foreign-born, and the percentages in Utica, Syracuse, and Rochester were also relatively high. Irish and Germans were the largest groups in each case but tended to change in proportion from east to west. Thus in the metropolitan area there were about 250,000 Irish and 130,000 Germans, while in Erie County (Buffalo) that ratio was somewhat more than reversed in favor of the Germans. The addition of several thousand English newcomers in each of these urban areas, of Scotch in most, Welsh in

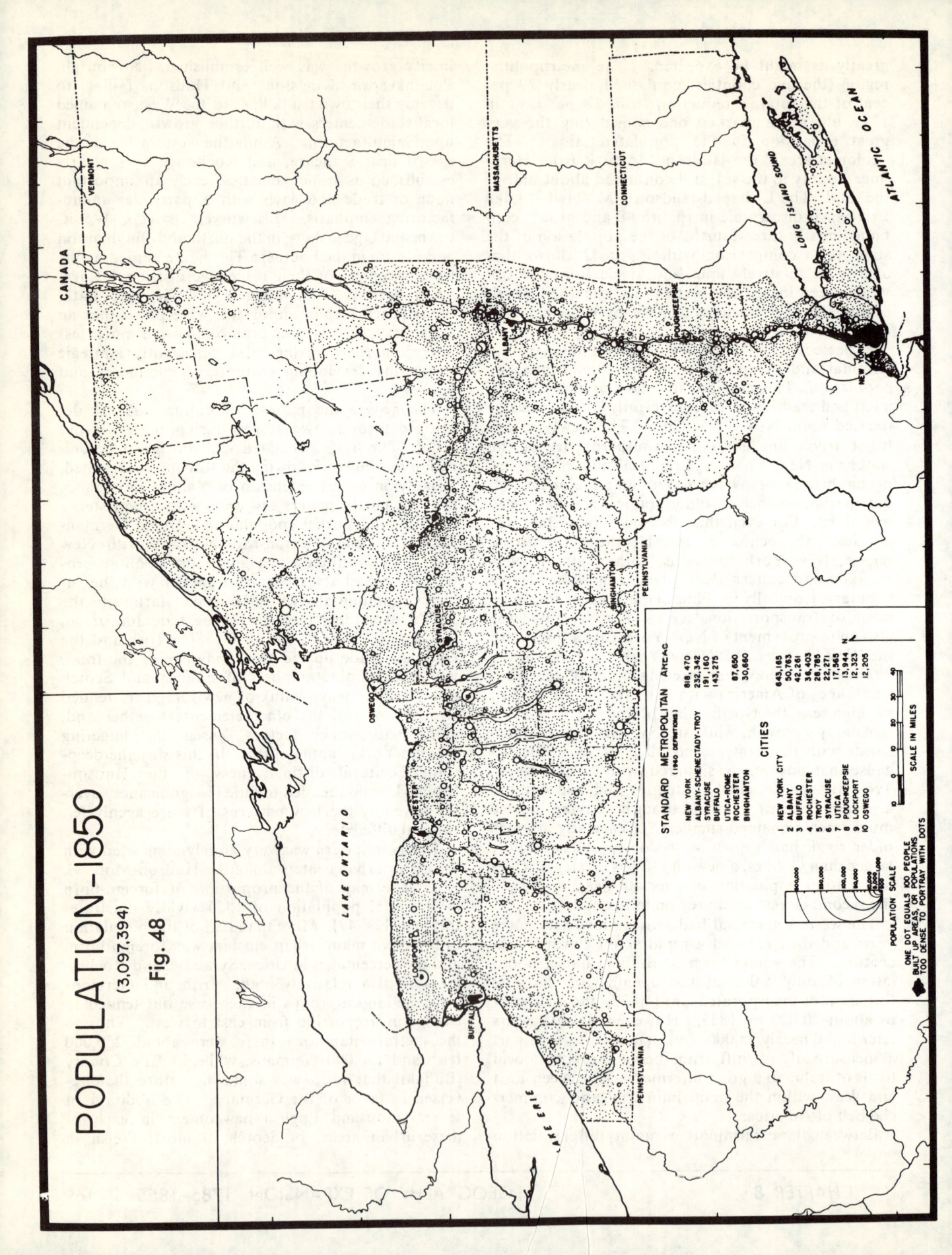

POPULATION—1850
(3,097,394)
Fig. 48

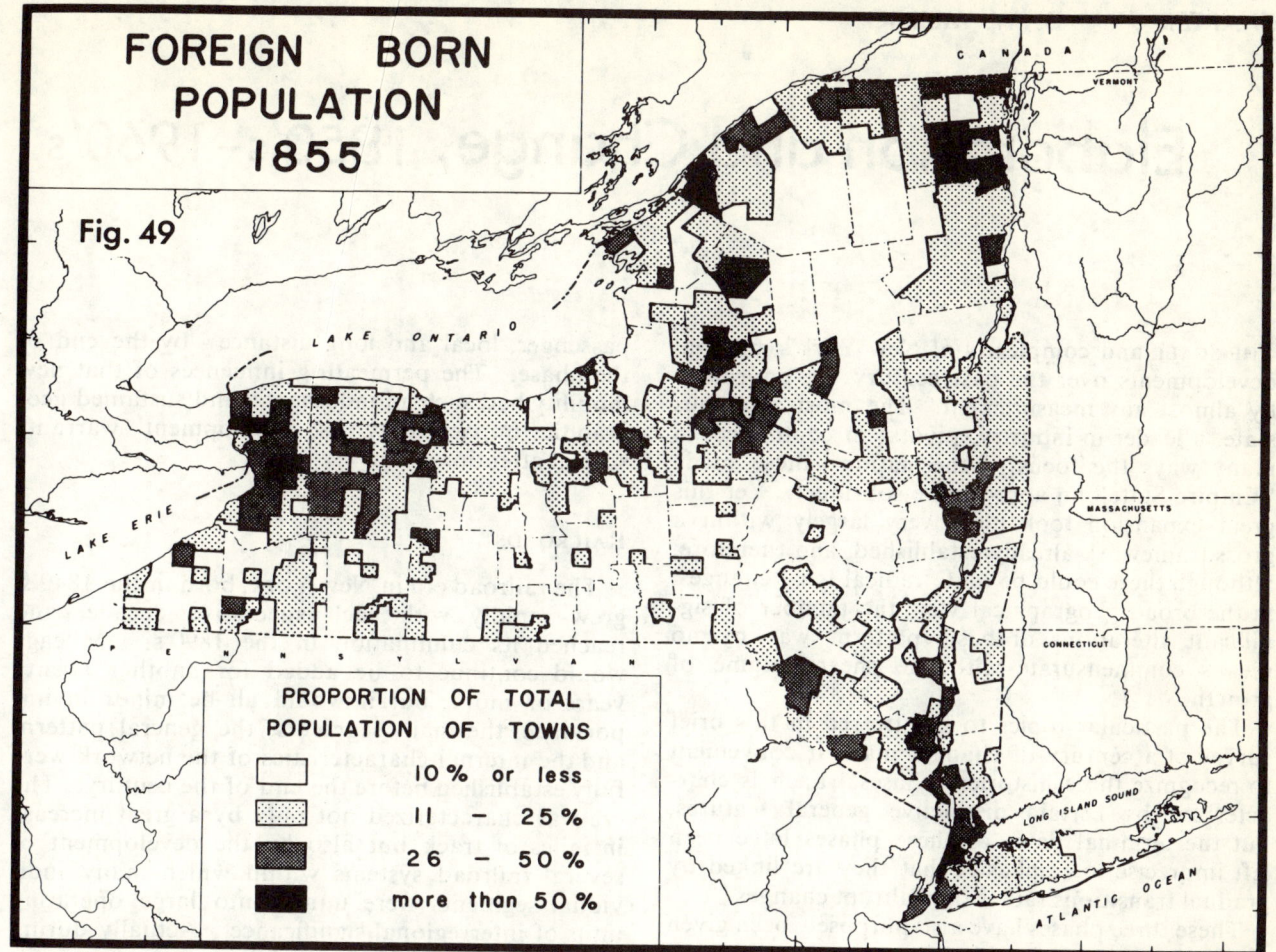

Utica, and in all a smaller sprinkling of other immigrants was clearly rapidly transforming these urban clusters into little regions of cultural diversity quite distinct from the relative uniformity of the rural areas and smaller cities and villages. The sharp contrast with the absence of any significant number of immigrants in the whole broad southern region from the Catskills to Dunkirk, even in the larger cities, is especially striking and is an important accentuation of new patterns of cultural regionalism. One other element in these new patterns deserves comment: the influx of Canadian-born persons into the northern border region, a movement which had gradually developed and would long persist, associated at this time with the lumber industry. Such portals as Oswego, Niagara Falls, and Buffalo, too, had important numbers of Canadians in their populations.

Emanating from these demographic developments were three important "geographic tensions," each on a different scale. Within the state as a whole, the chief tension was between the metropolis and the remainder, commonly expressed as between "the city" and "upstate." Within each of the regions of the state there were growing tensions between the by now relatively stable rural areas and the rapidly developing urban centers. Within those major urban centers, there was a sharp cleavage between the older "native" group and the new "foreign" elements. These tensions, each a distinct geographic pattern, were of major importance in the evolving political, economic, and social life of the state. Clearly discernible in the mid-nineteenth century, they would still be very much apparent in the latter part of the twentieth.

CHAPTER 9   D. W. MEINIG

Elaboration and Change, 1850's–1960's

THE SCALE and complexity of New York's material developments over the past century are enormous by almost any measurement. The most populous state, a leader in industry, pivotal in commerce, in many ways the focus of the nation—these made "Empire State" an appropriate sobriquet. Yet this great expansion took place very largely within a gross framework already established, and therefore, although there could be quite radical local changes, in the broad geographical view the number of significant alterations of basic patterns was by no means commensurate with the sheer volume of growth.

The particular topics to be stressed in this brief survey of a century of changes make it convenient to recognize three historical phases. Each is characterized by certain distinctive general features, but the terminal dates of these phases have been left imprecise to emphasize that they are linked by gradual transitions rather than abrupt changes.

These three phases have also purposely been given unequal treatment in this chapter simply because the general era grades into the present and thereby merges into the perspective of subsequent chapters. Throughout, only a few selected themes have been stressed and emphasis has been given increasingly to certain general trends rather than to local changes. Furthermore, the developments in this chapter are increasingly and intricately associated with broader patterns of change in the nation, but no more than a hint of these can be suggested.

Phase I: 1850's–1890's

This period of about forty years is a time of great growth in most activities. It is an era of rapid urbanization and of major increase in population growth, as well as in the quantity and diversity of manufacturing. Through most of the phase there is also some expansion in rural population and agricultural production. In many ways the most succinct and useful characterization of the phase is "the railroad age," for the railroad was sufficiently perfected at the beginning of this time to allow it to be rapidly extended and elaborated so as to dominate nearly all types of movement—freight and passenger, local and long-distance—by the end of the phase. The permeating influences of that new mobility had such a fundamental and sustained geographic impact that its development warrants special attention.

RAILROADS

The railroad era in New York, born in the 1840's, grew rapidly with each successive decade, and reached its culmination in the 1890's. Mileage would continue to be added for another twenty years or more, but it would all be minor in importance; the main segments, the general pattern, and the internal characteristics of the network were fully established before the end of the century. The era was characterized not only by a great increase in miles of track but also by the development of several railroad systems within which many individual segments were united into large operating units of interregional significance. Actually during these latter decades, as in earlier ones, most railroads were begun as local lines to connect nearby settlements. However, whereas amalgamation had only begun in the 1850's, by the 1890's the New York network was basically a set of large systems to which all short lines functioned as feeders even if nominally independent as corporations.

It was also an era characterized by wild speculations, extravagant competitions, and widespread fraud. Such features had some marked geographic results. The most obvious was the duplication of trackage along certain routes, a form of competition so costly as to be confined usually to routes associated with rivalries between major systems. Another feature was the construction of lines that had little prospect of economic success, the promoters of which hoped to make their fortunes out of the sales of stocks and bonds and from construction contracts. New York laws which allowed local communities to incur public debts in support of railroad construction greatly fostered this kind of speculation. In its most extreme form this could result in the building of railroads along devious routes, wandering across the countryside seeking those towns and cities which would tax themselves in the line's support and avoiding those which refused (see Pierce). Such deformations of basic geographic purposes of a transportation facility inevitably brought serious and often crippling financial distress. Yet it is not easy to evaluate the extent to which the railroad mileage exceeded the actual traffic needs. The fact is that the efficiency of the railroad was inherently so superior to rival forms of transportation at the time that even short lines through rural country serving only small settlements could have considerable traffic. It seems safe to suggest that most of the bankruptcies of the time were caused more by financial extravagance and mismanagement than by a traffic too small to support a properly operated railroad.

The over-all network in the 1890's was comprehensive (Fig. 50). No large settled area was without at least one railroad, few villages of more than hamlet size remained unserved (and these were of course destined by their lack of a railroad to remain small), and even the major mountainous regions had been not only tapped at their margins but completely transversed. The completion of the Adirondack line from Herkimer via Remsen and Tupper Lake to Malone in 1892 was the last lengthy construction in the state and opened the last extensive isolated area. The entire mileage in the state was an actual network in one new sense, for during the 1870's practically all of the trackage had been brought to a uniform "standard" gauge of 4 feet, 8½ inches. Nevertheless, in actual function this comprehensive pattern of lines was rather sharply segmented into various systems that markedly shaped the main flows of traffic and determined the degree of competitive service enjoyed by cities and regions. It is important to examine these systems as a means of understanding, not only the railroad geography of the time but also that of the present, for the railroad network of today, while considerably reduced in mileage, is but slightly altered in basic patterns from that of the 1890's.

Trunk lines were those with a major interest in through traffic between New York City (and in some cases New England) and Buffalo and points west (Fig. 51). Competition was involved at two different scales: for the traffic between New York

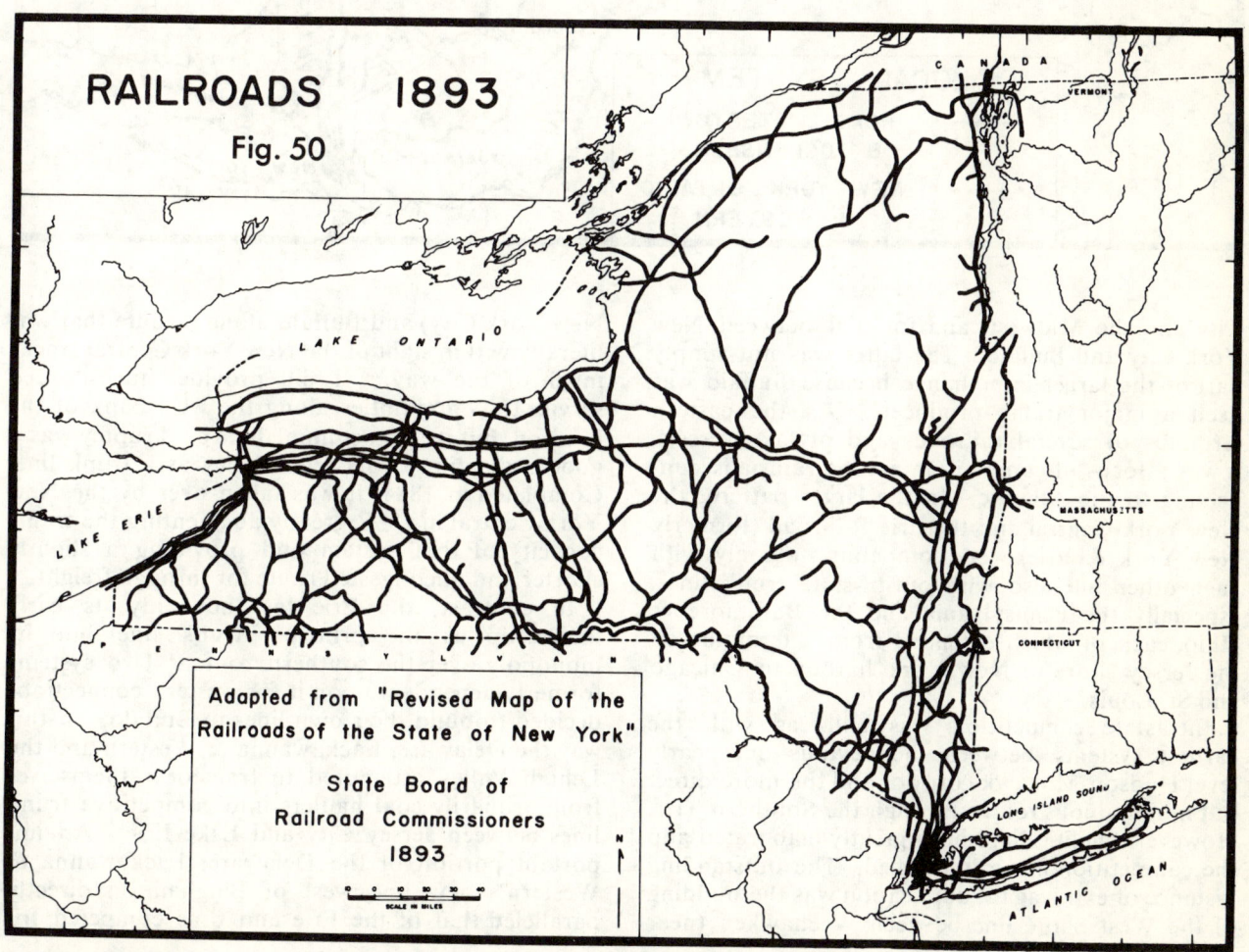

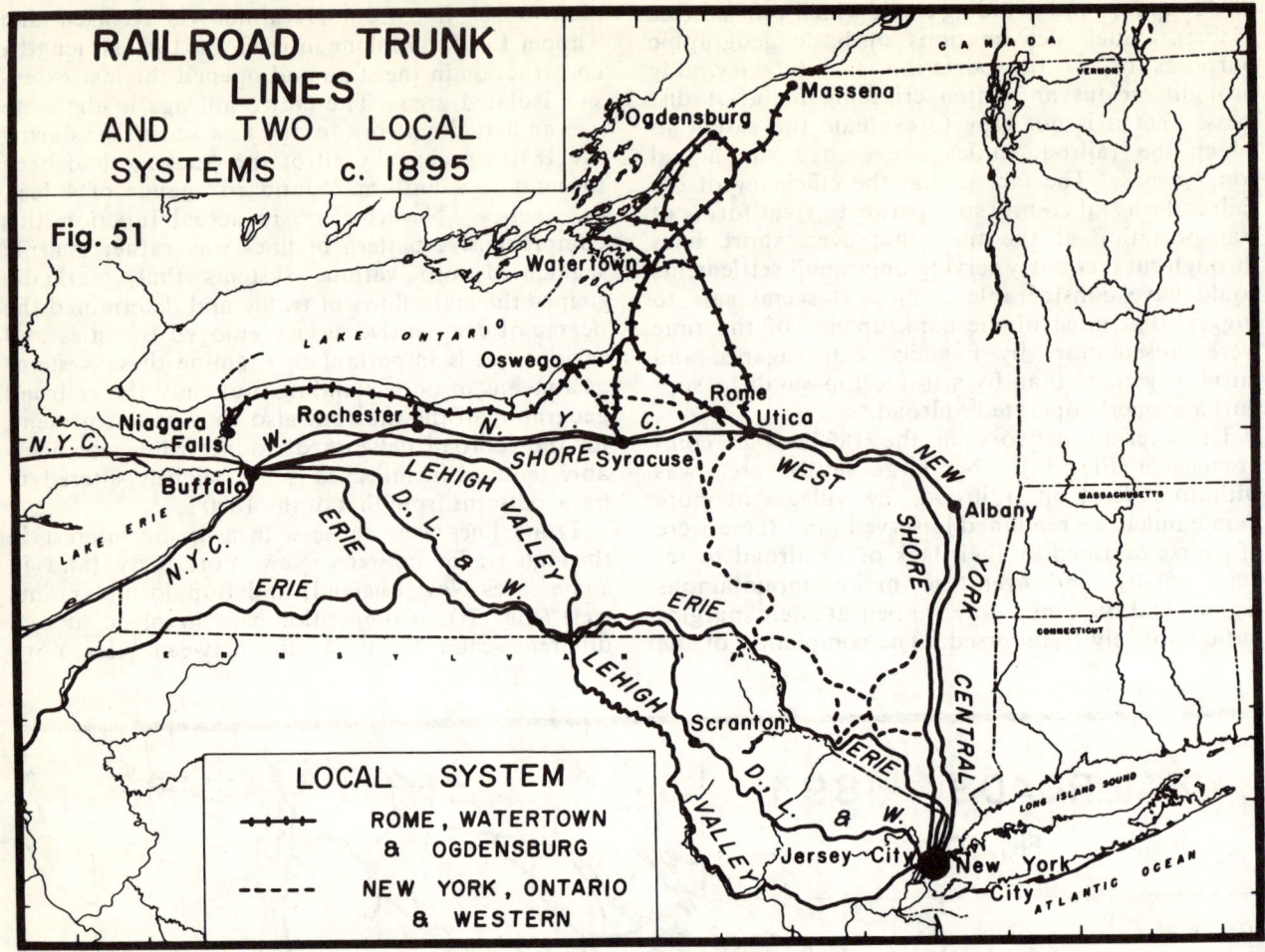

Fig. 51. RAILROAD TRUNK LINES AND TWO LOCAL SYSTEMS c. 1895

City and the Midwest, and for that between New York City and Buffalo. The latter was not simply part of the larger interchange because Buffalo was itself a major traffic producer. As the eastern terminus of several railroads and of lake vessels, it was a focus of competition among railroads connecting to the Atlantic. In the larger pattern, the New York Central and the Erie Railroad (formerly New York & Erie) were competing not only with each other but also with out-of-state trunk lines, especially the Pennsylvania and the Baltimore & Ohio, each of which extended from a terminus on the Jersey shore of New York harbor to Chicago and St. Louis.

Intrastate competition was still, as with the earliest systems, between the devious but nearly level Hudson-Mohawk corridor and the more direct but mountainous routes through the Southern Tier. However, facilities had been greatly elaborated and the competition much intensified. The outstanding instance of extravagant duplication was the building of the West Shore line between Weehawken (near New York City) and Buffalo along a route that was literally within sight of the New York Central tracks much of the way. It did provide much-desired service to communities along the west bank of the Hudson, but its almost immediate bankruptcy was a good indication of its redundancy as a trunk line. Completed in 1884, it was taken over by the New York Central in 1885, greatly augmenting the traffic capacity of that system and providing a slightly shorter and uncongested route for through freight.

In contrast, the Erie lost not only its early leadership as the principal trunk line but its monopoly over the southern route. Two systems formerly dependent upon it for western connections decided to build their own lines to Buffalo. In this way the Delaware, Lackawanna & Western and the Lehigh Valley attempted to transform themselves from primarily coal haulers into competitive trunk lines between Jersey City and Lake Erie. An important portion of the Delaware, Lackawanna & Western's new line west of Binghamton directly paralleled that of the Erie and thus competed for the local traffic as well. Although the statistics are incomplete, it appears clear that the New York Central was well ahead of the three southern lines combined in volume of through freight and passenger traffic. On the other hand, local tonnage was considerably greater on the Erie than on the New York Central.

Connections west were through the Buffalo and Niagara Falls gateways across southern Ontario to Michigan, or south of Lake Erie through northern Ohio. Buffalo was the one great link with lake service; Dunkirk, initially the main terminus of the Erie Railroad, was a minor port now relegated to a branch line of that system.

Connections east into New England were of a somewhat different character but equally important to the traffic patterns of New York (Fig. 52). The principal linkages were in four districts, and each was a focus of intense competition, though it was more a rivalry among New England systems than among those of New York. The New Haven route (*1* on Fig. 52) along the north shore of Long Island Sound funneled much of the Connecticut Valley traffic directly into New York City; however, for many years a combination rail and boat service via the Long Island Railroad competed for through traffic with Providence and Boston. The situation at the Albany-Troy gateway (*2* on Fig. 52) was drastically changed by the completion of the Hoosic Tunnel by the State of Massachusetts in 1875. This provided a second route through the Berkshires and made the Fitchburg (Massachusetts) line competitive with the long-established Springfield-Worcester route for the Boston traffic. Competition with the New York Central westward was provided via the Albany & Susquehanna (the present Delaware & Hudson) between Albany and Binghamton with connections to other trunk lines. In the far north the old Ogdensburg and Lake Champlain line (*3* on Fig. 52) continued as the western portion of a minor trunk line between Boston and Portland and the Great Lakes, and, despite the fact that its traffic persisted at a level far below the hopes of those interested in it, there was sporadic competition among Vermont railroads for its control.

The fourth area of linkage between the New York and New England networks was of a rather special character. The density of lines to the east and southwest of the Newburgh-Kingston area of the Hudson Valley (*4* on Fig. 52) reflected a special competitive situation. Some of the lines east of the river had originally been built to haul produce and iron ore to the Hudson, and some of those on the opposite side had been built for similar purposes and then later expanded to include the movement of coal from Pennsylvania to Hudson River docks. But, as railroad traffic mounted, systems began to

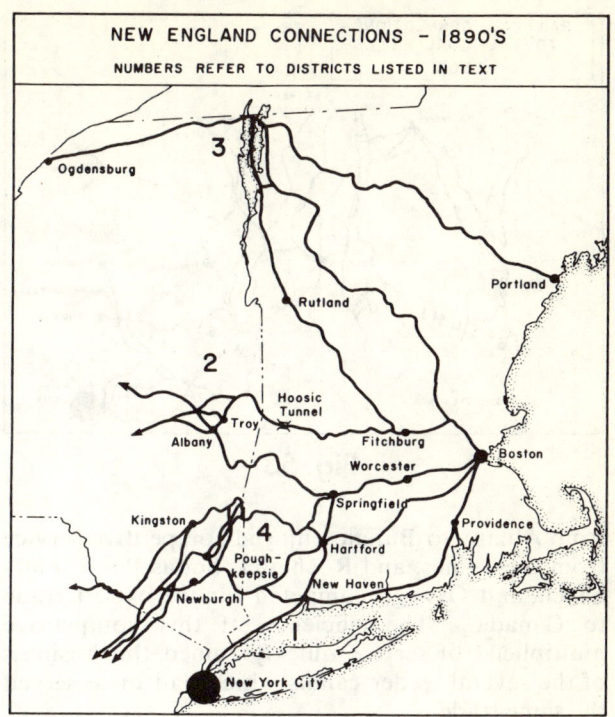

Fig. 52

be formed, and competition intensified, a bridge route between New England and Pennsylvania which bypassed the costly congestion of New York harbor became increasingly important. There ensued complex competition among various New England systems and among those of New York and Pennsylvania. Geographically much of the rivalry focused upon Newburgh, Poughkeepsie, and Kingston as to which would become the main gateway. By the 1890's Poughkeepsie had triumphed, primarily because the Hudson was bridged there, though the very elimination of the ferrying business considerably reduced the economic importance of the victory for the local community.

A striking feature of the railroad network was the density of north-south lines in central and western New York. Many of these had begun as short feeders to one of the east-west trunk lines, later expanding into lateral links between the two, but by the 1890's most had become primarily involved in the interstate movement of Pennsylvania coal northward to New York cities and ports (Fig. 53). The pattern of routes reflected the origination of such traffic from two distinct regions: one radiating from the anthracite area around Scranton and entering New York at various points from Waverly east; the other drawing from the much more dispersed bituminous coal areas through numerous portals west of Waverly. Major cities

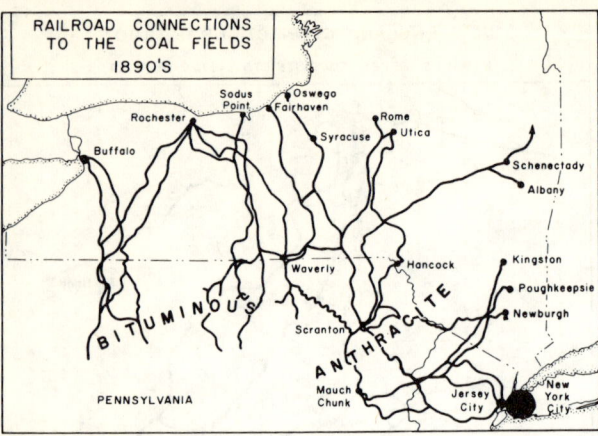

Fig. 53

from Albany to Buffalo enjoyed competitive service from these lines, and Rochester, Sodus Point, Fairhaven, and Oswego competed for the export trade to Canada. The efficiency of this competitive multiplicity of service quickly ruined the business of the several feeder canals which had once served the same trade.

Almost from its inception the New York, Ontario & Western became an infamous example of a railroad built less to serve an area than to make fortunes for its builders, and such fortunes were made not from its operation but from the public subsidies received for its construction. The generation of some enthusiasm for a new, shorter trunk line between New York City and Oswego, despite the obvious difficulty of the intervening Catskills and the persistently minor importance of Oswego, reflected an optimism about railroads as common at the time as was an ignorance of geographic realities. The line wound its way through the Catskills and across the Chenango hills, its pace of construction and direction of route determined by the availability of local public subsidies. For example, angling the line to the east and north of Lake Oneida was reputedly done to punish Syracuse for its refusal to offer financial assistance. That the New York, Ontario & Western was soon in financial difficulties was to be expected and its later melancholy corporate history became so well known as to obscure the fact that in some of its early years it made profits. It served a growing Catskill resort traffic cut through the rich Delaware and Chenango dairy area, and by building branches to Scranton and to Rome and Utica it garnered a small share of the anthracite trade. Yet it could never escape the burden of its physical and financial handicaps. That communities were willing to incur heavy debts to insure its construction was testimony to the common view that a railroad was absolutely vital. That even such a railroad as this occasionally made a profit was evidence that the railroad was indeed a singularly important instrument.

Compared to the New York, Ontario & Western, the Rome, Watertown & Ogdensburg (Fig. 51) was a model of success. The line between the first two towns of its name was the first to tap that part of the North Country, and it soon absorbed a neighboring company that extended to Potsdam and Ogdensburg. It was profitable from the first and soon sought further expansion. It added links to Syracuse and Oswego and in 1875 purchased the bankrupt Lake Ontario Shore Railroad between Oswego and Niagara Falls. At that time it encountered its first opposition. The Utica & Black River, an old but feeble expression of the chronic Utica-Rome rivalry, finally reached Ogdensburg and soon had branches to Clayton and Sackets Harbor. But in 1886 the two railroads joined and thus monopolized the traffic of this area. It was by then a rather impressive small system, with 643 miles of track, a main line from Niagara Falls to Massena connecting with Canadian lines at each end (and by ferries across the St. Lawrence at Morristown and Ogdensburg) and entry into every important city from Utica west except Buffalo— with plans to tap the latter. Such a local system greatly added to the tensions among larger rivalries. The addition of the Rome, Watertown & Ogdensburg to any one of the major systems (it was earlier briefly a part of the Delaware, Lackawanna & Western) could cause a major change in competitive situations. When it appeared that it was considering the possibility of building from Rome to Utica and down the Mohawk Valley to a connection with a New England system and thereby making itself a trunk-line competitor, the New York Central took action and quickly ran surveys into the North Country. Faced with the possibility of having its monopoly broken by so powerful a competitor, and weakened by the burden of the unproductive Niagara line, the Rome, Watertown & Ogdensburg chose to sell out at a good price to the New York Central in 1891.

This whole sequence of local lines, amalgamations, rivalries, further mergers, overextension, dreams of grandeur, and capitulation to the duress of a larger system is important not merely as a series of episodes in the history of the particular railroads described, but as an illustration of important processes common in the historical geography of American railroad development.

CANALS

The canal era, commencing before that of the railroad, continued during the latter half of the century to move through its development phases ahead of the latter. In terms of total tonnage moved the use of the canal system expanded steadily toward a peak during the years 1868–73, then gradually declined to a level about half as great in the 1890's. However, the impact of change was markedly different among various segments and the over-all geography of the system at the end of the century was quite different from that in the 1850's.

In 1862 the first comprehensive enlargement of the Erie Canal was completed. Carried out gradually over many years, the result was a deeper (7 feet instead of 4 feet), wider, straighter canal with fewer and larger locks, which considerably increased its capacity and efficiency. Larger boats were soon common, and the use of steam tugs to pull barge trains was begun. Traffic reached an annual volume of three to four million tons in the 1850's and stabilized. Grain and lumber persisted as the two great commodities. While the Erie did not suffer a general decline in grain tonnage, it did lose its proportionate dominance to the railroads; in the 1890's only about 28 per cent of the grain arriving at New York City came by canal boat.

The Champlain and Oswego canals were enlarged to the same general specifications. The former had a relatively stable annual tonnage, very largely forest products moved southward to the mills and market centers of the Glens Falls–Albany–Troy complex. The Oswego, however, reached a peak tonnage in the 1860's and declined to a tenth of that average in the 1890's. The principal factors were the rapid decline in the Onondaga salt trade and of local flour-milling, and the loss of what remained of the latter shipments to the railroads which paralleled it on either side. Traffic on the Cayuga and Seneca canal also declined very considerably.

Conditions on most of the other laterals became so bad as soon as local railroads began to compete that pressure for abandonment quickly gathered strength. While these canals had presumably benefited their localities, none had ever garnered sufficient income to cover maintenance costs. They were clearly not worth enlarging, but failure to do so accentuated their problems as it made them, in a sense, "narrow gauge," unable to accomodate the larger, more efficient carriers of the trunk line. Thus traffic declined further, structures deteriorated, and costs increased. And thus in the 1870's the state abandoned the Chenango, Chemung, Crooked Lake (Keuka), and Genesee canals, the Black River canal narrowly escaping the same fate. It had been costly to construct, but it did not as yet need expensive repairs, its traffic (very largely lumber) remained steady, and, probably most important, it was a critical water source for the Erie.

Thus the canal system, contracted in extent, remained to provide a particular kind of service for which it was well suited: the transport of heavy, low-valued bulk goods which did not require fast movement. The Erie was now primarily part of a waterways trunk line for the shipment of the western grain, and it was argued that its value must be measured not only by the tonnage moved but also by its service as a regulator of railroad charges for the same traffic. The sharp increase in railroad rates each winter upon the closing of navigation seemed to support that contention.

Agriculture

In agriculture the last half of the century was in general characterized by accentuation of changes already discernible in the 1850's. Wheat production, already declining at mid-century, revived somewhat during the Civil War and shortly after, then followed a gradual reduction toward a level of but half what it had been at its peak. It was, of course, a precipitous decline in national significance—from almost first in rank in 1850 to seventeenth in 1890. On the other hand, New York was still one of the most important barley producers, yet the peak output of barley, too, was past, more recently than that of wheat but just as irrevocably. Production had shifted from its former center in Onondaga to the Genesee (where it was the main replacement of wheat), a gradual westward movement which would soon take it right on out of the state as a major crop. This barley-growing area was the last of the grain specialty regions in New York.

As in the earlier grain districts, the shift was to more intensive activities, primarily dairying, but in some localities to some form of horticulture. New York still held national preeminence in milk, butter, and cheese production, a position further reflected by her leadership in hay acreage, and in the enhanced local significance of oats and corn. The main dairy region now extended farther west from the Mohawk into Onondaga and Oswego counties; the Watertown hinterland and the Appalachian Upland all along the southern border were important areas, too. The cheese district in southern Herkimer County remained the most intensively specialized subregion. The chief regional development in horticulture was the emergence of the Niagara-Ontario fruit belt behind the lake shores from Oswego to Erie County. Numerous smaller districts of specialization, such as the western Finger Lakes area and Chautauqua County with their wine grapes, had become well established too. The extension and development of commercial dairying and horticulture was of course made possible in large part by the spread of a railroad network that gave close contact with rapidly developing urban markets. A further concomitant of this kind of intensification was a steadily increased emphasis upon improvements in husbandry which greatly enhanced efficiency and quality. New York's continued leadership in 1890 in total investment in agricultural equipment and second rank in value of total farm products despite such a marked decline in the national rank of her volume of production were indications of this emphasis.

Furthermore, New York agriculture was not only changing in character but also contracting in areal extent. The over-all acreage of land in farms and of improved land was already declining in 1890. This reduction was state-wide. Only in a few towns, particularly in the Appalachian Upland where pastures were still expanding in the wake of lumbering, were increases reported, and these were minor in extent. The dominant pattern was the reverse: the abandonment of fields and in some cases whole farms, in the steeper, thin-soiled, or poorly drained uplands and in the gravelly and sandy lowlands. The greatest areal extent of farm land of census record in New York was in 1880, as it was in every state of the Northeast from Pennsylvania to Maine. The onset of a decline in farm land in that general region, while that in every other state in the union was still expanding, is a point of considerable historical and geographical significance. It was not simply a matter of age but of a new phase. It was not simply the result of competition from states to the west, though that was an important factor, but also the result of at least two generations or more of testing the land, of amassing a great deal of intimate empirical evidence as to what could be expected from the climate and soil of each locality. Such experience was as valuable as the emerging sciences and technology in forming the basis for the improvement of agriculture.

Manufacturing

New York's over-all industrial activity, already thriving and diversifying at mid-century, was given a powerful stimulus by the Civil War. The state was strategically situated to respond to the demands of the time: protected from any land or naval harassment and poised to ship goods with equal facility to the Potomac front or the Ohio Valley. The demands for such materials as textiles and garments, boots and shoes, arms and munitions, and transportation equipment were especially great. A war effort of this scale could not help but stimulate virtually every kind of industry.

In general manufacturing growth continued strong after the war. There were, however, some shifts in the relative importance of various activities and of localities. Lumber-processing, flour-milling, brewing, meat-packing, and tanning, all originally based upon local resources, were of declining national significance because of depletion of the local resources as well as the rise of newer regions in the Midwest. The impact of such changes, however, was often very unevenly felt within the state. For example, Rochester and Oswego lost their prominence as flour-milling centers, but the strategic position of Buffalo as the most convenient intermediary between lake shipments of Midwest grain and the Northeast market was steadily enhanced and it became firmly established as one of the three greatest milling centers of the nation (along with Minneapolis and Kansas City), a position it still holds today.

While total production in lumbering declined after 1860, output in the Adirondacks continued to expand and a high level of output was sustained between 1870 and 1890. The Albany-Glens Falls complex persisted as the most important focus, but other important clusters of sawmills flourished around the periphery of the highlands, especially in the north where railroads intersected log-driving streams, such as at Carthage, Gouverneur, Canton, Potsdam, and Norwood. By 1890 the end of that era of lumbering was clearly discernible, for only the higher and isolated sections of the Adirondacks remained untouched and the completion of the railroad across the highlands from Remsen to Malone would soon give better access to these last remnants. By that date the Albany-Troy market had already lost its leadership even within the state. Tonawanda was now second only to Chicago in the shipment of lumber. As the former as well as the latter drew upon the rich resources of Michigan and Wisconsin, it was a change comparable in type and pattern to that in flour-milling—another example of the Buffalo district's capitalizing upon its intermediate position between the resources and the market.

Despite such declines in certain activities, in 1890 New York had a sizable national leadership in the number of factory employees and in the total value of its manufactures. Yet, again, the state as a whole was much less intensively industrialized than southern New England. The New York City area accounted for about 60 per cent of manufacturing in the state, which made it by far the largest such center in America, and a half-dozen other major districts accounted for much of the remainder.

All of the major districts were to some degree diversified centers, although considerably varied in character and degree. The Albany district, for example, was composed of Albany, Troy, Watervliet, Cohoes, and Schenectady. Each was rather different in its manufacturing emphasis, and they together constituted a diversified economy based upon such major items as lumber, textiles, iron and steel products, transportation equipment, and a federal arsenal. Syracuse had shifted from its heavy dependence upon salt—an industry now suffering severely from depletion and midwestern competition—to a highly varied range of products including chemicals, pottery, steel, candles, shoes, agricultural machinery, typewriters, and other machinery. The decline in flour-milling in Rochester had been offset chiefly by an increase in textiles, but here too considerable variety had developed, including optics, instruments, and in the 1880's photographic supplies and cameras. Utica, principally a textile center, was the least diversified of the major areas, while New York City was by far the most diversified, with very few classes of industrial production not represented there.

Some smaller cities were flourishing on the basis of rather narrowly specialized activity: Corning had its glassworks, Jamestown its furniture factories, a number of others were dependent upon a large agricultural machinery works (e.g., Hoosick Falls) or a textile or carpet plant (e.g., Amsterdam). Probably during this period the greatest number and variety of small plants, though not the highest percentage of manufacturers, were scattered over the countryside in small cities and villages. For example, descriptions in the 1870's of Columbia County reported more than sixty factories, mostly in country villages, and it may be taken as fairly typical of the counties along the main Hudson-Mohawk axis. Products were principally cotton goods, paper (much of it from rye straw), and agricultural equipment. Employment in individual factories varied from three to three hundred, with the majority having a work force of about ten to twenty. This scale and variety of factories was partly a carry-over from an earlier era characterized by many small water-powered mills and partly of the newer era of larger steam-powered mills. Clearly the trend toward fewer, larger, more highly mechanized, and more favorably located plants in the larger urban areas was continuing, but such developments had not as yet destroyed the competitive position of smaller factories in the smaller communities.

POPULATION

New York's six million people in 1890 made it still by far the most populous state. Actually its rate of increase was lower than that of many others, but the size of its total was such that even a relatively low rate resulted in a large annual increment.

There were some important changes in distribution during the last part of the century. Figure 54 shows the situation in 1880, a date important because it marks the peak of rural density. Such an even spread of people over the state will not again occur. Between 1880 and 1890 nearly every rural town declined in population, while every settlement over 3,000 increased. Hence it was no longer appropriate to think of growth as a general phenomenon even along the main axis, but rather of growth confined to urban clusters, for all rural areas and even some of the large villages (of 1,000 to 2,500 size) within that axis were stagnant or declining.

There were some significant shifts in proportionate growth and position of the various urban areas, too. The New York City metropolitan region, with 45 per cent of the total in 1890, was rapidly rising to a proportionate position it had lost more than a century before. In the 1880's Buffalo surpassed the Albany-Troy cluster as the second largest, but Rochester, Syracuse, and Utica retained the same ranking along the Mohawk-Erie axis. In the south Binghamton had doubled its population since 1880, and Elmira, Corning, and Jamestown had also grown at a rapid rate. The latter two were typical of a considerable number of smaller industrial centers that had undergone sharp increases, such as Yonkers, Gloversville, Amsterdam, Glens Falls, and Tonawanda. The unevenness of the pattern of urban growth was accentuated by the relatively slight increases in such older cities as Kingston, Hudson, Oswego, and Canandaigua.

Because so much of the population increase was the result of foreign immigration, it might be expected that distribution of foreign-born people would be essentially identical in pattern to that of urban growth. But such was not quite the case. One-fourth of the total population in 1890 was foreign-born, and obviously the average proportion in the cities was greater. Yet in some of the smaller cities of most rapid growth, such as Binghamton, Elmira, Oneonta, Norwich, and Middletown, the proportion of new immigrants was much less than the state-wide average. A detailed examination of such disparities reveals that the influx of foreigners was rather sharply channeled into urban centers within the main Hudson-Mohawk-Lake Erie axis. Within that belt there were still significant differences. Some of the small but heavily industrialized centers such as Yonkers, Cohoes, Dunkirk, and Jamestown, received considerably larger numbers in proportion to their total population than such older and larger centers as Albany, Troy, Utica, and Syracuse.

The principal sources of immigration had remained much the same over the past forty years. Irish and Germans each accounted for about a third of the total and British sources for the biggest share

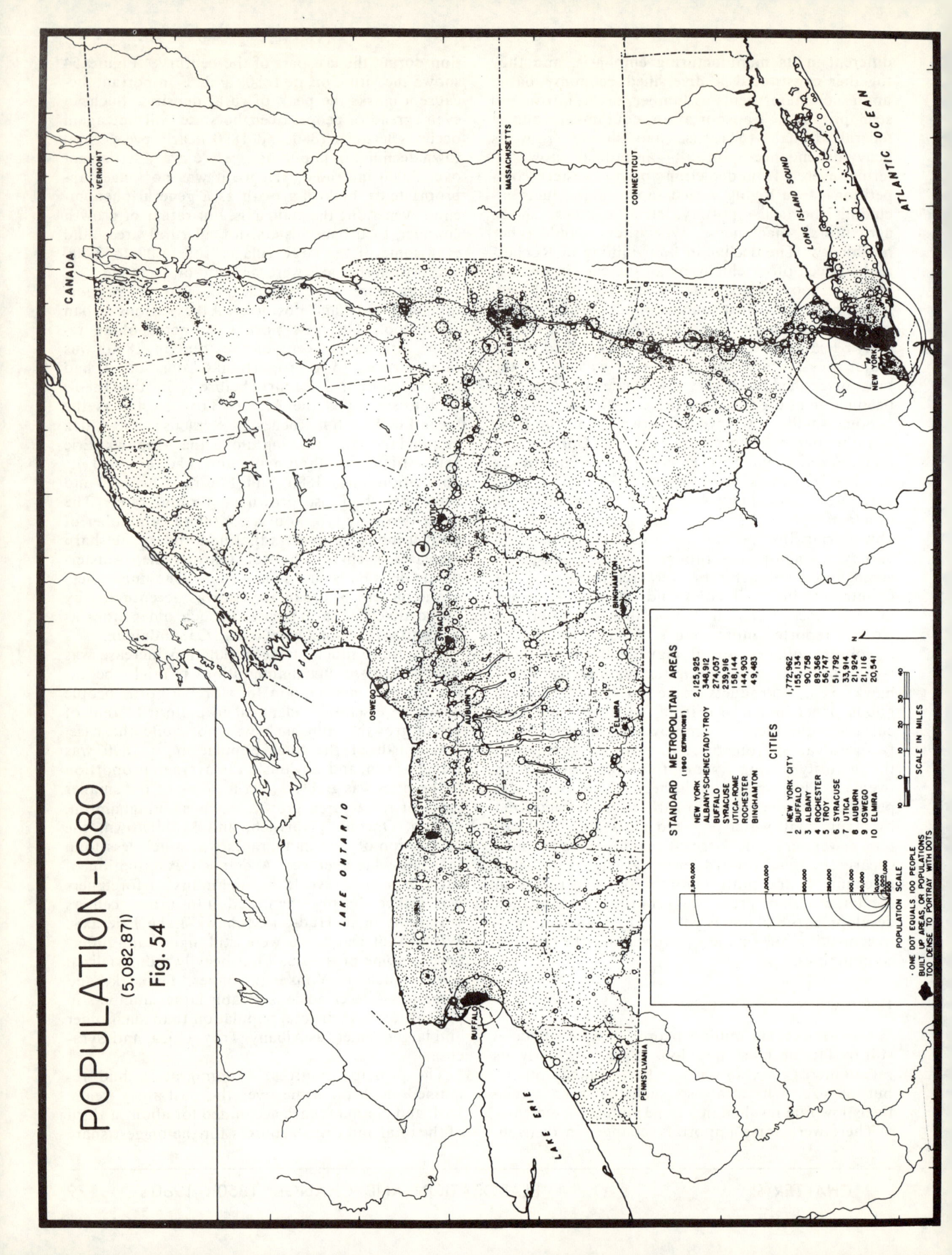

of the remainder. A hundred thousand Slavs from Eastern Europe and 64,000 Italians within the total foreign-born in 1890 were relatively new groups, increasing rapidly with each year but still constituting but a small share of the 1,500,000 total. In general, the proportionate distribution of these national groups was not significantly different among the various urban areas, although a few anomalies, especially the concentration of Swedes in Jamestown, Welsh in Oneida County, and Poles in Buffalo, were quite apparent.

Phase II: 1890's–1920's

The period spanning the opening years of the twentieth century is peculiarly important because so much of the present is directly rooted in that particular past. Many of the strands connecting these periods are related in some way to the perfection of the electric motor and of the internal combustion engine. The electric railway and the automobile each, in turn, had important impact upon the transportation facilities of the state. On the other hand, the general spatial patterns of manufacturing were not radically altered during these years, although a few localities did undergo exceptionally rapid growth. Certain new kinds of industry became prominent, too, thereby altering somewhat the particular character of various industrial districts. Overall, there was strong continuance of the trend toward concentration of industrial activities into fewer and larger urban areas, but now it was at the expense of hundreds of smaller centers.

Agriculture was characterized by a slow but persistent decline in nearly all of its gross measures: e.g., in total land in farms, in cropland, and in number of farms. Only those more intensive activities such as dairying and horticulture, which were developing in the preceding phase, showed much stability.

All of the above trends of course had some impact upon population distributions. Furthermore, the population of the state was not only undergoing some shifts in area but growing very rapidly and being profoundly modified in character by large numbers of immigrants from new sources overseas.

Transportation

The electric railroad was one of the most characteristic expressions of the transitional nature of these times between the older technology of the "coal-steam-iron" complex and the newer one just emerging. Its competitive advantage was clearly the product of the new technology: a relatively compact, efficient motor receiving continuous power which made possible fast, frequent service by one or a few passenger cars, allowing a flexibility of service quite uneconomic for steam railroads except in unusual circumstances. Yet, at the same time, it remained a railroad, with all the inherent rigidities in route, enormous investment in roadbed and maintenance, and hence high profitability only under conditions of mass movement.

Once perfected, the electric railroad was promoted with the same frenzied speculation that had characterized the initial stages of earlier transport forms. The first application was to streetcar operations, whose advantages in speed and economy over horsecars were immediately apparent. By 1900 practically every city over 5,000 population and some smaller ones had a streetcar system. Radiating from the central business district, these electric lines were a major factor in promoting elongated sector patterns in commercial and residential growth in the larger cities. They also allowed suburban commuting to become a normal pattern for a rapidly increasing proportion of the urban labor force, and, in fact, streetcar companies and real estate companies were often closely associated in the promotion of residential subdivisions at the outskirts of cities. It was also common for streetcar firms to develop resorts and amusement parks some distance beyond the city as a means of increasing traffic.

The electric interurban was in form, and often in fact, merely the extension of a streetcar system. The earliest ones were just short links between neighboring communities such as Olean and Allegany, Dunkirk and Fredonia, Middletown and Goshen, Newburgh and Walden. These were built in 1894/95, and within a very few years scores of companies had been formed and electric lines were being extended out of dozens of communities. By 1910 the boom was over and over 1,100 miles had been constructed. Figure 55 suggests several apparent networks, but in ownership and operations these were made up of many separate pieces. In fact one could travel from Oneonta to beyond Chicago by electric interurbans, but only by means of changing cars many times, including half a dozen changes in crossing New York. The interurbans made no attempt to compete with steam railroads for long-distance travel, but in many cases they were built directly parallel with steam railroad lines and were quite successful in garnering passenger traffic between nearby cities.

The prosperity of electric railroads was short-lived because the automobile was being perfected concurrent with the evolution of this new form of rail service. After 1900 the number of autos increased rapidly, and because of its size and wealth New York had a greater number than any other state. Yet it took time for the impact to be felt.

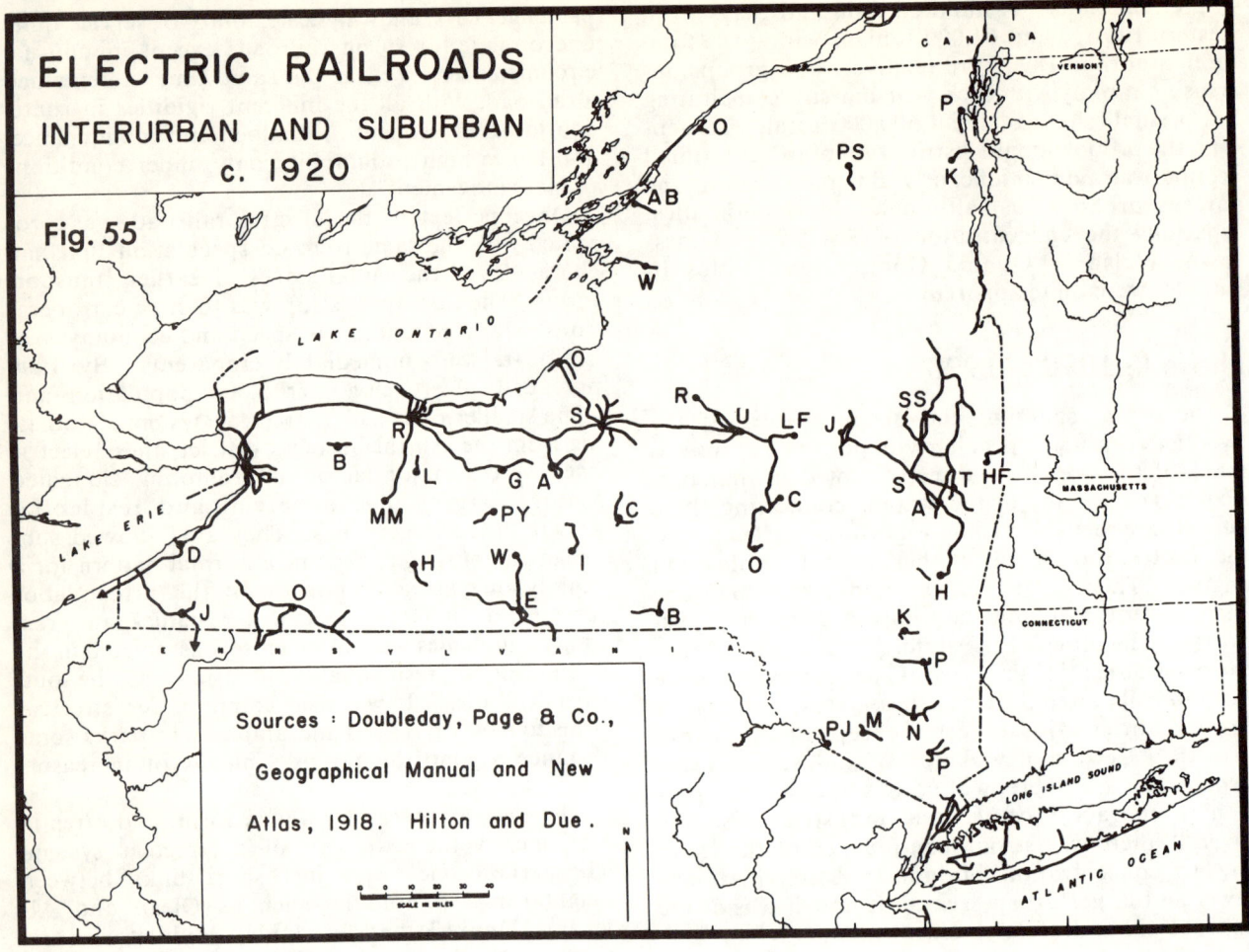

Fig. 55. ELECTRIC RAILROADS INTERURBAN AND SUBURBAN c. 1920

Sources: Doubleday, Page & Co., Geographical Manual and New Atlas, 1918. Hilton and Due.

By 1910 the efficiency and feasibility of the new machines within the cities were well demonstrated, and taxicabs had driven most of the horse-drawn hansoms from the streets. Within a few years city buses were beginning to capture a share of traffic from streetcars. But in the country, even between major cities, mud and snow made the auto a seasonal vehicle at best.

In 1916 the first federal highway construction program was launched. As a result of federal and state efforts, by 1926 New York had a comprehensive and unsurpassed network of paved roads (Fig. 56). In view of the fact that at that time most western and southern states had almost no paved highways and that pavement on the transcontinental routes ended at the Mississippi or shortly beyond, it was a remarkable development. The very rapid increase in the number of automobiles and buses concurrent with the building of such a road system put a quick end to electric interurban profits. Several smaller lines were abandoned in the 1920's, and the entire system collapsed in bankruptcy in the depression of the 1930's and eventually the trackage was abandoned.

During this same transition period, the canal system underwent its second major alteration with the creation of the New York State Barge Canal. Basically it was a widening and deepening of the Erie Canal and was completed in 1918. However, though it followed the same general route, there were numerous important local innovations. The old canal along the Mohawk was abandoned and replaced by the river itself, whose channel was deepened and straightened. Farther west the old route through Syracuse was abandoned and replaced by one via Oneida Lake; Syracuse was connected to the new route by way of Onondaga Lake and the Seneca River. The most compelling argument for investing in this enlargement was the belief in a continued need for water competition with the trunk-line railroads.

The Black River canal was finally abandoned.

The Champlain, the Oswego, and the Cayuga and Seneca canals have remained in operation, but unprofitably.

AGRICULTURE

The impact of the newer technology was as yet largely indirect upon agriculture. Evidence of this was the fact that the nearly 200,000 farms of the state reported fewer than 7,500 tractors in 1920 and probably only few of these were used as mobile power for field work. And although autos and trucks were developed more rapidly than tractors, the almost complete absence of improved roads in the rural districts strongly retarded the use of motor vehicles by farmers. Nor as yet had rural electrification projects brought the new convenient source of power and light to many farmsteads.

Yet the farming economy was undergoing some change. The steam and electric railroads and the widening influence of the urban centers were reflected in an ever greater commercial emphasis on crops. The family orchard, garden, and dairy cow steadily gave way to greater specialization. Primarily, such features were associated with the steady spread of commercial dairying. While the acreage in cash grain crops sharply declined and the over-all total land in farms and in cropland contracted, the acreage in corn and hay and the number of cows were not markedly affected. The total milk produced actually increased. Although special horticultural areas continued to exist, the general result was the development of greater similarity among the various agricultural districts of the state with emphasis on dairying.

MANUFACTURING

In general the impact of electricity was gradual and did not produce abrupt geographic changes. From the beginning of widespread use of electricity, steam-powered plants generated the greatest proportion of the total supply. However, it has always been the great hydroelectric plants that have been

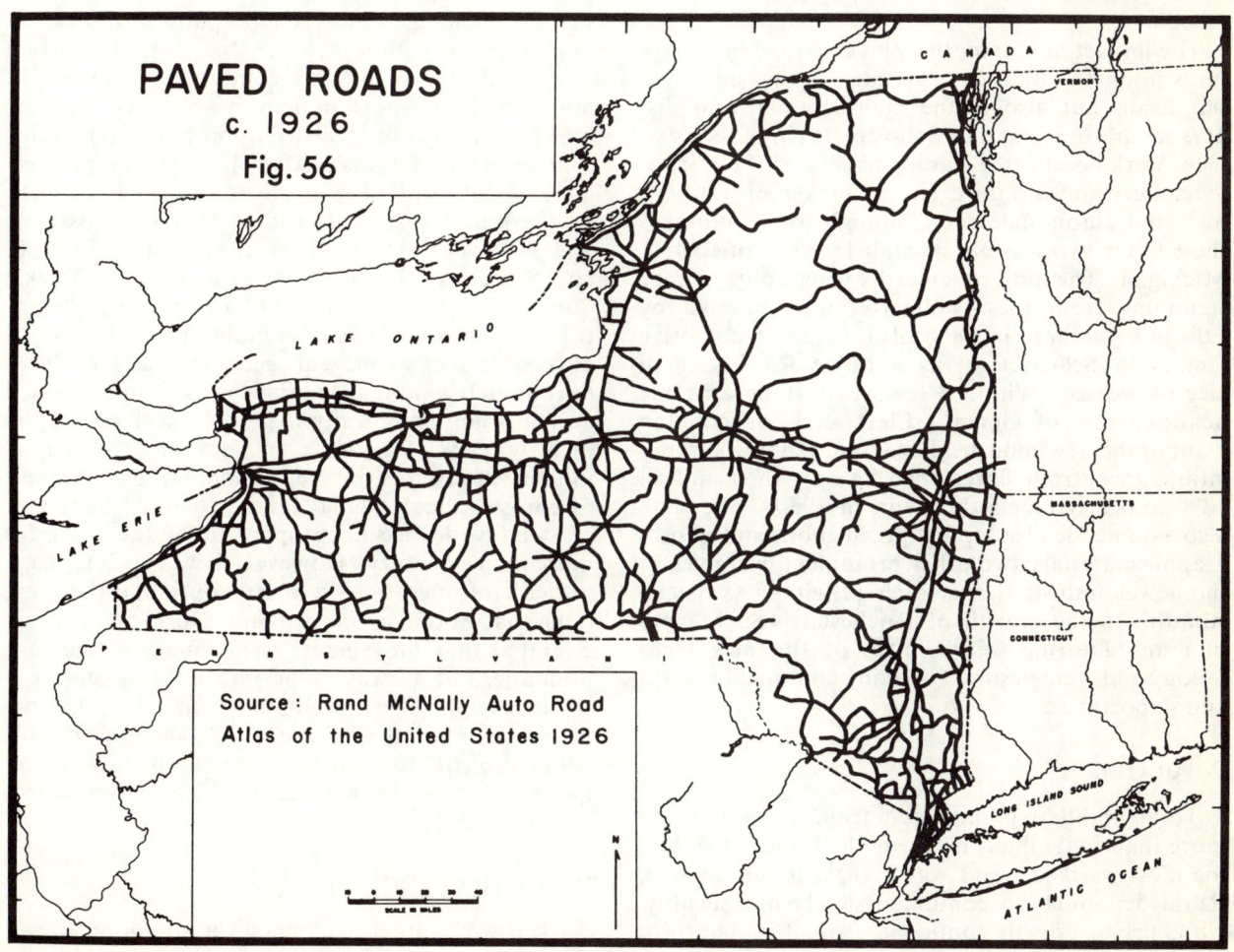

PAVED ROADS c. 1926
Fig. 56
Source: Rand McNally Auto Road Atlas of the United States 1926

CHAPTER 9   ELABORATION AND CHANGE, 1850's–1960's

most closely associated in the public mind with electric power production. Furthermore, because such power sites were often in localities quite different from older power sources, they are of special geographic interest. New York was of course blessed with the most spectacular single hydroelectric site in the nation.

The first generator at Niagara was installed as early as 1881, but it was another twenty years before the quantity produced was of importance and transmission lines were beginning to be extended over the western half of the state. The availability of such power soon transformed the city of Niagara Falls from purely a resort into an important industrial center as well. Between 1890 and 1900 its population nearly quadrupled, and by 1920 it had more than doubled again to a total of fifty-one thousand. There were many smaller power sites in the state and many were soon developed. However, because electricity can be transmitted over short distances quite efficiently, most of the power so generated was sent to established industrial and urban centers and thus the geographic changes associated with them were too small and too numerous to be detailed here.

The impact of automotive and electrical manufactures upon the industrial pattern of the state was significant but also in the main gradual and dispersed among existing manufacturing districts. New York became the leading producer of electrical machinery and the third largest maker of automobiles and automobile parts, though total output of these latter two was but a small fraction of that of Michigan. The most noteworthy geographic change stemming from these industries was initiated by Edison's purchase in 1886 of a vacant locomotive factory in Schenectady as a home for his small electric works. Within a few years it became the headquarters of General Electric, a burgeoning giant of the new industrial age, and the city's population grew from fewer than twenty thousand in 1890 to nearly ninety thousand in 1920. The state also assumed leadership in optical goods and photographic materials, two other prominent examples of the newer industrial age, each providing a special stimulus to the growth of Rochester. Such types of manufacturing were typical of the new technology and were destined to enjoy continued growth and importance.

POPULATION

Total population increased from six million to more than ten million between 1890 and 1920, and the most marked trend was a rural to urban one. Rural depopulation continued slowly but steadily, while urban growth continued strongly. The proportion of the state's total in urban centers rose from 65 to 82.7 per cent. The large cities generally at least doubled in size. Industrial and residential cities at the fringe of metropolitan centers, such as Yonkers and New Rochelle, Lackawanna and Tonawanda, clearly mirrored this growth. Among smaller cities there was much more unevenness. Schenectady and Niagara Falls were spectacular examples of increases, and there were a number of others, such as Binghamton, Jamestown, Corning, Geneva, and Watertown, which doubled or nearly doubled in size. And yet there were others that were very nearly stagnant, among them Troy, Cohoes, Oswego, Hudson, and Kingston. Quite clearly, industrial growth was no longer a general feature to be enjoyed almost as a matter of course by all; it was becoming increasingly associated with particular kinds of industries, and even, perhaps, with particular corporations within industry groups. Such variability in city trends was a sign of the maturity and complexity of the state's economy.

These decades marked the last great surge of essentially uncontrolled immigration and such newcomers continued to play an important role in sustaining the growth of the state. There was little significant change in the general pattern of location of immigrants within the state (Fig. 57). The urban centers of the major axis continued to receive a much greater proportion than those elsewhere. The big change was not in distribution but in the major sources of immigrants. After 1890 the preponderance quickly shifted from northwestern Europe to southern and eastern Europe. Italians, Russians, and Poles now replaced Irish, Germans, English, and Scotch as the principal groups, and Greeks, South Slavs, Hungarians, and Baltic peoples all contributed important streams of migrants. The spread of such peoples so unevenly geographically and proportionately among the various urban centers added an important complexity to the cultural geography of New York (Fig. 58). An additional ingredient in that variety was the Negro population. Present, of course, since colonial times, Negroes had been of steadily decreasing proportion of the total for more than a century. However, there had long been a trickle of migration from the South and the great industrial boom accompanying World War I accelerated that movement. The increase was not dramatic, but it was sufficient to bring about a slight reversal of a long-term trend. In 1920 the Negro population was about 200,000, less than 2 per cent of the total of the state; more than three-quarters of these lived in the New York City metropolitan area.

Phase III: 1920's–1960's

Clearly no other technological development of this century has had a geographic impact so impor-

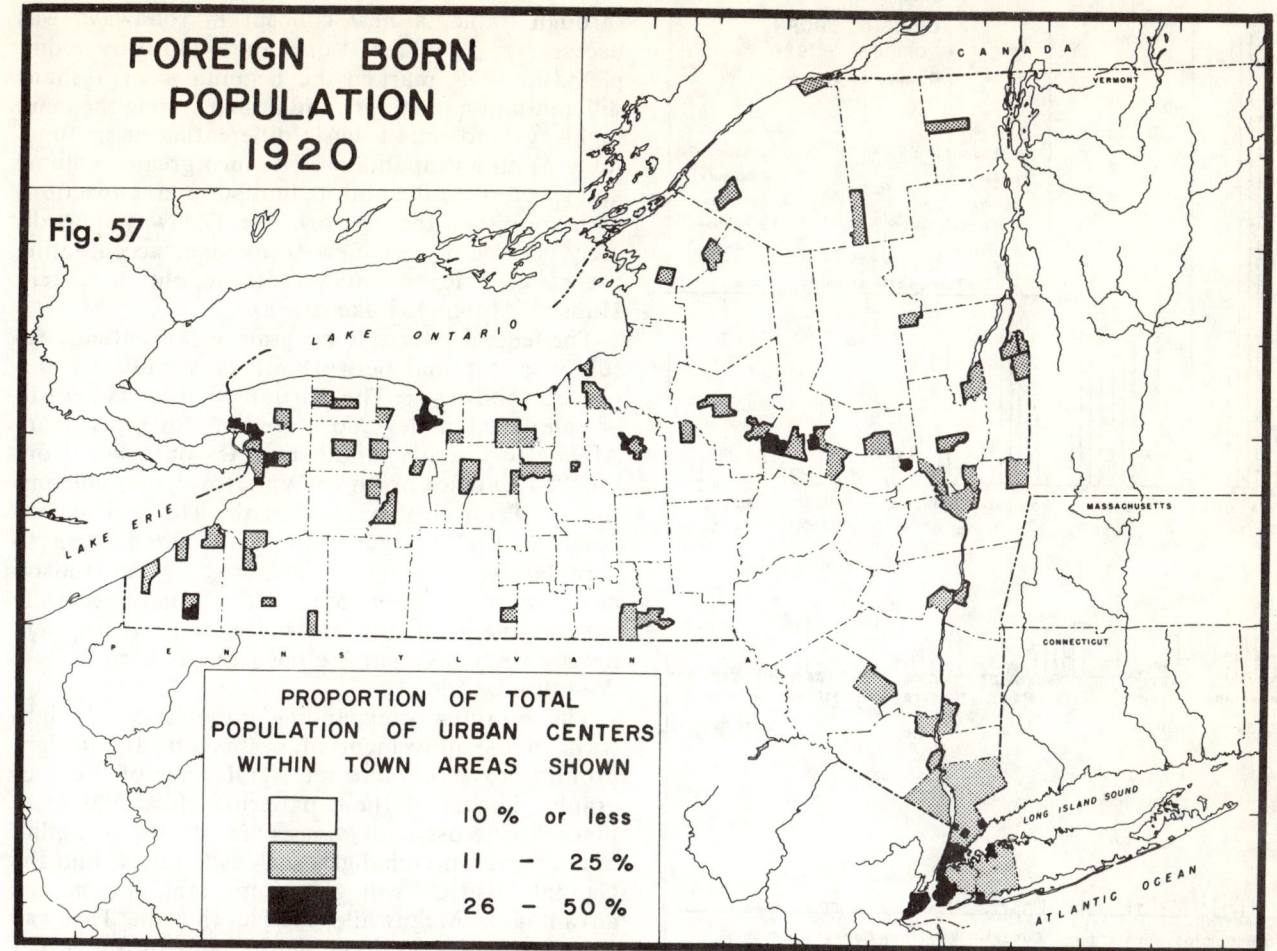

Fig. 57 FOREIGN BORN POPULATION 1920

PROPORTION OF TOTAL POPULATION OF URBAN CENTERS WITHIN TOWN AREAS SHOWN
- 10 % or less
- 11 − 25 %
- 26 − 50 %

tant and comprehensive as that of the motorcar and truck. An ever-accelerating increase in the quality and quantity of vehicles and roads brought a rather sudden burst of the automobile age in the 1920's, with promise of an even more radical impact than that of the railroad age nearly a century before. The application of the gasoline motor to air travel lagged about a generation behind. The airplanes and services of the 1930's were analogous in stage with the autos of 1910, and World War II would give an enormous stimulus to developments in aircraft, just as World War I did to those in land vehicles. In each case the new instrument emerged readied for competitive service in mass transportation.

The general characteristics and trends in agriculture and manufacturing since the 1920's are largely a continuation of those evident earlier, and they merge without sharp alteration into those of the present. Farms and farmers became fewer and dairying more dominant. Manufacturing gravitated more to the larger urban complexes and, in the latter half of the period, to the periphery of the larger urban places.

Similarly, the dynamics of the population geography of the state in the main reflect an accentuation of trends already well under way, although the automobile had and continues to have a special impact upon the detailed patterns of urban growth.

TRANSPORTATION

The ramifications of the new mobility of this era cannot be summarily stated, but the broad influence upon the state's general transportation complex must be indicated. The basic pattern of development of improved roads was not sharply different in type from that of the railroads; initially, local extensions of streets into the countryside and the linking of nearby cities, then expansion of the latter into trans-state routeways with an increasing number of interconnections among them until a complex network was created. Of course, the highway network became far more dense than that of the railroad,

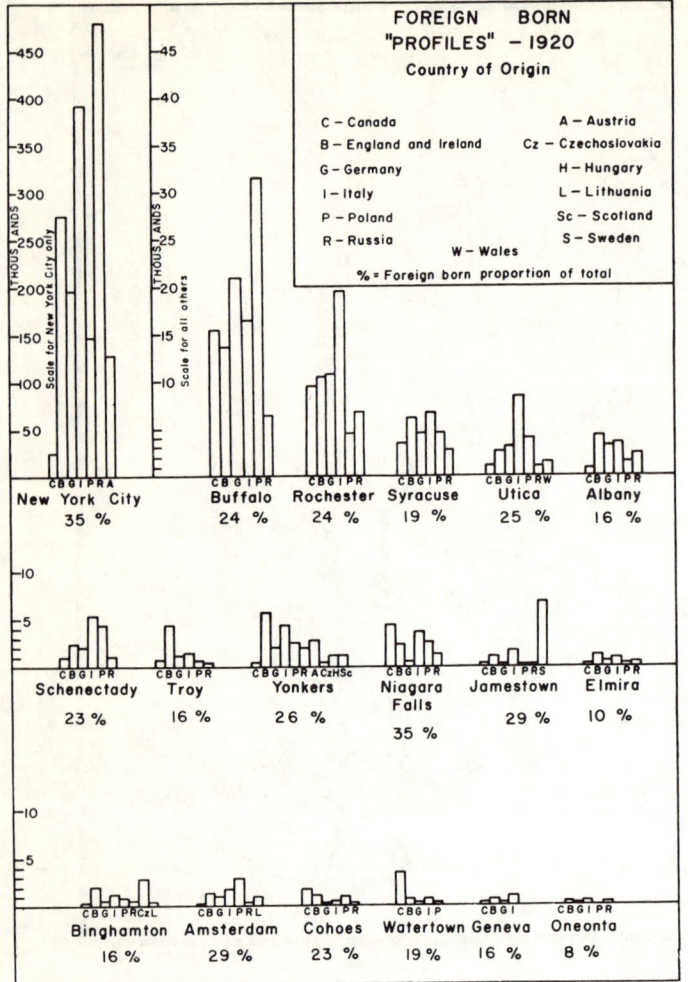

Fig. 58

and it was different in kind in that its every segment was equally open to every user (although scheduled bus and truck lines are restricted to franchised routes). Thus the highway system represents a new phase of far greater multiplicity of interconnections between areas, giving a far greater flexibility to spatial interactions.

Even within such a web there were of course great variations in the volume and type of traffic, as well as in dominant function of trunk lines and feeders. However, as long as even the biggest and best roads were open to local as well as long-distance traffic, and even local roads were paved and improved so as to be open to all kinds of vehicles, the trend was toward increasing flexibility of potential spatial interaction.

The construction of the New York Thruway marked a reversal of this trend. As the old system became increasingly unable to meet efficiently the demands of the constantly increasing volume of through traffic, a new concept in roadways was necessary. The New York State Thruway, completed in 1955, marked the beginnings of a sharp differentiation in quality and design among the main highways and thus a basic differentiation in function. With its capability of a much greater volume and speed of traffic and its limited interconnections with the rest of the network, the Thruway immediately became a great new trunk line, accentuating the geographic advantages of the old, persistent Hudson-Mohawk-Lake Erie axis.

The federal interstate program will eventually result in a national network of such radically improved roadways. The portions of that system to be completed in New York are the Northway, from Albany to Canada; another north-south link from the St. Lawrence north of Watertown to Binghamton and Pennsylvania, crossing the Thruway at Syracuse; and a "bridge route" from Connecticut to New Jersey and Pennsylvania, crossing the Hudson near Newburgh (Fig. 59). Other roads generally comparable in quality but not part of the system have been constructed elsewhere, and more will doubtless be added.

The continually expanding importance of highways in the movement of passengers and freight promises to accentuate the significance of the geographic impact of these patterns. It would seem that major crossroads areas enjoying the flexibility of numerous interchanges, such as Syracuse and the Capital District, will gain important commercial advantages. Meanwhile, the fact that the Thruway was completed so long before any of the other major routes must surely have already worked to the advantage of the cities along that old, persistent axis. Viewed at a different scale, the local impact of such superhighways can obviously be critical. Failure to obtain a local access road (interchange) to such routes will surely be almost as depressing to the commercial hopes of a community as was the failure to obtain a railroad in the previous transport era.

During this same period the airlines added still another mode to the state's transport complex. Inherently different from other forms, air service also has a particular geographic character. High costs make airlines dependent upon near-capacity use of planes, and thus air service began as exclusively trunk-line service between major cities and only very gradually extended its network to lesser ones. Nor is air service simply a matter of size of populations, for the cities must be spaced far enough apart to make air travel sufficiently advantageous over ground transport. A comparison of scheduled air routes in New York in 1933 and 1963 shows the increase of geographic coverage (Fig. 60), but of course gives little indication of the enormous in-

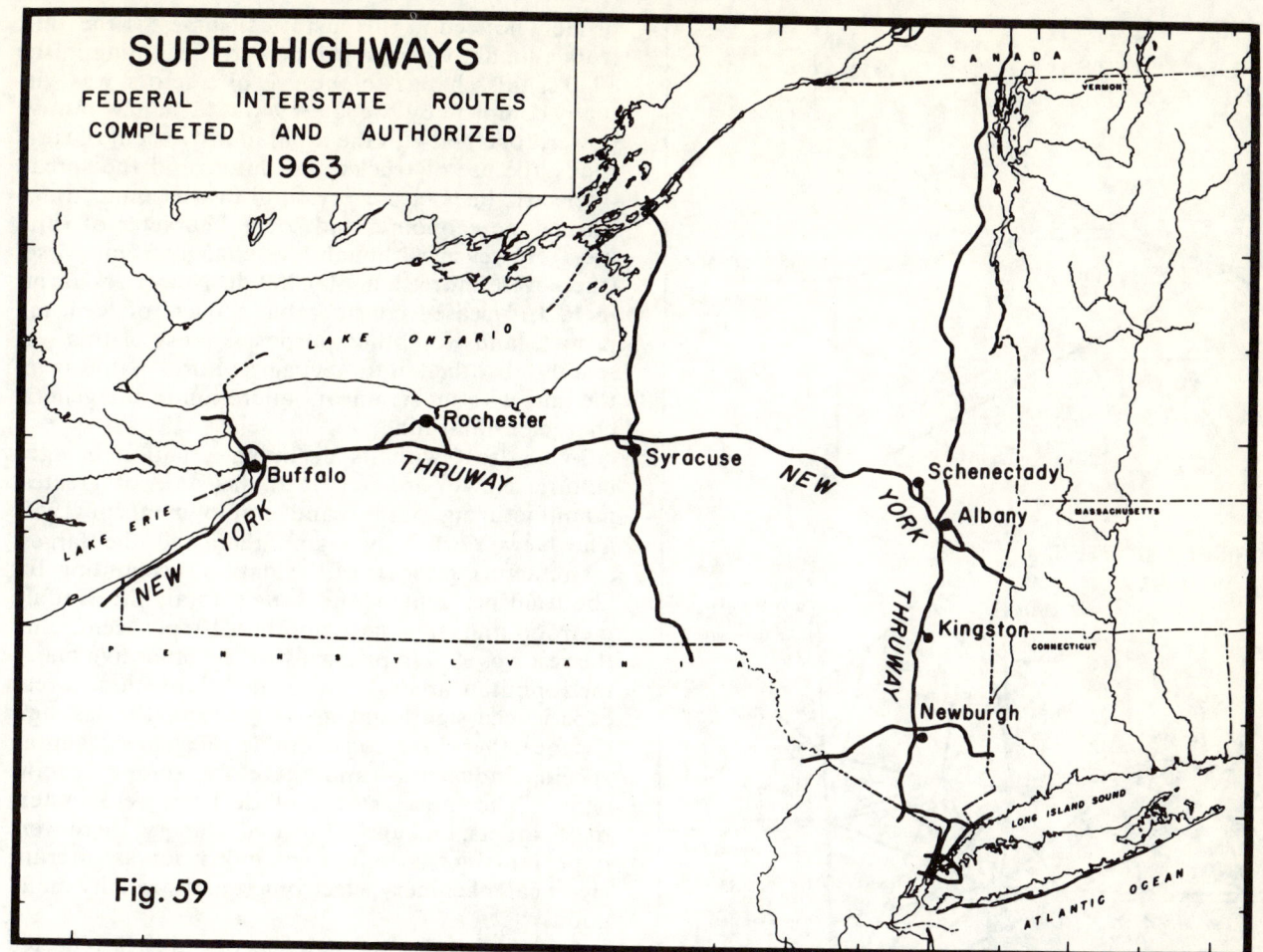

Fig. 59

crease in the volume of services offered between major centers. The geography of air service has been somewhat unstable; trial periods have shown insufficient traffic from some of the smaller cities. Further development of the superhighway system may well cut into some of the present air traffic so much as to make it unprofitable. On the other hand, technological advances may make it feasible to provide air service competitively even to smaller and more closely spaced cities than is now possible.

The impact of these newer transport mediums upon the older has been very heavy. The railroads lost virtually all their local intercity passenger and freight business to autos, buses, and trucks, most of their long-distance freight traffic to trucks. Thus just as competition from the railroads narrowed the transportation role of canals to certain types of commodities and movements in the nineteenth century, so the railroad was similarly, if not quite so drastically, narrowed in function in the middle twentieth. In addition to the impact of competition, the great reduction in the demand for coal has had an especially heavy effect upon railroad traffic. At the same time, the great increases in the use of petroleum products have been of little benefit, as seagoing tankers, barges, and pipelines completely dominate the long-distance movement of such commodities.

This reduction has inevitably led to a major contraction of the railroad network. Just as with the canal system, local branches and feeders (often separate railroad companies) were the first to succumb. Extensive abandonments began during the 1930's; World War II brought a temporary respite, followed by a rapid increase in the scale and rate of contraction. In addition to hundreds of miles of feeder lines, two small systems were completely abandoned: the Rutland, which operated the old Ogdensburg–Lake Champlain trunk line, and the long-struggling New York, Ontario & Western. Mergers, such as the Erie–Lackawanna (formerly the Erie and the Delaware, Lackawanna & Western), have allowed another means of contraction through the elimination of duplicate lines and facili-

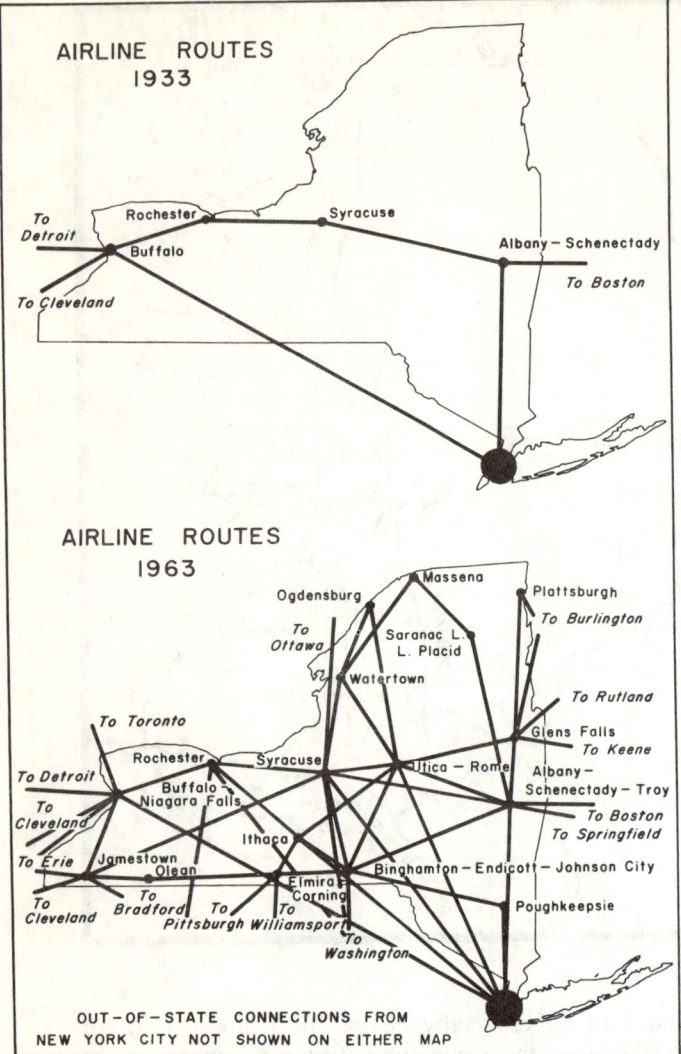

Fig. 60

ties. Other mergers are presently under consideration and if accepted may further alter the state's railroad pattern. The present network of about 6,300 miles (as compared with 7,870 miles in 1918) will almost certainly be further reduced. (See Fig. 85).

AGRICULTURE AND MANUFACTURING

New York's agriculture continued to contract in areal extent, in its significance to the state's over-all economy, and in national importance. First in rank in value of production in 1860, New York had slipped to thirteenth in the nation in 1960. The important aspects of these changes, as well as many characteristics of that agriculture, are described and analyzed in Chapter 10.

One feature, however, which can be most appropriately noticed in this historical survey is the shift from animal to mechanical power. Beginning in the 1920's, this change to the use of tractors was virtually complete by the late 1940's. The new motive power, together with the rapid improvement of rural roads, the use of trucks and autos, and the spread of electric lines into every rural district quite transformed the economic and social character of rural life. However, although the change from horses (there were more than 500,000 on New York farms in 1920) released considerable acreage of feed and pasture lands for other purposes, most of this was readily absorbed into the agricultural patterns of the time without significant alteration in the general character of farming.

In contrast with its declining situation in agriculture, New York remained the state of greatest manufacturing output and employment in 1960. The New York City region remained the largest manufacturing center of the nation, accounting for about 60 per cent of the state's total; the Buffalo metropolitan area had another 10 per cent; and the rest was shared primarily by the other five major metropolitan areas. Although all of these areas experienced significant growth during the last few decades, there was considerable unevenness among specific industries—and therefore among specific cities. The most significant declines were in textiles, carpets, and agricultural machinery; there were important increases in such industries as aircraft, electrical machinery, electronics, and specialty metal goods.

Geographically, the patterns of industry reflected other general trends of the times. On the one hand, there was greater concentration into the few major urban areas; on the other, there was considerable dispersal from the center to the periphery of those areas. Whereas late nineteenth-century trends were tending to bring a concentration of factories into tightly clustered industrial districts within cities, dictated strongly by power sites, sidings, and canal access, mid-twentieth-century trends were favoring the development of industrialization on cheaper and more plentiful land at the peripheries of large urbanized areas. Easy access to superhighways and arterial commuting routes had become a determining factor. Although there remain hundreds of factories in the smaller cities of the state, the competitive position of such localities for industrial expansion is in most instances deteriorating unless they are near enough to a major metropolitan area to partake of many of its advantages without sharing some of its disadvantages.

Such trends and characteristics, as well as other features related to the present geography of manufacturing, are examined in greater detail in Chapter 11.

POPULATION

In the 1920's New York's population total increased by more than two million, the greatest numerical growth in any decade of its history; in the next ten years, quite in contrast, the total gain fell below 900,000, less than any decade since 1880. In both of these spans New York was merely mirroring the national characteristics of the "Booming Twenties" and the "Great Depression." Since 1940, however, New York has lagged well behind the national growth rate; the increase during the most recent census period, 1950-60, was only about half the national average. Nevertheless, the base population is so great that even such a reduced rate meant that nearly two million people were added during that decade, bringing the total to 16,783,604 in 1960 (Table 14). Thus there were obviously some areas of vigorous expansion within the state.

The principal geographic characteristic of the period since 1920 was an ever greater concentration of growth into fewer and larger urban areas. Population expansion was much less a general urban characteristic and much more specifically a metropolitan one. In 1960 the New York City region held 63 per cent of the state's total, and each of the major urban agglomerations upstate had also increased its proportionate share of the state's people.

Examined at a larger scale, the principal geographic characteristic was clearly the "explosion" of the peripheries of the urban areas, with each city becoming surrounded by rapidly expanding rings of new suburbs, engulfing former country villages and often served by several satellite centers some distance away. This was, of course, one of the great geographic impacts of the radical new mobility of the automobile age. A corollary of this rapid expansion outward was the decline of population within the old central portion of the city and the rise of all the problems attendant upon such retrogression. In each of these cities, the new on the outside produced a desperate need for renewal on the inside.

Beyond the influence of these major clusters, the situation was in general quite different. Nearly every smaller city stagnated during the 1930's, few have had any significant growth since, and some have been slowly declining for thirty years (e.g., Amsterdam). Actual losses are usually associated with the closing of major manufacturing plants, and smaller cities heavily dependent upon the textile or carpet industry have been the hardest hit. The stagnation or very slow growth of the great majority of smaller cities is of course a reflection of the failure to attract new manufacturing or tertiary industries or to be selected as places of expansion by those manufacturing plants already there. In general, this reflects a national trend toward greater industrial concentrations in the larger metropolitan areas. However, that trend appears even stronger in New York, for some smaller cities in the South and West are expanding rapidly with new industries. Thus local patterns of growth as well as over-all state trends are reflecting not the general trends of the nation but those of the older industrialized Northeast.

In 1924 the United States reduced by law the annual volume of immigration and established a rigid quota for each country, heavily favoring those of northwestern Europe. Recent legislation has increased somewhat the annual quota and altered those for each country, but the general law remains restrictive and selective. In 1960 the foreign-born population numbered about 2,250,000, a proportion of the total (13.6 per cent) only about half what it was in 1920. The leading groups are listed in Table 15. Such a language classification of immigrants is particularly useful in showing the importance of some of the central and east European Jewish groups which otherwise become unidentified parts of the totals from such countries as Germany, Poland, and the U.S.S.R. On the other hand, this table obscures any geographic insight into the actual patterns of migration associated with the English group, which obviously includes large numbers from Canada and Eire as well as the United Kingdom and smaller trickles from many other parts of the Commonwealth; so, also, the Spanish have come from every part of Latin America as well as Spain, and the French are primarily from Canada rather than Europe.

In general, the more recent immigration pattern

TABLE 14
POPULATION, 1780-1960

	New York State Total Population	Absolute Increase	Percentage of Increase
1780	210,514		
1790	340,120	129,606	38.1
1800	589,051	248,931	73.2
1810	959,049	369,998	62.8
1820	1,372,812	413,763	43.1
1830	1,918,608	545,796	39.8
1840	2,428,921	510,313	26.6
1850	3,097,394	668,473	27.5
1860	3,880,735	783,341	25.3
1870	4,382,759	502,024	12.9
1880	5,082,871	700,112	16.0
1890	5,997,853	914,982	18.1
1900	7,268,894	1,271,041	21.0
1910	9,113,614	1,842,720	25.4
1920	10,385,227	1,271,613	14.0
1930	12,588,066	2,202,839	21.2
1940	13,479,142	891,076	7.1
1950	14,830,192	1,351,050	10.0
1960	16,783,304	1,953,412	11.6

TABLE 15
FOREIGN-BORN POPULATION CLASSIFIED BY "MOTHER TONGUE," 1960 NEW YORK STATE

Language	Number	Percentage of Total Foreign-born
Italian	430,843	18.8
English	406,128	17.7
German	317,578	13.9
Yiddish	275,308	12.0
Polish	139,591	6.1
Russian	92,928	4.1
Spanish	87,776	3.8
French	56,523	2.5
Hungarian	53,039	2.3
Greek	42,720	1.9

has been reduced somewhat in volume from its earlier heights and become more balanced in terms of the major historical source regions. The basic law favors northwestern Europe, but special regulations and political pressures have continued to bring in large numbers from southern and eastern Europe. As might be expected, such groups are most heavily concentrated in the central city areas of the major urban centers.

Insofar as New York City is concerned, since the 1940's the United States has become a greater source than foreign countries of distinct migrant groups. The Negro population of the state was nearly 1,500,000 in 1960, 8.4 per cent of the total, as compared with 2.1 per cent in 1920. About 90 per cent of these persons lived in the New York City area, where they constituted slightly more than a tenth of the total population. The second group is the Puerto Ricans, totaling perhaps 700,000 and almost entirely confined to New York City. These American migrants—the one distinguished by race, the other by language—constitute the latest in the long series of major groups of newcomers who have moved into the rather sharply compartmentalized cosmopolitan structure of the major urban centers to begin the slow and often agonizing processes of gaining social acceptance and assimilation into the main stream of American life.

Three and a Half Centuries in Geographical Perspective

The historical survey in Part Two is designed to provide a general review of the changing geography of New York State. It is hoped that every reader will recognize the modern relevance of many of these historical patterns and that students will discern therein many topics worthy of further geographical investigation. A summary of certain topics may provide a useful interpretation and conclusion.

POPULATION PATTERNS

A review of the evolving patterns of population distribution suggests four main phases since European settlement. The first is that of the colonial period, featuring the initiation by design of a principal center, the setting of a basic axis and inland terminal, and the long, slow filling in of that framework. The Revolution was a brief interval of retrogression, but it did not deform the basic geographical structure and it was quickly followed by the second phase, featuring the rapid expansion of colonists over the entire habitable area of the state. Broadly viewed, it was a great sweep of the moving frontier of settlement across the land; in detail, however, it was a complex, selective, uneven process governed by a host of factors, some common to the American frontier, some peculiar to New York. This second phase, the peopling of the western and northern lands, was of prime geographic importance because it set the framework, imprinting a basic pattern of distribution out of which all subsequent differentiations developed. The third phase emerged gradually out of the second, featuring the accentuation of differential growths among the many settlements. In this era the main axis was extended from the Mohawk to Lake Erie and along it and in every district urban centers of various sizes emerged, each representing a triumph over rival settlements in the same general locale. It was a phase of growth everywhere, but such growth was increasingly uneven geographically. The fourth phase, largely a feature of this century and still in progress, is characterized by the steady depopulation of rural districts, the stagnation of most smaller urban centers, and the heavy dominance of massive, complex metropolitan clusters of population.

It is not possible here to review more local details in this over-all perspective, but the accompanying graph (Fig. 61) illustrates one important historical pattern, the fluctuations in the proportionate size of New York City within the state's total population.

CULTURE GROUPS

While it is trite to say that New York—or America—is the product of many different peoples, the complexities of just where and how and to what degree each of these peoples has had its influence still remain a challenge for study. Even a simple geographic inventory of the distributions of these groups at different times is an essential basic step

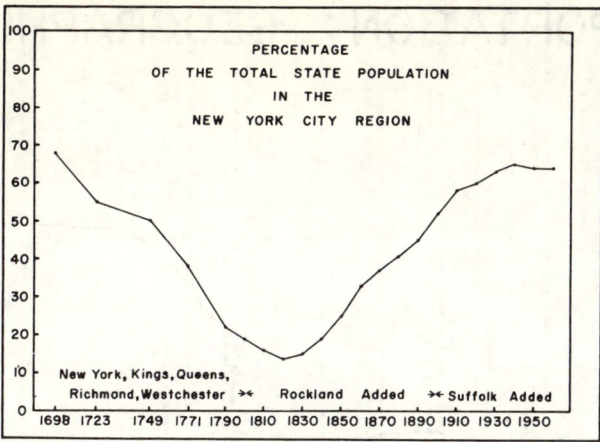

Fig. 61

yet to be taken in detail. Nevertheless, even this brief survey suggests important patterns.

The colonial period was characterized by competitive expansions of the "internal" population, that Dutch-German-English compound in the Hudson-Mohawk valleys which came to be thought of as "Yorker" and the "external" Yankee pressures from the East. The second phase was the Yankee triumph, the dominance of New Englanders over all of the newly opened lands of the state, and not only enveloping the Hudson-Mohawk area but establishing important footholds directly within it and in New York City.

Next came the influx of foreign immigrants into the cities. It was a movement so heavily urban as to be an important but often overlooked element in the widening cleavage between rural and urban interests and attitudes which was so characteristic of the nineteenth century. To the rural folk of New York State, city dwellers not only lived differently but also in many cases were very different people from very different backgrounds, speaking different languages, worshiping at different churches. Further, New York State provides a good measure of how complex such a commonplace general feature was in its detailed imprint, for the flow of immigrants was selective and the variation in volumes, proportions, and specific groups among the various urban centers was significant. Internally, of course, the compartmentalization of the cities into "foreign" and "native" areas, and of the former into Irish, German, Italian, and other such sectors, and the movements and successions of these various groups within that urban structure were also important geographic phenomena. The ramifications of these areal differentiations in the social, political, economic, and religious life of the communities involved were, and in many cases still are, immense.

In terms of these patterns, New York appears now to be in transition toward a new phase. Despite the fact that thousands of immigrants arrive each year, the volume is much reduced from that of half a century ago, and, instead of large blocs of a single national group, the sources are more diverse and thus the impact is much less. Barring some catastrophe which might send new waves out of Europe, it seems certain that this century-old source of cosmopolitanism will steadily dwindle in importance. Puerto Ricans have given fresh impetus to the long-established process in New York City, but the rather marked reduction in Puerto Rican immigration recently and the continued economic development of the source region suggest that this may not continue.

Negroes seem likely to have a much more important role in the future of this familiar drama. The American South is a source region capable of supporting a very large movement. Furthermore, the pressures seem sufficiently strong to sustain it indefinitely, and the problems of social assimilation in New York are so severe as to create highly localized "geographic tensions" that will certainly persist for a long time. The proportion of Negroes in cities outside the New York City region is smaller, but it has increased quite rapidly in recent years.

Increasingly the newcomers to New York will come from all of America itself, as part of the increasing mobility, drawing upon people of every economic class and from every region. It is a national process that is steadily redistributing and will inevitably bring about a greater uniformity of the American population, a process that will certainly gradually decrease the cultural variety, that rich cosmopolitanism which has so long characterized most cities in New York.

TRANSPORTATION PATTERNS

New York's development provides a good display of how the patterns of spatial interaction have been conditioned by the application of different technologies to a particular setting. The basic character of transport facilities has evolved through a sequence of eras, each dominated by a particular form and each having a particular geographic impact. The diagrammatic representation of these eras in Figure 62 is an attempt to emphasize the principal geographic characteristic of each in New York.

The dominance of the natural waterways lasted from the beginning of colonization until shortly after 1800. Throughout the colonial period only the Hudson-Mohawk sector was of major importance, but the rapid spread of settlement over the whole state after the Revolution brought all of the

Fig. 62 TRANSPORTATION: GEOGRAPHICAL

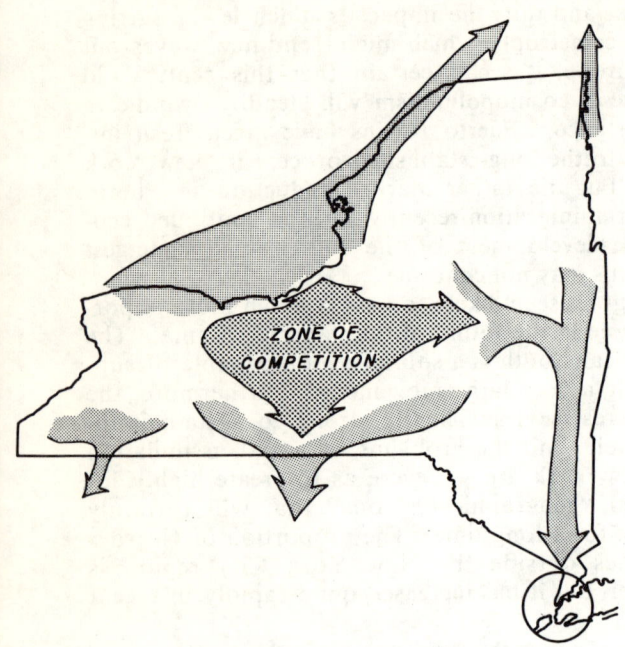

NATURAL WATERWAYS
DIVERGENCE

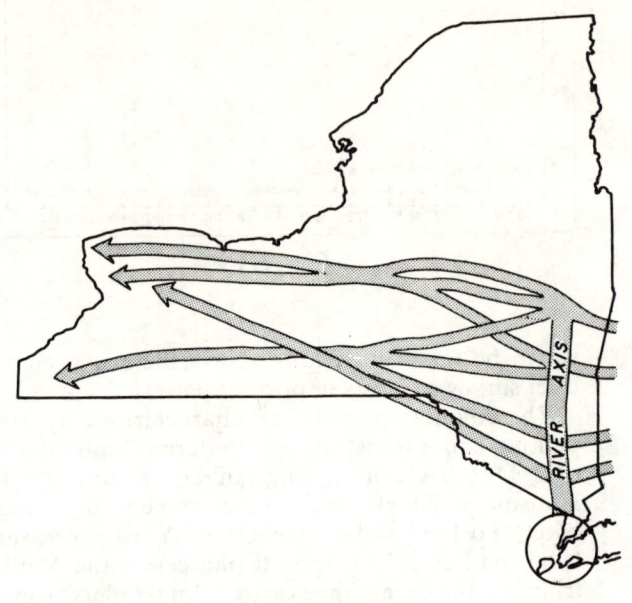

TURNPIKES
MULTIPLE TRUNK LINES (INCIPIENT)

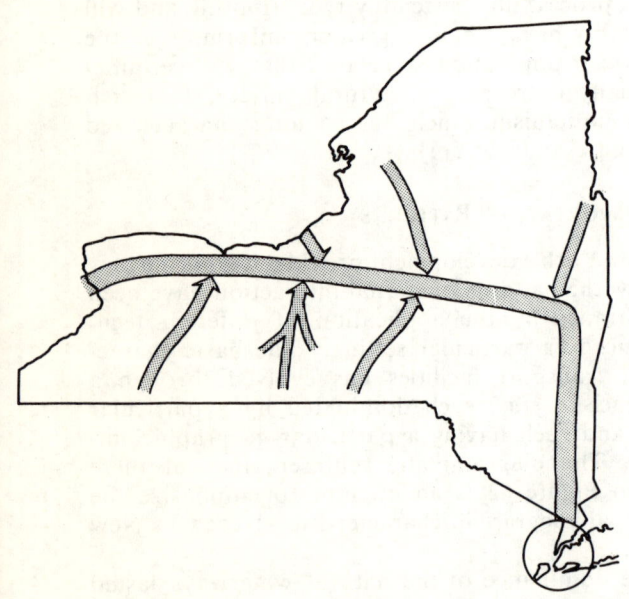

CANALS
SINGLE TRUNK LINE

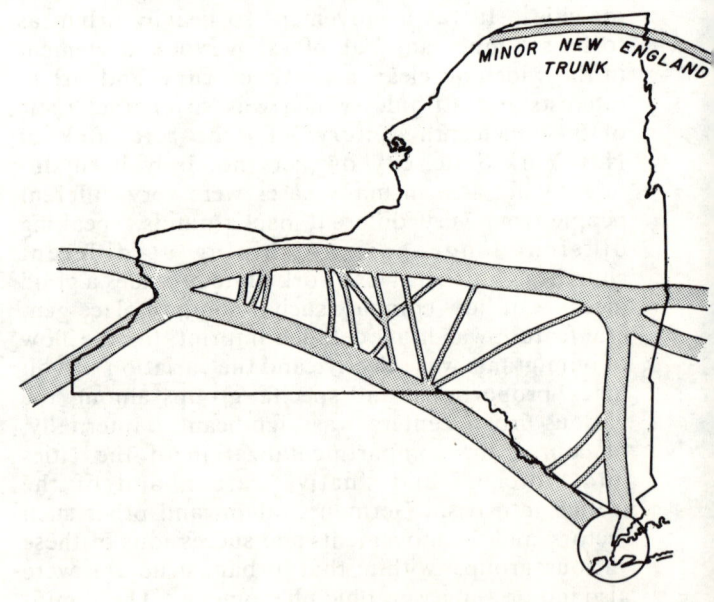

RAILROADS
TWO TRUNK LINES WITH NUMEROUS LATERAL LINKAGES

CHARACTERISTICS (DIAGRAMMATIC)

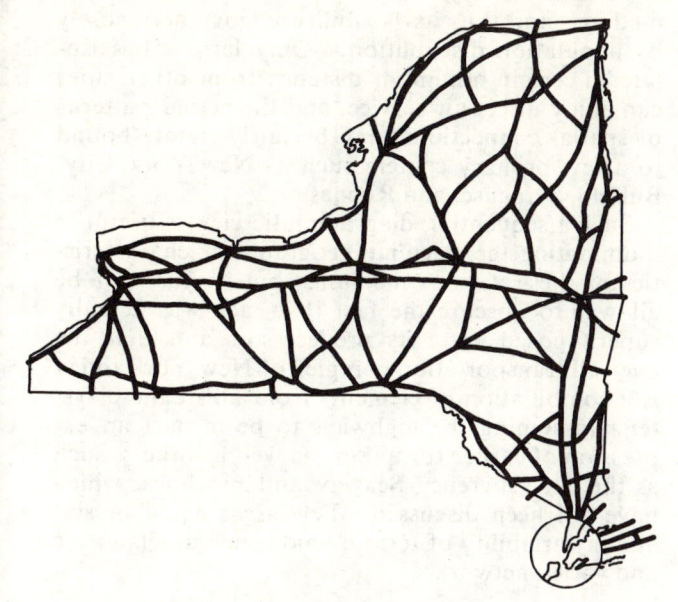

HIGHWAYS
DENSE WEB

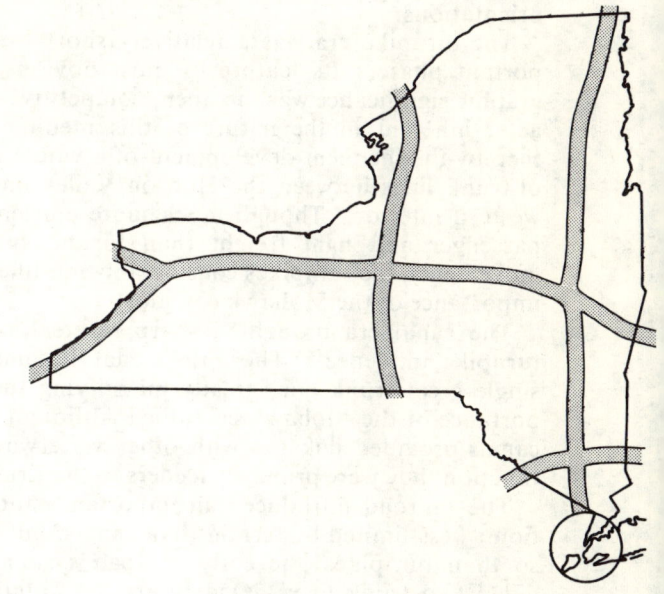

SUPERHIGHWAYS
TRUNK LINE GRID

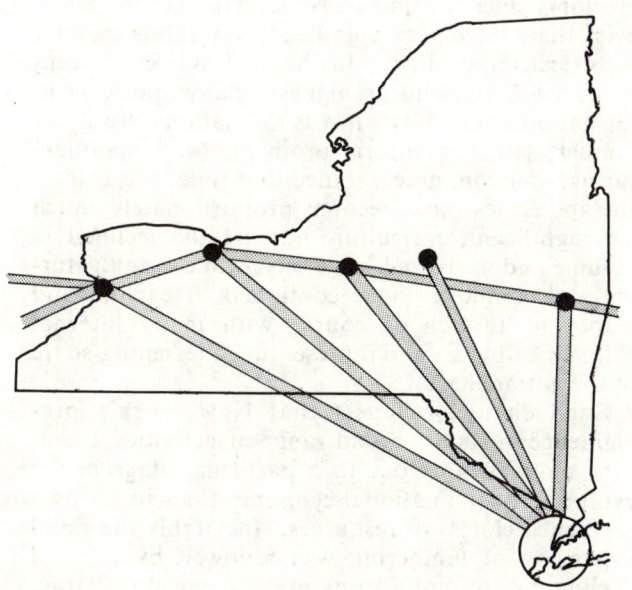

AIR SERVICE
MULTIPLE TRUNK LINES FOCUSED

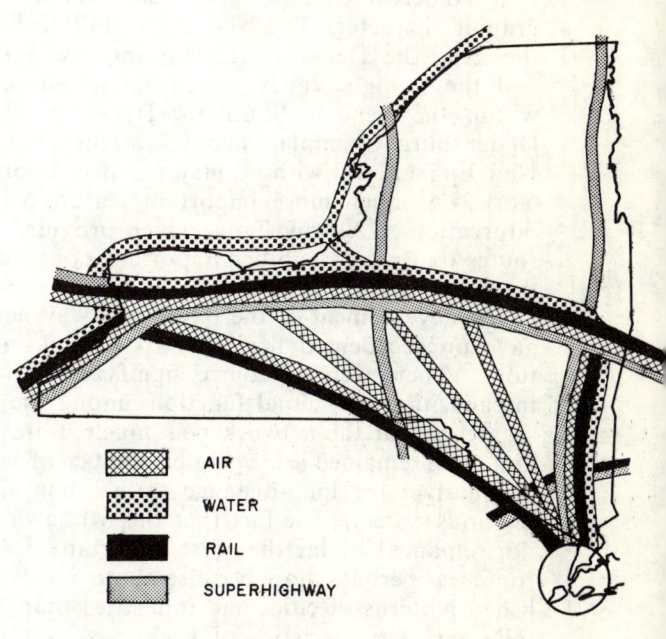

AIR
WATER
RAIL
SUPERHIGHWAY

TOTAL COMPLEX
TRUNK LINES OF FOUR MEDIA

CHAPTER 9 — ELABORATION AND CHANGE, 1850's–1960's

several systems into use. The basic geographic characteristic is one of marked divergence, with Montreal, New York City, Baltimore (principally), and New Orleans as the ultimate ocean ports. There were of course, as previously described, numerous factors that influenced the relative importance and the degree of competition among these various orientations.

The turnpike era was a relatively short but important phase. Its feature of most obvious geographic significance was the open, competitive character inherent in the nature of this medium and thereby the incipient development of a whole series of trunk lines between the Hudson Valley and the western interior. Though much more efficient for passenger and light freight than for heavy bulk movements, the turnpikes considerably modified the importance of the Mohawk corridor.

The canal era brought a sharp reversal of the turnpike influence. The Erie Canal became the single great trunk line, greatly magnifying the importance of the Mohawk corridor. Although other canals provided linkages with other waterways, in function they were primarily feeders to the Erie.

The railroad introduced an important modification. Less limited by terrain than canals, but more so than turnpikes, the early railroad system provided two trunk lines. Once more the dominance of the Mohawk route was somewhat reduced, although by adding a railroad to its canal it of course remained very significant. Later elaborations of the railroad network did not greatly alter its basic geographic character. The New York–Buffalo lines of the Erie, the Delaware, Lackawanna & Western, and the Lehigh Valley were merely alternatives within the general Southern Tier route. The Ogdensburg–Champlain line was a trunk route for New England but without major impact upon New York. A much more important feature was the bifurcation at Albany–Troy, which provided trunk routes to Boston, another important contrast with the patterns of the canal era.

The development of the paved highway network gave unprecedented flexibility to spatial interactions. There were of course significant variations in capacities and actual function among the many segments, but the network was much more dense and there remained a far greater number of feasible alternatives for long-distance traffic than in any previous system. The fact that the Mohawk corridor remained by far the most important east-west route was perhaps more because of previously established patterns of cities and industries than of the inherent characteristics of highways and motor transport. The completion of the Thruway and the present extension of other limited-access routes mark the re-establishment of a more rigid pattern of trunk lines, modifying the open, competitive nature of the highway system and thus accentuating the geographic importance of certain routes.

Theoretically the airplane, essentially freed of terrain restrictions, represents a maximum of flexibility. In fact it is restricted by its inherent high costs to serve very largely as a trunk-line passenger medium, and thus its flexibility is governed closely by population distribution. Only large cities isolated a certain minimum distance from other cities can enjoy direct air service, and the actual patterns of spatial connections may be fairly rigidly bound to a few primary centers such as New York City, Buffalo, Syracuse, and Rochester.

Such a sequential, diagrammatic view is useful in illuminating the peculiar geographical characteristics of these various mediums, but it must not be allowed to obscure the fact that each was actually superimposed upon its predecessors and that the over-all transportation complex of New York today is a combination of elements from all of these systems (assuming the highways to be in part an expression of earlier turnpikes) as well as others, such as the St. Lawrence Seaway and pipelines, which have not been discussed. Few areas equal in size and in variability of terrain enjoy such an elaborate and varied network.

ECONOMIC PATTERNS

Extractive, agricultural, manufacturing, and commercial activities have been a part of the New York economy since the time of New Netherlands. However, there have been significant variations in their historical importance. In the mid-nineteenth century New York held its highest relative position in the national economy: it was the national leader in lumber, salt, agricultural products, total manufacturing, and commerce. Since that time, the extractive industries have become proportionately much less significant, agriculture has greatly declined in volume and narrowed in variety, while manufacturing and commerce have continued to expand and elaborate, though of course with many internal shifts in emphasis and representing a lessening share of the national total.

Such changes suggest that New York's preeminence in such a broad range of activities a century ago was more due to a particular stage in the evolution of the national economy than to its own inherent richness of resources. Inevitably the rapid expansion of lumbering was followed by a rapid decline, but though forests are a renewable extractive resource, the natural conditions for renewal are much less favorable in New York than in many other forest regions of the nation. Furthermore, recreation has been given priority over exploitation in the use of these areas in New York. Agricultural leadership was possible only prior to the full realization of the much better inherent qualities of several other agricultural areas of the nation. The continued prominence of dairying and horticulture is probably more the result of a favorable location within a great market area than of especially favorable natural conditions. And in manufacturing, those industries based primarily upon local materials have mostly declined or disappeared, either because of depletion or of competition from more richly endowed areas. Furthermore the so-called soft-goods manufacturers often find New York and the rest of the Northeast a difficult area in which to compete.

Geographically there has been a steady narrowing of the most productive sections of the state for several decades. In the uplands deforestation was followed by farming, then abandonment; the amount of land in farms has been contracting for eighty years. Thousands of small factories and workshops in hundreds of villages have been abandoned; many of those in the smaller cities are no longer thriving. Thus, the over-all trend has been toward the concentration of an ever-increasing proportion of wealth-producing activities into the few great urban agglomerations.

Such a perspective leads to the conclusion that New York has in fact not been richly endowed with those raw material resources which have been most basic to economic development over the past century. However, it has been immensely favored by its general location within the nation and its facilities of connection with important regions. Furthermore, on closer examination it becomes clear that despite the importance of upstate industrial centers, the persistent significance of the Hudson–Mohawk axis, and the over-all density and high capacity of New York's transport network, it is primarily the enormous industrial production, singular location, and incomparable connections of New York City that have long given New York State its leadership in industry and commerce. Which is to say that New York deserves the title "Empire State" not so much because of the large population, resources, and wealth of the state as a whole, but simply because it is the seat of the *de facto* economic capital of the American "empire."

In 1963 a good deal of national attention was given to the fact that California had replaced New York as the most populous state in the union. Such a loss of leadership—held for nearly a century and a half—has certain political implications in a federal democracy, but other than that it is not in itself a matter of much importance, except perhaps to the pride and prestige of local citizens. Indirectly, however, that change is a reflection of very different economic growth rates—not just the remarkably rapid rate in California but the noticeably slowed rate of expansion in New York. This rather marked slackening of growth in New York is not new, nor is it common to all types of economic activities, and it is shared by all of the states in the Northeast. But it is an indication that some rural, industrial, and commercial interests are in economic difficulties. Such problems are rooted in the trends noted in this chapter. They reflect the impact of technological changes upon the geographic patterns and positions of New York.

For a century New York was an impressive leader not just in quantity but in quality and progress—in inquiry, innovation, and improvement. In the decades to come there remains the challenge to reassert its qualitative leadership in the solution of those problems which first arose in this oldest industrial part of the country but which are steadily spreading their affliction over an ever larger share of the nation. Drawing upon all of its past, upon its special geographical position within the nation and the world, and upon its diversity of resources in land and people, New York can best merit the retention of the title "Empire State" by leading the way toward solving such problems as urban blight, rural deterioration, destructive exploitation, and water pollution; by providing answers to such questions as how to diversify and adapt old industrial sectors to the new age, how to meet the political implications of decline and redistribution of population, how to effect a basic equality of treatment for highly diverse peoples.

Selected References

Baker, George P. *The Formation of the New England Railroad Systems.* Cambridge: Harvard Univ. Press, 1937.

Barck, Oscar T., and Hugh T. Lefler. *Colonial America.* New York: Macmillan, 1958.

Benson, Adolph B. (ed.). *The America of 1750; Peter Kalm's Travels in North America.* 2 vols. New York: Wilson-Erickson, 1937.

Bigelow, Timothy. *Journal of a Tour to Niagara Falls.* Boston: Press of J. Wilson & Son, 1876.

Bond, M. C. *Changes in N.Y. State Agriculture 1850–1950.* ("Cornell Agricultural Extension Bulletin," 917). Ithaca: New York State College of Agriculture, 1954.

Burr, David H. *An Atlas of the State of New York.* New York: By the author, 1829.

Cooper, William. *A Guide in the Wilderness.* Dublin: Printed by Gilbert & Hodges, 1810.

Cowan, Helen I. *Charles Williamson, Genesee Promotor—Friend of Anglo-American Rapprochement.* Rochester: Rochester Historical Society, 1941.

Darby, William. *A Tour from the City of New York, to Detroit, in the Michigan Territory.* New York: Kirk & Mercein, 1819.

Durant, Samuel W. *History of Oneida County, New York.* Philadelphia: Everts & Fariss, 1878.

Durrenberger, Joseph A. *Turnpikes.* Valdosta, Ga.: Printed by Southern Stationery & Printing Co., 1931.

Ellicott, Joseph. *Holland Land Company's Papers, Reports of Joseph Ellicott.* 2 vols. ed. Robert Warwick Bingham. Buffalo: The Buffalo Historical Society, 1941.

Ellis, David M., James A. Frost, Harold C. Syrett, Harry J. Carman. *A Short History of New York State.* Ithaca: Cornell Univ. Press in cooperation with the N.Y.S. Historical Assoc., 1957.

Ellis, Franklin. *History of Columbia County, New York.* Philadelphia: Everts & Ensign, 1878.

Evans, Paul D. *The Holland Land Company.* Buffalo: The Buffalo Historical Society, 1924.

Flick, Alexander C. (ed.). *History of the State of New York.* 10 vols. New York: Columbia Univ. Press under the auspices of the N.Y.S. Historical Assoc., 1933–37.

Fox, Dixon Ryan. *Yankees and Yorkers.* New York: University Press; and London: H. Milford, Oxford Univ. Press, 1940.

French, J. H. *Gazetteer of the State of New York.* Syracuse: R. P. Smith, 1860.

Friis, Herman R. *A Series of Population Maps of the Colonies and the United States 1625–1790.* ("American Geographical Society Miscellaneous Publication," No. 3.) New York: The Society, 1940.

Gordon, Thomas F. *Gazetteer of the State of New York.* Philadelphia: Printed for the author, 1836.

Hedrick, Ulysses P. *A History of Agriculture in the State of New York.* Albany: Printed for the N.Y.S. Agricultural Society, 1933.

Higgins, Ruth L. *Expansion in New York.* ("Ohio State University Contributions in History and Political Science," No. 14.) Columbus: The University, 1931.

Hilton, George W., and John F. Due. *The Electric Interurban Railways in America.* Stanford: Stanford Univ. Press, 1960.

Hungerford, Edward. *The Story of the Rome, Watertown and Ogdensburg Railroad.* New York: R. M. McBride & Co., 1922.

Jameson, J. Franklin. *Narratives of New Netherland.* New York: Scribner's, 1909.

Johnston, James F. W. *Notes on North America, Agricultural, Economical, and Social.* 2 vols. Edinburgh and London: Blackwood, 1851.

Knittle, Walter Allen. *Early Eighteenth Century Palatine Emigration.* Philadelphia: Dorrance, 1937.

Kouwenhoven, J. A. *The Columbia Historical Portrait of New York.* Garden City, N.Y.: Doubleday, 1953.

MacGill, Caroline E. *History of Transportation in the United States before 1860.* Washington: Carnegie Institution of Washington, 1917.

McNall, Neil A. *An Agricultural History of the Genesee Valley, 1790–1860.* Philadelphia: Univ. of Pennsylvania Press, 1952.

Mau, Clayton. *The Development of Central and Western New York.* Rochester: The DuBois Press, 1944.

Maude, John. *Visit to the Falls of Niagara in 1800.* London: Longmans, 1826.

New York State. *Annual Report of the State Engineer and Surveyor on the Canals of the State of New York.* Albany: Department of the State Engineer and Surveyor, various years.

O'Callaghan, E. B. (ed.) *The Documentary History of the State of New York.* 4 vols. Albany: Weed, Parsons & Co., 1849–51.

_____. (ed.). *Documents Relative to the Colonial History of the State of New York.* 5 vols. Albany: Weed, Parsons & Co., 1853–61.

_____. *History of New Netherland; or, New York Under the Dutch.* 2 vols. New York: D. Appleton & Co., 1845.

Pierce, Harry H. *Railroads of New York.* Cambridge: Harvard Univ. Press, 1953.

Rayback, Robert J. (ed.). *Richards Atlas of New York State.* 1st ed. Phoenix, N.Y.: Frank E. Richards, 1957–59.

Spafford, Horatio J. *A Gazetteer of the State of New York.* Albany: H. C. Southwick, 1813.

Stilwell, Lewis D. *Migration from Vermont.* Montpelier: Vermont Historical Society, 1948.

Sutherland, Stella H. *Population Distribution in Colonial America.* New York: Columbia Univ. Press, 1936.

Turner, O. *History of the Pioneer Settlement of Phelps and Gorham's Purchase, and Morris' Rreserve.* Rochester: W. Alling, 1851.

_____. *Pioneer History of the Holland Purchase of Western New York.* Buffalo: Jewett, Thomas & Co., G. H. Derby & Co., 1849.

Weaver, John C. *American Barley Production.* Minneapolis: Burgess Pub. Co., 1950.

Wertenbaker, Thomas J. *The Founding of American Civilization, the Middle Colonies.* New York and London: Scribner's, 1938.

Whitford, Noble E. *History of the Canal System of the State of New York.* Albany: Brandow Printing Co., 1906.

PART THREE

ECONOMIC ACTIVITIES TODAY

The Introduction contained some thoughts on development theory and categorized economic activities into three sectors—primary, secondary, and tertiary. Part Three contains four chapters, one on each of these sectors and a fourth on the strength of the state's economy. The three-way sector breakdown makes sense because each of the sectors exhibits strikingly different trends (see Fig. 6) and affects the geography of the state in such different ways.

The *primary sector,* which comprises agriculture, mining, lumbering, and fishing, was once of great relative significance to the state, but in the last hundred years it has not employed as much as half the labor force. Although a supplier of food, minerals, and wood products, it is insignificant today as an employer of men, accounting for but 2 per cent of the labor force. It is, on the other hand, a significant user of land. Encompassing the principal economic activities on the frontier, the primary sector—particularly agriculture—was instrumental in originally distributing people over the state; it has been the great disperser of people over the countryside. Although it gave rise to small market centers, it has not been able to induce growth of large cities.

The *secondary sector,* composed of manufacturing and construction, accounted for 30 per cent of the labor force in 1870 and has not fallen below that figure since. It reached its peak of relative significance in the economy in the period between 1900 and 1920, when immense numbers of European immigrants swelled the industrial labor force of the state as in no other part of the country, and when World War I drove the demand for manufactured products sharply upward. Quite in contrast to agriculture, manufacturing was a concentrator of people; instead of spreading them over the countryside, it brought them together, first in small towns which had favorable sites for gristmills, sawmills, textile mills; later largely in the metropolises. The secondary sector is a city-builder and depopulator of rural areas; it gives rise to transportation development; it triggers the trend toward unevenness in population distribution which so characterizes New York today.

The *tertiary sector,* once of little significance to the economy of the state, now dominates it, and will become increasingly significant as the years pass. It includes sales and service activities of all kinds. As our demands for education, government, and recreation increase, and as our automobiles, television sets, and electric driers break down in larger numbers, the tertiary sector must grow. The demand for these kinds of services is becoming almost insatiable; and perhaps this is good, for it appears as though most of the state's expanding labor force will have to be absorbed in the process of providing them. The tertiary sector, like the secondary, is a concentrator of people, a city-builder; but, most important of all, it is the major provider of jobs for the future.

Almost everyone these days knows what Gross National Product (GNP) is. It may be defined as the total market value of everything ("final" goods and services) the nation produces over a year's time —whether bought by the government, a foreign customer, a factory, or an individual. Because of computation complexities Gross State Products (GSP) are not calculated and published; however, the New York GSP can be approximated. Table 16 conservatively shows it to be 10.46 per cent of the GNP in 1960, a truly remarkable contribution for one state to make. Table 16 also shows the relative contribution of the three economic sectors to the GSP for the same year. The primary sector accounts for 1.5 per cent, the secondary sector 32.6 per cent, and the tertiary sector 65.9 per cent.

TABLE 16
ESTIMATED NEW YORK GROSS STATE PRODUCT, 1960

	1960 GNP (In Billions $)	N.Y.S. Employment as a % of U.S. Employment, 1960	GSP* Column 1 × Column 2 (In Millions $)	% of GSP
Agriculture Forestry, Fishing	$ 22.2	2.73	$ 606.1	1.1
Mining	11.0	1.76	193.6	.4
Primary Sector Total	33.2		799.7	1.5
Manufacturing	140.9	10.77	15,174.9	28.8
Construction	23.8	8.44	2,008.7	3.8
Secondary Sector Total	164.7		17,183.6	32.6
Wholesale & Retail Trade	89.1	10.20	9,088.2	17.3
Finance, Ins., Real Estate	61.3	15.19	9,311.5	17.8
Transportation	22.4	11.56	2,589.4	4.9
Communication	10.3	12.93	1,331.8	2.5
Pub. Utilities	13.0	9.56	1,242.8	2.4
All Services	54.3	10.88	5,907.8	11.2
Government & Gov't. Enterprise	52.9	9.80	5,184.2	9.8
Rest of World	2.3	—	—	—
Tertiary Sector Total†	305.5		34,655.7	65.9
Total GNP 1960	503.4			
Total GSP 1960			52,639.0	100.0
N.Y.S.'s Per Cent of GNP	$\frac{52.6}{503.4} = 10.46$			
Total GNP 1963	585.0			
N.Y.S.'s Est. 1963 GSP (585 × .1046)			61,191.0	

*Computation of GSP on this table assumes that the productive contribution of each N.Y. employee in a given category of the economy is the same as that for each U.S. employee in the same category. This assumption probably actually produces a conservative New York GSP figure.

†The tertiary breakdown shown is somewhat different from that commonly used by the census but is comparable in total to the usual tertiary designation.

Sources: GNP—Statistical Abstract of the United States, 1963. Employment—U.S. Census of Population, 1960.

CHAPTER 10   JOHN H. THOMPSON

The Primary Sector

As there was brief coverage of the relatively unimportant forest industries in Chapter 4 and because commercial fishing is of such minor significance in New York State, the principal emphasis in this chapter is on agriculture and mining. Table 17 shows the comparative significance of the four subdivisions (agriculture, lumbering and other forest uses, mining, and commercial fishing) of the primary sector for about 1954 and 1958. The primary sector, although an important user of land, employed but 2 per cent of the labor force and accounted for only 1.5 per cent of the GSP.

Before proceeding with the discussion of the components of the primary sector, three basic concepts relative to resource utilization should be introduced. They are: (1) the functional concept of resources, (2) the concept of renewability vs. nonrenewability, and (3) the concept of sustained yield.

It has already been pointed out that the way man uses the physical resource base at hand depends on attitudes, objectives, and technical skills. As human desires and abilities expand, parts of the resource base never used before suddenly become extremely useful. To put it another way, resources are dynamic in response to both increased knowledge and changing individual wants and social objectives. According to the functional concept of resources, *resources* are phenomena which expand and contract in response to human effort and behavior. The concept holds that something becomes a resource only when it satisfies a human want or need. New York State's petroleum was of no value to the Indian and so was not a resource to him. Only with the evolution of modern society, with its good roads and large buildings, have sand and gravel and cement become significant. New technology in the paint industry found important use for titanium, a New York mineral resource hitherto of little value. New fertilization and cultivation techniques have sorted out and made much more productive soils that respond well to these techniques. Technology permitting use of hardwoods in the pulp and plywood industries has given the widespread supply of maple, birch, and other hardwoods new meaning. Two exciting questions arise out of the functional concept of resources. What items in the physical environment not now considered resources will satisfy our needs in the future? How much can we expand our skill in making our current resources of greater value to us?

Some resources, such as minerals, when once used are gone forever. These are the nonrenewable resources. Their use and management require a full understanding of nonrenewability. Any area whose economy is based on nonrenewable resources must seek other means of support prior to exhaustion of

TABLE 17
THE PRIMARY SECTOR
New York State, 1954 and 1958

	Employment				Value of Products Sold (In Thousands of Dollars)			
	1954	% of Primary Total	1958	% of Primary Total	1954	% of Primary Total	1958	% of Primary Total
Agriculture	160,549	87.6	122,906 (1959)	86.8	668,488	79.2	755,410 (1959)	78.8
Lumbering & Other Forest Uses	9,100	5.0	8,100	5.7	38,250	4.5	47,200	4.9
Mining	11,517	6.3	9,657	6.8	128,393	15.2	147,603	15.4
Commercial Fishing (est.)	2,000	1.1	1,000 (1960)	.7	8,755 (1956)	1.1	8,904 (1960)	.9
Total	183,166	100.0	141,663	100.0	843,886	100.0	959,117	100.0

Sources: Census of Agriculture 1959, Census of Population 1960, Census of Mineral Industries 1958, Annual Report of the N.Y.S. Dept. of Conservation 1958 and 1961.

the resource or be doomed to economic deterioration.

Renewable resources, as the name suggests, can be renewed after use or depletion. For example, forests that have been cut may regrow on their own, or man may replant previously forested areas. Depleted fisheries under enlightened management may possibly be returned to their former, more productive state. Soils subjected for years to misuse and fertility depletion usually may be restored to more productive levels. Often the key question in renewing resources is whether the renewing process can be carried on economically or not. In many parts of New York there are areas where forests or soils could be renewed from the technical standpoint, but the costs would be so great as to make the endeavors unprofitable.

When using renewable resources man should always practice the concept of sustained yield if possible. He should use the resource in such a way as to insure future production. He should cut only as much timber as regrows; he should harvest only the quantity of fish that is naturally or artificially replaced; and he should see that the soil is as good for next year's crop as it was for this year's. Only through practice of sustained yield can our invaluable renewable resources supply posterity as they have supplied us.

Many years of exploitative rather than sustained-yield practice have left their mark on the resource base of New York and have produced obstacles to successful and smooth operation of the primary activities. Much has been learned, though, and more rational future use of resources in general should be anticipated.

Agriculture

Farms occupy about 44 per cent of New York's land, but less than 2 per cent of the state's population live on and are supported by these farms. Land in demand for urban or transportation uses ordinarily cannot be economically used for farming; similarly, land on which good farming profits can be made cannot be left in forests. This places agriculture in an intermediate position as regards land use intensity and suggests that farms will occupy land good enough to produce adequate farm profits but not in demand for the urban uses. As profit requirements rise with an improving economy, poorer lands often will no longer support farming; also, as urban and transportation uses take up more land, farmed acreages are bound to decrease. Thus, farming is squeezed from both sides, as it were.

During earlier periods of New York's history, much more of the land was used for farming than now. Higher percentages of the population were farmers, too. Paradoxically, the total value of farm produce and the quantity of many commodities produced are greater today than ever before. The fact that fewer farmers are producing more on less land can only be explained by improved farming methods. As a result of such a trend, the average farmer has been able to increase his income. However, as we shall see later, since this increase has in many instances not been commensurate with increases in urban enterprises, farmers tend to leave farms for the city. The area in all farms has declined at the rate of 250,000 acres per year from 1950 to 1960, a decline that is expected to continue. By census definition about 85 per cent of the farm land (over 11.5 million acres) could be classified as in commercial farms in 1959. This involved 56,760 farm units, a drop of over 30,000 in the last decade. These farms averaged about 200 acres in size, 82 acres of which were harvested cropland. Figure 63 portrays three aspects of the state's agricultural trends.

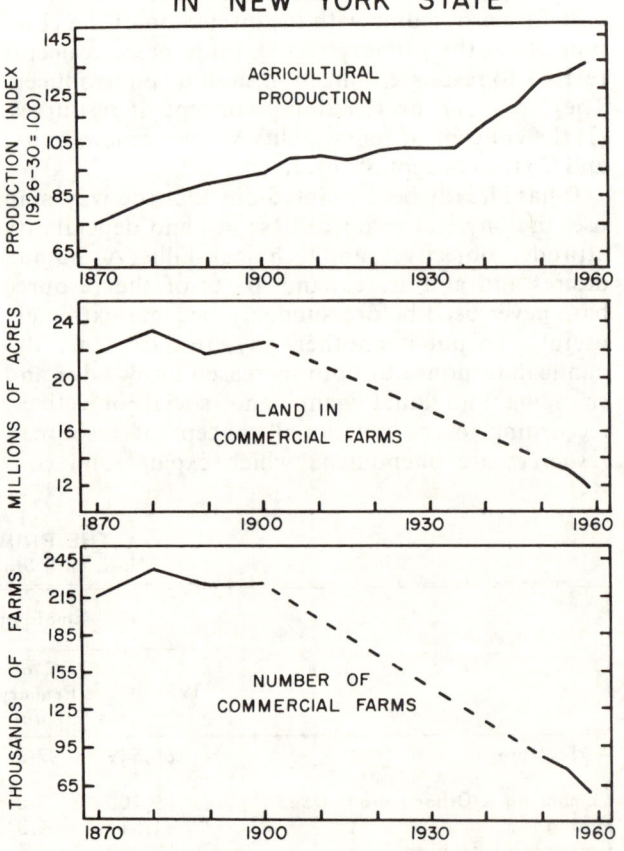

Source: U.S. Census of Agriculture (various years); and H.E. Conklin, "The Dynamics of Land Use in New York State," The Conservationist, State of New York Conservation Dept., 1964.

Fig. 63

In addition to the commercial farms there were some 25,000 units in 1959 on which some farming was practiced but not in a serious enough way for the units to be designated as commercial farms by the Census of Agriculture. These units, averaging about 75 acres in size and harvesting only about 15 acres of cropland each may be considered of little significance to the agricultural scene except as targets for early abandonment.

Thus, counting both commercial and noncommercial farms, there are about 82,000 units. It is interesting to note, however, that although 44 per cent of the state's land was in farms in 1959, only about 16 per cent was actually planted and harvested. This figure will fall in the years ahead. Dairying dominates agriculture, but there are areas, for example, where potatoes are important, or where fruit and vegetables are principal crops. Poultry, too, is significant in some localities. Such areal variations in agricultural production are due to variation in soil, climate, terrain, and markets, as well as to farmer preferences. Three maps in this chapter provide information on crop distribution; agricultural regions are outlined in another.

The Farm Economy Is Changing

In 1850 over 70 per cent of the state's population was classified as rural. A substantial, although unknown, portion of these people lived by farming. Today, as has been pointed out, less than 2 per cent of the state's people are on farms. Not only did the percentage of the state's population farming go down, but the actual number of these people dropped from perhaps two million in 1850 to about one-third million in 1960. The greatest amount of total land in all farms shows up in the census figures for 1880, and the greatest acreage of cropland was harvested in about 1900. Since then, there have been generally declining trends in both aspects of the agricultural scene. On the other hand, while the acreage harvested, the amount of farm land, and the number of farmers have decreased, the quantity of agricultural production has roughly doubled (Fig. 63), and the average farm has increased in size by 50 per cent. There has also been a healthy increased output per acre of land on the farm, reflecting higher efficiency and a more sound and well-established commercial type of agriculture.

When the state was first settled, farming was more or less of the subsistence type. That is, crops were produced largely for home consumption, and there was a minimal exchange of goods. The farm did not have to produce much to satisfy a family's needs, and so basically could have been a pretty poor farm and perhaps even run in a pretty poor fashion. Later, as jobs became plentiful in cities, and as demands for mechanical equipment and luxuries grew among farmers, the farms simply had to produce more to satisfy their operators. Between 1875 and 1960 nearly two-thirds of the state's farms (142,000 units in all) went out of business. They simply could not compete as production units in the changing economy. The reasons why they could not compete are many, but certainly the following were important: (1) the physical resource base of many (soils, terrain, climate, etc.) was poor, (2) some were too small, (3) others were poorly located with respect to transportation and markets, (4) in numerous instances there was unenlightened management and operation.

The problem of maintaining a farm as an operating and satisfactorily competing unit in the present economy can be illustrated with a few figures. Suppose Farmer Jones owns and operates a 160-acre dairy farm. Let us say his investment in land, twenty milkers, plus a few young stock, machinery, and buildings is $30,000. Let us say, too, that his net profit (income less expenses) from this operation is $3,000 a year. This is not spectacular, but it may be a livable income in the country, and after all, there are housing, garden produce, eggs, and milk which may not have been taken into consideration when figuring profits. Furthermore, being one's own boss may be worth something. Now let us suppose Jones is offered a factory job at $2 an hour in a nearby city and he is able to liquidate his investment in the farm. What would he do? If Jones looks at the money angle only, he would probably leave the farm, for he could invest his $30,000 at 5 per cent (producing $1,500 per year) and earn $4,160 a year in wages. This is a total of $5,660 as compared to the $3,000 earned on the farm. Even if housing, food, and other costs run $1,800 higher per year in the city, he would still have $3,860, or $860 more than the farm provided. Higher income taxes would take part of this, but he has two days off each week, perhaps does not have to get up so early or work so hard; and maybe his wife and children are glad to get off the farm.

This is a hypothetical case only and therefore does not fit exactly the situation of any specific farmer, yet it exemplifies the kind of condition which has brought, and will continue to bring, about a decline in the number of farms. It might be added that there are many farms in the state today showing profits of less than the $3,000 of Farmer Jones. These will disappear as full-time occupational enterprises in the years ahead. Animals will be slaughtered, or if good, probably purchased by neighboring dairy farmers. Some of the land, too, may be purchased and used by adjoining farms, but much will go unused and eventually return to forest from which it was originally hewed. These abandoned areas are the submarginal agricultural lands of the state. Perhaps in the future, when the population is much greater than now and when agricultural prices are higher, this land may be returned to production, but that time is a long way off.

There is a different, perhaps brighter, side to this story. The sequence of events just described has in essence freed large areas of land for other uses. Tens of thousands of families have satisfied their longing for the great out-of-doors by buying 10, 15, or 100 acres of abandoned farm land and either fixing up an old farmhouse or building a new home on the property. In all, New York has 11 million acres of obsolete farm land for such family living.

TYPES OF PRODUCTION

Three-fourths of the gross farm income for the state is derived from sale of livestock and livestock products, mostly in dairying. New York lies in the great American dairy belt which extends from New England to Minnesota. A relatively cool, moist climate and soils well suited to the growth of hay have aided in the development of this dairy industry. More important, however, is the state's location close to the huge fluid milk market of the middle Atlantic Seaboard, and New York City especially. New York City draws on many parts of the state for milk, including the faraway St. Lawrence Valley. Other urban areas in the state have substantial demands for milk in their own right and draw heavily on their surroundings. The area supplying milk to an urban center is ordinarily referred to as the *milk shed* for that center (in a sense the same concept being applied as in the use of the term watershed to designate the area from which a stream drains water).

Years ago, before transportation facilitated shipment of fluid milk to the great urban centers and when the urban centers were much smaller, much of the milk was converted into butter or cheese. The demand for fluid milk and the competition from margarines, as well as from butter and cheese production in Wisconsin and Minnesota, have resulted in a decline of these products in New York.

Today dairying is widespread, as Figure 64 shows. Only the Adirondacks, the Catskills, the Tug Hill Upland, and the more rugged portions of the Appalachian Upland show up as not supporting numerous dairy cattle. Production is relatively light in other places, where greater attention is given to fruit and vegetables. Forage crops, the mainstay of the dairy industry, occupy far more land than other crops. Included are clover, timothy, alfalfa, grass, and corn silage, and various combinations of these. Cereal crops such as corn, oats, wheat, barley, rye, and buckwheat are raised too and are commonly fed to dairy animals in one form or another.

Dairy products—fluid milk in particular—widely recognized as healthful and necessary foods, are consumed in large quantities in the populous northeastern United States. As the population of this

A GOOD DAIRY FARM along Route 20 in southern Oneida County. The size of the milking herd (about 75) is large enough to suggest continued successful operation, and the buildings exhibit recent capital investment. Many dairy farms in the state are less well endowed. *N.Y.S. Dept. of Commerce.*

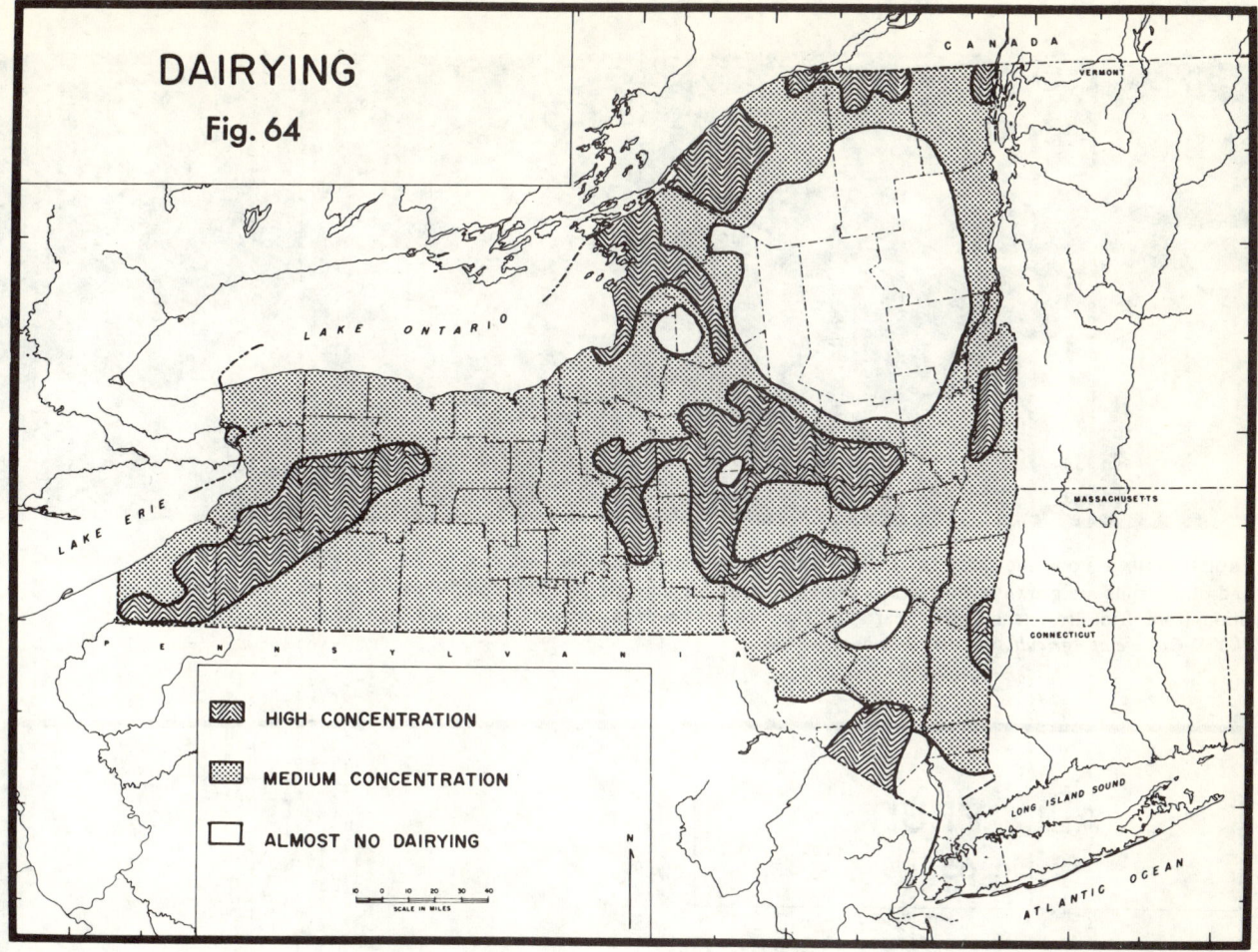

area rises, the demand for fluid milk will likely rise and the dairy industry should grow. However, two important trends may have substantial countering effects. One is the growing concern that there is a relation between the intake of saturated animal fats and heart disease. Such a relation is likely to be definitely proved or disproved sometime in the future. In the meantime, however, many people are cutting down on whole milk consumption, turning exclusively to skim milk or giving up milk completely. The second trend is related to the increased attention toward keeping slim and has resulted in tens of thousands of milk drinkers turning to skim milk. Others have given up milk entirely for this reason, too. In the long run these two trends could have sweeping effects on the size and general economic health of the state's dairy industry.

Certain forecasts can be made for the dairy industry: (1) there will be fewer producing units in the future and these will be found largely on the better lands, (2) the average producing unit will be larger and more scientifically operated, and (3) there may be a trend away from the family farm toward corporate enterprises run by specially trained managers and employing workers on forty-hour-week schedules.

Together, fruit and vegetables account for approximately 16 per cent of the value of agricultural sales. Potatoes and a variety of beans are most significant among the vegetables produced, and apples, grapes, cherries, and berries rank in that order as fruit. Lands just south of Lake Erie and Lake Ontario are significant because of moderating influences of the lakes, and the Finger Lakes district and the lower Hudson Valley, with their high number of growing degree months, are important, too. (Fig. 65).

The fruit acreage along Lake Erie and in the vicinity of the Finger Lakes is primarily in grapes, which require 80 per cent of the work on all fruit-raising in those areas. The largest fruit region lies just south of Lake Ontario, and in this area there is considerable diversification. Apples are by far the most widely grown, but cherries, peaches, pears, plums, and other fruit are important. The Hudson Valley between Newburgh and Albany also special-

FRUIT-FARMING along the shores of Lake Ontario near Sodus. Apples, peaches, pears, cherries, and other fruits are grown in this area because of favorable climatic and soil conditions. Competition from fruit-growers in other parts of the country and scarcity of labor are handicaps.
N.Y.S. College of Agriculture, Cornell Univ.

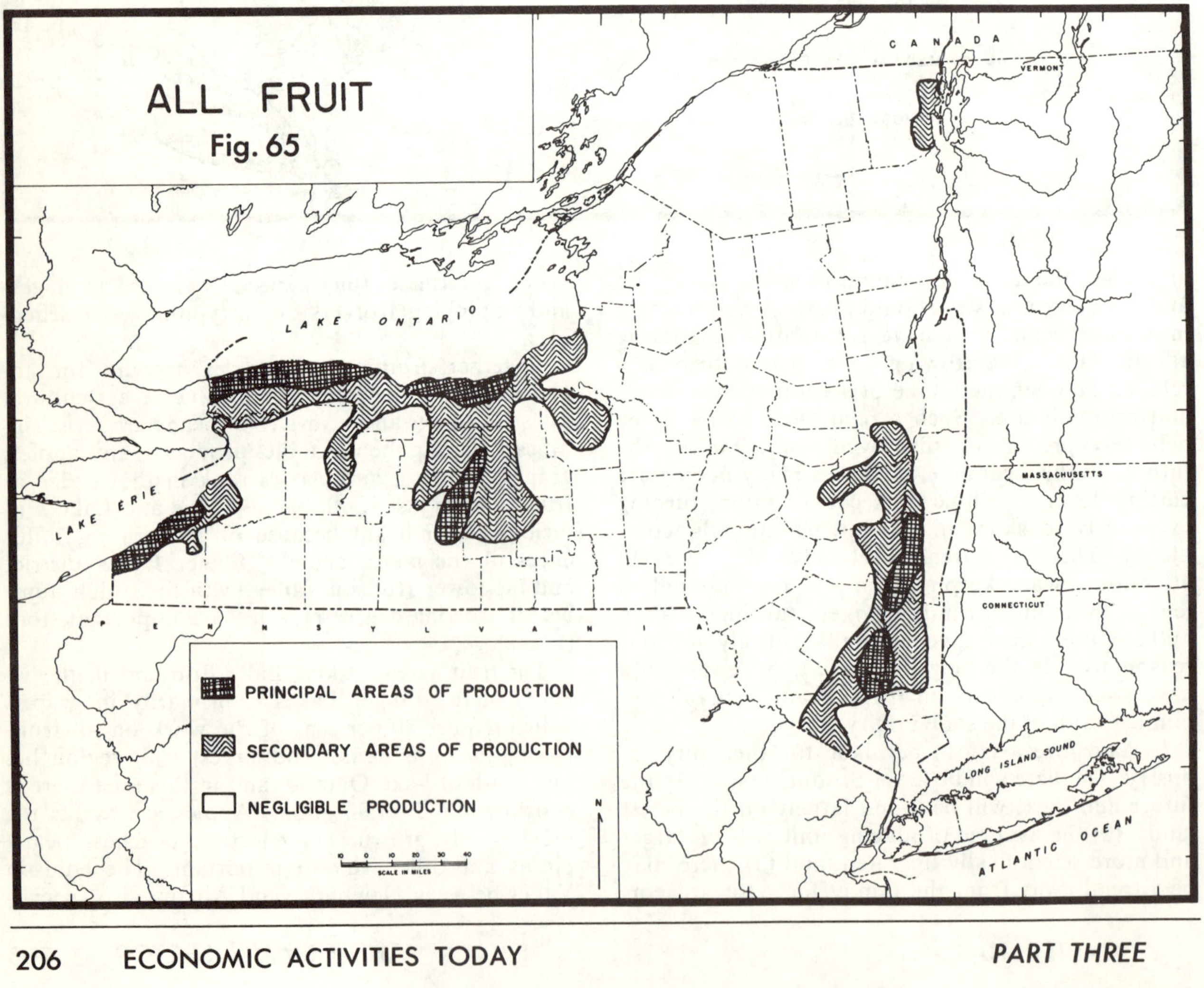

ALL FRUIT
Fig. 65

PRINCIPAL AREAS OF PRODUCTION
SECONDARY AREAS OF PRODUCTION
NEGLIGIBLE PRODUCTION

206 ECONOMIC ACTIVITIES TODAY   PART THREE

VINEYARDS, like these in the Finger Lakes area, make New York an important producer of grape juice and wine. *N.Y.S. Dept. of Commerce.*

izes in apples, as well as in peaches and other fruits. Raspberries and strawberries are commercially raised in many parts of the state. A climate free from sudden changes in temperature with long growing seasons and considerable energy receipt, and accompanying soils with good internal drainage favor fruit-raising. Those areas just south of the Greak Lakes, the Hudson Valley, and the Finger Lakes area have more growing degree months (130 to 155) than any other part of the state except the metropolitan New York and Long Island area.

Vegetable acreage is largely concentrated on well-drained soils along Lake Ontario and southward in valley bottoms toward the Southern Tier. Long Island, too, is important. Proximity to the ready markets of important urban areas is an asset which results in heavier than otherwise expected concentrations around cities. This is particularly noticeable in the vicinity of metropolitan New York and Buffalo. Fertile bottom lands in the Hudson Valley region and muck soils in scattered parts of the state have proved to be especially conducive to vegetable production. Included among the many commercial vegetables are sweet corn, lima and snap beans,

A MECHANICAL CHERRY-PICKER shakes ripe cherries onto a large catchment surface, from which they feed by gravity into a crating machine. This kind of equipment, as well as mechanical grape-pickers, will eventually lead to substantial reduction in labor needs at harvest time. *N.Y.S. College of Agriculture, Cornell Univ.*

POTATOES growing near Malone in Franklin County. Potatoes are produced successfully in many parts of the state from the St. Lawrence Valley to Long Island. *N.Y.S. College of Agriculture, Cornell Univ.*

beets, cabbage, onions, peas, tomatoes, and potatoes. New York is a very important producer of potatoes, often ranking close to Maine and Idaho. Long Island is by far the major growing area, but the Ontario Lake Plain and certain locations in the Appalachian Upland, especially in Steuben and Wyoming counties, are significant too. At the time of this writing sugar beet production has begun in the vicinity of Auburn and a refinery is under construction to process the anticipated produce.

The level to gently rolling fertile land in the Finger Lakes area and to the north and west from there is adapted to extensive cash crops such as wheat, corn for grain, and dry beans. The bulk of the production of these crops in the state comes from this area.

Chickens and other fowl are distributed widely over the state; however, concentrations occur in the Hudson Valley, on Long Island, and in the Southern Tier. Eggs account for 61.5 per cent of the value of all poultry products sales. Commercial poultry farms commonly produce little or none of their feed

BEANS AND ONIONS do well in New York State. Beans are widely raised on better soils throughout the state. Onions are commonly produced on muck lands. *N.Y.S. College of Agriculture, Cornell Univ.*

TABLE 18
VALUE OF AGRICULTURAL PRODUCTION AND SALES
New York State, 1954 and 1959

Product	Quantity Unit	1954				1959			
		Acres (1,000's)	Quantity (1,000's)	Value of Production ($1,000)	Sales ($1,000)	Acres (1,000's)	Quantity (1,000's)	Value of Production ($1,000)	Sales ($1,000)
Primarily Dairy-oriented Crops				220,986	35,832			196,857	33,430
Hay	tons	2,981	5,263	122,072	6,819	2,827	5,128	115,164	10,011
Corn	bu.	713	11,263	51,444	5,995	623	12,256	42,452	6,213
(silage only)	tons		3,850				4,002		
Sorghum				26	1	4		235	5
Oats	bu.	645	23,165	19,227	2,407	613	32,386	23,642	4,229
Buckwheat	bu.	40	701	701	388	18	291	321	218
Wheat	bu.	336	10,328	22,308	18,661	242	7,260	12,704	11,440
Barley	bu.	30	940	1,024	301	25	745	730	286
Rye	bu.	17	353	423	261	13	308	339	234
Other				3,711	999			1,270	794
All Fruit & Nuts				54,417	54,417			45,187	44,256
Apples	bu.		18,371	32,150	32,150		18,617	25,133	24,202
Cherries	lbs.		53,061	6,289	6,289		48,635	3,237	3,237
Grapes	lbs.		157,545	7,877	7,877		161,935	9,716	9,716
Peaches	bu.		1,040	2,339	2,339		693	1,802	1,802
Pears	bu.		295	841	841		584	1,344	1,344
Strawberries	qts.	4	7,892	2,841	2,841	3	7,110	2,488	2,488
Raspberries	qts.	3	2,690	1,130	1,130	2	1,557	623	623
Nuts & Other Fruits				950	950			844	844
All Vegetables		407		79,351	74,002	345		85,507	75,140
Irish Potatoes	cwt.	88	16,817	33,179	28,047	82	16,034	32,869	29,181
Beans (Soy, Field & Snap)		179				138			
Cabbage		14				10			
Sweet Corn		39				39			
Onions		10		46,172	45,955	13		52,638	45,959
Peas (Green)		14				11			
Tomatoes		18				15			
Other Vegetables		45				37			
Forest Products				5,028	5,028			6,132	6,132
Horticultural Specialties				37,758	37,758			42,104	42,104
Total Value of Crop Production				397,538				375,787	
Total Crop Sales					207,036				201,063
Dairy Products Sales					334,322				408,632
Poultry Products Sales					78,292				70,093
Other Livestock Products Sales					48,837				75,621
Total Value of Agricultural Products Sold					644,488				755,410

Source: Census of Agriculture, 1959.

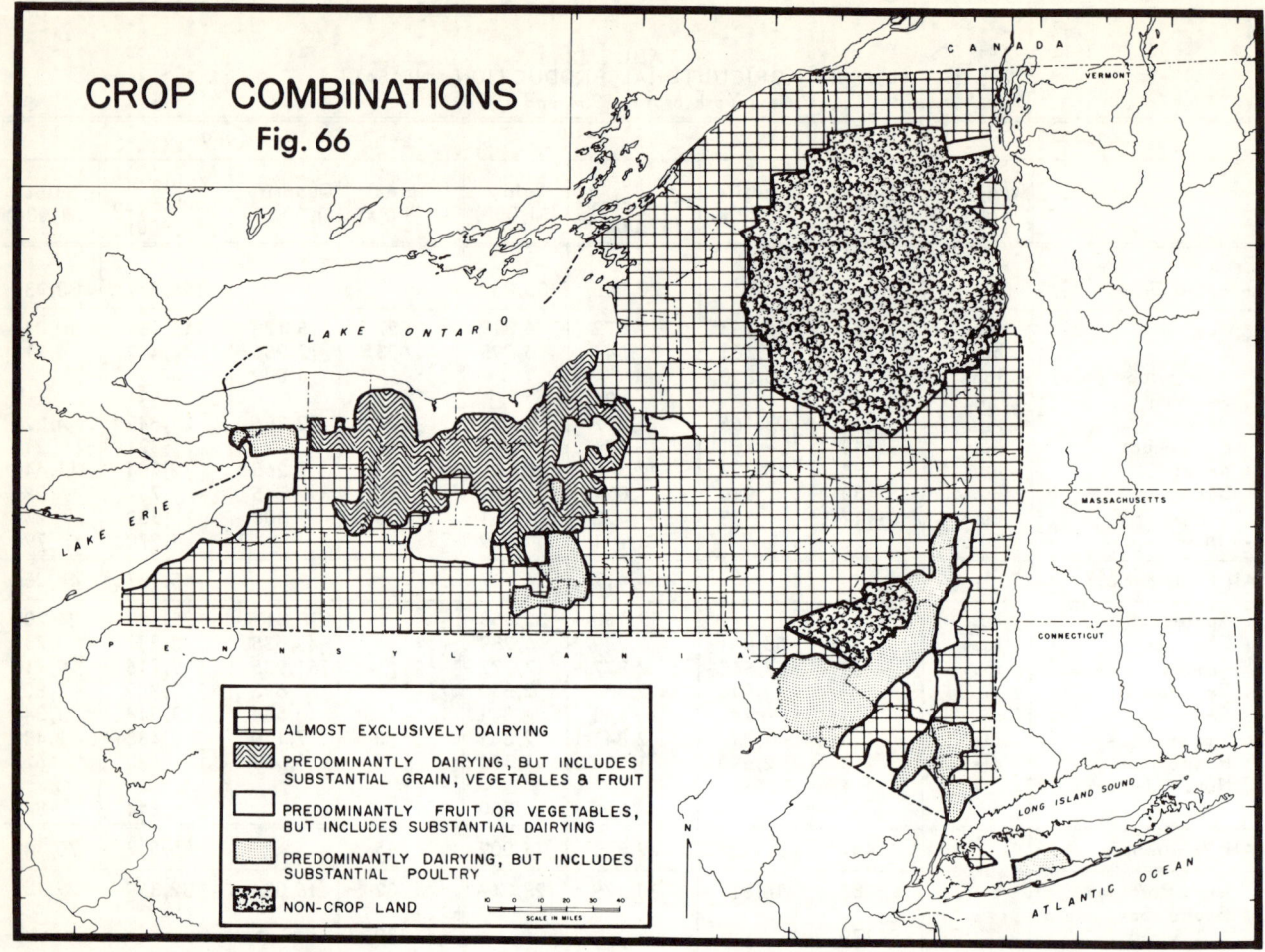

CROP COMBINATIONS
Fig. 66

needs, and so must buy feed, often from outside the state. The relative importance of agricultural production of different types is shown in Table 18.

Using town data as a basis and ignoring the existence of urban areas, one can map categories of crop combinations. Clearly, Figure 66 illustrates the overwhelming dominance of dairying but at the same time points up the versatility of New York as far as a wide range of agricultural production is concerned.

AGRICULTURAL REGIONS

Climate, soils, topography, and markets are important in their effects on types of agriculture. Those types of agriculture now carried on in the various parts of the state represent years of effort and experimentation by farmers to find optimums for each locality. This has resulted in fairly clearly identifiable areal regionalization of agriculture in terms of both crop emphasis and level of success.

According to the regional concept developed in other chapters, an *agricultural region* would be an area that exhibits considerable homogeneity in agricultural production, practices, and problems. For over twenty-five years agricultural economists from the College of Agriculture at Cornell University have been working toward the identification of such areas of agricultural homogeneity within the state, recognizing that a map of properly identified agricultural regions would be a great aid in thinking about, and planning for, the agricultural future of the state. The eight agricultural regions presented in Figure 67 are a synthesis and modification of the Cornell efforts.[1] Each one has certain important characteristics in common throughout and differs from the adjoining regions. Differentiating characteristics are as follows:

Appalachian Upland Dairy Region

This region occupies about 26 per cent of the state and approximately 45 per cent of the land could

[1] K. C. Nobe, E. E. Hardy, C. P. Grotto, "Agricultural Regions of New York State" (Preliminary), A.E. 1090, N.Y.S. College of Agriculture, Cornell Univ., 1958.

theoretically be tilled.[2] Farms average 155 acres in the central section, slightly less in the west, and slightly more in the east near the Catskills. Dairying predominates, but other types of production are important locally, such as potatoes in Steuben and Wyoming counties or wheat in the western portion. Except in valleys, soils, terrain, and climate are not especially attractive for other than forage crops. Because of relative proximity to New York City, market advantages seem to exist for those areas fringing the western Catskills. Agriculture in valleys is doing well but is subjected to urban pressure. Many hill farms have been declining and therefore abandoned. This trend will undoubtedly continue. Better lands, especially on the eastern and western ends of the region, are being improved.

---

[2] What could be tilled at any time depends on quality of the land as well as proper management practices and the state of competition (market conditions). The implication here is that land quality and market conditions are such that with proper management this per cent of the land could be cultivated.

Central Mixed Farming Region

Occupying 7 per cent of the state, this region includes land that is nearly 75 per cent tillable. Farms average 150 acres in size. A wide variety of fruit, vegetables, and grains are raised, even though more land is in dairy farms than any other kind. Except for drumlin and heavy clay areas, a combination of favorable soils, terrain, and warm climate make possible many alternatives in farming. The use of irrigation might be profitably expanded. This is a good area for farming, and certainly can be expected to be one of the most prosperous agricultural regions in the state in the years ahead.

Lake Ontario–Lake Erie Fruit and Vegetable Region

This region occupies 5 per cent of the state; 78 per cent of the land could be tilled. Farms average a relatively low 110 acres in size. Fruits, vegetables, and grain crops are very important. This is the principal fruit region in the state and is often referred to as the fruit belt. Dairying is carried on

APPALACHIAN UPLAND DAIRY REGION southeast of Syracuse. The more level and lower sites are still widely used for dairy crops; the steeper and higher locales show evidence of considerable abandonment. Generally the farm buildings in this photograph give little evidence of highly successful farm operations. *N.Y.S. Dept. of Commerce.*

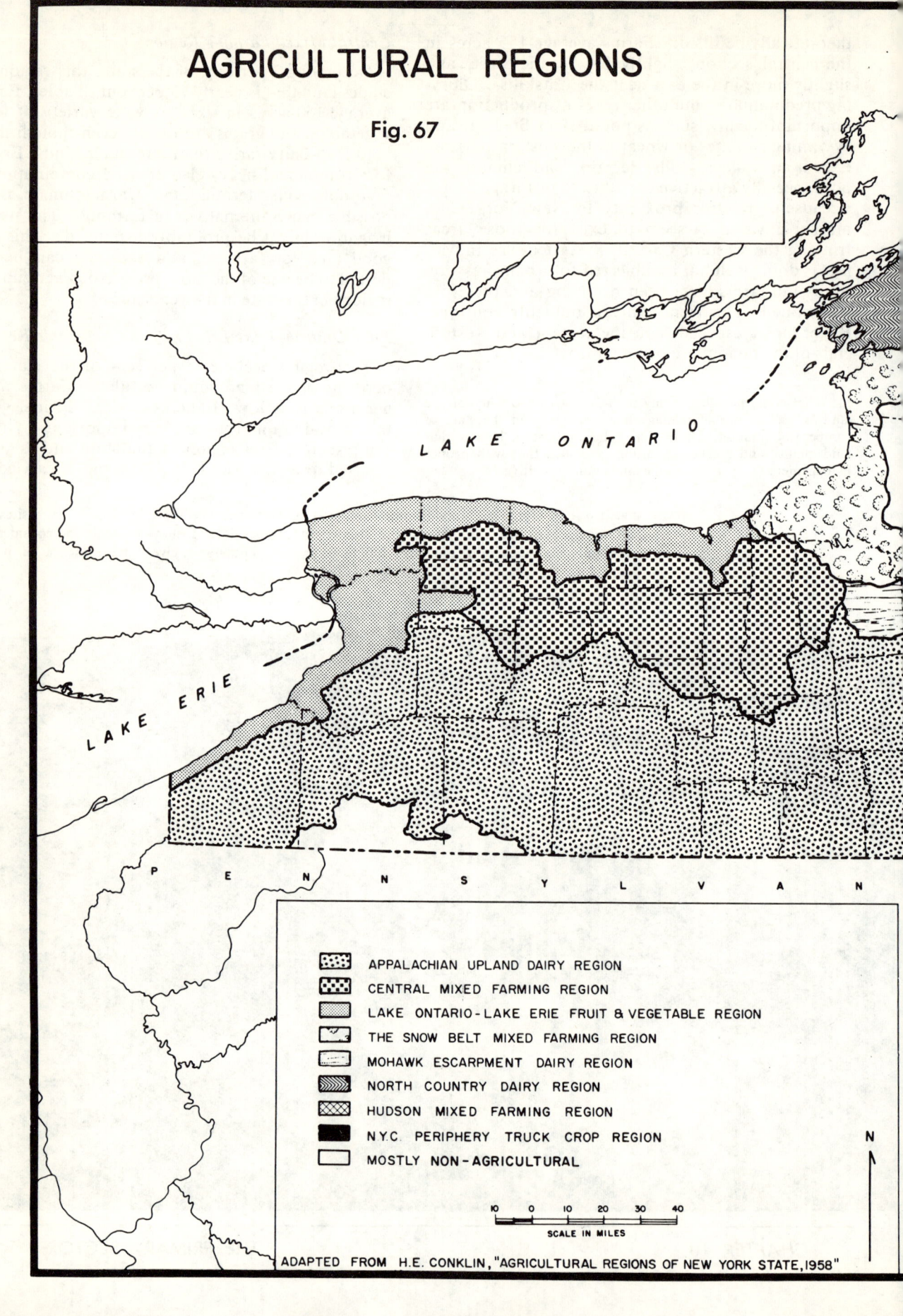

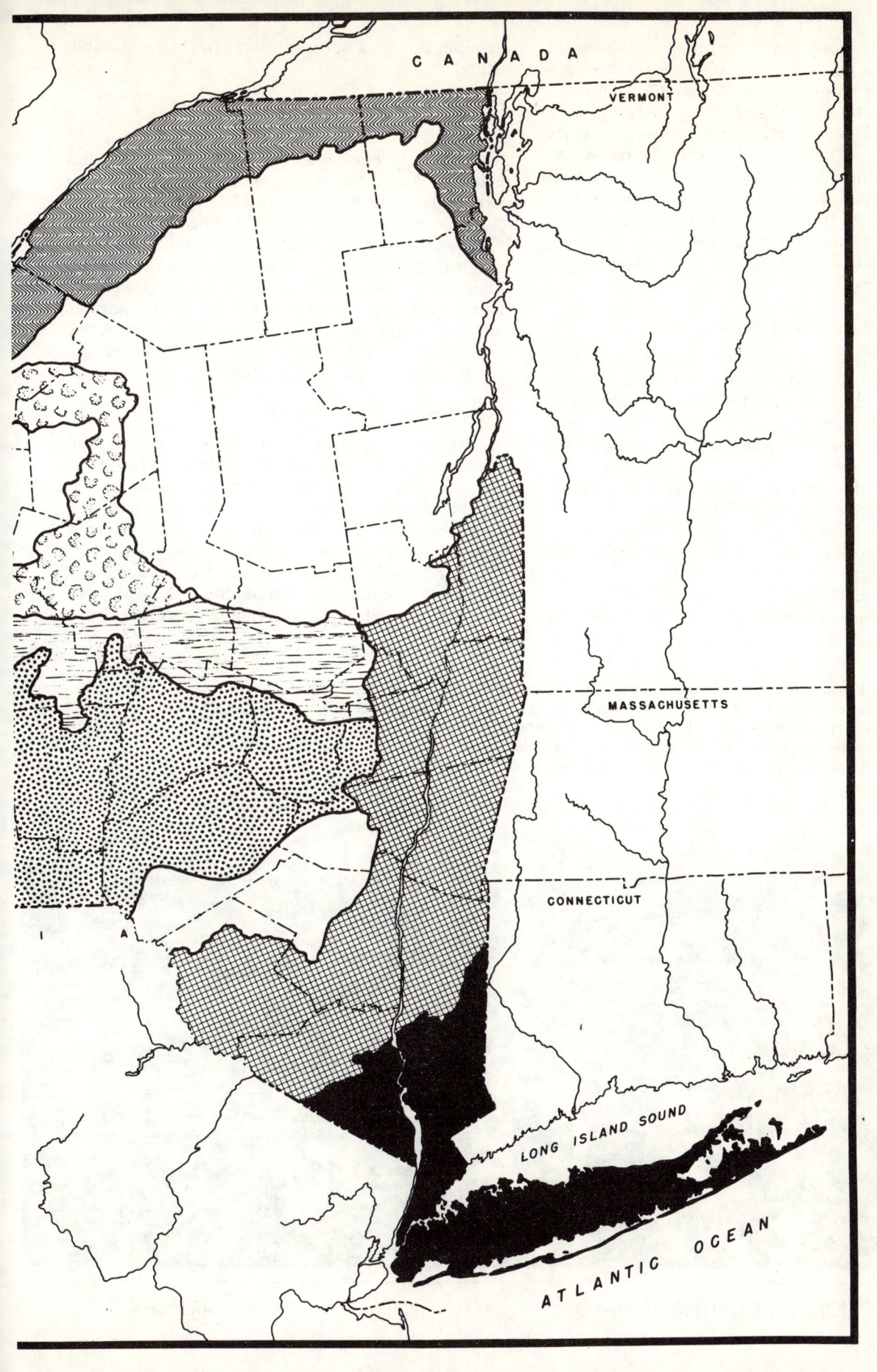

but is definitely secondary. The topography is good for agriculture, but soils vary extensively in capability from place to place. The Great Lakes modify temperatures and thus retard spring budding; they also prolong growth in fall by cutting down on early frost damage. The spring influence is more important than the fall effect. Rainfall is so low that irrigation is often desirable. This is an area of variable possibilities from place to place. It is subject to market and labor problems and to competition from other parts of the country. The use of land is relatively intense.

Snow Belt Mixed Farming Region

This region, which occupies 6 per cent of the state, surrounds the unproductive Tug Hill Upland. Fifty-five per cent of the land is arable. Farms average about 135 acres in size. The region is devoted largely to dairying, but some fruit and vegetables are grown on favorable soils. Although terrain is satisfactory, soils and drainage problems occur. Snows, averaging well above 100 inches a year, do not lend attractiveness to the region either. Many rural people work in cities. Basically, this is a poor agricultural region with few specific advantages. With a demand for more agricultural land it could be developed into a considerably greater producing area, but this does not seem likely in the foreseeable future.

Mohawk Escarpment Dairy Region

This region occupies 4 per cent of the state; 55 per cent of the land could be tilled. The region lies in the Mohawk Valley and along the northern escarpments of the Appalachian Upland. Farms average 165 acres in size and dairying is carried on with but few exceptions. There are some soil and terrain handicaps, but generally the resource base is good for forage crops, and thus dairying. It is well situated in relation not only to nearby urban markets but to fast transportation services into New York City as well. This has proved to be one of the strongest dairy regions of the state.

North Country Dairy Region

This region occupies 7 per cent of the state and 46 per cent of the land could be tilled. Farms average 185 acres in size and are almost exclusively devoted to dairying. Soils vary from good to bad, this range reflecting itself in variation in farm success. Climate and terrain are all right for forage crops, but a relatively great distance from consuming markets is a handicap. Although quantity of overall dairy production is surprisingly great, economic

CENTRAL MIXED FARMING REGION. In this area southeast of Rochester the level Erie–Ontario Lowland supports some of the most profitable agriculture in the state. Most farms emphasize dairying, but a wide variety of crops are produced. The Thruway in the center of the picture finds easy passage across this country and adds greatly to speed of travel, but it has resulted in parcelization or other disruption of agriculture on many farms. *N.Y.S. Thruway Authority.*

A Dairy Farm in the Central Mixed Farming Region. Three silos, a new house, and other buildings in good repair suggest success. Farms such as this are being substantially enlarged. *N.Y.S. College of Agriculture, Cornell Univ.*

Mohawk Escarpment Dairy Region. High lime, well-drained soils, and a good climate for alfalfa and corn silage make this kind of prosperous farm landscape possible. *N.Y.S. College of Agriculture, Cornell Univ.*

North Country Dairy Region near Malone in Franklin County. Level areas are widely used for hay crops. As roughness of terrain increases toward the south in the Adirondack foothills, profitable agriculture disappears. *N.Y.S. Dept. of Commerce.*

CHAPTER 10 THE PRIMARY SECTOR

IRRIGATING A POTATO CROP in the New York City Periphery Truck Crop Region. This land is well adapted to intensive agricultural use and in addition to potatoes produces fresh vegetables, horticultural specialties, and fowl. Perhaps more capitalization and specialization exists here than in any other part of the state. *N.Y.S. College of Agriculture, Cornell Univ.*

success on individual farms is spotty. Farms are comparatively large, but investments are low; many are marginal or submarginal.

Hudson Mixed Farming Region

With 53 per cent of its land tillable, this region occupies 12 per cent of the state. It includes the Hudson Valley, the highlands to the east, and most of Sullivan County to the west. Dairy, poultry, fruit, and vegetable farms are all common; they average 100 acres in size. Soil and terrain are not attractive, but growing seasons are long, energy receipts are high—especially at low elevations, and precipitation is plentiful. The best areas with greatest agricultural opportunities are found in the Hudson Valley, which has developed a noteworthy fruit and vegetable production. Requiring little land and usually depending on outside sources for feed, poultry enterprises can be, and have been, established widely in the region without concern for the local resource base. This is a region of varying success. There are many part-time farmers and much urban influence. Abandoned land in the higher and rougher areas is widespread.

NORTH COUNTRY FARMS traditionally have big barns, grass, hay, and permanent pastures. Soil drainage, modern machinery, extensive fertilization, and generally new techniques are needed for success. This farm apparently has not utilized enough of these things and has deteriorated. *N.Y.S. College of Agriculture, Cornell Univ.*

New York City Periphery Truck Crop Region

This region occupies 6 per cent of the state and 63 per cent of the land not currently used for urban purposes could be tilled. For a radius of forty or fifty miles around New York City land not already occupied by urban functions frequently can best be used for truck crops. This intensive use helps satisfy the almost insatiable demand for these products by the city and at the same time results in enough production from a given acre to warrant the longest possible continuation of the land in agricultural use. The region includes, for the most part, Long Island and Westchester, Rockland, and Putnam counties. Farms average only seventy-five acres.

Vegetables are dominant, but poultry-raising is very important, too, and dairy enterprises, many of them with very large numbers of milkers but little pasture, are numerous. Terrain is good on Long Island but limiting to the north. Soils are generally low in fertility, although they respond well to fertilization and thus can support intensive use. Adequate precipitation, the longest growing season, and the greatest amount of energy receipt in the state are distinct advantages. Unquestionably, though, the region's chief advantage is a market one. Farms are highly capitalized, specialized, and technically advanced. Success varies widely with prices and with skill of the operator.

Nonagricultural Areas

Twenty-seven per cent of the state is included in those areas shown in Figure 67 as nonagricultural. Included are the Adirondacks, Catskills, Tug Hill Upland, and Allegheny Hills. Previous discussions of the physical environment would lead the reader to the conclusion that the terrain, soils, and climate of these areas are not conducive to farming. This is certainly a correct conclusion. In fact, in these four areas only 2 per cent of the state's farms may be counted, and most of these are unproductive, marginal or submarginal, part-time units. Large portions of these areas have never been farmed, others once were used but were abandoned long ago. Their future use seems to be along lines of forestry or recreation, not agriculture.

Chapter 12, especially, considers how these lands can best be used so that the people of the state can get maximum benefits. In a sense it might even be argued that it is fortunate that these lands are not suited to agriculture, for as they are they contribute mightily to the aesthetic attractiveness of the New York landscape. Although some might still maintain that aesthetic attractiveness is of minor importance, growing pressures in a society moving at ever accelerating paces are likely to make more and more people aware of the relaxing values of such things as a remote mountain lake, a dark spruce forest, or a view from a wilderness mountain peak.

FARM PROBLEMS AND PROSPECTS

Obviously, the people of New York would like to see the agricultural resource base of the state used wisely. There is much land and it can produce a considerable amount of food and directly provide a livelihood for many farmers if economic opportunities and general living conditions are attractive. That farming be economically attractive is crucial. The future geography of agriculture should reflect largely areal variations in economic attractiveness of farming over the state.

As an area develops economically, the number of farmers, and often the amount of farmed land, may decrease. Furthermore, opportunities for deriving income ordinarily increase faster and attain higher levels in cities than in rural areas. In face of this a major problem of any society is to keep the brighter and most able younger farmers on the farm. This can be done only on economically successful farms where incomes are high enough to compare attractively with the amount of money that can be earned in nearby cities. The number of farms and farmers in New York has been dwindling, largely because of insufficient income. The next thirty years may see the number of farms cut by another third or half. Some of the land of these farms will be purchased and used by farmers who continue to farm; some will grow back to trees and be lost as farm land for a long time, at least; some of this will be used by rural nonfarmers for homes and recreation; some will be absorbed by growing urban centers, highways, and other similar uses and be lost forever as farm land. Change in the agricultural scene is bound to affect the trend in land use on farms. Figure 68 tells much about this changing land use situation.

Competition from other parts of the country is a problem. New York State apples must compete with those raised in the state of Washington, peaches with those in Georgia, grapes and vegetables with those in California, and dairy products with those in Wisconsin and Minnesota. Keeping abreast of technical advances, even staying in the vanguard of progress, is the best defense against this competition. Even then, losses may occur.

Prospects for profits on farms are being affected not only by rising labor costs but also by the scarcity of labor willing to work on a farm. Migratory workers are imported to harvest fruit and vegetables when thousands of unemployed people are but a short distance away. The wages that farmers can afford to pay—or do pay—to migratory labor commonly are not attractive even to the individual

218 ECONOMIC ACTIVITIES TODAY PART THREE

THE HYPOTHETICAL RANGE OF FARM STATUS. These six photos show that farming in New York State varies from high success levels to complete termination of operations. The first shows a high level of capitalization; the second, a rather average situation; the third, general farm failure but still attempt is being made probably as a secondary means of support; the fourth, farming has ceased although pastures are rented for grazing of young stock from a nearby farm; the fifth, buildings are falling down and weeds and brush are taking over the land; the sixth, buildings are gone and trees are beginning to reclaim the land they once dominated. *N.Y.S. College of Agriculture, Cornell Univ.*

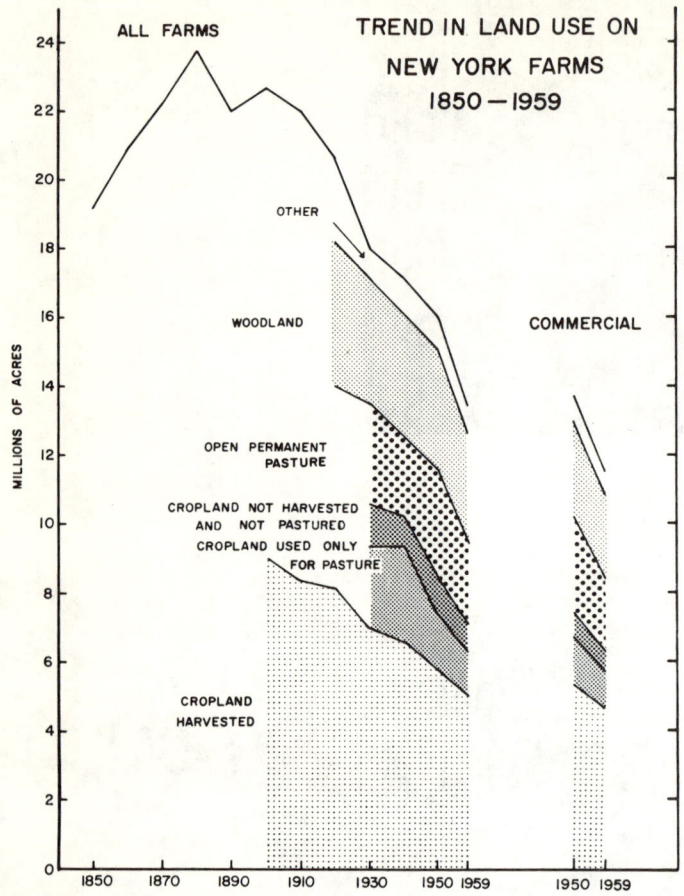

Fig. 68

who is drawing unemployment or welfare checks, to say nothing of the worker who is holding down another job.

Expensive specialized machinery has encouraged specialization in farming; it has made it possible for one man to do more work, to handle more acres, to get more from each acre, but it also requires much more investment in non-real estate items. The need for large cash outlays for such things as equipment and fertilizer makes it difficult for young men to get started in farming and easy for farmers to go broke in a hurry. On the brighter side of things, where soils are being wisely used by knowledgeable farmers, modern techniques of fertilization and cultivation of the last several decades have not only permitted sustained yield but have also increased yield of responsive soils beyond their original virgin levels.

Perhaps the most important geographic question has to do with which parts of the state will experience the most agricultural success in the years ahead and which areas will languish. All of the agricultural regions have some farms with a bright future and some that are bound to fail sooner or later. One part of a county may have the potential for an excellent future, while another part of the same county has already lost most of its farms. Good soils and relatively level terrain with good drainage are essential ingredients to success, but adequate capital and farm size, efficient, technically advanced management, and sound marketing opportunities may be even more important. Urban growth, too, has profound effects locally. Obviously, as city fringes spread, agricultural lands are taken over by residences, factories, roads, and other urban uses. But perhaps as important and less obvious to the casual observer is the land held for speculation. In and around the urban fringe much land is commonly purchased by speculators, taken out of farming, and held until it can be sold for urban uses. This procedure allows the holder of the land to take advantage of what economists call the unearned increment. Farmers themselves occasionally stop farming and also just keep their land for future sale for urban uses. The result is a belt around cities containing a good deal of land not used for farming or anything else, but waiting to be moved into higher value categories related to urban use.

Agricultural economists at Cornell University have made maps of considerable portions of the state, designating several classes of farm land ranging from land which can be expected to experience general farming success in the future to that in which farming cannot be expected to succeed. In setting up this classification, both physical qualities of the land and the general nature of the farming enterprises are taken into consideration. From these maps it may be deduced that none of the agricultural regions shown in Figure 67 will escape future loss of farms or loss of cultivated acreage, but that some will do much better than others. Four of these regions should have relatively bright futures: Central Mixed Farming Region, Mohawk Escarpment Dairy Region, Lake Ontario–Lake Erie Fruit and Vegetable Region, and Truck Crop Region of the New York City Periphery. Parts of other regions should do reasonably well, too, such as the western and eastern extremities of the Appalachian Upland Dairy Region, parts of the Hudson Valley, and a few sections in the North Country Dairy Region. Other areas in the state that still have farms probably will experience a strongly declining agricultural economy.

Mining

New York is not one of the leading mining states. In 1962 it ranked eighteenth nationally, and it generally contributes less than 2 per cent of the nation's minerals by value. Mining employed but .14 per cent of the state's labor force. On the other hand Figure 69 shows a surprising number of mining operations producing a wide variety of minerals. Almost every county has commercial mineral deposits of some kind or other. Actually, mining is somewhat more significant than statistics would indicate. Heavy, bulky products such as cement, stone, sand and gravel, and gypsum are needed locally in large quantities by a state with so much industry and construction, so large an urban population, and so many miles of good highways and streets. Fortunately, New York State has these products. If they had to be procured from a considerable distance, economic development would be more difficult and costly. Furthermore, the state contributes all, or substantial per cents, of the national production of certain items such as emery, garnet, and titanium concentrates.

THE MINERAL PRODUCTION MAP

Figure 69 (in back-cover pocket) shows five things: (1) location, size, and type of reported currently operating mining establishments; (2) location of abandoned or unreported mining establishments; (3) known mineralized sites which may or may not have formerly supported mining; (4) general localities underlain by salt, limestone and dolomite, gypsum, and petroleum and natural gas; and (5) the relative importance of the various counties (as designated in the inset). Refer to Table 19 for county production at different times.

Some 290 operating mining operations turning out twenty-five different products are recorded on the map. Three sizes of circles are employed to differentiate between large-, medium-, and small-scale operations. Size categories are relative only to producers of the same product in New York State. In other words, gypsum producers in New York State are classed as large, medium, or small as compared only with other gypsum producers in the state, not with garnet mines in the state or with gypsum quarries in Michigan. Letters within all circles indicate type of production. Clearly, sand and gravel operations are most numerous, followed in order by limestone and dolomite, clay, slate, and cement producers. Highway Department sand and gravel workings and small-scale gravel pits in general are omitted because of their transitory character and because they are so numerous that they would clutter the map.

Abandoned, or in some cases unreported, mining operations that are designated on the map by blue letter symbols are numerous. Their inclusion on the map contributes to the distributional picture of mineralization in the state and implies something of the state's mining history as well as its possible future. Those sites which have proven mineral occurrence, but for which no information is available that indicates production ever took place, make a similar contribution. They are designated by a tan (nonmetallic) or brown (metallic) symbol. Mining is changing so rapidly that some places of production shown in the map had already been abandoned by publication date. Two iron operations are cases in point. A few new producing units, especially of sand and gravel, usually open each year.

Extensive areas in the western and southern parts of the state are shown on the map as underlain by salt or gypsum. Salt is produced at seven places and gypsum at five. Dolomite outcrops and limestone occur in narrow bands and irregularly shaped areas in many parts of the state. Both the Adirondacks and the southern plateau country are almost completely surrounded by these outcrops, which are worked at many localities. Although oil and natural gas fields are scattered over the western half of the state, most of the production is in the southwestern-most counties. The pre-eminence of St. Lawrence, Essex, and Erie counties in total mineral production is strikingly portrayed in the inset map. Unfortunately, data limitations make it impossible to represent every county. In fact, 37 per cent of the mineral value comes from counties for which data are withheld for strategic or other reasons.

TRENDS IN MINERAL EXPLOITATION

The use of New York's minerals began with the Indians, who made weapons and tools out of stone and prized the salt in the vicinity of Syracuse. Salt has since been an important item in the development of Upstate New York. The first iron mine was opened about 1750 and over seventy million tons of iron ore have been mined since. Significant changes in the relative importance of mineral production have occurred in the last half-century. In 1904 clay products accounted for over 40 per cent of the total mineral value. Today they are insignificant. Mineral waters from such places as Saratoga were once significant; now they no longer appear in the statistics. Cement, stone, salt, sand and gravel, and iron ore have become the recent leaders, while petroleum and natural gas have decreased in importance. The value of mineral production has increased over the years, reaching 242 million dollars in 1962. Table 20 summarizes production for selected years since 1940. Large reserves of salt, gypsum, stone, sand and gravel, clay, titanium, iron ore, zinc, talc, wollastonite, and many other mineral products form a base for higher levels of pro-

TABLE 19
COUNTY OUTPUT OF MINERALS
(In Thousands of Dollars)

County*†	1955	1956	1961	1962	Leading Minerals by Value, 1962
Albany	$ 1,538.0	$ 1,416.2	‡	$ 2,435.9	Stone, cement, clays, sand and gravel
Allegany	9,061.0	10,665.3	$ 311.1	387.2	Sand and gravel
Broome	816.1	690.4	1,023.0	1,082.4	Sand and gravel, stone, clays
Cattaraugus	2,333.1	2,987.2	833.7	1,055.8	Sand and gravel
Cayuga	‡	‡	542.2	‡	Stone, sand and gravel
Chautauqua	423.0	339.3	160.8	154.8	Sand and gravel
Chemung	‡	‡	‡	‡	Sand and gravel
Chenango	‡	122.0	‡	‡	Sand and gravel
Clinton	4,852.9	5,311.8	‡	2,745.5	Iron ore, stone, sand and gravel, lime
Columbia	13,670.1	‡	‡	‡	Cement, stone, sand and gravel, clays
Cortland	92.3	‡	116.4	110.9	Sand and gravel
Delaware	439.9	681.7	‡	1,208.1	Stone, sand and gravel
Dutchess	‡	‡	‡	‡	Stone, sand and gravel, clays, gem stones
Erie	22,656.2	24,108.3	16,458.6§	15,970.9	Cement, stone, sand and gravel, gypsum, lime, clays
Essex	20,689.1	25,741.5	‡	‡	Ilmenite, iron ore, wollastonite, sand and gravel, garnet
Franklin	111.6	111.0	178.5	131.9	Stone, sand and gravel
Fulton	30.8	116.0	120.6	113.0	Sand and gravel
Genesee	3,293.7	3,316.7	2,746.9	2,699.0	Stone, gypsum, sand and gravel
Greene	‡	‡	‡	18,382.7	Cement, stone, sand and gravel, clays
Hamilton				‡	Sand and gravel
Herkimer	‡	‡	‡	‡	Stone, sand and gravel, gem stones
Jefferson	768.9	‡	1,172.1	1,085.9	Stone, sand and gravel
Lewis	29.7	‡	‡	‡	Stone, sand and gravel
Livingston	‡	‡	‡	‡	Salt, sand and gravel, stone
Madison	243.4	‡	387.3	490.1	Stone, sand and gravel, gem stones
Monroe	2,750.8	2,924.0	3,087.6	3,609.0	Stone, sand and gravel, gypsum
Montgomery	‡	‡	‡	485.4	Stone, sand and gravel
Nassau	5,979.1	6,250.7	6,957.7	8,142.2	Sand and gravel, clays
Niagara	1,243.7	‡	‡	3,941.5	Lime, stone, sand and gravel
Oneida	1,673.1	1,798.0	‡	2,732.0	Stone, sand and gravel, iron ore, gem stones
Onondaga	10,514.2	11,370.3	‡	18,418.2	Lime, salt, stone, cement, sand and gravel, clays
Ontario	668.8	‡	‡	1,410.4	Stone, sand and gravel
Orange	733.6	‡	1,172.5	1,037.9	Sand and gravel, clays, stone, peat
Orleans	‡	‡	‡	105.5	Sand and gravel
Oswego	285.3	226.0	‡	377.9	Sand and gravel
Otsego	62.6	‡	‡	166.4	Stone, sand and gravel, gem stones
Putnam	197.5	‡	‡	‡	Sand and gravel
Rensselaer	397.2	444.1	‡	903.4	Sand and gravel, stone, clays
Richmond			‡		Gem stones
Rockland	‡	‡	‡	7,144.5	Stone, sand and gravel
St. Lawrence	34,493.9	38,234.3	35,280.8	31,872.8	Iron ore, zinc, talc, stone, lead, sand and gravel, silver
Saratoga	662.5	606.7	‡	1,161.6	Stone, sand and gravel
Schenectady	‡	340.7	‡	256.2	Sand and gravel
Schoharie	‡	‡	‡	‡	Cement, stone, clays, sand and gravel
Schuyler	‡	‡	‡	‡	Salt, sand and gravel
Seneca	‡	‡	‡	‡	Peat, sand and gravel
Steuben	562.0	‡	‡	527.5	Sand and gravel
Suffolk	6,178.8	6,613.0	6,946.9	6,364.3	Sand and gravel
Sullivan	255.9	291.6	‡	‡	Stone, sand and gravel
Tioga	155.0	‡	350.8	296.4	Sand and gravel
Tompkins	‡	‡	‡	‡	Salt, stone, sand and gravel
Ulster	2,745.2	‡	‡	12,598.0	Cement, stone, clays, sand and gravel
Warren	‡	‡	‡	‡	Cement, garnet, stone, gem stones
Washington	1,476.3	1,074.9	789.6	632.9	Stone, sand and gravel
Wayne	‡	‡	‡	‡	Stone, sand and gravel
Westchester	808.6	570.3	730.3	779.8	Stones, sand and gravel, emery, peat
Wyoming	‡	‡	‡	‡	Salt, stone
Undistributed	63,949.4	89,989.9	154,465.3§‖	90,872.5	
Total#	216,907.0	237,016.0	233,833.0§	241,892.0	

* Bronx, Kings, New York, Queens, and Yates counties are not listed because no production was reported.
† Fuels, including natural gas and petroleum, not listed by counties; value included with "Undistributed."
‡ Figure withheld to avoid disclosing individual company confidential data; included with "Undistributed."
§ Revised figure.
‖ Includes natural gas and petroleum, some gem stones and sand and gravel that cannot be assigned to specific counties, and values indicated by footnote ‡.
Data may not add to totals shown because of rounding.

Sources: U.S. Bureau of Mines, *Minerals Yearbook*, 1956, 1962.

duction, providing market prices and competition from other areas are favorable.

Mining is a dynamic business. The old law of supply and demand, prices and price supports, the development of new technology, and many other factors can cause the fortune of any operation to rise and fall. A mine operating on a narrow margin of profit one day might well be forced to close down tomorrow if faced with reduced profits. On the other hand, improved techniques of mining and milling and changed market conditions may allow mineral exploitation which was previously uneconomical. For example, titanium-bearing ores were known to exist near Tahawus in Essex County for many years. During World War II titanium came into demand and new milling techniques which separated the titanium from the iron that occurred with it resulted in a new and vigorous mining enterprise. New exploration procedures also may turn up new mineral deposits of value. It was not until the airborne magnetometer, a magnetic device invented during the war to detect submerged submarines, was used that the magnitude of the iron ore deposit at Benson Mines was suspected. Today, Benson Mines has one of the largest open-pit magnetite (an iron mineral) mines in the world, and contributes substantially to the health of New York's mineral industries. New developments like these could conceivably take place in the future.

TYPES OF PRODUCTION

Mineral products may be categorized into three groups: metals, nonmetals, and fuels. In 1962 these represented 17, 79, and 4 per cent respectively of the value of the state's mineral production. Nearly 90 per cent of the income from mineral production is attributable to seven major items (Table 21).

TABLE 21
MAJOR ITEMS IN NEW YORK STATE MINERAL PRODUCTION

Item	Per Cent of State's Mineral Production as Measured by Value	
	1956	1962
Cement (est.)	21	25
Stone	15	19
Salt	11	13
Sand and Gravel	12	13
Iron Ore	17	10
Zinc	7	5
Petroleum	5	3

Metals

At least a dozen metals occur in the state in rich enough deposits to have been mined at one time or another. Most occur in the ancient crystalline rocks of the Adirondack region or around the Adirondack margin. Five of the more important ones are discussed below.

Iron Ore.—Iron ore of one kind or another occurs in numerous sections of the state, and mines have been worked many years. In recent decades production has been largely from the Adirondack area. Mines producing magnetite in Essex, Clinton, and St. Lawrence counties are most important. Open-pit operations at Benson Mines in St. Lawrence County dominate production, but Tahawus in Essex County is significant, too. At Tahawus iron ore is produced as a coproduct in the extraction of titanium concentrate. The magnetite ores occur in

TABLE 20
GROWTH OF NEW YORK STATE MINERAL INDUSTRIES BY QUANTITY AND VALUE*
(In Thousands)

Commodity†	Unit of Measure	1940		1945		1950		1956		1962	
		Quantity	Value	Quantity	Value	Quantity	Value	Quantity	Value	Quantity	Value
Cement (Portland)	376 lb. bl.	8,251.0	$11,687.0	5,578.9	$ 9,009.4	13,271.4	$ 30,895.2	NA	$ 50,000.0‡	NA	$ 60,000.0§
Clays	Short ton	NA	NA	NA	NA	1,443.1	1,155.6	1,234.8	1,507.7	1,397.0	1,618.0
Emery	Short ton	1.0	9.3	7.8	75.9	5.9	75.3	12.1	174.0	4.3	71.0
Gypsum	Short ton	798.2	1,037.1	557.9	1,262.9	1,280.1	3,876.1	1,140.1	4,817.3	601.0	3,122.0
Iron Ore	Long ton	874.3	4,031.0	1,965.6	15,136.1	2,917.2	27,914.8	3,188.2	41,093.6	2,099.0	24,953.0
Lead	Short ton	1.9	197.3	.8	148.2	1.4	400.6	1.6	504.9	1.1	196.0
Natural Gas	Million cu. ft.	12.1	8,246.0	6.0	4,395.0	3.3	837.0	4.0	1,160.0	4.2	1,198.0
Petroleum	42 gal. bl.	4,999.0	11,600.0	4,648.0	17,470.0	4,143.0	15,660.0	2,748.0	12,091.0	1,789.0	8,229.0
Salt	Short ton	2,117.6	6,523.7	2,862.2	10,327.0	2,806.9	14,405.3	3,872.7	27,544.9	4,456.0	32,236.0
Sand and Gravel	Short ton	13,225.1	7,639.6	7,477.6	5,049.9	21,778.0	18,075.2	27,814.9	28,721.7	29,447.0	31,346.0
Silver	Troy ounce	35.7	25.4	14.2	10.1	32.6	29.5	84.1	76.1	19.0	21.0
Slate	Short ton	NA	479.0	NA	NA	151.1	2,054.7	64.2	943.5	NA	NA
Stone	Short ton	9,782.1	10,398.4	7,900.5	9,133.7	15,729.0	22,529.0	22,805.4	36,134.7	27,589.0	47,256.0
Talc	Short ton	113.6	1,402.5	NA	NA	163.9	4,039.9	NA	NA	NA	NA
Zinc	Short ton	35.6	4,496.4	24.9	5,744.9	38.3	10,883.1	59.1	16,196.4	53.6	12,340.0
Other			8,346.2		12,522.4		3,697.0		16,049.7		19,183.0
Totals‡			76,119.5		90,286.0		156,529.0		237,016.0		241,892.0

NA: Figures are either unavailable or inconsistent in definition with those of other years.
*Figures are taken primarily from various issues of U.S. Bureau of Mines, Minerals Yearbook.
†Items that cannot be listed because they would disclose data for individual companies or because production figures are not available are garnet, lime, cement, talc, titanium concentrate, wollastonite, molding sand, graphite, feldspar, diatomite, strontium, pyrite, molybdenum, manganese, magnesium, limonite, arsenic ore, and beryl.
‡Estimated.

Pre-Cambrian gneisses and other ancient crystalline rocks. On the average the ores are not too rich and so must be concentrated and sintered before being shipped from the locale of the mine. The resulting enriched product is often shipped to steel mills near Buffalo or in western Pennsylvania. Table 20 shows that production increased from less than one million tons in 1940 to more than three million tons in 1956 and then dropped back to two million tons in 1962. During the latter year New York accounted for about 3 per cent of the domestic iron ore production. The iron mineral, hematite, is mined underground in Oneida County but is used only as red iron oxide pigment in metallic paints.

Zinc.—Zinc ore, sphalerite, occurs in St. Lawrence County and in the southeastern part of the state, but all production now comes from Balmat and Edwards in St. Lawrence County. Here the ore is found in a belt of rocks, including Pre-Cambrian crystalline dolomites and limestones, which extends for about eight miles from Sylvia Lake to Edwards. The Balmat mine is one of the largest individual zinc producers in the nation, and lead, silver, and pyrite are recovered as by-products of the zinc-mining. The future seems bright, for perhaps one-fifteenth of the zinc known to exist in the United States occurs in the state. Production levels have been reasonably steady for the last decade or so, with only Idaho and Tennessee producing more.

Lead.—Galena, the chief mineral for lead, is produced in conjunction with zinc-mining at Balmat in St. Lawrence County. The amount is small compared to zinc, and future production will probably be a reflection of the economic fortunes of zinc-mining at Balmat. Price levels can greatly affect the amount of both lead and zinc that are mined.

Silver.—Silver is a by-product of lead production at Balmat. About ten ounces of silver are recovered from each ton of galena ore treated. In the past small quantities of silver have been produced in various parts of the state, but no commercial operations now exist except at Balmat. Many people have thought that they had located silver and even gold in the state, but their claims have generally been erroneously based on mica or pyrite.

Titanium.—Titanium is one of the most interesting and important products turned out by New York State's mineral industries. It is produced from the mineral ilmenite, which contains both iron and titanium. Production occurs at Tahawus in the crystalline rocks of Essex County. Here the largest ilmenite mine in the nation supplies more than 50 per cent of the domestic production. Although attention has long been given to the use of titaniferous ores in the manufacture of titanium alloys, it was not until about 1908 that the possibilities of titanium dioxide as a source of white paint pigment was demonstrated. Production at Tahawus began in 1941. The use in paint contributed extensively to remarkable growth in production, until in 1956 nearly 400,000 tons of concentrate were produced at Tahawus. Although production data for 1962 are not disclosed, it may be assumed that production was well above that of 1956. Titanium pigments are also used in the manufacture of white and light-colored rubber, paper, asbestos shingles and siding, soaps, asphalt tile, linoleum, plastics, leather finishes, textile printing, wallboard, and white glue. The huge reserves of ilmenite in the Tahawus area will support production at high levels for many years.

Nonmetals

In total nonmetals are much more significant in the state than metals, accounting for 79 per cent by value of all mineral production. Figure 69 shows production to be extensively distributed over the state. Because of widespread stone and sand and gravel deposits, production locations commonly are reflections of market demand. That is, pits and quarries are developed near large cities or in response to highway needs.

Portland Cement.—The Portland cement industry now ranks first among the mining and quarry activities of New York State and accounts for nearly one-fourth by value of the state's mineral production. In 1962 New York ranked fifth in production behind Pennsylvania, California, Texas, and Michigan. Actually, cement is a manufactured product made from limestone, shale, and clay, and is included here under minerals only because it is so listed traditionally in New York state and Bureau of Mines publications. The mineral materials used in the cement industry are widely available, and so cement, being a heavy, bulky commodity, is produced at sites readily accessible to large markets. The Hudson River Valley, including portions of Greene, Columbia, Ulster, and Albany counties, is the major area of production. From here cement moves with ease to satisfy the almost insatiable market in and around metropolitan New York City. The Buffalo area, with its large local demand, and utilization of limestones brought in by lake transport, ranks second. Substantial production also occurs in Warren, Schoharie, and Onondaga counties.

Altogether there are twelve important producers of Portland cement and one producer of natural cement. Considerable expansion of plant facilities has taken place in recent years, and some shipments are made out of the state, particularly to New England. Two large cement plants recently opened in Albany County. The state cement industry can be expected to grow as the demand increases, for raw materials are available to support this growth almost indefinitely. In 1962 Portland cement manufacture consumed 2.9 million tons of limestone, 2.5 million tons of so-called cement rock, 285,000 tons of clay and shale, and 141,000 tons of gypsum. It is interesting to note that in 1962 shipments to customers were principally in bulk; two-thirds by truck, one-quarter by railroad, and the rest by waterway.

Stone.—A wide variety and huge quantity of stone is quarried in the state. Included under the heading of stone are sandstone, granite, limestone and dolomite, marble, trap, and quartzite. Together they represented about 19 per cent of New York's mineral production by value in 1962 and thus placed second behind Portland cement. Limestone and dolomite account for nearly 80 per cent of the stone quarried. These materials are used in road construction, concrete aggregate, certain chemical industries, lime manufacture, steel furnace flux, and fertilizers, as well as in cement. Traprock (basalt) is used extensively for road metal. Dimension stone of various varieties, used for flagging, curbing, repairing, fireplaces and other interior decoration, and building purposes, is produced in minor quantities.

Salt.—Salt has been an important mineral product throughout New York's development. In 1962 the state turned out over 16 per cent of the nation's supply and ranked third among the seventeen salt-producing states. About 6,000 square miles of the Appalachian Upland in southern and western New York are underlain by Silurian salt beds, and production occurs at localities in Livingston, Onondaga, Schuyler, Tompkins, and Wyoming counties. Salt is either mined as rock salt or produced as brine by forcing water into drill holes to dissolve the salt at depth. The largest rock salt mine is at Retsof in Livingston County, and the most important area of wells is at Tully in Onondaga County. These wells supply brine to a large chemical plant in Syracuse. Chemical industries probably use over 70 per cent of New York's salt production for the manufacture of such things as chlorine, soda ash, and dyes, and, considerably to the consternation of the state's automobile owners, most of the rest is used for ice control on streets and highways. No more than 3 per cent of the state's salt is used for food seasoning and preservatives.

Sand and Gravel.—Sand and gravel are fourth in value among New York's mineral commodities, and production has been substantially increased in recent years. Ninety per cent of the sand and gravel is washed and screened for use in structural and paving concrete. Fill, industrial sands, and railroad ballast account for most of the remainder.

Fortunately, sand and gravel deposits occur widely over the state. Most of the deposits are related in some way to continental glaciation, usually lacustrine deposits. Some 140 enterprises are shown in Figure 69, in spite of the fact that small producers and highway department gravel pits are omitted. If all small gravel pits were shown, there would be several hundred more.

With the deposits being as widely distributed as they are, production is pretty much in response to markets, with the urban impact obvious. Clusters of sand and gravel operations appear in Figure 69 near the larger cities. Sand and gravel are heavy and bulky but of low unit value so cannot stand the costs of long-distance shipping.

Gypsum.—New York State is the only major gypsum producer in northeastern United States, and the increasing demand for plaster, wallboard, and similar products made of gypsum in the building trades would suggest a bright future for production. Excellent quality and enormous reserves enhance future production possibilities. On the other hand, increasing competition from Canadian imports may limit production. Figure 69 shows a long, narrow area underlain by gypsum extending from Buffalo to east of Syracuse. All five of the state's mining operations are situated between Rochester and Buffalo.

Talc.—New York is the chief talc producer in the United States, accounting for over 25 per cent of the domestic production. Talc occurs in a belt of Pre-Cambrian rocks near Gouverneur in St. Lawrence County and is mined in that locality by two companies. It is used in the manufacture of paints, ceramics, paper, insecticides, various building materials such as composition floor and wall tile, and miscellaneous fillers. Production has climbed appreciably in recent years, but figures have been withheld from publication since 1953.

Clay and Shales.—New York clays are used primarily in brick manufacture and lightweight aggregates, the Hudson Valley being the most important producing area. The very large market for clay products in metropolitan New York can easily be served from there. Buffalo, Syracuse, and Binghamton, among other cities, obtain at least a part of their clay requirements from local deposits. Considerable tonnages of clay and shale are also used in the manufacture of cement and for abrasive wheels and shapes. Most of the clay deposits were laid down by glacial melt waters and occur at or near the surface of the ground in various parts of the state. Usable shale is largely of Devonian age and occurs in the Hudson Valley and elsewhere.

Garnet.—New York can boast of having the world's largest garnet mine at North Creek in Warren County. Here, and in Essex County, garnet occurs as an important mineral in the Pre-Cambrian crystalline rocks of the eastern Adirondacks. New York garnet is among the harder varieties and has the added quality of breaking with very sharp edges. It is used principally in the manufacture of abrasives—garnet-paper (sandpaper) and glass-polishing powder. Reserves are large, but the future of production will depend extensively on the amount of competition from artificial abrasives.

Emery.—The entire domestic supply of emery comes from two mines near Peekskill in Westchester County. These mines have been operating for over sixty years and have contributed most of New York's production. Emery is a natural mixture of a variety of very hard minerals. Of these, corundum and spinel are most significant to its quality. Emery is used for grinding wheels, emery paper, polishing pastes, and other abrasives. A recent important use is for nonskid surface coatings for concrete floors and steps. Although competition, particularly from Turkish and Greek imports, exists, New York's production has increased in recent decades.

Slate.—Slate production in the state has suffered as a result of competition from artificial products, unfavorable freight rates, and a failure of operators to modernize. Production comes from the general vicinity of Granville in Washington County, where fine deposits of green, purple, black, gray, and variegated slates occur. Slate is used largely for roofing granules, slate flour, and flagstone. Waste materials from the mines are now sometimes marketed as lightweight aggregate. The future of the slate industry, in its fight to withstand competition from artificial products, does not appear too bright. Recently a number of mines have had to shut down.

Peat.—Peat is ordinarily thought of as a fuel, but production in New York is not for fuel. Instead peat is sold as peat moss for lawn and garden cultivation. It is taken largely from bogs in Seneca County, but recently production began at Armonk in Westchester County.

Others.—Although a number of other minerals of the nonmetal type occur in the state and are shown in Figure 69, they either are not mined today or are mined only in relatively small quantities. These include calcareous tufa, diatomite, feldspar, graphite, marl, wollastonite, and vermiculite.

Fuels

Both petroleum and natural gas have been produced in New York for many years. Production, though, has been declining for some time, and the position of the state in the national picture has been declining at even a more rapid rate. Between 1956 and 1960 their portion of the total value of mineral production in the state fell from 6 to 4 per cent.

Petroleum.—In spite of declining production, the high quality of New York's paraffin-base petroleum has kept it among the money producers in the state. Figure 69 shows that the oil fields are confined largely to Cattaraugus, Allegany, and Steuben counties. There are over seven hundred individual producers with perhaps twenty times that many wells. Production per well is small. Proven reserves of 23 million barrels in 1962 will last about thirteen years at current production levels. Production in 1962 was only about three-fifths that of 1956 and the long-range future is not bright. Virtually all of the New York crude oil is treated in Pennsylvania refineries.

Natural Gas.—Figure 69 shows that natural gas occurs in various places in western New York State, but the largest production now comes from Erie, Chautauqua, and Cattaraugus countries. The demand for natural gas in the Northeast is tremendous, but competition from the West and local production costs must be contended with. The estimated 41-billion-cubic-foot reserve in 1962 would last but ten years at current production levels. Many wells have been developed for storage of gas imported from western states, and there is over twice as much of this now stored in the wells as there is native gas remaining.

The Future

New York certainly will never become a Texas, a California, or a Pennsylvania as far as minerals are concerned, but total value of mineral production has been rising some and probably will continue to do so. Exploration and research are the best assurance of continued vitality in the mineral industries, and certainly the brightest future seems to be in nonmetals, not metals or fuels. The fact that the nonmetals accounted for 67 per cent of the total mineral production by value in 1956 and 79 per cent in 1962 exemplifies this trend.

One thing that should be kept in mind is that minerals are nonrenewable resources, that is, resources which when once used will not re-form, regrow, or otherwise replenish themselves as, for example, forests may do. The West is full of ghost towns which once flourished as mining centers, but which found no means of support once the mineral deposits were depleted. Whole districts within states suffered economic deterioration as minerals were used up. While some communities in the Adirondacks, St. Lawrence County, the southwest oil area, and elsewhere have suffered the same fate, or may in the future, most of New York is not sufficiently dependent on minerals to have to worry. Furthermore, the important production of cement, stone, gypsum, sand and gravel, and even salt is based on deposits so extensive and widespread that they may be considered almost inexhaustible. Future production of these commodities in the state is likely to be limited only by demand and competition, not by supply. Production of the heavy, bulky nonmetals will be geographically related to the demand centers. Road-building, construction, and manufacturing in and near the great urban agglomerations will be "magnets" to much of the state's mineral industries of the future.

Lumbering and Other Forest Uses

Although lumber and pulp production has been relatively significant in the past and although nearly half of the state is covered with forest at the present time, forests contribute surprisingly little to the state's economy (see Chapter 4). Employment in primary forest activities is even less than in mining, accounting for but .12 per cent of the state's labor force. Is this because the large acreage of existing forest resource is of poor quality, or is it because man has not been able to organize his activities in such a way so as to utilize it efficiently? In part the currently low production levels are due to both of these causes. The forests, though extensive in total area, are composed of second-growth stands, commonly dominated by low-quality hardwoods which are in limited commercial demand. Furthermore, much of the forest land is in small private holdings which are difficult, if not impossible, to develop economically; other large acreages in the state Forest Preserve have been reserved from cutting by law. The real crux of the problem is that forests cannot be commercially used unless at a profit, and commercially using the fragmented second-growth timber stands in New York at a profit in the usual lumbering and pulping sense is not easy. Using them in less obvious, and at best only semicommercial ways, for watershed protection and recreation is, and will prove to be, rational, whether profitable from a dollar-cents standpoint or not.

While New York's forests produce only about 1 per cent of the nation's lumber they do support numerous sawmills, pulp and paper mills, and a substantial number of wood-using furniture and veneer plants. Altogether, these activities add up to but a fraction of 1 per cent of the productive capacity of the state. Descriptive accounts sometimes show employment of well over 100,000 in New York's forest-dependent industries, but these figures exaggerate the importance of the New York forest resource base because much lumber and other forest products are brought into the state from the outside and processed here. In fact four-fifths of New York's timber and wood needs must be supplied from other parts of the country, Canada, and elsewhere.

Forest Potentials

Is New York forest exploitation currently carried out on a sustained-yield basis? Certainly it was not during the heyday of lumbering in the nineteenth century. Now the annual growth is more than twice the annual cut, but much of the growth is on trees of small diameter, poor form, or low-value species. The rate of removal of favored species, sizes, and grades exceeds the rate of natural replacement. Thus, the state is easily on sustained yield as far as "cubage" is concerned but not as far as economically usable saw timber and pulpwood are concerned.

With 12 million acres of commercial forest land and 14 billion cubic feet of wood growing on these acres, it would seem that there ought to be some method for getting the state on a profitable sustained-yield basis as far as saw timber and pulpwood are concerned. About three-fourths of the saw timber is hardwood, chiefly sugar maple, beech, and yellow birch. Proper management of these species, as well as a growing demand for them, must precede any substantial rise in the relative importance of New York's forest industries. There is much small-dimension stock that could physically be used for pulp but is too far from pulp mills or otherwise not available to them.

During the past fifty years both public and private endeavors have been oriented toward improvement of forest land and timber potentials. Those endeavors have taken the form of improved fire, insect, and disease protection, reforestation, encouragement in wiser weeding, thinning, pruning, and cutting procedures, and the development of new processes such as use of hardwoods for pulp and the manufacture of pressed boards and insulation from logging and milling residues.

Reforestation has been a particularly interesting, if not wholly successful, development. Well over half a million acres of softwoods (largely pine, spruce, and fir) have been planted in New York. Most of the trees used (over one billion in all) have been produced in state nurseries and distributed free or at low cost. The first plantings occurred just after the turn of the century, and for twenty-five or thirty years there was considerable enthusiasm for the economic success of the reforestation process. Trees grow slowly, however, and small plantations, poorly managed during growing periods, have turned out to be anything but "golden eggs." Quality is less than was hoped for in many instances, and serious marketing problems, especially for pole and pulp sizes, exist. Furthermore, rising labor costs (for thinning, pruning, etc.) which made management increasingly expensive over the years have dulled the prospects for ready profits. Interest has turned in recent decades to Christmas tree production, both

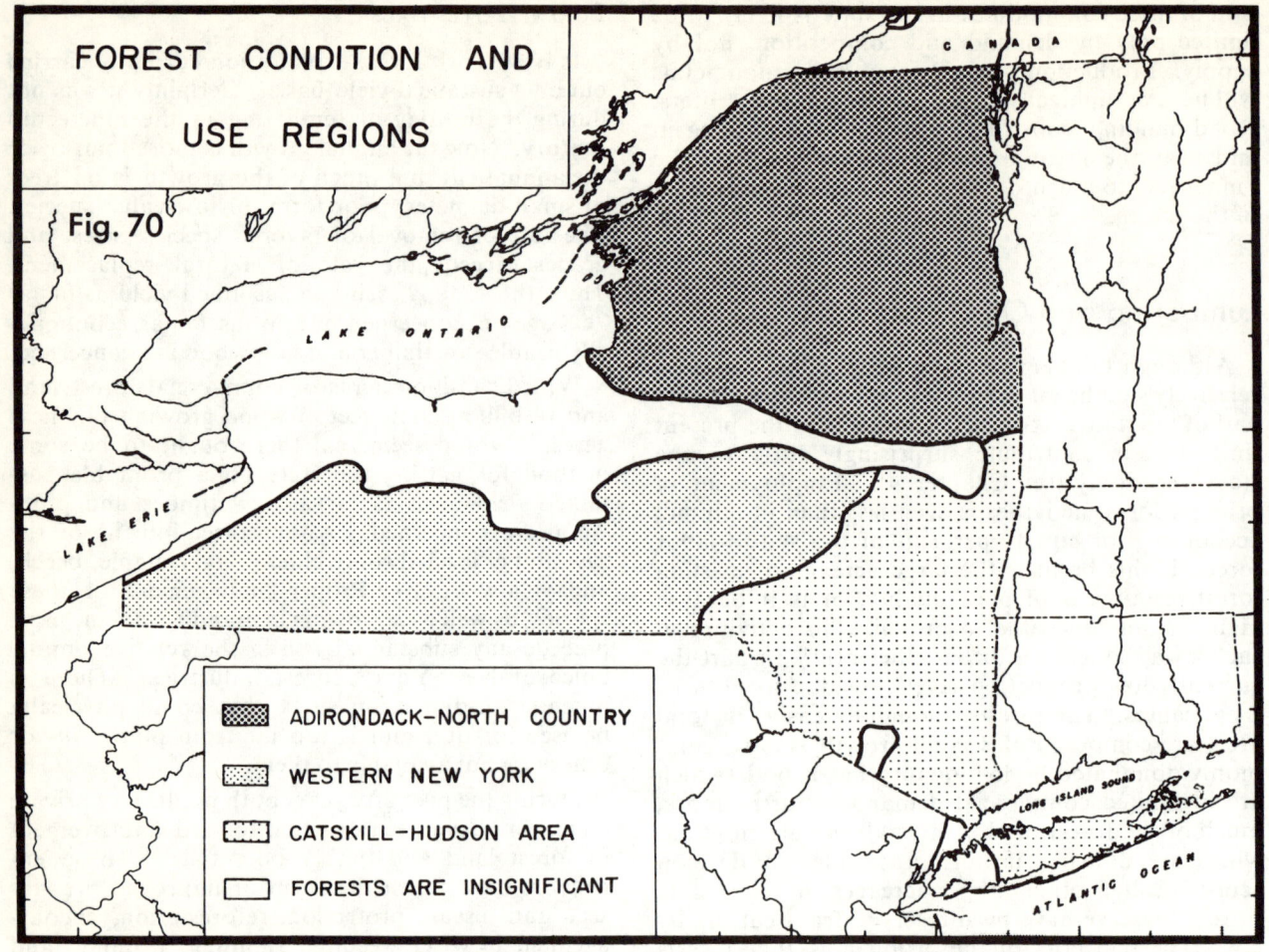

Fig. 70. FOREST CONDITION AND USE REGIONS

as part of reforestation for lumber where Christmas trees are cut as part of the thinning process and as a business in itself. Depending on species, Christmas trees are ready for marketing in from six to twenty years from the time they are set out as seedlings, and as the population rises and the supply of trees from natural forests diminishes it would appear that their production makes sense. However, they too, have to be shaped and trimmed, and there is a growing indication that supply will outrun demand. Recent studies show that big profits from Christmas tree production seldom are attained. Net returns of about eighteen to twenty dollars per acre per year are normal in New York. If it were not for the fact that most land in Christmas trees is abandoned farm land that would, if left alone, be growing back largely to worthless weed trees, the eighteen-to-twenty-dollar figure would look very low. It might be added that a great many people, both farmers and nonfarmers, plant trees not to make a profit but to provide watershed protection and wild-life habitat or just to improve the appearance of the land.

FOREST CONDITION AND USE REGIONS

Regionalizing the state in a very general way as to forest condition and use is of value to the understanding of the spatial forest resource potential. Without getting into the matter of extent of forest cover shown in Figure 30, the state can be divided into three forest condition and use regions: (1) the Adirondack-North Country, (2) Western New York, and (3) the Catskill-Hudson area (Fig. 70).

The Adirondack-North Country contains 5.4 million acres of commercial forest land, 2 million of which are in Forest Preserve land in Adirondack State Park. The latter, by law, will never be cut again. Essentially all of the accessible portions of the region were lumbered for the softwoods they contained between 1820 and 1920. Forest fires, especially in 1903 and 1908, destroyed huge acreages, and the Great Blowdown of November 25, 1950, wrought havoc on the second-growth stands. Most of the large private and industrial forest properties are in the white areas within the park boundaries shown in Figure 71, or just outside the park bound-

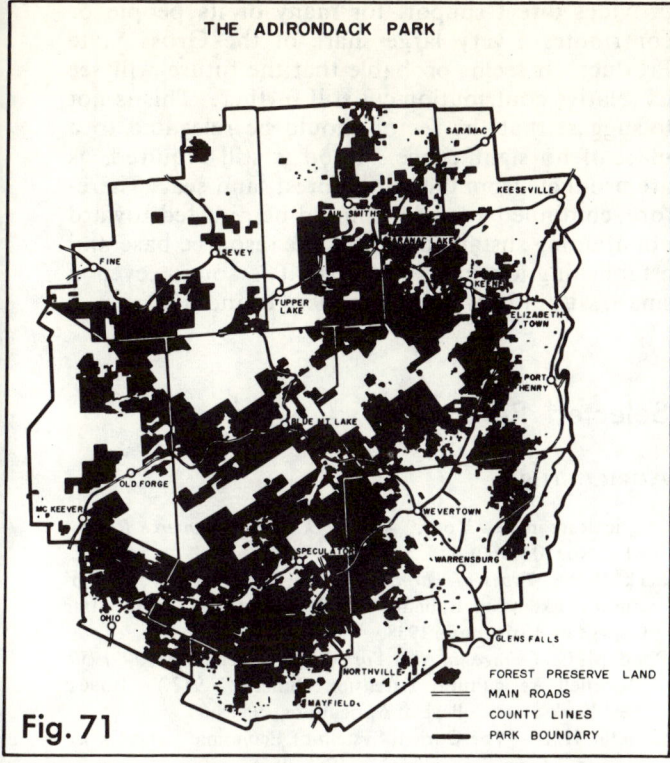

Fig. 71

From the N.Y.S. Conservationist, courtesy of the N.Y.S. Dept. of Conservation.

aries. Several are over 50,000 acres in size. Virtually all of the state's pulp mills and many of its larger sawmills are in this forest use region. In terms of total available commercial forest acreage, and especially in terms of total softwoods and large tract ownership, this region would seem to offer considerable possibility for wood industries while supplying a recreation environment destined to greater future significance.

Commercial forests in the western New York region are largely confined to the Appalachian Upland west of the Catskills. In all there are a surprising four million acres of forest land here designated as commercial. Most of it, however, is in relatively small private holdings or in fragmented state forests, both evolving out of widespread land abandonment and subsequent natural succession or reforestation. Hardwood types predominate, especially maple and beech, but ash, cherry, and oak are fairly common too. Excellent stands on some of the better sites offer some of the best opportunities for economical intensive wood-lot management in the state. The region supports a number of small sawmills and wood-using plants. Many Christmas tree plantations have been established. Fire has never been a serious problem, but grazing of cattle in wood lots is.

The Catskill-Hudson area contains 2.6 million acres of commercial forest land, about 200,000 acres of which are in the Catskill State Forest Preserve. The forests are composed of many species but support few primary wood-using plants. A large proportion of the forest land is in estates, club ownerships, or public recreation areas. It would seem that population distribution, terrain, and forest conditions all support the rationale of this kind of use.

Commercial Fishing

Commercial fishing is relatively unimportant in New York. In recent years there have been no more than a thousand to two thousand people employed full time as commercial fishermen, although vastly larger numbers are engaged in handling, canning, selling, and distributing fish products. The retail value of the fish may approach five times the dockside value, which in 1960 was just under nine million dollars (see Table 17).

MARINE FISHERIES

Marine fisheries are in the vicinity of Long Island, where thirty-eight species of fish and twelve species of shellfish are taken. Fish, including menhaden, fluke, flounder, porgies, sea bass, cod, and bluefish, account for 92 per cent of poundage and 60 per cent of value, while shellfish, including clams, oysters, bay scallops, and mussels add up to only 8 per cent of the weight but 40 per cent of the value. The catch, even though not large, is distributed nationally.

The production from marine fisheries used to be much larger than it is now. Storms, natural predators, shortages of oyster seed, overfishing, and pollution have all taken a toll. The last two are of major importance. The Indians, and then the Dutch, are known to have taken substantial quantities of oysters from the lower Hudson, but by 1885 these beds produced essentially nothing. Long Islanders have long cultivated extensive underwater oyster acreages around their island, but since World War II production has fallen off drastically there too. In addition, hard clams, fluke, and other types of catch have been much reduced in recent years, although the future of the hard clam harvest seems to offer the best possibilities. Sanitation programs need to be improved, and predators which attack young shellfish will have to be dealt with. A number of techniques employed in Japan's fisheries are now being studied in terms of their applicability. About two-thirds of the total fish poundage taken from New York's marine fisheries now is low-value menhaden.

FRESH-WATER FISHERIES

New York has over 2.5 million acres of fresh-water fisheries in the Great Lakes and major rivers such as the Niagara and Hudson, yet the commercial catch had a dockside value of only $165,000 in 1960 and has been dropping significantly for many years. As recently as 1952 the catch was valued at $276,000, about two-thirds of which came from Lake Erie and most of the rest from Lake Ontario and the Hudson and Niagara rivers. Lake Erie has always been the best fish producer of New York's two Great Lakes because of its shallowness. Whitefish, never very important in Lake Ontario, has until recently been the mainstay of Lake Erie fisheries. The 1960 catch of whitefish, however, was negligible. Other species, caught in both lakes, include blue pike, smelt, yellow perch, muskalonge, and alewives. Net fishing, formerly widely used, especially in Lake Erie, now is little employed in that lake and seldom used in Lake Ontario either, except at Chaumont Bay.

The sea lamprey is a major problem in Lake Ontario, but is being fought by both the United States and Canada. Some stocking in the Great Lakes of such species as lake trout and Atlantic salmon has taken place, but those lakes are rather large for the assurance of major success of any stocking methods now employed.

A SHIFT FROM COMMERCIAL TO SPORT FISHING

It is difficult to make direct comparisons, but it seems likely that, more and more, sport fishing will outrank commercial fishing in both monetary contribution to society and number of people involved. There were in 1964 nearly 350,000 private motorboats registered in the state, not counting rowboats, canoes, sailboats, and commercial vessels, and in 1962/63 over 700,000 fishing licenses of all kinds were issued. These included, by the way, 37,000 to nonresident fishermen. In addition, perhaps 250,000 youngsters under sixteen years of age are fishing without licenses. Large lakes and small ones, the major rivers and the small trout streams, the waters of Long Island Sound, and even farm ponds are being used increasingly by a growing New York population that finds sport fishing enjoyable. All of this sets up a chain reaction of substantial increase in sporting goods store sales, increase in motel, hotel, and camp rentals, and vigorous development of boat liveries and marinas.

Outlook

The primary sector in New York State involves use of a good deal of the state's lands, but no longer provides direct support for many of its people or contributes a very large share of the Gross State Product. It seems probable that the future will see its relative contribution cut still further. This is not to suggest that the sector should be relegated to a place of no significance. Food is still required, as are products from the mine, forest, and sea. Therefore, continued attention should be directed toward maintaining sustained yield of the resource base and establishing a primary sector that is sound, even if small, within New York's modern economy.

Selected References

AGRICULTURE

"Agriculture in New York," *New York State Commerce Review*, XIV, No. 10 (1960), 1–9.

Beck, R. S. *Types of Farming in New York*. ("Cornell Agricultural Extension Bulletin," 704.) Ithaca: New York State College of Agriculture, 1938.

Bond, M. C. *Changes in New York State Agriculture 1850–1950*. ("Cornell Agricultural Extension Bulletin," 917.) Ithaca: New York State College of Agriculture, 1954.

Conklin, H. E. "The Cornell System of Economic Land Classification," *Farm Economics,* No. 198 (1955).

———and I. R. Starbird. *Low Incomes in Rural New York State*. Ithaca: New York State College of Agriculture, Cornell University, 1958.

Fippin, E. O. *Rural New York*. New York: Macmillan, 1921.

Hedrick, U. P. *A History of Agriculture in the State of New York*. Albany: Printed for the New York State Agricultural Society, 1933.

Higbee, Edward. *Farms and Farmers in an Urban Age*. New York: Twentieth Century Fund, 1964.

Nobe, K. C., E. E. Hardy, and H. E. Conklin. *The Extent and Intensity of Farming in Western New York State, 1960*. "Cornell Economic Land Classification Leaflet," 7.) Ithaca: New York State College of Agriculture, 1961.

———, ———, and C. P. Grotto. "Agricultural Regions of New York State" (Preliminary). ("Cornell Agricultural Extension Bulletin," 1090.) Ithaca: New York State College of Agriculture, Cornell University, 1958.

Tobey, J. S. *Changes in New York State Agriculture*. ("Cornell Agricultural Extension Bulletin," 101.) Ithaca: New York State College of Agriculture, 1960.

Warren, S. W., and W. McD. Herr. *Location of Agricultural Production in New York State*. ("Cornell Agricultural Extension Bulletin," 952.) Ithaca: New York State College of Agriculture, 1954.

LUMBERING

Armstrong, G. R., and J. C. Bjorkbom. *The Timber Resources of New York*. Upper Darby, Pa.: Northeast Forest Experiment Station, U.S. Department of Agriculture, 1956.

Deckert, R. C., and R. J. Hoyle. *Wood-Using Industries of New York*. Syracuse: State University College of Forestry at Syracuse University, 1962.

Dinsdale, E. M. "The Lumber Industry of Northern New York: A Geographical Examination of Its History and Technology." Unpublished Ph.D. dissertation, Department of Geography, Syracuse University, 1963.

Fox, W. F. *A History of the Lumber Industry in the State of New York.* ("Bureau of Forestry Bulletin," 34.) Washington, D.C.: U. S. Department of Agriculture, 1902.

Hair, Dwight. *The Economic Importance of Timber in the U.S.* ("U.S. Department of Agriculture Miscellaneous Publication," 941.) Washington, D.C.: Govt. Printing Office, 1963.

Linton, R. E. *A Look at Economics in the Production of Christmas Trees in New York State.* ("Cornell Agricultural Extension Res.," 142.) Ithaca: New York State College of Agriculture, 1964.

New York Forest Facts. Washington, D.C.: New York Forest Industries Committee with cooperation of American Forest Products Industries, Inc., 1962.

Stillinger, J. R., and R. J. Hoyle. *Wood Using Industries of New York: A 1946 Census Including Comparisons with Surveys of 1912 and 1926.* ("State University College of Forestry Bulletin," XXII, No. 2.) Syracuse: State University College of Forestry, 1949.

Fishing

New York State Department of Conservation. *Annual Report.* Albany: The Department, various years.

———. *The New York State Conservationist.* Albany: The Department, various dates.

Minerals

Graham, John A. *The Mineral Industries of New York State, 1949–50.* ("New York State Museum and Science Service Circular," 41.) Albany: New York State Museum, 1955.

Hartnagel, C. A., and John G. Broughton. *The Mining and Quarry Industries of New York State, 1937 to 1948.* ("New York State Museum Bulletin," 343.) Albany: New York State Museum, 1951.

International Salt Company, Inc. *Salt.* Scranton, Pa.: The Company, 1951.

Kreidler, Lynn W. "1957 Gas and Oil Developments in New York," *Bulletin of the American Association of Petroleum Geologists*, XLII, No. 6 (June, 1958), 1143–46.

Lieberman, Percy. "Mining Laws of New York," *The New York State Conservationist*, IX, No. 4 (February–March, 1955), 13–15.

New York State Department of Commerce. *The Mineral Industries of New York State.* Albany: The Department, 1950.

Otte, Herman F. *The Expanding Mineral Industry of the Adirondacks.* Albany: New York State Division of Commerce, 1943.

Prucha, John J. "Mining and Prospecting in New York," *The New York State Conservationist*, IX, No. 4 (February–March, 1955), 12–13.

U.S. Bureau of Mines. "The Mineral Industry of New York." Reprint from *Minerals Yearbook 1962.* Washington, D.C.: Govt. Printing Office, 1962.

———. *Minerals Yearbook.* Washington D.C.: Govt. Printing Office, various years.

CHAPTER 11   JOHN H. THOMPSON

The Secondary Sector

The secondary sector, dominated by manufacturing but including construction, has been extremely significant in New York's economy for many years. As a matter of fact, for about 120 years New York has led all states in manufacturing and has ranked high in construction.

In contrast to agriculture, manufacturing and construction are small users of land and large employers of men. In 1960 the secondary sector employed 35 per cent of the state's labor force and accounted for just under 33 per cent of the Gross State Product. It is a concentrator of people, a city-builder.

Manufacturing

Although all kinds of manufacturing are carried on in New York, production of certain types of goods is outstanding. With *value added by manufacture* (value of products produced less cost of materials and contract work) as an indicator, over one-third of the nation's apparel, nearly as high a portion of scientific instruments, and one-fourth of printed and published products are turned out here. Relative employment figures are almost as impressive. Table 22 and Figure 72 show the importance of the major groups of manufacturing (employing Standard Industrial Classification Code numbers) and New York State's share of national production since World War II.

Basis for Industrial Development

Why has New York State become the manufacturing giant that it has, and how has it maintained so important a position within the fast-growing economy of the United States? Ordinarily a large manufacturing economy develops as a result of a favorable combination of industrial location factors such as raw materials, market, transportation, power, labor, and capital. Although some raw materials are produced in the state, it has no outstanding advantage in this respect at the present time. While this would seem to handicap industrial development, in fact most factories use materials previously processed in other factories rather than raw materials from mines, forests, or farms. Thus many factories attract other factories, and New York has had many factories for a long time. There are many people available to work in factories, but, on the other hand, labor surpluses and probably equal labor skills exist in other states. The state ranks high as a power producer, but just as cheap or cheaper power is available elsewhere. The Buffalo and Massena areas have profited most because of power advantages.

Actually, proximity to a huge market, a long tradition of favorable transport service, the existence of a number of large urban systems, and the location of the nation's financial center in New York City have been most significant in building and sustaining manufacturing in New York State. Favorable location with respect to heavily populated northeastern United States (the state and areas within 300 miles of its borders contain about 90 million people), and to the so-called American Manufacturing Belt (the national core, or principal economic axis, as designated in Fig. 1), which occupies a parallelogram-shaped area between Portland, Maine, Milwaukee, Wisconsin, St. Louis, Missouri, and Baltimore, Maryland, has been of great importance. New York thus not only is well situated in relation to consumer markets, but is literally in the middle of the greatest manufacturing region in the world. To some factories of this manufacturing region it can sell goods, and from others it can handily buy materials which it further processes.

Much has already been said about the importance of the Hudson-Mohawk corridor as a concentrator of transportation arteries and economic activities. Through it passed the Erie Canal, an early stimulant to manufacturing growth. Later came the railroad, the New York State Barge Canal, good highways, and finally a limited-access superhighway. These connected the Great Lakes and the Midwest with New York City. Other important transport lines from Pennsylvania, New York, and New England focused on New York City. With the aid of superb harbor and port facilities, New York City early became the transportation and commerical hub of the east. This fact is well known. What

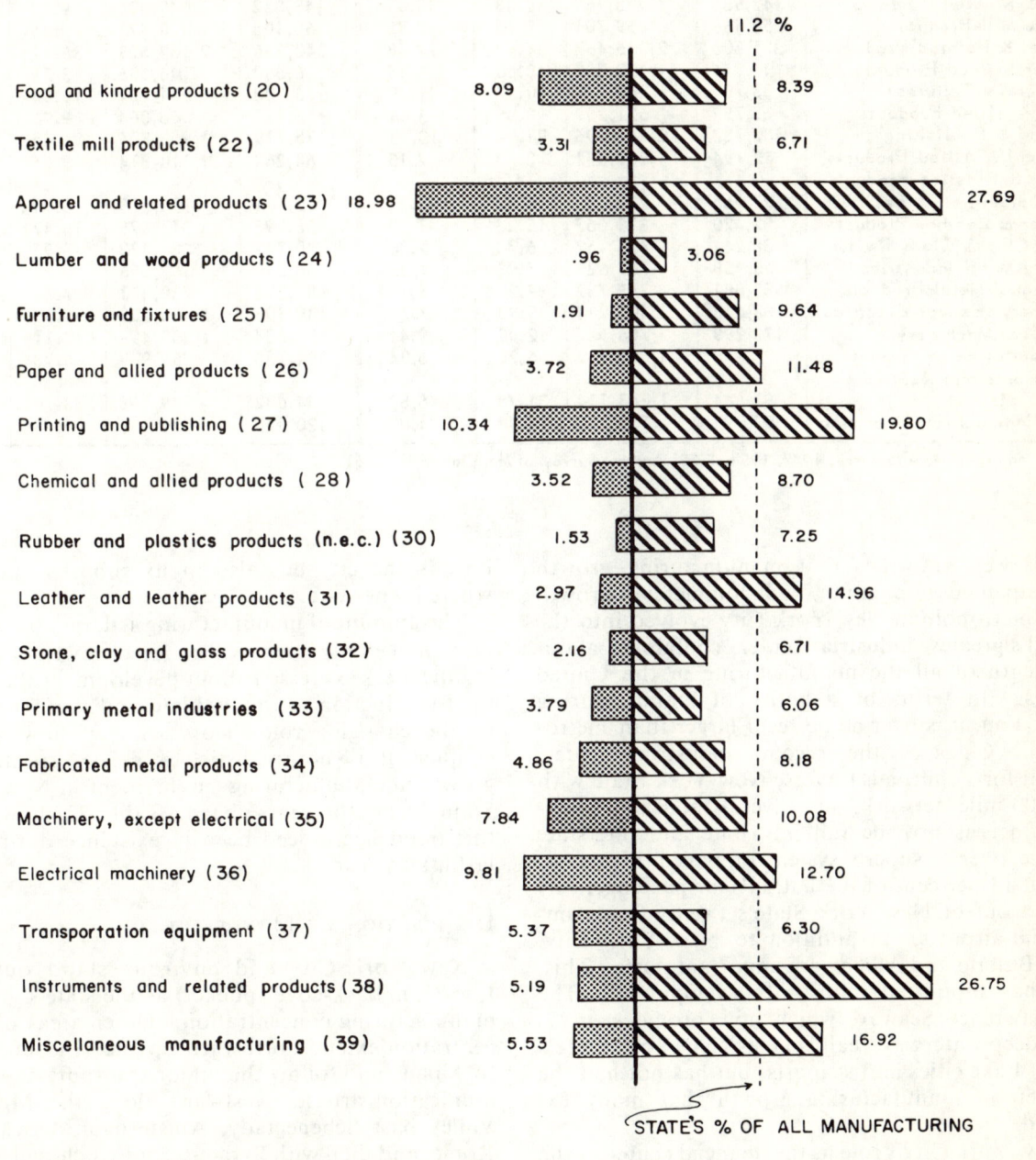

Fig. 72

TABLE 22
MANUFACTURING PRODUCTION

Code	STATE AND INDUSTRY	1961				1958			
		All Employees	Adjusted Value Added ($1,000)	Value Added as % of U.S. Industry Value Added	Industry Value Added as % of State Total Value Added	All Employees	Adjusted Value Added ($1,000)	Value Added as % of U.S. Industry Value Added	Industry Value Added as % of State Total Value Added
	New York Total	1,767,360	$18,039,754	10.98	100.00	1,914,258	$15,891,767	11.23	100.00
20	Food & Kindred Products	142,988	1,903,769	9.43	10.55	148,832	1,649,400	9.33	10.38
22	Textile Mill Products	58,466	459,701	8.19	2.55	60,105	410,471	8.45	2.58
23	Apparel & Related Products	335,505	2,305,467	34.42	12.78	342,286	2,182,529	36.31	13.73
24	Lumber & Wood Products	16,946	122,382	3.60	.68	16,560	106,506	3.31	.67
25	Furniture & Fixtures	33,701	268,772	10.57	1.49	36,233	271,401	11.53	1.71
26	Paper & Allied Products	65,792	621,257	9.35	3.44	65,723	568,364	9.96	3.58
27	Printing & Publishing	182,787	2,369,206	24.96	13.13	175,119	1,955,806	24.64	12.31
28	Chemical & Allied Products	62,194	1,289,371	8.73	7.15	64,261	1,110,814	9.05	6.99
30	Rubber & Plastics Products (not elsewhere classified)	27,055	220,686	5.62	1.22	25,401	193,518	5.91	1.22
31	Leather & Leather Products	52,420	311,067	15.23	1.72	58,795	311,675	16.42	1.96
32	Stone, Clay & Glass Products	38,247	426,157	6.73	2.36	40,742	384,423	6.95	2.42
33	Primary Metal Industries	66,968	767,625	5.98	4.26	68,340	667,696	5.72	4.20
34	Fabricated Metal Products	85,881	745,737	7.25	4.13	88,272	738,471	7.84	4.65
35	Machinery, Except Electrical	138,613	1,405,214	9.93	7.79	139,103	1,180,700	9.53	7.43
36	Electrical Machinery	173,299	1,706,527	12.40	9.46	138,684	1,258,499	12.11	7.92
37	Transportation equipment	94,947	1,089,912	6.20	6.04	104,735	959,939	6.28	6.04
38	Instruments and Related Products	91,644	1,243,114	31.78	6.89	84,632	979,398	33.70	6.16
39	Miscellaneous Manufacturing	97,792	730,776	14.43	4.05	120,948	924,136	19.43	5.82

Sources: U.S. Census of Manufactures, 1947, 1954, 1958; Annual Survey of Manufactures, 1961.

is less well known is that manufacturing growth accompanied population and commercial growth until metropolitan New York City evolved into the world's greatest industrial center, containing about one-tenth of all the manufacturing in the United States. In terms of quantity of manufacturing carried on, it is over 60 per cent larger than metropolitan Chicago, the second industrial center. About forty railroads traverse New York State with a 6,900-mile network, and several thousand trucking concerns provide full intrastate and interstate service over a superb system of paved highways. About 13 per cent of the nation's air passengers fly in and out of New York State's twenty-four commercial airports. In addition to New York City, both Buffalo and Albany are important ports. This, too, has supported industrial development. The St. Lawrence Seaway, which now provides a 27-foot-deep waterway, really makes New York State's Great Lake cities into seaports, but has not had the impact on manufacturing growth that many expected.

New York City's role as the financial center of the United States and its ever expanding roster of head offices for large manufacturing corporations serve as a constant stimulant to industrial success not only in the city but also in its suburbs and elsewhere in the state.

The amount of manufacturing is large, but future growth in employment will proceed slowly. Following the ideas expressed about development theory in the Introduction to this volume and the comments on the changing role theory which follow in this chapter, it is not realistic to expect spectacular growth in manufacturing employment in New York State during the years ahead. Table 22 shows that this trend has indeed been in existence during the last fifteen years.

Distribution of Manufacturing

New York City and environs stand out (see Fig. 73, in back-cover pocket) as the state's greatest manufacturing concentration. Other areas of concentration extend northward up the Hudson River to Albany and follow the major transport and communication arteries westward along the Mohawk Valley past Schenectady, Amsterdam, Utica, and Rome, and then with Syracuse and Rochester as the major centers, spread out over the Ontario Lake Plain. The east-west belt across Upstate New York ends with the Buffalo metropolitan area at the east-

TABLE 22
MANUFACTURING PRODUCTION (Continued)

1954				1947					
All Employees	Unadjusted Value Added ($1,000)	Value Added as % of U.S. Industry Value Added	Industry Value Added as % of State Total Value Added	All Employees	Unadjusted Value Added ($1,000)	Value Added as % of U.S. Industry Value Added	Industry Value Added as % of State Total Value Added	Employment Growth 1947–61	Code
2,008,072	$14,140,524	12.09	100.00	1,773,138	$9,655,859	12.97	100.00	−5,778	Total
152,442	1,365,436	10.19	9.66	128,624	920,179	10.20	9.53	14,364	20
66,184	416,048	8.76	2.94	90,126	421,225	7.89	4.36	−31,680	22
379,925	2,012,415	39.10	14.23	380,873	1,998,314	44.97	20.70	−45,368	23
17,872	92,459	2.90	.65	16,957	69,421	2.78	.72	−11	24
38,158	250,320	12.73	1.77	37,389	186,159	13.51	1.93	−3,688	25
64,671	476,173	10.39	3.37	64,369	368,998	12.83	3.82	1,423	26
168,091	1,583,450	25.28	11.20	166,125	1,123,621	26.55	11.74	16,662	27
67,752	872,140	9.24	6.17	67,366	585,020	10.90	6.06	−5,172	28
9,891	68,566	3.60	.48	8,764	46,909	3.60	.49	18,291	30
61,364	274,670	16.77	1.94	68,997	290,804	18.97	3.01	−16,577	31
39,140	316,646	8.29	2.24	39,216	209,104	9.07	2.17	−969	32
76,394	624,129	6.40	4.41	77,490	369,908	6.42	3.83	−10,522	33
98,360	694,884	9.15	4.91	88,382	458,260	9.31	4.75	−2,501	34
142,549	1,169,041	9.47	8.27	137,940	687,471	8.80	7.12	673	35
125,278	938,632	12.68	6.64	101,328	456,410	11.72	4.73	71,971	36
141,131	1,188,463	8.53	8.40	89,803	408,482	6.96	4.23	5,144	37
79,999	700,389	32.90	4.95	87,320	426,050	39.44	4.41	4,324	38
161,624	1,055,735	23.60	7.47	105,600	495,225	23.69	5.13	−7,808	39

ern end of Lake Erie and along the Niagara River. In the Southern Tier there are a number of important centers extending from Binghamton in the east to the Corning area in the west. Elsewhere over the state little other than isolated centers can be found.

Approximately 98 per cent of New York State's manufacturing occurs in the 450 urban centers shown in Figure 73. Although the map is based on factory employment figures of about a decade ago, current employment concentrations are largely the same. As the map legend indicates, the quantity of manufacturing employment in the various centers is proportional to the area of the circles. A list of the leading manufacturing communities has been provided for quantitative and comparative examination. The communities named are located at the centers of the proportional circles. (The use of proportional circles to represent manufacturing centers provides more precise information than a portrayal of manufacturing by counties. Compare Figs. 73 and 76.)

It should be pointed out that the circumferences of the circles commonly extend beyond the area of manufacturing concentration. Furthermore, it will be noted that the larger metropolitan areas actually are composed of a number of urban centers, so care must be taken in the visual comparison of these areas. For example, the circle for Rochester is slightly larger than that for Buffalo because the city of Rochester has the larger number of factory workers. On the other hand, the metropolitan area of Buffalo is considerably more important for manufacturing than metropolitan Rochester (see Table 23), as becomes clear if manufacturing employment figures for Niagara Falls, Tonawanda, Lackawanna, and other communities in the Buffalo metropolitan area are added to those for Buffalo. Besides the city of Rochester, the Rochester metropolitan area includes only a few very small suburban or satellite industrial centers.

In summary, then, the map permits quantitative comparisons of the state's manufacturing centers. Not only can separate and isolated centers be compared, but the internal distributional picture of the large metropolitan areas can be seen. (Note the internal complexity of metropolitan New York City.) By showing the distribution of manufacturing centers over the state, the map points up the areas of heavy concentration in comparison with those areas devoid of manufacturing. Thus, one may easily compare areas of manufacturing concentration shown in the map with geographic patterns such as population and transportation shown in other maps in this book. The population map (Fig. 2) exhibits concentrations in approximately the same localities as the manufacturing map. This is as expected, for, as has been said, manufacturing is a concentrator of people. Transportation maps in Chapters 9 and 12 shows strong correlations with Figure 73, too.

TABLE 23
MANUFACTURING IN METROPOLITAN AREAS
1947, 1954, 1958, 1961*

SMSA	1947				1954			
	Employees	Value Added ($1,000)	Employees as % of N.Y.S. Employees	Value Added as % of N.Y.S. Value Added	Employees	Value Added ($1,000)	Employees as % of N.Y.S. Employees	Value Added as % of N.Y.S. Value Added
New York State	1,773,138	$ 9,655,859			1,895,982	$14,140,524		
Albany-Schenectady-Troy	78,785	324,346	4.4	3.4	79,997	584,708	4.2	4.1
Binghamton	35,769	159,879	2.0	1.7	40,502	268,720	2.1	1.9
Buffalo	183,876	1,023,231	10.4	10.6	200,801	1,677,778	10.6	11.9
New York	1,027,001	5,984,484	57.9	62.0	1,120,638	8,133,224	59.1	57.5
Rochester	107,493	514,543	6.1	5.3	111,385	919,779	5.9	6.5
Syracuse	67,639	275,257	3.8	2.9	67,697	478,545	3.6	3.4
Utica-Rome	50,975	232,985	2.9	2.4	42,941	328,206	2.3	2.3

SMSA	1958				1961			
	Employees	Value Added ($1,000)	Employees as % of N.Y.S. Employees	Value Added as % of N.Y.S. Value Added	Employees	Value Added ($1,000)	Employees as % of N.Y.S. Employees	Value Added as % of N.Y.S. Value Added
New York State	1,782,380	$15,891,767			1,767,360	$18,039,754		
Albany-Schenectady-Troy	63,275	640,902	3.6	4.0	55,153	608,624	3.1	3.4
Binghamton	38,057	242,601	2.1	1.5	37,519	273,557	2.1	1.5
Buffalo	169,877	1,715,627	9.5	10.8	158,703	1,805,755	9.0	10.0
New York	1,074,561	9,388,523	60.3	59.1	1,081,734	10,801,941	61.2	59.9
Rochester	100,876	1,072,709	5.7	6.8	101,989	1,281,213	5.8	7.1
Syracuse	66,647	679,910	3.7	4.3	66,693	751,033	3.8	4.2
Utica-Rome	40,251	344,834	2.3	2.2	38,455	404,617	2.2	2.2

*Value-added figures on this table are not in constant dollars, so one year cannot be compared with another year.
Sources: U.S. Census of Manufactures, 1947, 1954, 1958; Annual Survey of Manufactures, 1961.

Manufacturing Regions

Six manufacturing regions may be identified. They are in order of their importance: (1) Metropolitan New York, (2) Ontario Lake Plain, (3) Niagara Frontier, (4) Mohawk Valley, (5) Southern Tier, and (6) Mid-Hudson. Figure 74 shows that in most cases these regions are not widely separated from each other. In fact, all but the Southern Tier are a part of a greater industrial belt which extends from New York City to Buffalo. The status of these manufacturing regions can perhaps be better understood if it is realized that the geography of manufacturing in any large, advanced nation like the United States evolves into a sort of hierarchy of concentrations ranging from something like the American Manufacturing Belt at the top, down through regions such as referred to here, then on to single metropolitan concentrations, followed by city concentrations, districts within cities, and finally the factory itself. As one works downward through the hierarchy, the size of the area decreases but the per cent of the land devoted to manufacturing increases. It will be useful to consider the six regions separately, for each has certain distinct characteristics, problems, and growth trends.

Metropolitan New York.—Only that part of metropolitan New York which is in New York State is of concern here. Even so, this portion contains slightly over one million factory workers, which is 61 per cent of the state's total. It also contains 40,000 individual manufacturing establishments and involves 80 urban centers outside of New

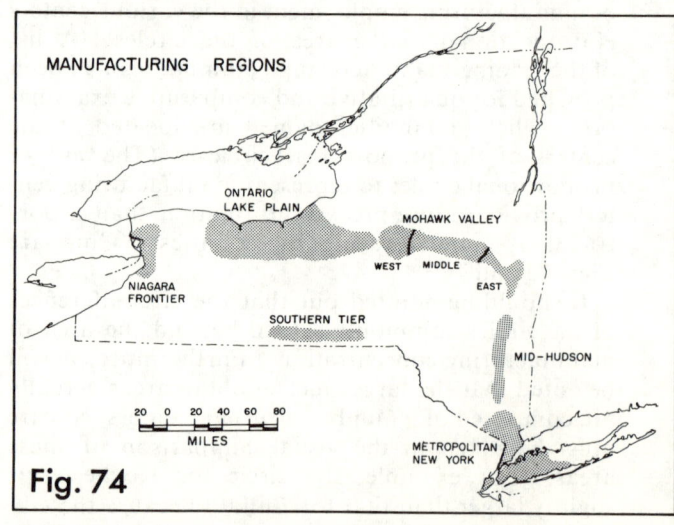

Fig. 74

York City proper. Between 1947 and 1961 manufacturing employment in this area increased by 4.65 per cent, providing a total of 48,000 new jobs in manufacturing (Table 24). Figure 75 shows a peak in employment in 1954 and lower levels since then. This region has done better than the state as a whole and probably can expect to continue this trend, as manufacturing growth on the periphery slightly more than makes up for losses from congested parts of New York City. Every major kind of manufactured product is turned out, including both durable and nondurable goods.[1] Most important are apparel, printing and publishing, machinery, and the food industries, but leather products, furniture, transportation equipment, fabricated metals, instruments, pharmaceuticals, toys, jewelry, and many other products are produced in great quantity. Small-scale manufacturing operations are characteristic of New York City and are relatively more significant there than in most other parts of the country. Thousands of these small firms carry on production in rented quarters such as lofts in multistory multitenant buildings. This is particularly

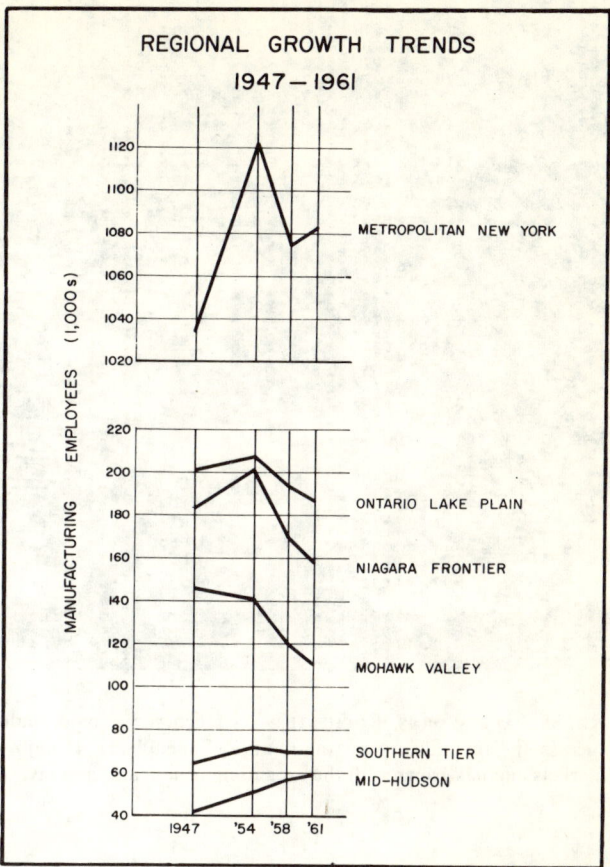

Fig. 75

---

[1] A number of rather imprecise pairs of terms describing types of manufacturing are in common usage and appear in the regional descriptions below as well as in various other parts of this volume. They are: (1) *durable* and *nondurable*—the former refers to goods generally lasting over 3 years, the latter to those used up or worn out in less than 3 years. (2) *soft* and *hard* goods—this is an attempt to distinguish between items such as apparel, textiles, rugs, etc., on one hand and those generally made of metals such as machinery, appliances, and transportation equipment on the other. (3) *light* and *heavy*—here a differentiation between lightweight small items such as toys, optical equipment, or small machinery, and heavy or bulky items such as steel, chemicals, cement, and large machinery. (4) *small-, medium-,* and *large-scale* refer to size of factory as usually measured by number of employees. A factory with fewer than 100 workers is considered a small-scale operation; one with from 100 to 1,000, medium-scale; and one with more than 1,000, large-scale.

characteristic of the apparel industry of Manhattan but also occurs in other industries and in other parts of the metropolitan area. The availability of the lofts enables small firms, particularly new enterprises, to avoid large outlays of capital for plant sites and structures. Expansion is frequently accomplished by renting additional space—space which

TABLE 24
THE MANUFACTURING REGIONS

Region	Employment 1961	Employment 1947	Percentage Growth 1947–61	Absolute Growth or Decline 1947–61	Percentage State Manufacturing Employment 1961	Principal Industrial Types*
Metropolitan New York	1,081,734	1,033,634	4.65	48,100	61.21	23,27,36,20,34
Ontario Lake Plain	192,220	201,442	− 4.58	− 9,222	10.87	36,38,35,20,28
Niagara Frontier	158,703	183,876	−13.69	−25,173	8.98	33,37,20,35,28
Mohawk Valley	111,106	146,113	−23.96	−35,007	6.29	
Eastern	55,063	70,839	−22.27	−15,776	3.12	36,23,20,26,32
Middle	26,263	39,531	−33.56	−13,268	1.49	33,22,23,20
Western	29,780	35,743	−17.68	− 5,963	1.68	33,34,35,39,20
Southern Tier	71,414	64,908	10.02	6,506	4.04	35,31,38,32,20
Mid-Hudson	58,238	42,057	38.47	16,181	3.30	35,23,22,20,27
Total for State	1,767,360	1,773,138	− .33	− 5,778	100.00	23,27,36,20,35

*Standard Industrial Classification codes; rank by employment.

KODAK PARK WORKS, ROCHESTER. As Rochester's major industry, Kodak is a good example of successful light industry on the Ontario Lake Plain. It enjoys nation-wide as well as world-wide markets and has grown with the expanding American economy. *Eastman Kodak Company.*

formerly may have been occupied by nonmanufactural activities.

It should not be construed that large firms are absent in metropolitan New York. Many have existed there for years; others have been recently established. Space, transport congestion, and high land values are serious problems, however. This is particularly true of New York City itself. Suburban areas such as Long Island and Westchester and Rockland counties have attracted large-scale factories, both from the city and from other parts of the United States. The trend for manufacturers to move from congested parts of the city to the suburbs is noteworthy and is likely to continue. It should be added that new and thriving firms in the suburbs strengthen the economy of New York City, on which they may depend for business services and markets. In the last two decades considerable expansion has occurred in the fields of aircraft and ordnance equipment, radio and electronics, and computer manufacture.

The importance of manufacturing in the New York metropolitan area can be attributed to a combination of the following factors: (1) superb location in relation to ocean, overland, and air transport, (2) proximity of a huge market for both consumers' and producers' goods, (3) existence of extensive local business services, including numerous central administrative offices, (4) long tradition of a large labor supply bolstered by a continued influx of immigrant workers, and (5) an early start in the scheme of New York State and of American economic development.

In years ahead metropolitan New York City may not increase in manufacturing at rates equal to some smaller and newer industrial regions in the state, but every indication points to the fact that it will continue to grow and will not only remain dominant in the New York State manufacturing picture but will not be surpassed in the Foreseeable future by any metropolitan area anywhere.

Ontario Lake Plain.—The Ontario Lake Plain region extends from the Syracuse area on the east to Genesee and Orleans counties west of Rochester. It contains just under 11 per cent of New York State's manufacturing workers and about 2,200 factories. Although manufacturing employment decreased by 4.58 per cent between 1947 and 1961 and about 9,000 factory jobs were lost as a result, value added by manufacture increased substantially and adjustments in the direction of the most successful types of production took place. The Syracuse and Rochester metropolitan areas, each quite distinct from the other in kinds of production, dominate the region. Rochester has always been famous for production of photographic and optical goods, electrical machinery, instruments, and apparel. Syracuse, on the other hand, is primarily a machinery and metal-working center, with over 60 per cent of the industrial labor force employed in these lines. Chemicals and automotive parts are important, too. The numerous small centers of this region produce a variety of light manufactures, but foods, machinery, and high-value specialty products are most significant.

This region borders on Lake Ontario, but the lake has had but limited effect on industrial development and is little used by factories in the area for transportation. A few mineral resources such as salt, limestone, and gravel have supported industrialization, and the establishment of some large factories by entrepreneurs who by chance chose this locale has contributed to growth. By far the region's greatest assets, however, have been its location astride the main east-west transport arteries across the state and the development of the two large urban systems of Rochester and Syracuse, which have functioned as growth foci for manufacturing.

Expansion in production of machinery, automotive parts, photographic and optical goods, instruments, and fabricated metals has been noteworthy since 1940. Expansion in hard-goods lines particularly is likely to continue, although substantial increase in the manufacturing labor force seems unlikely.

Niagara Frontier.—The Niagara Frontier region might, with justification, be called metropolitan Buffalo, for that is in essence what it is.

As shown in Figure 73, most of the manufacturing in this region is in Buffalo, Niagara Falls, Tonawanda, Lackawanna, Lockport, and North Tonawanda. Nine per cent of the state's manufacturing employees work here in about 1,800 factories. In addition to the major centers cited above there are about twenty-five minor ones. The region has had its employment problems. Between 1947 and 1961 the number of factory workers dropped nearly 14 per cent for a loss of 25,000 jobs. The loss is even more spectacular when it is realized that factory employment fell from a peak in 1954 of over 200,000 to less than 160,000 in 1961. Automation, competition from production in other areas, and termination of government contracts have all taken their toll.

This is a heavy industry region specializing in durable goods. Transportation equipment, primary metals, chemicals, and machinery are most important. Buffalo is also perhaps the world's largest flour-milling center. The industrial scene is dominated by two large steel mills, an aircraft plant, automotive parts and assembly works, electrochemical plants, and the great flour mills.

A fine combination of water and rail transport services and a good supply of hydroelectric power are the principal factors behind the industrial development. Access to Great Lakes traffic gives the area all of the advantages of any of the other

NEW INDUSTRIAL SITE. Although most factories tend to associate themselves with the large urban systems, some move well out to the periphery of those systems, or even beyond into rural settings where land is cheap and a surplus farm population is allegedly seeking work. This is the Smith-Corona-Marchant Plant near Cortland. *N.Y.S. College of Agriculture, Cornell Univ.*

The Lackawanna Plant of the Bethlehem Steel Company shown here is not only the largest of its kind in New York but the largest industrial employer in the Niagara Frontier Manufacturing Region. With water, rail, and highway transport facilities available, this concern occupies certainly one of the finest locations for heavy industry in the state. The Buffalo harbor and CBD are visible in the background. *Buffalo Area Chamber of Commerce.*

Great Lakes ports. This is particularly important to the steel mills, automotive plants, and flour mills, which depend extensively on lake shipment of industrial materials. The old Erie Canal and later the New York State Barge Canal, which gave water access to the Hudson and New York City, played important roles in industrial development, as have the New York Central Railroad and major highways. It is worth noting that industrial expansion has been going on near the Thruway interchanges. Power installation at Niagara Falls has provided large supplies of hydroelectric power, a factor particularly attractive to chemical industries.

Most successful industrial activity in recent years has occured in the metals, electrical machinery, chemicals, and motor vehicles lines. These industries seem to have the most promising future. Being a durable-goods region with considerable dependence on government contracts and the general demand for steel, this area is affected rather extensively by government military orders and the vigor of nation-wide steel-using industries such as the automotive industry. Recent drops in employment should not be looked upon as a forecast of industrial doom. The Niagara Frontier can be expected to remain very important in the manufacturing picture of the state, even increasing its industrial output in the years ahead. Substantial increase in the region's manufacturing labor force, however, will not come easily, barring major military entanglements or the opening of unforeseen markets in the steel and chemical lines.

Mohawk Valley.—The Mohawk Valley region extends from the Albany-Schenectady-Troy metropolitan area on the east to the Utica-Rome metropolitan area on the west and includes some fifteen smaller centers along the Mohawk River in between. Slightly over 6 per cent of the state's factory

I.B.M. Plant and Laboratory at Kingston. This concern not only exemplifies the new type of industrial structures and the large amount of land required by these structures and parking lots, but it is the kind of high-value-added industry which competes well in the industrial environment of the Northeast. It also might be added that it does not contribute to air pollution or ground dirt or generate objectionable amounts of noise. *N.Y.S. Dept. of Commerce.*

workers are employed by about 1,550 establishments in the region.

Employment statistics for the postwar period portray this to be the state's least successful manufacturing region. Between 1947 and 1961 its manufacturing labor force dropped 24 per cent from 146,000 to 111,000. Production of soft goods, which could not compete well with similar production in newer manufacturing areas elsewhere in the country or which did not have national market growth trends commensurate with most manufactured goods, was unfortunate. Even some individual hard-goods producers contributed to the region's dilemma by cutting back local labor forces or moving branches to distant locations.

In Figure 74 the region is subdivided into eastern, middle, and western sections. Both the eastern and western sections now have their economy heavily based in durable goods, particularly electrical and other kinds of machinery. Moreover, each has as its core a large urban system. They have experienced manufacturing employment decline between 1947 and 1961 of 22 and 18 per cent respectively, which is substantial but much less than the middle section. They probably will roughly sustain their current manufacturing employment levels in the near future. The Middle Mohawk, lacking large urban systems and having a history of soft-goods production, has become perhaps the number one industrially depressed area in the state

Contrasts Between the Old and the New in the vicinity of Utica. The photograph on the left shows the old-style multifloor textile plant so characteristic of Mohawk Valley cities. It has lost its original tenant but fortunately has been refurbished for the electronics industry. Many such buildings stand empty, and the communities that depended upon them are experiencing economic distress. The plant is surrounded by older urban construction and the original tie to rail transport is obvious. The photograph on the right shows the new type single-floor structures, located where land was available in the suburbs and oriented toward highway transport rather than railroads. They produce machinery. *Sidley K. MacFarlane.*

CHAPTER 11

THE SECONDARY SECTOR

despite its fine location in the Mohawk corridor. Manufacturing employment in the Middle Mohawk declined by one-third in the 1947–61 period, and it is quite possible that further decline will occur unless hard-goods manufacturers can be enticed to locate in the small cities of the area.

Before World War II the Mohawk region as a whole was largely known for such products as textiles, rugs, gloves, and apparel, but competition from the South has forced a shift in emphasis toward the hard-goods lines of machinery, metal fabrication, and transportation equipment. For example, in the Utica-Rome area textile employment dropped from 10,200 in 1948 to 2,100 in 1955. On the other hand machinery employment grew from 6,700 in 1950 to 11,100 in 1955. Rome has long been an important copper city, its mills producing about 10 per cent of the nation's copper and brass. Schenectady, the largest manufacturing center, is particularly well known for industrial electrical machinery.

Albany is an important river port and the Mohawk Valley has always had easy access by water, rail, and highway to eastern markets. An early start provided momentum which has resulted in continued but sporadic growth. Those centers which have undergone an extensive changeover from nondurable (soft) to durable (hard) goods have had greatest success. Where the changeover has been slow or nonexistent, as in some of the smaller centers of the Middle Mohawk, substantial decline has produced economic crises. A few large firms dominate the industrial employment scene, and their decisions to expand or contract operations could well determine the industrial future of the whole region.

Southern Tier.—The Southern Tier region includes the Binghamton-Endicott area in Broome County and the Elmira-Corning area to the west. In all there are fewer than fifteen urban centers involved. The zone contains 4.4 per cent of the state's factory workers and about 525 establishments. Shoes and business and photographic equipment from the Binghamton area, glass from Corning, and typewriters from Elmira are famous throughout the nation and abroad. Since 1940 there has been a trend toward production of machinery, ordnance, and transportation equipment, with such items as computers, compressors, valves, engines, aircraft, machine tools, and structural steel being important. Growth in employment between 1947 and 1961 amounted to 10 per cent, and some additional growth may occur. There is little in the way of favorable location factors, however, that would indicate that this region would reach the size and importance of any of those discussed above.

Mid-Hudson.—The Mid-Hudson region extends from Orange County in the vicinity of Newburgh in the south up the Hudson Valley nearly to the Albany and Rensselaer county lines. It is the least important of all the regions, employing only about 3.3 per cent of the state's factory workers, but it is growing at the highest rate. Between 1947 and 1961 this region grew by a spectacular 38 per cent, adding over 16,000 factory jobs. There are about 750 establishments. Poughkeepsie is by far the leading center, followed in order by Newburgh, Kingston, and Hudson.

Machinery production, of high-value specialty types, dominates manufacturing in the region. Apparel, textiles, leather goods, aircraft parts, explosives, and cement are also turned out in large quantities. A number of plants have been built in recent years. This region has the advantage of excellent accessibility to the metropolitan New York and Mohawk Valley regions. Good connections in both directions can be made via cheap Hudson River transport, railroad, or the Thruway. Local limestones, clays, and gravels make this area particularly suitable for those kinds of manufacturing that are based on these items. Cement production is a typical response.

It is sometimes suggested that the steep, hilly sides of the Hudson Valley offer limited opportunity for industrial development. This may be a minor handicap, but there is plenty of room for many hundreds of new factories. Proximity to, yet severance from, New York City should spell better than average success in the years ahead, particularly in the durable- or hard-goods fields.

Isolated Smaller Cities

Neither isolation nor smallness seems to be attractive to manufacturing in the Northeast at the present time. Those communities which turned out to be far removed from the evolving major transport arteries and distant from the faster-growing larger cities generally fell behind in growth. Some of them had a measure of manufacturing success over the years, but most have recently found it difficult to maintain existing status as manufacturing centers.

Among the more significant isolated cities are: Jamestown, which has specialized in furniture, metal products, and yarns; Massena, important for aluminum; Dunkirk, noted for steel and transportation equipment; Olean, for machinery, tile, and cutlery; Sidney, for aircraft components and machinery; Cortland, for machinery, apparel, primary metal products, and transportation equipment; and Watertown, for transportation equipment, machinery, and paper.

Some of the smaller and medium-sized cities

within the manufacturing regions are rather isolated, too, at least in terms of distance from the great urban systems, if not in terms of position in relation to the major transport arteries. These places, even though they are within the manufacturing regions, have many of the same kinds of problems of attracting and holding manufacturing that plague the isolated cities outside the regions. Examples of cities in this category are Ilion, Auburn, Seneca Falls, Geneva, Oswego, Oneida, Little Falls, and Le Roy. However, as the metropolitan systems grow and their influence spreads, the isolated cities within the regions, unlike their counterparts beyond the regional boundaries, will be quicker to lose their handicaps of isolation and actually begin to profit from proximity to the nearby urban system. For example, Fulton probably is better off economically because of its proximity to the Syracuse Urban System. The same may be said for Lockport's relation to Buffalo; and, of course, a host of communities around metropolitan New York City have been doing very well from a manufacturing standpoint because they have recently become part of that urban system.

The handicaps of the isolated smaller manufacturing center are further elaborated upon in the discussion of the theory of industrial agglomeration below.

Empty Areas

Nearly 95 per cent of New York State's manufacturing is carried on in the six regions that have been described. On the other hand, some parts of the state have little or no manufacturing. It would seem to follow that in these latter parts disadvantages for manufacturing outweigh advantages. Perhaps it would be more accurate to say that manufacturing could be carried on in many of these empty areas, but that opportunities for profits are less than in the existing industrialized areas. More factories could operate in Essex County, but opportunities for profits are less there than in the existing industrialized areas. Onondaga County, for example, seems to offer better opportunities than Essex County for profits. Corporations seeking new places for plants evaluate and compare numerous locations, and more often than not it turns out that well-established industrial districts, especially the large urban systems, prove more attractive than areas without factories or large cities. Thus, there is the tendency for the industrial regions to account for higher and higher percentages of the state's manufacturing and for the nearly empty areas to grow relatively more empty of manufacturing.

There are three outstandingly empty areas: (1) the Adirondacks and the North Country, (2) the Catskills, and (3) the plateau country lying south of the Mohawk Valley and Ontario Lake Plain regions but north of the Southern Tier region. These three areas all suffer in varying degrees from relatively poor access to major transport arteries and market centers. Also, generally rough terrain, snowy, cold winters, and lack of large cities with attendant local markets and industrial consciousness do not encourage industrialization. Even the St. Lawrence Seaway and Power Projects have not provided the added impetus necessary for extensive industrial expansion in the North Country. A few new factories have been established near Massena because of hydroelectric power, but because of this area's "off-the-beaten-track" location, the New York side of the St. Lawrence Valley has not experienced any cataclysmic growth of manufacturing and probably will not in the future.

Manufacturing Over the Years

The industrial distributional picture which has been described has evolved over a period of three hundred years. Yet most of the growth has occurred in the last century. This growth, however, has been geographically uneven. Some areas exhibit fairly continuous and rapid development, while others seem to lack economic vitality. Some areas have exhibited more industrial growth than the state for a period of years, then have stagnated while other areas have moved forward. Although historical trends in manufacturing were dealt with in Part Two, a few summarizing thoughts are appropriate to the discussion here.

The first manufacturing in the state was carried on largely in homes, where cloth and shoes were produced. This was accompanied by the development of small factory establishments such as gristmills, sawmills, brick-making concerns, shipbuilding concerns (New York City), distilleries, and tanneries. New York City and the lower Hudson Valley were first settled and had their first manufacturing back in the seventeenth century.

Both foreign and domestic political policies have affected industrial growth. Early application of Great Britain's mercantile policy restricted colonial manufacture except for items that were in demand by the British such as naval stores. But New Yorkers, by developing a triangular trade with Britain and the West Indies, managed to make some progress. After the French and Indian wars, British tax policies which affected imports actually spurred manufacturing, and there was considerable expansion around 1770. Again most of this took place near New York City and the lower Hudson Valley. By 1810 tanneries, distilleries, sawmills, gristmills, and clotheries were widespread, probably dis-

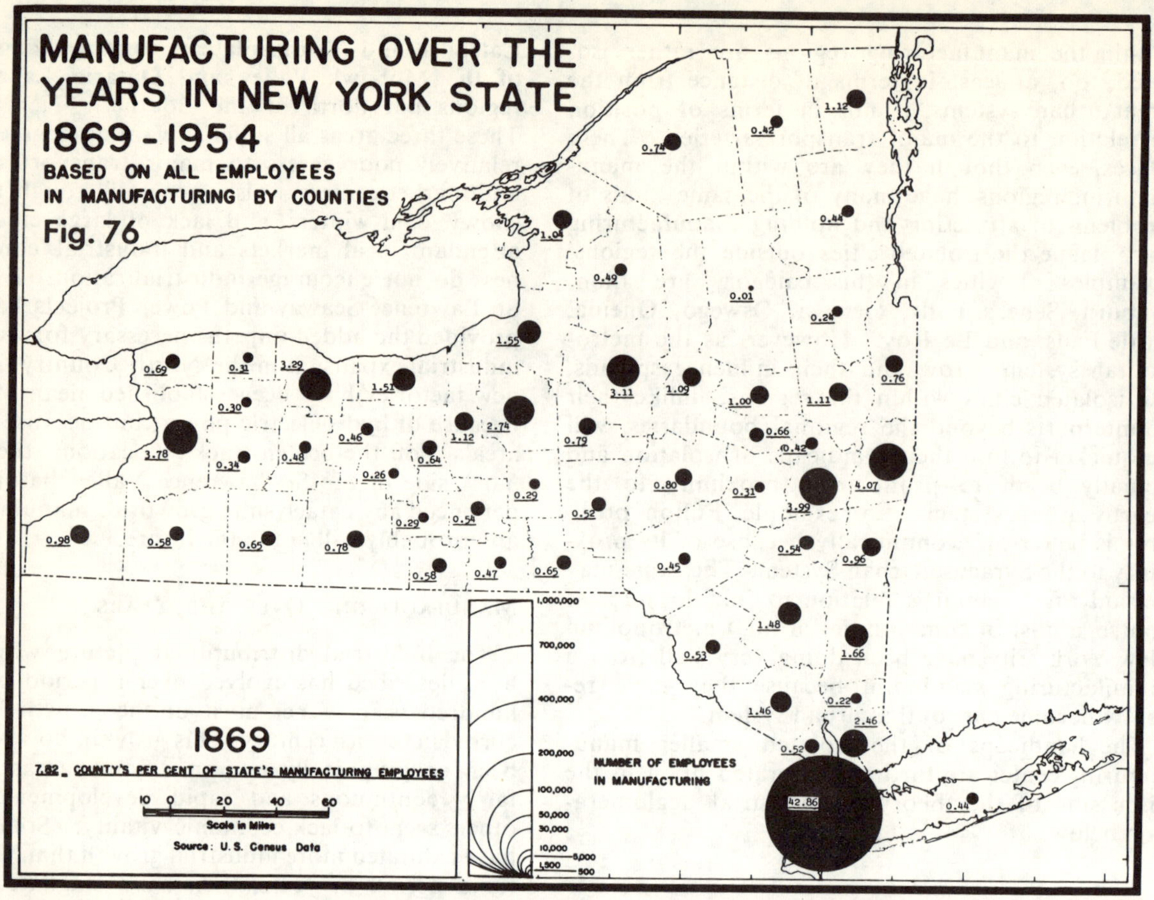

MANUFACTURING OVER THE YEARS IN NEW YORK STATE 1869-1954
BASED ON ALL EMPLOYEES IN MANUFACTURING BY COUNTIES
Fig. 76

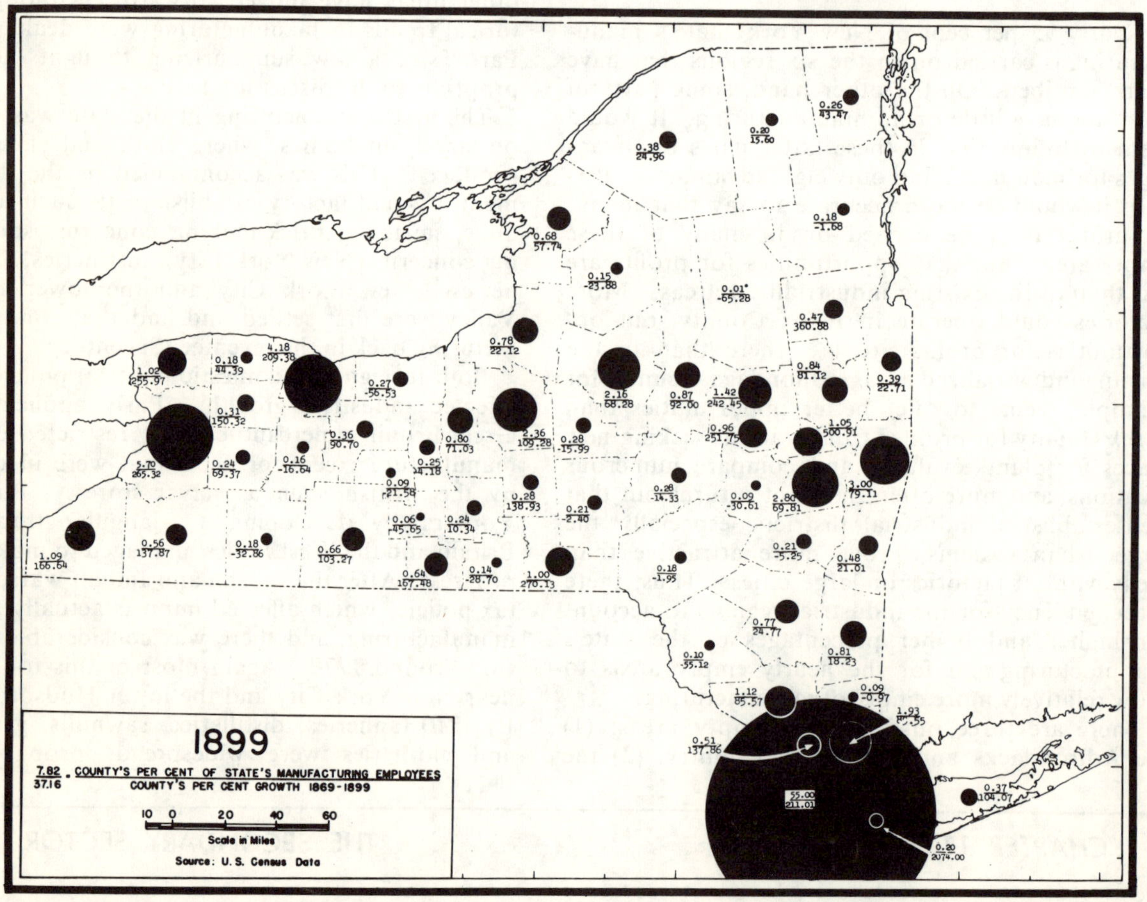

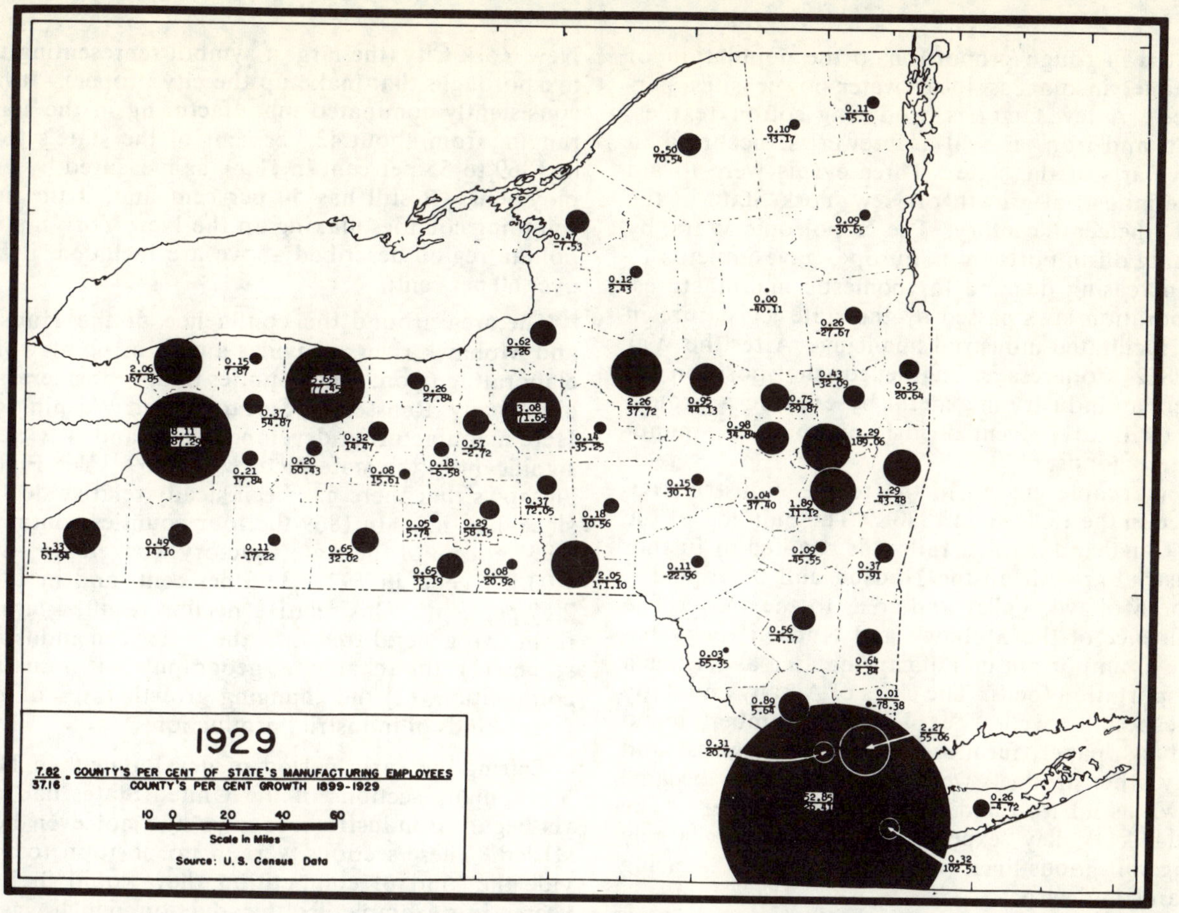

1929

7.82 — COUNTY'S PER CENT OF STATE'S MANUFACTURING EMPLOYEES
37.16   COUNTY'S PER CENT GROWTH 1899-1929

Scale in Miles: 10 0 20 40 60

Source: U.S. Census Data

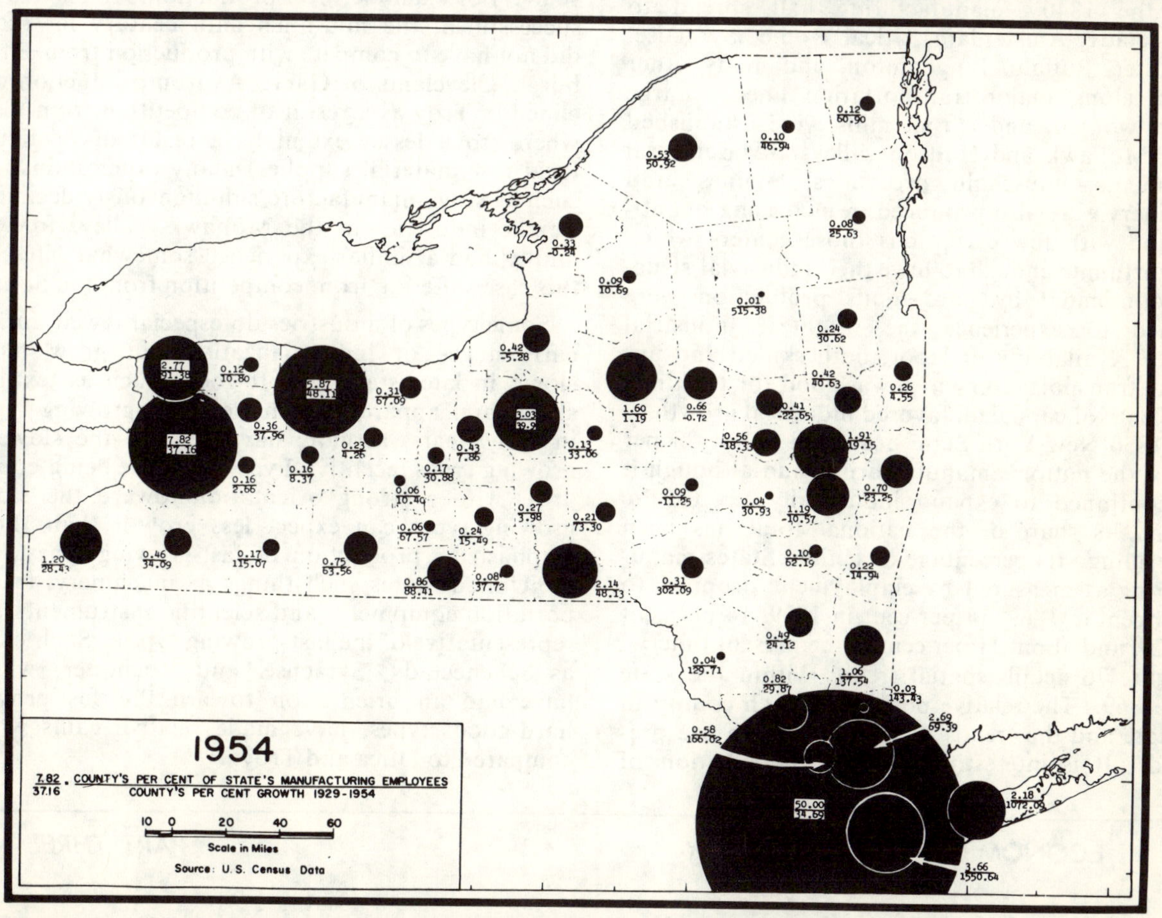

1954

7.82 — COUNTY'S PER CENT OF STATE'S MANUFACTURING EMPLOYEES
37.16   COUNTY'S PER CENT GROWTH 1929-1954

Scale in Miles: 10 0 20 40 60

Source: U.S. Census Data

tributed in rough proportion to the population of the time, insofar as local water-power sites permitted. A few factories producing cotton textiles, paper, and iron, as well as breweries, occurred in many parts of the state. Three events were to aid in the industrial growth of New York State in the early nineteenth century. The Napoleonic Wars, by shutting off imports from Europe, gave impetus to the increasing demand for domestic manufactures. Corporation laws passed by the state legislature in 1811 facilitated industrial financing. After the War of 1812 Congress strengthened the position of American industry in general by erecting a protective tariff to prevent British dumping of manufactured goods.

Remarkable growth in the textile industries occurred in the 1820's and 1830's. The building of the Erie Canal and the first railroads resulted in further industrial growth up the Hudson and the opening of the Mohawk Valley and areas to the west. The confluence of the Mohawk and Hudson rivers became an important manufacturing area as well as a transportation focus. The cities of Albany and Troy did especially well. Textiles, iron, lumber, metal products, paper, furniture, brick, flour, glass, and many other products were produced. Troy became known as an iron and iron products center. The Mohawk Valley experienced expansion largely along soft-goods lines, particularly textiles, wearing apparel, and carpets.

By the 1840's manufacturing had spread to Schenectady, Amsterdam, Utica, Rome, Syracuse, Rochester, Buffalo, Binghamton, and many other centers along major transportation lines. Large cotton, woolen, and carpet mills were established in the Mohawk and Hudson valleys. Agricultural implements, household necessities, engines, and machinery were also produced in increasing quantities, and with few exceptions those centers which were fortunate enough to have their industrial structure dominated by hard-goods production were destined to experience the brightest industrial future. Availability of labor, both skilled and unskilled, transportation which was good for the time, and plenty of capital all favored industrial growth.

In 1850 New York State possessed about 23 per cent of the nation's manufacturing, and although it has continued to expand industrially up to the present, its share of the national total has been diminishing. Its percentage of United States manufacturing as measured by employment dropped to 20 per cent in 1869, 18 per cent in 1899, 14 per cent in 1929, and about 11 per cent at the present time.

Figure 76 details spatial trends within the state since 1869. The relative position of each county in the state and the trends of county growth are presented. It is interesting to note the position of New York City (the largest symbol representing the five boroughs that make up the city proper). It has consistently dominated manufacturing in the state, ranging from about 43 per cent of the state's total in 1869 to 55 per cent in 1899, as measured by employment. It still has 50 per cent and, if the surrounding counties making up the New York metropolitan region described above are included, it has over 60 per cent.

The area around the confluence of the Hudson and Mohawk rivers provides a case in point worth elaborating upon. The counties under consideration are Albany, Rensselaer, Saratoga, and Washington. Here manufacturing developed early and was quite significant in the state picture by 1869. Yet Figure 76 shows that there has been steady relative decline since that time. In 1869 the four counties contained 9.93 per cent of the state's factory workers; in 1899, 7.03 per cent; in 1929, 3.96 per cent; and in 1954, 2.57 per cent. This relative decline resulted largely from two general trends in the American industrial scene: (1) the changing geography of industrial competition, (2) the changing growth rates of different kinds of industrial production.

During the early period of development in New York, many sections of the United States had not yet begun to industrialize; some had not even been settled. These sections were in no position to provide the kind of competition they would in later years. For example, iron production, which was so successful in the mid-nineteenth century in Troy, did not have to compete with production from Pittsburgh, Cleveland, or Gary. As iron production declined in Troy as a result of competition from elsewhere (to a lesser extent as a result of declining local raw material supplies) many other industries such as stove manufacture simultaneously declined. Textile industries in the Mohawk Valley, for example, had a similar experience somewhat later, in this case suffering from competition from the South.

Some types of industries do especially well during early stages of industrialization but grow more slowly in later stages. Soft goods such as textiles, shoes, and apparel, which were fast-growing types in early years, actually have become the slowest-growing types recently. Everything else being equal, areas with a strong orientation toward the slow-growing types can expect less growth than those emphasizing production of fast-growing types. In the United States such things as machinery, transportation equipment, and scientific instruments are representative of the fast-growing types. Such cities as Schenectady, Syracuse, and Rochester, which have had an orientation toward the fast-growth hard-goods types, have made relative gains when compared to Utica and Troy.

THE MAJOR THEORIES OF
MANUFACTURING GEOGRAPHY

After a discussion of the present manufacturing geography of the state and a glimpse of its evolution, a treatment of the major theories of manufacturing geography will help the reader attain a more thorough understanding of why things are as they are today. There are five major theories: (1) changing role theory, (2) industrial cycle theory, (3) least cost point theory, (4) agglomeration theory, and (5) redistribution theory.

Changing Role Theory

The changing role theory has already been alluded to as part of general development theory. In essence it stipulates that in early periods of economic development manufacturing is of little significance. Later, after the industrial revolution gets well under way, the role of manufacturing becomes pre-eminent, cities grow, and transportation facilities greatly expand in importance. Manufacturing becomes the great *basic* economic activity, i.e., the part of the economy that brings money in from the outside. As time goes on manufacturing can be expected to grow at a declining rate and lose relative significance as the tertiary sector overtakes and passes it as an employer and as a producer of GNP or GSP. Although the crossing of the secondary and tertiary lines did not occur at the same time in all parts of the state, it should be understood by anyone concerned with economic growth that the crossing of those lines has long since occurred in most communities, and, even though manufacturing may continue to be important, its period of pre-eminence is past. Development and planning effort should proceed accordingly. Areas or communities which are not receiving their proportionate share of tertiary growth today but instead are leaning on manufacturing alone for economic sustenance are likely to experience relative economic declines. Or, communities which attempt to solve their economic problems by attempting to attract only manufacturing establishments are likely to find the competition severe.

Industrial Cycle Theory

Just as the role of manufacturing in the economy changes in time, so also do the problems confronting manufacturing in a particular area. The *industrial cycle theory* holds that a manufacturing area once established goes through an aging process with an attendant changing set of problems that strongly affect the competitive position of the area and the types of production.

In analyzing individual manufacturing establishments, it is apparent that some are young and vigorous, some are mature and stable, others are old and declining, and although they are changing at varying rates of increase or decrease they all tend to exhibit a common course of development involving periods of experimentation, rapid growth, diminished rate of growth, and stability or decline. This growth tendency, known as the law of industrial growth, is approximated in Figure 77. Among the factors responsible for it are: changes in technical progress, market demand competition, regional and local cost advantages, and management vigor.

Manufacturing areas commonly exhibit a growth curve similar to that expressed by the law of industrial growth for individual establishments. The causes are regional or local in nature rather than internal to an enterprise, but they often are quite similar. Recognition that industrial areas go through an aging process and as a result experience a changing set of problems is vital to industrial regional analysis and adequate development work.

Manufacturing in the United States had its start in the Northeast, first in New England and a little later in New York State, A rapidly growing economy, spurred on by the heavy demands of the Civil War, made manufacturing prosper. No other part of the United States could as yet compete with this vigorous, industrial Northeast. Even Europe, because of the disadvantage of distance, had difficulty selling manufactured goods in the United States. The most favorable localities in New York were experiencing unusual manufacturing growth, and large cities began to evolve. Transport nets into the Midwest and South were being developed, but a manufacturer in Troy or Utica did not have

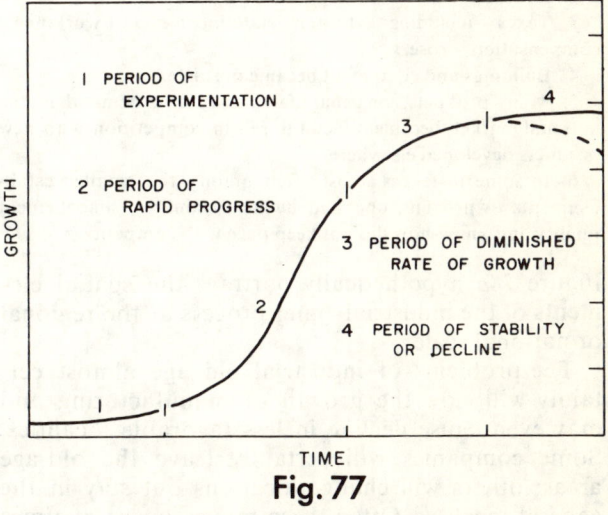

Fig. 77

to contend much yet with competition from Illinois, Ohio, South Carolina, or Mississippi. The wave of industrialism was moving forward, but it had not reached these areas. Still largely beyond the "industrial frontier," they were nonindustrialized. Few places had the locational advantages of the Mohawk Valley or most other New York State localities. New York was in the stage of vigorous industrial youth.

In the ensuing years, as manufacturing continued to grow rapidly in the state, entrepreneurs who had learned their business here began to be "exported" to other parts of the country. Big northeastern companies established branches elsewhere and new industrial areas opened. Population shifts occurred toward the West and South, and accompanying these shifts was a change in the geography of markets. World War I found New York a strong, virile manufacturing state, but one which was in the midst of a competition battle with more rapidly expanding industrial regions elsewhere. It might be said the state was in the stage of industrial maturity.

By the 1920's and 1930's population and market shifts had been extensive, the transportation net was efficiently serving the entire country, and new sources of industrial raw and processed materials became available. Many of New York's clear earlier locational advantages in the geography of markets, transportation, and industrial materials had disappeared. While all of this was going on, the following major problems, symptoms of industrial old age, began to appear in New York and the rest of the Northeast:

1. There was a rise in unionism and an increase in labor costs. Older experienced individuals who made up a large part of the factory payrolls demanded more pay, shorter working hours, and less input of work.
2. Land occupied by factories became surrounded by the urban land uses of cities. It cost more and there was little available space for expansion.
3. Taxes—including real estate, state income, and workmen's compensation—rose.
4. Buildings and equipment became obsolete.
5. Many products, long manufactured and in demand in the national market, became difficult to sell in competition with new products developed elsewhere.
6. In some instances, most often among the smaller establishments owned and operated by local families, management quality and know-how did not keep pace with competition.

Figure 78*a* hypothetically portrays the spatial elements of the industrial aging process at the regional or national scale.

The problems of industrial old age almost certainly will slow the growth of manufacturing and may even cause decline in less favorable localities. Some companies will certainly leave the old-age areas; others will change locations but stay in the general locality. Often these moves occur at times

THE INDUSTRIAL AGING PROCESS
A. NATIONAL OR REGIONAL SCALE

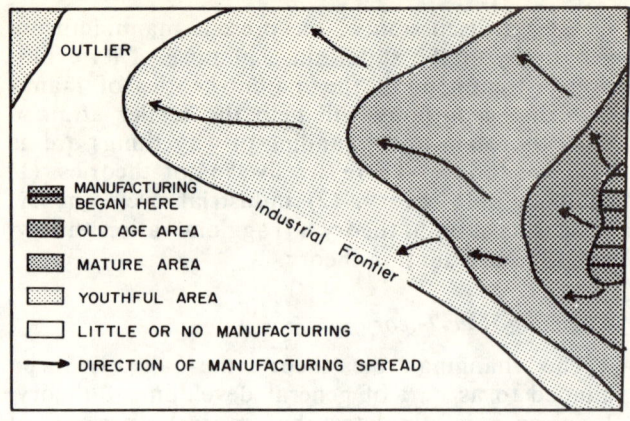

B. METROPOLITAN SCALE

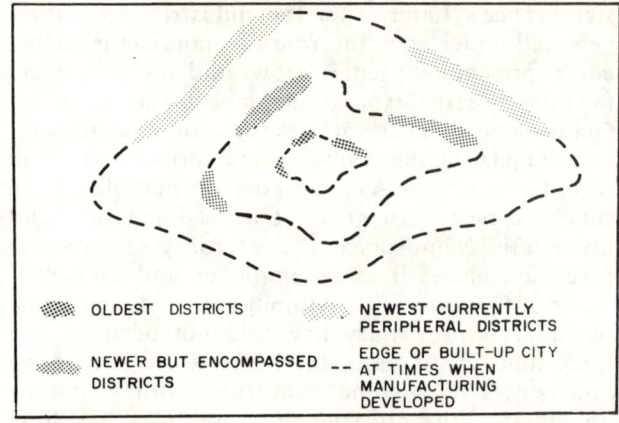

Fig. 78

when plant and equipment are deemed obsolescent and lack of room for expansion makes arguments for staying untenable. New industries are wary about coming in.

Certain kinds of manufacturing are especially subject to difficulties arising from the six problems of industrial aging and therefore are least able to survive in old-age areas. In general these include the soft-goods industries, especially certain kinds of textiles, shoes, and apparel manufacture, and, to a lesser degree, carpets, furniture, and ceramics. These are generally *low-value-added* types (value added by manufacture is a small percentage of the total value of the product produced), where high percentages of the jobs are of the unskilled variety. The kinds of manufacturing that do best in old-age areas—in fact in some cases may do better there than anywhere else—are those that require maximum skills and are *high-value-added* types. These include electrical machinery and appliances, precision instruments and computers, and complex parts for transportation equipment. Often they are the things that society buys more of as it becomes increasingly complex and affluent. Research and development divisions of large nation-wide manufacturing concerns do especially well too.

The aim of manufacturing development efforts in an old-age industrial region should be to effect a satisfactory manufacturing economy within the framework of the problems at hand. To do this the problems of industrial aging must be understood and dealt with intelligently. The alternative is almost certain: eventual, if not immediate, economic distress.

Big interregional waves of industrial aging spread across regions and whole nations and smaller interregional waves of metropolitan scale are usually simultaneously occurring. In this case it is a matter of obsolescence of buildings and equipment, congestion of traffic, lack of expansion space, and high land and local tax costs which produce most of the difficulties. These things show up in the oldest manufacturing districts of cities, which inevitably lose relative significance to newer peripheral districts where the above-mentioned difficulties are less severe. The result is that factories normally are an important part of the rapidly expanding doughnut ring which is to be found in different localities at different times around most large urban systems (Fig. 78b). A factory forced to move from a congested city center has two choices: to move to the periphery of the city or to move outside the general region, say from New York State to the South. Developers often negotiate in favor of the former choice and save the industry for the local area even though it is lost to the city center.

Least Cost Point Theory

Manufacturing concerns tend to locate where chances for profit are greatest. Ordinarily this is where the sum of the costs of assembling materials to be used, processing the product, and distributing the product to market is least. There is for any given manufacturing concern a theoretical place—known as the *least cost point*—where assembly, processing, and distribution costs are at a minimum. The location of this least cost point depends on the *cost structure* of the concern, i.e., how much of the concern's cost of operation is devoted to such things as raw material, power, transportation, and labor. A good location for a paper mill might be a poor one in terms of cost for a steel mill or a watch factory. Nevertheless, certain areas have such a generally favorable array of locational advantages for manufacturing that they support many kinds of factories; others, with few advantages, may have no manufacturing at all.

Many things go into giving an area a favorable or unfavorable cost advantage. They are usually referred to as industrial *location factors*. Four are so commonly significant that they will be called here the "big four": market, materials, transportation, and power and fuel. Two others of great significance, which tend to be highly fluid and therefore easily moved to locations where the "big four" are in favorable combination, are labor and capital. Among the additional location factors that may in some instances affect cost of production are: cost of living, climate, and tariff and trade conditions. All of the location factors, functioning in complex combination, make some areas attractive to manufacturing and others unattractive. Relative attractiveness or unattractiveness can change with time as the regional and national geography of the factors themselves changes.

Irrespective of the degree of attractiveness of a general area, for example metropolitan New York City, there are some precise sites within that area that have cost advantages. The factors that make one precise spot cheaper for a concern to use than another are called *site factors*. They include local tax differences, availability and cost of land or buildings, availability and cost of utilities such as water and waste disposal, access to railroad sidings or major highway interchanges, suitability for on-premises advertising, and nature of topography and subsurface bearing materials.

If from the standpoint of assembly, processing, and distribution costs a factory is built on the best site in the best general locality, it is situated at the least cost point and should have the optimum competitive possibilities. Actually, there are for many concerns a number of precise localities so close in their cost advantages that it would make little difference which is used. Nevertheless, approximating the least cost point in location makes sense for any concern and is the essential aim of most industrial location analyses.

There are also some rational but non-cost situations that result in the locating of many manufacturing establishments. For example, a factory may be located in a given place simply because the entrepreneur happened to live there or otherwise liked the place. Once located, industrial inertia tends to keep the plant where it is even though costs may be unfavorable. Strategic considerations also fall in the category of non-cost location factors, as may government distribution of contracts for manufactured goods. A zoning ordinance might be considered a non-cost site factor, too.

Industrial Agglomeration Theory

Because of the importance of the least cost point to manufacturing location, and because that point is so often found in or near large cities, an advanced economic society motivated by the desire for profits tends to experience more and more concentration of manufacturing as its stage of economic development advances. The specific result is ever increasing amounts of manufacturing in large urban systems—particularly in the doughnut ring—at the relative expense of smaller central places and rural areas.

In New York State it appears that there is a threshold size of approximately 100,000 for substantial manufacturing success. A smaller urban system and rural areas seem to have difficulty in competing for most kinds of manufacturing establishments with the newly emerging industrial districts at the peripheries of urban areas with over 100,000 in population. In analyzing this it is worth remembering that for both market and material reasons factories attract factories. Furthermore, as equipment and processes become more complicated, advanced service facilities are most likely to be available in the big cities, where they are in greater demand. Also, air connections and other transport facilities between large urban systems are usually superior to those of smaller centers. Finally, there has been a labor and "brain" drain away from smaller places and rural areas toward the large cities for years, and the peripheries of the large urban systems provide plenty of space at reasonable cost. What is being described here is a situation in which the big get bigger and the small have difficulty. Concentration of manufacturing in the large urban systems of course attracts tertiary activities and people. The result is that the few very large urban systems will get very much larger. This is the essential element in the growth pattern of New York State. It may be that someday, if largeness and congestion in cities become overpowering, men may try to induce deagglomeration processes which will again spread people more evenly over the land. Success in this direction, however, will be costly, probably requiring a forced flow of capital from large urban areas to smaller centers and rural districts.

Redistribution Theory

The *redistribution theory* stipulates that industrial development efforts or other forces which induce locational change succeed primarily in redistributing manufacturing rather than in increasing the total amount. Suppose, for example, a vigorous development effort in one state or community succeeds in luring a manufacturing concern from another locality or in attracting a new factory that might have gone to the other locality. The state or community that lured the factory would enlarge its economy at the expense of the other locality. In this case it is largely the geography of production that is changed, not the amount. No argument is being made here to discourage development efforts; in fact, they should be encouraged, for they probably sharpen competition, which may be valuable between areas in the same way it is valuable between individuals or businesses.

Ironically, the depressed community is often very busy with development efforts only to find that its outlay of development capital and talent cannot match that of the more successful larger community. The net result is that the depressed community loses ground in the competitive struggle with the large urban system.

Development effort often results in redistributing manufacturing via a net flow of manufacturing out of old-age industrial regions, particularly away from city centers and smaller communities in those regions. This has been the nature of the flow of considerable soft-goods manufacture from New York to various locations in the South, of industrial moves from the centers of large cities to their peripheries, and of losses of smaller centers to larger ones. In the latter two instances particularly, development effort is actually one of the forces producing agglomeration.

Construction

The amount of construction carried on in any area is roughly related to the size of the labor force but also to the area's general rate of economic growth. This means that construction activity is concentrated in the great urban systems, especially in their fast-growing peripheries. Nassau and Suffolk counties, which are true periphery areas to metropolitan New York, had nearly 7 per cent of their total nonagricultural labor force in construction activities in May, 1964, and Westchester County had over 6 per cent. On the other hand, the Utica-Rome metropolitan area, with a much less rapidly growing periphery and very largely static central cities, had only 2.7 per cent of the nonagricultural labor force in construction at that same month. Among upstate metropolitan areas, economically vigorous Syracuse exhibited the highest relative orientation toward construction activities in May, 1964. Recent urban renewal activities have expanded the need for construction workers near the centers of the great cities.

Because of New York's very large labor force and relatively good economic health, the state ranks second nationally (behind California) and contains about 9 per cent of the nation's construction workers. Although part of the secondary sector, construction is greatly overshadowed in New York State by manufacturing and employs only about 5 per cent of the state's labor force as compared to manufacturing's 30 per cent. It is a type of activity that can vary considerably from time to time in any one area, often dropping or increasing by 5 or even 10 per cent in a month. These changes can be due to a variety of things from seasonal difficulties and weather adversities to dating of major contracts. Such variations are likely to result in substantial unemployment.

The Future

The leadership in the secondary sector which New York State has enjoyed will probably remain unchallenged for a long time, even though growth in manufacturing employment has not occurred in recent years. Other states are in the process of catching up, but they have a long way to go.

Perhaps too much of the discussion in this chapter concerning manufacturing trends has been related to employment. Automation and efficiency are constantly on the rise, which means that a single worker produces more now than he did a few years ago and will produce more a few years hence than he does now. As has been pointed out previously, value added by manufacture is an excellent measure of the actual size of manufacturing activity and can be used for trend analysis providing figures are expressed in dollars of constant value. While manufacturing employment in the state experienced little change between 1947 and 1961, Table 25 shows that the value added by manufacture, expressed in constant dollars, has increased nearly 50 per cent. This suggests that labor's efficiency has increased by about the same amount and that there is roughly 50 per cent more money accruing to the state as a result of manufacturing changes in this period. It is interesting to note, too, that over half of all value added by manufacture is expended in salaries and wages. Table 25 also shows the relative strength of the Rochester and Syracuse metropolitan areas of the Ontario Lake Plain region.

All evidence points to the fact that most future investment in manufacturing, both for expansion and relocation, will take place in the peripheries of the larger urban systems. The Ontario Lake Plain and Mid-Hudson regions may be expected to exhibit the most favorable trends, but absolute gains in these regions will be well behind those in metropolitan New York.

Figures 79 and 80 tell much about the kinds of manufacturing which have done well and badly in New York State in recent years and offer some indications of what might be expected in the way of type structure success in the future. The trends of eighteen of the twenty major industry groups are shown on Figure 79, the two omitted being insignificant in the state. From this figure it is apparent that recent manufacturing growth has not been in soft-goods lines such as apparel, textiles, and leather goods, but instead has been strongly dominated by electrical machinery. Printing and publishing, rubber goods, transportation equipment, and instruments have exhibited growth, too. If it can be assumed that recent trends will continue, Figure 79 provides useful hints as to what is likely to happen in the years ahead.

The major industry groups, however, are a bit too general, for each contains a wide range of manufacturing types. For example, Group 37 (transportation equipment) includes both fast-growing types such as aircraft parts and less successful types such as railroad equipment and motor vehicles and equipment. For this reason Figure 80, which uses a finer breakdown of manufacturing, is included here. On this figure aircraft and parts, communication equipment, scientific instruments, and office machines stand out as the growth industries in the state, while women's outerwear, men's and boys' clothing and furnishings, and nonrubber footwear are the principal loss types. This continues to support the thesis that the future of manufacturing in the state will be increasingly tied to the high-value-

TABLE 25
TRENDS IN VALUE ADDED BY MANUFACTURE FOR NEW YORK STATE AND SELECTED AREAS, 1947-61
(In Constant Dollars*)

	1947 ($1,000)	% of State Total	% Ave. Ann. Growth 1947-54	1954 ($1,000)	% of State Total	% Ave. Ann. Growth 1954-58	1958 ($1,000)	% of State Total	% Ave. Ann. Growth 1958-61	1961 ($1,000)	% of State Total
New York State	$9,655,859	100.0	4.3	$12,535,926	100.0	.5	$12,764,471	100.0	4.3	$14,408,749	100.0
New York City	5,984,484	62.0	2.9	7,210,305	57.5	1.2	7,540,982	59.1	4.8	8,627,748	59.9
Niagara Frontier	1,023,231	10.6	6.5	1,487,392	11.9	-2.0	1,378,014	10.8	1.5	1,442,296	10.0
Rochester & Syracuse SMSAs	854,680	8.8	7.5	1,306,277	10.4	1.9	1,407,726	11.0	5.1	1,623,200	11.3

*The wholesale price index (1947-49=100) for all manufactured goods published by the U.S. Bureau of Labor Statistics is used to establish a constant dollar value.
Source: U.S. Census of Manufactures, 1947, 1954, 1958; Annual Survey of Manufactures, 1961.

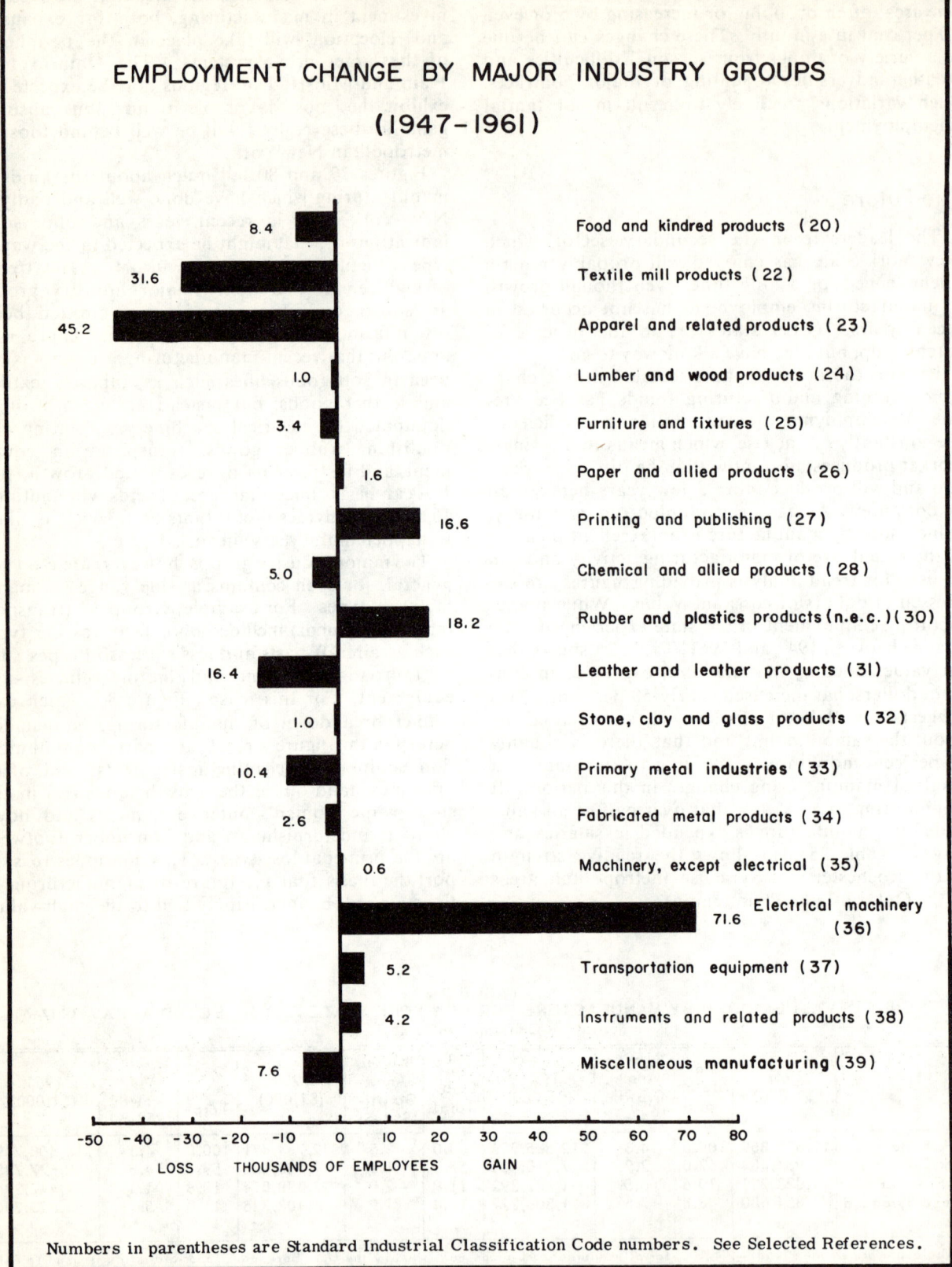

Fig. 79

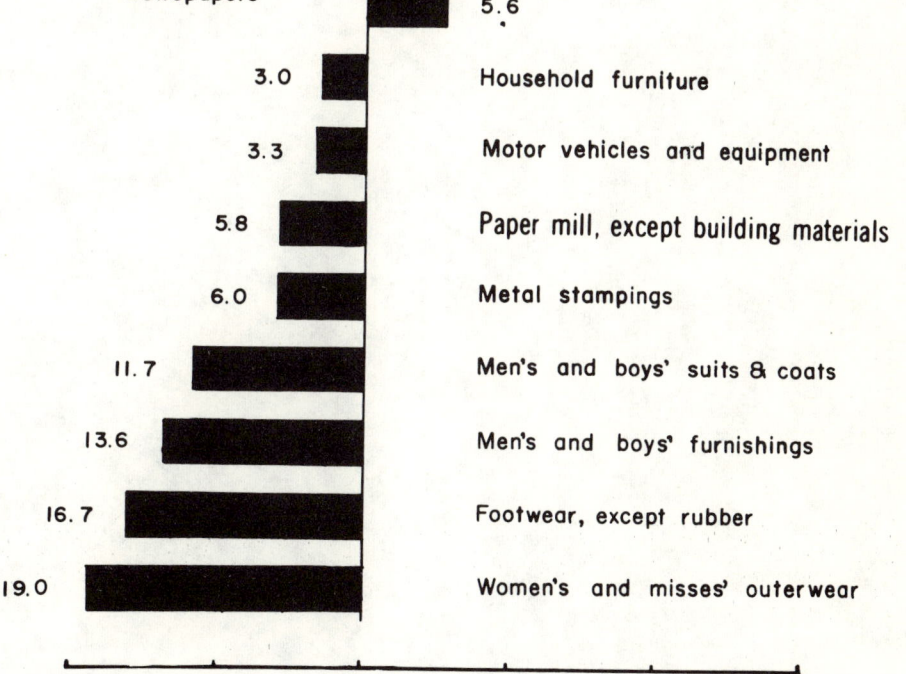

Fig. 80

CHAPTER 11 THE SECONDARY SECTOR

added hard goods, which require a high level of skill in their manufacture. The position of the aircraft industries in New York is indicative of a national trend that has made them the largest employers of factory workers in the United States, exceeding even the employment levels of the automotive and steel producers. The future of the aircraft industries depends extensively on the international situation and the size of governmental defense spending. Production in electronic and computing device lines, also vigorously expanding in the state, reflects both consumer and government demands for these products.

Trends in construction employment will be relatively more favorable than those in manufacturing. It is probable that expansion of construction activities in the years ahead in New York will roughly parallel or fall slightly behind the general growth of the labor force. Thus, it can be expected that construction will account for between 4 and 5 per cent of the state's labor force in the foreseeable future.

As a concentrator of people and a city-builder, the secondary sector will remain critical to New York's future. It is doubtful, however, that it will account for as much of the state's labor force in the future as it does now (35 per cent in 1960).

Selected References

Fuchs, V. R. *Changes in the Location of Manufacturing in the U.S. Since 1929*. New Haven and London: Yale Univ. Press, 1962.

Greenhut, M. L. *Plant Location in Theory and Practice*. Chapel Hill: Univ. of North Carolina Press, 1956.

Macfarlane, Sidley K. "The Characteristics, Problems and Potential of Manufacturing in the Utica Area." Unpublished Ph.D. dissertation, Department of Geography, Syracuse University, 1960.

Miller, E. W. *A Geography of Manufacturing*. Englewood Cliffs, N.J.: Prentice-Hall, 1962.

New York State Department of Commerce. *Industrial Directory of New York State*. Albany: The Department, 1958.

_____. *New York State Commerce Review*, I–XIV. Albany: The Department, 1946–60 (ceased publication).

Thompson, J., and J. Jennings. *Manufacturing in the St. Lawrence Area of New York State*. Syracuse: Syracuse Univ. Press, 1958.

U.S. Bureau of the Budget. *Standard Industrial Classification Manual*. Washington, D.C.: Govt. Printing Office, 1957.

U.S. Bureau of the Census. *Annual Survey of Manufactures: 1961*. Washington, D.C.: Govt. Printing Office, 1963.

_____. *Census of Manufactures:1958*. Washington, D.C.: Govt. Printing Office, 1961.

_____, *County Business Patterns: 1st Quarter, 1962*. Washington, D.C.: Govt. Printing Office, 1962.

Yaseen, L. *Plant Location*. New York: American Research Council, 1956.

CHAPTER 12   JOHN H. THOMPSON, *with a section on* Recreation and Conservation
by HENRY G. WILLIAMS, JR. *and* ROGER THOMPSON

The Tertiary Sector

THE TERTIARY SECTOR employs 63 per cent of New York's labor force and accounts for an even higher percent of the GSP. Furthermore, its already overwhelming significance will increase as the economic development of the state advances. It is not unreasonable to expect that 75 to 80 per cent of all new jobs formed in the state in years ahead will be tertiary jobs.

The sector ordinarily is considered to be made up of the following eight parts: (1) transportation, communication, and other public utilities, (2) retailing and wholesaling, (3) finance, insurance, and real estate, (4) business and repair services, (5) personal services, (6) entertainment and recreation services, (7) professional and related services, and (8) public administration. In computing the GNP the government uses a slight modification of the above as exhibited in Table 16. This chapter, after dealing with the tertiary sector as a whole, will be divided into sections dealing with transportation, sales activities, service activities, trade and service centers and trade areas, power, and recreation and conservation. Except for the latter the tertiary activities, like the secondary ones, tend to agglomerate people and thereby are strong city-builders. Transportation provides the link or possibility of interaction between places, while wholesale and retail activities as well as services congregate at points of maximum transport advantage. The result is a trade center. Each trade center, in competition with trade centers around it, dominates its own trade area. The trade areas in turn produce a functional, spatial fabric representing the interaction patterns of society.

The Sector as a Whole

Census data do not provide complete production or value information for the tertiary sector and its parts, but Table 16 gives some indication of its dollar significance. Table 26 compares the various parts of the sector and shows the trends from 1950 to 1960.

Wholesale and retail trade and professional and related services are the largest employers, both in New York State and nationally. Together they account for somewhat over half of all tertiary employment. Between 1950 and 1960 the largest gain by

TABLE 26
THE TERTIARY SECTOR NEW YORK STATE

	1950		1960		Absolute Gains or Losses 1950–60
	Employment	% of Sector Total	Employment	% of Sector Total	
Transportation, Communication, and Public Utilities	522,237	14.6	508,572	13.0	− 13.665
Wholesale and Retail Trades	1,240,961	34.7	1,202,643	30.8	− 38,318
Finance, Insurance, and Real Estate	336,789	9.4	409,243	10.5	+ 72,454
Business and Repair Services	179,631	5.0	210,799	5.4	+ 31,168
Personal Services	387,106	10.8	358,087	9.2	− 29,019
Entertainment and Recreation Services	76,315	2.1	63,539	1.6	− 12,776
Professional and Related Services	566,650	15.8	842,400	21.5	+275,750
Public Administration	270,532	7.6	313,773	8.0	+ 43,241
TOTAL	3,580,221	100.0	3,909,056	100.0	+328.835

Source: U.S. Census of Population.

far in employment occurred in professional and related services. This particularly reflects gains in activities related to education and medicine. Also growing were the finance, insurance and real estate, public administration, and business and repair services. Both wholesale and retail and personal services categories showed substantial losses. These figures reflect recent trends as well as likely future ones. Not made clear from the data in Table 26 are the size and growth trends related to recreation. Recreation and the activities it induces have been expanding, but the census categories are not so constituted as to show this because the recreation impact extends through most of the categories. As a matter of fact, probably the three most rapidly expanding individual items within the tertiary sector are education, recreation, and public administration.

The Tertiary Surplus and Deficit Concept

It can be hypothesized that any state, county, or city has specific needs in terms of sales and service functions. It might be further hypothesized that some places produce more of these functions than they need and others less than required. If this be so, tertiary activities set up a basis for economic interchange between areas.

In 1960 the United States had about 36 million tertiary workers providing sales and service functions for a population of 179 million. As a ratio, this is very near 1:5. If this ratio is established as a norm, the assumption may be made that any place not having at least one tertiary worker for every five people has a *tertiary deficit* and depends on other places for some of its services; on the other hand, any place having more than a 1:5 ratio, say, 1:4, has a *tertiary surplus* and is in a position to serve people beyond its limits. Obviously there are problems in assuming that the 1:5 ratio is normal everywhere, for undoubtedly city people use more services than country people and New Yorkers in general use more than Mississippians. Nevertheless, it is a useful model against which any area can be compared.

Tertiary potential for any area is calculated through multiplying the tertiary employment by the national ratio figure of five and represents the number of people that the tertiary labor force can serve at the national norm. A comparison of this figure with the actual population shows whether a theoretical surplus or deficit exists. The percentage of the tertiary potential which is surplus is referred to as the *basic percentage*; the percentage of the population exceeding the tertiary potential is called the *deficit percentage*. These two percentages imply the extent to which the tertiary sector brings in capital from the outside or the extent to which the area must export capital to the outside in order to have the services not locally available.

New York State, with 4,747,000 people employed in the tertiary sector, can, according to the model, provide enough tertiary functions for 23,736,000 people (the tertiary potential). When comparing this figure with the state's population of 16,782,000 it can be said that New York has a tertiary surplus of about seven million; or, to put it another way, the state provides services enough to satisfy its own needs plus those of about seven million people beyond its borders. This is over three times the surplus of any other state and reflects primarily the tremendous tertiary role of the nation's primate city, metropolitan New York. The surplus of the New York Standard Metropolitan Statistical Area alone is 6.4 million, so the rest of the state, with less than 600,000 surplus, is less impressive as an exporter of tertiary functions.

The Concept Applied to Counties and Central Places

Spatial variation in degree of self-sufficiency in tertiary function can illuminate much about economic geography. Table 27 and Figure 81 portray the geography of tertiary activity by county. Of the sixty-two counties in the state, only twenty-two have surpluses and only nine have substantial surpluses. There is a strong correlation between population and tertiary employment, with the big-city counties having the most tertiary employment and ordinarily the largest tertiary surpluses as well. Exceptions to the population-surplus relationship are encountered in Bronx and Kings counties, which both have large deficits. These deficits can be explained by the dependence of these areas on Manhattan for services. Albany County on the other hand, because of its public administration function, has an abnormally large surplus.

The same calculations can be carried out for urban centers. Table 28 and Figure 82 show the results for the urbanized areas of major cities and for the other larger centers of the state. Figure 82 points up the tremendous role of New York City but also significantly sets off Buffalo, Rochester, Syracuse, and Albany as being in a class by themselves. Of these, Syracuse has by far the largest surplus for its size as shown by its basic percentage. This suggests an unusually wide dissemination of that city's service function over Upstate New York; and, in view of the general trend in the growing importance of the sector, this speaks well for the possibility of economic growth in the Syracuse area.

Figure 83 suggests some aspects of the likely distribution of the Syracuse surplus. From the figure it can be seen that the central city has a surplus of 211,000. This means that the central city theoretically provides the services it needs plus enough for 211,000 people beyond its boundaries. Proceeding,

TABLE 27
TERTIARY ACTIVITIES IN NEW YORK STATE COUNTIES

County	Tertiary Employment	Tertiary Potential	Population 1960	Surplus or Deficit	Basic or Deficit Percentage
State Total	4,747,248	23,736,240	16,782,304	6,953,936	29.30
Albany	98,560	492,800	272,926	219,874	44.62
Allegany	7,444	37,220	43,978	− 6,758	−15.37
Bronx	171,694	858,470	1,424,815	−566,345	−39.75
Broome	46,701	233,505	212,661	20,844	8.93
Cattaraugus	17,047	87,235	80,187	7,048	7.87
Cayuga	13,216	66,080	73,942	− 7,862	−10.63
Chautauqua	28,400	142,000	145,377	− 3,377	− 2.32
Chemung	21,431	107,155	98,706	8,449	7.88
Chenango	7,835	39,175	43,243	− 4,068	− 9.41
Clinton	12,661	63,305	72,722	− 9,417	−12.95
Columbia	8,625	43,125	47,322	− 4,197	− 8.87
Cortland	7,411	37,055	41,113	− 4,058	− 9.87
Delaware	6,756	33,780	43,540	− 9,760	−22.42
Dutchess	36,498	182,490	176,008	6,482	3.55
Erie	253,011	1,265,055	1,064,688	200,367	15.84
Essex	7,253	36,265	35,300	965	2.67
Franklin	9,097	45,485	44,742	743	1.63
Fulton	8,923	44,615	51,304	− 6,689	−13.04
Genesee	10,263	51,315	53,994	− 2,679	− 4.96
Greene	6,808	34,040	31,372	2,668	7.84
Hamilton	1,127	5,635	4,267	1,368	6.53
Herkimer	10,600	53,000	66,370	− 13,370	−20.14
Jefferson	19,095	95,475	87,835	7,640	8.00
Kings	468,704	2,343,520	2,627,319	− 283,799	−10.80
Lewis	3,272	16,360	23,249	− 6,889	−29.63
Livingston	8,348	41,740	44,053	− 2,313	− 5.25
Madison	8,765	43,825	54,635	− 10,810	−19.74
Monroe	131,607	658,035	586,387	71,648	10.89
Montgomery	10,371	51,855	57,240	− 5,385	− 9.41
Nassau	257,008	1,285,040	1,300,171	− 15,131	− 1.16
New York	1,794,502	8,972,510	1,698,281	7,274,229	81.07
Niagara	39,502	197,510	242,269	− 44,759	−18.47
Oneida	62,183	310,915	264,401	46,514	14.96
Onondaga	113,018	565,090	423,028	142,062	25.14
Ontario	12,004	60,020	68,070	− 8,050	−11.83
Orange	40,957	204,785	183,734	21,051	10.28
Orleans	5,839	29,195	34,159	− 4,964	−14.53
Oswego	12,966	64,830	86,118	− 21,288	−24.72
Otsego	10,431	52,155	51,942	213	.41
Putnam	5,545	27,725	31,722	− 3,997	−12.60
Queens	342,893	1,714,465	1,809,578	− 95,113	− 5.26
Rensselaer	27,531	137,655	142,585	− 4,930	− 3.46
Richmond	45,191	225,955	221,991	3,964	1.75
Rockland	22,224	111,120	136,803	− 25,683	−18.77
St. Lawrence	20,750	103,750	111,239	− 7,489	− 6.73
Saratoga	13,878	69,390	89,096	− 19,706	−22.12
Schenectady	33,085	165,425	152,896	12,529	7.57
Schoharie	3,821	19,105	22,616	− 3,511	−15.52
Schuyler	2,314	11,570	15,044	− 3,474	−23.09
Seneca	5,380	26,900	31,984	− 5,084	−15.90
Steuben	18,214	91,070	97,691	− 6,621	− 6.78
Suffolk	107,708	538,540	666,784	−128,244	−19.23
Sullivan	13,293	66,465	45,272	21,193	31.89
Tioga	5,481	27,405	37,802	− 10,397	−27.50
Tompkins	20,823	104,115	66,164	37,951	36.45
Ulster	22,297	111,485	118,804	− 7,319	− 6.16
Warren	12,264	61,320	44,002	17,318	28.24
Washington	6,429	32,145	48,476	− 16,331	−33.69
Wayne	16,390	81,950	67,989	13,961	17.04
Westchester	202,566	1,012,830	808,891	203,939	20.14
Wyoming	5,896	29,480	34,793	− 5,313	−15.27
Yates	3,342	16,710	18,614	− 1,904	−10.23

*Tertiary employment on this table is estimated on the basis of place of work and therefore differs from that shown in Table 26, which is based on place of residence. Out-of-state commuters come largely from New Jersey and Connecticut.

Source: Richard T. Lewis, "The Measurement of Tertiary Activity," unpublished Master's thesis, Department of Geography, Syracuse University, 1964.

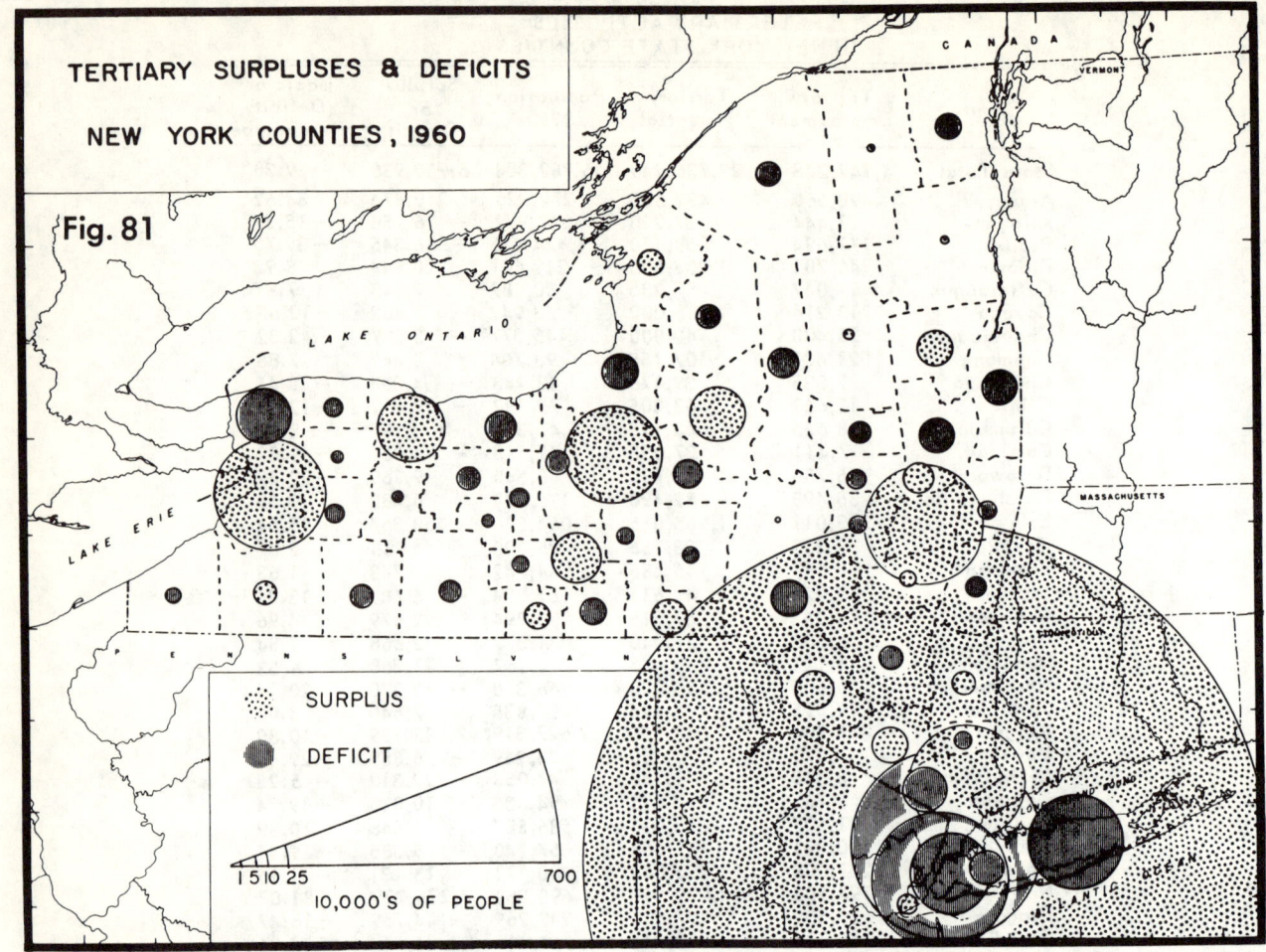

Fig. 81 — TERTIARY SURPLUSES & DEFICITS NEW YORK COUNTIES, 1960

Figure 83 shows the urbanized area to have a surplus of 159,000, or 52,000 less than the central city. This is reasonable, for areas just outside central cities are always dependent on the central city for a wide array of services. In this case 52,000 of the city's surplus is used up to take care of deficits in the area between the city boundary and the urbanized area boundary. Similarly, another 17,000 of the surplus is used between the urbanized boundary and the county boundary and 32,000 is absorbed in the rest of the Standard Metropolitan Statistical Area. There still remains a surplus of 110,000 not required in the SMSA. It is safe to assume that some of this is distributed widely over the state and nation, but most goes for servicing the North Country and other parts of Upstate New York. This example of Syracuse illustrates that examination of the tertiary surpluses provides some idea of the "reach" of central places and suggests something about spatial interaction.

It has often been implied that Utica and Binghamton are in the same general city, or central place, category as Syracuse. Table 28 and Figure 82 would indicate that this is not so, for in both tertiary employment and tertiary surplus they are very much smaller and so have very much less central place function. In fact, residents of either city will commonly confirm that they occasionally drive to Syracuse for certain tertiary functions.

TERTIARY STRUCTURE

The question of what parts, if any, of the tertiary sector dominate the tertiary structure of any place can be ascertained, and it is also possible to compare the tertiary structure of any central place to the average structure of all urban areas in the United States. Table 29 does this for selected places. Notice the extent to which professional and related services dominate the tertiary structure of Ithaca, actually amounting to nearly 55 per cent of the tertiary labor force when the national average is only 21 per cent. This reflects the importance of Ithaca's educational function, which supports much of the professional and related services employment there. Table 29 also points up the high-level em-

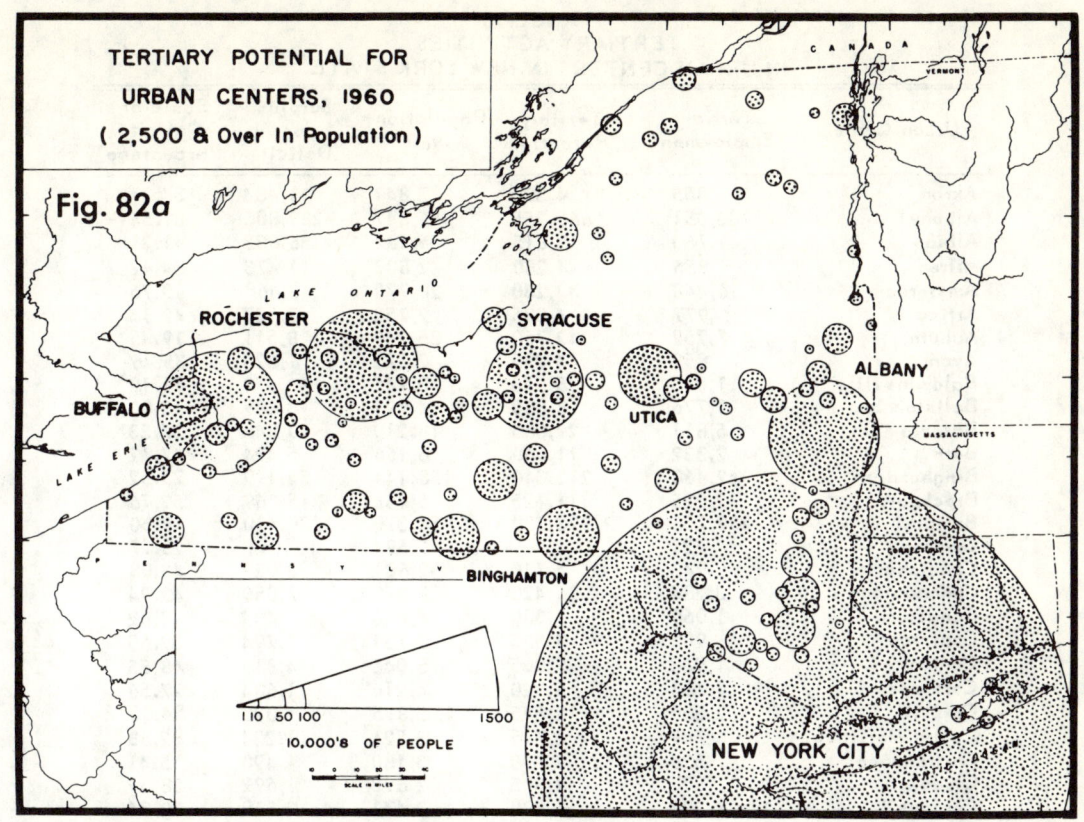

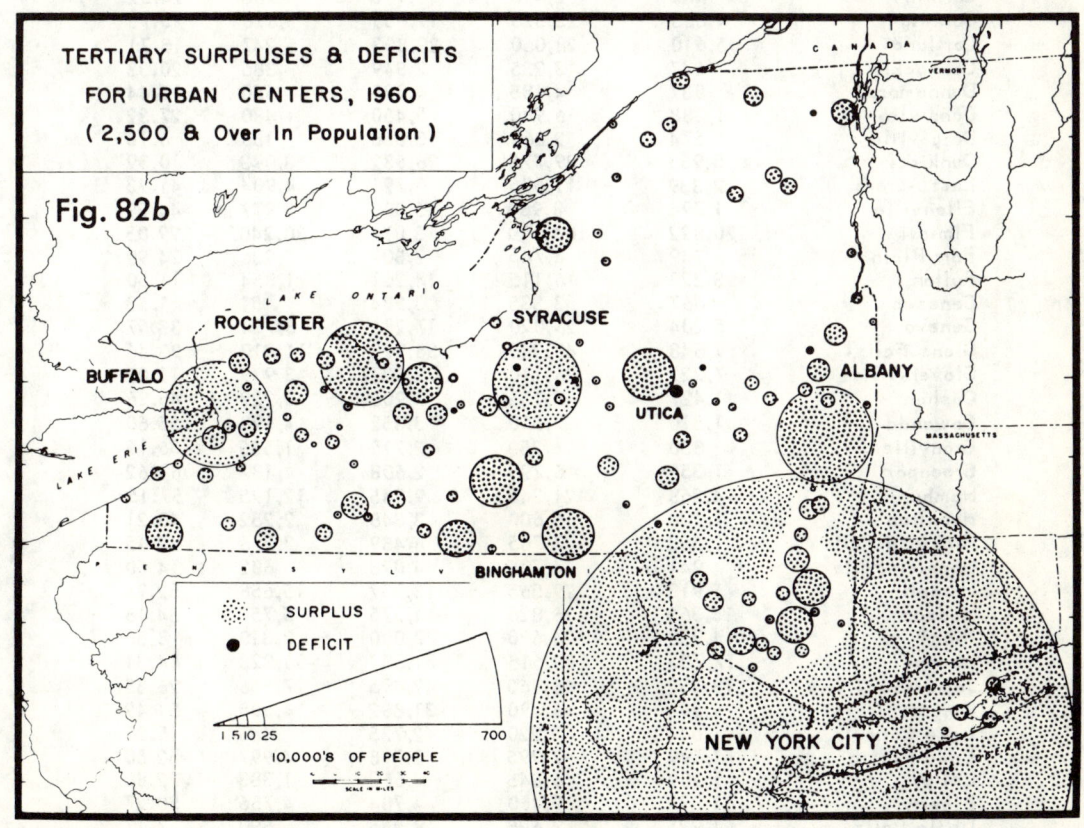

CHAPTER 12 THE TERTIARY SECTOR 259

TABLE 28
TERTIARY ACTIVITIES
IN URBAN CENTERS IN NEW YORK STATE

Urban Center	Tertiary Employment	Tertiary Potential	Population 1960	Surplus or Deficit	Basic or Deficit Percentage
Akron	865	4,325	2,841	1,484	34.31
Albany†	133,051	665,255	455,447	209,808	31.54
Albion	1,763	8,815	5,182	36,633	41.21
Alfred	856	4,280	2,807	1,473	34.42
Amsterdam	6,448	32,240	28,772	3,468	10.76
Attica	977	4,885	2,758	2,127	43.54
Auburn	8,752	43,760	35,249	8,511	19.45
Avon	859	4,295	2,772	1,523	35.46
Baldwinsville	1,063	5,315	5,985	− 670	−11.19
Ballston Spa	1,776	8,880	4,991	3,889	43.80
Batavia	5,811	29,055	18,210	10,845	37.33
Bath	2,332	11,660	6,166	5,494	47.12
Binghamton†	42,468	212,340	158,141	54,199	25.52
Brockport	2,225	11,125	5,256	5,869	52.76
Buffalo†	258,756	1,293,780	1,054,370	239,410	18.50
Camden	721	3,605	2,694	911	25.27
Canajoharie	942	4,710	2,681	2,029	43.08
Canandaigua	3,284	16,420	9,370	7,050	42.94
Canastota	1,060	5,300	4,896	404	7.62
Canisteo	905	4,525	2,731	1,794	39.65
Canton	1,973	9,865	5,046	4,819	48.85
Carthage	1,164	5,820	4,216	1,604	27.56
Catskill	2,569	12,845	5,825	7,020	54.65
Center Moriches	749	3,745	2,521	1,224	32.68
Chittenango	538	2,690	3,180	− 490	−15.41
Clyde	877	4,385	2,693	1,692	38.59
Cobleskill	1,398	6,990	3,471	3,519	50.34
Cooperstown	1,559	7,795	2,553	5,242	67.25
Corinth	482	2,410	3,193	− 783	−24.52
Corning‡	4,665	23,325	19,655	3,670	15.73
Cortland‡	5,610	28,050	22,803	5,247	18.71
Coxsackie	647	3,235	2,849	386	20.33
Dannemora	837	4,185	4,835	− 650	−13.44
Dansville	1,388	6,940	5,460	1,480	27.32
Dolgeville	574	2,860	3,058	− 188	− 6.15
Dunkirk‡	5,955	29,775	26,682	3,093	10.39
East Aurora	2,339	11,695	6,791	4,904	41.93
Ellenville	1,796	8,980	5,003	3,977	44.29
Elmira‡	20,822	104,110	73,870	30,240	29.05
Fort Plain	749	3,745	2,809	936	24.99
Fulton	3,223	16,115	14,261	1,854	11.50
Geneseo	667	3,335	3,284	51	1.53
Geneva	5,204	26,020	17,286	8,734	33.57
Glens Falls‡	9,648	48,240	36,923	11,317	23.46
Gloversville‡	7,222	36,110	32,131	3,979	11.02
Goshen	1,427	7,135	3,906	3,229	45.26
Gowanda	1,510	7,550	3,352	4,198	55.60
Granville	850	4,250	2,715	1,535	36.12
Greenport	1,359	6,795	2,608	4,187	61.62
Hamburg	4,268	21,340	9,145	12,195	57.15
Hamilton	1,120	5,600	3,348	2,252	40.21
Highland Falls	1,553	7,765	4,469	3,296	42.45
Hoosick Falls	941	4,705	4,023	682	14.50
Hornell	5,913	29,565	13,907	15,658	52.96
Hudson	3,365	16,825	11,075	5,750	34.18
Ilion‡	4,938	24,690	27,000	− 2,310	− 8.56
Ithaca‡	17,703	88,515	31,587	56,928	64.31
Jamestown‡	13,332	66,660	49,495	27,566	26.35
Kingston‡	9,298	46,490	31,882	14,608	31.42
Lake Carmel	644	3,220	2,735	485	15.06
Lake Placid	1,599	7,995	2,998	4,997	62.50
LeRoy	1,209	6,045	4,662	1,383	22.88
Liberty	1,892	9,460	4,704	4,756	50.27
Little Falls	1,925	9,625	8,935	690	7.17
Lockport	7,029	35,145	26,443	8,702	24.76
Lowville	1,162	5,810	3,616	2,194	37.76

TABLE 28 (Continued)

Urban Center	Tertiary Employment	Tertiary Potential	Population 1960	Surplus or Deficit	Basic or Deficit Percentage
Lyons	1,027	5,135	4,673	462	20.68
Malone	3,072	15,360	8,737	6,623	62.65
Massena	4,304	21,520	15,478	6,042	28.08
Mechanicville	2,044	10,220	6,831	3,389	33.16
Middletown	8,114	40,570	23,475	17,095	42.14
Monroe	1,334	6,670	3,323	3,347	50.18
Monticello	2,485	12,425	5,222	7,203	57.97
Mount Morris	1,287	6,435	3,250	3,185	49.49
Newark	9,288	46,440	12,868	33,572	72.29
Newburgh‡	16,821	84,105	54,551	29,554	35.14
New Paltz	1,268	6,340	3,041	3,299	52.03
New York City*	3,412,490	17,062,450	10,694,633	6,367,817	37.32
Norwich	3,351	16,755	9,175	7,580	45.24
Ogdensburg	3,506	17,530	16,122	1,408	8.03
Olean	6,771	33,855	21,868	11,987	35.41
Oneida‡	3,215	16,075	14,599	1,476	9.18
Oneonta	4,684	23,420	13,412	10,008	42.73
Orchard Park	1,215	6,075	3,278	2,797	46.04
Oswego	4,971	24,855	22,155	2,700	10.86
Owego	1,597	7,985	5,417	2,568	32.16
Palmyra	776	3,880	3,476	404	10.41
Penn Yan	1,699	8,495	5,770	2,725	32.08
Perry	961	4,805	4,629	176	3.66
Plattsburgh	6,976	34,880	20,172	14,708	42.17
Port Jervis	2,865	14,325	9,268	5,057	35.30
Potsdam	2,386	11,930	7,765	4,165	34.91
Poughkeepsie‡	17,548	87,740	58,204	29,536	33.66
Riverhead	2,041	10,205	5,830	4,375	42.87
Rochester†	126,465	632,325	493,402	138,923	21.97
Salamanca	2,363	11,815	8,480	3,335	28.23
Saranac Lake	2,370	11,850	6,421	5,429	45.81
Saratoga Springs	5,610	28,050	16,630	11,420	40.71
Saugerties	1,416	7,080	4,286	2,794	39.46
Seneca Falls	1,633	8,165	7,439	726	8.89
Sidney	1,141	5,705	5,157	548	9.61
Silver Creek	1,120	5,600	3,310	2,290	40.89
Skaneateles	1,124	5,620	2,921	2,699	48.02
Southampton	1,819	9,045	4,583	4,463	49.34
Springville	1,081	5,405	3,852	1,553	28.73
Syracuse†	98,491	492,455	333,286	159,169	32.32
Ticonderoga	1,124	5,620	3,568	2,052	36.51
Tupper Lake	1,250	6,250	5,200	1,050	16.80
Utica†	48,438	242,190	187,779	54,411	22.47
Walden	1,361	6,805	4,851	1,954	28.71
Walton	874	4,370	3,855	515	11.78
Wappingers Falls	1,307	6,535	4,447	2,088	31.95
Warsaw	1,341	6,705	3,653	3,052	45.52
Warwick	1,083	5,415	3,218	2,197	40.57
Waterloo	973	4,865	5,098	−233	−4.57
Watertown	11,549	57,745	33,306	24,439	42.33
Watkins Glen	1,035	5,175	2,813	2,362	45.64
Waverly	1,463	7,315	5,950	1,365	18.66
Webster	1,152	5,760	3,060	2,700	46.87
Wellsville	2,214	11,070	5,967	5,103	46.10
Westfield	1,073	5,365	3,878	1,487	27.72
Whitehall	1,288	6,440	4,016	2,424	37.64

*SMSA.
†Urbanized Area.
‡Combined Urban Places (satellite and suburban places added to larger places to give better representation of true central function)—Corning: Painted Post; Cortland: Homer; Dunkirk: Fredonia; Elmira: Elmira Heights, Elmira Southwest, Horseheads, Victory Heights, West Elmira; Glens Falls: Fort Edward, Hudson Falls, South Glens Falls, West Glens Falls; Gloversville: Johnstown; Ilion: Frankfort, Herkimer, Mohawk; Ithaca: Cayuga Heights; Jamestown: Falconer, Lakewood; Kingston: Port Ewen; Newburgh: Beacon, Cornwall, Cornwall Southwest, New Windsor; Oneida: Sherrill; Poughkeepsie: Arlington, Fairview, Highland.
Source: Lewis, "The Measurement of Tertiary Activity."

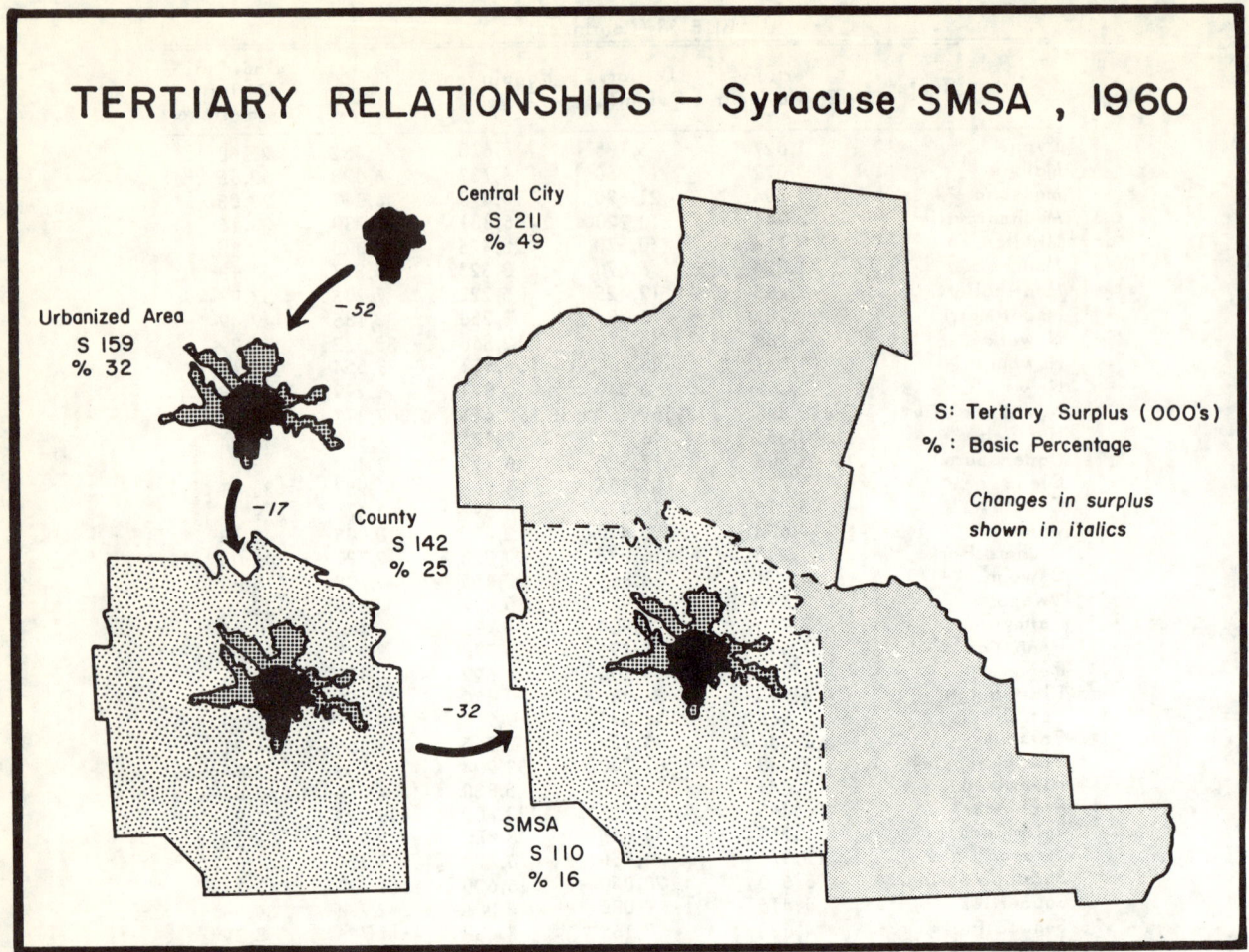

Fig. 83

phasis on public administration in Albany, professional and related services in Binghamton, Rochester, and Syracuse, finance, insurance, and real estate in New York, and transportation, communication, and other public utilities in Hornell. In each of these cases the special tertiary emphasis has a substantial impact on the general economy and the over-all structure elucidates the functional nature of the city.

Transportation

Chapters 8 and 9, particularly, have treated the evolution of transportation in the state. At different times different mediums have dominated. There was the turnpike era, the short period when plank roads were important, the canal era, the railroad heyday, the highway era, and now the new period when the role of the limited-access highway is strongly affecting area development. Following is a brief discussion of transport network development theory and of the extent to which transportation affects the state and its various sections today.

Transport Network Development Theory

Man at different stages in his development uses the environment differently to solve his transport requirements, but at all times there are certain principles which determine the geography of transportation and interaction between places. These are the principles of complementarity, intervening opportunity, and transferability.

A transportation route develops as a result of demand for connection between two traffic generating points, say points A and E in Figure 84. This demand for connection or the need of one area for the produce or services of another is called *complementarity*. When point A has a surplus of a commodity (e.g., flour or shoes) that is in short supply but in demand at point E, a transport route theoretically should develop between points A and E so that commodity flow can take place. It would be helpful to development of the transportation route if E had a surplus of wheat or leather which was in demand at A for manufacture into flour and shoes. In any case it is apparent that complementarity is a function of both natural and cultural areal differentiation. Or, to put it another way, it may be differences in the physical resource base between A and E or differences in man's attitudes and productive capacities, or a combination of the two which produces complementarity and thus the rise of transportation routes. Resource and production differences between New York City and the lower Hudson on one hand and western New York and those regions bordering on the Great Lakes on the other gave rise to a need for the Erie Canal, for example. The thoughtful reader might try to weigh the relative importance of the kinds of factors that instigated development of any given railroad line, or even the new limited-access highway system within the state.

The second fundamental principle affecting transportation network development is *intervening opportunity*. It may be explained in the following way: If point B, which lies between A and E, could provide the flour and shoes needed by E, then E would buy less from A, giving A less exchange to trade with E, and there would be far less complementarity and thus less need for transport facilities connecting A and E. On the other hand, intervening opportunity of another kind more often creates interaction between distant potentially complementary areas by making construction of routes profitable. It is possible that complementarity between A and E would hardly be great enough to induce established transport routes between them if there were no intervening opportunities for business at B, C, and D. Note, too, that the locations of B, C, and D affect the precise location of the routeway between A and E. The precise location of the routeway may also be affected by terrain and other factors. The great transport artery from New York City to Buffalo and beyond profits from intervening opportunities for business at Albany, Utica, Syracuse, Rochester, and a number of other places. Thus complementarity and intervening opportunity work hand in hand in pushing forward the transport route in an

TABLE 29
TERTIARY STRUCTURE FOR SELECTED CENTRAL PLACES, 1960*

	All Urban Areas of U.S.	Urbanized Areas						Urban Places	
		Albany	Buffalo	Syracuse	Binghamton	Rochester	New York†	Ithaca‡	Hornell
Transportation, Communication and Public Utilities	12.6	11.7	14.7	12.5	10.3	9.7	13.8	4.8	34.2
Wholesale and Retail Trades	32.7	29.8	35.1	33.0	34.5	34.3	30.6	16.4	23.7
Finance, Insurance, and Real Estate	8.1	6.1	7.0	9.0	7.0	7.8	12.0	4.0	2.9
Business and Repair Services	4.5	3.6	3.9	4.0	3.5	4.1	6.1	2.3	2.0
Personal Services	10.6	7.2	7.4	8.3	8.7	8.6	8.9	12.6	10.2
Entertainment and Recreation Services	1.5	.9	1.4	1.2	1.3	1.6	1.7	1.5	.5
Professional and Related Services	20.9	23.0	22.6	24.7	27.3	26.6	19.4	54.6	21.7
Public Administration	9.1	17.7	7.8	7.4	7.4	7.3	7.5	3.9	4.8

*The data in this table are not exactly comparable to those in Table 26 because Table 26 includes rural as well as urban areas. Figures are percentages of total tertiary employment.
†Including some areas in northeastern N.J.
‡Including Cayuga Heights.
Source: U.S. Census of Population, 1960.

area as it is settled and affect the volume of goods and people moving over it as long as it exists.

Transferability is the third factor in an interaction system. If the distance between the market, E, and the supply, A, is too great and shipment costs too high to overcome, interaction will not take place in spite of existing complementarity. In such a case there will be a transfer or substitution in goods used at E to those which can be secured from the local area or at least at points closer than A. Transferability undoubtedly was much more important in the state in early years when people were forced to make do with local products because of poor transport facilities and high transport costs from distant points.

After settlements were established around the mouth of the Hudson, it was not long before complementarity with the interior developed. Routes of least terrain resistance and maximum intervening business opportunity were followed by early lines of transportation called lines of penetration. Feeders or spurs branched from the line of penetration as soon as business warranted. In time complementarity expanded as economic growth occurred at both New York City and interior points. Specialization started to replace self-sufficiency. Feeders grew into lines of lateral interconnection as distant points and secondary routes became connected with the original line of penetration. Very heavy traffic began to develop on main routes until they entered the category of trunk lines. Where transport routes cross, central places come into existence. The more important the routes, the greater the likelihood that the central places will grow.

The dominant transportation mode along any routeway may have varied from time to time from canal to railroad to highway, and in some cases a trunk rail line may be paralleled by a canal or roads, which are trunk lines in their own right. Such paralleling and congestion of trunk lines produces what may be referred to as a transportation axis, or artery. The Hudson-Mohawk axis, or artery, has long been made up of all three of the above modes of transportation and the selection of it for the route of the New York State Thruway reflects the effective forces of both complementarity and intervening opportunity.

Transportation Patterns Today

New York has enjoyed relatively good transport facilities for many years. It was a leader in American canal construction and railroads early criss-crossed the state, providing service to rural areas and major central places alike. Since the 1920's the road and its counterpart, the automobile, have provided stiff competition to earlier canal and rail systems, but have at the same time provided a valuable elaboration of the over-all transport network. Air service between major centers, and particularly between major centers and New York City, has become especially important to businessmen and other clientele for whom "time is money."

Water Transportation

A quick look at a physical map of New York would suggest that numerous opportunities for water transportation exist. The Great Lakes, the St. Lawrence River, the Hudson River, and Lake Champlain provide obvious opportunities for water traffic.

Certainly the most publicized and exciting recent development in water transportation has been the construction of the St. Lawrence Seaway. For many years those sections of the St. Lawrence which had fast, shallow water made navigation difficult. Early canalization on the Canadian side enabled small vessels to bypass the rapids, but not until 1959 did a cooperative venture between Canada and the United States open a major (27-foot-deep) channel and lock system that enables large ocean vessels to reach the Great Lakes. Considerable resistance by competing transport and dock facilities and other groups had held back construction of the seaway for years, but the gradual depletion of domestic iron ore reserves in Minnesota and the discovery of large iron ore bodies in Labrador broke resistance, and the seaway was built.

Some people expected much of the seaway in the way of promoting economic expansion in northern New York. However, although a good deal of traffic passes through the seaway and past northern New York, not much originates or terminates in the state. Thus New York communities on the seaway, and even on the lakes, have profited less than their boosters had hoped. The seaway does provide the state with a "third coast," and much interest is still alive in communities such as Oswego, Ogdensburg, Watertown, and Massena. Ways and means are being explored by development groups in these communities to find out how the seaway might pay off more extensively for them.

As might be expected, a large part of the seaway traffic is bulk type goods including iron ore from Labrador, wheat and other grains from the interior of the continent, and fuel oil coming in from the Atlantic. Bulk commodities account for about 90 per cent of the total traffic by weight. Direct shipments of general nonbulk cargoes between United States Great Lake ports and foreign countries have increased since preseaway days, but the amount of this kind of trade has been disappointingly small. It probably can be forecast that shipments on the seaway will gradually rise, that the average size of vessels using it will increase, and that it will become increasingly vital to the American iron and steel industry (some of which occurs in the Buffalo area) as larger tonnages of iron ore from Labrador are used. It probably cannot be expected, though, that the St. Lawrence Valley in New York will profit much by all of this because it is, as previously indicated, off the principal economic axis of the United States. As far as the seaway is concerned, northern New York is a *transit region*, one by which goods will pass but in which few will originate or terminate. The Canadian side opposite New York will profit relatively more because it lies astride the principal economic axis of Canada.

Commercial traffic occurs on the Hudson River between New York City and Albany, and limited quantities of goods move between lake ports on the Great Lakes. Elsewhere present-day water transportation would have to be described as insignificant. Canals everywhere are declining in use for commercial purposes or have been abandoned altogether. They may, however, continue to be used, perhaps even in increasing amounts, by pleasure craft and other kinds of recreation. Currently the New York State Barge Canal, which roughly parallels the course of its predecessor, the Erie Canal, allows water traffic to move from the Hudson to Buffalo and provides water transportation service to many centers en route. Although the Barge Canal has always been a disappointment in terms of the volume of traffic it carried and the revenue it produced, until fairly recently substantial amounts of petroleum products, sand and gravel, cement, grain, stone, and other heavy, bulky commodities were shipped on it. Pipelines, rail, and bulk truck-carrying facilities have all but brought about its demise. The St. Lawrence Seaway has not helped either. From the Barge Canal water routes lead into several of the Finger Lakes as well as into both of the Great Lakes. A canal is still maintained between the Hudson and Lake Champlain, and the Richelieu Canal leads from Lake Champlain to the St. Lawrence River below Montreal. So, even though many miles of canals have been abandoned in past years, if navigable mileage in rivers, lakes, and the remaining maintained canals is totaled, there is in and around New York State one of the most extensive and interesting waterway systems to be found anywhere. Although failing as a boon to commercial traffic in recent years, the system has become a paradise for a growing number of pleasure craft. Interesting public discussions are likely to be carried on in the state in the future to determine whether these waterways will be maintained for noncommercial purposes alone. Or, the so-called noncommercial waterway uses, because they do generate business, may be looked upon as a new kind of commercial use.

Railroad Transportation

More than any other transportation medium, the railroads laid down the economic base for the state, determined the lines of interaction, and directed the pattern of land use. There are over thirty railroads operating in the state with almost 6,300 miles of road. Most of the larger central places are still served. However, as in much of the Northeast, rail transportation in New York is suffering a slow but inexorable decline. Financial difficulties, reduction of passenger and freight service, and outright abandonment of lines, stations, and allied railroad enterprises have marked operations in recent decades.

That problems for the railroads exist and that lines have been and are being abandoned is not surprising. The continued process of urbanization and rural depopulation in the state over the past half-century, without complementary modifications in the railroad network, produced an outmoded physical pattern for the railroads. The major part of New York's railroads were built to service a rural and agrarian economy which has experienced an absolute as well as relative decline. Population increases and economic growth have been associated almost solely with the larger urban systems along the Hudson-Mohawk-Ontario Lake Plain transport artery and a few areas of the Southern Tier. The actual railroad mileage servicing these principal urban areas amounts to less than half of the still existing trackage. Add this "lack of match" between the spatial patterns of lines and the spatial patterns of modern traffic demand to the changing availability of competition from other transport mediums, and reasons for railroad difficulties are relatively clear. Specifically, these difficulties have been, or are, taking the following forms: (1) abandonment of certain lines of declining or low traffic generation, (2) abandonment of multiple trackage for the purpose of reducing right-of-way taxation and maintenance, (3) removal of passenger service from all short-haul (less than 200 miles) roads, (4) consolidation or merger of railroad systems with accompanying abandonment of duplicating trackage and operations, (5) general reduction of freight or passenger service on all roads so that fewer trains will haul more economically, and (6) abandonment of allied railroad enterprises such as railway express agencies, passenger terminal facilities, and freight

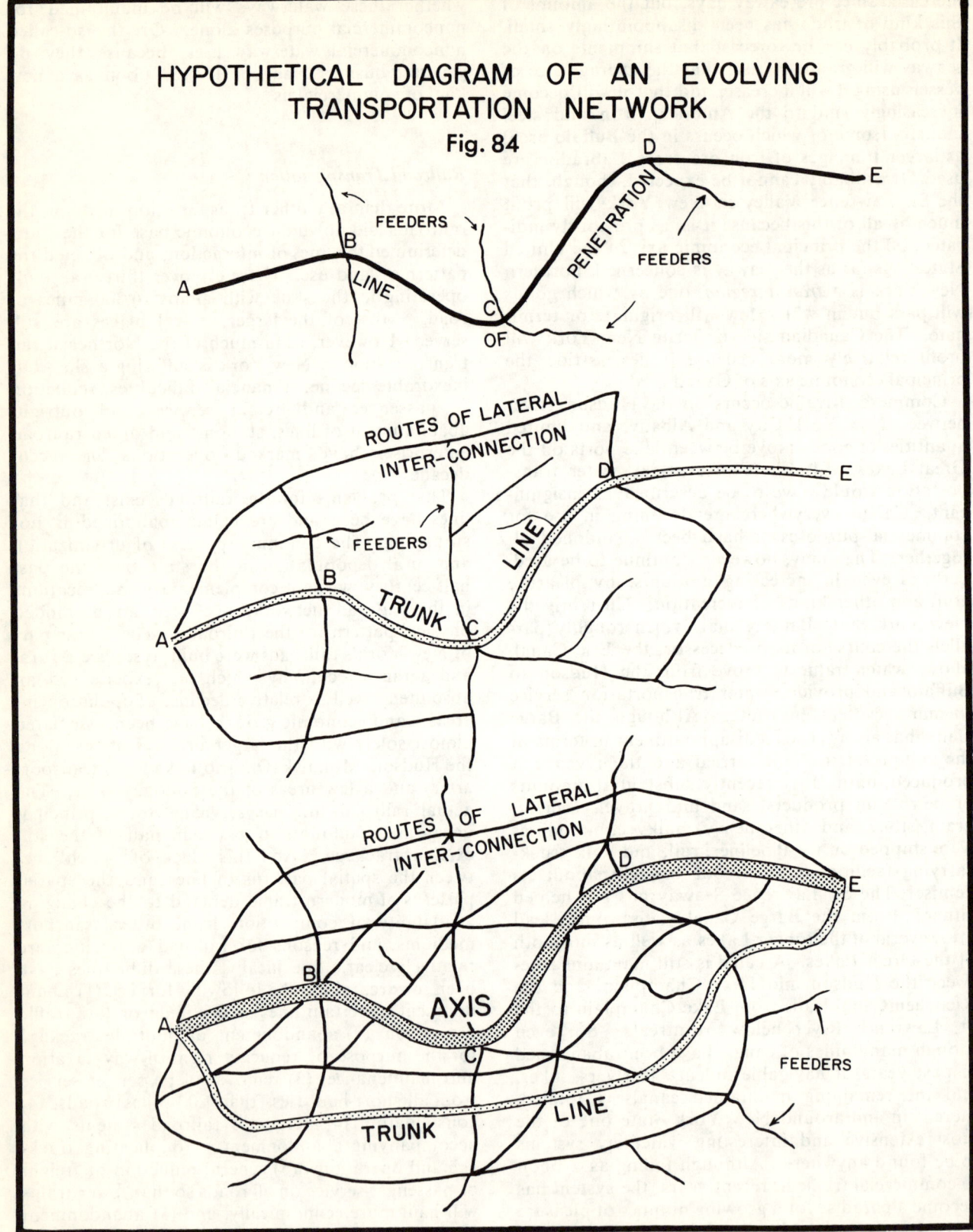

HYPOTHETICAL DIAGRAM OF AN EVOLVING TRANSPORTATION NETWORK
Fig. 84

houses. The private automobile, bus, truck, and airplane have all cut into the business on the state-wide traffic system once so dominated by railroads.

Figure 85 shows most of the modern railroad network and designates lines that have been abandoned. One thing this map does not show is the relative amount of traffic moving over the various lines. Data are not available for construction of such a map. If they were, the Hudson-Mohawk-Ontario route would stand out and the Erie-Lackawanna main line between New York City and Buffalo would appear easily second in significance.

Abandonment of trackage has been going on since the 1920's. Figure 85 excludes the nearly 1,500 miles of interurban electric trackage abandoned between 1920 and 1945, but still shows the tremendous extent to which the original steam railroad system has been affected by abandonment. The process is still going on as this chapter is being written. It has been largely the north-south feeder roads and the several lines of lateral interconnection which have been unable to compete successfully with other transport modes. By 1962, 30 to 35 per cent of all the 1913 trackage had been abandoned, and this included at least 20 per cent of the state's first steam trackage.

A railroad map for the year 1980 will still show some operating facilities in the state, but these will be primarily the major arteries and trunk lines connecting the large urban systems, lines which competitively provide these urban systems with a continued transport service for large volume, bulky, and heavy goods which must be moved considerable distances. Passenger service on railroads, unless there are attractive innovations, will be largely a thing of the past and the familiar railroad stations in cities and villages of all sizes will become rare indeed.

Most observers feel that railroads' success in future competition for traffic will depend largely on two trends: (1) their inclination and ability to innovate and thereby improve and extend service, and (2) the taxing and depreciation policies on railroad right-of-way and facilities. Should these two trends exhibit unfavorable characteristics, it may become progressively difficult for railroads to be maintained as privately operated enterprises. That they should be maintained as carriers could scarcely be denied by anyone. It would seem physically impossible for the transportation needs of the state to be served without the railroads. Furthermore, they play other less obvious but still important roles in the economy of the state. In 1963 alone capital improvements on the state's railroad network totaled almost 30 million dollars, and railroad purchases, primarily from manufacturers, amounted to 43 million dollars. They employ nearly 50,000 workers, and payrolls total more than 330 million dollars. This kind of contribution by the railroads still touches, in a small way at least, many parts of the state.

Highway Transportation

Highways, and the motor vehicles that use them, have come to dominate New York State traffic. At the close of 1963 there were 5.8 million motor vehicles registered in the state, or about one for every three people. Of these, 4.9 million were passenger cars, 570,000 were trucks, 57,000 were buses, and the remainder were special-purpose vehicles. Fifty-two billion vehicle miles were traveled during that year. Not only have the roads of the state become principal lines of interconnection, but along them and at junction points where two or more come together have evolved thousands of business establishments that serve road-users or are themselves served by the road. These include service stations and garages, restaurants and motels, warehousing and retailing establishments, and even factories. The road has indeed become a major locator of economic activity.

Not including city and village streets, there are nearly 90,000 miles of roads in the state, or 1.8 miles per each square mile of area. This may be compared with a density of 0.6 miles per square mile in sparsely settled Maine or 2.4 in congested New Jersey. Seven-eighths of the New York mileage is surfaced and can be used in all weather. Parkways, limited to noncommercial traffic, are becoming more common, especially around metropolitan New York. In 1954 the Thruway opened, providing a high-speed, limited-access link between major urban centers along the entire Hudson-Mohawk-Lake Plain axis. Any of the better road maps available at service stations will show distribution of United States highways, limited-access highways, and state and major county roads, and designate such things as number of lanes. Because the ordinary road map is a source of information about roads and because it contains so much other information as well, it can be a valuable reference tool for the reader of this book. If more detailed road data are needed, county road maps published by county highway departments or topographic quadrangles may be used.

Except for the Adirondacks, the Tug Hill Upland, and portions of the Catskills, it would be difficult to find any place in the state more than a few miles from a good road. Even though snowfall in New York is heavy, winter snow removal is remarkably good. Any school bus route is given early snowplow attention, and, because of the centralized school system in the state, this means that most rural roads, even though they have but a single year-round residence on them, will be well maintained.

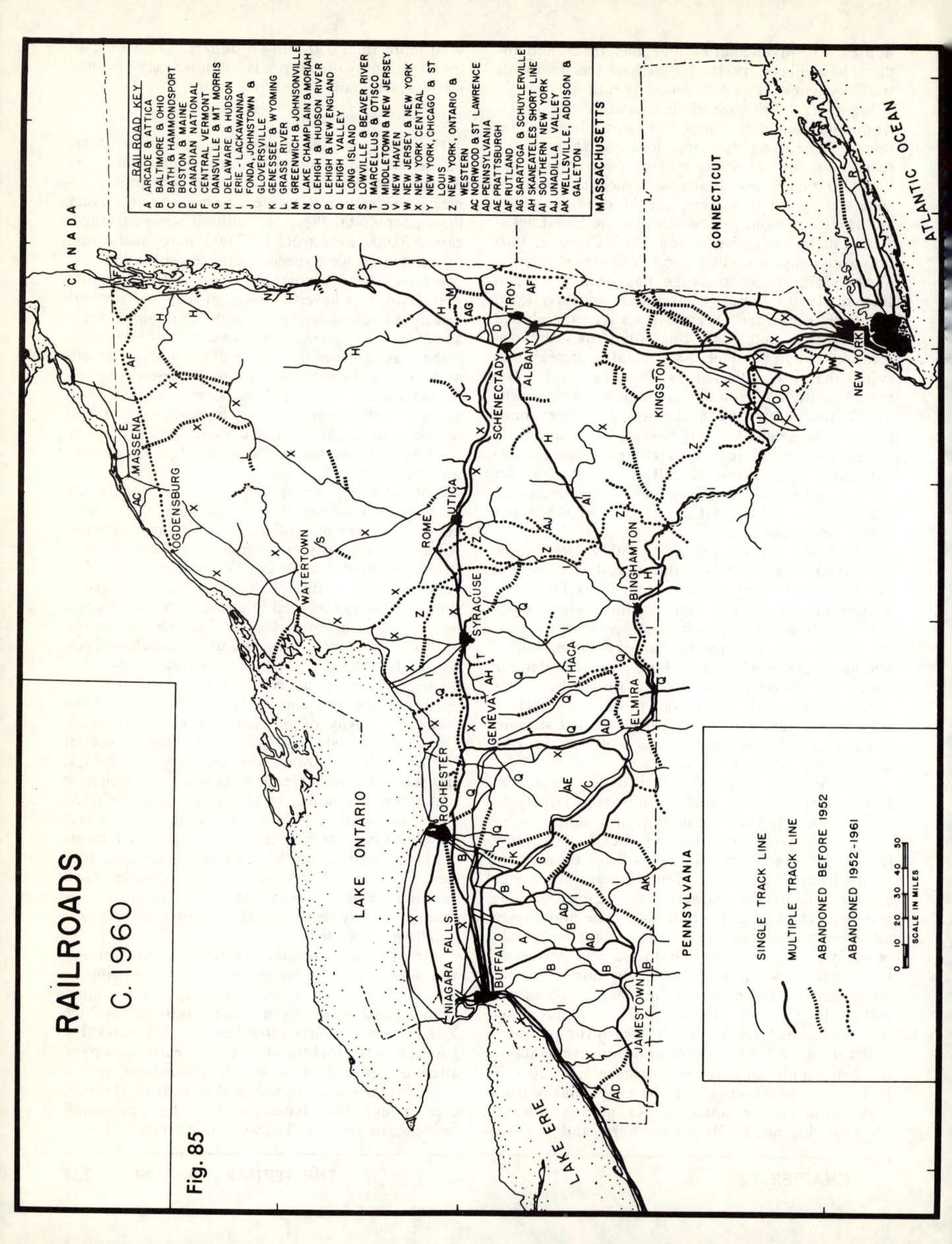

Fig. 85 RAILROADS C. 1960

Figure 86 functions in a way that Figure 85 and an ordinary road map cannot in that it shows comparative traffic flow. The degree of interaction between places, insofar as it is reflected by highway traffic, is effectively shown by this kind of map. The geographic position of the major urban systems on the flow pattern signifies extensive possibilities for business generation from the transport function in these systems, but remote localities may be assumed to profit little. This substantiates the thesis that the largest centers will become relatively larger. The Hudson–Mohawk–Lake Plain axis is accentuated by the great flow of highway traffic along it. The heaviest line on the map along this route is the Thruway. If the figure showed traffic for town and county roads, there would be a filling by light traffic routes of most of the state except for the Adirondack, Tug Hill, and Catskill areas.

The impact of the highway transport function at various points is roughly implied by the size of the black areas in and around central places in Figure 86. Those which appear most significant are the same centers which enjoyed large tertiary surpluses (Fig. 82b). In the years ahead, although traffic will tend to increase generally, substantial relative changes in the route and central place importance shown in Figure 86 are not likely. There are a few exceptions, however. For example, when Interstate Route 81 is completed south from Syracuse, traffic in much larger volume will flow southward from Syracuse to New York City via Binghamton. In giving Syracuse even more of a "crossroads" location than it now enjoys, this will contribute to substantial economic development of that urban system. Binghamton should profit some, too.

Limited-access highways, which include the Thruway, the Interstate System, and some of the parkways, have a somewhat different geographic influence than the ordinary highway in that their impact occurs largely at the interchanges. At the interchanges of the limited-access highway a nodal type of economic growth tends to take place. The ribbon type, which commonly occurs along a busy highway, will not materialize. The extent of nodal

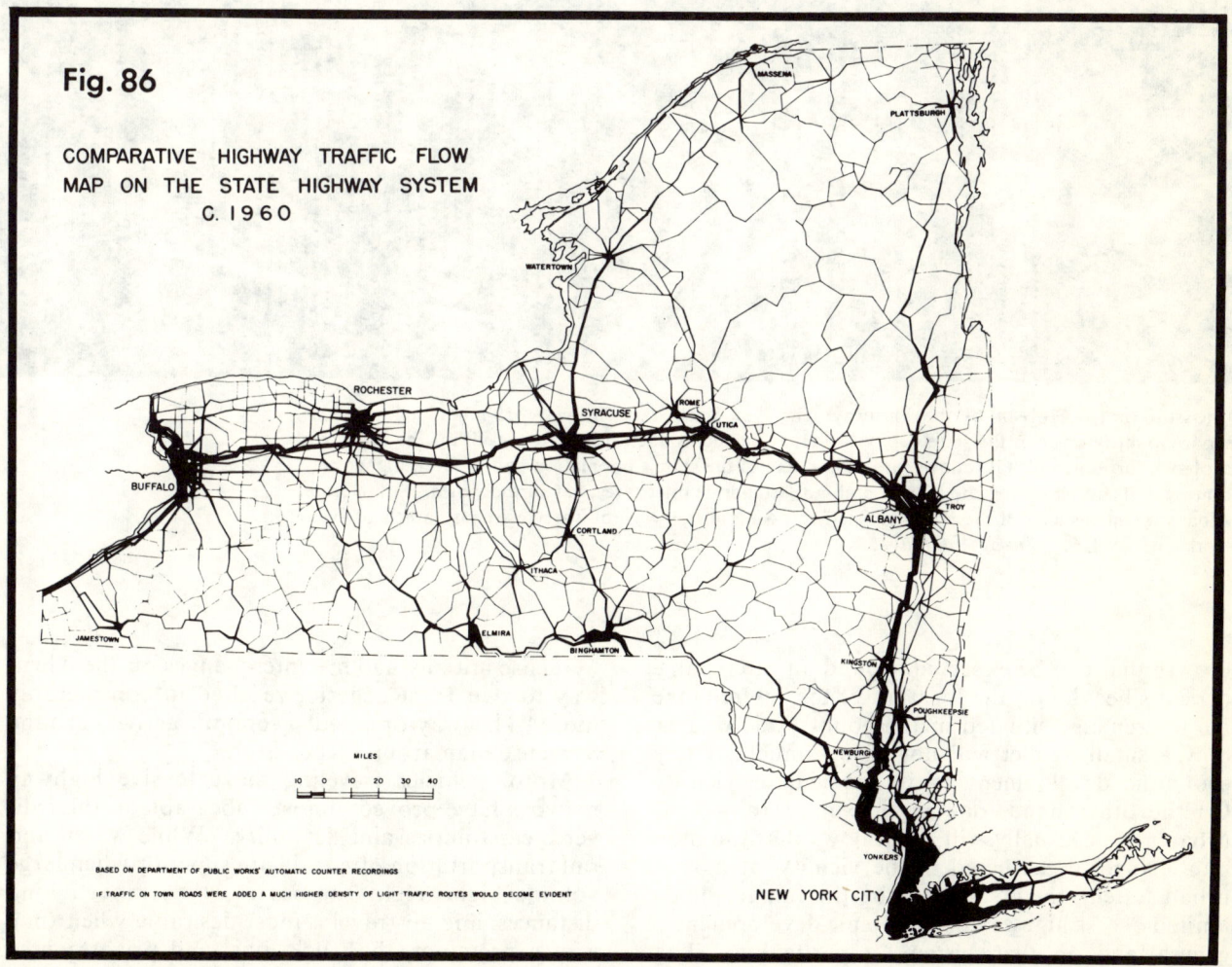

Fig. 86 COMPARATIVE HIGHWAY TRAFFIC FLOW MAP ON THE STATE HIGHWAY SYSTEM C. 1960

CROSSING OF THE HUDSON by the Thruway. The New York State Thruway, 559 miles long, provides service to thirty-seven of the state's sixty-two cities, including the nine largest. Within the Hudson–Mohawk corridor, which it follows, are 80 per cent of the state's population and an even higher percentage of its business. It is the longest and most elaborate toll expressway in the world. South from Albany it remains west of the Hudson, finally crossing in this view of Tarrytown just north of New York City. *N.Y.S. Thruway Authority.*

growth that can be expected around an interchange depends heavily on the location of the interchange. An interchange situated in the countryside or near only a small hamlet will not induce much, if any, economic development in its immediate vicinity. On the other hand, one in proximity to a large urban system usually will. It follows that the more interchanges constructed in the vicinity of a large urban system, the greater will be the impact of the limited-access highway on economic development in general and on that system in particular. That Syracuse initially had five interchanges on the Thruway to two for Rochester resulted in considerably more "Thruway-oriented" economic activity around Syracuse than around Rochester.

Motor vehicles traveling the extensive highway network have proved almost unbeatable in their diverse capabilities and flexibility. While water and rail transportation often still are superior when large volume–high weight cargoes must be moved long distances, and air travel is most desirable when time is at a premium, the truck, bus, and private automobile will probably continue to exhibit their diversity and flexibility by taking over increasing proportions of the total transportation business. Door-to-door pick-up and delivery, small lot handling, relatively good speed, lower crating and handling costs, and good roads are the advantages of the highway era in transportation.

Air Transportation

New York State makes extensive use of air transportation. More than 7.6 million passengers boarded scheduled airlines for domestic destinations in 1963 and another 1.8 million, mostly out of New York City, embarked for foreign destinations. In addition, there were probably about 45,000 non-scheduled domestic passengers and a substantial amount of air freight movement. Twenty communities provide scheduled air service, and over 300 additional landing strips are scattered throughout the state.

The high cost of time in the United States makes it sensible for men in responsible, high-paid positions to travel by air. Much of the air traffic is of this kind. The possibility of leaving Syracuse or Rochester in the morning, conducting business in New York City, Chicago, Washington, or Cleveland, and returning the same day is attractive to

THE MOHAWK TRANSPORTATION CORRIDOR (looking east). Between the villages of Mohawk (right) and Herkimer (left) the elements of this transportation corridor are clearly visible. The New York State Barge Canal in the center, and the Thruway, Route 5, and the New York Central Railroad on the left are most important. Secondary facilities south of the canal exist, too. The northern edge of the Appalachian Upland is clearly visible at upper right. Although on good transportation axes, small central places in the Middle Mohawk do not easily attract new business. Some good farm land occurs both in the valley and on the upland. *N.Y.S. Dept. of Commerce.*

many. Also, as more college students travel longer distances to attend college and as more people with adequate financial resources take vacations in distant places, perhaps renting a car after they get to their destination, air travel is bound to flourish. That over 1,000 seats for departing aircraft are sold in an average day in a city the size of Syracuse indicates the extent to which people have turned to air travel. It also helps explain the difficulties the railroads are having in attracting sufficient passenger business to stem the tide of service decline.

Origination of air passenger traffic is shown in Figure 87. As might be expected, it is the urban systems with the large tertiary surpluses which stand out as air traffic generators. Albany is not quite as significant as might be expected, and this can be explained by the fact that it is close enough to New York City and other eastern portions of the state for the private automobile to compete effectively with the airplane.

The overwhelming position of New York City with John F. Kennedy International, La Guardia, and Newark airports reflects both its traffic-generating capacity as a business and manufacturing center and its function as the principal origination and destination focus for overseas air travel to and from the East Coast.

Is there sufficient demand for air traffic at other

THE MOHAWK CORRIDOR near Fultonville (looking west). Here the principal elements of the corridor are all back in the narrow valley bottom. Note lock on the canalized Mohawk River and the tree-covered steep sides of the valley. *N.Y.S. Thruway Authority.*

LITTLE FALLS AND THE MOHAWK RIVER (looking east). Crowded into the narrowest part of the Mohawk Valley, Little Falls shares limited space along the canalized Mohawk River with roads and railroads. At this point the Thruway is several miles away in the hills to the south, its engineers finding little room or reason to keep it in the valley bottom. Old textile mill buildings, many now empty, line the waterfront. Considerable "ribbon" development has occurred between Route 5 and the New York Central Railroad tracks in the foreground. Tree-covered slopes on the left and right are very steep. *N.Y.S. Dept. of Commerce.*

points in the state to support scheduled service? A few additional commercial airports may be established in the years ahead, but it is doubtful that many more can be supported because ordinarily anything other than a fairly large urban place just cannot generate enough traffic demand to warrant establishment of commercial service. A recent survey indicated substantial traffic demand at a mid-Hudson location somewhere between Albany and Poughkeepsie. Money also is currently being spent on airport construction in Herkimer County and at Indian Lake in the Adirondacks. Air travel from most existing airports may be expected to rise faster than the population of those centers.

THE TRANSPORTATION FUNCTION OF
NEW YORK'S CITIES

The degree of transportation function of urban places may in part be evaluated by examining the size of the labor force engaged in transportation-oriented activities. The transportation labor force of one place can be compared with that of another or against a national norm in a fashion similar to that used above for identifying tertiary surpluses or deficits. The transportation surplus or deficit of a place may be expressed either in number of workers or, as the tertiary surplus has been expressed, in number of people these workers serve according to the national norm. Figure 88 shows both transportation potential and transportation surpluses and deficits. These and other data are shown in Table 30. New York City overshadows all other areas in the state. In fact, it has a transportation potential of nearly 10 million and a surplus of more than 8 million. This means it theoretically provides enough transport function for itself plus 8 million people over the state and nation. That New York City is so significant in transport function should not be

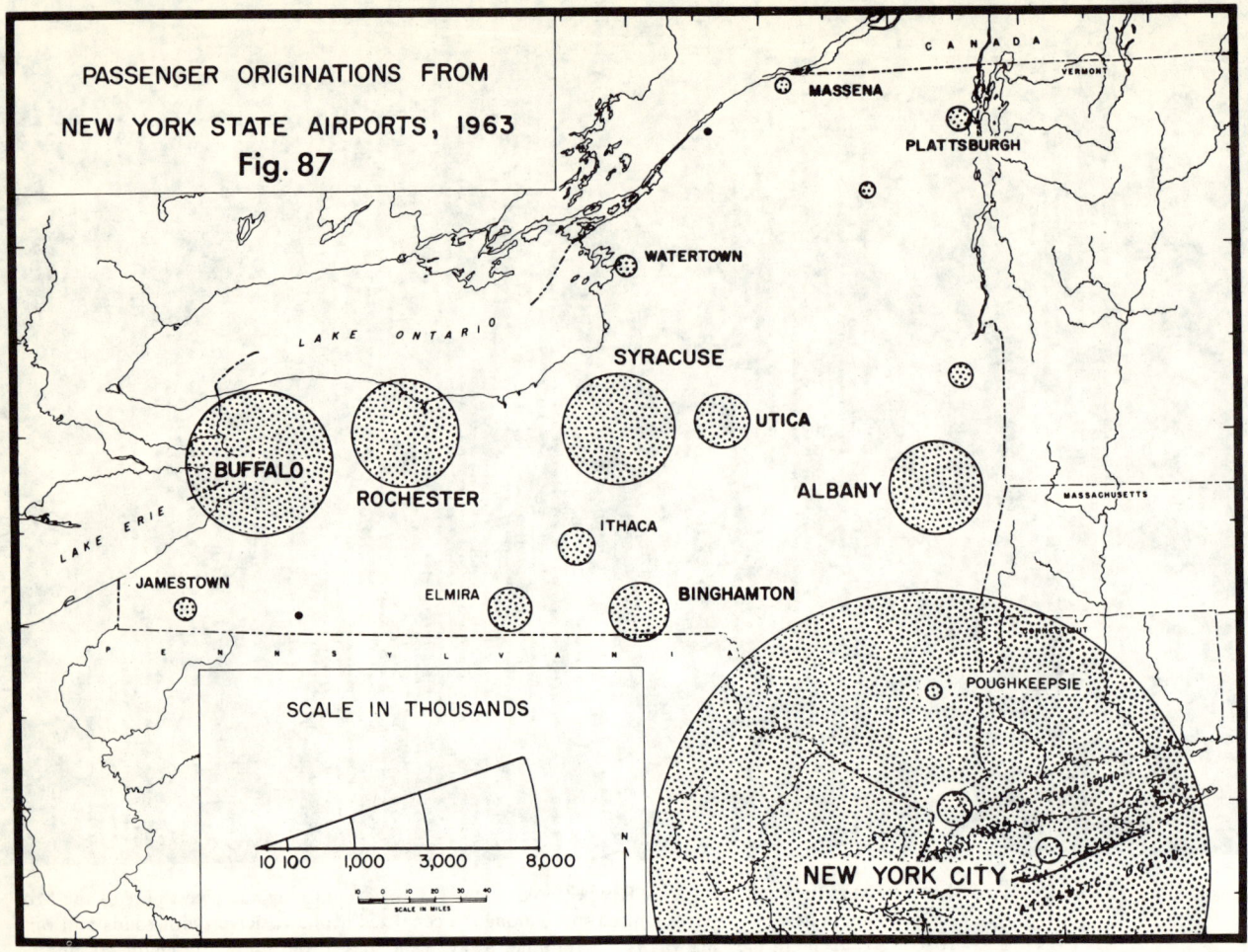

too much of a surprise. After all, it ranks very high as a rail and air traffic focus and is the world's greatest port. Passengers and freight passing through it are destined for the far corners of the earth. The more evidence brought out concerning this city's economic status, the more clear it becomes what a truly remarkable central function it has.

The state's second transport center is Buffalo, with a potential of 1.6 million and a surplus of 560,000. This is small by New York City standards but very large compared to any other central place in the state. Buffalo, situated on major rail, highway, and air lines at the eastern end of the Great Lakes and with access to the St. Lawrence Seaway, is the great western gateway to New York. Through it pour a large volume and wide range of products, as well as much passenger traffic from the interior of the continent. In it are produced manufactured goods that must be transported to markets around the nation.

Behind Buffalo in transportation surplus is Albany, with its favorable situation at the confluence of the Hudson River and the Mohawk route west. Syracuse, as a major upstate distribution center, is significant, too. Of considerable interest is Hornell's high ranking in transportation surplus. This is a case of a relatively small central place having concentrated in it a large transport function based in railroad activity. Of interest, too, is the fact that Rochester actually has a theoretical transportation deficit, suggesting that some substantial part of the transport service needed by that city is provided by people working elsewhere—perhaps in Buffalo, New York City, or Syracuse. Many of the smaller places along the Hudson–Mohawk–Ontario Lake Plain route exhibit deficits indicating the low affinity of smaller places for transport function even when on a major transport artery. This artery across the state has New York City, Albany, Syracuse, and Buffalo as its principal transport nerve centers.

Significant transportation patterns include individual lines and routeways, flow volumes, and activity nodes. The precise way in which these fit

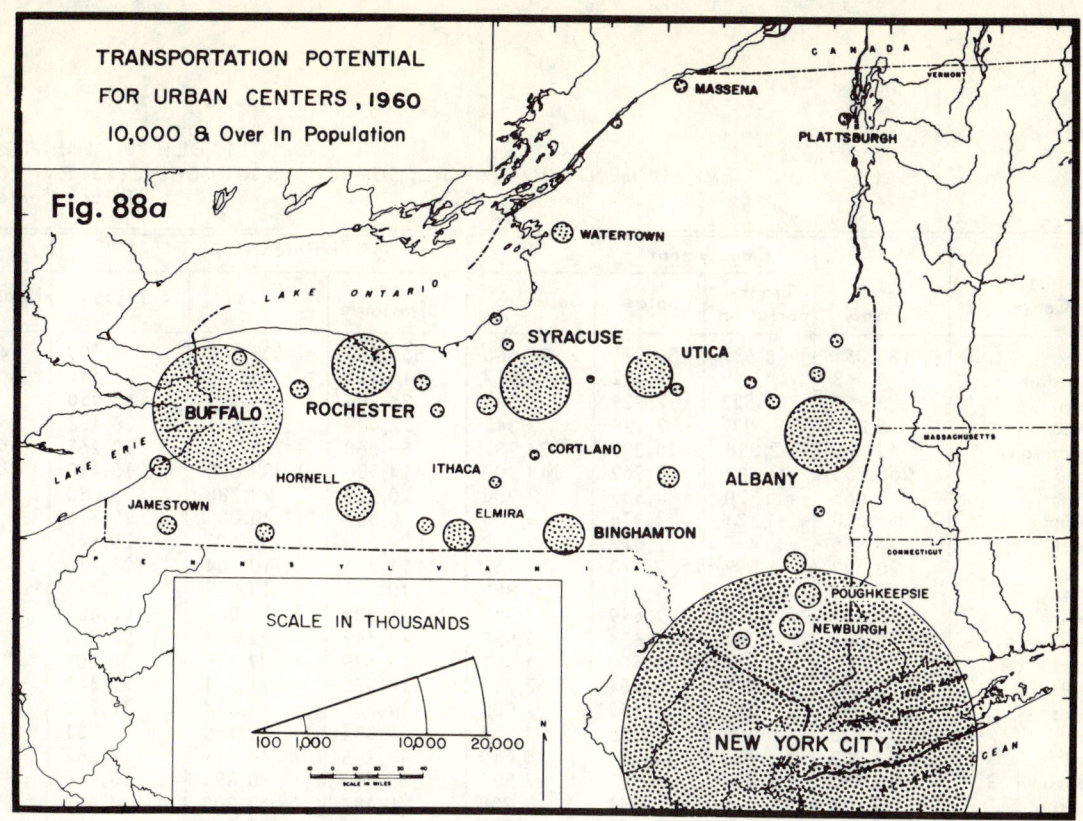

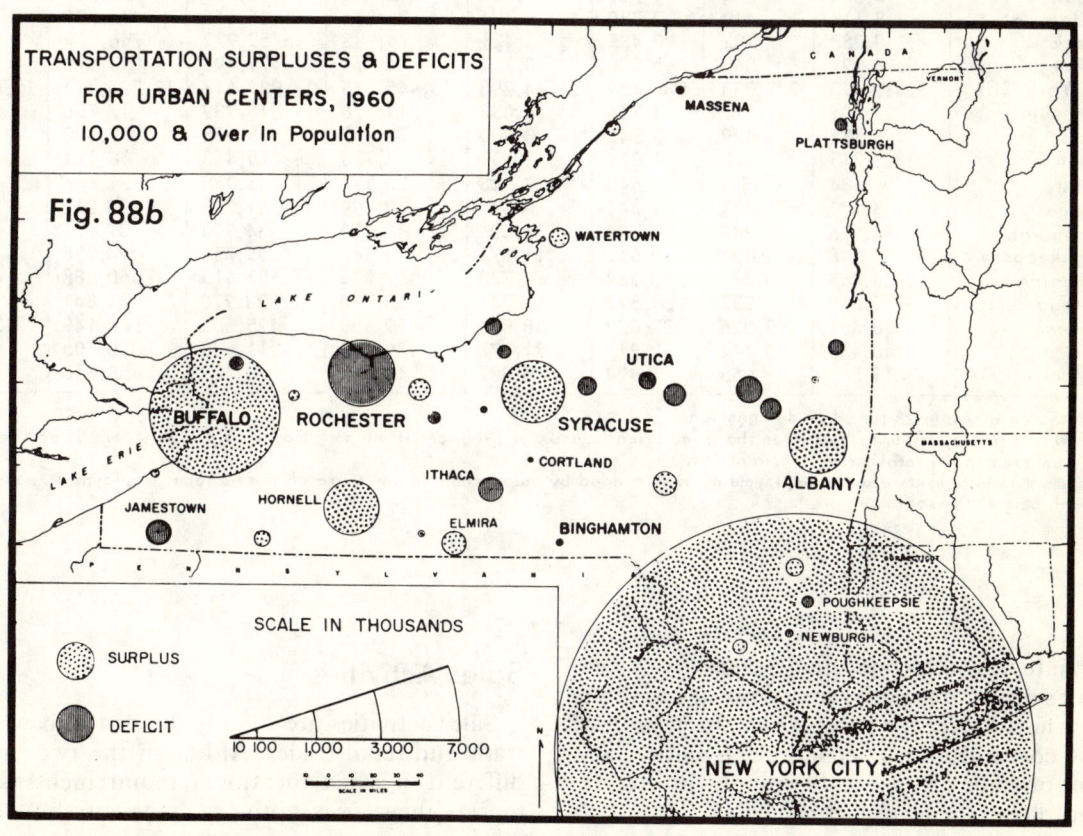

CHAPTER 12 THE TERTIARY SECTOR 275

TABLE 30. THE TERTIARY
EMPLOYMENT, POTENTIAL, SURPLUSES AND DEFICITS, EMPLOYMENT
(Urban Centers over

Urban Center*	Total Tertiary	Employment			Potential†			Population
		Transportation	Sales	Services	Transportation	Sales	Services	
1. Albany	133,051	8,632	39,819	84,600	564,533	605,249	719,100	455,447
2. Amsterdam	6,448	249	2,382	3,817	16,285	36,206	32,445	28,772
3. Auburn	8,752	523	2,929	5,300	34,204	44,521	45,050	35,249
4. Batavia	5,811	337	2,129	3,345	22,040	32,361	28,433	18,210
5. Binghamton	42,468	2,394	15,337	24,737	156,568	233,122	210,265	158,141
6. Buffalo	258,756	24,687	90,962	143,107	1,614,530	1,382,622	1,216,410	1,054,370
7. Corning	4,665	320	1,565	2,780	20,928	23,788	23,630	19,655
8. Cortland	5,610	131	1,976	3,503	8,567	30,035	29,776	22,803
9. Dunkirk	5,955	447	2,079	3,429	29,234	31,601	29,147	26,682
10. Elmira	20,822	1,397	7,270	12,155	91,364	110,504	103,318	73,870
11. Fulton	3,223	154	1,219	1,850	10,072	18,529	15,725	14,261
12. Geneva	5,204	210	1,649	3,345	13,734	25,065	28,433	17,286
13. Glens Falls	9,648	267	3,454	5,927	17,462	52,501	50,380	36,923
14. Gloversville	7,222	148	2,603	4,471	9,679	39,566	38,004	32,131
15. Hornell	5,913	1,749	1,404	2,760	114,385	21,341	23,460	13,907
16. Hudson	3,365	187	1,143	2,035	12,230	17,374	17,298	11,075
17. Ilion	4,938	197	1,655	3,086	12,884	25,156	26,231	27,000
18. Ithaca	17,703	188	2,896	14,619	12,295	44,019	124,262	31,587
19. Jamestown	13,332	487	5,322	7,523	31,850	80,894	63,946	49,495
20. Kingston	9,298	619	3,441	5,238	40,483	52,303	44,523	31,882
21. Lockport	7,029	320	2,450	4,259	20,928	37,240	36,202	26,443
22. Massena	4,304	204	1,292	2,808	13,342	19,638	23,868	15,478
23. Middletown	8,114	410	2,330	5,374	26,814	35,416	45,679	23,475
24. Newark	9,288	382	3,485	5,421	24,983	52,972	46,079	12,868
25. Newburgh	16,821	862	5,230	10,729	56,375	79,496	91,197	54,551
26. New York City	3,412,490	282,038	986,159	2,144,293	18,445,285	14,989,617	18,226,491	10,694,633
27. Ogdensburg	3,506	177	1,246	2,083	11,576	18,939	17,706	16,122
28. Olean	6,771	429	2,542	3,800	28,057	38,638	32,300	21,868
29. Oneida	3,215	70	1,014	2,131	4,578	15,413	18,114	14,599
30. Oneonta	4,684	513	1,648	2,523	33,550	25,050	21,446	13,412
31. Oswego	4,971	211	1,441	3,319	13,799	21,903	28,212	22,155
32. Plattsburgh	6,976	248	2,302	4,426	16,219	34,990	37,621	20,172
33. Poughkeepsie	17,548	831	5,642	11,075	54,347	85,758	94,138	58,204
34. Rochester	126,465	5,473	43,264	77,728	357,934	657,613	660,688	493,402
35. Saratoga Springs	5,610	287	1,577	3,746	18,770	23,970	31,841	16,630
36. Syracuse	98,491	7,026	32,624	58,841	459,500	495,885	500,149	333,286
37. Utica	48,438	2,744	13,917	31,777	179,458	211,538	270,105	187,779
38. Watertown	11,549	668	3,994	6,887	43,687	60,709	58,540	33,306

*See footnotes in Table 28 for urban designations.
†The national ratios, or norms, based on the Population Census of 1960, are: transportation 1:65.4, sales 1:15.2, and services 1:8.5, as compared to the total tertiary ratio of 1:5.
‡Employment requirements are the employment sizes needed by each urban center to match the national employment/population ratio in each part of the sector.

together into an interconnecting transport system contributes heavily to spatial variations in economic success or health over the state. Thus, the growth of cities, the concentration of economic activities, and the extent of trade areas are all strongly affected by the geography of the transport system.

Sales Activities

Sales activities are usually divided into wholesale trade and retail trade. Although the two are quite different in exact locational requirements and the market they serve, both are active city-builders and

SECTOR DISSECTED:
REQUIREMENTS OF POPULATION, AND EMPLOYMENT SURPLUSES AND DEFICITS
10,000 in Population, 1960)

Surpluses or Deficits			Employment Requirements‡			Employment Surpluses or Deficits§			
Transportation	Sales	Services	Transportation	Sales	Services	Transportation	Sales	Services	
109,086	149,802	263,653	4,964	29,964	53,582	3,668	9,855	31,018	1
− 12,487	7,434	3,673	440	1,893	3,385	− 191	489	432	2
− 1,045	9,272	9,801	539	2,319	4,147	− 16	610	1,153	3
3,830	14,151	10,223	278	1,198	2,142	59	931	1,203	4
− 1,573	74,981	52,124	2,418	10,404	18,605	− 24	4,933	6,132	5
560,160	328,252	162,040	16,122	69,366	124,044	8,565	21,596	19,063	6
1,273	4,133	3,975	301	1,293	2,312	19	272	468	7
− 14,236	7,232	6,973	349	1,500	2,683	− 218	476	820	8
2,552	4,919	2,465	408	1,755	3,139	39	324	290	9
17,494	36,634	29,448	1,130	4,860	8,691	267	2,410	3,464	10
− 4,189	4,268	1,464	218	938	1,678	− 64	281	172	11
− 3,552	7,779	11,147	264	1,137	2,034	− 54	512	1,311	12
− 19,461	15,578	13,457	565	2,429	4,344	− 298	1,025	1,583	13
− 22,452	7,435	5,873	491	2,114	3,780	− 343	489	691	14
100,478	7,434	9,553	213	915	1,636	1,536	489	1,124	15
1,155	6,299	6,223	169	729	1,303	18	414	732	16
− 14,116	−1,844	−769	413	1,776	3,176	− 216	− 121	− 90	17
− 19,292	12,432	92,675	483	2,078	3,716	− 295	818	10,903	18
− 17,645	31,399	14,451	757	3,256	5,823	− 270	2,066	1,700	19
8,601	20,421	12,641	487	2,098	3,751	132	1,343	1,487	20
− 5,515	10,797	9,759	404	1,740	3,111	− 84	710	1,148	21
− 2,136	4,160	8,382	237	1,018	1,821	− 33	274	987	22
3,339	11,941	22,204	359	1,544	2,762	51	786	2,612	23
12,115	40,104	33,211	197	847	1,514	185	2,638	3,907	24
1,824	24,945	36,646	834	3,589	6,418	28	1,641	4,311	25
7,750,652	4,294,984	7,531,858	160,987	692,666	1,238,649	127,151	269,349	860,693	26
− 4,546	2,817	1,584	247	1,061	1,897	− 70	185	186	27
6,189	16,770	10,432	334	1,439	2,573	95	1,103	1,227	28
− 10,021	814	3,515	223	960	1,718	− 153	54	413	29
20,138	11,638	8,034	205	882	1,578	312	766	945	30
− 8,356	− 252	6,057	339	1,458	2,606	− 128	− 17	713	31
− 3,953	14,818	17,449	308	1,321	2,373	− 60	981	2,053	32
− 3,857	27,554	35,934	890	3,829	6,848	− 59	1,813	4,227	33
−135,468	164,211	167,286	7,544	32,461	58,047	−2,071	10,803	19,681	34
2,140	7,340	15,211	254	1,094	1,956	33	483	1,790	35
126,214	162,599	166,863	5,096	21,927	39,210	1,930	10,697	19,631	36
− 8,321	23,759	82,326	2,871	12,354	22,092	− 127	1,563	9,685	37
10,381	27,403	25,234	509	2,191	3,918	159	1,803	2,969	38

§Employment surpluses and deficits differ from regular surpluses and deficits in that the former measure the number of workers available to serve outsiders while the latter are the number of outsiders who can be served.
Source: Compiled from data in U.S. Census of Business, 1958, and U.S. Census of Population, 1960.

seem to do especially well in the large urban centers. Both are extensive users of transportation and contribute much to determining the areal reach or trade area of a central place. About 19 per cent of the state's labor force and 17 per cent of the GSP are attributable to sales activities.

WHOLESALE TRADE

New York's role as the nation's leading distribution center is verified by a wide variety of census data, and wholesale trade is no exception. In 1958 more than 40,000 wholesale trade establishments registered 55 billion dollars in sales. This volume

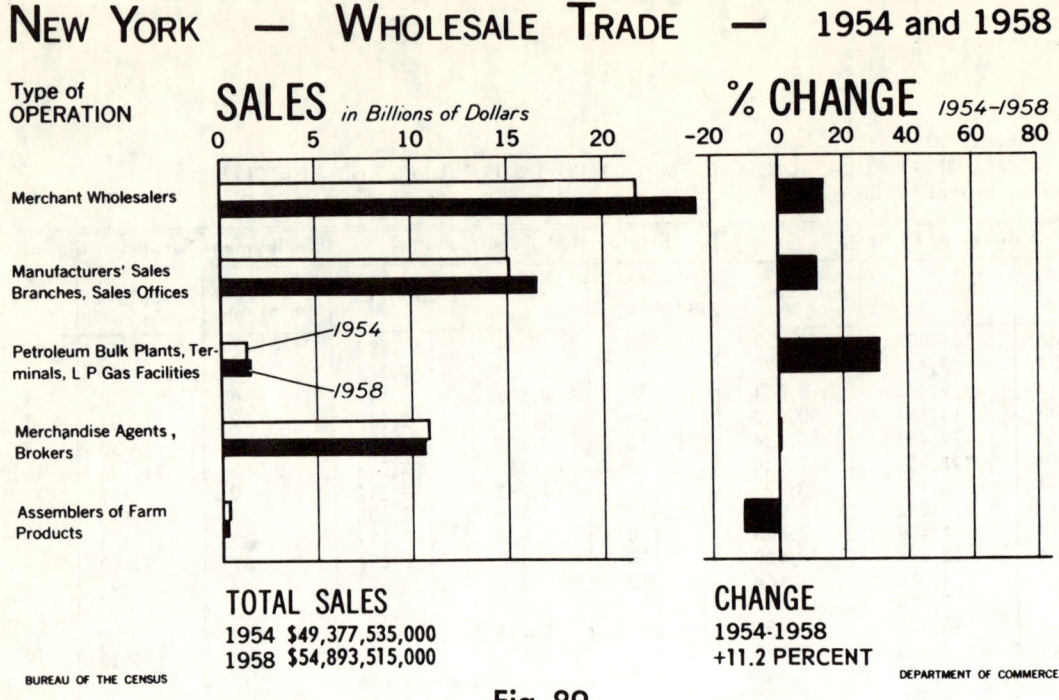

Fig. 89

was more than double that of second-ranking California and well over three times that of nearby Pennsylvania. Over 415,000 persons were employed in the same year. The census identifies the five major types of wholesale operations shown in Figure 89. By far the most important of these is merchant wholesalers, which account for 45 per cent of the sales volume and two-thirds of the employment. Merchant wholesalers are those who purchase, take title to, and, where customary, store and handle goods made by others. They sell this merchandise chiefly to retailers or to industrial, institutional, or commercial users. Included are distributors, exporters, importers, cash-and-carry wholesalers, and warehouses.

The geography of wholesaling shows an overwhelming concentration in metropolitan New York, where between 80 and 90 per cent of the state's wholesaling is carried on. Almost 70 per cent occurs in Manhattan alone. No equal concentration can be found in any other urban area. Much of this is oriented toward overseas markets. Upstate, Buffalo is most significant, followed in order by Syracuse, Rochester, the Albany area, Utica, and Binghamton. Because of its very nature, wholesaling gravitates toward the larger urban systems. A wholesaler or warehouse operator will want to be situated where he can serve effectively and competitively the maximum retail market, other wholesalers, industrial markets, and the like. Quick, cheap access to these markets is important. An isolated location in a small town ordinarily will not permit this quick, cheap access except perhaps to the limited number of smaller retailers in that town. So, once again, as the big city becomes bigger, its relative attractiveness to business such as wholesaling increases. Smaller, more sparse rural and village retail markets do have to be served, but this is usually done by regional wholesale centers with distribution radii of perhaps several hundred miles. It is in this particular capacity that Buffalo, Syracuse, and the Albany area function for Upstate.

Retail Trade

Over 183,000 retail establishments registered nearly 21 billion dollars in sales in New York in 1958. Eight hundred thousand persons were employed in the same year. The census identifies eleven major types of retail operations, which are shown in Figure 90. Food stores account for well over one-fourth of the retail sales. Automotive dealers and eating and drinking places follow in importance. With the rapid increase in number of motor vehicles, it is not surprising that among the eleven types of retail sales, gasoline service stations enjoyed the greatest increase in business between 1954 and 1958.

Retailing spatially correlates with population more nearly than does wholesaling and thus exhibits less concentration in New York City and somewhat more in most other parts of the state. For example, metropolitan New York accounts for 65 per cent of the retail business in the state but 87 per cent of the wholesale business. Buffalo has 7.3 per cent of the retail trade as compared to 4.4 per cent of the wholesale trade. Syracuse generates 3.2 per cent of the state's retailing trade but only 2.2 per cent of the wholesaling. The above areas have 64, 7.8, and 3.4 per cent of the population respectively. A rural area such as Yates County has .0009 per cent of the state's retail trade, .0003 per cent of the wholesale trade, and .001 per cent of the population. It may be fairly safely assumed that the future geography of retailing and population will be closely related.

Every central place, from the small hamlet to the large city, has its retail establishments. As a result of surrounding populations coming to trade at these establishments, patterns of tributary trade areas and divides between trade areas evolve. The "pull" of a retail complex, i.e., the distance people will travel to take advantage of it, depends on its diversity and size, the degree of access to it, and competition from other retail complexes. All but two (Ilion and Oswego) of the central places in the State with populations over 10,000 exhibit a surplus in over-all

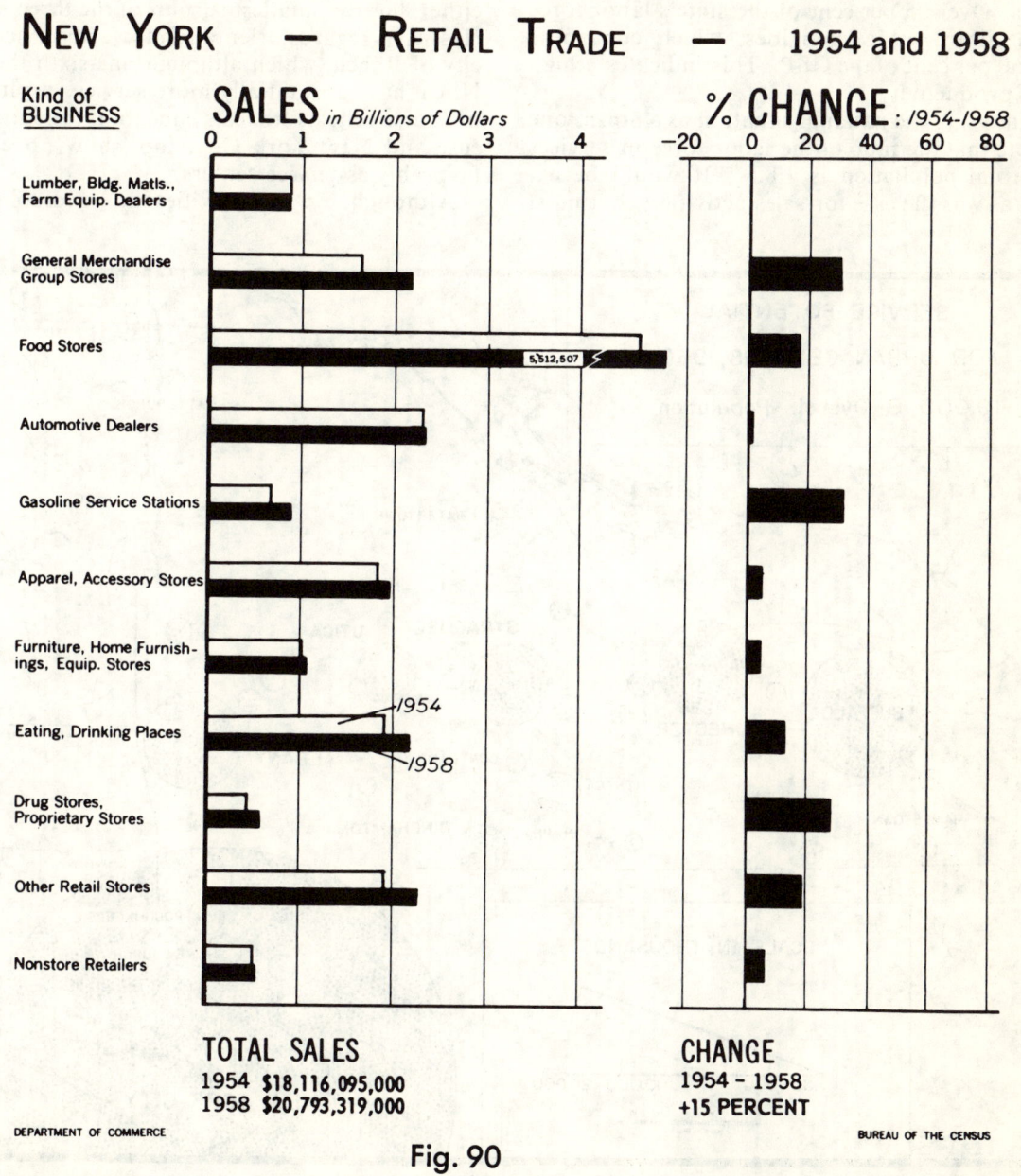

Fig. 90

CHAPTER 12 THE TERTIARY SECTOR 279

sales function when compared to the national norm. Because of the tie of sales to population and central places, this is not surprising.

Service Activities

Service activities include finance, insurance and real estate, business and repair services, personal services, entertainment and recreation, professional and related services, and public administration. That each person ordinarily demands more in the way of these things as society develops and becomes more complex is undeniable. That most of the service activities promote city development is fairly obvious. Over 35 per cent of the state's labor force is engaged in service activities, which contribute nearly 50 per cent of the GSP. This indicates a high level of productivity.

Again, using the national situation as a norm, one may note that the ratio of the labor force in services to the total population is 1:8.5. It would be expected, as was the case for sales activities, that most central places would show a surplus in service function, the surplus services being required by the people living outside the central places. Figure 91b, shows this to be the case. While large centers generally have more service potential and service surplus than small centers, there is much less correlation between service and population than there is between retailing and population. Concentrations of educational institutions, government services, or hospital facilities with wide reaches beyond the boundaries of certain central places will result in very large service surpluses for these places.

For example, Albany, with its important function as state capital, is smaller than Rochester but has a much greater service function. Buffalo, larger than either, has the smallest surplus of the three. Perhaps classic, as regards orientation toward services, is the city of Ithaca, which although one-sixth the size of Utica, has substantially more service surplus. This of course reflects Ithaca's education function. Syracuse and New York City, too, show up especially favorably as service centers.

Although service activities do tend to gravitate

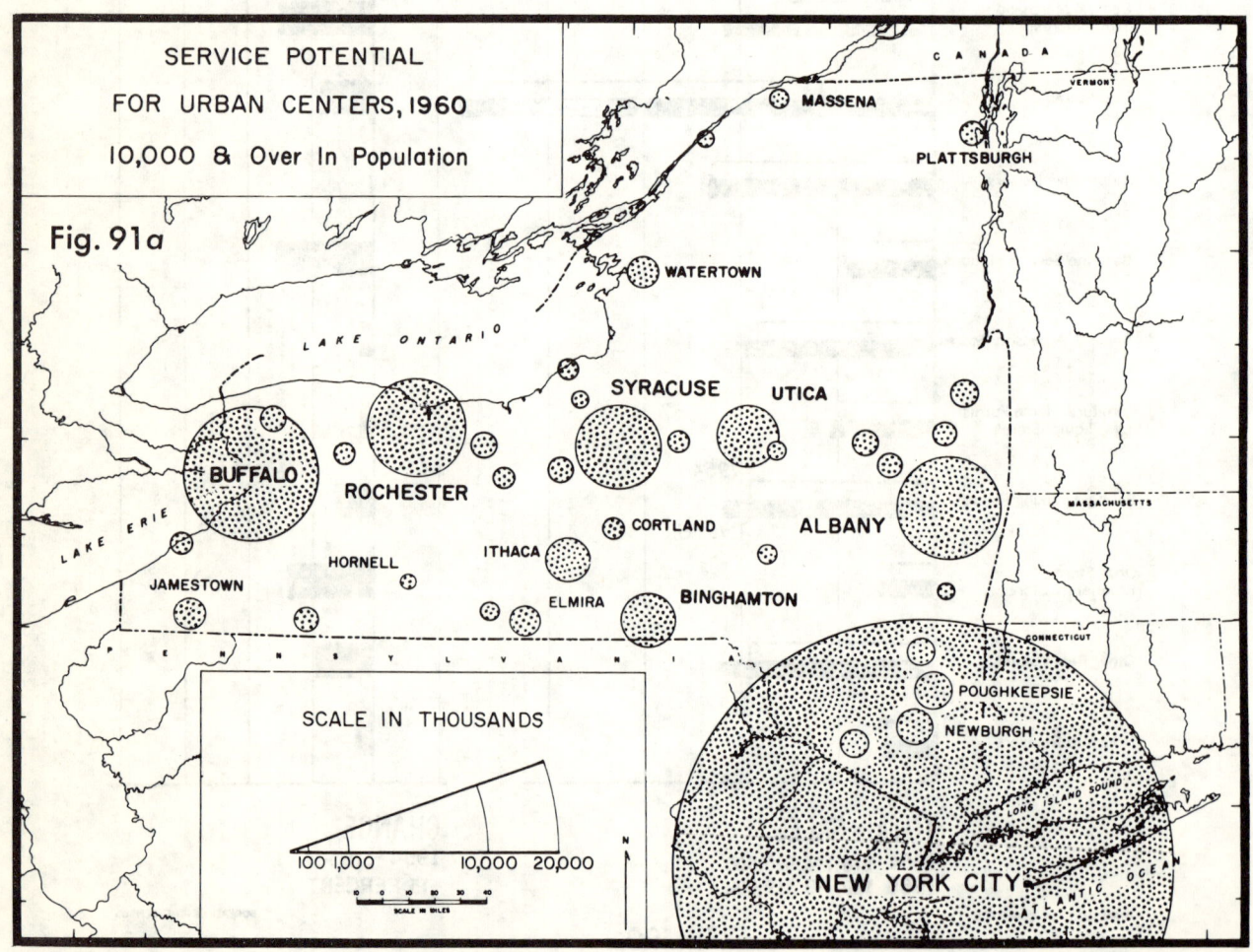

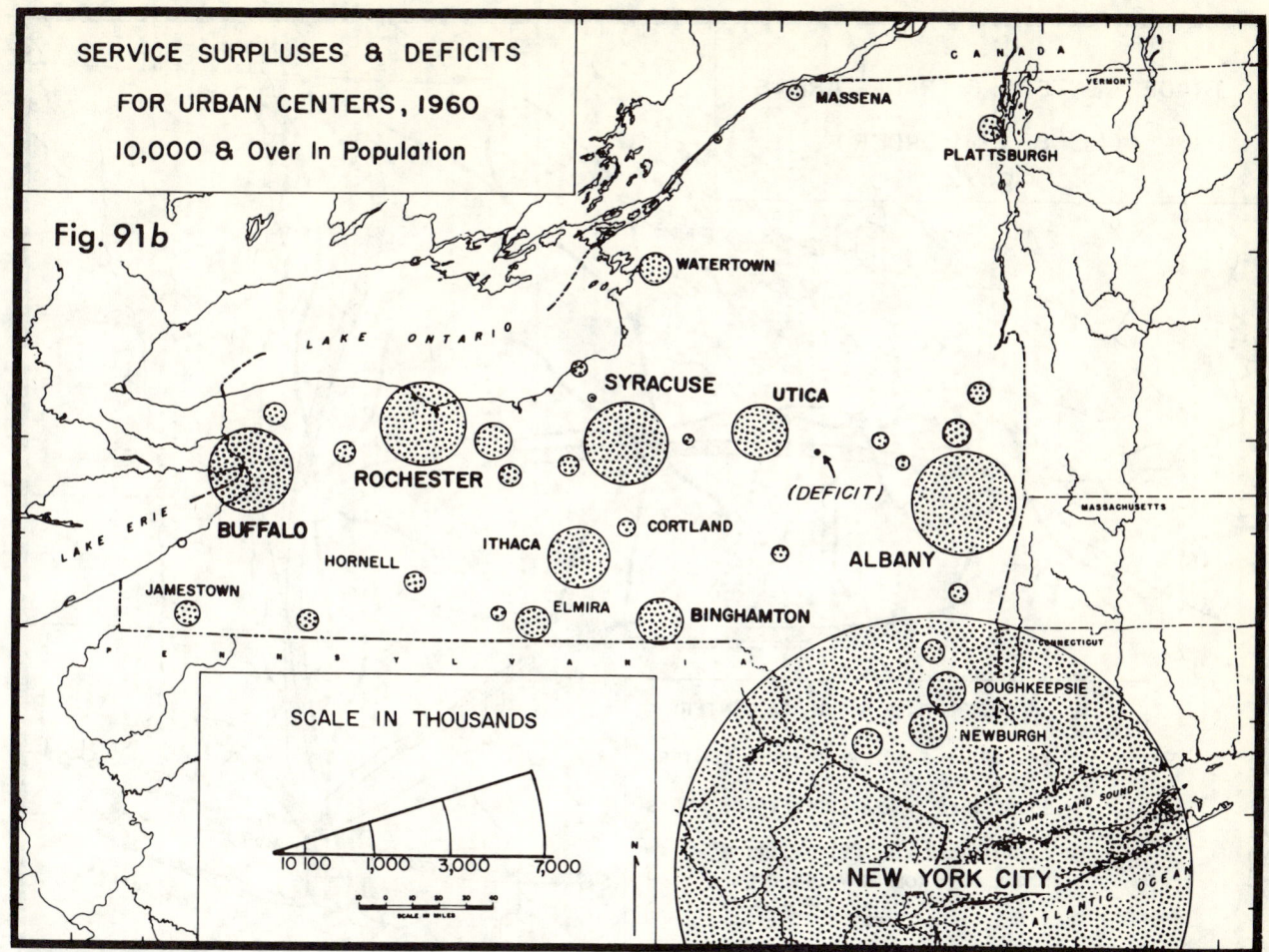

toward major urban systems and even though it can be expected that these major systems will attract the lion's share of expansion in the service function in the state, it probably is safe to say that smaller central places have a better chance of competing today with larger ones in certain of the service categories than in any other type of economic activity. If the economic health of central places is to be maintained, special attention by development groups might be directed toward identifying those service enterprises which have the highest affinity for smaller cities and encouraging their establishment in the smaller places.

Trade and Service Centers and Trade Areas

Most of what has been said in this chapter points to the fact that cities and villages function as nerve centers or focal points for a trade and service system. They are in one sense a concentration of trade and service facilities to which surrounding populations are attracted. Together with their pattern of tributary areas they form a connectivity "fabric" and if mapped exhibit much about the areal functional organization of a given region.

ORDERS

Different orders of trade centers and trade areas may be identified. New York City and its trade area is of a very high order, while a crossroads hamlet is of a very low order. At least five orders may be identified in New York State. Each order, except for the transitional third order, completely blankets the state with certain trade and service functions. Any center of a given order (e.g., second order) functions also as a lower-order center (e.g., as a third-, fourth-, and fifth-order center). New York City, being of great size and generally recognized as the primate city of the United States, is clearly of the *first order*. Its trade area extends over the entire state and even over the nation. It has certain business, cultural, and entertainment functions not precisely duplicated elsewhere and thus draws a certain trade and service clientele from all parts of the state. All people within a first-order trade area, however, can not be expected to make use of the highly specialized trade and service function.

CHAPTER 12   THE TERTIARY SECTOR

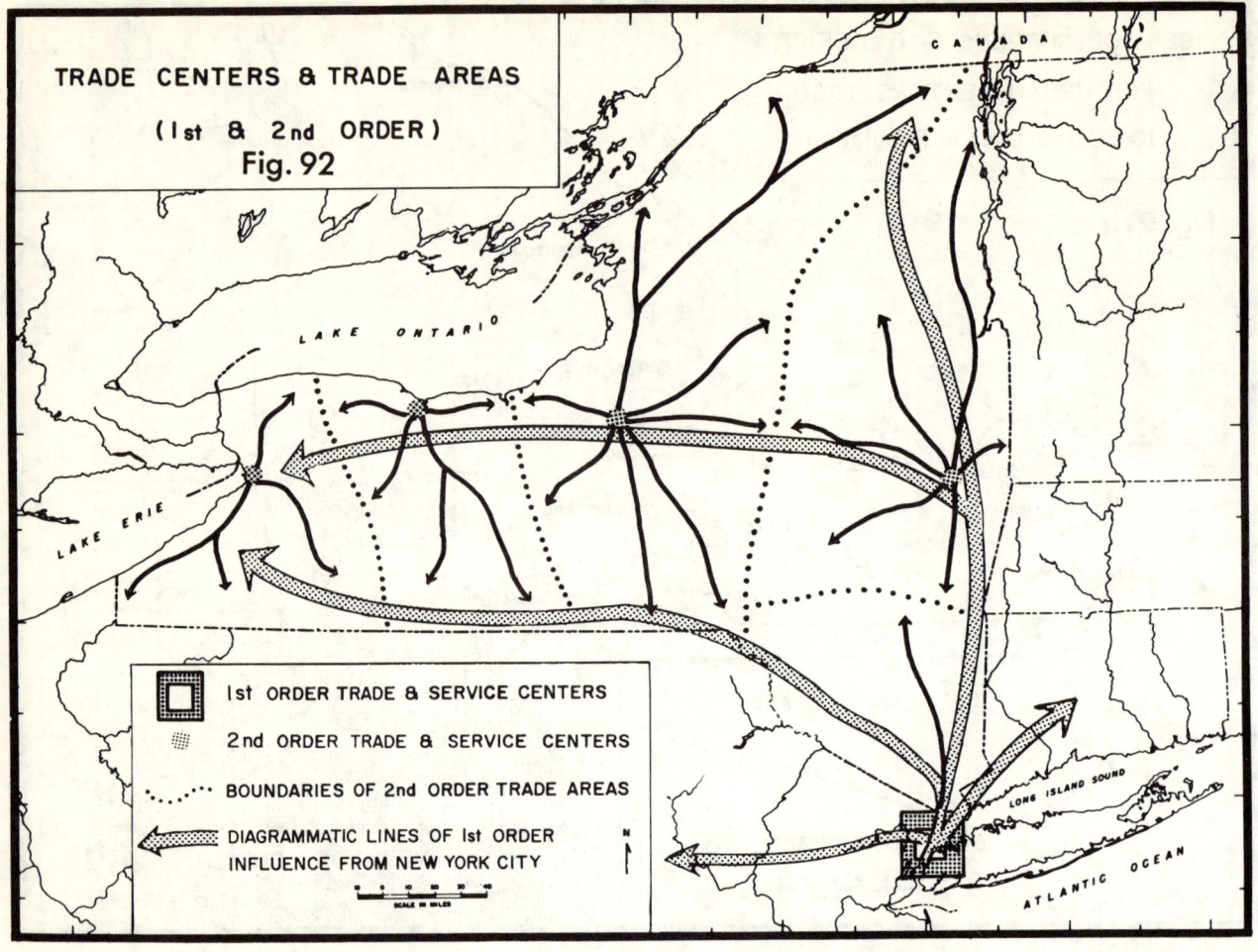

Fig. 92. TRADE CENTERS & TRADE AREAS (1st & 2nd ORDER)

Legend:
- 1st ORDER TRADE & SERVICE CENTERS
- 2nd ORDER TRADE & SERVICE CENTERS
- BOUNDARIES OF 2nd ORDER TRADE AREAS
- DIAGRAMMATIC LINES OF 1st ORDER INFLUENCE FROM NEW YORK CITY

Buffalo, Rochester, Syracuse, and the Albany area must be categorized as second-order trade centers with second-order trade areas. New York City, although of first order, of course has second-order functions as well. Together these five places extend their trade and service to "reach" all parts of the state, and even beyond. By definition a *second-order* trade center is one that provides specialized trade and service functions to essentially all people. This statement suggests that everyone in New York State at some time (usually as a minimum several times a year) turns to the second-order centers for purchases of items and services of special kinds that are not ordinarily found in lower-order centers. Involved may be things and services ranging from certain kinds of machinery to the latest style of clothing, from repair of a complicated instrument to receiving specialized hospital care, from attending a zoo to enjoying a big-time sporting event. Second-order places commonly function as regional wholesaling and warehousing centers.

Figure 92, based on statistical data, sample interviewing, and consultation with the experts who prepared succeeding chapters on the large urban systems, shows the approximate trade areas of the second-order centers and diagrammatically illustrates lines of trade and service influence radiating from these centers and from first-order New York City. The Albany area and New York City provide second-order functions to the eastern part of the state; Syracuse, to much of the central and northern areas; and Rochester and Buffalo, to western New York. It must be emphasized that for lack of detailed information the boundaries of the second-order trade areas should be considered as being only in approximately the right place. A farmer or even a community on one side of the Syracuse–Rochester boundary line, for example, might properly be on the other side of the line. In fact, these boundaries at best are zones rather than precise lines. Also, it is quite likely that the position of the trade area boundaries can be expected to move from time to time in response to improved or deteriorating trade and service functions in one of the centers or as a result of changing quality of access.

A number of readers from the Utica and Binghamton areas are going to question the omission of their communities from second-order designation. When research on this particular subject was first begun, it seemed apparent that they were somewhat smaller, but of the same general order as Syracuse. As investigations proceeded, however, it became increasingly clear that they were of a different order. Substantial numbers of people who live in these two cities, for example, depend on Syracuse and the Albany area for second-order trade and service functions. Data on tertiary potentials and surpluses place Utica and Binghamton in a different category than Syracuse (see earlier figures in this chapter). Are these places on their way up to becoming second-order places, or were they perhaps second-order places some years ago and now on their way down to a lower order? The answer to this question is tied up in many things including greater mobility of the population and differential growth rates of the tertiary sector. It seems probable, however, that instead of moving closer to second-order designations, Utica and Binghamton are "suffering" from improved transportation and more rapid expansion of trade center attractions in the second-order places and are moving further away from second-order character. In any case, a special *third-order* category is reserved for these cities. They do not have the specialized draw of second-order places, but they have more than the day-to-day trade and service functions of fourth-order places. They, together with perhaps Watertown and one or two other places which are not too different in their tertiary functions, do not produce a categorical cover or blanket for the state as do the places of the other orders.

Fourth-order trade centers and areas are those which provide all of the day-to-day trade and service needed by a population. They have good stores but not the array of products for sale found in retail establishments of second-order places. They have some wholesaling, but it is quite limited, and second-order places must be turned to for considerable amounts of this function. They have doctors and hospitals, but services of outstanding specialists and large first-class hospitals ordinarily must be sought in the second-order places. They have much

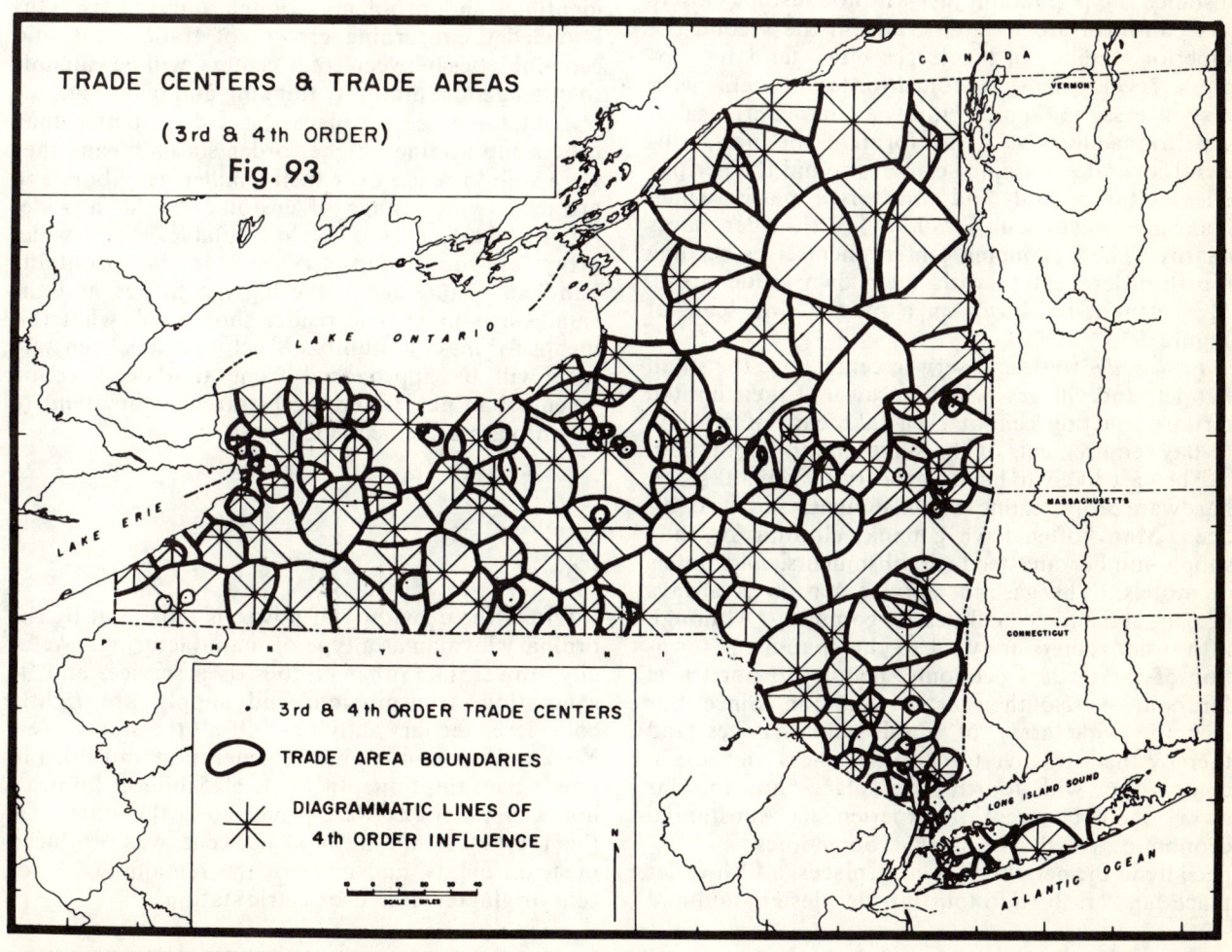

in the way of repair services, but for special problems and special parts the customer has to go elsewhere. They have good transportation and some warehousing but do not usually function as regional distribution centers. These are the central places ranging in population from perhaps a thousand or two on the small side to several tens of thousands in the upper range. It should be repeated that higher-order places, or even peripheral shopping centers within them, perform fourth-order functions for their immediate surroundings.

Figure 93 exhibits the third- and fourth-order centers and their theoretical trade areas. The boundaries of the trade areas were determined mathematically, using retail functions and distance relationships between centers. This technique had proved useful in other areas and without doubt roughly approximates the real situation in New York. Actual boundaries could be identified only by time-consuming field studies. It is noticeable that fourth-order trade areas are generally larger where: (1) populations are extremely sparse and (2) where second- or third-order places are really providing the fourth-order function. In the first instance the population just will not sustain closely spaced fourth-order centers, and in the second the superior facilities of the larger cities tend to suppress development of fourth-order centers with easy access to those cities. Characteristically, superior facilities take the form of the new, efficiently operated shopping centers around the peripheries of the second- and third-order places which make growing difficult for older, fourth-order places nearby. These shopping centers themselves become fourth-order centers with their own trade areas. Most cannot be shown on a map of the scale of Figure 93.

Fifth-order trade and service centers are the small hamlets and villages, which because of their limited tertiary function cannot even take care of all day-to-day requirements of the surrounding population.

They are likely to have a grocery or general store, hardware store, eating establishment, and post office. Many often have a bank, clothing or shoe shops, lumber and feed establishments, and hotels or motels. The gasoline service station is always present and garage facilities are common. Although fifth-order centers are vital to the spatial organization of the tertiary economy, people do not travel far to make use of their tertiary function. Since they lack the wide array of goods and services and thereby the attractive force for business and are in competition with fourth-order places not too far away, their chances of experiencing substantial economic growth are small. Sometimes, and especially in the periphery of large places, a fifth-order place can "crash" into fourth-order designation and greatly extend its trade area by adding a wide range of new retail facilities. Many of the larger suburban shopping centers have just this experience.

A Program for Local Study

Whatever the order of a trade center and its trade area, knowing about its characteristics means being better able to solve local problems and plan for the future.

To accumulate information concerning fourth- and fifth-order places and their trade areas is a practical and useful project for a community study group to undertake. Identifying such things as the extent of rural mail delivery and school districts, talking to store-owners and bankers about the area they serve, making sample checks of trade habits by asking people where they do their everyday trading and many other endeavors would provide contributory information. As each bit of the information becomes available, it should be recorded on maps and used in overlay analysis (see Introduction). Eventually the details of a trade area and the special strengths and weaknesses of a trade center can be identified and problems pointed out. This, plus knowledge concerning growth of trade areas and economic ties between trade centers, will be valuable to intelligent community thinking and planning.

That some central places have done better than others and attained higher-order status means they were able to reach over their smaller neighbors and attract a wider range of customers. As a wider range of customers is made available, a still wider array of goods and services can be profitably handled. Thus, again, the big get bigger and the small stay small. The reader should ask what this means for his community. Much that has been said here will be approached from another direction when urban places and the hierarchies they tend to fall into are discussed in Chapter 14.

Power

Often, the providing of power is looked at by the ordinary layman as a type of manufacturing. Actually, power, like other utilities, is a service, and its generation, transmission, and supply are rightly considered tertiary activities. Of all the states, New York ranks second in power generation and third in power consumption. In 1964, 68.5 billion kilowatt hours of electricity were produced in the state. Of this total 50.5 billion, or 74 per cent, was produced in steam plants, and most of the remaining 26 per cent originated at hydroelectric stations.

It may be said that in the mid-twentieth century there are three main types of power-generating regions: resource regions, high-demand regions, and low-demand regions. In the case of resource regions local advantages in water power or fuel are so great that power is generated whether or not a demand exists. Surpluses are transmitted to distant demand points. High-demand regions may have no particular resource advantages, but huge demands result in local generation nevertheless, usually with imported fossil fuels. Low-demand regions have neither the resources nor the demand to justify a great deal of local generating capacity. As a result they are likely to be supplied by smaller local plants or nearby resource regions. In addition to power-generating regions there are service regions, that is, regions serviced by single suppliers.

HISTORICAL DEVELOPMENT

Early gristmills and sawmills located at potential power sites on streams and used direct water power. Later steam engines provided direct power, too. Conversion of water and steam power into electricity and the transmission of this electricity through wires was technologically a great step forward. In essence, power-users were released from the previous geographic control of the power-producing site.

Developments in the field of electric power transmission progressed to the point where in 1896 the mayor of Buffalo closed a switch, and the streets of the city were lighted for the first time with electricity generated by the waterfall of the Niagara River. Great utility companies evolved during the next forty years, and with the encouragement of President Hoover and the establishment of the Rural Electrification Administration (REA), lines were extended to almost every potential user in the state. Large producers and long transmission distances became common. Today seven major corporations generate about 75 per cent of the electrical power in the state, and another 23 per cent comes from installations owned by the Power Authority of the State of New York (PASNY). (See Fig. 94.) Power can be transmitted 400 miles or more if need be, a transmission distance equal to that between Niagara Falls and New York City. Generating stations and the hundreds of miles of transmission lines are coordinated into a vast integrated power system called a *grid*, which serves all of New York State as well as areas beyond its borders. About 160 individual generating plants and all three types of power regions are involved. The expressed purpose of the grid is to satisfy demand at all times and in all places within its jurisdiction regardless of demand levels at any time and in any place. So completely integrated is the whole grid that it is virtually impossible to identify the exact source of any given kilowatt of power.

THE SERVICE REGIONS

The geography of power cannot be grasped without examining the areas served by the major electric utility companies. Figure 94 shows that four utilities serve most of Upstate New York. Of these Niagara Mohawk Power Corporation's territory is the largest, extending in a broken fashion from the westernmost portion of the state to the eastern border. Rochester Gas and Electric Corporation serves Rochester and portions of surrounding counties as well as smaller areas in the Cattaraugus Hills to the southwest. A major part of the Appalachian Upland and a small area of northeastern New York get their power from New York State Electric and Gas Corporation, while the Mid-Hudson area is supplied by Central Hudson Gas and Electric Corporation.

Metropolitan New York is serviced by Consolidated Edison Company of New York, Long Island Lighting Company, and, to a lesser degree, by Orange and Rockland Utilities Inc. These are the great suppliers of electricity for home and industry, the key elements in the power grid that serves New York. Figure 94 also shows how the facilities of these corporations are interconnected at the state boundaries with other utilities, forming a grid that extends over a far larger area than New York State.

In addition to the seven major utilities, PASNY has a large generating capacity at Niagara Falls and at Barnhart Dam on the St. Lawrence. Most of this power is either sold direct to major users near the generating source or sold to one of the major corporations, which subsequently sells and distributes it. Forty-two municipal power companies operate in the state too. A few of these generate their own power; most buy from the major corporations.

POWER-GENERATING REGIONS

Table 31 and Figure 94 show the size and location of all generating units. Concentration of generation capacity occurs at the Niagara Frontier, in metropolitan New York, between the St. Lawrence River and the Adirondacks, and near other of the larger urban systems.

Clearly the Niagara Frontier and the St. Lawrence Area are resource power regions oriented toward favorable hydro sites. Metropolitan New York is a high-demand region utilizing imported fossil fuels, while large sections of the state, particularly portions of the Appalachian Upland, are low-demand regions. The large urban systems, including Buffalo, all have nearby steam plants.

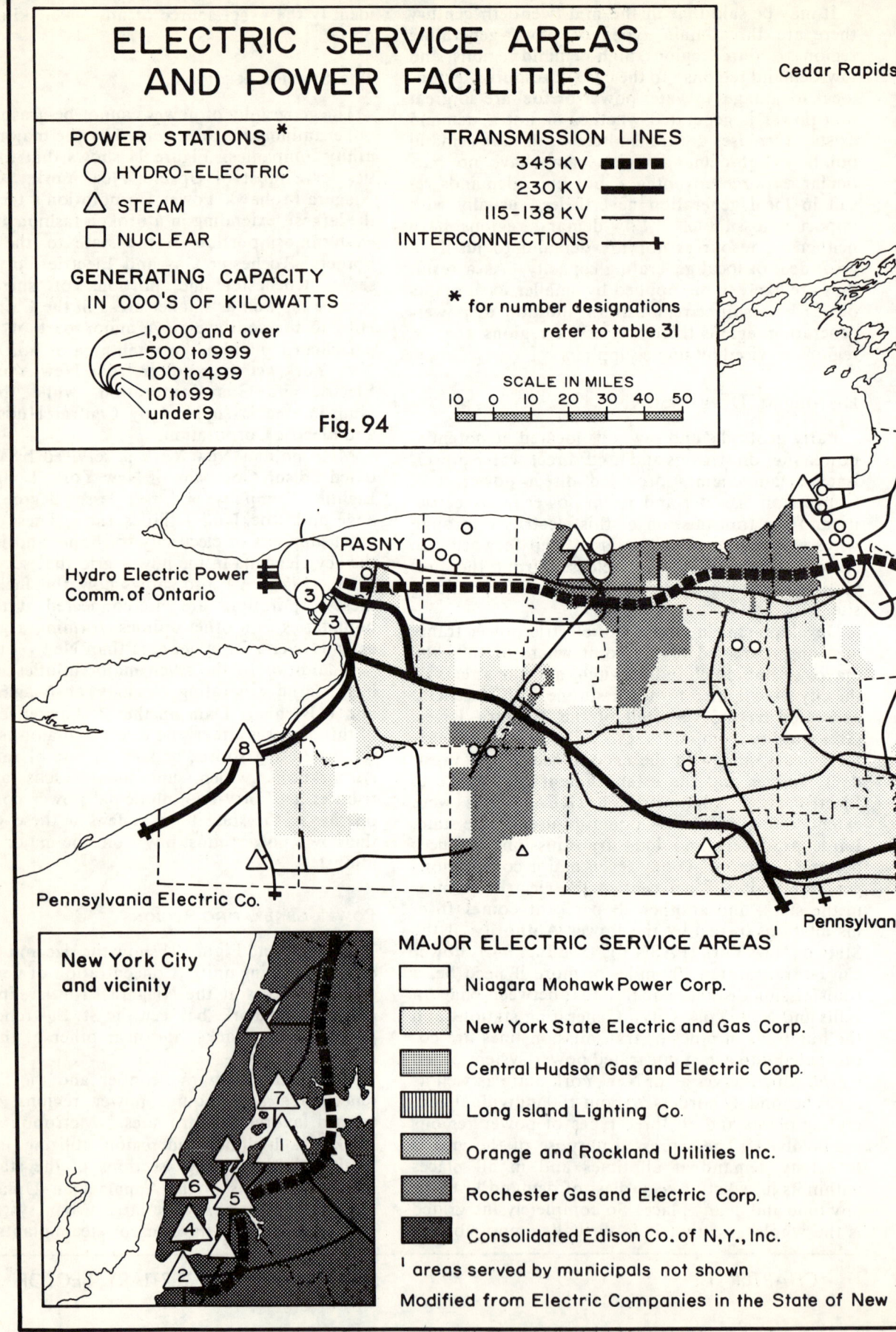

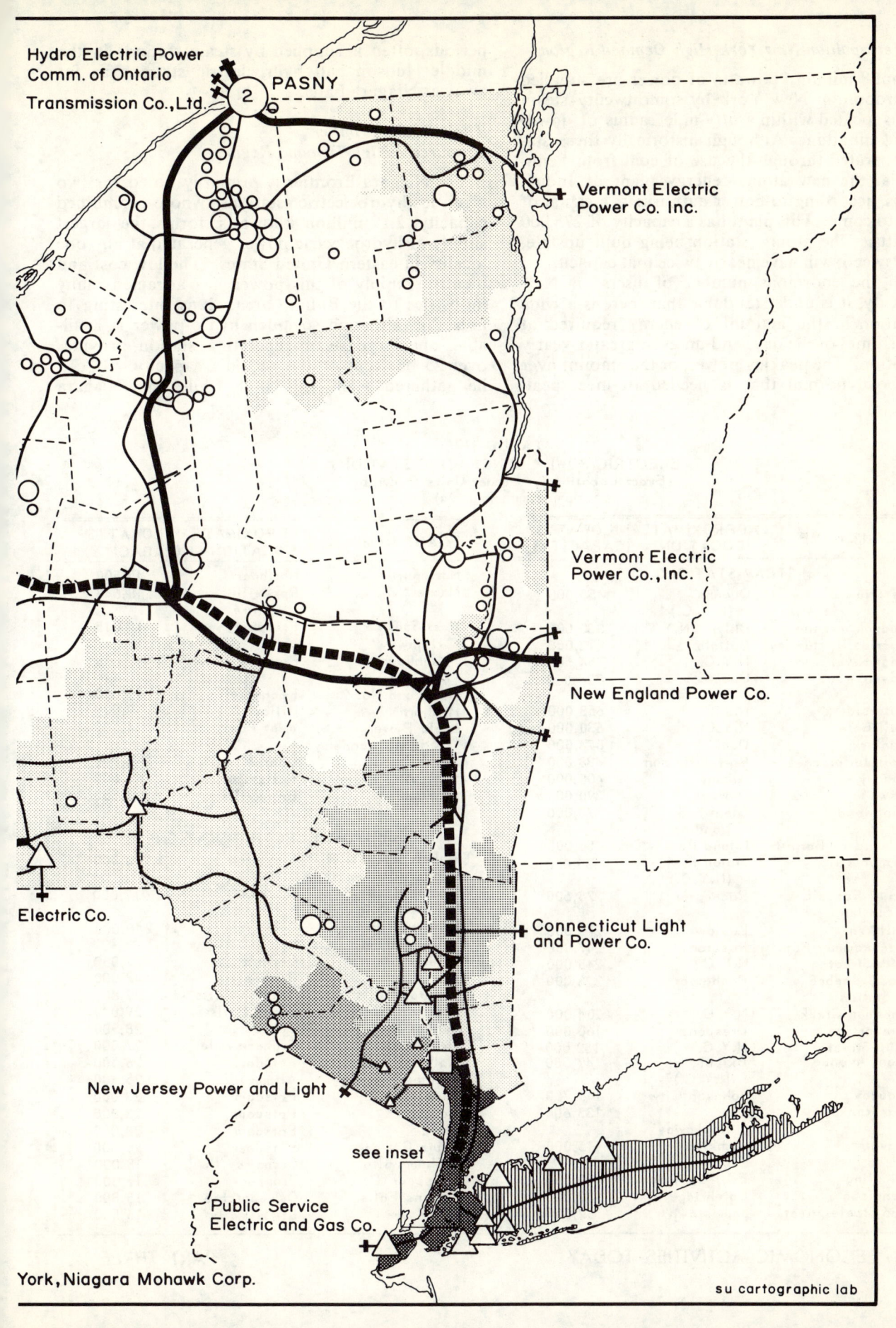

The Metropolitan New York High-Demand Region

About 8 million kilowatts of power are supplied to metropolitan New York by some twenty steam stations located within a fifty-mile radius of downtown Manhattan. Although historically these stations operated through the use of coal from Pennsylvania, the new atomic energy plant at Indian Point is, according to power authorities, a signal of things to come. This plant has a capacity of 275,000 kilowatts. The atomic station being built upstate, near Oswego, will have nearly twice that capacity.

With the enormous number of users in New York City, it is understandable that there is a wide fluctuation in the amount of energy required at various times of the day, and an even greater yearly fluctuation. The peaking power, or the amount over and above normal that is needed to meet peak periods, often is supplied by steam stations in the middle Hudson and hydroelectric stations as far away as Niagara Falls.

*The Niagara Frontier
Resource—High-Demand Region*

The Niagara Frontier is primarily noted for two gigantic hydroelectric stations whose combined capacity, 2.19 million kilowatts, forms the largest supply of hydroelectric power generated at any one locality in eastern United States. The low cost and plentiful supply of this power has attracted many industries to the Buffalo area. It is interesting to note that although so much hydro-power is available, one large steam-generating station provides over 25 per cent of the region's capacity. It can be gathered from this that while the Niagara

TABLE 31
ELECTRIC POWER GENERATING STATIONS
(Exact Location for Larger Units Shown by Numbers in Fig. 94)

PLANT NAME	APPROXIMATE LOCATION	KILOWATTS (CAPACITY)
a) STEAM STATIONS		
1 Astoria	Queens (N.Y.C.)	1,455,000
2 Hudson Avenue	Bklyn. (N.Y.C.)	830,000
3 Charles R. Huntley	Buffalo	812,000
4 East River	N.Y.C.	754,500
5 Ravenswood	Queens (N.Y.C.)	726,000
6 Waterside	N.Y.C.	658,000
7 Hell Gate	N.Y.C.	630,000
8 Dunkirk	Dunkirk	624,000
Port Jefferson	Port Jefferson	438,000
Albany	Albany	408,000
Oswego	Oswego	390,000
Glenwood	Glenwood Landing	370,000
Edward F. Barrett	Island Park	350,000
Arthur Kill	Staten I. (N.Y.C.)	335,000
No. 7 Russell Station	Rochester	275,500
Milliken	Ludlowville	270,000
Danskammer Point	Roseton	247,000
74th Street	N.Y.C.	245,000
No. 3 Beebee Station	Rochester	223,000
Sherman Creek	N.Y.C.	204,000
Greenridge	Dresden	160,000
50th Street	N.Y.C.	152,000
Kent Avenue	Bklyn. (N.Y.C.)	147,500
Goudey	Johnson City	145,750
Far Rockaway	Far Rockaway	133,600
Lovett	Tomkins Cove	115,000
Hickling	East Corning	70,000
Jennison	Bainbridge	60,000
136 Steele Street	Jamestown	52,500
Northport	Northport	45,500
Maple Ave.	Rockville Centre	18,648
West Sunrise Highway	Freeport	12,015
Poughkeepsie Elec. Sta.	Poughkeepsie	12,000
No. 8 and No. 9	Rochester	11,000
Hillburn	Hillburn	9,000
Light Power—U.S. Military Academy	West Point	4,500
Monroe	Monroe	3,000
Wellsville Plant	Wellsville	1,500
Lawrence	Bronxville	1,350
b) HYDROELECTRIC STATIONS		
1 Robert Moses—Niagara Power Plant	Lewiston	1,950,000
2 Robert Moses Power Dam	Massena	912,000
3 Reservoir Pump-Generating	Lewiston	240,000
Spier Falls	Corinth	49,000
No. 5	Rochester	42,000
Trenton Falls	Trenton Falls	29,800
Sherman Island	Glens Falls	29,000
School St.	Cohoes	28,500
Neversink	Grahamsville	27,000
Colton	Potsdam	26,500
Bennett's Bridge	Altmar	25,400
Five Falls	Potsdam	23,000
Rainbow	Potsdam	23,000
Stark	Potsdam	23,000
South Colton	Potsdam	20,000
Grahamsville	Grahamsville	18,000
Prospect	Prospect	17,300
Browns Falls	Oswegatchie	15,800
Beardslee	Little Falls	15,500

TABLE 31 (Continued)

PLANT NAME	APPROXIMATE LOCATION	KILOWATTS (CAPACITY)	PLANT NAME	APPROXIMATE LOCATION	KILOWATTS (CAPACITY)
Soft Maple	Croghan	15,400	Ft. Edward	Ft. Edward	2,800
Sturgeon Pool	Rifton	15,000	Rainbow Falls	Ausable Chasm	2,640
Schaghticoke	Schaghticoke	14,600	Union Falls Plant	Hawkeye	2,400
High Falls	Moffitsville	14,100	Cadyville	Cadyville	2,400
Blake	Potsdam	14,000	Eel Weir	Ogdensburg	2,400
Stewart's Bridge	Hadley	13,500	Parishville	Potsdam	2,300
E. J. West	Hadley	13,500	Cadyville	Cadyville	2,280
Deferiet	Deferiet	11,000	Franklin Falls Plant	Bloomingdale	2,265
No. 2 and No. 26	Rochester	10,600	Mill C	Cadyville	2,250
Rio	Lumberland	10,000	Piercefield Plant	Piercefield	2,000
Moshier	Croghan	8,800	Sewalls Island	Watertown	2,000
Lighthouse Hill	Altmar	8,500	Baker Falls	Hudson Falls	2,000
Seneca Falls	Seneca Falls	8,000	Norwood #2	Norwood	2,000
High Dam	Oswego	8,000	Raymondville	Norwood	2,000
Varick	Oswego	7,700	Keuka	Keuka	2,000
Hannawa	Potsdam	7,600	Waterloo	Waterloo	1,920
Minetto	Minetto	7,000	Belfort	Croghan	1,900
Black River	Watertown	6,900	High Falls	High Falls	1,800
Swinging Bridge No. 1	Forestburg	6,750	Port Henry	Port Henry	1,800
Swinging Bridge No. 2	Forestburg	6,750	Elmer	Croghan	1,600
Beebee Island	Watertown	6,600	Glenwood	Medina	1,500
Inghams	Little Falls	6,500	Walden	Walden	1,475
Kents Falls	Cadyville	6,400	Schuylerville	Schuylerville	1,400
Flat Rock	Oswegatchie	6,000	Theresa	Theresa	1,300
Higley	Potsdam	5,700	Diamond Island	Watertown	1,200
Marble St.	Watertown	5,400	Fulton	Fulton	1,100
Eagle	Croghan	5,300	Victory Mills	Victory Mills	1,100
Moreau	Hudson Falls	5,000	No. 170	Wiscoy	1,080
Ephratah	Ephratah	5,000	Macomb	Malone	1,000
Kamargo	Black River	5,000	Middle Falls	Greenwich	900
Dashville	Rifton	4,800	Riverside	Fulton	900
Taylorville	Croghan	4,600	Heuvelton	Heuvelton	800
Mechanicville	Mechanicville	4,500	Oswegatchie	Edwards	800
Norfolk	Norfolk	4,500	Chateaugay	Chateaugay	720
Herrings	Herrings	4,300	Baldwinsville	Baldwinsville	700
Allen's Falls	Potsdam	4,200	Deerland	Long Lake	688
Sugar Island	Potsdam	4,200	Hogansburg	Massena	500
Chasm	Malone	4,100	Springville	Springville	500
Waterport	Medina	4,000	Whittlesey	Malone	400
Mongaup Falls	Forestburg	4,000	Yaleville	Norwood	400
Colliers	Colliersville	3,810	No. 160	Rochester-Mt. Morris	340
Greenport	Greenport	3,637	Lake Placid #2	Lake Placid	332
South Glens Falls	Glens Falls	3,500	Canal	Medina	300
Johnsonville	Johnsonville	3,400	Oak Orchard	Medina	300
East Norfolk	Norfolk	3,400	Lake Placid	Lake Placid	300
Miller Street	Plattsburgh	3,236	Keese's Mills Plant	Paul Smith's	245
South Edwards	Edwards	3,100	No. 172	Mills Mills	220
Effley	Croghan	3,100	Greene	Greene	200
Granby	Fulton	3,000	Champlain	Cohoes	100
Hydraulic Race	Lockport	3,000	Philadelphia	Philadelphia	80
Stuyvesant Falls	Stuyvesant Falls	3,000			

Source: *Electric Utilities Directory—1965* (New York: McGraw-Hill, 1965).

Frontier has the attributes for being a resource power region, it also is a high-demand region.

The St. Lawrence Valley Resource Region

The St. Lawrence Valley, including the western and northern fringes of the Adirondacks, is a typical resource power region, based exclusively on the hydro sites of the St. Lawrence, Raquette, Black, and other rivers. The PASNY facilities on the St. Lawrence and many smaller installations have a total capacity of over one million kilowatts. Early hydro-power potential in this area attracted the aluminum industry.

TABLE 32
VARIATION IN ELECTRIC POWER COSTS

Utility Company	Location	Typical Electric Bill (Residential)	
		250 KWH	1,000 KWH
Niagara Mohawk Power Corporation (cost uniform in central and eastern divisions)	Albany	7.18	18.76
	Dunkirk	7.18	18.76
	East Syracuse	7.18	18.76
	Glens Falls	7.18	18.76
	Gouverneur	7.18	18.76
	Buffalo	6.90	18.52
	Massena	7.18	18.76
	Niagara Falls	6.90	18.52
New York State Electric and Gas Corporation	Alfred	8.10	22.05
	Auburn	8.10	22.05
	Binghamton	8.10	22.05
	Cooperstown	8.10	22.05
	Elmira	8.10	22.05
	Walden	8.49	22.44
Long Island Lighting Company	Amityville	8.19	20.42
	Babylon	8.19	20.42
	Hicksville	8.19	20.42
	New Hyde Park	8.19	20.42
	Syosset	8.19	20.42
	N.Y.C.: Queens	8.19	20.42
Central Hudson Gas and Electric Corporation	Arlington	8.93	21.98
	Cornwall	8.93	21.98
	Newburgh	8.93	21.98
Consolidated Edison Co.	Mount Kisco	8.75	24.19
	N.Y.C.: Bronx	8.83	24.43
	N.Y.C.: Queens	8.83	24.43
	Ossining	8.75	24.19
Rochester Gas and Electric Corporation	East Rochester	8.53	21.25
	Geneseo	8.53	21.25
	Rochester	8.53	21.25
Orange and Rockland Utilities, Inc.	Port Jervis	8.90	20.26
	Suffern	8.90	20.26
	Warwick	8.90	20.26
Cost Range for Large Utilities			
High		8.93	24.43
Low		6.90	18.52
Sample Municipal Power Companies	Akron	4.71	14.46
	Boonville	5.63	15.00
	Patchogue	7.80	20.43
	Ellenville	8.00	19.00
	Fairport	5.62	15.74
	Jamestown	5.43	16.68
	Mohawk	5.71	15.09
	Plattsburgh	3.89	11.20
	Tupper Lake	7.88	24.75
	Watkins Glen	5.40	13.40

Source: Federal Power Commission, *Typical Electric Bills 1964* ("Federal Power Commission Publication," R-65 [1965]).

The Appalachian Upland Low-Demand Region

Figure 94 shows that throughout much of the Appalachian Upland there are few generating stations. Still, power is available at every village and hill farm which requests it in this whole area. Source of the power used varies from time to time. It may come from steam plants in Binghamton, from Niagara Falls, from Rochester, or from the Hudson Valley.

POWER COSTS

Residential costs for 250 kilowatt hours of power in the state vary from $3.89 to $8.93, with the greatest fluctuation occurring among small municipal producers (see Table 32). Basically the rate schedules within New York State are currently set up in such a way as to allow producers a return in the order of 6 ½ per cent (the level established by the New York State Public Service Commission). Variations in the electricity bills received by residential, commercial, and industrial users throughout the state are therefore largely a function of the differing expenses incurred by the producers, the efficiency of the equipment and company organization, taxes paid, the amount of capital each company has invested, and the total sales.

High rates in New York City reflect in part high costs associated with installations and maintenance of underground cables. Some large utility companies elsewhere such as Niagara Mohawk or New York State Electric and Gas have both high-cost and low-cost installations making intermediate charges possible throughout their service area.

The wide spread of rates among municipal companies results from many local factors. Low rates occur when no taxes are paid by the municipals or when they purchase power at discount from PASNY or one of the large corporations and thus have no generators to install or maintain. High rates often result because of antiquated generation or transmission facilities.

THE GREAT BLACKOUT OF 1965

At 5:16 P.M. on November 9, 1965, the lights went out all over the Northeast. According to theories supporting the use of integrated grid systems, this should never have occurred. The needed power for all portions of a system ought to be available even when there is a sudden loss of power in one area. Somehow, somewhere, some link failed. At this particular moment, and in the confusion of the following days, nobody seemed to know the exact answer. It was thought that perhaps a switching device had failed or that a giant generator had gone out of phase.

Two weeks later the pieces of the puzzle were still being put together and answers were being deduced. As the result of a disconnection of the Sir Adam Beck #2 hydro station (Queenston, Canada) from its network to the north without loss of generation, well over one million kilowatts of power were suddenly fed back into the grid system covering Upstate New York, causing, in effect, a chain reaction of overloaded circuits and subsequent disturbance and automatic tripping out of lines in many areas. The sudden dropping of a significant amount of load in one direction from an interconnected system without prior loss of generation, and therefore overloading of circuits in other directions, is extremely rare. But it happened at the Canadian station. As a result, about 80,000 square miles and nearly thirty million people in the northeastern portion of the United States were without power for periods of time ranging from a few seconds to many hours.

Could this happen again? The experts say that with the present system of automatic and semi-automatic devices it could. Probably the best thing that can be said for the great blackout is that it is causing the electrical power industry to re-examine the structure of the power grid and to take necessary measures to prevent a recurrence of the disaster.

Perspectives on Recreation and Conservation

Recreation, which represents one of the most rapidly growing aspects of the tertiary sector, is an important user of land and will demand increasing acreages in the years ahead. Thus this chapter on the tertiary sector has a significant section devoted to recreation. In addition, because of its obvious close relation to conservation—a vital subject of the modern age—the two are examined together under the joint heading of perspectives.

The elevation of a large mass of the American public to a position of comparative affluence and leisure is a relatively recent phenomenon. It is a situation with enormous potential—for good or evil—for the human personality. But whatever the qualitative aspects of "free time," its quantitative aspects have had an enormous impact on both recreation and conservation.

Impressive postwar increase in outdoor recreation, amounting to perhaps 15 per cent a year, has stimulated the demand for additional areas to be devoted to varying recreational interests ranging from wilderness hiker to family camper. .It has accentuated the need for including recreation as a normal and wise use of land. A forest, for example, might well provide recreation not merely as an accidental by-product of the material production of wood but as a rightful and equally valuable service to be demanded of the forest environment.

The American conservation movement, which began gathering headway during the latter part of the nineteenth century and developed irresistible momentum under Theodore Roosevelt and Gifford Pinchot, has always benefited from the political popularity accorded the efforts to preserve nature for aesthetic enjoyment. But in most of the United States lumber and scenery have not enjoyed equal status in practice. Marketable values have generally held priority over the subjective nonmarketable values which recreation represents.

New York State is an interesting exception to this latter generalization. Well endowed with forest land, and an enormous consumer of forest products, it has for many years directed the bulk of its efforts in conservation toward satisfaction of the subjective values of recreation rather than the commercial values inherent in the forest.

The Forest Preserve

Of New York's 30.7 million acres, about 14.5 million are forest, and of this, about 22 per cent, or 3.5 million acres, are owned by the state. Of the 3.5 million publicly owned acres, 80 per cent is devoted to wilderness use and enjoys an interesting and unique legal status. These wilderness-dedicated lands are such lands as the state owns or may subsequently acquire within the twelve Adirondack counties and the four Catskill counties: Clinton, Essex, Franklin, Fulton, Hamilton, Herkimer, Lewis, Oneida, Saratoga, St. Lawrence, Warren, Washington, Delaware, Greene, Sullivan, and Ulster. They are known as the Forest Preserve counties, and most state-owned land within them is known collectively as the Forest Preserve. By constitutional law, lands acquired for the Forest Preserve assume an inalienable status and may not be used for any purpose deemed incompatible with that of a wilderness park. The significance of this lies in the interpretation of the wording of the New York State constitution as an operational mandate within the geographical and political context of the Forest Preserve.

The constitution states that the land comprising the Forest Preserve "shall be forever kept as wild forest lands. They shall not be leased, sold or exchanged, or be taken by any corporation, public or private, nor shall the timber thereon be sold, removed or destroyed." This provision has been described as at once the most loved and most hated portion of the constitution. The reason for this is fairly straightforward. By prohibiting the disturbing of any portion of the forest cover by public or private corporations, the constitution baffles even the executive branch of the state government as to just how to pursue its operational mandate in the face of so adamant a sovereign declaration. The Department of Public Works, for example, cannot use Forest Preserve lands for new road construction without a specific amendment to the constitution. The state Conservation Department, which is charged with the administration of the Forest Preserve, has overtly expressed its bewilderment as to whether the constitution permits the department to perform such functions as cutting trees for public sites or manipulating the forest environment for game production.

A large measure of the difficulty stems from the fact that the Forest Preserve holdings are not a solid block but are rather a shotgun pattern of publicly owned wilderness interspersed with private holdings of a varied character and superimposed over the existing political structure of local government. (See Fig. 71.) Thus, for example, the demands of the local economy may suggest that the straightening of a road would be beneficial and yet find that this is impossible because parcels of Forest Preserve land are involved.

However, despite the operational difficulties imposed by the Forest Preserve's constitutional status, since its inception in 1885 the preserve has served as a symbol capable of arousing powerful political sentiments among the people of the state. The Forest Preserve has had a stormy and sometimes bitterly controversial existence, yet it has not only survived essentially unchanged but has actually increased in size. It is a central fact in understanding New York State's position in the conservation-recreation spectrum. It is perhaps a monument to an urban population's desire for the beauty and quiet of the forest environment.

Since 1951 the New York State Joint Legislative Committee on Natural Resources has been struggling with the delicate political problem posed by the Forest Preserve. Treading gingerly amid the emotion generated by the symbol of the preserve, it has sought to make the administration of the preserve a more rational affair. Among its proposals has been that of wilderness zoning. This would delineate solid blocks of truly wild land and thus ease the administrative difficulties posed by the random imposition of parcels of legal wilderness over a well-established local economy and local units of government.

Public Administration of Recreation

Although the Forest Preserve constitutes the bulk of the area of state-owned land, much public recreation takes place in state parks outside the preserve. An excellent recreation map is available from the Conservation Department in Albany. It shows all lands administered by the Conservation Department: state parks, camp sites, fish hatcheries, public fishing streams, and the like. Figure 95 is a simpli-

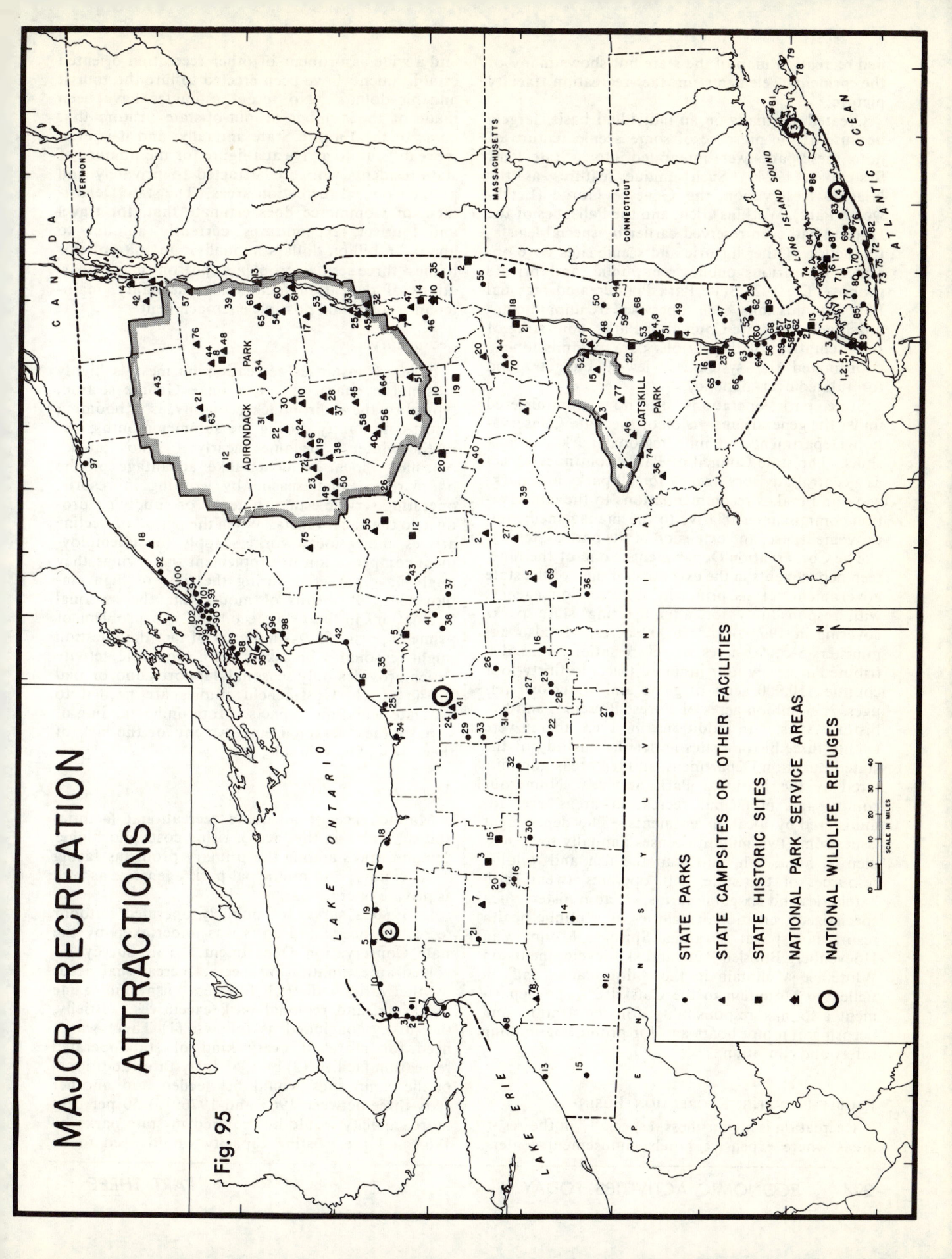

Fig. 95 MAJOR RECREATION ATTRACTIONS

fied recreation map of the state but shows many of the principal elements in the recreation facility picture.

Created originally on an individual basis, largely because of the presence of some scenic feature of note, these parks were integrated into a State Park System in 1924. Such unique features as the Niagara Reservation, the Genesee Gorge (Letchworth Park), Watkins Glen, and the Palisades of the Hudson had been reserved earlier by special legislative action. Other historic and scenic sites were operated by various public, semipublic, and private groups. The 1924 State Park Plan created regional Park Commissions, each supervised by unpaid commissioners. A state Council of Parks, consisting of representatives of the various park commissioners, coordinated the system, reviewed budgets, and formulated over-all policy.

State park operations are now administered under the general supervision of the state Conservation Department and nine regional park commissions. The state Council of Parks continues to act as a central advisory agency for all parks and parkways and makes recommendations to the conservation commissioner relative to the management, improvement, use, and extension of the park system.

The Conservation Department is one of the nineteen departments in the executive branch of the state government. It is primarily a recreation agency, with responsibility for administering state parks covering in 1963 over 216,000 acres. It also administers 585,000 acres of reforestation areas distributed in nearly four hundred tracts in thirty-four counties, 50,000 acres of game and waterfowl refuges, 2.56 million acres of Forest Preserve, and five historic sites. In addition, in 1963 there were twenty-three historic sites under the custody of the state Education Department, and ten areas administered by the National Park Service. Numerous county and municipal recreation areas are administered by local governments. The department issues some two million licenses annually to sportsmen to hunt, fish, and trap the fish and wildlife resources of the state. It operates twenty fish hatcheries and six game farms. It administers such specialized recreational facilities as a public health resort and spa at Saratoga Springs, Mount Van Hoevenberg Bobsled Run, and ski developments on Whiteface Mountain in the Adirondacks and on Belleayre Mountain in the Catskills. The department also has responsibility for registration and licensing of motor boats and for promoting boating safety and education.

Problems in the Recreation Business

Recreation is big business, especially in the resort areas where expensive hotels, amusement centers, and a wide assortment of other recreation-oriented establishments have been erected to lure the tourist and his dollars. No recent estimates have been made of the number of out-of-state visitors that come to the Empire State annually; and it is even more difficult to arrive at a figure for the number of state residents who are attracted to privately and publicly owned recreation areas. The state Department of Commerce does estimate that, for travel and tourism, expenditures currently amount to about 2.6 billion dollars annually. In spite of this bigness three serious problems confront many operations of recreation facilities: seasonality, maintenance costs, and possible encroachment.

Seasonality

In many instances recreation business is highly seasonal in character. The Lake George section, typical of the Adirondack economy, is a booming resort community during the summer months; but in the winter it becomes nearly a "ghost area." Normally, local residents take advantage of the boom recreational season by working in resorts, operating service establishments, or supplying produce to the vacationists. When the period of decline arrives, many local workers apply for unemployment compensation to supplement the savings they might have gathered during the time of high employment. A means of moderating the seasonal fluctuations in the resort economy is a problem of prime importance. As a matter of fact the question might reasonably be asked: Is an economic activity which provides only seasonal support good or bad for an area? Cost-benefit studies are needed to evaluate economic impacts. More understanding of social values, seasonal employment, or the lack of them, would be valuable, too.

Maintenance Costs

Maintenance of adequate recreational facilities and standards in the face of rising costs and higher demand ranks among the primary problems facing state, county, and municipal park agencies, as well as private operations.

In 1960 a survey of publicly operated outdoor recreation needs and costs was undertaken by the state Conservation Department. This survey involved an estimation of specific recreational needs by the various district foresters, fish and game managers, and regional park executives. Briefly, the survey concluded as follows: (1) There was a need for more of every kind of state-operated recreation facility; (2) by 1965 forty-three additional public camp sites would be needed and another sixty-three between 1965 and 1976; (3) 50 per cent more capacity would be required in state parks by 1965 and the existing capacity would need to be

TABLE 33
RECREATIONAL DATA
I. State Parks—Acreage and Attendance*

Map No.†		Acres	Attendance		
			1963	1958	1953
1	Niagara Reservation State Park	433	4,458,700	3,804,820	2,830,800
2	Whirlpool State Park	109	582,075	360,800	222,500
3	Devil's Hole State Park	42	280,250	209,650	138,400
4	Fort Niagara State Park (inc. Old Fort and 4 Mile Creek Campsite Annex)	536	372,605	362,965	235,555
5	Golden Hill State Park	378	3,900	Undev.	Undev.
6	Beaver Island State Park	918	630,990	564,950	567,075
7	Buckhorn Island State Park	896	115,400	107,750	38,850
8	Evangola State Park	733	212,960	202,145	Undev.
9	Lower Niagara River State Park	260	Undev.	Undev.	Undev.
10	Wilson Harbor State Park	260	Undev.	Undev.	Undev.
11	Reservoir State Park	134	Undev.	Undev.	Undev.
12	Allegany State Park	60,220	689,100	638,300	549,800
13	Lake Erie State Park	240	88,500	77,000	67,000
14	Cuba Reservation State Park	650	67,500	73,000	86,250
15	Long Point State Park	320	Undev.	Undev.	Undev.
16	Letchworth State Park	14,021	610,000	496,000	418,000
17	Hamlin Beach State Park	1,118	288,500	241,500	229,000
18	Braddock Bay State Park	2,115	21,000	11,000	Undev.
19	Lakeside Beach State Park	642	Undev.	Undev.	Undev.
20	Silver Lake State Park	484	Undev.	Undev.	Undev.
21	Darien Park Site	1,472	Undev.	Undev.	Undev.
22	Watkins Glen State Park	605	443,397	364,590	451,900
23	Buttermilk Falls State Park	675	77,838	70,805	62,300
24	Cayuga Lake State Park	188	201,189	141,370	107,100
25	Fair Haven Beach State Park	860	268,768	246,530	232,900
26	Fillmore Glen State Park	857	65,983	53,810	37,500
27	Newtown Battlefield Reservation State Park	330	40,966	39,240	32,600
28	Robert H. Treman State Park	1,020	148,147	133,830	112,800
29	Seneca Lake State Park	141	71,093	Undev.	Undev.
30	Stony Brook State Park	554	135,941	131,360	84,900
31	Taughannock Falls State Park	608	232,061	286,905	215,500
32	Keuka Lake State Park	554	Undev.	Undev.	Undev.
33	Sampson State Park	1,507	Undev.	Undev.	Undev.
34	Chimney Bluffs State Park	596	Undev.	Undev.	Undev.
35	Battle Island Park	235	38,067	30,660	29,398
36	Chenango Valley State Park	983	286,469	252,449	210,748
37	Chittenango Falls State Park	123	60,624	62,186	51,926
38	Clark Reservation State Park	228	38,555	44,948	38,450
39	Gilbert Lake State Park	1,569	121,870	87,006	97,791
40	Glimmerglass State Park	593	Undev.	Undev.	Undev.
41	Green Lakes State Park	925	469,708	477,996	302,781
42	Selkirk Shores State Park	980	151,402	151,595	105,069
43	Verona Beach State Park	1,355	192,664	236,100	121,755
44	John Boyd Thacher Park	1,138	468,000	363,646	256,750
45	Lake George Beach and Battlefield Parks	97	148,048	150,753	226,112
46	Saratoga Spa State Park		NA	NA	NA
47	Clarence Fahnestock Memorial State Park	5,890	60,025	63,311	47,345
48	Clermont State Park	396	Undev.	Undev.	Undev.
49	James Baird State Park	590	168,921	190,529	130,177
50	Lake Taghkanic State Park	1,568	150,336	113,542	77,644
51	Margaret Lewis Norrie State Park	323	124,675	98,617	33,253
52	Mohansic State Park	822	96,353	5,197	Undev.
53	Ogden Mills and Ruth L. Mills State Park	419	32,118	23,116	14,613
54	Taconic State Park	4,570	53,180	117,300	91,002
55	Grafton State Park	382	Undev.	Undev.	Undev.
56	High Tor State Park	492	Undev.	Undev.	Undev.
57	Hook Mountain State Park	661	Undev.	Undev.	Undev.

*Figures taken from Annual Reports of New York State Department of Conservation. Acreages are as of January 4, 1965.
† Numbers correspond to those in Figure 95.
NA: Not Available.

TABLE 33 (Continued)

Map No.†		Acres	Attendance		
			1963	1958	1953
58	Nyack Beach State Park	61	80,200	29,300	28,540
59	Rockland Lake State Park	1,036	97,500	Undev.	Undev.
60	Stony Point Reservation State Park	45	34,500	25,985	17,280
61	Storm King State Park	1,142	Undev.	Undev.	Undev.
62	Tallman Mountain State Park	687	137,900	116,825	147,335
63	Bear Mountain State Park	4,933	2,475,000	2,483,000	1,962,000
64	Harriman State Park	45,828	2,617,000	2,213,000	1,456,000
65	Highland Lakes State Park	809	Undev.	Undev.	Undev.
66	Goosepond Mountain State Park	1,559	Undev.	Undev.	Undev.
67	Blauvelt State Park	590	Undev.	Undev.	Undev.
68	Haverstraw Beach State Park	73	Undev.	Undev.	Undev.
69	Bayard Cutting Arboretum	690	157,600	134,500	Undev.
70	Belmont Lake State Park	459	494,900	573,800	511,000
71	Bethpage State Park	1,477	526,700	458,100	355,200
72	Captree State Park	298⎫	12,412,600	9,851,500	Undev.
73	Jones Beach State Park	2,413⎭			7,805,200
74	Caumsett State Park	1,476	Undev.	Undev.	Undev.
75	Gilgo State Park	1,223	Undev.	Undev.	Undev.
76	Heckscher State Park	1,657	631,700	606,900	597,400
77	Hempstead Lake State Park	867	432,400	539,500	406,400
78	Hither Hills State Park	1,755⎫	279,800	222,300	255,700
79	Montauk Point State Park	724⎭			
80	Massapequa State Park	596	9,500	7,700	7,500
81	Orient Beach State Park	357	75,600	70,700	64,400
82	Robert Moses State Park	1,000	134,400	104,000	80,000
83	Southside Sportsmen's Club Park Site	3,473	Undev.	Undev.	Undev.
84	Sunken Meadow State Park	1,266	1,232,400	1,109.900	471,400
85	Valley Stream State Park	97	229,000	286,600	374,300
86	Wildwood State Park	699	268,400	175,300	172,000
87	Wyandanch Club Park Site	543	Undev.	Undev.	Undev.
88	Burnham Point State Park	12	17,552	16,243	9,994
89	Cedar Point State Park	48	80,383	56,484	43,381
90	Wellesley Island State Park	2,636	135,420	56,688	1,315
91	Grass Point State Park	27	50,780	62,350	40,214
92	Jacques Cartier State Park	461	53,303	26,969	Undev.
93	Keewaydin State Park	179	16,540	Undev.	Undev.
94	Kring Point State Park	41	48,611	43,118	23,340
95	Long Point State Park	23	13,839	10,618	10,385
96	Sackets Harbor Battlefield State Park	6	50,600	39,280	49,600
97	Robert Moses State Park	3,115	198,216	Undev.	Undev.
98	Westcott Beach State Park	319	154,341	107,313	74,600
99	Canoe-Picnic Point State Park	70	14,192	10,635	1,408
100	Cedar Island State Park	10	3,151	2,900	4,900
101	Mary Island State Park	13	4,821	3,025	4,275
102	DeWolf Point State Park	13	7,798	10,095	6,474
103	Waterson Point State Park	6	5,794	2,900	4,575
	GRAND TOTAL	200,670	35,334,271	30,030,400	23,062,736

*Figures taken from Annual Reports of New York State Department of Conservation. Acreages are as of January 4, 1965.
†Numbers correspond to those in Figure 95.
NA: Not Available.

II. Campsite or Other Facility‡

Map No.		Total Acreage	Visitor Days		
			1963	1958	1953
1	Ausable Point Campsite	47	Undev.	Undev.	Undev.
2	Bear Waller Multiple Use Area	NA	Undev.	Undev.	Undev.
3	Belleayre Mtn. Ski Center	300	207,280	181,911	33,302
4	Beaverkill Campsite	72	46,195	40,064	37,588
5	Bowman Lake Campsite	11,275	40,571	Undev.	Undev.
6	Brown Tract Pond Campsite	50	23,176	Undev.	Undev.
7	Carlton Hill Multiple Use Area	1,694	Undev.	Undev.	Undev.
8	Caroga Lake Campsite	60	58,563	62,084	95,965
9	Cascade Lake Campsite	556	Undev.	Undev.	Undev.
10	Cedar River Campsite	602	Undev.	Undev.	Undev.
11	Cherry Plains Day-Use Area	43	36,365	Undev.	Undev.
12	Cranberry Lake Campsite	144	35,774	26,595	20,397
13	Crown Point Reservation Campsite	351	53,779	83,398	51,284
14	Cumberland Bay Campsite	68	171,258	197,730	202,889
15	Devil's Tombstone Campsite	24	14,898	15,124	12,376
16	Dryden Lake Multiple Use Area	190	Undev.	Undev.	Undev.
17	Eagle Point Campsite	16	33,882	29,511	37,301
18	Eel Weir Campsite	8	13,079	19,576	33,643
19	Eighth Lake Campsite	175	49,822	45,818	41,396
20	Featherstonehaugh Lake Mult. Use Area	682	Undev.	Undev.	Undev.
21	Fish Creek Ponds Campsite	121	156,328	191,981	152,593
22	Forked Lake Campsite	127	25,147	21,345	14,445
23	Fourth Lake Access & Picnic Area	10	Undev.	Undev.	Undev.
24	Golden Beach Campsite	140	81,707	86,851	79,272
25	Hearthstone Point Campsite	99	137,911	124,018	111,869
26	Hinckley Reservoir Campsite	335	Undev.	Undev.	Undev.
27	Hunt's Pond Campsite	1,710	Undev.	Undev.	Undev.
28	Indian Lake Islands Campsite	315	11,380	Undev.	Undev.
29	Kehr Multiple Use Area	374	Undev.	Undev.	Undev.
30	Lake Durant Campsite	125	16,496	15,506	12,693
31	Lake Eaton Campsite	130	55,052	56,826	45,586
32	Lake George Battleground Campsite	10	30,740	29,488	20,896
33	Lake George Islands	NA	119,260	66,999	45,451
34	Lake Harris Campsite	28	10,064	Undev.	Undev.
35	Lake Lauderdale Campsite	117	Undev.	Undev.	Undev.
36	Lake Superior Campsite	1,367	Undev.	Undev.	Undev.
37	Lewey Lake Campsite	140	66,870	59,644	26,015
38	Limekiln Lake Campsite	150	17,999	Undev.	Undev.
39	Lincoln Pond Campsite	744	Undev.	Undev.	Undev.
40	Little Sand Point Campsite	35	28,444	25,986	17,356
41	Long Point Multiple Use Area	95	Undev.	Undev.	Undev.
42	Macomb Reservation Campsite	151	54,816	66,199	Undev.
43	Meacham Lake Campsite	125	97,584	60,532	68,784
44	Meadowbrook Campsite	14	24,683	18,730	15,274
45	Moffitt Beach Campsite	125	96,276	141,475	93,247
46	Mongaup Pond Campsite	1,086	Undev.	Undev.	Undev.
47	Moreau Lake Campsite	689	Undev.	Undev.	Undev.
48	Mount Van Hoevenberg Bobsled Run	44	7,386	6,396	10,395
49	Nelson Lake Campsite	2,408	Undev.	Undev.	Undev.
50	Nick's Lake Campsite	3,747	Undev.	Undev.	Undev.
51	Northampton Beach Campsite	103	130,512	69,672	Undev.
52	North Lake Campsite	472	120,701	96,881	103,229
53	Palmer Pond Campsite	244	Undev.	Undev.	Undev.
54	Paradox Lake Campsite	6	18,344	14,591	17,666
55	Pixley's Falls Campsite	652	16,239	14,707	26,951
56	Point Comfort Campsite	15	28,894	31,799	23,052
57	Poke-O-Moonshine Campsite	2	7,741	13,828	16,133
58	Poplar Point Campsite	15	14,316	12,937	18,860
59	Pulver Multiple Use Area	715	Undev.	Undev.	Undev.
60	Putnam Pond Campsite	62	21,060	Undev.	Undev.
61	Roger's Rock Campsite	186	126,743	111,282	82,576

‡Other facilities include outdoor recreation areas operated by the Division of Lands and Forests, New York State Department of Conservation. Multiple Use Areas provide outdoor recreation compatible with other resource uses.
NA: Not Available.

TABLE 33 (Continued)

Map No.		Total Acreage	Visitor Days		
			1963	1958	1953
62	Rohner-Townsend Multiple Use Area	905	Undev.	Undev.	Undev.
63	Rollin's Pond Campsite	100	80,120	56,094	9,402
64	Sacandaga Campsite	35	44,390	57,495	38,989
65	Schroon River Campsite	134	Undev.	Undev.	Undev.
66	Sharp Bridge Campsite	19	16,312	17,223	20,019
67	South Lake Campsite	869	Undev.	Undev.	Undev.
68	Taconic-Hereford Multiple Use Area	883	Undev.	Undev.	Undev.
69	Tarbell Farms Campsite	2,186	Undev.	Undev.	Undev.
70	Thompson's Lake Campsite	51	54,094	Undev.	Undev.
71	Toe Path Mountain Campsite	70	9,638	Undev.	Undev.
72	Treasure Island Campsite	5	Undev.	Undev.	Undev.
73	Valcour Island Campsite	477	Undev.	Undev.	Undev.
74	Waneta Lake Campsite	225	Undev.	Undev.	Undev.
75	Whetstone Gulf Campsite	531	26,688	13,343	19,812
76	Wilmington Notch Campsite	9	41,582	31,720	31,868
77	Woodland Valley Campsite	18	17,624	14,545	13,056
78	Zoar Valley Multiple Use Area	3,474	Undev.	Undev.	Undev.

III. State Historic Sites

Map No.		Acres	Location
1	Baron Von Steuben Memorial	50	Remsen, Oneida County
2	Bennington Battlefield	171	Walloomsac, Rensselaer County
3	Boyd-Parker Memorial	4	Cuylerville, Livingston County
4	Clinton House	1	Poughkeepsie, Dutchess County
5	Fort Brewerton	1	Brewerton, Oswego County
6	Fort Ontario	20	Oswego, Oswego County
7	Grant Cottage	1	Mt. McGregor, Saratoga County
8	John Brown Farm	160	Lake Placid, Essex County
9	John Jay Homestead	30	Bedford, Westchester County
10	Johnson Hall	18	Johnstown, Fulton County
11	Knox Headquarters	50	Newburgh, Orange County
12	Oriskany Battlefield	5	Oriskany, Oneida County
13	Philipse Manor	1	Yonkers, Westchester County
14	Saratoga Battle Monument	3	Schuylerville, Saratoga County
15	Sullivan Monument	1	East Groveland, Livingston County
16	Temple Hill	1	Newburgh, Orange County
17	Walt Whitman House	1	Huntington, Nassau County
18	Fort Crailo	7	Rensselaer, Rensselaer County
19	Guy Park	1	Amsterdam, Montgomery County
20	Herkimer Home	143	Little Falls, Herkimer County
21	Schuyler Manson	1	Albany, Albany County
22	Senate House	1	Kingston, Ulster County
23	Washington's Headquarters	6	Newburgh, Orange County

IV. National Park Service Areas

Map No.		Acres	Attendance, 1963	Location
1	Castle Clinton National Monument	1	NA	New York City
2	Federal Hall National Memorial	.45	58,800	New York City
3	General Grant National Memorial	.76	304,100	New York City
4	Home of Franklin D. Roosevelt National Historic Site	93.69	371,600	Hyde Park, Dutchess Co.
5	St. Paul's Church National Historic Site	6.09	NA	New York City
6	Sagamore Hill National Historic Site	75	NA	Oyster Bay, Nassau Co.
7	Theodore Roosevelt Birthplace National Historic Site	.11	NA	New York City
8	Vanderbilt Mansion National Historic Site	211.65	172,200	Hyde Park, Dutchess Co.
9	Statue of Liberty National Monument	10.38	848,000	New York City
10	Saratoga National Historical Park	5,500	97,100	Stillwater, Saratoga Co.

NA: Not Available.

V. National Wildlife Refuges

Map No.		Acres	Location
1	Montezuma National Wildlife Refuge	6,776	Cayuga County
2	Iroquois National Wildlife Refuge	3,144	Orleans County
3	Elizabeth Morton National Wildlife Refuge	187	Suffolk County
4	Wertheim National Wildlife Refuge	11	Suffolk County

doubled or trebled by 1976; (4) there existed a definite need for wilderness areas, winter sports areas, multiple purpose areas, fishing and hunting rights or easements and boating facilities, particularly launching sites; (5) there was a need for local parks and recreational facilities in the state's municipalities and counties. In order to meet these needs, a program of land acquisition involving a bond issue in the amount of 100 million dollars was approved and the process of acquisition by the state and numerous municipalities has begun. Approval of the bond issue has been a big step forward, since 50 million dollars of the total issue has been made available to municipalities and counties under a cost-sharing schedule. The costs of desirable sites have increased greatly with increased demand—sometimes to levels which public agencies are unable to pay.

Protection from Encroachment

Another important problem relates to the protection of the state's recreational resources from encroachment by incompatible land uses, including garish billboard signs, roadside taverns, junk yards, and, perhaps most important, unrestrained commercialism involving recreation itself. Some form of zoning control has been advocated as a solution, especially in the Adirondack and Catskill regions, but little has been accomplished by way of implementation.

RECREATION RESOURCES

The variety of New York's recreational resources is tremendous. Few states can boast of such an array of physical and cultural phenomena so attractive to the traveling and recreating public. Geologic and topographic features, the flora and fauna, water and sand, historic and cultural sites, and urban attractions are all part of this array.

Geologic and Topographic Features

Geologic and topographic features (some of which were described in Chapter 1) range from extensive areas of sand and beaches on the Atlantic Ocean, Lake Erie, Lake Ontario, and Long Island Sound to rugged mountain peaks and forested wilderness-like areas in the Adirondacks and Catskills. Throughout much of the state local relief is sufficient to provide topographic interest, and in many areas very pleasing vistas of sweeping valleys exist. Long slopes and heavy snowfall combine to form a vigorous and growing winter ski industry. The Appalachian Upland particularly exhibits many interesting examples of terrain modified by the effects of continental glaciation or running water. Features fascinating to tourists include the Finger Lakes, the Palisades, Niagara Falls, and world-renowned caverns and chasms.

Flora and Fauna

The varied topography and climate and the existence of much sparsely used land in New York are conducive to the support of a wide variety of animals and plants. More than four thousand species of plants and nearly eight hundred different kinds of animals are found.[1] While such a variety of animals and plants are useful for general interest and in biological research, they also provide unexcelled opportunities for hunting, fishing, and just collecting. The Conservation Department operates several game management areas, and in 1963 the state acquired the perpetual public fishing rights on over 880 miles of some of the finest trout-fishing waters. A survey conducted in 1955, in connection with a national survey of fishing and hunting, revealed that 24.1 per cent of the state's population in the twelve-eighteen age group fished and 2.3 per cent hunted, 4.5 per cent of the eighteen-twenty-five age group fished and 3.4 per cent hunted, 9.7 per cent of the twenty-five-forty-four age group fished and 4.4 per cent hunted, and 6.3 per cent of the forty-four and over age group fished and 1.8 per cent hunted. One in every eleven persons twelve years of age and over was a fresh-water fisherman. Annual expenditures for fishing amounted to an estimated $169,365,000 in 1955, of which 31.5 per cent was accounted for by the cost of gear and equipment. Expenditures for hunting were somewhat less at $54,512,000.

Water

Rivers and streams are important elements of the over-all state recreation resource picture. In addi-

---

[1] The figures given are for so-called valid forms; sometimes these are of species rank, sometimes of lesser rank, such as subspecies or races or varieties.

WATKINS GLEN (left) AND TAUGHANNOCK FALLS (215 feet high, right) are two examples of the many attractions in the Appalachian Upland which combine geologic or topographic and water resources. Both are bases for state parks. *N.Y.S. Dept. of Commerce.*

tion to the value of water bodies for fishing purposes, water-related activities, especially boating and swimming, are becoming increasingly popular. The rapid expansion of motorboating as a recreational activity in the state has led to the creation of a new administrative unit within the state Conservation Department to supervise the licensing of motorboats, provide launching sites, and promulgate boating safety regulations. The Division of Motor Boats estimates that there are more than half a million motorboats (about 80 per cent of which are of the outboard type) in the state; and the number of boats has increased about 12 per cent annually between 1950 and 1960. In state and county park planning, the provision of suitable facilities for swimming has become a prerequisite to public acceptance. There are only a few large parks in the state which do not provide swimming pools, whether natural or constructed.

Major rivers, such as the Hudson, St. Lawrence, Mohawk, Susquehanna, and Genesee, together with countless smaller streams, are recreationally valuable. Hundreds of lakes, many of them small, are located in the Adirondack Region; and the Finger Lakes, such as Skaneateles, Cayuga, and Seneca, are well known for their scenic qualities. Other large lakes include Chautauqua Lake, where a famous cultural center is operated in the summer;

WATER FOR RECREATION. Whether it be for swimming at congested Jones Beach (left) or canoeing on remote Indian River (right), water makes for recreation. *N.Y.S. Dept. of Conservation.*

Otsego Lake, the site of summer theatres and places of historic interest; Oneida Lake, a popular fishing place; Lake George, a center for intensive summer tourist activity, amusement centers, dude ranches, and similar entertainment; and Lake Champlain, which vacationists from both New England and New York can rightly call their own. Several large reservoirs have been constructed for flood control, water supply, or hydroelectric power purposes. Some of these, such as Sacandaga Reservoir in Fulton and Saratoga counties, are used intensively for public recreational purposes.

Recreationally valuable water resources of a more spectacular character include Niagara Falls, Taughannock Falls, Buttermilk Falls, Genesee Gorge (known as the "Grand Canyon of the East"), and Ausable Chasm.

The New York State Barge Canal, a state-owned and -operated waterway system having an aggregate length of 522 miles, includes the Erie Canal (Waterford to Tonawanda), Oswego Canal (Syracuse to Lake Ontario), Cayuga-Seneca Canal (Montezuma to Watkins Glen and Ithaca via Seneca and Cayuga lakes), Champlain Canal (Waterford to Whitehall) and connections with canal harbors at Utica, Syracuse, and Rochester. Commercial traffic on the canal has been meager for many years, but recreational use, as measured by the number of permits issued to pleasure boats, has increased from 1,940 in 1952 to about 7,800 in 1963. At the general election in 1960, the voters of the state approved an amendment to the state constitution authorizing the legislature to lease, sell, or transfer the canal system to the federal government. Subsequently a joint legislative committee was appointed, and after appraising the various uses of the canal and how such uses would be affected by a lease or transfer, it recommended to the legislature that the canal be retained in state ownership. It is anticipated that a comprehensive study of the canal system and its future uses will be made before definite action is taken by the legislature.

The potentialities of the canal for public recreational use have hardly begun to be realized. Marinas and boat service facilities are becoming increasingly popular, and municipalities or private companies add many new units each year on leased land. The state, however, could do much more to develop the canal as a major recreational resource. Some suggestions have been advanced to provide parks, picnic areas, and similar facilities at scenic or historic sites on the canal. Moreover, attention has been directed to the need for controls to regulate adjacent private uses such as docks, boathouses, and the like, which might interfere with the recreational function of the canal.

Historic and Cultural Sites

Major themes in the state's historical sites relate largely to the colonial and revolutionary periods. Excellent examples of colonial architecture remain in New York City, the Hudson and Mohawk valleys, and the Catskill Region. Places of significance during the Revolutionary War, such as Saratoga Battlefield, have been preserved for the enjoyment of future generations by the National Park Service. Altogether, the National Park Service administers eight such facilities in the state, five of them in New York City.

In addition to well-preserved places of the colonial and revolutionary periods, relics and camp sites associated with early Indian cultures are maintained for public inspection and study by the state, county, and municipal governments, or private groups. Finally, some of the places of historical or scientific interest owned by the state are administered by the state Education Department. In the past, the state erected highway historical markers to describe significant events which took place throughout the state, but this practice has been discontinued.

Urban Attractions

Although the following regional descriptions of the state's recreational resources are oriented primarily toward outdoor activities, the story of recreation in the Empire State would be incomplete if consideration of the recreational opportunities offered by the larger cities were omitted. Unfortunately, adequate statistical data for the extent of use of urban attractions are not available. As a result, even an approximation of the total number of recreational visitors, economic values of recreation, and similar useful information, is difficult to ascertain. There is little question, however, that the place of urban attractions in the total recreational picture is great and probably accounts for more recreational visitors annually than all other types of recreational use combined.

New York City, of course, is an unexcelled urban recreational magnet for the entire world. But notwithstanding the naturalistic qualities that characterize some of the metropolitan parks, such as famous Central Park, the type of recreation which is available—and which people seek in New York City—is highly specialized and unique. Sightseeing, night clubs, amusement centers, and cultural facilities such as theaters, art galleries, and museums are examples of the types of recreation resource which the large metropolitan center can provide. It has been estimated that about six million persons visit New York City annually for recreational purposes and these visitors spend upward of 500 million dollars.

Similar recreational opportunities, at a reduced scale, are offered by other urban places such as Buffalo, Rochester, Syracuse, and Albany. Cultural attractions such as libraries, museums and art galleries, concerts, lectures and exhibitions, and an assortment of other intensive recreational pursuits are available in most of these cities. Moreover, many of the cities in the state maintain official park and recreation agencies. In some places the park and recreation functions are separated, while in other places these two functions are the responsibility of the same agency. Some cities, which formerly possessed adequate parks, have allowed the pressure of burgeoning populations and other needs to overshadow the requirements of their park systems. This is a pity and must be remedied. In many places municipal expenditures have been customarily utilized by other agencies, and park maintenance and new construction have been retarded. New school developments, highways, and other public improvements have usurped scarce municipal park land.

In cities the types of outdoor recreation provided normally range from the small ornamental park or square to the large park where natural influences predominate. Intermediate in this range are playgrounds, well-organized athletic fields, and specialized facilities such as ice-skating rinks and golf courses. Some of the metropolitan counties also provide integrated recreational systems through an officially designated county park board or department. Less than a dozen of New York's counties, however, have such administrative units. In some counties whatever park facilities are offered are developed and maintained by the Highway Department.

RECREATION REGIONS

Variations in recreational resources from place to place in New York State and differences in driving time from population centers strongly affect the way a given area is actually used for recreational purposes. Some sections of the state attract large numbers of recreational visitors as a result of the uniqueness or quality of physical or historical features, or the presence of developed facilities such as camp sites, picnic grounds, and boating areas. Other sections are characterized by very high recreation use intensity owing primarily to their proximity to urban centers; and still other areas may be identified which have unrealized potential for recreational use.

Using type of recreational attractions, current use characteristics, and potential future use as differentiating criteria, six recreation regions can be identified in the state (see Fig. 96). Each of the regions has unique and distinct recreational attractions and the volume of recreational use is high compared with areas lying outside the regional units. The large part of the state which has not been assigned specific regional identity, however, is not completely lacking in recreational significance. The Hudson-Mohawk Lowland, for example, offers scenic and historical features which have interest from a recreational point of view; and the Appalachian Upland west of the Catskills may have potential for future recreational development.

The Adirondacks

Perhaps the most widely known of all the recreational areas in the state, the Adirondack region is characterized by rugged, mountainous terrain and innumerable lakes and streams. The high-peak area concentrated in Essex County has more than forty mountains that rise higher than 4,000 feet above sea level, including Mount Marcy, whose summit at 5,344 feet marks the highest point in the state. Elsewhere in the region, dense forest covers more than 80 per cent of the land. Low temperatures and heavy snowfall in the winter months and cool summers contribute to the popularity the region enjoys for both winter and summer recreation. The abundance of such game species as white-tailed deer, black bear, varying hare, ruffed grouse, woodcock, and beaver attracts hunters; and the region's varied water resources provide excellent opportunities for fishing, canoeing, and motorboating.

A great diversity of recreational facilities, both summer and winter, have been introduced. The state has constructed over four hundred miles of forest trails, lean-to shelters, and public camp sites on Forest Preserve lands. For the winter-sports enthusiast, extensive skiing facilities have been provided at Whiteface Mountain, and the Mount Van Hoevenberg Bobsled Run is of Olympic quality. The state Conservation Department has also provided boat-launching sites for the convenience of the public. In addition to recreational facilities which have been constructed by the state, the region contains hundreds of privately owned recreational developments, including private camps and clubs, hotels, dude ranches, organization camps, amusement centers, and privately operated ski slopes.

Both winter and summer recreational activities are carried on. Although no reliable estimates have been compiled regarding the total number of visitors at any season, it is evident that summer is by far the principal recreation season. Transportation facilities within the region are limited, partly as a result of the "forever wild" mandate of the state constitution, which restricts highway improvements and new construction on state Forest Preserve lands. Peripheral highways, completed or under construction, may be expected to enhance access. The area does draw recreationists from all parts of the state and many from the heavily populated urban areas along the Atlantic Seaboard beyond the state's borders.

The potential for additional development and use is related in substantial measure to future legislative enactment of regulatory procedures and public policy regarding the use of Forest Preserve lands. Relaxation of the restrictions governing use of such land may influence more intensive, but not necessarily more desirable, use. Continuation of the trend toward greater use for recreational purposes may be anticipated in the ensuing years, but unquestionably this region will remain as close to a wilderness area as any part of eastern United States.

Great Lakes–St. Lawrence Shore Line

This region extends along the shore line of Lakes Erie and Ontario and the banks of the Niagara and St. Lawrence rivers. The principal physical attractions consist of these major water resources themselves and the sand beaches and dune formations adjacent to the water. In addition to opportunities for water-related recreation and scenic enjoyment, completion of the spectacular engineering accomplishments of the St. Lawrence Seaway and Power Project and the Niagara River Power Project have provided new interests for the sightseer. Ni-

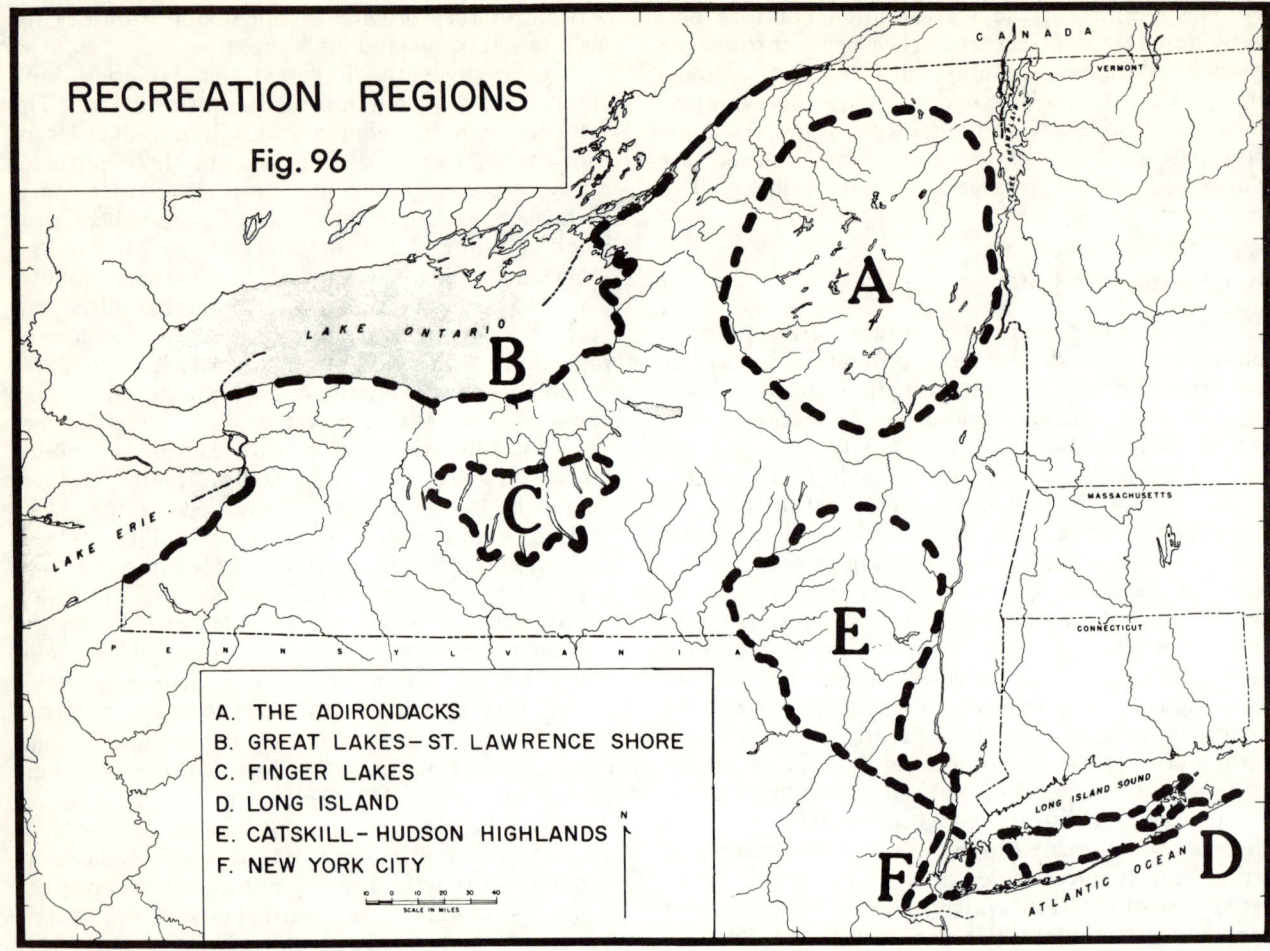

RECREATION REGIONS
Fig. 96

A. THE ADIRONDACKS
B. GREAT LAKES–ST. LAWRENCE SHORE
C. FINGER LAKES
D. LONG ISLAND
E. CATSKILL–HUDSON HIGHLANDS
F. NEW YORK CITY

agara Falls and the Niagara River Gorge have always been a world-famous attraction.

Recreational facilities include state parks and camp sites, private summer homes, resorts, and tourist-oriented recreational businesses. Except for some ice-fishing, recreational use is confined almost exclusively to the summer season. The state has constructed some boat-launching sites, and more are planned in order to provide for the continuing demand.

The volume of recreational use is high and trending upward. Private recreational use of shore line sites for summer residential purposes and commercial establishments has already occupied the better sites, and the cost of shore frontage has become high. The state has expanded its park facilities in the Thousand Islands area, which has sustained the greatest recent percentage increase in attendance of any of the state parks. In 1963, 855,341 persons visited the units of the Thousand Islands State Parks alone.

The opportunities for potential development of recreational opportunities are somewhat limited as a result of the relatively high intensity of existing use which is already characteristic. The state plans to improve the Ontario Lake Parkway, which presently extends west from Rochester, and to construct new state parks. Since private interests have already occupied much of the shore line, the possibility of obtaining adequate sites for state park construction is limited. Acquisition of wetlands along the shore of Lake Ontario and the St. Lawrence River has already been undertaken by the state Conservation Department. Wetlands have value in recreational resource development because they are important to migratory waterfowl and because they provide habitat for wildlife, fish, and shellfish.

Finger Lakes

Because of their common glacial origin and finger-like shape, six large lakes—Skaneateles, Owasco, Cayuga, Seneca, Keuka, and Canandaigua—and five smaller ones—Otisco, Honeoye, Canadice, Hemlock, and Conesus—may correctly be called Finger Lakes. These picturesque lakes offer excellent opportunities for boating, swimming, and fishing. The Appalachian Upland rises sharply from them, producing almost fiord-like scenery. Historical landmarks and other natural features, notably waterfalls, add to the recreation resource base of the region. Facilities attracting the recreational visitor to the region include many state parks and camp sites, summer hotels and resorts, and numerous recreational cottages.

Recreational use is heavy, particularly in close proximity to the lakes. Usable sites for shore line development such as summer cottages and commerical tourist facilities have already been largely claimed. The cities of Rochester and Syracuse are the principal sources of customers, contributing thousands of day-use and week-end visitors, but people from Binghamton, Ithaca, and Elmira, too, enjoy the Finger Lakes' recreational attributes. Access is good, with the main roads running north–south, in conformity with the terrain, to connect with the Thruway and Route 20 on the northern edge of the region. A few miles south of Watkins Glen, located at the southern tip of Seneca Lake, Routes 15 and 17 provide convenient access to Elmira, Binghamton, and more distant places.

Some of the best sites have been used for public

SACANDAGA CAMP SITE in the Adirondacks. Camping provides an opportunity to get away from the hustle and bustle of everyday life at a minimum cost, and the widespread use of most of the state's seventy camp sites attests to its growing popularity. In fact, demand for camp sites seems to keep ahead of new construction. *N.Y.S. Dept. of Conservation.*

CHAPTER 12 THE TERTIARY SECTOR 307

WHITEFACE SKI CENTER. Skiing as a form of recreation is expanding rapidly and new ski facilities are being opened each year, both as private enterprises and state-managed facilities. Heavy snows and good slopes occur so widely that many suitable sites still remain. More and more it is being recognized that summer resorts can enhance their chance of economic success by encouraging nearby winter sports such as skiing. *N.Y.S. Dept. of Commerce.*

or private recreational developments, but new parks are being added at Keuka and Seneca lakes and more growth in recreational facilities can be expected.

Long Island

This region owes its unique functional character to its long shore line and ready access to the immense populations in and around New York City. Recreational resources are dominated by water and sand, as evidenced by such well-known recreational places as Jones Beach and Sunken Meadow state parks. Aside from the water- and sand-related activities, including sun-bathing, swimming, sailing, and deep-sea fishing, hunting is a major recreational activity because Long Island is on the flyway of, and a wintering ground for, a large waterfowl population.

Recreational facilities include a well-organized system of state and county parks, parkways, and private developments. Private clubs, golf courses, race tracks, shooting preserves, resorts, and amusement centers provide recreation for millions of persons.

Nearly 17 million visitors to recreation facilities on Long Island were recorded in 1963, with more than 12 million being accounted for by Jones Beach and Captree state parks alone. Facilities have been expanded in the state parks system to prevent overcrowding, but the rapid growth of the New York metropolitan area continues to produce almost insatiable demands for recreation.

The potential for additional recreational developments is inhibited by rapid conversion of open space to residential and commercial uses. Some expansion opportunity exists, however, in Suffolk County, which is in the process of organizing a county park plan.

Catskill-Hudson Highlands

One might consider this region as two separate areas: the more remote mountainous portion, and the intensively used resort area centered in Orange, Ulster, and Sullivan counties. The recreational attractions in the mountain section include rough, hilly, heavily forested land. Slide Mountain, the highest peak, reaches an elevation of 4,204 feet. Ready access, scenic beauty, and moderate summer temperatures make the place highly attractive to New York City residents. Considerable snow at higher elevations supports winter recreation. Encompassing an area of nearly a million acres, the Catskill high-peak section includes some of the more rugged topography in the state. Forests cover a large portion of this area, with the state owning about 238,000 acres as Forest Preserve lands both inside and outside the Catskill State Park boundary. This acreage is subject to the same constitutional protection as applies in the Adirondack region.

South of the high-peak area the topography is less varied, but woodlands and the feeling of open country are important recreational attractions that have fostered the development of a wide variety of intensively used resort and private recreational establishments.

Public facilities include state camp sites and trails and the state Ski Center at Belleayre Mountain. Private facilities such as luxurious summer hotels, private clubs and estates, and shooting preserves are numerous.

Current recreational use of the region is great. In 1963 about seven million persons visited the Palisades Interstate Park alone. Although adequate statistics are not available, it is obvious that the volume of visitor use of the various recreational facilities provided by private enterprise is great, too.

Opportunities are possible for continued development in the region. The availability of choice sites for recreational purposes is annually becoming more restricted, however, and the need for some form of land-use control is evident.

New York City

The recreational attractions that contribute to the identification of New York City as a recreation region include, in some degree, all the manifold components of this great city. Places of historical and cultural interest, amusement centers and theaters, sports events, and the over-all character and impression of the city provide foci for the individual's enjoyment. And the opportunities for variety of recreational experience are nearly as broad as the attitudes and objectives of the observer.

While it is impossible to indicate precisely the number of persons who visit New York City for recreational purposes, the estimate of six million visitors annually, mentioned above, is probably low. Moreover, the potential for future recreational visits is limited only by the ability of the city to provide attractive facilities and accommodations.

The Sector in Summary

The tertiary sector, which includes the whole range of sales and service activities, has been dealt with here in greater depth than is customary in most geographic analyses, yet there are many other kinds of investigations and interpretations needed before a complete geographic understanding of the sector can be achieved. For example, its further dissection into parts and the analysis of growth trends of the various parts under different locational environments would be an aid to detailed spatial forecasting of the sector's behavior.

Attempts have been made here to provide a limited theoretical frame as well as a factual background for transportation and sales and service activities. In so doing, some light is shed on the functional organization of economic activity, especially as elucidated by the trade area concept. The tertiary sector is now the chief employer in the state, and in this function its importance will increase in the years ahead. It is a trade center or city-builder and a trade area determinant. Associated with it are many of the factors necessary for explaining the geography of connectivity.

Recreation activities represent a fast-growing element in the sector and are increasingly important users of land. Interesting, but not yet fully understood, patterns of connectivity exist between recreation facilities on one hand and the areas from which their customers come on the other. How will the lines of connectivity change? Which recreation areas will expand most rapidly and why? What will be the variation in the impact of recreation on different kinds of areas? These are all important questions for which complete answers are not yet available.

If knowledge concerning the geographic structure of the tertiary sector is added to that of the secondary and primary sectors dealt with in previous chapters, the geography of economic health and prospects for regional development may become better understood. It is this subject with which the next chapter deals.

Selected References

Alexander, John W. "The Basis-Nonbasic Concept of Urban Economic Functions," *Economic Geography*, XXX (July, 1954), 246–61.

Brush, John E., and H. E. Bracey. "Rural Service Centers in Southwestern Wisconsin and Southern England," *Geographical Review*, XLV (October, 1955), 559–69.

Burton, Ian, and Robert Kates (eds.). *Readings in Resource Management and Conservation*. Chicago: Univ. of Chicago Press, 1965.

Getis, Arthur. "The Determination of the Location of Retail Activities with the Use of a Map Transformation," *Economic Geography*, XXXIX (January, 1963), 14–22.

Godlund, S. *The Function and Growth of Bus Traffic Within the Sphere of Urban Influence*. ("Lund Studies in Geography," Series B., No. 18.) Lund, Sweden: Department of Geography, The Royal University of Lund, 1956.

Harris, Chauncy D. "A Functional Classification of Cities in the U.S.," *Geographical Review*, XXXIII (January, 1943), 86–99.

———. "The Market as a Factor in the Localization of Industry in the U.S.," *Annals of the Association of American Geographers*, XLIV (December, 1954), 315–48.

Johnson, Lane J. "Centrality Within a Metropolis," *Economic Geography*, XL (October, 1964), 324–36.

Jones, Victor. "Economic Classification of Cities and Metropolitan Areas," *Municipal Yearbook*, XXI (1954), 62–70.

Lewis, Richard T. "The Measurement of Tertiary Activity." Unpublished Master's thesis, Department of Geography, Syracuse University, 1964.

Mayer, Harold M., and Clyde F. Kohn. *Readings in Urban Geography*. Chicago: Univ. of Chicago Press, 1959.

Nelson, Howard J. "A Service Classification of American Cities," *Economic Geography*, XXXI (July, 1955), 189–210.

New York State Department of Commerce. *New York State Commerce Review*, I–XIV. Albany: The Department, 1946–60 (ceased publication).

New York State Department of Conservation. *The New York State Conservationist*. Albany: The Department, 1946 to date.

New York State Office of Transportation. *Annual Report, 1963*. Albany: The Office, 1964.

Outdoor Recreation Resources Review Commission. *Outdoor Recreation for America. A Report to the President and the Congress*. Washington, D.C.: Govt. Printing Office, 1962.

Philbrick, Allen K. "Principles of Areal Functional Organization in Regional Human Geography," *Economic Geography*, XXXIII (October, 1957), 299–336.

Siddall, William R. "Wholesale-Retail Trade Ratios as Indices of Urban Centrality," *Economic Geography*, XXXVII (April, 1961), 124–32.

Stafford, Howard A., Jr. "The Functional Basis of Small Towns," *Economic Geography*, XXXIX (April, 1963), 165–75.

Thompson, Roger C. "The Doctrine of Wilderness: A Study of the Policy and Politics of the Adirondack Preserve-Park." Unpublished Ph.D. dissertation, State University College of Forestry at Syracuse University, 1962.

Ullman, Edward L. "The Role of Transportation and the Bases for Interaction," in *Man's Role in Changing the Face of the Earth*, ed. William L. Thomas. Chicago: Univ. of Chicago Press, 1956. Pp. 862–77.

U.S. Bureau of the Budget. *Standard Industrial Classification Manual*. Washington, D.C.: Govt. Printing Office, 1957.

Vance, J. E. "Labor-Shed, Employment Field, and Dynamic Analysis in Urban Geography," *Economic Geography*, XXXVI (July, 1960), 189–220.

Van Cleef, E. *Trade Centers and Trade Routes*. New York: Appleton-Century, 1937.

CHAPTER 13   SIDNEY C. SUFRIN *and* JOHN H. THOMPSON

Strength of the State's Economy

THE LAST THREE chapters have given some indication of New York's contribution to the national economy and analyzed the geography of the three economic sectors. They have demonstrated that the state has an advanced type of economy based largely on secondary and tertiary activities, with these two sectors employing about 98 per cent of the labor force. Because secondary and tertiary jobs are found largely in and around cities, during recent decades urban growth has occurred while rural and village sections have diminished in importance. In the years ahead it is overwhelmingly likely that most new jobs will be of tertiary variety and will be located in or near the seven major urban systems. Somewhat in excess of 80 per cent of the people of the state already live in a 40-mile-wide band which follows the Hudson–Mohawk–Lake Plain axis. As the state's population doubles in the future, this narrow band, perhaps grown slightly wider, will continue to house and employ at least 80 per cent of the population.

This chapter introduces the concept of social organization significance and then analyzes the strength of the state's economy in terms of external comparisons and internal spatial variations.

Social Organization and the State's Economy

The over-all character of an area—and this includes economic strength—depends to a large extent upon what man wants to do, how he does it, and what technical skills are at his command. As is true for any economic society, the strength of New York's economy depends heavily on its people and on the social organization, or institutions, they have developed in order to make the state a going concern. There is no intention to degrade the importance of geographical location, natural resource base, or man-made resources such as transportation or power, but these things in themselves will not assure an economy of success and a high standard of per capita output and income unless a skillful people have developed an effective social organization. Social organization is effected through many kinds of institutions such as those related to government, education, health, religion, philanthropy, and similar noneconomic but economy-influencing institutions.

Certainly among the above-mentioned elements of social organization, governmental structures and activities are of great significance. Governmental significance becomes apparent when one stops to realize the extent to which institutions today are established by government, at one level or another, to administer a wide range of things from education to health facilities. Because of its importance, government will be used here to illustrate the significance of social organization for the development of human skills and economic success.

GOVERNMENT, TAXES, SOCIAL ORGANIZATION, AND INCOME

Money is required if government is to make an efficient contribution to social organization. A state's income, that is, the amount of taxes or loans it can secure, is related to the total income produced, and the amount of income commonly reflects the effectiveness of the social organization. Thus, the more income generated or otherwise accruing to the population, the more public revenue possible. Revenue gives government (state and local) the opportunity to develop more effective social organization, which results in more income—and around and around the cycle goes. It may be hypothesized that taxes may turn out to be not a drain on or handicap to development but rather a reflection of income and economic success and an aid to the development of effective social organization. This is somewhat in opposition to the commonly held view that taxes—especially high taxes—tend to slow up industrial growth and new investment. High taxes instead may well provide high-quality services and high-quality social organization. Elaboration of this idea follows.

State and local taxes are a proper charge to business as a cost, and so can be considered deductible items when the businessman computes his federal corporation taxes. This reduces the importance of state and local tax costs in business calculations. It

TABLE 34
AVERAGE TOTAL TAX BILLS FOR THREE HYPOTHETICAL
CORPORATIONS IN SELECTED STATES, 1957

States	Corp. A (Rank)*	Corp. B (Rank)	Corp. C (Rank)
Michigan	(1)	(1)	(1)
1. Average total state-local taxes	$235,579	$241,100	$234,455
2. Unempl. & Workmen's Comp. taxes	49,140	56,160	56,160
3. Total, all taxes	284,719	297,260	290,615
Maryland	(2)	(2)	(2)
1. Average total state-local taxes	224,694	209,531	227,375
2. Unempl. & Workmen's Comp. taxes	53,900	61,600	61,600
3. Total, all taxes	278,594	271,131	288,975
Indiana	(5)	(3)	(3)
1. Average total state-local taxes	189,433	200,720	197,249
2. Unempl. & Workmen's Comp. taxes	48,160	55,040	55,040
3. Total, all taxes	237,593	255,760	252,289
New York	(3)	(4)	(4)
1. Average total state-local taxes	184,308	115,120	139,210
2. Unempl. & Workmen's Comp. taxes	89,950	102,800	102,800
3. Total, all taxes	274,258	217,920	242,010
New Jersey	(6)	(5)	(6)
1. Average total state-local taxes	130,120	111,261	111,423
2. Unempl. & Workmen's Comp. taxes	89,900	101,600	101,600
3. Total, all taxes	219,020	212,861	213,023
Pennsylvania	(4)	(6)	(5)
1. Average total state-local taxes	170,548	119,581	145,716
2. Unempl. & Workmen's Comp. taxes	67,935	77,640	77,640
3. Total, all taxes	238,483	197,221	223,356
Illinois	(7)	(7)	(7)
1. Average total state-local taxes	109,988	105,909	102,777
2. Unempl. & Workmen's Comp. taxes	45,710	52,240	52,240
3. Total, all taxes	155,698	158,149	155,017

*A ranking of (1) refers to highest tax-paying position, (7) to lowest.
Source: Adapted from Pennsylvania Economy League Study 1957; Associated Industries of N.Y. State.

does not, however, completely remove them as costs, for the tax rates and bases of different places vary widely. Also, the services paid for by taxes and rendered to business and individuals vary so greatly from place to place that high-tax areas may be low-cost areas, or vice versa. Only when high-tax areas remain high-cost areas can economic difficulties be expected.

Tax levels and rate of tax increase are ultimately determined by what the people want in the way of services. In general, it appears that the higher the per capita income of a community, the more services the people of that community demand from public sources. In communities that make great demands for schools, hospitals, roads, and other services, taxes are apt to be higher than in communities where demands are less. For example, good municipal water and sewage systems, good road maintenance, or high-quality public schools, all of which require tax money, should relieve the corporation or individual from having to provide these things through out-of-pocket costs. What anyone might legitimately ask and what the businessman should want to know about any given area is not only whether taxes are high but also whether the tax burden is higher or lower than what competitors are paying in other areas and whether it increases or decreases costs. This complex statistical-economic question is difficult to answer, for it is almost impossible to find situations of sufficiently similar character which occur in widely separated communities with different tax structures.

A 1957 comparison between conditions for three hypothetical corporations situated in New York State and six other industrialized states is attempted in Table 34. Although there are widely different taxing policies and structures in the seven states, the table does provide a rough idea of the position of New York in the tax bill hierarchy. Clearly it ranks about in the middle, and the implication from the foregoing discussion is that therefore the service range and the cost outlay functions of the services should be about medium. It is difficult to prove, but most experts feel that this service range, at worst, is medium.

EDUCATION, A KEY ELEMENT IN SOCIAL ORGANIZATION

Education is perhaps one of the best uses for tax money. It is generally recognized that it is the tool which converts an unskilled population into a skilled one, the tool which makes the good mind, the capable person, efficient, the tool which allows the individual to compete successfully in an increasingly complex society. Nearly half (46 per cent) of the New York State budget is devoted to education, and when this expenditure is added to that of local school districts more money is probably spent to educate an elementary or secondary school pupil in New York than in any other state in the union. This sounds impressive, but state and local public school revenue as a per cent of personal income amounts to but 3.9, giving New York a "middle" national ranking of twenty-ninth. The state government expressly takes the position that no young man or woman with the desire and capacity will be denied a college education for lack of personal financial means or insufficient facilities. To further this goal, the state supports a very large program (at a cost of 198 million dollars in 1964/65) of scholarships and fellowships, is accelerating expansion of the State University, and gives financial assistance to numerous city- or community-sponsored colleges throughout the state. All of this is a reflection of meritorious intent and substantial effort, but as yet results have not been spectacular. For example, the median number of school years completed by people over twenty-five years of age is 10.7, which gives New York an unimpressive rank of twenty-eighth among the fifty states. Other indicators of educational success and national ranking are as follows: eighth graders who graduate from high school (74.1 per cent—rank eighteenth), population with less than five years of schooling (7.8 per cent—rank thirty-second), selective service registrants failing mental test (22.4 per cent—rank thirty-fourth), and adults with four years of college (8.9 per cent—rank twelfth).[1]

If the above measurements are at all indicative, it can be concluded that New York is cognizant of the need for improved education and is spending a large sum per pupil, but as yet is perhaps only a bit above medium in attainment and has ample room for improvement.

DIFFERENT ROUTES TO IDENTICAL OBJECTIVES

Again it is appropriate to return to the idea that states may choose different routes to implementing the strength of their economies. While some may hope to stimulate private industrial development by keeping taxes low, others may seek to do so by providing a high level of public facilities and services, reflecting the view that the newer high-value-added electronic and chemical industries, for example, place considerable significance on availability of good-quality school, hospital, library, higher educational, water, and sanitation facilities. Investments in these facilities in New York would seem to be justified, as it has been suggested in Chapter 11 that high-value-added industries have the brightest futures in New York.

Some states, of course, have large fiscal capacity or taxing base as compared to others. The extent to which states and their local governments use the fiscal capacity available to them is called tax effort. Theoretically, if a state is interested in providing excellent services, it will put forth a strong tax effort. Table 35 contains indexes of tax effort which show the position of all states in relation to the (theoretically) average state. In the left-hand column, where fiscal capacity is taken to mean personal income, New York ranks eighteenth among the states and is 13 per cent above the average state. In the right-hand column, fiscal capacity is assumed to be the sum of average base components of the tax system of all the states. This is called here the yield under a representative tax system. Because New York makes use of certain kinds of taxes not employed, or less effectively employed, by many other states, it exhibits a very high tax effort, ranking second among the fifty states and 36 per cent above the average state. Its personal and corporate state income tax and New York City's gross-receipts tax are largely the cause of the state's high rank.

SUMMARY

Tax burdens of New York State are not especially high as compared to other industrial states, or to incomes. The total tax base and the total taxes collected are high, however, and the tax effort is vigorous. If tax money is at least as wisely spent here as in other states, it should follow that New York provides better than average social organization and services, and in so doing is in a position to compete rather favorably for those kinds of business establishments which seek a wide service range.

---

[1] These data are taken from report of National Committee for the Support of Public Schools, summarized in *Changing Times, The Kiplinger Magazine* (November, 1964).

TABLE 35
TAX EFFORTS

Rank	Personal Income as a Measure of Fiscal Capacity		Yield Under Representative Tax System		Rank
	State	Per cent of U.S. Average	State	Per cent of U.S. Average	
1	South Dakota	140	Hawaii	155	1
2	North Dakota	136	New York	136	2
3	Vermont	132	Vermont	130	3
4	Louisiana	126	Maine	126	4
5	Hawaii	124	Massachusetts	121	5
6	Arizona	121	Alaska	116	6
7	Montana	121	Washington	114	7
8	Mississippi	120	Mississippi	113	8
9	Maine	119	Oregon	113	9
10	Minnesota	119	Rhode Island	112	10
11	Utah	117	Michigan	110	11
12	Idaho	116	Wisconsin	110	12
13	Kansas	116	California	109	13
14	Wyoming	115	Louisiana	106	14
15	California	114	Maryland	106	15
16	Colorado	114	South Carolina	106	16
17	Iowa	113	Minnesota	105	17
18	New York	113	Arizona	104	18
19	Oregon	113	Georgia	102	19
20	Washington	113	West Virginia	101	20
21	Nevada	109	Colorado	100	21
22	Wisconsin	108	Utah	98	22
23	Oklahoma	105	New Jersey	97	23
24	Michigan	104	Kansas	96	24
25	New Mexico	104	North Carolina	96	25
26	Florida	103	Pennsylvania	96	26
27	Massachusetts	103	New Hampshire	95	27
28	South Carolina	103	Connecticut	94	28
29	New Hampshire	101	Oklahoma	94	29
30	Arkansas	100	Nevada	93	30
31	North Carolina	99	Tennessee	93	31
32	Georgia	97	South Dakota	92	32
33	West Virginia	97	Alabama	91	33
34	Rhode Island	96	Iowa	91	34
35	Tennessee	95	North Dakota	91	35
36	Nebraska	94	Ohio	91	36
37	Indiana	92	Arkansas	90	37
38	Maryland	92	Florida	90	38
39	Texas	92	Idaho	89	39
40	Alabama	91	Illinois	88	40
41	Ohio	88	Delaware	87	41
42	Illinois	85	Indiana	87	42
43	New Jersey	85	Montana	86	43
44	Pennsylvania	85	District of Columbia	85	44
45	Kentucky	83	New Mexico	84	45
46	Connecticut	82	Virginia	84	46
47	Virginia	82	Kentucky	80	47
48	District of Columbia	79	Missouri	76	48
49	Missouri	75	Wyoming	73	49
50	Delaware	73	Nebraska	72	50
51	Alaska	71	Texas	67	51

Source: "Measures of State and Local Fiscal Capacity and Tax Effort," in *Staff Report M-16* (Washington, D.C.: Advisory Commission on Intergovernmental Relations, 1962), p. 82.

These kinds of establishments tend to employ high-income people and often are associated with high standard of living.

That the relation of income, taxes, service range, social organization, and business climate be kept clearly in mind by state officials and that the tax structure of New York be periodically re-examined with an eye to keeping pace with the dynamism of national business climate is vital. In the end it becomes a matter of evaluating the service range level desired and then establishing a tax structure, so that maximum advantage to the taxpayer will accrue per dollar of taxes levied.

External Comparisons

In comparison to other states New York has been given a better than average grade for social organization—which, it has been implied, means general business climate. There are many other criteria that will aid in shedding light on the relative strength or health of the state's economy. These range from population characteristics, through income and employment, to those which are indicative of general business conditions. Nearly thirty criteria in at least four broad categories have been selected for the following analysis. For each criterion New York is compared to the nation as a whole, the neighboring states of Vermont, New Jersey, and Pennsylvania, the depressed states of West Virginia and Arkansas, and the economically younger and faster-growing states of Florida, Arizona, and California. (Data are shown in Table 36.) The analysis assumes that level of economic conditions and trends in economic conditions are two different but essential elements in the concept of economic health or strength. For example, increase in per capita income between 1950 and 1960 measures economic health, but in a different way than the level of per capita income at any one time. As has been explained previously, New York, being a part of the older Northeast, should not expect such favorable trends as states in the West and South. On the other hand, it may be ahead of these states in level of economic attainment.

Population

Population trends, especially, are valuable indications of economic health. Both total population and population in the most productive age group are examined below.

Total Population

Over 9 per cent of the nation's population in 1960 was in New York, and its increment of nearly two million in the 1950–60 decade was nearly one-fifteenth of the national gain. It did not grow nearly so fast as the Florida-Arizona-California group, and ranked behind New Jersey among its neighbors; but it was better off in this respect than the depressed states, which showed actual declines in population. That neighbor New Jersey ranked above New York in per cent gain reflects in part at least the economic expansion of metropolitan New York City, which affects New Jersey. New York can be expected to add substantial increments in the years ahead, but its growth rate will be slightly behind the national trend, and well behind that of California in both percentage and absolute gains. A lag behind the nation and California is no reason for alarm, and, as a matter of fact is quite as should be expected considering the state's geographic location in the earlier developed Northeast.

Population in the Most Productive Age Group

It is assumed that people aged twenty to forty-nine best represent the productive working population. Areas with a good job history and opportunities attract people in this productive age group, while areas with poor employment opportunities experience relative losses as people seek jobs elsewhere.

Table 36 shows that in both 1950 and 1960 New York ranked slightly above the United States and Pennsylvania, slightly below New Jersey, and well above Vermont in this productive age group. It can be assumed that Vermont is exporting substantial numbers of people to New York and adjoining states. In fact, it ranks with the depressed states by this as well as many other measurements. That New York ranks above Arizona and Florida and only slightly below California is a reflection of the relatively large numbers of retired persons in the population mix of these states. All in all, New York appears very competitive in its ability to attract and hold the twenty to forty-nine age group.

Income

Income is a standard measurement of economic conditions. Both level of income and trend in income are highly indicative. Per capita income, per cent of families in low-income brackets, and per cent of families in high-income brackets are utilized in this analysis. New York ranks well by all three measures.

Per Capita Income

Neither the nation nor any of the rapidly growing states shown in Table 36 have a per capita income quite as high as that of New York State, although California's is only slightly lower and may well

TABLE 36
STRENGTH OF NEW YORK'S ECONOMY
(External Comparisons)

	POPULATION				INCOME					EMPLOYMENT							GENERAL BUSINESS		
										Manufacturing			Tertiary				Bank Assets		
	Total (Thous.)	Per Cent of U.S.	Per Cent Change (1950-60)	Per Cent in 20-49 Age Group	Per Capita (Dol.)	Per Cent Change (1950-60)	Per Cent of Families with under $3,000	Per Cent of Families with over $10,000	Number of Families with over $10,000 (Thous.)	Total (Thous.)	Per Cent of U.S.	Per Cent Gain 1950-60	Total (Thous.)	Per Cent of U.S.	Per Cent Gain 1950-60	Unemployment Rate	Total (Mil. 1950 Dol.)	Per Cent of U.S.	Per Cent Change 1950-60
Column	1	2	3	4	5	6	7	8	9	10	11	12	13	14	15	16	17	18	19
United States																			
1960*	179,323	100.0	18.5	38.4	2,223	49.1	21.4	15.1	6,814	16,730	100.0	16.4	37,559	100.0	19.2	5.1	256,987	100.0	34.3
1950	151,326	100.0		43.6	1,491		48.4	3.1	1,188	14,370	100.0		31,511	100.0		4.8	191,317	100.0	
New York																			
1960	16,782	9.4	13.2	39.4	2,789	48.2	13.8	19.9	863	1,957	11.7	12.7	4,747	12.6	18.8	5.2	65,806	25.6	32.3
1950	14,830	9.8		45.9	1,882		38.2	4.8	185	1,737	12.1		3,996	12.7		6.0	49,724	26.0	
Neighbors																			
Vermont																			
1960	390	.2	3.2	34.8	1,859	56.5	23.1	8.9	8	37	.2	−10.8	80	.2	6.7	4.5	502	.2	66.2
1950	378	.2		39.6	1,188		60.4	1.7	2	41	.3		75	.2		5.5	382	.2	
New Jersey																			
1960	6,067	3.4	25.5	40.4	2,665	48.9	11.4	22.0	343	814	4.9	13.4	1,235	3.3	24.7	4.6	8,161	3.2	38.3
1950	4,835	3.2		46.3	1,790		34.1	4.7	59	718	5.0		990	3.1		5.1	5,899	3.1	
Pennsylvania																			
1960	11,319	6.3	7.8	38.8	2,266	44.7	16.9	13.9	403	1,449	8.7	2.5	2,410	6.4	7.0	6.2	15,975	6.2	20.9
1950	10,498	6.9		44.6	1,566		45.3	2.7	71	1,413	9.8		2,253	7.1		5.4	13,216	6.9	
Depressed States																			
West Virginia																			
1960	1,860	1.0	−12.3	36.1	1,674	52.5	32.6	8.4	39	117	.7	−6.0	326	.9	10.1	8.3	1,247	.5	19.9
1950	2,006	1.3		41.3	1,098		60.2	1.6	8	124	.9		296	.9		4.8	1,040	.5	
Arkansas																			
1960	1,786	1.0	−6.9	34.0	1,341	66.2	47.7	5.5	24	103	.6	45.1	292	.8	17.7	6.0	1,223	.5	34.5
1950	1,910	1.3		39.3	807		78.2	1.3	6	71	.5		248	.8		4.7	909	.5	
Growth States																			
Florida																			
1960	4,952	2.8	78.7	37.8	1,988	54.5	28.4	11.1	145	193	1.2	121.8	810	2.2	35.7	5.0	4,634	1.8	115.7
1950	2,771	1.8		44.8	1,287		61.7	2.8	20	87	.6		597	1.9		4.5	2,148	1.1	
Arizona																			
1960	1,302	.7	73.6	38.6	2,011	55.3	21.3	14.4	45	48	.3	220.0	252	.7	81.3	5.3	1,222	.5	142.5
1950	750	.5		42.6	1,295		52.6	2.8	5	15	.1		139	.4		7.6	504	.3	
California																			
1960	15,717	8.8	48.5	40.6	2,741	49.0	14.1	21.8	870	1,296	7.7	76.6	3,553	9.5	40.5	6.1	23,184	9.0	52.6
1950	10,586	7.0		46.2	1,839		37.0	3.8	107	734	5.1		2,529	8.0		7.9	15,193	7.9	

Sources (by column): 1, 5, 17, 30, 33, 35, 38—Statistical Abstract of the U.S., 1952, 1962; 4, 7, 8, 9, 16—U.S. Census of Population 1950—60; 10, 25—Annual Survey of Manufactures, 1950, 1960; 13—Census of Business, 1948, 1958, and U.S. Census of Population, 1950, 1960; 20—U.S. House of Representatives, Subcommittee on Science, Research, and Development; 22, 23, 24—U.S. Census of Manufactures, 1954, 1958, and Annual Survey of Manufactures, 1955, 1956, 1957, 1959, 1960; all remaining columns computed from present data.

Indexes: 17, 35—Wholesale Price Index, all commodities; 25—Wholesale Price Index, manufactures; 30—Consumer Price Index, all commodities.

show relative improvement in the immediate future. In 1960 New York's per capita income was $125 above that of New Jersey, more than $500 above Pennsylvania, and $900 above Vermont, and over twice that of Arkansas. Low incomes in Florida and Arizona reflect once again the situation of the retired population, and possibly the relatively less developed secondary and tertiary sectors of those states.

It is interesting, but not alarming, that the per cent gain in per capita income between 1950 and 1960 in New York was less than for the United States as a whole or for any other of the states shown except Pennsylvania. This simply reemphasizes the point that higher per cent gains are easier to achieve when the base or starting point is low. Arkansas, for example, which had a per capita income of $807 in 1950 increased to $1,341 in 1960.

TABLE 36 (Continued)

GENERAL BUSINESS					EMPLOYEE EFFICIENCY														
Government Contract Awards for Research and Dev. (1963)		Average Annual Expenditure for New Plant and Equipment (1954–60)			Manufacturing					Retail Trade					Wholesale Trade				
Total (Mil. Dol.)	Per Capita (Dol.)†	Average Annual Expenditure (Mil. Dol.)	Per Capita (Dol.)	Per Manufacturing Employee (Dol.)	Total Value Added (Mil. 1950 Dol.)	Per Cent of U.S.	Per Cent Gain 1950–60	Per Employee (Dol.)	Per Cent Change 1950–60	Total Sales (Mil. 1948 Dol.)	Per Cent of U.S.	Per Cent Gain 1948–58	Per Employee (Dol.)	Per Cent Change 1948–58	Total Sales (Mil. 1948 Dol.)	Per Cent of U.S.	Per Cent Gain 1948–58	Per Employee (Dol.)	Per Cent Change 1948–58
20	21	22	23	24	25	26	27	28	29	30	31	32	33	34	35	36	37	38	39
9,864	55.01	9,531	55.70	571	135,406	100.0	51.0	8,094	29.7	177,786	100.0	36.2	19,081	26.6	250,199	100.0	32.6	84,367	17.5
					89,676	100.0		6,241		130,521	100.0		15,072		188,689	100.0		71,829	
917	54.64	664	40.69	336	14,911	11.0	41.9	7,619	26.0	18,450	10.4	26.1	18,448	19.1	48,068	19.2	11.4	113,187	15.9
					10,506	11.7		6,048		14,627	11.2		15,496		43,140	22.9		97,699	
5	12.82	14	35.71	383	236	.2	14.6	6,378	27.0	393	.2	16.3	18,916	15.5	202	.1	20.2	51,100	3.4
					206	.2		5,024		338	.3		16,382		168	.1		49,426	
289	47.63	439	77.56	537	7,090	5.2	45.7	8,710	28.5	6,455	3.6	44.1	19,521	20.4	7,423	3.0	106.4	80,600	43.7
					4,867	5.4		6,779		4,479	3.4		16,214		3,597	1.9		56,072	
349	30.83	770	69.94	523	10,679	7.9	31.5	7,370	28.2	10,933	6.1	20.6	17,384	23.6	14,279	5.7	29.7	78,054	19.6
					8,124	9.1		5,749		9,069	6.9		14,061		11,011	5.8		65,275	
5	2.69	151	80.17	1,221	1,160	.9	31.5	9,915	39.4	1,426	.8	10.7	16,620	13.3	1,157	.5	7.4	58,470	.9
					882	1.0		7,113		1,288	1.0		14,671		1,077	.6		57,925	
17	9.52	55	31.11	616	623	.5	94.1	6,049	33.8	1,364	.8	25.9	16,644	18.3	1,019	.4	19.6	61,467	19.5
					321	.4		4,521		1,083	.8		14,071		852	.5		51,452	
248	50.08	127	29.75	800	1,493	1.1	239.3	7,736	53.0	5,182	2.9	121.5	18,130	28.2	4,827	1.9	142.4	57,536	54.4
					440	.5		5,057		2,340	1.8		14,137		1,991	1.1		37,276	
140	107.53	23‡	20.20	594	380	.3	196.9	7,917	−7.8	1,253	.7	90.4	18,358	13.4	1,063	.4	153.1	61,481	29.2
					128	.1		8,533		658	.5		16,190		420	.2		47,592	
3,808	242.29	706	49.44	586	11,471	8.5	124.0	8,851	26.9	17,700	10.0	60.6	20,146	21.0	22,685	9.1	70.9	78,625	38.3
					5,121	5.7		6,977		11,020	8.4		16,643		13,271	7.0		56,831	

*In columns 30–39, data are for 1948 and 1958 Business Census years rather than 1950 and 1960.   †Per capita figures based on 1960 populations.
‡Arizona's average expenditure is based on six years; 1957 data are not available.

The per cent change was 66; the absolute was $534. New York, which had a per cent change of only 48, however, had an absolute change of $907, leaving Arkansas further behind in actual income dollars in 1960 than it was in 1950. None of the states shown in Table 36 equaled New York's absolute gain.

Any simple discussion of per capita income cannot really bring out departures from a normal curve which the distribution of income may exhibit. For example a few cases of very high income may partially cover up the degree of lowness of income earned by the majority of the population. This problem may be set aside by examining the distribution of incomes at different levels.

Percentage of Families with Low Incomes

A $3,000 annual family income has recently been designated by the federal government as the minimum necessary to escape poverty status in the United States. Obviously $3,000 will "go" much further in southern rural environments than in northern cities, but comparisons of New York and

CHAPTER 13 STRENGTH OF THE STATE'S ECONOMY 317

other states in terms of the percentage of families earning less than $3,000 will give some indication of the relative strength of the states' economies.

Only New Jersey, among the states shown in Table 36, had in 1960 a lower percentage of its families below $3,000 than New York. The nation as a whole is not so well off as New York in this respect. Attention might be drawn to the very high percentages of low-income families in the depressed states of Arkansas and West Virginia (32.6 per cent). Clearly, if the figure $3,000 means anything, the "war on poverty" will have to be waged vigorously in these states.

Percentage of Families with High Incomes

Twenty per cent of New York families had incomes over $10,000 in 1960. This is well above the national average, and higher than in any of the states compared except New Jersey and California. Neighboring Pennsylvania and Vermont are far behind, the latter again comparing fairly closely with the depressed states. With more than 860,000 families earning above $10,000 in 1960, New York ranks with California and far ahead of the other states in affluence, as measured by this criterion.

EMPLOYMENT

Total employment tends to follow fairly closely the population trends, and so need not be referred to here. However, because modern economies are so closely related to conditions in manufacturing and tertiary activities, these can be used to draw comparisons between New York and other areas. Unemployment is a useful indicator too.

Manufacturing Employment

New York, with 12 per cent of the nation's factory workers, outranks all other states. Although New Jersey, among its neighbors, experienced a greater per cent gain between 1950 and 1960, New York's addition of 220,000 factory workers in that decade exceeds by far the combined absolute gains of New Jersey and Pennsylvania. Vermont and West Virginia experienced losses.

That New York did not keep pace with national growth rates or trends in Florida, Arizona, and California in the 1950–60 decade was to be expected. California, with a growth in manufacturing employment of over half a million, left all other states far behind. Arizona's 220 per cent gain amounted in absolute jobs to but a seventh of New York's.

Tertiary Employment

Chapter 12 emphasizes the importance of the tertiary sector as a current and future employer. New York State with over 12 per cent of the nation's tertiary employment in 1960 but less than 10 per cent of the population, exhibits a strong tertiary orientation. As pointed out earlier, this is largely a reflection of New York City's great sales and service function. While Pennsylvania, New Jersey, and Vermont together added 407,000 tertiary jobs in the 1950–60 decade, New York added 751,000. Only California added more. New York very nearly held its own in percentage of the national tertiary workers, which is rather remarkable for a northeastern state.

Unemployment

It may be nearly impossible for any state to lower its unemployment rate below about 4 per cent, for those who simply are not skilled enough to hold jobs plus the short-term unemployed make up at least that portion of the labor force. No state in Table 36 either in 1950 or 1960 shows unemployment rates below 4.5 per cent. In 1960 New York was about at the national average and ranked in the middle of the states shown. Interestingly enough Vermont had the lowest unemployment rate in 1960. This may be explained in part at least by the fact that much unemployed farm labor does not get counted and that numerous Vermonters find jobs outside the state. It is interesting, too, that the rapidly growing states on the whole are faring a little less well than New York as far as the unemployment problem is concerned. Of the depressed states, West Virginia shows up badly.

GENERAL BUSINESS INDICATORS

Many indicators of general business might shed light on New York's comparative economic strength. The four used below are (1) bank assets, (2) government contract awards for research and development, (3) expenditures for new plant and equipment, and (4) certain employee efficiency indexes.

Bank Assets

Twenty-five per cent of the bank assets of the United States are in New York, concentrated in New York City. Although the per cent growth in bank assets between 1950 and 1960 in New York was slightly less than for the nation as a whole, the additions amounted to as much as 16 billion dollars, which was twice that of California and ten times that of Florida. Neighbors Pennsylvania and New Jersey each added between 2 billion and 3 billion dollars. This outstanding position in bank assets indicates the great role of New York City as the seat of the nation's banking activities.

Government Contract Awards for Research and Development

The distribution of government funds for research and development is highly changeable. It probably reflects both the potential of an area to do the kinds of things the government wants done and the effectiveness of political pressures and lobbies. Data for 1963 indicate that a handful of states dominate the booming government program known as research and development. California received 38.6 per cent of the total outlay and New York was second with 9.3 per cent. It is reasonable that these two populous and industrially advanced states should rank high but surprising that California should do roughly four times as well per capita as New York. That nearly 1 billion dollars in this kind of federal spending occurred in New York in a single year, however, speaks well for the kind of industrial situation the state enjoys. From the 1963 figures, and based on population, it probably can be assumed that the state should receive in the years ahead about a tenth of the federal outlays for research and development. This is bound to do much for its economy because research and development activity attracts related activities.

Investment in New Plant and Equipment

Investment in new plant and equipment reflects faith in business climate and thus says something about prospects for the future. Also, it may indicate something about the general industrial success recently experienced. On first observation New York appears very strong in this regard, ranking third behind Pennsylvania and California in total investment for the 1954-60 period. In terms of investment per capita, however, it ranks well behind the nation and outranks only Vermont, Arkansas, Florida, and Arizona among the states compared. Perhaps the investment in new plant and equipment per manufacturing employee (which implies investment per given quantity of manufacturing) is a more useful criterion of comparison. By this measure New York unfortunately appears in last place in Table 36. In other words, although New York is the greatest of our industrial states, the lack of proportional new investment would not support the contention that the state will hold its own. Although it might not be expected to compare favorably with the young, rapidly growing state in new investments per manufacturing employee, it might be assumed that it should do as well as the neighboring states of New Jersey and Pennsylvania. In spite of West Virginia's generally depressed condition, it experienced the highest apparent effort in this kind of new investment. It would be unfortunate if most states waited until depressed conditions set in to expend large investment effort.

Employee Efficiency

That one employee produces more than another because of the kind of activity in which he is engaged, or because he has better equipment with which to work, or because he is more energetic and conscientious, is to be expected. Attempts are made here to compare New York's employee efficiency with that of other states in terms of: (1) value added by manufacture per employee in manufacturing (2) volume of retail trade per employee in retail trade, and (3) volume of wholesale trade per employee in wholesale trade.

Value Added by Manufacture per Employee in Manufacturing.—New York is the nation's greatest manufacturing state, but in terms of value added by manufacture per employee it ranks below the national average as well as below all of the other states shown in Table 36 except Vermont, Pennsylvania, and Arkansas. That its position is no stronger is a reflection, in part at least, of the state's emphasis on certain low-value-added industry types such as garments. It seems desirable that more of its industrial base should shift to high-value-added types in the future to insure a competitive position in the over-all industrial structure of the nation. This point is particularly pertinent because New York is in an old-age industrial region, where low-value-added industries have more than their share of difficulties. The peculiar market, labor, and plant space advantages of New York City for the garment industry will tend, however, to maintain that industry in the face of what would seem to be major obstacles.

In terms of gain in value added by manufacture per employee between 1950 and 1960, the nation and all states shown except Arizona outstripped New York. This may be more serious than its low rank.

Volume of Retail Trade per Employee in Retail Trade.—Although New York's efficiency level as measured by volume of retail sales per employee in retail sales appears somewhat better than when measured by value added per manufacturing employee, its position is still not overly favorable by this measure. The state ranks below the national average and behind New Jersey and California but ahead of the rest of the states shown in Table 36. In both per cent gain and absolute gain in efficiency between 1950 and 1960, it ranks ahead of only Vermont, Arizona, and the depressed states of West Virginia and Arkansas.

Volume of Wholesale Trade per Employee in Wholesale Trade.—New York exceeds by far all states in volume of wholesale sales and is more than 40 per cent above any other state in the volume of wholesale sales per employee. This prodigious position of the state reflects the efficient wholesaling functions of metropolitan New York as well as that of a number of upstate cities such as Syracuse.

From the standpoint of per cent gains it does not compare well, but here again is the case of the state with a high level of attainment in the base year not being able to achieve great per cent gains. Absolute gains in volume of wholesale sales per employee in New York between 1950 and 1960 amounted to over $15,000. This was more than the national average and above that of all states shown in Table 36 except New Jersey, Florida, and California.

SUMMARY

There are many other criteria that might be used to present New York State's comparative status in economic strength, but the above indicators, including fifteen individual measures of level and fourteen measures of trend, probably provide a reasonably comprehensive evaluation. Table 37, which ranks New York, the United States as a whole, and the other eight states according to each measure referred to in Table 36, shows that New York ranks favorably in level, being above the United States as a whole and all the states compared except California and New Jersey, which has the same rating as New York. New York enjoys a relatively stronger position than neighbor states Pennsylvania and Vermont. The trend indicators, as might be expected for an older area with a large economic base, paint a less favorable picture for New York. Measured by these kinds of indicators, it ranks behind the United States average and behind Florida, California, New Jersey, and Arizona. It still looks good when compared with neighbors Pennsylvania and Vermont.

With the fifteen level indicators weighted equally, New York as shown in Table 37 has a mean rank of 3.3. This is well above a theoretical middle ranking of 5.5. Its mean trend rank is exactly the middle ranking of 5.5. If the trend means and level means are given equal weight and again averaged, New

TABLE 37
RANK* OF UNITED STATES AND NINE STATES IN STRENGTH
OF ECONOMY AS MEASURED BY SELECTED INDICATORS

	United States	New York	Vermont	New Jersey	Pennsylvania	West Virginia	Arkansas	Florida	Arizona	California
LEVEL INDICATORS†										
Per Cent of Population Age 20–49	6	3	9	2	4	8	10	7	5	1
Per Capita Income	5	1	8	3	4	9	10	7	6	2
Per Cent of Families with under $3,000 Income (reverse order for rank)	6	2	7	1	4	9	10	8	5	3
Per Cent of Families with over $10,000 Income	4	3	8	1	6	9	10	7	5	2
Number of Families with over $10,000 Income	5‡	2	10	4	3	8	9	6	7	1
Per Capita Value of Federal Contracts	3	4	8	6	7	10	9	5	2	1
Manufacturing Employment	5‡	1	10	4	3	7	8	6	9	2
Tertiary Employment	5‡	1	10	4	3	7	8	6	9	2
Rate of Unemployment (reverse order for Rank)	4	5	1	2	9	10	7	3	6	8
Bank Assets	5‡	1	10	4	3	7	8	6	9	2
Average Annual Investment in New Plant & Equipment	5‡	3	10	4	1	6	8	7	9	2
Average Annual Investment in New Plant & Equipment per Manufacturing Employee	6	10	9	7	8	1	3	2	4	5
Value Added per Manufacturing Employee	4	7	9	3	8	1	10	6	5	2
Retail Sales per Employee in Retail Sales	3	5	4	2	8	10	9	7	6	1
Wholesale Sales per Employee in Wholesale Sales	2	1	10	3	5	8	7	9	6	4
MEAN	4.5	3.3	8.2	3.3	5.1	7.3	8.4	6.1	6.2	2.5

TABLE 37 (Continued)

	United States	New York	Vermont	New Jersey	Pennsylvania	West Virginia	Arkansas	Florida	Arizona	California
TREND INDICATORS										
Per Cent Change in Population, 1950–60	5	6	8	4	7	10	9	1	2	3
Absolute Change in Population, 1950–60	5‡	3	8	4	6	10	9	2	7	1
Per Cent Change in Per Capita Income, 1950–60	6	9	2	8	10	5	1	4	3	7
Per Cent Gain in Manufacturing Employment, 1950–60	5	7	10	6	8	9	4	2	1	3
Absolute Change in Manufacturing Employment	5‡	2	10	4	6	9	8	3	7	1
Per Cent Gain in Tertiary Employment, 1950–60	5	6	10	4	8	7	3	1	2	
Absolute Change in Tertiary Employment	5‡	2	10	3	6	9	8	4	7	1
Per Cent Change in Bank Assets, 1950–60	7	8	3	5	9	10	6	2	1	4
Absolute Change in Bank Assets, 1950–60	5	1	10	6	3	9	8	4	7	2
Per Cent Change in Value Added per Employee, 1950–60	4	9	7	5	6	2	3	1	10	8
Per Cent Change in Retail Sales per Employee, 1950–60	2	6	8	5	3	10	7	1	9	4
Absolute Change in Retail Sales per Employee, 1950–60	1	6	8	5	4	10	7	2	9	3
Per Cent Change in Wholesale Sales per Employee, 1950–60	7	8	9	2	5	10	6	1	4	3
Absolute Change in Wholesale Sales per Employee, 1950–60	7	4	9	1	6	10	8	3	5	2
MEAN	4.9	5.5	8.0	4.4	6.3	8.6	6.5	2.4	5.2	3.1
Rank If Level and Trend Means Are Averaged	4.7	4.4	8.1	3.9	5.7	8.0	7.5	4.3	5.7	2.8
Cumulative Rank	5	4	10	2	7	9	8	3	7	1

*A rank of 1 is top, one of 10 is bottom.
†All level indicators are for the year 1960 or as designated.
‡Where the criteria are not appropriately applied to the U.S., they are given a medium rank of 5.

York shows a cumulative mean rank of 4.4 among the ten units compared. This is rather encouraging for a northeastern state and speaks relatively well for the strength of the state's economy. That California and Florida outrank New York is to be expected, and that New Jersey does as well as indicated above can be explained in part by its proximity to metropolitan New York.

Of those criteria examined in which New York ranks low, certain ones appear to be critical. Among these are: (1) average annual investments in new plant and equipment per manufacturing employee, (2) value added by manufacture per manufacturing employee, and (3) percentage change in volume of retail sales per retail sales employee, 1950–60. If, through major effort, something could be done to strengthen New York's position in reference to these three indicators, then the situation relative to others such as change in per capita income and absolute change in retail sales per employee would be automatically improved.

Internal Spatial Variations

New York's economy is of above average strength as compared to the United States and selected states. Of equal, if not greater, significance to this geographic analysis of New York is the spatial variation of economic strength within the state. In any area of the country some sections will be doing much better than others, and generally explanations for this unevenness can be discerned. New York State has some of the most economically vigorous sections in the United States as well as some of the country's more lagging sections. Many factors contribute to the variations in economic health or strength, but basically it is now a matter of the location of the large urban systems. Almost without exception the doughnut rings around the great urban systems have the best economic health, central cities are doing less well, and strictly rural areas are doing most poorly. This situation has been in effect for many decades and is likely to continue in the foreseeable future. The fast-growing doughnut rings are expanding in width, ever outward from the cities that spawn them.

It is quite possible to demonstrate spatial variation in economic health within the state by using much the same techniques employed in arriving at New York's external comparisons. In this case county and central place data replace national and state statistics. Two such studies, already completed, provide the basis for the following analysis.[2]

[2]Sidney C. Sufrin, John H. Thompson, Marion A. Buck, and Arland E. Charlton, *The Economic Status of Upstate New York at Mid-Century, with Special Reference to Distressed Communities and Their Adjustments* (Syracuse University: Business Research Center, College of Business Administration, August, 1960); and John H. Thompson, Sidney C. Sufrin, Peter R. Gould, and Marion A. Buck, "Toward a Geography of Economic Health: The Case of New York State," *Annals of the Association of American Geographers*, LII, No. 1 (1962).

TABLE 38
RANK OF NEW YORK COUNTIES FOR NINE INDICATORS OF ECONOMIC HEALTH*

	Increase in Per Capita Income, 1950–57	Per Cent Population Growth, 1950–56	Per Cent Population in 20–49 Age Group	Per Cent Growth Employment, 1947–56	Per Capita Income, 1958	Per Cent Growth Retail Sales, 1948–58	Per Cent Average Unemployment, 1948–1958	Per Cent Increase Value Added by Manufacturing, 1947–54	Increase Average Weekly Earnings, 1950–58
New York City (considered as one statistical unit)	5	58	1	45	2	53	19	51	31
Nassau	27	2	2	1	3	1	5	2	19
Rockland	12	6	6	8	8	4	2	4	10
Suffolk	54	1	14	2	5	2	25	1	4
Westchester	1	8	5	11	1	5	21	30	11
Broome	6	17	12	26	14	27	4	25	26
Chenango	41	36	51	22	43	9	26	12	31
Delaware	44	44	43	3	34	42	16	5	29
Otsego	51	40	47	29	38	43	52	27	50
Cattaraugus	33	42	31	35	37	51	14	49	44
Chautauqua	18	26	27	33	23	45	32	34	28
Erie	16	14	7	30	15	22	23	29	12
Niagara	3	7	9	15	9	17	8	23	5
Albany	24	32	10	25	10	34	3	43	25
Rensselaer	36	37	15	52	39	48	48	47	40
Saratoga	39	15	21	48	47	31	18	45	17
Schenectady	4	10	4	42	4	38	6	7	2
Schoharie	52	35	54	14	49	40	55	58	49
Warren	32	25	32	44	22	36	42	36	53
Washington	31	51	39	24	32	46	1	17	20
Allegany	35	47	46	55	48	50	29	13	9
Chemung	22	9	19	41	29	41	22	18	27
Schuyler	43	28	56	28	51	58	30	37	33
Steuben	20	34	48	17	27	33	17	9	21
Tioga	37	19	50	31	42	24	15	24	1
Tompkins	9	3	3	20	18	32	7	41	23
Columbia	29	48	42	38	24	14	40	48	48
Dutchess	9	4	13	12	13	21	12	8	7
Greene	40	56	52	9	33	30	34	20	47
Orange	21	11	17	18	17	26	32	31	41

TABLE 38 (Continued)

	Increase in Per Capita Income, 1950–57	Per Cent Population Growth, 1950–56	Per Cent Population in 20–49 Age Group	Per Cent Growth Employment, 1947–56	Per Capita Income, 1958	Per Cent Growth Retail Sales, 1948–58	Per Cent Average Unemployment, 1948–1958	Per Cent Increase Value Added by Manufacturing, 1947–54	Increase Average Weekly Earnings, 1950–58
Putnam	13	5	20	16	7	16	28	3	34
Sullivan	26	43	24	48	21	23	20	19	55
Ulster	9	39	23	4	26	12	27	11	6
Fulton	49	46	34	54	46	55	56	57	54
Hamilton	57	57	35	57	58	8	NA	22	58
Herkimer	38	50	33	58	45	39	53	35	56
Montgomery	58	54	16	56	35	56	45	56	57
Oneida	2	24	18	40	11	29	36	39	16
Clinton	56	22	22	19	56	7	47	53	24
Essex	30	49	45	36	50	10	24	46	36
Franklin	53	55	53	51	55	52	57	15	53
Jefferson	25	53	49	34	30	44	39	38	37
Lewis	50	52	58	46	53	54	11	42	43
St. Lawrence	14	16	40	5	44	6	37	19	3
Genesee	17	38	30	27	19	11	33	26	30
Livingston	34	41	29	37	25	20	46	33	15
Monroe	7	18	11	39	6	25	9	16	8
Ontario	45	23	28	43	28	18	38	52	39
Orleans	28	29	44	7	41	13	51	10	35
Seneca	15	30	57	10	16	15	10	6	18
Wayne	19	21	38	6	21	3	44	28	38
Wyoming	23	45	25	14	31	28	54	44	22
Yates	48	27	55	21	54	57	50	32	46
Cayuga	42	32	26	49	36	47	41	55	42
Cortland	46	20	37	50	40	49	35	50	45
Madison	47	13	36	32	52	37	49	40	51
Onondaga	11	12	8	23	12	19	13	21	13
Oswego	55	33	41	53	57	35	43	54	32

NA: Not Available.
*The counties in this table (and Table 39) are listed in a regional rather than alphabetical order. For example, the New York City suburban counties of Nassau, Rockland, and Westchester are grouped together, as are the Buffalo area counties of Cattaraugus, Chautauqua, Erie, and Niagara. This regional order is employed in statistical tables published by the New York State Department of Commerce, and is an attempt to group counties into what that department considers to be "economic areas."
Source: John H. Thompson, Sidney C. Sufrin, Peter R. Gould, and Marion A. Buck, "Toward a Geography of Economic Health: The Case of New York State," *Annals of the AAG*, LII, No. 1 (1962).

The idea is continued that level of economic conditions and trends in economic conditions are two different but essential elements in the concept of economic health or strength.

The County Analysis

Because of the wide variety of economic data available for counties, they are useful statistical units. Per capita income, unemployment, and population in the twenty to forty-nine age group are used as indicators of level of economic health; whereas increase in per capita income, growth of population, growth in total employment, growth in retail sales, increase in value added by manufacture, and increase in average weekly earnings are used to show trends in economic health. These indicators, selected from a larger number representing the spread of available data, seem most appropriate.

Table 38 shows the rank order of counties for each of the nine indicators. There are fifty-eight counties (counting New York City's five as one unit). Again, a ranking of 1 is an expression of first place or highest level of economic health, 58 indicates last place or lowest level. It might be noted that New York City ranks first in per cent of population in the twenty to forty-nine age group, implying that there is an in-migration of people in this, the most vigorous age level. On the other hand, it ranks forty-fifth in per cent growth of employment. Nassau County, known for its recent rapid growth because of its location on the periphery of New York City, ranks first in per cent growth of employment but twenty-seventh in increase in per capita income. These and most other variations in rank order among the indicators make sense when the situations are examined in detail. The important point is that a wide variation in rank order occurs. Table 39 shows the divergencies in actual figures, i.e., per capita income in dollars, population growth in per cent, and so forth.

Tables 38 and 39 provide considerable evidence of spatial variation in economic health and suggest that each indicator must measure a somewhat different facet. For Figure 97 the indicators within the level and trend categories are weighted equally to provide level and trend rankings. It is not surprising that on this map New York City exhibits a very high ranking (fourth) among the fifty-eight counties in level of economic development but a low one (forty-sixth) in trend; or that St. Lawrence County, a relatively poorly developed county, which profited from construction of the St. Lawrence Seaway and Power projects during the 1950's, should rank low (forty-third) in level but high (fourth) in trend. More recent trend data would not indicate such a favorable situation in St. Lawrence County.

If the equal weighting procedure is carried one step further, all indicators, both level and trend, may be combined to produce a single ranking of economic health. In such a case the single ranking is derived by adding the ranks of the individual indicators and reranking the totals. Such is the basis for Figure 98. The result shows Rockland County ranked first and Fulton County in the distressed middle Mohawk region ranked last (fifty-eighth). A complex statistical analysis suggested that the equal rating procedure produced acceptable results.[3]

---

[3] County rankings based on Factor 1 of a factor analysis correlated almost exactly with the rankings derived from equal weighting and shown in Figure 94.

TABLE 39
STATUS OF SELECTED NEW YORK COUNTIES ACCORDING TO NINE INDICATORS OF ECONOMIC HEALTH

	Increase in Per Capita Income 1950-57 (Dollars)	Per Cent Population Growth 1950-56	Per Cent Population in 20-49 Age Group	Per Cent Growth Employment 1947-56	Per Capita Income, 1958 (Dollars)	Per Cent Growth Retail Sales 1948-58	Per Cent Average Unemployment 1948-58	Per Cent Increase Value Added by Manufacturing 1947-54	Increase Average Weekly Earnings 1950-58 (Dollars)
New York State	$662	9.00	45.93	8.92	$2,609	16.66	8.36	21.94	$27.70
Upstate					2,086	17.99	8.70	36.24	27.81
New York City	761	1.90	48.20	1.18	2,906	1.26	8.33	1.63	27.95
Nassau	452	66.00	47.42	151.13	2,811	136.32	5.82	300.14	26.69
Rockland	655	14.70	45.34	41.55	2,406	61.46	4.76	116.56	29.98
Suffolk	310	70.30	43.48	150.10	2,571	122.58	8.82	486.34	34.64
Westchester	818	13.54	45.61	36.64	3,456	42.89	8.57	30.54	29.42
Broome	730	10.24	43.92	12.88	2,202	17.13	5.25	39.95	24.75
Chenango	392	7.22	38.13	18.97	1,650	36.24	8.99	54.32	22.97
Delaware	377	6.21	39.16	65.45	1,731	8.11	7.75	110.54	24.10
Otsego	332	7.03	38.66	12.00	1,697	8.03	17.11	39.02	17.84
Cattaraugus	425	6.34	40.47	7.48	1,720	3.79	7.07	3.56	20.52
Chautauqua	538	8.42	40.96	10.88	1,952	6.50	10.15	23.41	24.46
Erie	573	10.85	45.14	11.34	2,188	21.83	8.66	34.30	29.03
Niagara	770	13.97	44.77	33.01	2,401	25.08	6.14	42.73	33.01
Albany	472	7.79	44.40	13.67	2,373	13.64	5.14	13.45	24.76
Rensselaer	419	7.21	43.07	−.09	1,695	4.80	15.49	8.22	21.57
Saratoga	394	10.77	41.87	−.02	1,616	14.39	8.21	10.93	26.78
Schenectady	762	12.83	46.32	4.13	2,641	11.66	5.85	105.52	40.45
Schoharie	323	7.27	37.92	33.33	1,586	9.28	19.03	−24.59	18.04
Warren	430	8.49	40.45	3.13	1,965	11.89	14.16	17.73	17.76
Washington	434	3.80	39.78	15.63	1,741	5.92	3.84	48.38	25.76
Allegany	420	5.10	39.03	−.15	1,615	4.36	9.87	50.50	30.32
Chemung	482	13.07	42.45	4.48	1,777	8.56	8.58	48.09	24.70
Schuyler	383	8.33	37.04	12.50	1,523	−12.57	9.95	17.49	22.87

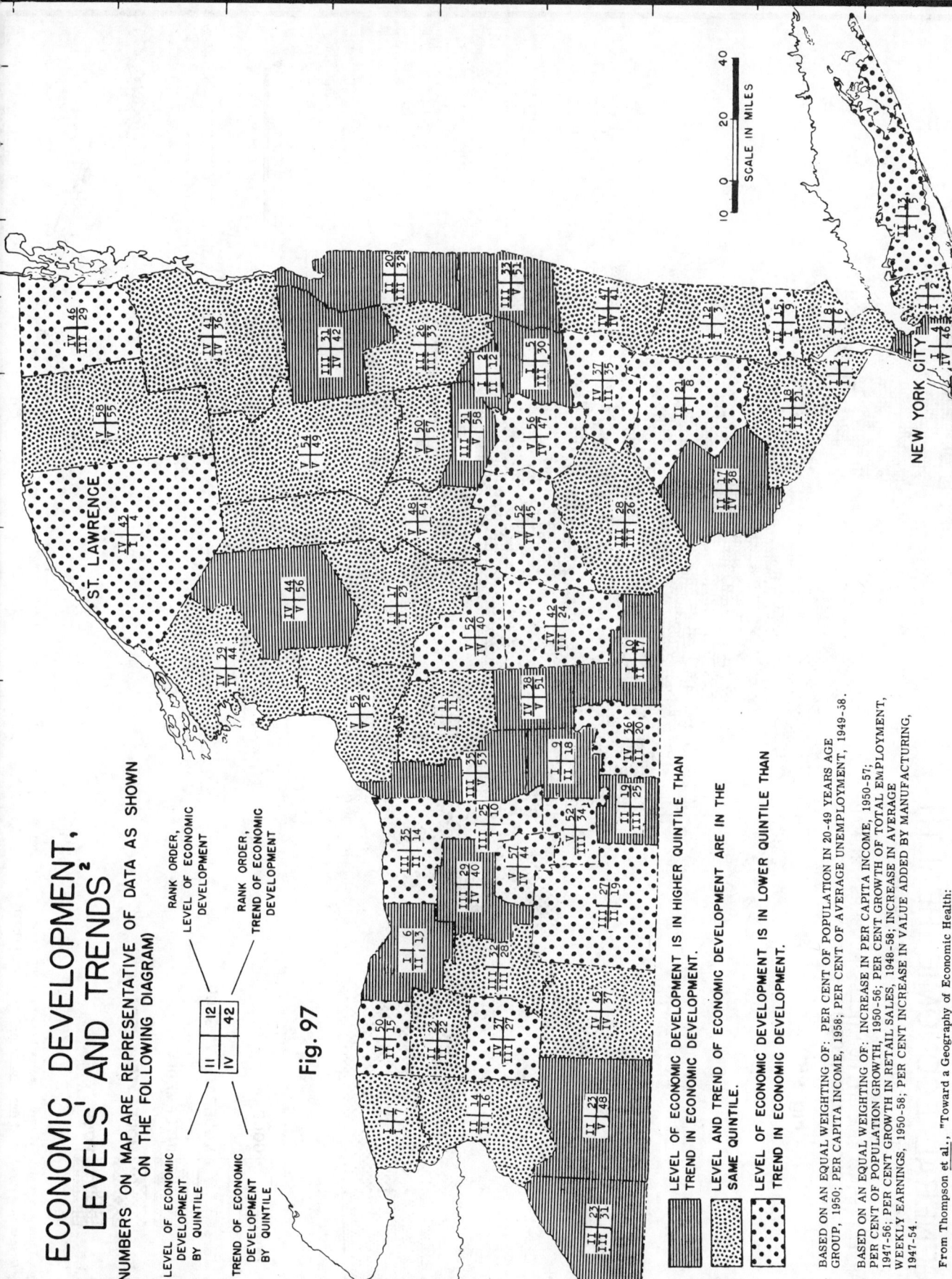

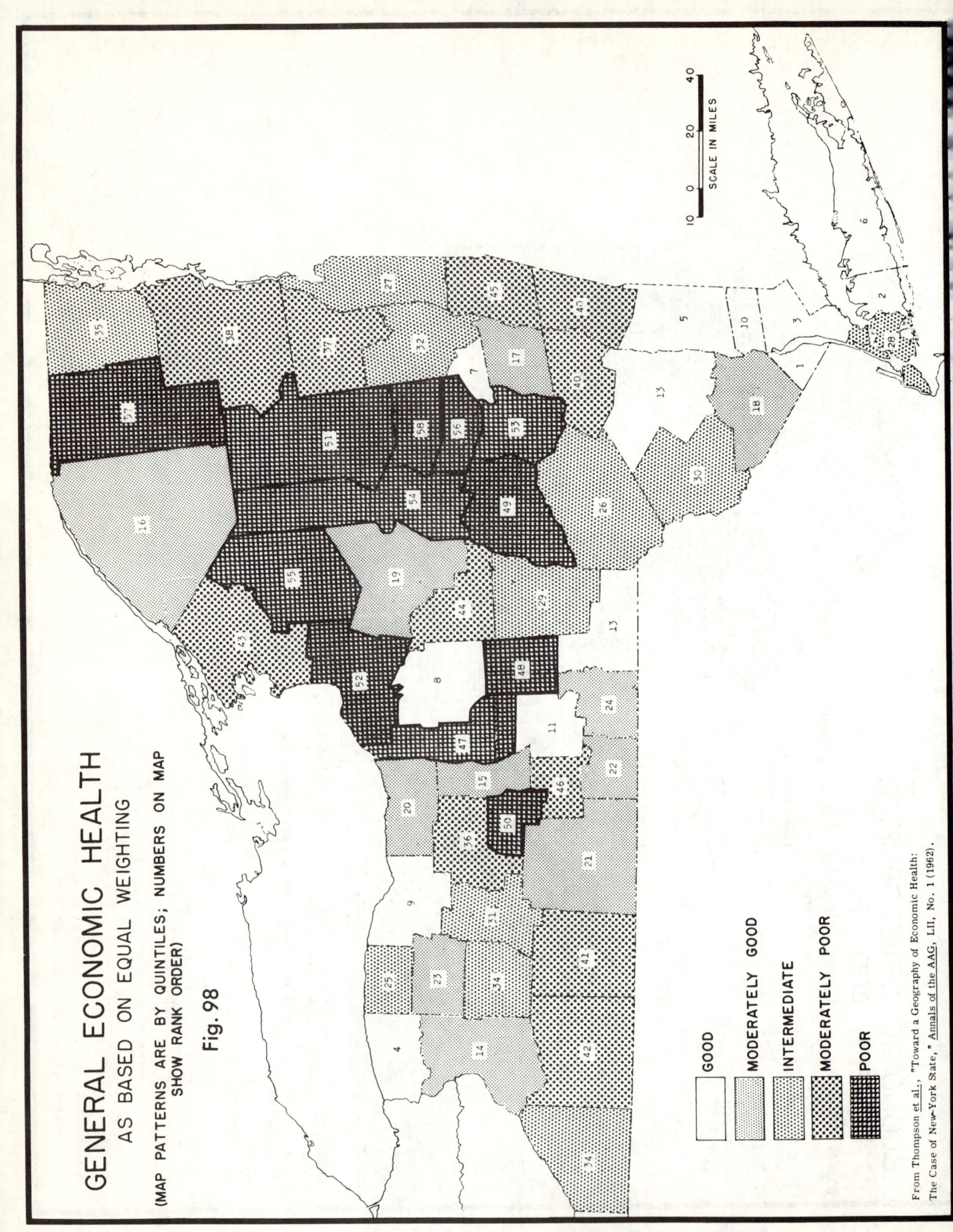

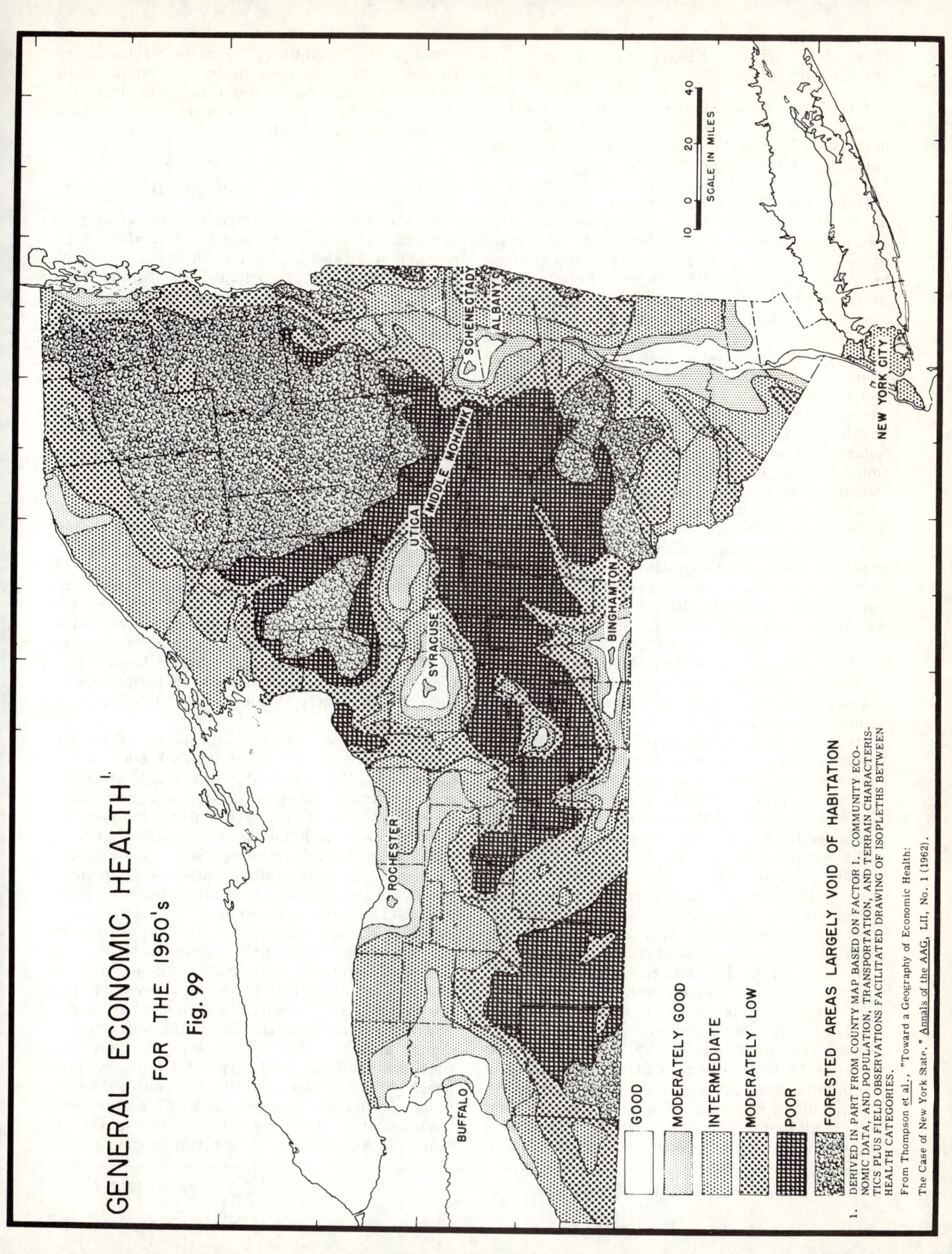

Making Geographic Reality out of the County Maps

A county map such as Figure 98 does not portray accurately the spatial variations of economic health in the state because the county statistical units are much too large. For example, not all of St. Lawrence or Albany County is doing equally well. One part may be urbanized, another part may be an agricultural area, and a third part may be forested wilderness. Thus, to satisfy the geographically trained mind and to keep from confusing potential users, Figure 98 must be converted into a more accurate isopleth map through use of central place data and other larger-scale information.

The following considerations were found pertinent to the isopleth drawing process: (1) Large metropolitan areas such as Rochester, Syracuse, and Binghamton generally have better economic health than smaller centers or rural sections. (2) Some urban areas, particularly medium-sized and small ones, are under considerable economic stress. Examples include places in the middle Mohawk region (between Utica and Schenectady), Auburn, or Oswego. Access to a variety of data on trends and levels for practically all cities of the state aided in placing them in their proper context in relation to the surrounding county or counties. (3) Within large metropolitan areas there is a doughnut effect, with poorer economic health in the old city than in its peripheral zones. Town and smaller central place data clearly indicate this. (4) Areas dominated by farming are not experiencing a proportionate share of economic growth and in some instances are actually declining. (5) State-owned lands such as parts of the Adirondacks and Catskills, which by law have been set aside as "forever wild" areas, cannot be expected to show high levels of economic health when measured by the above indicators. (6) Areas of principally hilly or rough terrain are not conducive to economic development and are generally becoming poorer and poorer. (7) Land largely in forests reflects a low intensity of land use. (8) The density of transportation networks of different kinds, although not always conclusive, gives some evidence concerning economic progress. (9) Population density (Fig. 2) is a valuable aid.

Maps based on data related to the above considerations, when overlaid and compared with Figure 98, served as guides for drawing isopleths. Field checks further facilitated the isopleth drawing process. The result is Figure 99, a map of general economic health in New York State. It is interesting to note that some counties that were well up in the top quintile in Figure 98 actually have part of their areas in all quintiles from the highest to the lowest in the isopleth map (e.g., Albany County). Others, although in one quintile in Figure 98, may in reality have most of their areas in another quintile. In such cases, small but economically and demographically significant parts of a county dominate the whole county statistically.

Contributions of the Economic Health Map

Figure 99 further substantiates a number of generalizations that have been made in earlier chapters and will function as a predictive aid in subsequent discussions. Clearly, the periphery of New York City, the Albany-Schenectady district, and the Syracuse, Buffalo, Rochester, and Binghamton areas stand out as being in relatively good economic health. They fall generally into the top two quintiles. These are the large urban areas, the places getting the largest share of new industrialization, expansion in tertiary activities, and residential construction. They have the size, the market, the services, and the general dynamism to attract economic activities, and they compete favorably with growing urban areas elsewhere in the country. They will provide most of the additional jobs of the future and absorb most of the state's population growth.

Rural areas tend to be in the lower quintiles of economic health in Figure 99. They have been going downhill relatively for a long time and are likely to continue to do so. Little of the increasing population in the decades ahead can be expected to be supported in the areas shown in the lowest quintiles, except as the rapidly growing peripheries of the great cities may spread outward to encompass them.

Although there are some exceptions, urban systems with population of less than 100,000 are not doing well. When places of this size do show exceptional economic vigor, it usually is because of proximity to one of the larger metropolitan systems, which actually is inducing the economic growth, rather than the smaller place itself. Some of the smaller urban places are in the lowest quintile. Standouts are those in the middle Mohawk region between Utica and Schenectady.

Figure 99, then, much more closely approximates the geography of economic health in New York than does Figure 98. The data are largely for the 1950's, and although some changes have modified the pattern in recent years, those changes are minor and will probably remain so in the years immediately ahead. Figure 99 reflects, in part, the economic evolution of the past and provides some insight into the future. Further, it mirrors something of what is happening in all of northeastern United States. It represents a starting point for inquiry into causes and, as such, can become a basis for geographic and economic analyses as well as for regional thinking and planning. It will enhance the analyses in Part Four of this volume.

Conclusions

These conclusions are aimed not only at government officials and agencies but also at the intelligent citizen who is consciously interested in seeing that New York State and its parts remain permanently competitive in the national economic scene.

Economic strength or health of any area commonly reflects the effectiveness of the social organization which is established. That the right kind of institutions exist for dealing with the problems, and that they are staffed with individuals who remain constantly cognizant of the changing nature of the problems, is vital. This statement would seem to be valid whether local or state-wide problems are under consideration.

If New York can maintain what appears to be an over-all economy a bit better than the national average, it will have to compete with other parts of the country which unquestionably are ambitious to improve their lot. "Soft spots" such as the apparent small effort at investment in new plants and equipment in the state need to be examined and remedied. There may be numerous other soft spots not brought out by the data used in this analysis. Searches for them should continue.

The tremendous difference in economic potential between the peripheries of the great cities and other parts of the state must be accepted as a fact of life today. This is not to say that society should give up on city centers and rural areas, but these kinds of locations must be recognized for what they are—mid-twentieth-century trouble spots. Urban renewal and other efforts are aimed at keeping city centers competitive, and excellent progress is being made in many of the cities of New York State. Studies are also being carried out by state, municipal, and university groups to determine what the best uses for the extensive rural lands will be in the future.

Perhaps relevant to this whole subject are the following questions: To what extent is the wave of urbanism, with attendant concentration of ever greater amounts and percentages of populations and economic activity, detrimental to society? Or, is it detrimental at all? If it is detrimental, to what extent should efforts be directed toward reversing the trend? These are not questions for which economics or geography necessarily pretends to have the best answers. They are questions, however, which seem reasonably to emerge from the analysis of the internal spatial variations of economic health in the state. They are also questions with which state and regional decision-makers eventually must wrestle.

Selected References

"How's Your State Doing in Education?" *Changing Times, The Kiplinger Magazine* (November, 1964), pp. 22–23.

"How Things Look in a Big New U.S. Industry (Research and Development)," *U.S. News and World Report* (October 19, 1964), p. 119.

Jones, Barclay, and Burnham Kelly. *Long Range Needs and Opportunities in New York State, a Series of Working Papers.* Ithaca: College of Architecture, Cornell University, 1962.

New York State Executive Department. *Budget Summary 1964–65.* Albany: The Department, 1964.

New York State Temporary State Commission on Economic Expansion. *Steps Toward Economic Expansion in New York State.* Albany: The Commission, 1960.

Resources of the Allegheny Plateau. Syracuse: State University College of Forestry at Syracuse University, 1962.

Sufrin, Sidney C., John H. Thompson, Marion A. Buck, and Arland E. Charlton. *The Economic Status of Upstate New York at Mid-Century.* Syracuse: Syracuse University Business Research Center, 1960.

Thompson, John H., Sidney C. Sufrin, Peter R. Gould, and Marion A. Buck. "Toward a Geography of Economic Health: The Case of New York State," *Annals of the Association of American Geographers*, LII, No. 1 (March, 1962), 1–20.

U.S. Advisory Commission on Intergovernmental Relations. *Measures of State and Local Fiscal Capacity and Tax Effort.* Washington, D.C.: Govt. Printing Office, 1962.

U.S. Bureau of the Census. *State Government Finances in 1963.* Washington, D.C.: Govt. Printing Office, 1964.

U.S. President. *Manpower Report of the President.* Washington, D.C.: Govt. Printing Office, 1965.

PART FOUR

LANDSCAPES
AND REGIONS

Significant aspects of the physical, historical, and economic geography have been analyzed in the preceding chapters. Whenever appropriate, the regional concept has been employed to aid in the analysis, with the result that a large number of maps showing both uniform and nodal regions have been presented. These maps—usually each regionalizing a single phenomenon such as land forms, agriculture, or trade—provide background for meaningfully differentiating one part of New York State from another. They are examined, compared, and overlaid in Part Four in an attempt to discover order in earth space and to provide a comprehensive general regional map which will be a useful predictive, and perhaps planning, tool.

Clearly, the greatest contrasts in the general geography of the state today are between urban and rural areas. In densities of economic activity, trends in development, conditions of economic health, and outlook for the future, rural and urban areas are distinctly different. So strong is this difference that it might be said New York has two landscapes, one urban and one rural. The word *landscape* has been used by geographers in a number of ways, but here it refers to the visible scene and the invisible forces that shape it. Reading a landscape, that is, seeing and understanding it, is one of the experiences the trained geographer most thoroughly enjoys and profits from and one in which the careful reader of this volume, even though not a geographer, may become interested.

Part Four contains three chapters, one on the urban landscape, one on the rural landscape, and a third on the problem of applying regional theory to the establishment of planning and development regions.

CHAPTER 14   DAVID J. DE LAUBENFELS, JOHN H. THOMPSON, AND JOHN E. BRUSH

The Urban Landscape

THE most valuable and most intensively used portions of New York are the urban areas, even though they occupy only a small fraction of the state. Modern cities have become complex functioning urban systems which have evolved characteristic qualities easily discerned as an urban landscape. The land is occupied by dwellings, stores, factories, churches, schools, warehouses, and many other types of buildings, together with parks, lawns, parking areas, and streets—modified surfaces intended for regular and intensive use by man. By contrast, rural landscapes, which account for the bulk of the state's area, are dominated by land uses that are more extensive, including cropland and forest. Urban systems exist because only where people and the buildings in which they carry on their affairs are concentrated can the complex industrial, commercial, and social activities that have become a part of modern advanced society be successfully performed. The people of New York State, as much as those anywhere else in the nation, are now dependent upon the urban habitat for their way of life.

The degree of dominance of urbanization in New York State is revealed by population data. Not until 1840 had the urban population exceeded 20 per cent, but by World War I it had reached an amazing 80 per cent. Currently it is well over 85 per cent. The current definition of *urban population* by the Census Bureau includes all people in places, incorporated and unincorporated, with at least 2,500 inhabitants plus the densely settled fringes of urbanized areas. *Urbanized areas* are densely built-up areas with an incorporated central city of at least 50,000 inhabitants. Over half of the almost 15 million urban people in the state reside in the five boroughs of New York City, and another 20 per cent live outside the boroughs but within the urbanized area of New York City. The other urbanized areas of the state in order of size and with populations in thousands (1960) are: Buffalo (1,054), Rochester (493), Albany-Schenectady-Troy (455), Syracuse (333), Utica-Rome (188), and Binghamton (158). Together they contain nearly 20 per cent of the state's urban population. The remaining 10 per cent are found in urban places with populations between 2,500 and 50,000.

The concept of an urban landscape means different things to different people. Formerly, the adjective urban denoted a way of life that was refined, polished, and sophisticated, while rural denoted that which was coarse, unpolished, and naive. Cities by their very nature offered more in education, exchange of ideas, opportunities, and goods. However, the revolution in transportation of the twentieth century has given rural people access to the cities and all of their advantages. At the same time, the overwhelming and growing impersonality of big cities, obsolescence, and congestion have allowed the accumulation in parts of the urban areas of whole communities of economically and educationally disadvantaged people. Cities thus not only have come to be exceedingly complex spatially, but have induced some of the most serious problems facing twentieth-century New York. The following examination of the urban landscape concerns itself first with the internal geographic structure of cities, then with their distributions, and finally with urban problems and their solutions.

Internal Structure

Urban systems have a spatial organization with distinct and important elements. Various kinds of neighborhoods can be identified, each with its characteristic location with respect to the rest of the urban area. There are concentric elements nested around the central focus, there are sectors radiating outward like spokes of a wheel, and there is a dispersion of subsidiary foci analogous to the many poles that help the center pole hold up a circus tent. Binding together all the parts of the city are linkages such as shopping trips, deliveries, conferences, service calls, removal, and all those other connections between the parts that make an urban system function. The spatial organization the system develops results from the never-ending drive to make it work better; it is a sum of all the little decisions to minimize cost, reduce time loss, increase flow, maximize profit. There are infinite numbers of unique things about individual cities, many of which are described in later chapters. This chapter is concerned primarily with generalizations about New York urban systems as a whole.

CONCENTRIC PATTERNS

Urban systems commonly exhibit a more or less regular sequence of land use patterns outward from their centers. This concentricity is the product of relative access to the central focus and of the stages of areal growth. There lies at some central position in any city a point of maximum accessibility for all dependent territory, including not only the surrounding countryside but the parts of the city itself. In the competition for access, the various land uses become sorted by relative position with respect to this point of maximum accessibility in terms of their requirements and their relative ability to pay the cost. The kinds of land use which appear and the weight of each contributing response vary immensely with the stage of the economy and the culture orientations with which the city is associated. However, much of the variation in structure is eliminated when an essentially uniform sample such as the cities of New York is being considered. Thus the concentric structure of an idealized New York urban system can be described.

Central Business District

At the center of an urban system, its point of maximum accessibility, is the most concentrated construct of man, the central business district (CBD), or downtown, as it is often called (see Fig. 100a). The *CBD* is a landscape of retail stores, offices, depots, and theaters, of buildings in which the activities require the support of the entire urban region and can pay the premium price for being located at the center. It is a busy place where large numbers of people come together to transact all kinds of business, to work, and to play. Even so small an area as the CBD, perhaps but half a dozen to a dozen blocks in its longest dimension, has distinct zonation within itself. The very heart of the city (the inner zone of the CBD) is all stores, of limited variety at that. For the most part they are apparel stores—clothing, jewelry, shoe, and department stores. Mixed in are a few top-quality variety establishments, drug stores, five-and-tens, and perhaps a book and stationery store, as well as establishments to serve the passing shopper such as cafés, candy counters, newsstands, and the like. Not quite so central (the outer zone of the CBD) is a combination of variety stores, theaters, banks, appliance stores and other bulky goods outlets, hotels, and all sorts of personal services and eating places. Lower-price apparel stores, too, may be in this outer zone of the CBD.

Frame

The central business district is bracketed by a *frame* which serves the CBD in many ways. Here are the parking lots and parking garages with such train and bus depots as still operate near the center, all facilitating the movement of people. Here, too, are warehouses and wholesale outlets which supply the goods to the CBD. Mixed in are many small manufacturing and repair shops which characteristically have the whole city as their market but which cannot pay the price of CBD locations. Included are such regularly encountered enterprises as newspapers, auto repair and print shops, laundries, foundries, and bakeries. Stores supplying paint, linoleum, plumbers' supplies, lumber, and so on to contractors are located here. Also within the frame are land uses requiring relatively more space, such as institutions (church and government), certain kinds of recreational facilities, automobile dealers, gas stations, and sometimes hospitals. The frame may be thought of as a zone of transition because expansion of the CBD is felt here and because land uses of the neighboring zones may be mixed in. As time goes by the frame gets more and more set in position and character as more structures of the above-described types are built. Institutions are especially hard to move, as are the railway yards which normally back up the warehouses and factories. In recent years a new factor is crystallizing the frame yet more. Inner rings of the superhighways are being built in all the major cities to speed the movement of people and goods into and out of the central business district, often replacing blighted or slum portions of the frame. Often the frame turns out to be a zone of transition because on its inner margin the pressure from the expanding CBD is felt and on its outer edge intermixing with residential uses is bound to occur.

Middle City

Much of an urban system's area is a great zone of residences which completely surrounds the CBD and the frame, the *middle city*. The middle city extends outward a little more than a mile in Utica and Binghamton, between two and three miles in Syracuse, Rochester, and Albany, about four miles in Buffalo, and at least fifteen miles in New York City. Characteristically, it is part of the incorporated city and is densely populated, a natural result of the lack of open space. The density takes two forms. The first form includes the older housing units which are likely to be crowded, or even overcrowded, with occupants in terms of their original design. The second involves multiple-unit housing such as apartment buildings. Apartment areas have barely taken form in the smaller urban centers, but they cover

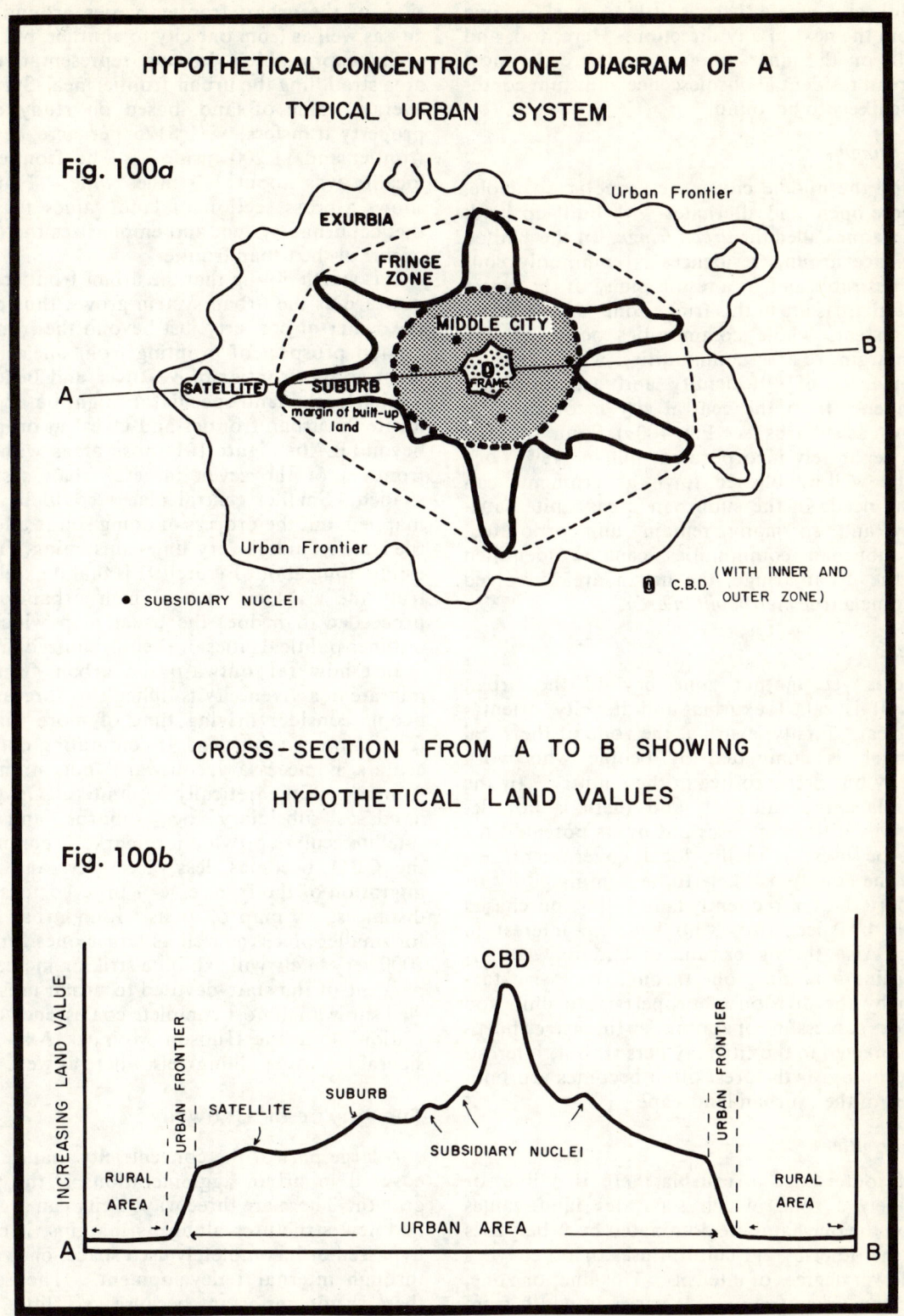

block after block in New York City. The impersonality of the city has its greatest expression in the middle city, where there is little to mark off one area from the next in any direction. Here, too, and especially on the inner margins, is where the most widespread residential obsolescence and slum conditions are likely to be found.

Urban Fringe

Beyond the middle city more space is available, and where open land alternates with built-up land, there is a zone called the *urban fringe*. In the United States space around residences is commonly considered desirable and, as a result, many of the better residential areas are in this fringe zone. In the larger urban systems whole communities occur on the urban margin. These communities, which enjoy a large measure of self-identity and are aided by independence from the central city incorporation, are known as suburbs (see Fig. 100a). Some suburbs may be separately incorporated, but in New York State the well-established town governments can serve the needs of the suburban areas quite satisfactorily and so many remain unincorporated. When suburban communities can be identified within the urban fringe, the urban area involved has become a true *metropolitan area*.

Exurbia

There is yet another zone beyond the urban fringe. It is called exurbia and its city orientation is clear. Firstly, exurbia, the zone of the rural non-farmer, is dominated by people who work in the city but prefer to live in the country. In the second place the value of land there is not determined by its farming uses but by its potential for urban-type uses. Finally, local government, especially the county, is likely to be dominated by its urban part. Not infrequently land will be purchased and held by speculators who have no interest in farming. Also, the higher value of land may actually force ordinary farming out through excessive taxation or by the division of property into units too small for successful operation. Intensive horticulture directed to the city may persist, but, interestingly, second-growth forest often becomes a prominent part of the exurban landscape.

Urban Frontier

At the outer edge of exurbia there is a line, or more likely a zone, which separates land values which on the one hand are dominated by urban uses and on the other by agricultural uses or other types of still lower degrees of intensity. This line, or zone, is called the *urban frontier*. It varies in width from a fraction of a mile to as much as ten miles, although zones as wide as ten miles are not found in New York State except in a few places around New York City. The difference in land values on the two sides of the urban frontier varies around a single city as well as from one city to another, but it is generally appreciable. In one representative sample area straddling the urban frontier near Syracuse the average value of land, based on study of actual property transfers, was $125 per acre beyond the frontier and $1,200 inside it. The frontier at this locality was about 1½ miles wide. Figure 100b shows a cross section of land values for a hypothetical urban system and emphasizes the drop that occurs at the urban frontier.

It is worth noting that the urban frontier migrates outward as the urban system grows, thus providing the owner of property just beyond the frontier with a good prospect of profiting from unearned increment as the frontier moves over and beyond him. Similarly, the land speculator might be expected to locate the urban frontier and invest in property just beyond it. In Figure 101 those areas within urban frontiers of the seven largest urban systems are shaded. Smaller central places could be similarly mapped, but the process of doing so for all places in the state would be very time-consuming. The interesting thing about Figure 101 is that it exhibits more truly the areal extent to which urbanization has proceeded than does the usual map which simply outlines political cities or their urbanized areas.

Just how far outward the urban frontier will migrate in a given city is difficult to forecast. Most people consider driving time of more than about forty minutes a burden. If commuting daily to city centers is necessary, outward migration of the frontier will theoretically be limited. On the other hand, as subsidiary foci, suburbs, and growing satellite centers provide job markets, commuting to the CBD becomes less necessary and outward migration of the frontier can proceed to surprising distances. A map of areas within urban frontiers for smaller places as well as larger ones for the year 2000 very likely will exhibit a striking increase in the per cent of the state devoted to urban uses. It may well show an almost complete coalescence of urbanization along the Hudson–Mohawk Axis and considerable "valley-filling" elsewhere as well.

THE IMPACT OF GROWTH

A large part of the concentricity that can be discovered in urban agglomerations is the result of growth. There are three ways a city can grow: it can add new structures along its margins, it can fill in bypassed or incompletely used space, or it can grow through internal redevelopment. The second of these is not important because very little available space is left within the central built-up area. As an urban area expands outward, vacant land is quickly

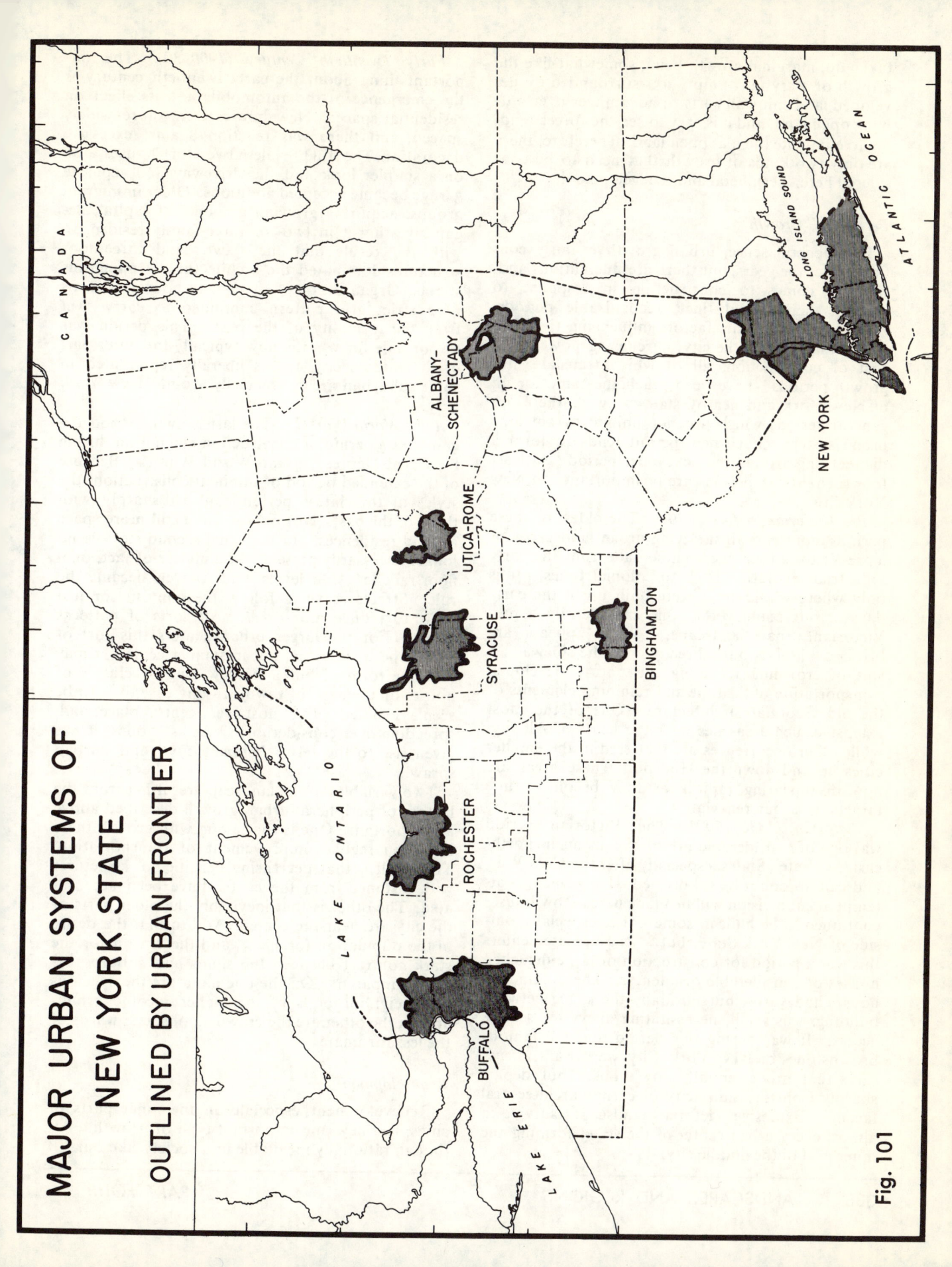

Fig. 101

taken up, marginal growth being somewhat like the growth of a crystal. Empty lots surrounded by developed land enjoy all sorts of advantages over outlying open land, and it is easy to see the forces tending to eliminate them as open land. Therefore, there are only two kinds of growth that need to be considered here: peripheral and redevelopment.

Peripheral Growth

In a general sense, urban growth is fairly continual over the years, but there are fluctuations and, when it comes to construction in response to growth, there are distinct, recognizable periods. Because the economic factors influencing the birth rate—migration to the city, purchasing power, and cost of construction, all of which translate into growth periods—have been much the same for all of New York and nearby states as well, there are general periods which are not unique to any one urban system. Each new period tends to form a distinct ring around the next older period. At least four such growth periods are of importance to New York cities.

Pre-Victorian (before 1830).—The oldest of these periods represents all surviving urban landscapes of a pre-Victorian vintage. These date from the early industrial revolution back to colonial times. It is only where whole areas were built up in the early days of our country that one can identify a pre-Victorian urban landscape. What appears are brick row houses built directly onto the sidewalk, a sort of crowding reflecting the poorly developed transportation of the time and retaining elements of the old Georgian architecture. One of the most extensive such areas lies near the heart of Albany, while other occurrences are scattered in the smaller cities up and down the Hudson. These areas are now disappearing rapidly as they become prime targets for urban renewal.

Victorian (1830–1900).—The Victorian period was one of considerable urban expansion in northeastern United States, especially after the Civil War, and can no doubt be subdivided. However, no attempt at subdivision will be made here. Row houses continued to be built in some cities, especially outside of New York State, but in the upstate centers this was a period for construction of large detached houses of considerable ornateness. These detached houses had yards, but normally the space between buildings was small, necessitating alleys for service access. It was during this period that the famous brownstones of New York City were built, structures that are essentially row houses but depart sharply from the simplicity of earlier architectural facades. Extensive Victorian landscapes survive in almost every urban center of the state, forming the inner part of the middle city.

Early Twentieth Century (1900–30).—The important thing about the early twentieth century is the emergence of the automobile and its effect on residential space. New houses were more widely spaced, and there was reaction against excessive decoration. The still prevalent two-story houses took on a simpler look and the driveway and separate garage became accepted additions. Older immigrant groups, acquiring growing amounts of capital, began investing it in two- or three-family residences, with the result that these two- and three-floor structures dominated the architecture of numerous streets. Organization of streets and houses in terms of a strict grid pattern continued to carry over from the formality of the past. This period was responsible for what is now typically the outer part of the middle city and is liberally represented in some suburban areas, especially around New York City.

Post-World War II.—The last growth period began after a period of construction stagnation during the Great Depression and World War II. Because of the extended period of minimal construction, the styles of this latest period contrast sharply with those of the past. Not only is there still more space around residences, but the usual second story is no longer necessarily present. As a matter of fact, one-floor ranches, split levels, and trilevels became the rule. Streets curve to follow the contour, or just curve to reduce monotony. All sorts of patterns occur. For the larger urban centers this sort of landscape *is* suburbia, but in smaller centers it may take the form of but a narrow ring or cluster of houses on the edge of town. In any case, it is to be seen associated with almost every central place, and, spreading over considerable space as it does, it has given rise to the terms "spread city" and "urban sprawl."

Two variables of the landscape resulting from the history of peripheral urban growth described above are important. One is that along with growth there has been regular improvement of transportation. The result is that decreasing densities of structures with distance from the center have been accentuated. The other is that inevitably the older parts of the city are near the center. Add to that the desire of the population for space, and the reason for the stark contrast between the slums and suburbs becomes apparent. On the one hand are the spacious new neighborhoods of the higher income groups and on the other are the crowded, obsolete wards of the less fortunate.

Redevelopment

Redevelopment, especially in the inner parts of cities, is a new and important type of growth. Although rationally inevitable in a society like ours, it has been slow to take place for two reasons. First, the ever improving transportation has worked against increasing demand for central positions, and second, rising wages and land values as well as rental demands for old structures have made it expensive to clear away old buildings to make way for the new. Sooner or later, however, most old structures, worn out and obsolete or inadequate, must disappear. Larger stores, office buildings, and parking garages rise in and around the central business district. More intensive residential land use takes the form of multiple-unit housing. Long ago apartment buildings began replacing the brownstones and older houses in New York City. This new wave of construction has been overdue in the lesser urban centers but is gathering momentum now with the assistance of urban renewal programs. The frame and inner portions of the middle city are becoming a landscape of apartment houses and housing projects and many central business districts are experiencing substantial change. One might wonder how long it will be before the outer rings will have to be redeveloped, too.

SECTOR PATTERNS

Just as urban systems exhibit concentric patterns, so they also have sector patterns. *Sectors* characteristically are wedges, belts, or broad zones of similar land use extending outward from the center of the city. Sectorial differences in land use occur because: (1) the physical environment sets limitations to land use, (2) axes of transportation concentrate certain kinds of land use, or (3) affinities occur between certain types of land use.

Effect of Physical Environment

The physical environment occupied by a city is seldom uniform from place to place. Rather there are marked contrasts in such things as terrain and drainage. Large uniform areas do exist, but cities have a way of being located along rivers, on waterfronts, or in natural channels for transportation. From the contrasts in site qualities comes a differentiation of urban areas into sectors, i.e., wedges of land extending from the center separated from one another by some physical barrier or dominated by different kinds of land use in response to a different surface condition. Several types of such sectors can be identified.

Water bodies near the center of a city can prevent growth in the directions where they lie. Thus, a city may have a semicircular shape as does Buffalo. When the water body is a major river or a lesser river with a hard-to-use floodplain, there may, in addition to the semicircular shape, be a sector of growth beyond the barrier. This pattern appears in many small cities and most large ones such as Albany, Utica, Niagara Falls (in Canada), and Schenectady, while it is even important in the layout of New York City. Small lakes, bays, and swamps cut smaller wedges into the urban areas, as in Syracuse and New York City. Sometimes drainage or filling operations can reclaim former water bodies, but the land use which follows is likely to differ from the surrounding land even so.

Small rivers or ravines cut through nearly every urban area. The number of sectors resulting may depend on how many tributaries join near the center of town. Buffalo, Rochester, and Albany have some spectacular ravines within the built-up area. Broad valleys alternating with hilly zones, and intersections of natural channels and the traffic they carry, are natural sites for a city even in the face of the need for drainage operations. Such is the case of Syracuse and, to a lesser degree, of Utica and New York City. Some of the differences in land use related to such terrain contrasts are: better residential development on hilly sectors, poorer quality residences in lowlands and on sites of difficult drainage, transport arteries down natural channels, and manufacturing on flat or reclaimed land.

Effect of Axes of Transportation

There are a number of major transportation routes leading into and out of any given city. The organization of the parts of a city with respect to these major routes produces distinct structural sectors. Each part of a city is influenced in a particular way by the increased accessibility along major streets which cut through it. The central business district tends to be elongated in the direction of the streets that carry the heaviest traffic. The business activities of the frame zone cluster along these same streets. On out into the middle city there are nearly continuous strings of commercial and industrial land use, again along major streets. Crowding on or near the major lines of circulation, too, are other of the more intensive land uses such as apartment districts and concentrations of offices. On the other hand, these same streets, with their attendant congestion, are generally considered poor places for detached residences. Along the external highways, which are commonly extensions of the major streets, arms of urban growth may penetrate even into the rural landscape.

As is apparent from this discussion, transportation is a common cement for the structural sector, but there are, and have been, different kinds of transportation, each with its own effect. Early cities were walking cities, and walking favors compactness of both size and shape. Streetcars, an early form of rapid transit, were not very rapid but were flexible enough so that more distant parts of cities could be served. Although they did not foster too much dispersion, some outward projection along principal streets occurred. A major change in the patterns of accessibility attached to the commuter train and the interurban electric car. Though much faster than earlier forms of transportation, they were less flexible. The fact that their special roadbeds were fixed along a limited number of specific routes required development to be concentrated along these routes for efficient service. And development did cluster along their routes because they offered distinctly better access to the urban center than land located away from them. Because rapid transit must limit the number of stops or cease to be rapid, the tendency was for a series of commuter suburbs to develop something like beads on a string, and the metropolitan area became shaped like a star with long thin arms. This phenomenon became especially well developed around New York City.

More recently there have been two revolutions of transportation involving the private automobile and its effect on urban shape. When the automobile became a common means of transportation, it destroyed the earlier exclusiveness of rapid transit. Many more paths from the city could now be traversed with equal ease, even if at somewhat higher cost. Urban areas, therefore, spread out—some say exploded—as it became more readily possible to satisfy the desire for space. It was not just the advent of the motorcar but also the ability of the average American family to have one that made the difference. But there is still a differential in accessibility to automobiles because traffic normally moves faster along major arterials. The auto-age city remains star-shaped, but the arms are much broader than before, and more numerous.

The second revolution with respect to the private car is going on now. In the larger cities more cars have accumulated than can be accommodated efficiently on ordinary streets. The solution, short of prohibiting cars from the congested zones, is to construct expressways, and that is being done. Expressways tend to be exclusive, as were rapid transit commuter trains earlier, in that movement is relatively much faster than on other roads and there are a limited number of access points, the interchanges. The tendency of the expressway is to lengthen the arms of the urban stars. Places forty miles away or more may be within reasonable automobile commuting time. This is something to think about when forecasting the possible extent of future urban expansion.

Effect of Affinities Between Land Uses

It is common for like kinds of land use to exhibit a high degree of affinity and cling together, actually forming a sector of like use. All categories of land use are involved and all cities have such sectors, but the specific position of each sector is the result of historical accident and certain site preferences. Once a grouping of similar units of land use becomes established it will direct the addition of new units of the same kind to the periphery of that particular sector as the over-all urban system grows. The linkages producing the clustering of like land use units are, broadly speaking, functional, social, and imposed. Functional linkages include use of the same services and suppliers or the transaction of business between like units. Social linkages include the tendency for people with like backgrounds, interests, or activities to associate. Imposed linkages result from zoning regulations or the limited availability of suitable sites. In passing, it should be pointed out that some kinds of establishments are ubiquitous and do not exhibit affinities between their kind. By their nature they gain from dispersal. Examples include drug stores, eating places, personal service establishments, parking lots, and gas stations.

The Mix of Sector and Concentric Patterns

The sectors that develop in the typical urban system as a result of the combined influence of physical conditions, axes of transportation, and affinities between similar kinds of land uses tend to vary from one concentric zone to another. There are broad, more or less homogeneous sectors, and there are narrow sectors literally tied to but the two sides of a major routeway. Some extend from the center of the city to perhaps nearly the urban frontier, others are fragmentary or clustered in shape and perhaps should be called simply districts rather than sectors.

Sectors in the CBD.—Various sectors appear even within the central business district. The larger the city, the more pronounced they tend to be. The very grouping of apparel stores at the center of the CBD results from the same forces that cluster other categories of business. For business the power of comparison shopping to bring like stores together is immense. Thus, book stores and office supply stores, to name a couple, will tend to form districts or sectors of kind. In like manner there will be theater districts and hotel districts with related service and supply facilities. Financial districts and office districts result from the need for mutual business as much as from any other factor. The differentiation of specialized districts has reached a high level in New York City, the center of the world's largest metropolitan agglomeration, but most of these sectors of the CBD will be at least weakly developed in any of the metropolitan centers of the state.

Sectors in the Frame.—The frame is roughly divided into two sectors, one for commodities, the other for community functions (see Fig. 102). Access of bulk goods toward the center is through rail yards and freight depots, both train and truck, as well as through ports where water transports are available. For various reasons, including transport focus, mutual linkage, and zoning requirements, these facilities tend to concentrate in the frame. Closely bound to freight and to each other, is the constellation of warehousing, wholesaling, and light manufacturing which often forms a well-defined sector adjacent to the central business district. More than a wedge-shaped sector is involved. Because of the importance of these supporting functions to the center, they cluster around, nest-like, partially enclosing the business district and forming, as one or more commodity sectors, a good half of the frame zone.

By contrast, a group of functions frequented by the general public tends to concentrate on another side of the urban center. This sector includes such things as government buildings, religious institutions, and recreation facilities. Civic leaders are now prone to advocate an integrated community plaza in this sector composed of city, county, state, and even federal halls, a coliseum, museum, library, formal park, and churches. It is not unusual for all these to be close together, even without planning, with the YMCA, YWCA, and other nonprofit organizations, and a bowling alley added. This community sector of public buildings, and of office workers and the general public in circulation, contrasts sharply with the commodity sector with its heavy traffic, numerous laborers, and warehousing facilities.

Within the frame there are smaller subsectors, or special districts. In a city of some size there may be differentiated a furniture district, a second-hand district, and an area called "skid row." In a smaller center farm supply districts may be characteristic. The variety of special districts that has developed in Manhattan's frame is fantastic. While the frame has two principal sectors, each is divided into a myriad of localized special districts, which, in their own right, exhibit considerable homogeneity in land use.

Sectors in the Middle City.—Extending outward from the frame are certain other narrow sectors of very particular characteristics. Well known is automobile row, where new car sales, used car lots, automobile repair, and various supply centers form a

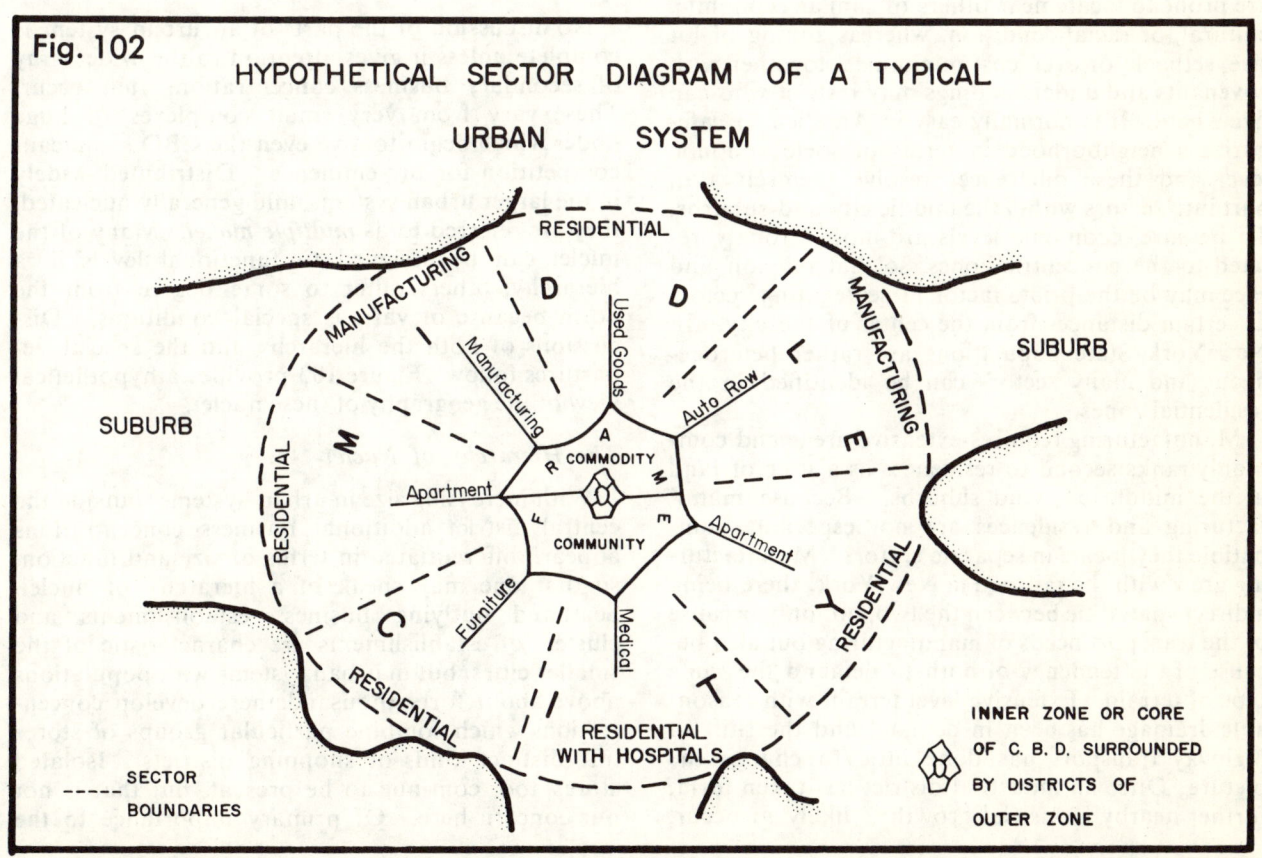

Fig. 102. HYPOTHETICAL SECTOR DIAGRAM OF A TYPICAL URBAN SYSTEM

CHAPTER 14   THE URBAN LANDSCAPE   341

continuous line along both sides of some major street for as much as a mile. Developing rapidly is the medical center, where hospitals, clinics, doctor's offices, and medical supply stores group together. Within both of these types of sectors, there are numerous cross-linkages which yield great advantages to grouping. In some cities a college or colleges may occupy a similar kind of location, with supply stores and eating places nearby. Already mentioned above was the apartment district which may concentrate along a single street or two. One can visualize each major street that emerges from the frame and penetrates the middle city as becoming the life line of a single sector. Certain types of sectors such as those dominated by apartments or manufacturing are more common than other types. Choking of access to the frame and CBD constitutes a serious problem which is solved by the construction of arterial expressways. What effect these will have on configuration of the commodity and community sectors of the frame and on the various sectors of the middle city remains to be seen.

Because the middle city and suburbs are heavily residential, sector development, except immediately along the above-described major routeways, may reflect classes of residences or some kind of differentiation of the people who live there. The separations may be voluntary or imposed, in that people are prone to locate near others of similar economic, cultural, or racial condition, whereas zoning of lot size, setback, or even cost minimums, together with covenants and understandings may restrict who can live where. It is normally easy in America to categorize a neighborhood in terms of socioeconomic level and these differences resolve themselves in part into sectors within the middle city and suburbs. To be sure, economic levels are more strongly related to the concentric zones, so that religion and race may be the prime factor in segregating sectors at certain distances from the center of the city. In New York State populations are rather heterogeneous and many sectors can be identified in the residential zones.

Manufacturing requires extensive areas and commonly ranks second to residences as a user of land in the middle city and suburbs. Because manufacturing and residences are not especially compatible they locate in separate sectors. Manufacturing grew with the railroad in New York, there being a direct spatial tie between the two, not only because of the transport needs of manufacturing but also because of the tendency of both to demand the same type of terrain. Extensive level terrain with reasonable drainage has been in demand and the shift to highway transport has done little to change the picture. Once an industrial district has taken form, further nearby industrial growth is likely to occur, both because of the need to do business with existing plants and established suppliers and because of severe zoning limitations.

Figure 102 shows the hypothetical middle city divided into seven broad sectors, five residential and two manufacturing. Penetrating from the frame into the middle city and into these seven broad sectors are the narrow routeway sectors. Again, hypothetically two of these are shown as apartment-oriented, one as manufacturing, one as used goods, one as auto row, one as furniture, and one as medical.

Sectors Beyond the Middle City.—The middle city sectors tend to expand outward with the growth of the city for the same reasons that sustained them earlier. Changes, however, take place through time, and for several reasons a given sector is not likely to extend outward indefinitely. Large, old, expensive residences lose their desirability and become converted into apartments, smaller homes get cleared away for urban renewal or parking lots. Zoning, or changes in surface conditions, may discourage continued outward growth of the same kind of land use. Finally, some other kind of land use may become established beyond an earlier sector or may engulf it, cutting off further outward growth.

MULTIPLE NUCLEI

No discussion of the parts of an urban system is complete unless it gives attention to the wide array of secondary business concentrations that occur. These vary from very small complexes to huge nodes which begin to give even the CBD significant competition for pre-eminence. Distributed widely in the larger urban systems, and generally nucleated, they are referred to as *multiple nuclei*. Many of the nuclei can be assigned to functional levels in a hierarchy; others differ to some degree from the norm because of various special conditions. Discussions of both the hierarchy and the special departures follow. Figure 103 provides a hypothetical view of the geography of these nuclei.

The Hierarchy of Nuclei

With increasing size in urban systems, outside the central district additional business concentrations appear, differentiated in terms of size and function, so that one may speak of a hierarchy of nuclei. Scattered outlying business establishments and clusters of establishments are characteristic of the smaller cities, but in urban systems with populations above about forty thousand there develop concentrations which combine particular groups of stores into distinct kinds of shopping districts. Isolated stores, too, continue to be present, but that is not our concern here. Of primary importance to the

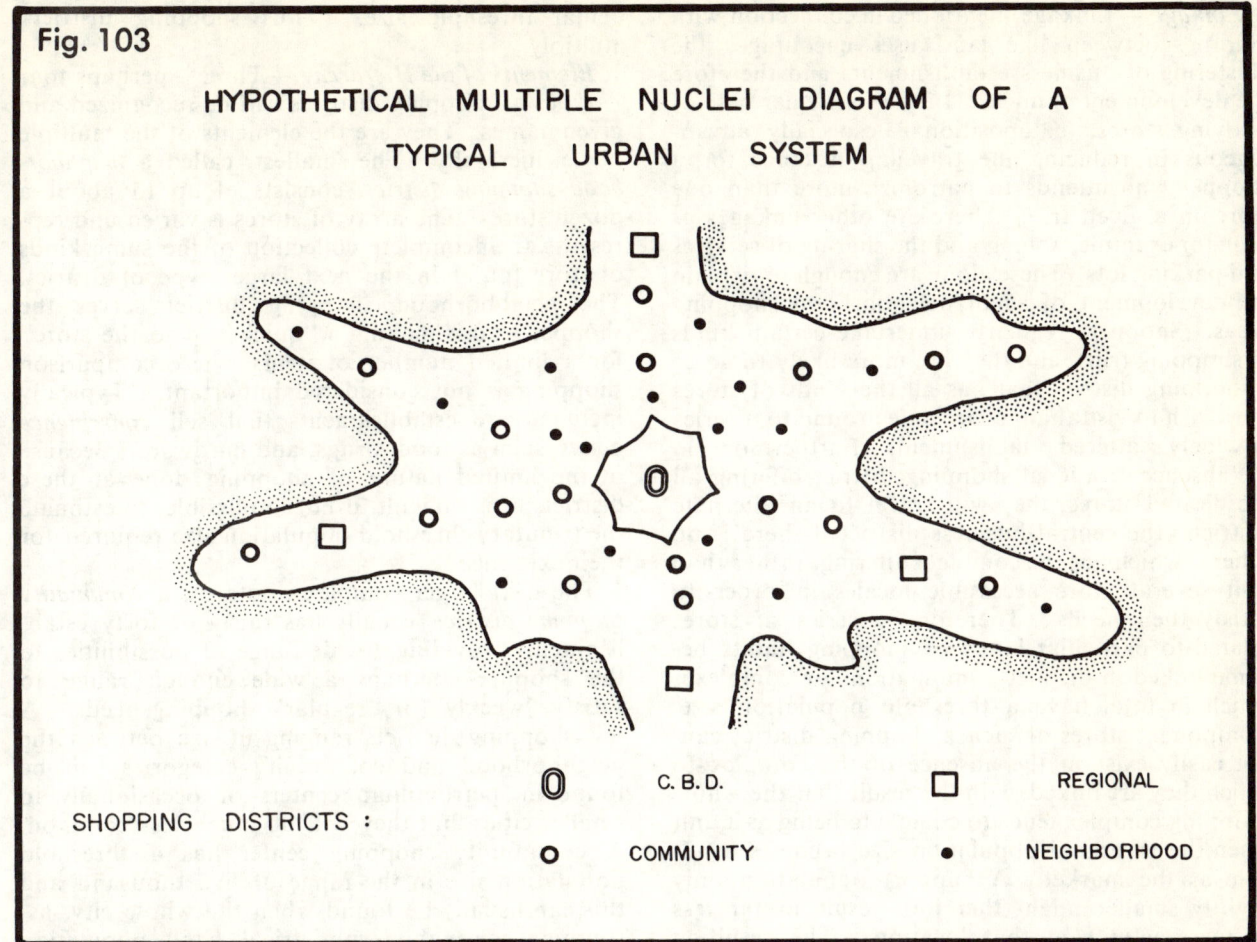

Fig. 103 HYPOTHETICAL MULTIPLE NUCLEI DIAGRAM OF A TYPICAL URBAN SYSTEM

SHOPPING DISTRICTS: ⓪ C.B.D. ▫ REGIONAL ○ COMMUNITY • NEIGHBORHOOD

business structure of urban areas are the forces that make possible concentrations of outlying stores and bring about the particular combinations of stores that can be classified into hierarchies. Of equal importance to the student of urban geography is a knowledge of what kinds of shopping districts result from these forces.

Threshold Concept.—Any store has a minimum volume of business necessary for its survival in its particular milieu, a level of operation called the *threshold business size*. When income from the goods and services being offered is at least equal to the sum of the fixed costs and the variable costs, theoretically the threshold has been achieved. The fixed costs include rent, heating, salaries, insurance, and all other items that are necessary for the operation but do not change with the volume of business. Normally a reasonable return on the investment is a part of the fixed cost. The variable costs include wholesale prices and handling—costs that vary with the amount of business transacted. Income is in some direct way related to the number of customers, which in turn is related to population. Therefore, for any store there is a threshold population size that it must draw on in order to realize success. When in some part of a city a point can be located which is more accessible than the central business district to a population equal to or larger than the threshold population of a given kind of store or group of stores, then such a store or group can, and very often will, be located there. It must not be imagined that the economic factors being described operate in an absolutely rigid way, because marketing conditions, variation in individual entrepreneurial skills, as well as variable and varying income and habits of the population all play a part. There may in some localities also be an actual absence of suitable entrepreneurs. What happens in the usual state of affairs is a large measure of trial and error. In any case, in America, where there are ordinarily plenty of investment capital and numerous individuals and corporations looking for investment opportunities, few locational possibilities are permanently overlooked.

CHAPTER 14 THE URBAN LANDSCAPE 343

Linkage.—Linkage, mentioned in connection with affinities between like land uses encourages the clustering of business establishments and therefore the development of nuclei. In the particular case of outlying stores, juxtaposition is especially advantageous in reducing the traveling distance for a shopper who intends to patronize more than one store on a given trip. There are other linkages as well: for example, zoning and the sharing of services and parking lots. These alone are enough to explain the development of small neighborhood shopping areas. Shoppers regularly undertake certain kinds of shopping trips, and they are more likely to go to a shopping district that has all the kinds of stores they wish to visit than to circulate around to a series of widely scattered establishments. Furthermore, in the absence of a local shopping district offering all the desired stores, they will travel to an alternate district—the central business district if there is no other—which has the complete offering, rather than visit several more accessible locales in order to satisfy their needs. Therefore, a series of stores related to particular kinds of shopping habits become linked in distinct combinations or complexes which in total have a threshold population size. Component stores of such a shopping district cannot easily exist in the absence of the complex to which they are linked, with the result that the whole shopping complex tends to come into being as a unit when the threshold population size becomes available as the market. A support population only slightly smaller might therefore result in far less business activity at that location. The resulting distinct groupings in both the size and the function of shopping nuclei become the elements for a hierarchy of shopping districts.

The same forces that bring distinct kinds of shopping districts into being work against their indefinite growth in size. To be sure, comparison shopping makes desirable more than one store of the same kind in a shopping center. The optimum number of comparisons more frequently than not apparently lies between two and three, and it is normal for there to be two or three of the dominant type of store in a given shopping district. As a city grows, considerations of symmetry require that at about the same time a threshold population accumulates in one part of the city it will appear in others as well; three, four, or even more shopping districts of a certain level in the hierarchy will develop almost simultaneously. With further growth the supporting population in each part may sooner or later become more than double the threshold size. In that event, it becomes likely that some new districts will come into existence between or beyond the established shopping districts at locations where superior accessibility is realized for a population of their particular threshold size. Thus shopping districts multiply.

Elements of the Hierarchy.—Three—perhaps four—types of shopping districts can be recognized and given names. They are the elements of the multiple nuclei hierarchy. The smallest, called a *neighborhood shopping district*, consists of up to about a dozen stores. The array of stores is varied and represents an incomplete collection of the same kinds of store found in the next larger type of district. The neighborhood shopping district serves the shopper who is making a "quick trip to the store" for a limited number of items where comparison shopping is not considered important. Typically included are establishments that sell *convenience goods,* such as food, drugs, and hardware. Because of the limited nature of shopping done at these districts, it is difficult, if not impossible, to estimate the tributary threshold population size required for their existence.

The next larger complex of stores, a *community shopping district*, usually has thirty or forty establishments providing a wide range of possibilities to the shopper—perhaps a wide enough range to satisfy "weekly" or "regular" shopping needs. A few shopping districts ranging in size between the neighborhood and community categories can be found in metropolitan centers or occasionally in smaller cities, but they will be much less common. A community shopping center has a threshold population size in the range of five thousand and this can usually be found when the whole city, assuming it is roughly symmetrical, has a population of about forty thousand. Dominating the typical community shopping district are two supermarkets. The majority of stores sell convenience goods, as did the neighborhood stores, but it must not be imagined that weekly or regular shopping needs are restricted to the limited range of stores that characterize neighborhood districts. As much as a third of the stores in a community district will be of a nonconvenience, i.e., shopper's, category, including restaurants and establishments selling furniture and appliances. Some time ago it was common for a community district to contain many specialty stores, each with its limited range of merchandise. This pattern still persists, but in the last generation there has been a vigorous growth of the *integrated store,* a store that carries under one roof all the lines that might be desired by a customer on one trip. Thus we have the supermarket, which strives to offer many of the products and services that might be expected to be found in individual stores, expanding from food lines into drugs, hardware, variety goods, and even to commonly used apparel, furniture, and personal services. The integrated store may take the form of a discount store too, although this phenomenon more characteristically associates itself with larger districts. The following empirically derived list of establishments is typical of the community shopping district: two supermarkets, liquor store, small grocery, two specialty food shops, drug store, hardware store, three cafés, two restaurants, specialty eating shop, five-and-ten-cent store, three specialty variety stores, barbershop, beauty shop, laundry, laundromat, two dry cleaners, two specialty apparel shops, appliance store, furniture store, bank, recreation establishment, office building.

The seven largest urban systems in New York have reached sufficient magnitude so that shopping districts of still higher class can be supported outside the CBD. For this higher class the name regional shopping district is applied. It used to be that threshold population in excess of 100,000 was necessary for the support of this larger complex, but trial and error have shown the investors who must underwrite the keystone establishment, the department store, that as few as 40,000 may be sufficient. Around the one or more department stores there will be at least seventy-five other business establishments. Only in the New York City and Buffalo urban systems can regional shopping districts be numerous. Places the size of Syracuse have but a few.

As indicated by the dominant position of the department stores, the emphasis in regional shopping districts is on shopper's goods, but all convenience goods are available, too. The customer makes infrequent trips to such a district for major purchases of clothing, furniture, and appliances, or for the purchase of any one of many special goods. A limited selection of all these items is available in smaller shopping districts but without the choice so necessary for more than routine purchases. In the regional shopping district approximately half the stores, among which are a majority of the larger kinds, are in the shopper's category. In fact the very continuance of an important number of convenience goods outlets stems from the simple fact that a regional district is at the same time also a community district for its immediate vicinity. Added to this is the linkage factor, whereby customers for the shopper's goods combine purchases at convenience stores in the same trip. Therefore, all of the stores listed above for the community district will be found in a regional district, although individual stores will tend to be larger. Services and eating places will normally exist in approximate proportion to the total size of the district, while there will be additional specialty food outlets such as candy stores, ice cream centers, and delicatessens. The greatest increase will be in the apparel category where, in addition to the department store that concentrates on apparel while offering the whole range of shopper's items, there will appear numerous women's, men's, and children's clothing stores, shoe stores, jewelry shops, and so on. Close behind apparel is an array of variety stores that handle sporting goods, toys, cameras, flowers, stationery, gifts, and many other items. Furniture stores, recreation establishments such as theaters, banks, and offices are present, too, but there are likely to be no more than two or three of the first three of these. Generally there are also insurance, real estate, dental, and legal offices, and a branch post office. The regional shopping district, because of its wide range of goods and services, gives the CBD real competition.

In the largest urban systems of America, metropolitan giants which encompass more than two million people, there is emerging yet another and higher class of business district for which there is no standard name. In such a nucleus there is a higher level of specialization in shopper's goods, but it is the office function which sets it apart. Instead of a few common types of offices, there are whole office buildings both public and private. Within the state it is only in the New York City area that such nuclei are to be found. Because this type of district has not been properly defined so far, a more detailed description of it will not be made here.

Departures from the Hierarchy of Nuclei

For various reasons individual shopping districts will differ more or less from the typical ones described above. There are variations in shape due to the conditions under which the district took shape. Instead of being nucleated there may be an orientation along a route of movement. This is often called a shopping strip. Sometimes small CBDs of peripheral cities become engulfed; in other instances planned districts which vary from the hypothetical basis described above emerge. The local geography of income, ethnic background, education levels, and even predominant age levels can all modify somewhat the make-up of shopping districts.

Shopping Strips.—As a great many potential customers make trips along major traffic arteries—generally radial routes out from the CBD—business establishments locate along such streets in order to be accessible to these customers. The result is shopping strings or strips. A linkage is involved with either the regular trip to work or the occasional expedition to some other business area, usually the central business district. Aside from its linear instead of nucleated conformation, the shopping strip differs in two ways from other secondary shopping districts. First, because its customers are largely passers-by, especially the working population, there will be a difference in the proportions of stores. There are fewer food stores and apparel shops in proportion to total numbers but more services and eating places. Imbedded in the strips may be lower-level nucleated centers. Stores which look to a wide area, perhaps the whole city, tend to locate closer to the center of the city, where shopping strips are most likely to occur. Such stores stand shoulder to shoulder with stores whose intended market is only local. Thus major appliance establishments, furniture stores, discount houses, contractor suppliers, and recreation establishments aimed at a city-wide market are mixed in with grocery stores and dry cleaners serving only the immediate vicinity.

Engulfed Central Business Districts.—Some shopping districts have grown slowly, may even have been a central business district of a community which has been engulfed by the expansion of a nearby larger urban system. Such nuclei are necessarily old and without consistent content or plan. There is chronically inadequate space for parking and the buildings are not only aged but poorly designed for present-day use. Both an inadequate complement of stores and a lack of compactness often accrue to old districts from lack of planning and the rigidity of alternate land uses already established. They therefore tend to be less desirable places in which to shop and in extreme cases present slum-like problems.

Planned Districts.—Along with increased size and rapid growth of an urban system, whole new shopping centers can be economically planned and constructed at one time, using every advantage of experience gained from study of problems and successes of existing districts. Plenty of parking area can be provided along with modern, conveniently arranged stores. The planner in selecting the stores to be included tends to encourage the large money-making store and pare down the number of small and marginal operations. More is seen of supermarkets, big drug stores, five-and-tens, department stores, and other large establishments. On the other hand there are fewer services and far fewer eating places. The usual result of the planned district is a lowering of the total number of stores for any given category of shopping district and the increasing of the relative importance of shopper's goods.

Special Orientations.—Each area of a city has its own peculiarities which become reflected in the make-up of shopping districts. For example, shopping districts located in neighborhoods characterized by lower income levels have more than the usual per cent of food markets and eating places (bars and cafés) but fewer apparel and appliance stores. Sometimes used furniture or even used clothing stores may be found. Different ethnic groups have particular shopping needs. Even more specialized in their needs are university or medical populations, who on the one hand require bookstores, services, and eating places and on the other need pharmacies and medical supplies. Areas with young populations have children's stores or maternity shops, while older apartment dwellers require more eating places and services.

External Patterns

There are a great many urban places, big and small, in an area of the size and importance of New York State. Each urban place has a location and a personality. It is not our intention at this point to pursue the uniqueness of any urban place, but rather again to clarify the regularities that can be established among the many diverse cities involved. These regularities resolve themselves into two types, one dealing with relative location or spacing of cities, and the other concerned with the classification of cities into hierachies based on function or growth tendencies.

RELATIVE LOCATION OR SPACING

The spacing of urban systems is highly complex. To some degree and in some places many cities are evenly spaced; in contrast distinct groups of cities can be identified. Between these two extremes a degree of randomness in urban distribution appears as a result either of specialized functions that apply to certain cities individually or as a result of the influences of site considerations.

Even Spacing

Inasmuch as cities serve their surrounding territory they tend to repel one another, each city being more accessible than any other to the area where its supporting population lives. Any other comparably sized city in the vicinity must necessarily divide the available support, thus limiting the size of the trade area of both. Eventually one will probably take trade from the other, enlarging its own trade area, unless there is some special reason why this should not be so, such as the existence of a river barrier running between them or a political boundary separating them. What is involved here is called the central place function, and the result is that central places offering comparable services tend to be evenly spaced.

Taking the leading urban systems of New York State and nearby areas there is a repeated spacing of approximately 150 miles. This is true for the pairs Albany-Boston, Albany-New York City, Albany-Burlington, Syracuse-Albany, Syracuse-Scranton, and Syracuse-Buffalo. Of course, several distortions can be quickly demonstrated. Cleveland is nearly two hundred miles from Buffalo. Burlington is much smaller than any of the other places mentioned, although it does stand alone and supreme in its region. Finally, Springfield, Massachusetts, is every bit as big as Albany and stands halfway between that city and Boston, while Rochester is half again the size of Syracuse and is located about midway between Syracuse and Buffalo. Nevertheless, there is no intermediate city whose size exceeds both cities that bracket it. Furthermore, nearly all of the cities of the state somewhat smaller than those already mentioned lie between fifty and seventy-five miles from one of the larger centers. Included are Binghamton south of Syracuse, Utica east of Syracuse, Watertown north of Syracuse, Elmira southwest of Syracuse, and Jamestown southwest of Buffalo. Several small cities alternate on either side of the Hudson between Albany and New York City, the leading ones being Newburgh, Poughkeepsie, and Kingston. Similar exercises demonstrating the validity of a tendency for the even spacing of cities can be carried out with smaller centers in many parts of the state or with similar centers in other parts of the country. Distances within a given area will differ from other areas and exceptions will always be found, but an element of regularity in spacing is commonly present. This regularity would most likely become nearly perfect in a uniform milieu, such as a featureless plain with equal access to all of its parts.

Grouping

That New York State does not present a uniform milieu has been abundantly demonstrated, so much so that one might well marvel at the degree of even spacing that can be found. The alternative to equal spacing takes the form of grouping of cities in certain areas and lack of cities in certain other areas. It should be mentioned that suburbs are considered here as a part of their larger metropolitan complex and do not enter this discussion.

Although it has been emphasized in earlier chapters that the cities of New York are sharply concentrated within limited parts of the state, re-emphasizing the point here in the way of illustrating the basis for urban grouping is worth-while. Several zones of better agricultural land or greater attractions to manufacturing, which are at the same time open to easy transportation, are connected by transport corridors to form the heavily traveled and often mentioned Hudson-Mohawk axis. Better land and better accessibility have concentrated cities along this axis and caused them to grow. Well over half of those in the state with at least ten thousand population are involved. A secondary route of travel crossing the Southern Tier of counties strings together half a dozen more cities of the ten thousand-plus size. Localities near Lake Ontario, in the St. Lawrence River lowland and along Lake Champlain, support some sizable central places, but cities of ten thousand or more in population for all the rest of the state are few and far between.

Several areas in and around the state are rugged and little populated. They are barriers to transportation and city development in general. Much of the state north of the Mohawk corridor consists of the Adirondack Upland, while to the south of the corridor the Appalachian Upland extends nearly across the state. In these areas of the state there are few cities of any size, each highland block forming a conspicuous disruption to the near-even spacing of central places which might otherwise have tended to develop.

Randomness

If even spacing might be expected to occur on a featureless plain with uniform access to all its parts and distinct grouping develops when nonuniformity in the milieu exists, there are still other conditions which produce largely a random pattern of cities. One such condition occurs when a highly specialized city serves a much wider area than its immediate vicinity; another condition occurs when the attributes of a particular site attract urbanization with such strength that the tendency for regularity in distribution is outweighed.

The typical central place function, which means providing trade and service facilities for an immediate hinterland, is not dominant in some cities. There may be instead an advanced degree of specialization. For example much of New York State lies within the manufacturing belt of the United States, a concentration of manufacturing in a zone most accessible to the over-all national market, raw materials, labor supply, and other requirements of industry (see Fig. 1). Each plant in this area is favorably located in reference to the national market, and thus there can be far more urban development than local trade area conditions alone would warrant. Manufacturing is important in a majority of the cities in New York and allows them to be relatively larger or crowds them closer to other cities than might be expected.

Other specialized functions contribute to randomness in city spacing. The educational function serving a national market makes Ithaca what it is, but has little to do with placing the city in a regularized locational scheme. Mining, lumbering, and recreation often stand out as city functions, as in certain Adirondack and Catskill communities. In these cases only small cities are involved.

The exact location of any city will be influenced by site conditions, the very best sites usually being occupied with disregard for any regularized spacing scheme. And what is the very best site at one time in history may not be the best in another.

It is no exaggeration to say that cities of today owe their particular site to the fact that in earlier periods there was a city there. The importance of this geographical persistence or inertia has been repeatedly referred to in other contexts. Sites were chosen for reasons that may not be important today, but the already existing city provides sufficient impetus to focus new growth and activity at the old site. Transportation junctions, such as those at Binghamton and Utica, and lake or river ports, such as those at Oswego and Ogdensburg, were favorite sites for developing central places. A mineral resource, such as salt at Syracuse, can localize a budding city. Some early locational factors continue in importance today, such as the water power at Niagara Falls and the excellent harbor at New York City.

The continuance of urban growth at a particular place can depend on fortuitous circumstances; the construction of the state buildings at Albany, the emergence of General Electric Company at Schenectady, and the development of Eastman Kodak at Rochester have insured these cities roles as leading urban centers. There are many other sites that would serve as well, or even better, for most urban places of today, but once the site is fixed it tends to persist because of the investment in buildings, streets, and the social and property structure. All of these special site conditions, both modern and outdated, contribute to randomness in city locations.

Hierarchies

Just as nucleated shopping districts can be arranged into a systematic sequence or hierarchy of types, so can urban places in general be ordered, either on a functional basis, very much like the shopping centers, or on the basis of growth tendencies.

Functional

There are many theories and some empirical evidence suggesting that cities can be grouped into levels of a hierarchy of increasing size and complexity. When population alone is considered there is no support for such theory. If central places of any area are ranked by population in order from the largest to the smallest, the ranking produces a continuum rather than a series of groupings (see Fig. 104). An urban hierarchy, therefore, must be based on discrete levels of the functions performed by the central places. Groupings by function are identifiable. This is much the same idea that was expressed in Chapter 12 in the discussion of trade centers and trade areas.

Any city or urban system provides a group of functions to a population which has two parts, the city or urban system population on one hand and the tributary trade area population on the other. Logically, therefore, the size and complexity of the urban functions will depend on the sum of the two supporting populations and not on the city population alone, and for any given level of urban functions there must be a threshold population size below which this level cannot successfully exist. Because these levels are not directly related to the size of the city itself, and the tributary population is difficult to calculate, the relation of functional level to population size must be approximate.

Urban functions, and particularly those serving the tributary population, which are usually referred to as the central place functions, are lodged primarily in the central business district. The nature of a hierarchy of business districts has already been developed and the same principles can be applied to a hierarchy of central places. There are certain differences, however, between central business districts and other business districts which should be clarified before proceeding.

A central business district is shielded from competition with other central business districts of equal or greater size by a distance factor much larger than that which operates between multiple nuclei within one urban area. This partial monopoly based on the friction of distance allows many kinds of functions to operate relatively inefficiently in the CBD. Compare, for example, a shopping district within a large city supported by a population of five thousand and having other shopping districts nearby, with an isolated town whose entire support population is also five thousand but with the nearest comparable competition being at least fifteen miles away in another town. The central business district in the isolated town would have several times as many business establishments, more in every category except perhaps food stores. There would be twice as many eating places, drug stores, variety stores, and recreation establishments. There would be several times as many appliance, hardware, and apparel stores, as well as beauty and barber shops. Two other categories of establishments make their appearance, not being found in the secondary shopping district of the large city. One is a set of shops and suppliers dealing in such things as fuel, repair, and building materials, totaling over thirty establishments. The other is a complex of office buildings and hotels. There will be several hotels or motels

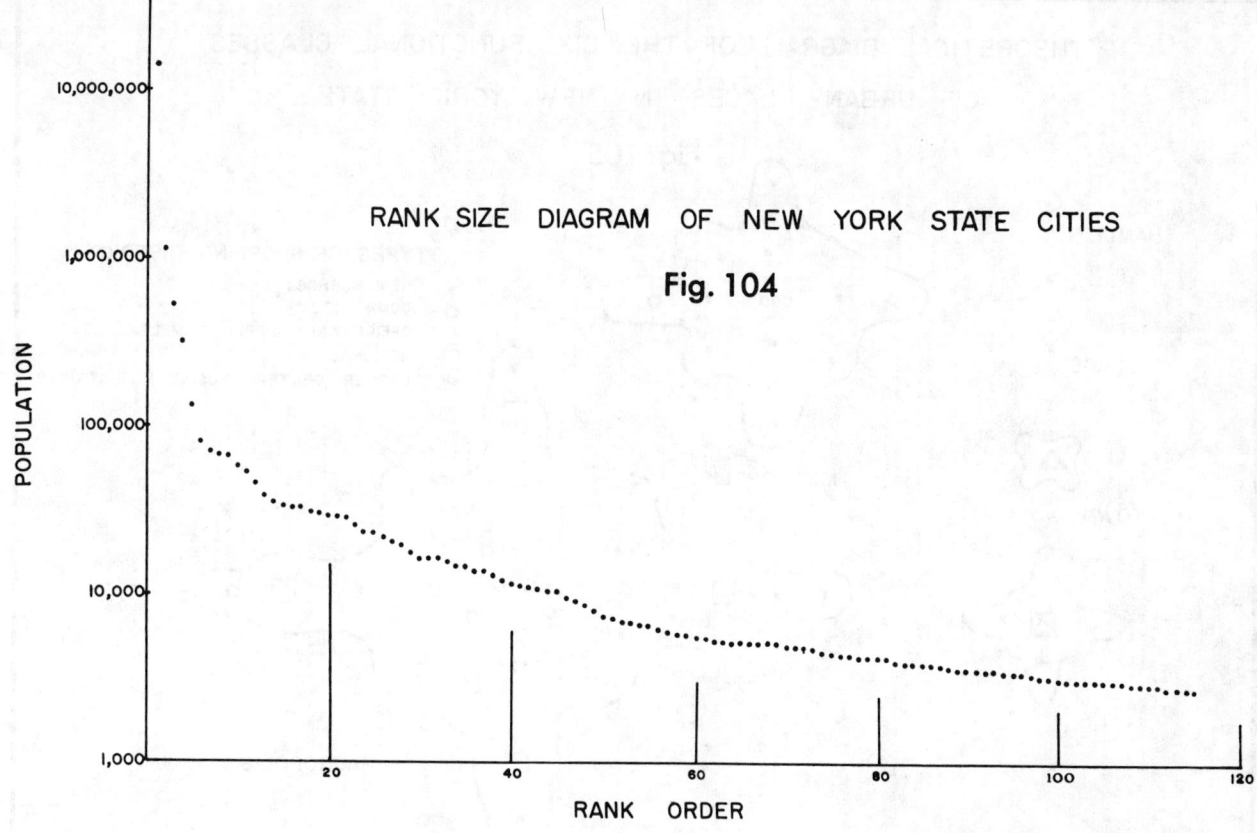

RANK SIZE DIAGRAM OF NEW YORK STATE CITIES
Fig. 104

and two dozen or more offices (occupied by doctors, dentists, optometrists, veterinarians, lawyers, accountants, insurance agents, and real estate brokers).

Aside from New York City, which clearly stands by itself, the urban centers of the state can be grouped into six functional classes (see Fig. 105). The smaller three operate functionally through a central business district only and have been designated by commonly used names. The larger three are more complex and have secondary shopping districts in addition to the central business district. For each of these a rough lower population size will be suggested in order to establish their general magnitude even though population is not the determining factor, as was made clear above.

The first two classes of central places, hamlets and villages, correspond roughly to the fifth-order trade and service centers of Chapter 12. The *hamlet* is identified as having from one to a few trade and service functions and is somewhat comparable to the neighborhood shopping center of a city. The emphasis on trade and service function eliminates from consideration here a cluster of residences unaccompanied by such function. As a usual minimum the hamlet contains a general store with an emphasis on grocery sales. Filling stations, schools, and churches appear in slightly larger hamlets, and as the size increases further various stores of convenience category appear. General stores, and other stores in larger hamlets, sell small amounts of many lines of merchandise which can be sold more cheaply or with better choice in special stores of larger central places, but the alternative of shopping in the city is not readily available to the clientele of the hamlet.

A *village* is a central place of several hundred people where there is sufficient support for the development of shopping facilities, corresponding in a general way to a small community shopping district of a city. It differs from the community district in displaying fewer food stores and shopper's goods outlets. Added are half a dozen suppliers ranging from oil dealers to lumber yards and auto parts dealers. Studies have shown that the number and variety of stores in an isolated village increase rapidly when the population surpasses about three hundred. This demonstrates a sort of threshold population size for the business complex that characterizes a village. Where the whole population is

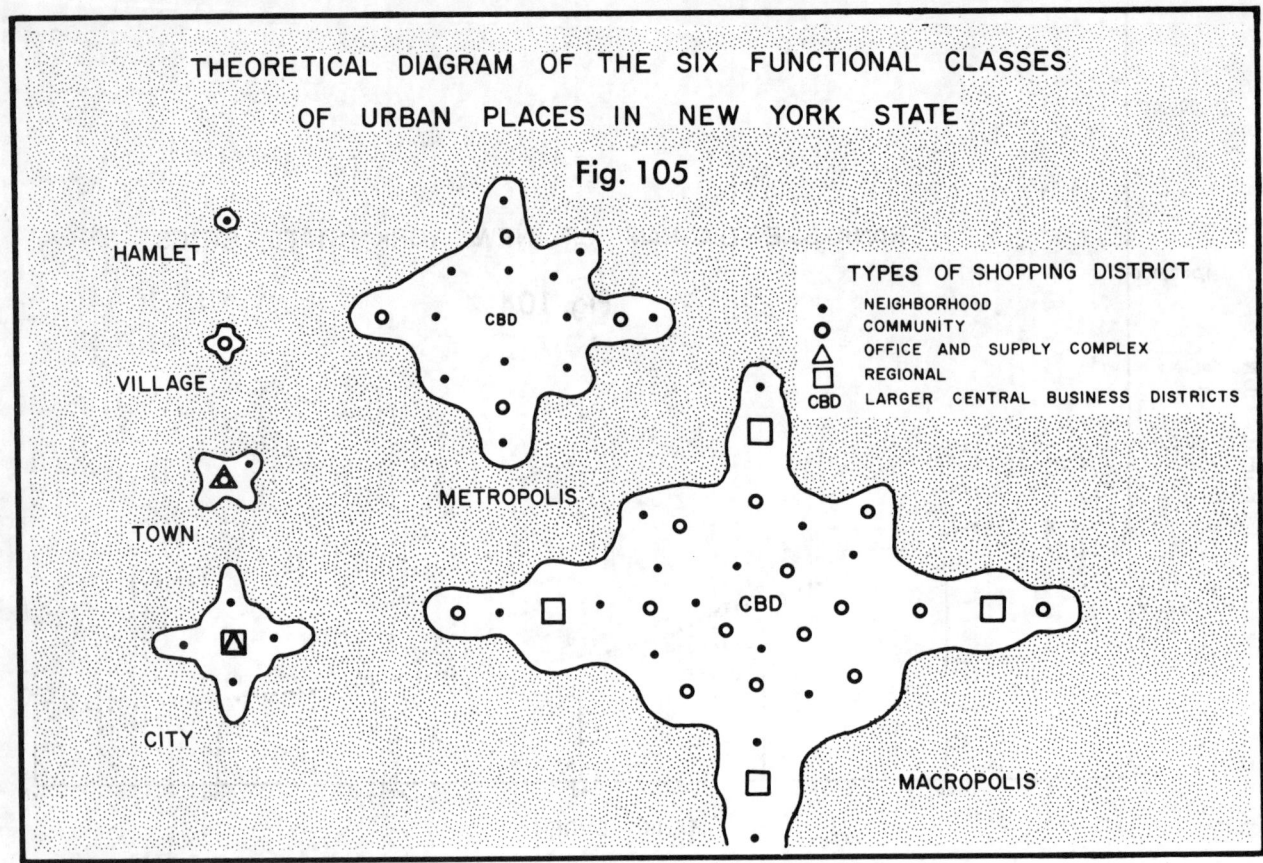

more largely rural, the village itself can be smaller and still the same complex of functions will be supported.

When a central place reaches a population of about twelve hundred, it is usually referred to as a *town*. A town by this definition is in the lower range of the fourth-order trade and service centers of Chapter 12. It should not be confused with the political unit called a town in New York State. The town business district emphasizes convenience goods as does the community district of a city. It does, however, contain a complex of offices, hotels, shops, and suppliers, which are largely lacking in the community shopping district. There are also more stores in general in the town, and there may be outlying stores or occasional groups of stores in the residential district.

The next three classes of central places are generally referred to as cities. Here, the smallest will be named simply city; the next larger, metropolis; and the third and largest, macropolis. Their distribution is shown in Figure 106. The city is in the upper range of the fourth-order trade and service center class of Chapter 12; the metropolis compares roughly with the third order; the macropolis with the second order. A *city* has a minimum population of about five thousand, and stores emphasizing shopper's goods dominate the CBD, making it correspond to the regional shopping district of the larger urban system. Offices are numerous; in fact, there are likely to be several office buildings of various sizes. It is normal for neighborhood shopping centers to exist, distributed about the residential areas. In a number of New York cities community shopping centers have been built at peripheral locations. These are designed to compete directly with the central business district for shoppers from the whole city, offering convenient parking and modern buildings. One or both of the two competing business areas are likely to suffer from the contest.

Places that can be classed as cities but not metropolises (excluding those which are suburbs) number not more than fifty in the entire state and have three characteristic kinds of location. Some, relatively close to the larger urban systems, are satellites strongly influenced by the larger city. Others line up along major transportation routes between the largest urban centers. The last group, where the central place function is dominant, are remote, often being roughly midway between two larger centers.

The *metropolis* category begins at a population of roughly forty thousand and is characterized by the appearance of community shopping districts. A threshold population in several sectors of the metropolis becomes available for community shopping districts when the total urban population reaches about this size. The same threshold may also be achieved in an odd sector of a very asymmetrical city of a smaller size. With the appearance of community shopping districts close to the homes of most of the residents, the nature of the central business district undergoes a notable change. No longer is it the focus of the regular shopping trip, and thus the convenience store ceases to be important. Rather, there is now an emphasis on shopper's goods with department stores being prominent. There is specialization in the apparel and variety type of shop. Large office buildings and hotels form the skyline of the CBD. The frame is developed including large areas of parking lots and even a few parking garages. Metropolises often have suburbs, but it is not clear whether there is a direct relation to function in this development. There is good reason to suspect that an outlying community shopping district in fact is the genesis of a true suburb. At any rate, those cities with suburbs merit the designation "metropolitan."

The few urban systems in New York State whose populations surpass a quarter of a million, can be called *macropolises,* being differentiated from metropolises because at about the quarter-million size regional shopping districts develop. The movement of many stores specializing in shopper's goods out to these regional shopping districts clearly alters the CBD. High levels of specialization continue to exist in the CBD, where the volume of trade will be some ten times the size of the largest regional shopping district. The office function of the whole urban system, growing at a rate greater than that of the population, will now rival the retail business function in the CBD in such things as the space it occu-

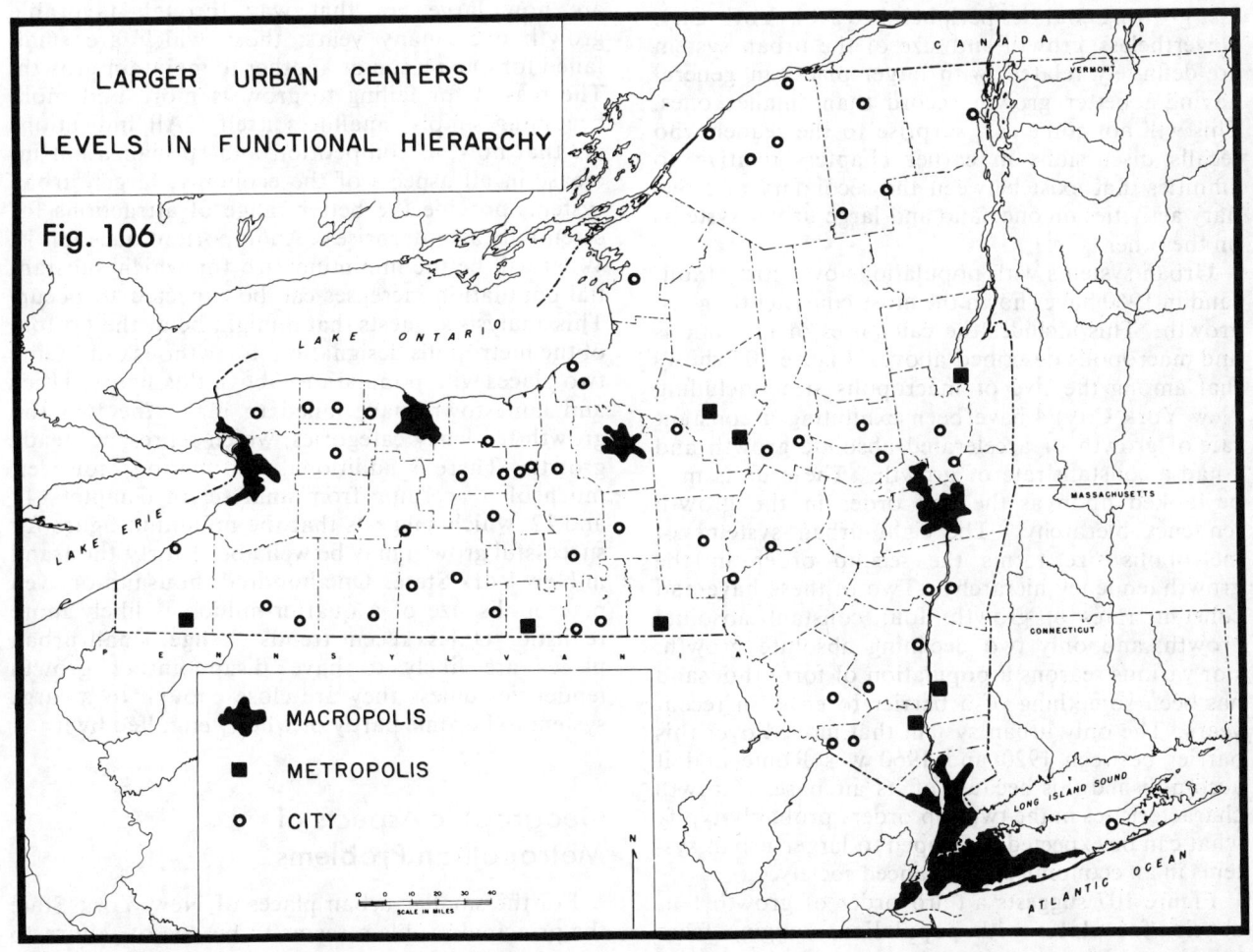

CHAPTER 14   THE URBAN LANDSCAPE   351

pies and the number of people it attracts. It is significant that many office buildings but few stores are being built in the center of the modern macropolis.

Growth Tendency

A study of the long-time growth tendencies for agglomerations of populations around and in all urban places in the state which had over one thousand in 1960 reveals that it is possible to classify urban systems into six growth-trend categories, namely, those places that: (1) have experienced a constant growth rate, (2) have a roughly constant absolute growth, (3) have declining absolute growth, (4) at one time stagnated but experienced renewed growth, (5) are exhibiting no growth (approximate stagnation), and (6) have declined. This is simply a refinement of the second and third types of population trends described in the Introduction.

There are, of course, many things that make central places grow, ranging from favorable sites on major transport axes to the establishment of a military base. Regional disadvantages and advantages are important, too, as in the case of the economically depressed cities of the middle Mohawk or the fast-growing counties peripheral to New York City. Nevertheless, growth and size of the urban system are definitely related, with larger places in general having a better growth record than smaller ones. This will not come as a surprise to the reader who recalls discussions in earlier chapters relative to affinities that exist between the secondary and tertiary activities on one hand and large urban systems on the other.

Urban systems with populations over forty thousand in 1960 have shown the most consistently good growth. This includes the categories of metropolis and macropolis described above. Figure 107 shows that among the five of macropolis size (including New York City) 4 have been exhibiting a constant rate of growth or accelerated absolute growth and 1 had a constant rate of growth. These units may be looked upon as the first order in the growth tendency hierarchy. The eight urban systems of metropolis size forms the second order in the growth tendency hierarchy. Two of these have had constant rates of growth, four constant absolute growth, and only two declining absolute growth. For various reasons a population of forty thousand has been something of a barrier to cross in recent years. The only urban system that moved over this barrier between 1920 and 1960 was Rome and it accomplished this because of its air base. Growth characteristics in the two top orders probably typify what can be expected to happen to larger urban systems in an economically advanced society.

Figure 107 suggests a third order of growth tendencies for places with populations ranging from fifteen thousand to forty thousand in 1960. Of this group of sixteen units, only one could boast of a constant rate of growth, five had constant absolute growth, and three had declining absolute growth. Six had no growth at all and one declined. As a whole this is distinctly a less favorable performance than enjoyed by places in the first two orders, and some of the better performances in this third order can be traced to installations of military bases or other nonpermanent and perhaps what might be called nonnormal stimuli.

The fourth order comprises the rest of the places shown in Figure 107 ranging in size from twenty-five hundred to fifteen thousand. While some thirty of these exhibit favorable growth trends (and there are usually special explanations for these cases such as proximity to larger places), the group as a whole performed unimpressively, with twelve actually declining.

It can be seen from the foregoing that places already large have been getting the lion's share of new growth. There is a general decline in population size from cities with constant absolute growth to cities with little or no growth. Those places which are now large got that way through favorable growth over many years; those which are small failed for one reason or another to maintain growth. The reason for failing to grow is more and more becoming simply smallness itself. All indications are that now, as competition and specialization increase in all aspects of the economy, larger urban systems provide the better range of attractions for essentially all enterprises. An important question is: What will be the minimum size for which substantial population increases can be expected to occur? This analysis suggests that it might be at the bottom of the metropolis designation, forty thousand. Only two places with populations above this figure, Utica and Jamestown, have failed to make the two top growth tendency categories, which represent steady growth. There is additional information, however, much of it accruing from analyses in Chapters 11 and 12, which suggests that the minimum figure for successful growth may be well above forty thousand in New York State. One hundred thousand or even macropolis size of a quarter-million is likely more realistic. Unless recent trends change, small urban places are likely to have disappointing growth tendencies unless they are close enough to a large system to become partly or wholly engulfed by it.

Geographic Aspects of Metropolitan Problems

For the smaller urban places of New York State the principal problem seems to be one of failure to

Fig. 107

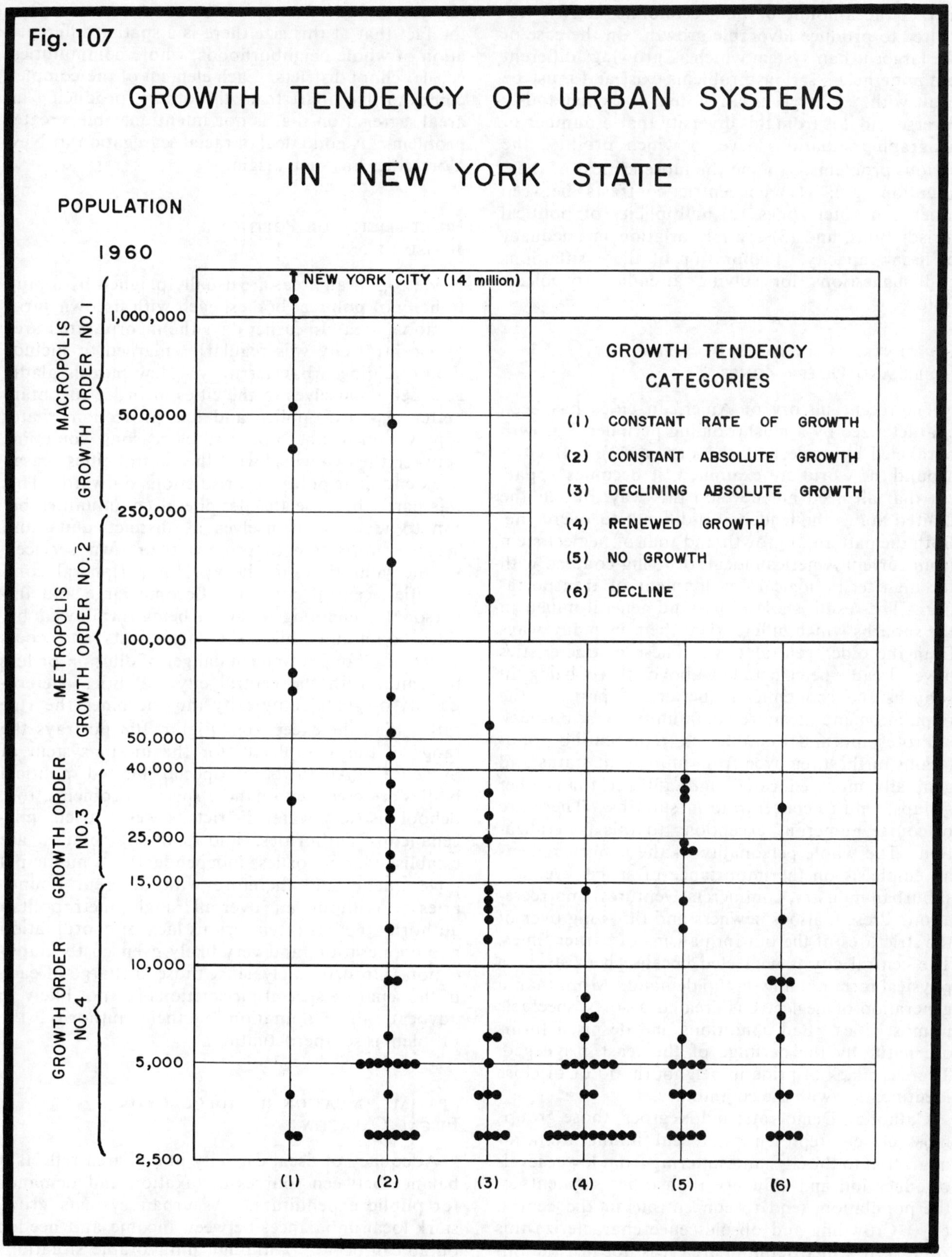

GROWTH TENDENCY OF URBAN SYSTEMS IN NEW YORK STATE

POPULATION 1960

NEW YORK CITY (14 million)

MACROPOLIS — GROWTH ORDER NO. 1
METROPOLIS — GROWTH ORDER NO. 2
GROWTH ORDER NO. 3
GROWTH ORDER NO. 4

GROWTH TENDENCY CATEGORIES

(1) CONSTANT RATE OF GROWTH
(2) CONSTANT ABSOLUTE GROWTH
(3) DECLINING ABSOLUTE GROWTH
(4) RENEWED GROWTH
(5) NO GROWTH
(6) DECLINE

CHAPTER 14   THE URBAN LANDSCAPE   353

attract the amount of new economic activity required to produce favorable growth. In the case of the larger urban system which is growing, different but nonetheless serious problems exist and must be dealt with. A large urban system is tremendously diverse and it is from this diversity that a number of geographic situations develop which produce the serious problems. Among the more critical of these situations are: (1) concentric contrasts between inner and outer zones, (2) multiplicity of political jurisdictions, and (3) spatial variation in adequacy of fiscal capacity. Elaboration of these situations and suggestions for solving attendant problems follow.

CONCENTRIC CONTRASTS BETWEEN INNER AND OUTER ZONES

The recent history of American cities has been characterized by a mushrooming peripheral growth paralleled by neglect of the urban core. If cities around the world are examined, it becomes apparent that all need not develop this way, but in the United States the tendency is difficult to overcome. Both the pattern of growth and that of neglect stem from current American ideals of living coupled with the high technological development of transportation. The result is a land use and general milieu in the suburbs which differentiate them in many ways from the older central city. These characteristics have already been detailed above, there being in suburbs the economically better off part of the population and the more community consciousness.

Protestants and Republicans, representing populations furthest removed from immigrant status and generally more educated and affluent than other groups, tend to concentrate in suburbs. There are of course numerous exceptions to this generalization. The whole personality of the suburb reflects an emphasis on the importance of space, even for suburban industry, commercial ventures, and recreation. Present also is newness and throwing over of the strictures of the old formalisms of earlier times. The central city is neglected because it exists in a physical form not now in top demand. More than a generation of neglect has created a sorry spectacle in most cities. Bad conditions and despised forms are naturally the heritage of the least privileged. There is a lack of pride in neighborhood or of close identification with the community.

Catholics, Democrats, and Negroes, those groups most closely representing recent immigration, or migration to the city, and suffering from lower levels of education and affluence than other segments of the population, tend to concentrate in the central city. Crowding and obsolescence characterize this zone. For a metropolitan area it is not just the differentiation of inner and outer zones, however, but the fact that at this size there is a spatial differentiation of whole neighborhoods, whole communities, whole school districts. Each element of the complex comes to live apart from the others, producing an areal segregation that is not intentional but creates problems. A good deal of racial segregation in New York cities has such origin.

MULTIPLICITY OF POLITICAL JURISDICTIONS

Metropolitan areas are usually plagued by a multiplicity of political bodies, each with its own jurisdictional area. In earlier days the incorporated area of a central city was regularly enlarged to include the expanding urban territory. New parcels gladly attached themselves to the cities in order to obtain better services: police and fire protection, water supply, and so on. For various reasons, sometime between the two world wars the central cities ceased to extend their political jurisdictions outward. This was partly because the peripheral communities began to identify themselves as distinct units and became more able to provide their own services. At the same time, in the eyes of peripheral communities, central cities had become large and impersonal, something to avoid being swallowed by. Each community has its own interests and goals which are thought to be in danger of dilution or loss by union with the central city. Political heterogeneity of great complexity thus becomes the rule rather than the exception. Figure 108 portrays the range of this complexity for the urban system of Syracuse. All sorts of special limited political bodies are created to handle specific problems, too. School districts, water districts, sewer districts, miscellaneous authorities, and planning bodies are established more or less independently of municipal governments and therefore with different boundaries. Without an over-all single metropolitan authority there is often serious lack of coordination and cooperation, and very likely even costly duplication. Unfortunately, since those in charge of each of the separate special jurisdictions are not likely to advocate the elimination of their authority, the problem is self-perpetuating.

SPATIAL VARIATION IN ADEQUACY OF FISCAL CAPACITY

Adequacy of fiscal capacity in any area reflects a balance between sources of taxation and demands for public expenditure. As urban systems grow, stark local imbalances between income and needed outgo can occur. Although unfavorable situations

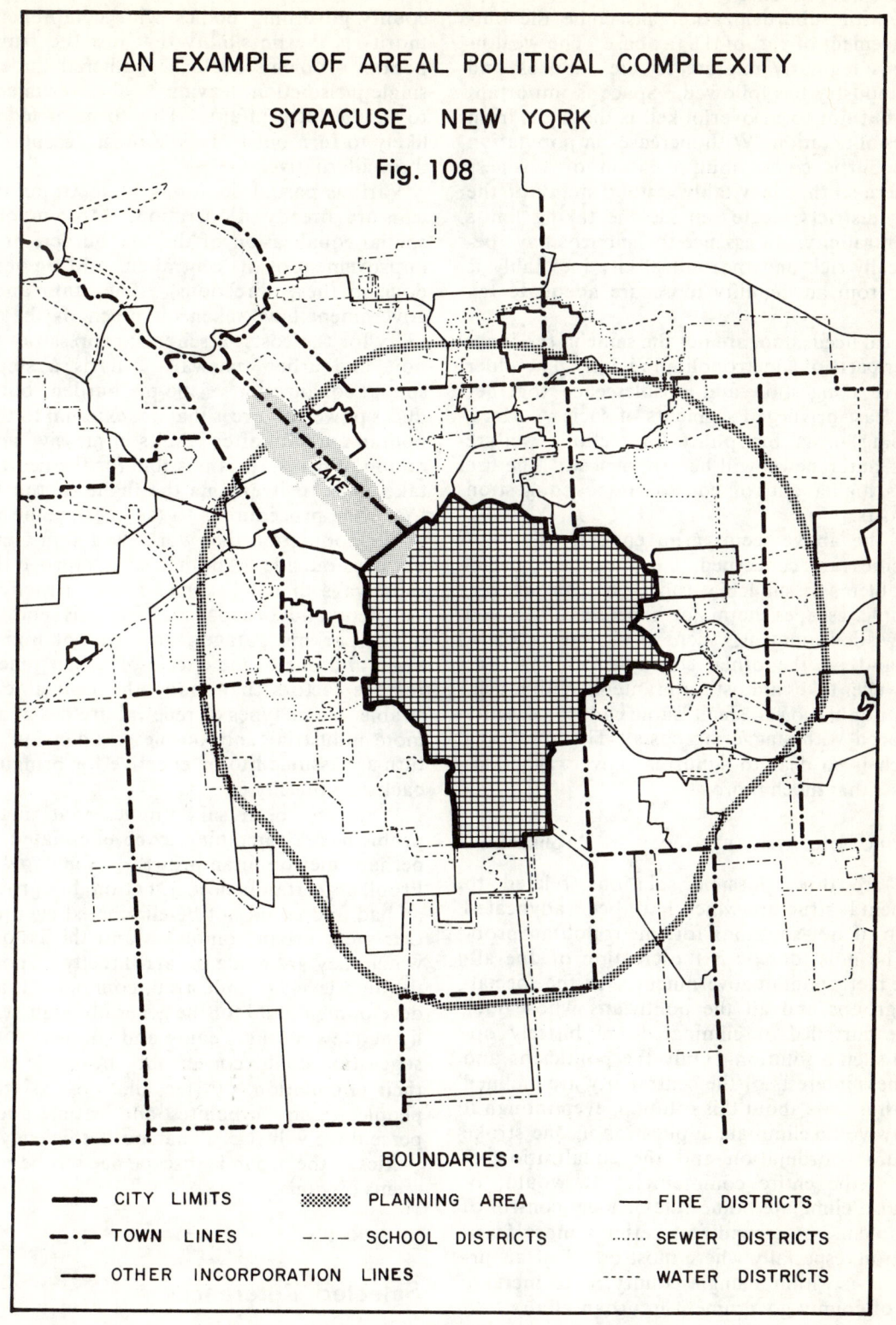

are more common in the central city, suburbia and exurbia are not immune.

Along with suburbanization has come the outward movement of part of the tax base. The wealthier citizens regularly migrate to the suburbs and recently industry has followed. Space is important to both, but not to be overlooked is the flight from high rates of taxation. With increases in population in the suburbs comes multiplication of business nuclei there so that inevitably a good many of the shopping districts locate outside the taxing limits of the central city. In essence the suburbs have become fiscally rich and the central city inevitably is suffering from an inability to secure adequate tax incomes.

The fiscal needs, too, are not the same in the inner and outer parts of a metropolitan region. The older areas have congestion and obsolescence together with the least privileged segments of society. Here, then, rising costs for police protection, welfare demands, and renewal will be experienced. The fire hazard is higher and of course traffic congestion creates costs.

All of the above seems grim enough as far as central cities are concerned, but there are special fiscal problems in suburbia, too. The central city, for example, escapes the major expenses associated with new development, such as schools, roads, sewers, and all the other conveniences that the urban dweller takes almost for granted. With mushrooming population a small suburban community can be faced with staggering costs. The spreading out which is so dear to suburbia only exaggerates these costs that much more.

SOME SOLUTIONS TO METROPOLITAN PROBLEMS

There are three possible solutions related to governmental structure which have been advocated as aids to, if not solutions for, metropolitan problems. The most drastic is the creation of one all-inclusive metropolitan government. All the special-interest groups and all the politicians whose jobs might be curtailed or eliminated are bitterly opposed to such a solution. Only the politicians and the business interests of the central city are inclined to be enthusiastic about this solution, even though it obviously would eliminate duplication in one stroke and insure coordination and the equalization of costs over the entire community. It would, of course, also eliminate home rule or local control of local problems. A second and perhaps more feasible solution, especially where most or all of an urban system is within a single county, is to increase the role of county government in urban affairs. At one extreme this could produce an effective county metropolitan government and at the other just one more big hand in an overstirred kettle of soup. Less drastic than having centralized metropolitan or county governing bodies with comprehensive authority is the possibility that just the truly metropolitan problems could be gathered under such a single jurisdiction, leaving local governments intact to handle local affairs. This form of federalism is likely to turn out to be the most acceptable of the three alternatives.

Various partial solutions to metropolitan problems are already in operation. Most involve either spatial equalization of the tax burden or physical improvement of the central city. More needs to be done in these directions. The state and federal government have taken on much of the responsibility for the costs of schools, expressway construction, and urban renewal. This is a step toward spatial equalization of the tax burden, but it introduces powerful forces that are external to the metropolitan unit and often results in massive noncoordination of plans. It is possible for the central city to take steps to overcome the flight of the tax base, too. One procedure is to tax the income and sales of the population that works and shops in the city but lives outside. Another is to improve the city in such ways that it becomes more attractive for investment and occupance. To this end shopping malls, parking garages, and arterial highways are built, and one of the effects of urban renewal is to remove factors that made the central city undesirable. Some types of renewal are designed to promote industrial and business expansion; these in turn are assumed to be effective for bringing people back into the city.

What has been said implies that metropolitan problems are something to be eliminated, and that perhaps metropolitan growth is somehow bad too. Problems certainly need attention, but growth is not all bad and could not be eliminated even if it were. Our great urban complexes and the landscapes of which they are made up are directly responsible for the high levels of industrial, commercial, and social development, and for the generally high standard of living New Yorkers enjoy and cherish. No serious suggestion could conceivably be made advocating their elimination. After all, most of the state's people are now urbanites, and as time goes on that percentage will rise. That both efficiency and aesthetics in the urban landscape need to be maximized seems obvious.

Selected References

Bartholomew, Harland. *Land Uses in American Cities.* Cambridge: Harvard Univ. Press, 1955.

Berry, Brian J. L., and Allen Pred. *Central Place Studies: A Bibliography of Theory and Applications.* Philadelphia: Regional Science Research Institute, 1961.

Christaller, W. *Die zentralen Orte in Süddeutschland.* Jena, Germany: Gustav Fischer Verlag, 1963.

Dickinson, Robert E. *City and Region.* London: Kegan Paul, 1964.

Duncan, O. D., *et al. Metropolis and Region.* Baltimore: Johns Hopkins Univ. Press, 1960.

The Editors of Fortune. *The Exploding Metropolis.* Garden City, N.Y.: Doubleday, 1957.

"The Future Metropolis," *Daedalus, Journal of the American Academy of Arts and Sciences* (Special issue, Winter, 1961), 1–193.

Garrison, William L., *et al. Studies of Highway Development and Geographic Change.* Seattle: Univ. of Washington Press, 1959.

Gottmann, Jean. *Megalopolis: The Urbanized Northeastern Seaboard of the United States.* ("A Twentieth Century Fund Study.") Cambridge: M.I.T. Press, 1961.

Harris, Britton (ed.). "Urban Development Models: New Tools for Planning," *Journal of the American Institute of Planners,* XXXI, No. 2, (May, 1965), 90–172.

Higbee, Edward. *The Squeeze: Cities without Space.* New York: Morrow, 1960.

Hoover, Edgar M., and Raymond Vernon. *Anatomy of a Metropolis: The Changing Distribution of People and Jobs within the New York Metropolitan Region.* Cambridge: Harvard Univ. Press, 1959.

Horwood, Edgar M., and Ronald R. Boyce. *Studies of the Central Business District and Urban Freeway Development.* Seattle: Univ. of Washington Press, 1959.

Klove, Robert C. "The Definition of Standard Metropolitan Areas," *Economic Geography,* XXVIII (April, 1952), 95–104.

Mayer, Harold M., and Clyde F. Kohn (eds.). *Readings in Urban Geography.* Chicago: Univ. of Chicago Press, 1959.

Norborg, Knut (ed.). *Proceedings of the IGU Symposium in Urban Geography, Lund 1960.* ("Lund Studies in Geography," Series B, No. 24.) Lund, Sweden: Department of Geography, the Royal University of Lund, 1962.

Tunnard, Christopher, and Henry H. Reed. *American Skyline.* New York: Mentor, New American Library, 1956.

Vernon, Raymond. *Metropolis 1985: An Interpretation of the Findings of the New York Metropolitan Region Study.* Cambridge: Harvard Univ. Press, 1960.

CHAPTER 15   JOHN H. THOMPSON

The Rural Landscape

The Urban Landscape is the milieu in which over 85 per cent of the state's people reside. On the other hand, it is the rural landscape which makes up over 85 per cent of the state's area. Here in the open country where uncrowdedness and relative quiet prevail, the hand of nature at least equals the hand of man in characterizing the landscape.

If one drives outward far enough from cities, suburban homes and commercial structures largely disappear and there remains little evidence of an urban orientation. The urban frontier has been crossed. In places farmsteads and cropped land are widespread, and agriculture is doing relatively well; in others it is evident that farming was once common but has become less important. Here land abandonment is pronounced. In still other parts of the state close observation reveals that farms never occupied the majority of the land. Except where recreational, lumbering, or mining enterprises have induced settlement, the land has always been fairly largely empty of human beings. Thus, rural New York exhibits three quite different landscapes: the first is dominated by farms and will be referred to as the *farm landscape,* the second is called the *mixed landscape of farms and abandoned farm land,* and the third, a landscape essentially empty of settlement, appropriately is designated the *empty landscape.* Few large sections of the state exhibit exclusively one of these three, yet it is quite possible to outline in a general way those areas which are predominantly of one type or another. This generalized breakdown of the state into three rural landscapes is shown in Figure 109.

Census statistics, both county and town, support field observations in making the differentiations. Of particular value are: (1) per cent of total land in farms, (2) per cent of land harvested, (3) per cent of total land in pasture or woods and idle land, (4) per cent of total land not in farms, and (5) per cent loss of harvested acreage between 1925 and 1960 (see Table 40).

Farm Landscape

The principal area of farm landscape occurs on the Erie-Ontario Lowland and in the Finger Lake area but extends along the northern edge of the Appalachian Upland and into the Hudson Valley. Smaller areas are found elsewhere in the state where conditions for farming are reasonably attractive. Attractive conditions for farming include fairly level land, good, well-drained soils, not too cold and snowy a climate, and favorable locations with respect to marketing.

Roughly 75 per cent of the total area designated as farm landscape is in farms, and 35 per cent is harvested (see Table 40). Harvested land includes fields from which hay is cut, or grain, fruit, vegetables, and other crops are harvested, but excludes pasture. Some whole counties in the area of farm landscape, such as Genesee, harvest crops from nearly 40 per cent of the land. This is high when compared to a figure of only about 16 per cent for the state as a whole. Forty per cent, the difference between the amount in farms and that harvested,

TABLE 40
LAND USE PERCENTAGES FOR THE THREE RURAL LANDSCAPES AND NEW YORK STATE*

Land Use	Farm Landscape	Mixed Landscape of Farms & Abandoned Land	Empty Landscape	New York State
Total Land in Farms	75	55	Under 10	44
Total Land Harvested	35	18	Under 1	16
Total Land in Pasture–Woods–Idle Land	40	37	Under 9	28
Total Land Not in Farms	25	45	At Least 90	56
Decline in Harvested Acreage, 1925–60	30	41	Most Lost Before 1925, but over 75	40

*Estimates based on county census data, U.S. Census of Agriculture.

FARM LANDSCAPE near Cayuga Lake. Fairly level land and good soils encourage use of most of the land for farming. Dairy-oriented crops predominate. The large area in trees suffers from poor drainage. Successful farming is likely to continue. *N.Y.S. College of Agriculture, Cornell Univ.*

may be assumed to be in a pasture–wood lot–idle land category. The 25 per cent of the total area not in farms is devoted to urban, institutional, and transportation uses, including state, county, or municipal parks, and the like.

Occupying the inherently best agricultural land, as would be expected, the farm landscape has experienced relatively less decline in harvested acreage than other areas. Actually, the decline between 1925 and 1960 was about 30 per cent as compared to 40 per cent for the state as a whole.

These land use statistics clearly give character to the farm landscape, but there are also certain identifying features which can best be observed in the field. Outstanding among these is a relatively dense network of good, paved, farm-to-market roads. These roads are well maintained in summer and winter, even in heavy snowfall areas upstate where drifts above car tops may make canyons out of the roads during January or February. Farmsteads, situated along roads, ordinarily vary in number from three to seven per square mile. Residences occupied by nonfarmers are usually less numerous than farm-houses and are almost entirely absent in some sections.

Because of the widespread emphasis on dairying, farms commonly exhibit large dairy barns and silos as well as the usual complement of other structures. A casual observer can tell whether activity around the barn, and condition of the barn and silo, point to active dairying or whether the farmstead has been abandoned as a farm and is now being lived in by a nonfarmer. Most farmhouses in New York were originally built in a fairly substantial fashion but are not as large on the average as those in the better farming areas of the American Corn Belt in the Midwest.

Fenced hayfields, pastures, and lesser acreages of corn and small grains dominate the over-all view, but where slopes are steep or bad drainage condi-

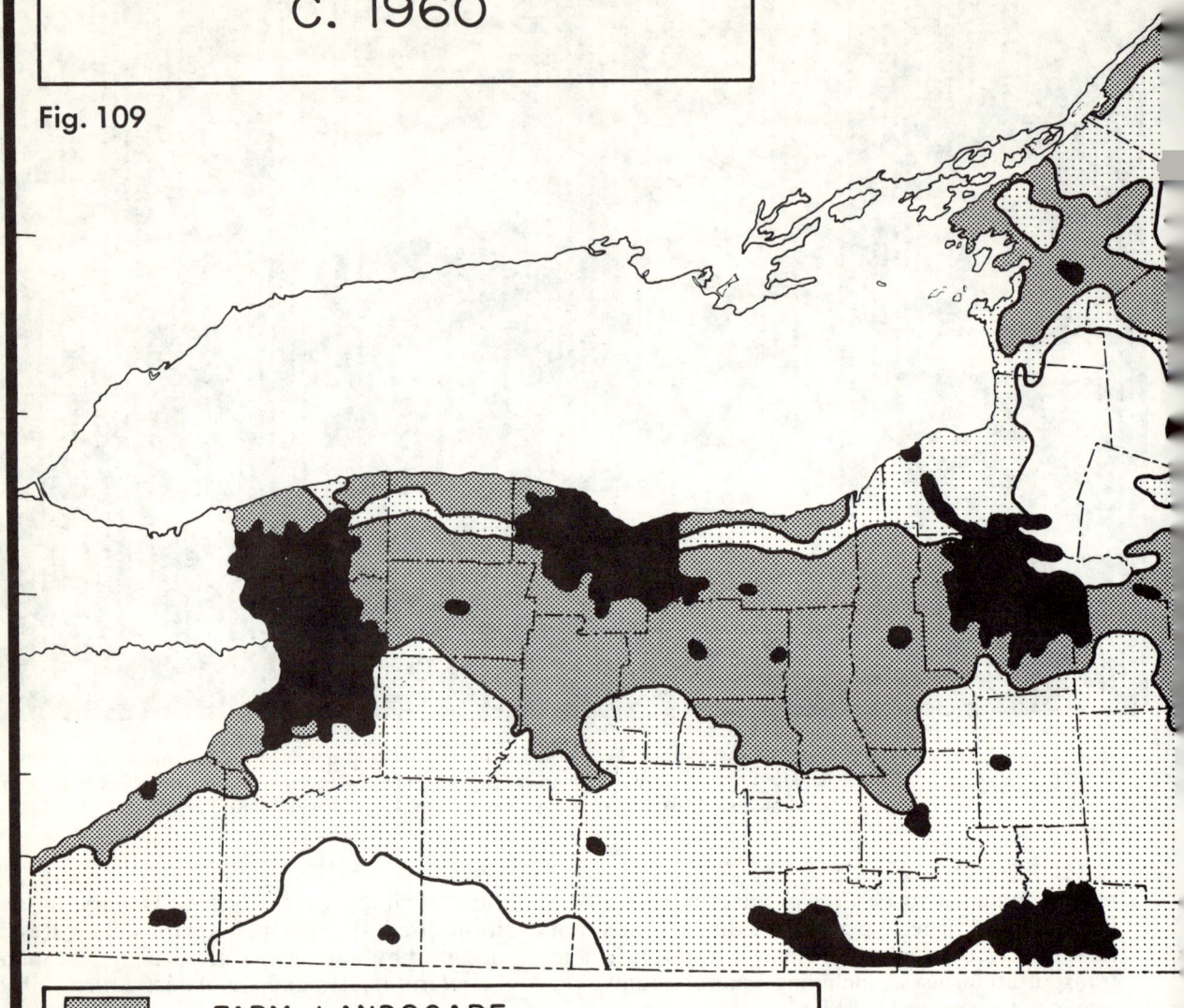

Fig. 109

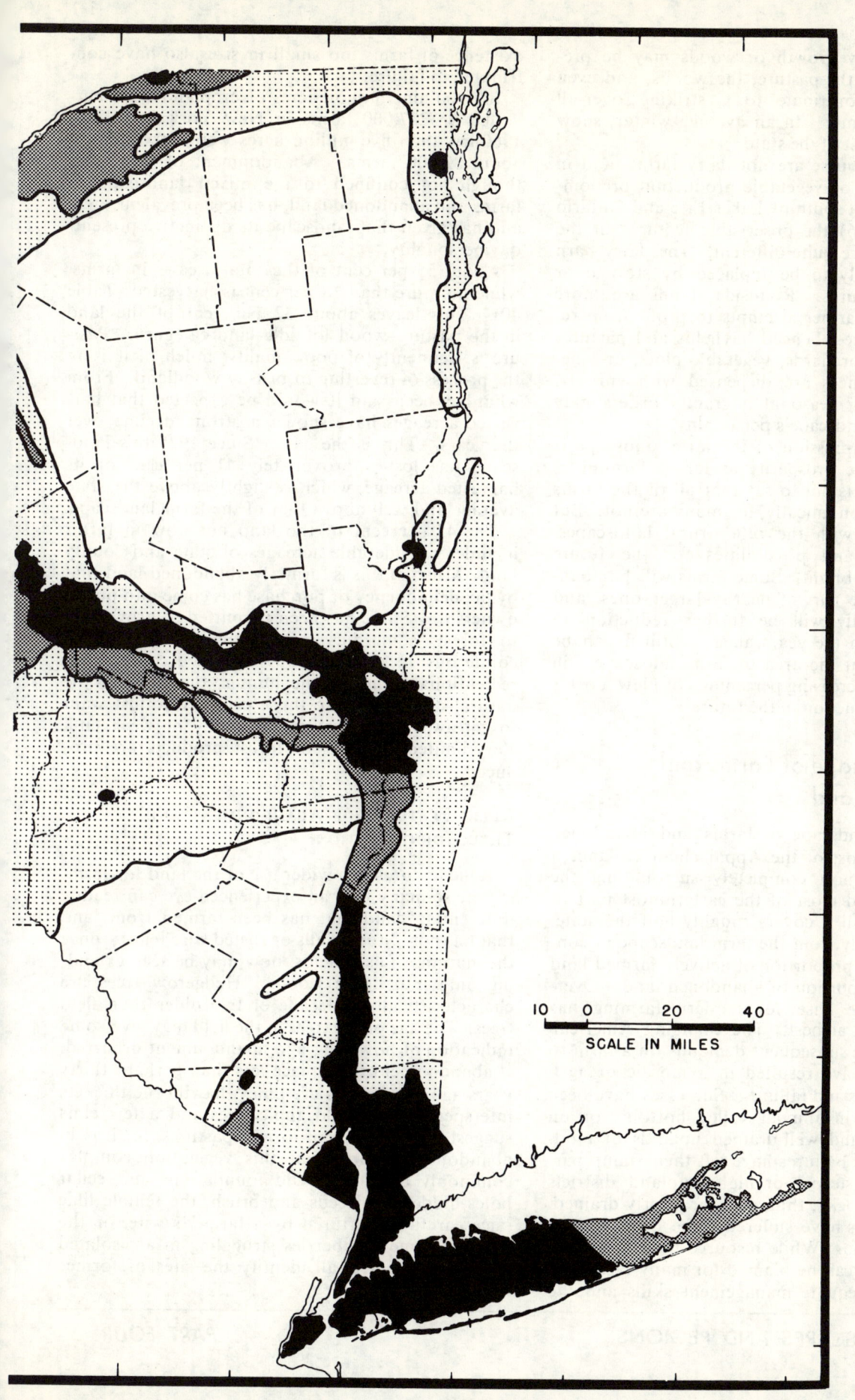

tions exist brushy growth or woods may be prevalent. The hay, the pasture, the woods, and even the grainfields contribute to a striking over-all greenness in summer. In an average winter, snow lies deep over most of the state.

All farms of course are not dairy farms, and in areas where fruit or vegetable production predominates, such as just south of Lakes Erie and Ontario (see Fig. 65 or 66) the observable features of the farm landscape are quite different. The dairy barn and silo are likely to be replaced by storage or processing structures. Roadside stands are more common as the farmer attempts to profit from retailing his produce. Fenced hayfields and pastures are replaced by orchards, vegetable plots, or vineyards. Often fences are dispensed with entirely. Drab housing for seasonal migrant workers may also add to the landscape's personality.

The general impression of the farm landscape is one of reasonable prosperity as far as farming is concerned. This is not to say that all of the farms are doing well economically, for many are not. But when compared with the other rural landscapes, current success and possibilities for the future appear relatively bright. Some farms will be abandoned or become part of nearby larger ones, and there undoubtedly will be further reduction in harvested land in the years ahead. Still it can be safely forcast that the area of farm landscape will account for an increasing percentage of New York's agricultural production in the future.

Mixed Landscape of Farms and Abandoned Land

The mixed landscape of farms and abandoned land occupies most of the Appalachian Upland, a narrow zone almost completely surrounding the Adirondacks, and most of the easternmost part of New York. In all it covers roughly half the state. It differs primarily from the farm landscape in containing a smaller proportion of actively farmed land and a larger proportion of abandoned land. Overall, the resource base for modern farming has proved marginal at best. The changing American economy and the subsequent demands on a farm to produce more have resulted in a mosaic of agricultural successes and failures. Successes have been most numerous in alluvial valley bottoms or on relatively level and well-drained uplands of moderate elevations. Failures have left their stamp particularly in the steeper or higher upland districts where soils are rocky, thin, acid, or poorly drained. These same areas have suffered some from relative inaccessibility too. While resource base and location deficiencies can be blamed for many farm failures, lack of adequate management skills and the existence of farms too small in size also have contributed to failure.

In 1900 nine million acres of cropland were harvested on 227,000 farms in New York State. In 1960 less than five million acres were harvested on about 82,000 farms. Abandonment of farms, although not confined to the mixed landscape of farms and abandoned land, has been prevalent there and has given that landscape its distinctive present-day personality.

About 55 per cent of the total area is in farms, while no more than 18 per cent is harvested (Table 40). This leaves about 37 per cent of the land in the pasture–wood lot–idle land category. Pasture is frequently of poor quality; much of it is in the process of reverting to brushy woodland. From what has been said it would be expected that harvested acreages have shown a strong decline over the years. This is the case. Since 1925 this landscape has lost approximately 41 per cent of its harvested acreage, which is slightly above the state average and well above that of the farm landscape.

The 45 per cent of the land not now in farms includes considerable acreages of state and county land. Much of this is formerly abandoned land that by tax delinquency or purchase has come into public ownership. Some has been permitted to grow back to forest; some has been actually replanted to coniferous trees. Such rough wooded country has recreational value to the hunter, the camper, the naturalist, as well as to those people who just want to get away from crowds for a day or week end. Very limited amounts of forest products are produced.

Relic Features of Abandonment and Their Identification

Relic features of a wider use of the land for farming are so many that an experienced eye can readily differentiate land that has been farmed from land that has not. Stone walls or rusted wire fences, once the margins of used fields, now may be seen extending only into dense woods. Hedgerow trees are characteristically a decade or two older than field trees. The age of the trees in the field may give some indication of how long ago abandonment occurred. If abandonment took place less than perhaps thirty years ago and if soils are limey, hawthorne thickets interspersed with wild apple trees and a few elms suggest the existence of former pastures, for it is in abandoned pastures that this vegetation complex commonly develops. Old foundations and cellar holes hidden by weeds and brush, the telltale lilac once carefully nurtured by a farmer's wife, or the patch of tame raspberries struggling in an isolated and untilled place, all identify the sites of former farmsteads.

MIXED LANDSCAPE OF FARMS AND ABANDONED LANDS in Chemung County. Essentially all of this land was once actively farmed. Upland areas and poorly drained areas have been abandoned and various stages in the abandonment process are visible. Eventually most of the farming will become unprofitable and the open areas will grow back to trees. *N.Y.S. Dept. of Commerce.*

One of the most interesting ways to locate abandoned farms is to compare an old, large-scale topographic map with a recent one of the same area. Old maps are frequently available in libraries or local historical association files, and new ones can be purchased at stationery or book stores. If these maps have rural structures on them for the dates on which they are drawn, overlaying the two maps will reveal the extent of abandonment and permit the user to go to abandoned sites with ease. Air photos, too, are useful tools. They have been taken intermittently for some parts of the state since the early 1930's. Even though only a few decades of time lapse are involved, study of old and new photos may provide startling evidence of the march of land abandonment.

ABANDONMENT, DEPOPULATION, AND REOCCUPANCE

Land abandonment and population shifts in rural areas have been alluded to in the Introduction and Chapter 10, as well as elsewhere in the text, yet here under the subject of the mixed landscape of farms and abandoned land once more these critical and fascinating processes deserve attention. Limitations in physical environment, small farm size, and questionable management skills, and the necessity of farms to produce better returns as the American economy advances have resulted in a loss of 145,000 farms in New York State in the last sixty years. If it is assumed that each of these farms supported a family of six, then over 850,000 people have been directly affected. Some departing farmers found better farms elsewhere, but most became a part of the great migration from rural to urban areas where jobs were more numerous and incomes were higher and allegedly more steady. Extensive depopulation of some rural areas has been fairly pronounced. This is apparent if the density of dots in Figure 54 (population map for 1880) and Figure 2 (population map for 1960) is compared for rural areas in the Appalachian Upland.

Once a farm is abandoned the house and other structures are not always doomed to disappearance. If the structures are of good line and in reasonable condition, city people frequently buy and modernize them for summer use, or perhaps for year-round living if the location is within reasonable commuting distance to urban areas. The advent of the automobile, the improvement of roads and snow removal facilities, and most important, the desire for a home in the country, so widespread among middle- and upper-income urbanites, have brought about this kind of use of abandoned farm structures. Sometimes, too, new homes are built on attractive sites on abandoned land. The new occupant, who earns his income from a city position, looks on the steep, rocky slopes around him not as deterrents to his future, as the farmer before him must have done, but as aesthetically attractive phenomena contributing to his attempt to relax in a countryside far from the noise and pressure of city environment. As was pointed out in the Introduction, people living in the

CHAPTER 15 THE RURAL LANDSCAPE 363

countryside but not gaining a livelihood from farming are referred to by the United States Census as rural nonfarmers. Two categories may be designated: (1) the *productive,* or those earning a living somewhere, and (2) the *unproductive,* those not earning their own way. Seldom will the spread of productive rural nonfarmers into an abandoned farm area result in a population as large as that on the earlier farms unless, of course, the urban frontier spreads beyond the area.

The case of the unproductive rural nonfarmer may be quite different. Indeed, rural populations considerably larger than original farm populations can occur. Unlike the productive rural nonfarmers, who enjoy adequate income from urban enterprise, the unproductive rural nonfarmers receive little or nothing from their own current toils. These people usually fall into one of three categories: (1) persistent recipients of welfare payments, (2) recipients of unemployment compensation for substantial parts of each year, and (3) retired people living on old-age pensions. After a farm is abandoned, and if the property is not attractive for one reason or another to a productive rural nonfarmer, both the buildings and land are likely to come on the market at very low prices. Sometimes properties can be bought for payment of back taxes alone or rented for very low rates. This means that people with little means may occupy an old farmhouse and, without remodeling or in any way making improvements, live in it at the lowest possible cost. In other cases an acre or two of abandoned land may be secured for a small sum of money which permits construction of a shack-type dwelling. House trailers are even occasionally used without title to the land on which they are parked, although unproductive rural nonfarm people cannot usually afford trailers.

If conditions are just right, an original farming population may be almost completely replaced by unproductive rural nonfarmers, most of whom are likely to be indigenous to the immediate vicinity. Unpainted, deteriorating, and slovenly over-all appearance of property is commonly, although not always, an indication of unproductive rural nonfarm occupance. Also, dilapidated barns and sheds, junk cars in yards, roofs and sidewalls patched with tar paper, storm windows made of plastic, or poorly constructed and maintained shacks are common indicators. This type of occupance may occur fairly near large urban centers, but rising costs due to suburbanization spread are likely to drive it beyond the urban frontier, to land not wanted by suburbanites or people of the productive rural nonfarm type.

Studies have shown that once they have become members of the unproductive rural nonfarming group, people tend to make this kind of life more or less permanent. Very often parents of welfare recipients were themselves on welfare rolls. Many were farmers forced out of business because of submarginal properties, or were descendants of former farmers. A few came from the city seeking cheaper living in the country. Families tend to be large, resulting in an ever increasing population probably destined to be satisfied with this kind of life. Unquestionably, when numerous, or accounting for a large portion of a population, they do induce local economic problems. Productive enterprises engaged in by others must provide some of their support and funds from distant parts of the state or nation will be needed by them. They require school and road maintenance but contribute little, if anything, to support them; it is possible that they even bring about changes in, or even control, local politics.

No implication is intended here that the unproductive rural nonfarmers live only on abandoned lands. As a matter of fact they are found in substantial numbers in good farming areas and of course their city counterparts are common in urban environments. But, because they are so prevalent, and indeed sometimes dominant, on the abandoned lands of rural New York, discussing them as a phenomenon of the mixed landscape of farms and abandoned land lends understanding to the processes at work and the changes taking place in that landscape.

The sequence of land use that has produced the mixed landscape of farms and abandoned land varies a bit from place to place as to kind and speed of change, but has the same general characteristics almost everywhere in the state. Figures 110 through 114 portray the general characteristics of land use change from pioneer settlement to occupance by a productive rural nonfarmer. They tell much about the economic geography and history of rural New York.

In Figure 110 the early pioneer of western New York and his family have barely succeeded in pushing the forest away from a hurriedly built log cabin and a small shelter for livestock before winter sets in. Even the chimney seems missing from the log cabin. The date might be 1790–1805.

Perhaps six months later (Fig. 111) the pioneer has cleared an additional few acres and probably planted crops among the stumps. A chimney has been added to the house, crude wooden fences have been built, and there is a log bridge over the small stream. Neighbors are aiding in removing recently felled logs from the field in the background.

Substantial acreage has been cleared a decade later (Fig. 112), but stumps in the background indicate that the clearing process is continuing. The house, although still simple, has been expanded and improved; a barn has been built; fields are enclosed, in this case with zigzag and straight rail fences, but

Fig. 110. In the Beginning

Fig. 111. The First Summer

CHAPTER 15 THE RURAL LANDSCAPE 365

Fig. 112. A Decade Later

in localities where boulders are common the stone wall would be present; an orchard has been planted and it may be surmised that fields (behind the barn), contain corn, small grains, and hay. A better bridge serves a road which passes in front of the house; a one-room schoolhouse appears at upper left.

Figure 113 shows the situation perhaps forty to one hundred years after pioneer settlement. Almost all of the land as far as the eye can see has been cleared, and farms are numerous. Only the distant ridge top and a wood lot or two remain in trees. Prosperity has smiled on the farm unit we have been following; the house has been rebuilt to almost pretentious proportions, there are several barns, and a railroad serves a small hamlet at upper right. The farm economy is perhaps approaching its zenith; certainly it has tamed the land.

Figure 114 is the post-World War II scene. The American economy has changed, and sometime over the past years the same farm unit became insufficiently productive to survive that change. The farmer found greater remuneration in a city job. Abandonment of the farm probably took place between 1920 and 1943. Fields quickly grew up to brush and the farm buildings deteriorated.

The house, however, was architecturally and structurally sound and in the late 1940's or early 1950's attracted a buyer who earns his livelihood in a nearby city. The drawing shows the house to have been completely renovated not with money made on the property but with funds coming from urban sources; the barn is kept in shape to house riding horses wanted by the new occupant's children; except for limited pasture land and perhaps a hayfield the rural nonfarmer makes no pretense at keeping the fields cleared. In fact he much prefers that they return to the wild. He is likely to help this process along by planting evergreen seedlings. The yard is manicured and outfitted like that of a suburban dwelling and includes a patio, foundation planting, the typical blue spruce and white birch clump, and a tarvia drive leading to a two-car garage. Tax payments are likely to at least equal those that would have accrued if the property had remained an operating farm.

Had the buildings been of poorer quality and unattractive to the productive nonfarmer, it is quite likely that by 1960 they either would have fallen into ruin or have been occupied by an unproductive rural nonfarmer. A third alternative would have been for the farm to have survived as an operating unit. In that case a dairy barn and cleared and improved fields, not a meticulously groomed residence, would have dominated the foreground.

An important part of future planning for the state will have to be related to use of abandoned lands.

Fig. 113. The Land Is Tamed

Fig. 114. The Productive Rural Nonfarmer Takes Over (*drawn by Sylvia S. Wyckoff*)

CHAPTER 15 THE RURAL LANDSCAPE

Their potential, as has been pointed out many times, has changed over the years, and it can be expected to change again as the economy of the state changes. Whether fifty years from now they are used for forests, farming, rural nonfarm occupance, or just for occasional housing will depend on the state of the economy and on man's awareness that all lands should be used optimally at all times. Except for favorable farming locations such as valley bottoms, where farms may be as prosperous as any in the state, the general status of the mixed landscape of farms and abandoned land is one of underuse. It is impressive that so much land is obviously contributing so little to the state's economy. It is apparent, too, that the landscape as a whole is on the economic downgrade; that only where secondary and tertiary activities are being carried on in and around urban places is there substantial economic growth. It seems safe to forecast that future economic success of the mixed landscape of farms and abandoned lands will depend on the extent to which it can be rationally used by productive nonfarm populations.

Empty Landscape

That parts of populous New York are essentially empty of human beings is not new to the reader of this volume. Figure 2 (1960 population) clearly shows this, and Figures 30, 34, and 67, as well as many of the other maps, indicate strong limitations in the resource base of those sections designated as empty landscape. The empty landscape occurs in four major areas: (1) Adirondacks, (2) Catskills, (3) Tug Hill, and (4) Allegheny and Cattaraugus hills. Most of the land in these areas was never farmed, and that which was has been largely abandoned. The few remaining farms are mostly part-time enterprises, the owners receiving supplementary income from seasonal jobs, part-time work, or welfare checks.

Less than 10 per cent of the land is in farms, and only a fraction of 1 per cent is harvested. Most of the latter is low-grade hay land. Although most harvesting ceased before 1925, land in crops has declined by over 75 per cent since that time. These areas actually were not even considered good by the Indians, who stayed pretty much out of them, and white man was late to try farming there. A few farming attempts for short periods of time gave ample evidence of the submarginal nature of the resource base. The forests were largely cut off in the nineteenth century. In 1885 farsighted men conceived of, and set aside, the New York State Forest Preserve in the Adirondacks and Catskills as forever wild areas. In both places considerable private property existed at that time, so these are not solid blocks of state-owned land today but rather a mosaic of public and private property.

The empty landscape seems to have three potentials. It represents a source of wood products and minerals and has an environment attractive to certain recreational activities. Although essentially all of it is forested, only limited revenues accrue today from forest industries. The forever wild areas, by law, are not open to lumbering and pulp-cutting, and many of the rest of the forests are not productive because they are currently submarginal for lumbering or because sound management policies are not being applied. There are those, however, who feel that New York's forests in general, and particularly those occupying what is here called the empty landscape, represent a large and valuable future resource. Better management and better market conditions will be required if this resource is to become significant. As has been pointed out in Chapter 10, minerals are produced in some quantity, especially in the northern Adirondacks. Undoubtedly such deposits will be important in shaping the economy of certain places. Never, though, will large numbers of people gain their livelihood from mineral industries.

The empty landscape's future probably will be most significantly tied up with recreation. Public lands are used for recreation, but it is on the private lands that most such development has taken place. Here are the resort towns, the lake shore settlements, and the isolated private camps. There are factions of people in the state who would open up the forest preserves by building more roads and other facilities so that more people could use them. There are also those who would adhere more strictly to the forever wild concept and limit access and facility construction. (This difference in opinion is elaborated upon in Chapter 12.)

In any case, the untold millions of additional people who will become part of New York State's population in decades ahead will need more recreational lands and facilities. The empty landscape can satisfy this need if properly managed, and in so doing can make an important contribution to society even though increase in permanent population and permanent job opportunities within the landscape may be slow in coming. Because of its very emptiness it is attractive to growing numbers of individuals who enjoy relaxing away from the crowded cities. Large urban populations are close enough to the empty landscapes to reach them in a reasonably short time, and by virtue of this closeness they take on great importance not attributable to more remote portions of the United States. The forests, lakes, and mountains have a widespread aesthetic appeal. The hiker on a wilderness trail is perhaps most keenly aware of this appeal, but the

EMPTY LANDSCAPE of the Adirondacks. This view, showing Lake Placid, is representative of hundreds of square miles of New York's empty landscape. Few attempts at farming were ever made here and those that were, were doomed to early failure by adverse environmental and market conditions. Only recreation flourishes. *N.Y.S. Dept. of Conservation.*

hunter and fisherman, the camper, the cottage occupant, the two-week vacationist at Lake Placid, and even the Sunday driver viewing a spectacular display of fall color are all aware of it, too. New York's empty landscape is a priceless heritage equaled today in few of our other eastern states. That it is not filled with farms or cities, that it is not contributing to the state's economy like the Ontario Lake Plain or Manhattan, that it does not offer many year-round jobs may in the long run be an asset, not a liability.

Selected References

The Appalachian Region. Annapolis: Maryland Department of Economic Development, May, 1960.

Barlowe, Raleigh. *Land Resource Economics.* Englewood Cliffs, N.J.: Prentice-Hall, 1958.

Chenango Area Development Study. A Report for the Joint Legislative Committee on Natural Resources. Syracuse: State University College of Forestry at Syracuse University, 1962.

Conklin, Howard E., and Irving R. Starbird. *Low Incomes in Rural New York.* Ithaca: Cornell University, 1958.

Fippin, Elmer O. *Rural New York.* New York: Macmillan, 1921.

Hedrick, Ulysses. *A History of Agriculture in New York State.* Albany: New York State Agricultural Society, 1933.

Higbee, Edward. *Farms and Farmers in an Urban Age.* New York: Twentieth Century Fund, 1963.

New York State Department of Conservation. *The New York State Conservationist.* Albany: The Department, 1946 to date.

The Resources of the Allegheny Plateau. A Report for the Joint Legislative Committee on Natural Resources. Syracuse: State University College of Forestry at Syracuse University, 1962.

Tobey, J. S. *Changes in New York State Agriculture.* ("Cornell Agricultural Extension Bulletin," 101.) Ithaca: New York State College of Agriculture, 1960.

Zelinsky, Wilbur. "Rural Population Dynamics as an Index to Social & Economic Development: A Geographic Overview," *The Sociological Quarterly,* IV, No. 2 (Spring, 1963), 99–121.

CHAPTER 16   JOHN H. THOMPSON

Planning and Development Regions

As man becomes more sophisticated, he begins to wonder whether he is using his environment in the best way. He concerns himself about trends and becomes curious about the long-range effects of these trends on his surroundings. He begins to develop the idea that planning for future development is not only worth-while but essential. He sees the development program as the medium to highest and best use, and he may indeed define highest and best use in terms of suitability in the over-all economic and aesthetic scheme of things, rather than just as production for the highest monetary return. He will understand that a plan well suited to one location is not applicable to a second location which differs strongly in basic characteristics and problems. He will realize, therefore, that in an area as large and diverse as New York State significant spatial variations will justify breaking up the state into planning and development regions with similar characteristics and problems. Once such regions are defined, planning and development councils may be established for each region and the process of attaining rational use and development is under way.

Identification of planning and development regions might well be expected as a by-product of geographic regional analysis. Involved are the application of the regional concept and the synthesis of regional information to determine its over-all meaning in a development context. The planning and development region is thus a product of regional addition, in essence a composite region. If well designed, it probably ought to be based on the concepts of uniformity and nodality and incorporate political reality and statistical practicality.

Concept of the Composite Region

Calling a piece of the earth's surface which approaches homogeneity in terms of specified criteria or which possesses functional cohesion a region, and not an area, is traditional in geography. Regionalizing for the purpose of examining areal differentiation in New York State, of finding similarities between parts of the state, and of revealing patterns of interconnection has been widely employed in this volume. It has been demonstrated that phenomena are organized in such a way as to produce a mosaic of regions, and land form, climate, soil, vegetation, agriculture, manufacturing, and trade regions, to mention a few, have been identified, mapped, and described. That this regionalizing process has elucidated New York's geography there can be little doubt. Does it lead, however, to significant over-all interpretations of the state? Or, to aim the question pointedly at the subject at hand, does it aid in defining rational planning and development regions? It is hypothesized at this juncture that the answer to this question is yes, but before elaborating it will be useful to review briefly the regionalizing range that should be used in definition of something as complex as planning and development regions.

THE REGIONALIZING RANGE

There are both uniform (homogeneous or formal) regions and nodal (functional) regions. Either type may be defined in terms of a simple, single phenomenon, or in terms of less specific but much more complex multiple phenomena. This range from simple to complex might be made clearer by identifying regions of three degrees of complexity.

Least complex are those defined in terms of a single phenomenon such as a slope or temperature region, both of which are uniform types, or a trade region defined simply in terms of department store deliveries, which is a nodal type.

A category of regions of intermediate complexity includes those defined in terms of multiple phenomena. Variation will occur from: (1) those based on closely related features operating under similar sets of processes such as climate, soil, and forest regions, to (2) those based on less closely related phenomena and different processes such as agricultural regions, trade areas, and economic health regions, to (3) those which are based on all the relevant elements of the physical environment or on all the relevant elements of the cultural environment and the attendant loosely related processes. Natural regions or cultural regions result from this kind of regional analysis. Although implications have been made from time to time concerning physical or cultural regionality, no identification of such natural or cultural regions has been attempted in this volume.

The most complex form of regionalizing must wrestle with the problem of defining regions in terms of the totality of human occupance, involving interrelated natural and cultural phenomena as well as interrelated physical and cultural processes. To identify such complex regions is not easy, and considerable subjectivity must be employed. It is a worth-while goal, however, for the past, present, and future status of New York State reflects, and will continue to reflect, the results of man's creative choices of endeavor in the total spatial setting of resources and culture. In geographic literature these most complex of regions have been variously referred to as total regions, compages, natural-societal regions, geographic regions, or composite regions. The last designation has been selected for use here. *Composite regions*, if of uniform type, are likely to exhibit similarities in resource base, history of use, current characteristics and problems, and development potential; if of nodal type, they will exhibit functional organization. Both composite uniform and composite nodal regions, because of their nature, should be useful, if not vital, to the definition of planning and development regions.

Tools and Procedures for Defining Composite Regions

The principal tools available for defining composite regions in New York are the many maps that have appeared in earlier chapters and the statistical data on which they are based. It should be reiterated that these maps and statistics have been evolved from the value judgments of numerous informants and hundreds of on-the-spot field observations, as well as published data. All of the maps in previous chapters were examined, overlaid, and their patterns compared and analyzed. No one aspect of either the physical or cultural milieu was so significant as to dominate the composite regionalizing process, but the following fifteen maps were perhaps the most useful: Figures 2 (Population, 1960); 3 (Per Cent Change of Population, 1950–60); 9 (Land Form Regions); 24 (Climatic Regions); 30 (Extent of Present Forest); 34 (Agricultural Potential); 43 (Spread of Settlement); 67 (Agricultural Regions); 73 (Manufacturing in New York State); 82 (Tertiary Potential; Surpluses, and Deficits for Urban Centers); 85 (Railroads, c. 1960); 86 (Comparative Highway Traffic Flow, c. 1960); 92 (Trade Centers and Trade Areas); 99 (General Economic Health for the 1950's); 109 (Rural Landscapes, c. 1960).

If, as asserted above, tne planning and development region ought to be four-dimensional in that it is based on the concepts of (1) nodality and (2) uniformity, but incorporates (3) political reality and (4) statistical practicality, then the following steps must be taken: First, composite uniform and nodal maps must be prepared; second, these two maps must be overlaid, analyzed, and combined; and third, a fitting of this combination to county or other political boundaries must be made so as to facilitate administrative and data-gathering procedures.

Composite Uniform Regions

Eight quite distinct composite uniform regions are shown in Figure 115. As is the case of most regions, each has a core area where regional characteristics are best exemplified. The core area of one region differs noticeably from core areas of adjoining regions. Regional boundaries usually are not the sharp lines shown on the map but rather transitional zones that partake of the character of neighboring cores. Each of the regions is described below in terms of general terrain and subsurface conditions, climate, vegetation, soils, settlement and population density and trends, economic health, and economic potential.

Adirondack Wilderness

This is perhaps the most readily discernible composite uniform region in the state; in fact, it is one of the finest such regions to be identified at this scale in the world. Regional characteristics have been accentuated by state law, which established much of it as a wilderness area.

1. An attractive isolated upland composed of hills, mountains, and numerous lakes; igneous and metamorphic rocks predominate; slopes generally excessive.
2. Severe climate with very cold, snowy winters and very cool, wet summers; energy for plant growth lowest in the state.
3. Original forest of the spruce-fir-northern hardwoods type; land almost completely covered with second- and third-growth forest today.
4. Shallow, poorly drained, acid soils on glacial till and steep terrain; farming next to impossible except on locally favorable sites.
5. Most of the region never permanently occupied; limited settlement after 1800; population densities very low and mostly associated with recreation activity; population stagnation or decline in recent decades.
6. Economic health very poor by state standards.
7. Agriculture not feasible; mining occurs and may expand, but will not employ many people; manufacturing expansion seems unlikely; greatest fortune would seem to be in recreation, for which the region has many attributes; will remain indefinitely but sparsely settled even in a generally rapidly rising state population trend; has an empty rural landscape designation.

Appalachian Country

Except for a few Southern Tier manufacturing cities, of which Binghamton is the largest, the Ap-

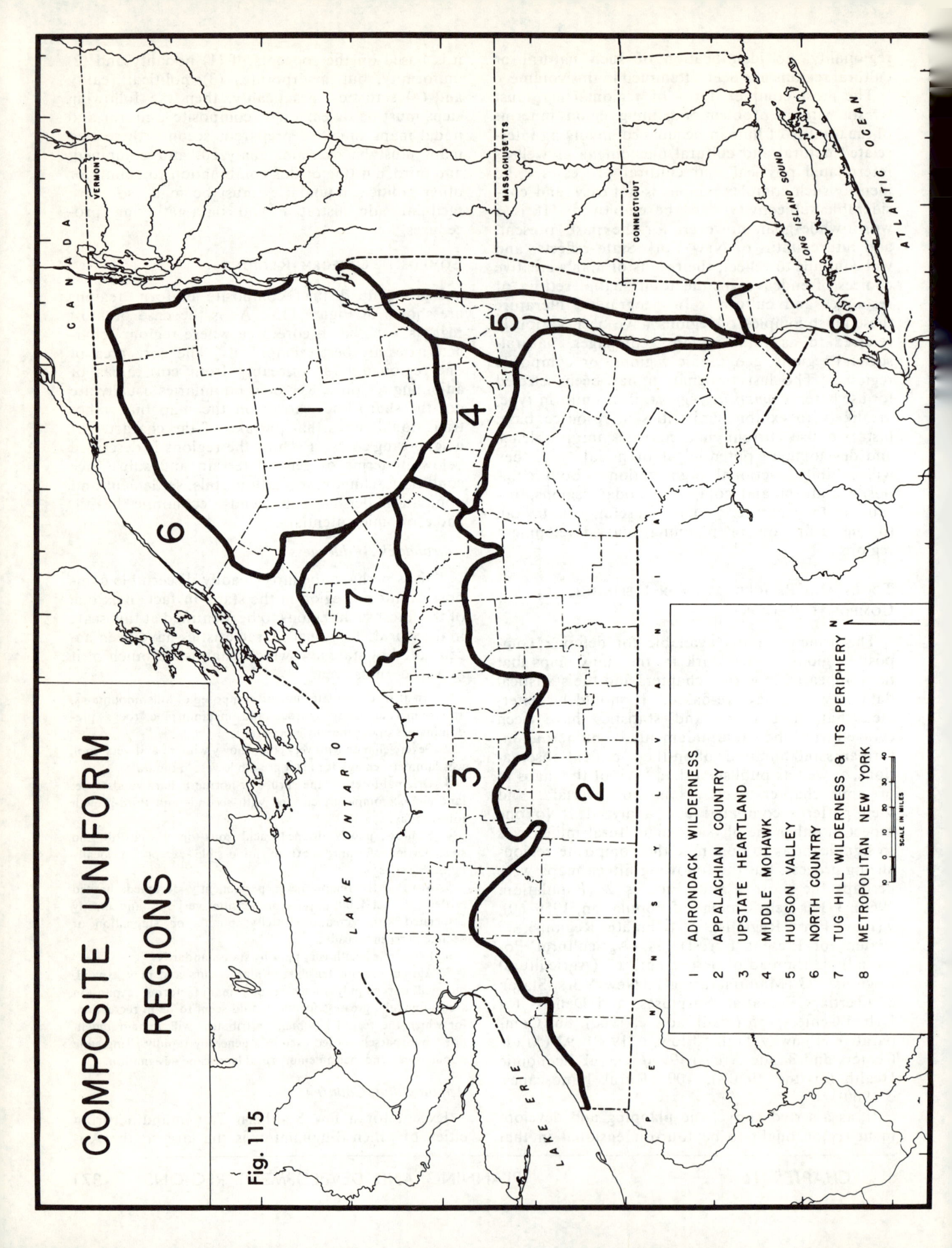

Fig. 115 COMPOSITE UNIFORM REGIONS

1 ADIRONDACK WILDERNESS
2 APPALACHIAN COUNTRY
3 UPSTATE HEARTLAND
4 MIDDLE MOHAWK
5 HUDSON VALLEY
6 NORTH COUNTRY
7 TUG HILL WILDERNESS & ITS PERIPHERY
8 METROPOLITAN NEW YORK

palachian Country stands out as an agricultural area fraught with numerous difficulties. What might be done with this, the state's largest composite uniform region, in the way of making it economically viable has aroused much speculation. Answers are not easily found. Most of the Appalachian Upland land form region and the Wallkill Valley are included.

1. Largely a dissected plateau composed of deeply incised flat-bottomed valleys and steep-sloped uplands, the latter greatly dominating in total area; sedimentary rocks throughout; few lakes; Catskills in the east scenically attractive and well located with respect to New York City; southern Cattaraugus County only part of state not glaciated.

2. Higher portions with cold, snowy winters and cool, wet summers; valleys and lower central section warmer and drier in summers; energy supply in valleys suitable for considerable agricultural versatility.

3. Original forests of oak and northern hardwoods; much land abandonment has resulted in widespread forest or scrub brush cover but generally not in large, continuous tracts.

4. Best soils, alluvial in nature, occur in valley bottoms; most of the uplands are covered with acid soils formed from glacial till; where the till contains limy material, soils are better, as is generally the case in the northern margins of the region.

5. Settled after the Revolutionary War, eastern portions first, then best valleys to the west, and finally the poor uplands in the west; farming population reached peak near end of nineteenth century and has been declining since; except around a handful of the industrial cities there has been little tendency for population growth; many sections have experienced losses in population.

6. Economic health moderately poor to poor except around larger cities.

7. Agriculture on the way out, although favorable valleys exhibit greatest staying power; mining and forestry do not amount to much; manufacturing potential low; interest in expanding recreation exists but possibilities limited; population growth not expected generally; rural areas to decline; the rural landscape designated as a mixed one of farms and abandoned land.

The Upstate Heartland

This region is the best agricultural area in New York, and, with its several large urban systems and many smaller cities lying astride major transport routes, it deserves the designation "economic heartland of upstate."

1. Rather level lowland in north; strongly rolling and hilly in south as extension into Finger Lakes area occurs; sedimentary rocks covered with heavy mantle of limy till; drumlins and lake sediments common.

2. Cold, snowy winters and warm, dry summers; climate reasonably suitable for wide range of crops from fruit and vegetables to grain and hay; could profit in places from irrigation.

3. Original forest of elm-red maple-northern hardwoods; currently least wooded part of state, with forests which do exist being in small tracts or on poorly drained land.

4. Soils rich in lime and situated on suitable terrain contribute strongly to making this the state's best farming area.

5. Mostly settled between 1790 and 1810, although limited earlier penetration in the Finger Lake Hills; current population density high, with the four major urban systems—Buffalo, Rochester, Syracuse, and Utica—dominating; population growth very rapid in, and adjacent to, these four urban systems.

6. Economic health generally good, again especially in the periphery of the large urban systems; some economic stagnation in smaller central places and rural areas.

7. Less farm land abandonment than elsewhere in state and generally more prosperous farming conditions; agriculture competes fairly well in northeastern markets; rural area designated as farm landscape; manufacturing production rising, but employment remaining about same; tertiary sector in the large urban systems is growing and will continue to grow rapidly, providing most of the new jobs needed in the region; population to double in next thirty-five or forty years.

Middle Mohawk

This region has been described as New York's most distressed area, and it is general economic difficulties which most noticeably set it off from lowland neighboring regions to the west and east.

1. The narrow lowland corridor between the Adirondacks and Appalachian Upland; largely underlain by sedimentary rocks; locally steep slopes, but generally only moderately rolling or valley bottom land.

2. Climate mild by comparison to adjacent uplands but not unlike the cold, snowy winter and warm summer type of most of lowland upstate.

3. Original forests of northern hardwoods; land today largely in farms or urban and transport uses.

4. Soils vary greatly from good to bad, depending on lime content, slope, and drainage conditions.

5. First settled prior to Revolutionary War; population density has been high for a long time, with most of the people living in middle-sized and small cities; recent population growth light or absent; early economic growth related to transportation advantages.

6. Economic health has become progressively poorer as transportation advantages faded, old factories closed, and new factories failed to be easily attracted.

7. Lacking large cities; substantial economic development and population growth not likely; principal asset remains the high degree of accessibility.

Hudson Valley

Focusing on the Capital District (Albany-Schenectady-Troy) and extending southward to the Hudson Hills, this is a relatively vigorous, rapidly growing region quite different from its surroundings on the north, east, and west. To the south, immediately adjacent to the Hudson River, it comes in contact with the metropolitan New York region. Favorable location, transport advantages, the political function of Albany, and manufacturing, to mention a few of the region's attributes, should keep it moving ahead.

1. Lowland lying generally between the Appalachian and New England uplands; narrow in the south and widening to about twenty miles at Albany; underlain with sediments; a substantial proportion in steep slopes, but level areas are widespread

enough to support considerable agriculture; south from Albany forms one of the most beautiful areas in America.

2. Climatically transitional between the more severe conditions of upstate and the milder New York City area; energy and moisture supply advantageous in south; droughty in the north.

3. Original forests of oak–northern hardwood type; over half of the land now in trees.

4. Good glacial lake soils in middle portion; sandy in north; acid in south.

5. Settled before the Revolutionary War and contains some of the finest examples of geographical persistence from this early period; population density is high and generally rising rapidly, although increases at specific localities have been meager.

6. Economic health is moderately good to good.

7. Expansion in both the secondary and tertiary sectors will result in continued growth; eventually an almost continuous urban landscape will develop along the Hudson River with certain urban systems such as Beacon-Newburgh, Kingston, and Poughkeepsie experiencing very rapid growth.

North Country

Lying in the far north of the state and a bit off the principal economic axis of the United States, this region has shown disappointing economic performance. It was hoped by some that the St. Lawrence Seaway and Power projects would stimulate the region enough to make it into one of the more prosperous sections of the state. This, however, turned out to be no more than wishful thinking. Substantial population growth and economic progress do not come easily here, although some variation from place to place in potential does exist.

1. Generally a relatively level lowland, being tributary to Lake Ontario, the St. Lawrence River, and Lake Champlain; limestones and sandstones covered with till predominate.

2. Very cold winters and cool, sunny summers; receives less energy than any other lowland in the state but supports dairy crops.

3. Original forest varied considerably from one part of the region to another as does the amount of land now in forest; better dairy areas in Jefferson and St. Lawrence counties have as little of their area in trees as does the Upstate Heartland, but elsewhere the region is more like the Appalachian Country in having about half its area in forests.

4. Like the vegetation, soils vary, being good in places and poor in others; thus are the variations in farming success largely explained.

5. Settled in 1790–1810 period, but population densities have remained low; little growth in recent decades except around Plattsburgh and that seems explainable in terms of military installations there.

6. Economic health intermediate at best; much of area falls below that designation.

7. Agriculture is widely carried on; although dairying predominates, there is some fruit production near Lake Champlain; farm abandonment is common and will continue except in the most suitable sections; manufacturing, centered in Massena and Watertown, will show little tendency to expand, and it is doubtful that tertiary activities will increase at the state average; region may be expected to drop in the per cent of the state's population it contains.

Tug Hill Wilderness and Its Periphery

Years ago the Adirondacks used to be referred to as the Great Wilderness and the Tug Hill as the Lesser Wilderness. Today the term "wilderness" is perfectly suitable for the central portion of Tug Hill. Although its periphery is by no means uninhabited, it exhibits little potential for economic growth.

1. Outlier of the Appalachian Upland, the Tug Hill Upland rises 1800 to 2000 feet above surrounding lowlands; the lower periphery, composed of rolling plains and swamplands, slopes away in all directions; sandstones and shales predominate; slopes not generally excessive.

2. Severe climate with cold, very snowy winters and cool, wet summers; low energy receipt a disadvantage.

3. Original forest of spruce-fir-northern hardwoods type; although all originally logged off, most of the area now in second- and third-growth forest.

4. Poorly drained acid soils generally unsuited for agriculture.

5. No permanent settlement in central Tug Hill; periphery settled late and extensively abandoned; periphery now generally stagnant or declining.

6. Economic health very poor.

7. Agriculture essentially gone; manufacturing limited largely to wood-using establishments, declining; lacks large central places, thus tertiary sector shows little tendency to grow; like the Adirondacks but with fewer physical attractions, recreation seems to offer brightest future; use for camps, hunting lodges, skiing already notable.

Metropolitan New York

Except for easternmost Long Island this region is completely urban. This is not to say there is no open country, but most of the region is within the urban frontier of metropolitan New York City. In the years immediately ahead the open places will surely disappear as this region's population rises.

1. Focused on perhaps the finest natural harbor in the world and connected to the interior by the Hudson-Mohawk corridor, this region was destined from the first to support a major urban system and become a focal point of not only New York State but the United States; ancient crystalline rocks in Manhattan, sandstones in Rockland County, and glacial moraines on Long Island provide a varied environment; Rockland and Westchester counties sufficiently rough to provide the varied terrain and pleasant views prized by suburban communities; Long Island, with irregular coastline and relatively smooth surface supports all kinds of urban activity ranging from manufacturing to exclusive suburbs to recreation.

2. Mild, wet winters, warm, humid summers, and the greatest energy receipt in the state favor intensive agriculture; even with the advance of urbanization, agricultural success continues to be considerable, especially on Long Island.

3. Original oak forests now heavily depleted by urbanization.

4. Some alluvial soils on Long Island of reasonably good quality.

5. First area settled in the state; population density now

highest of all the regions and growing rapidly in localities peripheral to New York City.

6. Economic health in peripheral locations best in state; some problems in older city districts.

7. Manufacturing and tertiary center of state and nation; will continue to contain about two-thirds of the state's population for the foreseeable future; serious economic and social problems evolving in older districts of New York City.

Composite Nodal Regions

In Chapter 12 trade areas of various orders were identified. These actually are composite nodal regions in that they were delineated on the basis of over-all functional organization. Although difficult to prove, it may be satisfactory to assume here that second- order trade centers and second-order trade functions are the most basic orders as far as the spatial organization of an area the size of New York State is concerned. Certainly the second-order trade centers are the most economically successful and the leading growth-generating elements in the economic geography. Therefore, these second-order trade centers and their trade areas (Fig. 92) are used as significant composite nodal regions. Five in number, they are named according to the major urban system on which they focus.

Buffalo

The Buffalo nodal region, comprising 5,676 square miles, contains 11.5 per cent of the state's area. Including both Lake Ontario and Lake Erie frontage, it occupies the western extreme of the state. Roughly 32 per cent is in the Upstate Heartland composite uniform region; the rest is in the Appalachian Country, including the very heavy snowfall area of the Western Appalachian Upland. The Buffalo urbanized area, second largest in the state, contains over one million people and has a tertiary surplus of 240,000, some of which no doubt is oriented toward Canada.

Rochester

The Rochester nodal region contains 6,118 square miles and accounts for 12.3 per cent of the state. It extends from Lake Ontario to the Pennsylvania line, including the western portion of the Finger Lakes area. Forty-eight per cent is in the Upstate Heartland and 52 per cent in the Appalachian Country. The Rochester urbanized area, third largest in the state, has a population of just under 500,000 and a tertiary surplus of about 140,000.

Syracuse

Containing 16,673 square miles or 33.6 per cent of the state's area, the Syracuse nodal region ranks first in size. Furthermore, in addition to Syracuse it contains two trade centers, Utica and Binghamton, which, as pointed out in Chapter 12, fall just below second-order designation. It extends 250 miles from the Canadian border to Pennsylvania and includes parts of five composite uniform regions; 19.6 per cent is in the North Country, 25.5 per cent in the Adirondacks, 10.8 per cent in the Tug Hill and its periphery, 19.0 per cent in the Upstate Heartland, and 25.2 per cent in the Appalachian Country. It is the most geographically complex of the state's nodal regions. The Syracuse urbanized area, with a population of about 330,000, has a tertiary surplus of 160,000, one-seventh larger than that of Rochester.

Albany Area

This is nearly as large and complex as the Syracuse nodal region and occupies 13,966 square miles in the east-central portion of the state. It contains 28.2 per cent of the state's total area. Again parts of five composite uniform regions are involved. In the northeasternmost part of the state, near Lake Champlain, 3.6 per cent of its area is found in the North Country. Coming southward 45.3 per cent is in the Adirondack Wilderness, 11.0 per cent in the Middle Mohawk, 10.6 per cent in the Hudson Valley, and 29.5 per cent in the Appalachian Country. The Albany urbanized area has a population of 455,000 and a tertiary surplus of 210,000, some of which, because of the political function of Albany, reaches all parts of the state.

New York

The New York nodal region contains 7,149 square miles and 14.4 per cent of the state's area. The area includes Long Island and the southeastern corner of the state; 63.5 per cent is in the Appalachian Country, 7.6 per cent in the Hudson Valley, and 28.9 per cent in metropolitan New York. Metropolitan New York has over 10.5 million people and a tertiary surplus of about 6.3 million. It has, of course, a nation-oriented first-order trade function as well as a second-order trade function in New York and adjoining states.

Figures 85 and 86 provide substantial information concerning location and amount of interconnection between centers. Other maps elsewhere in the text, such as those showing tertiary surpluses, also contribute to the understanding of interconnection.

Combining Composite Uniform and Composite Nodal Regions

Available now, then, is a map showing composite uniform regions, one showing approximately composite nodal regions, and several exhibiting patterns of interconnection. If sensible planning or development regions are to be designed using this material, an early decision has to be made as to whether uniformity or nodality should be the basic regionalizing element. The former commonly takes precedence in geographic analyses, but in this case it will be argued that the latter is more critical. This argument rests largely on the realization that most of the people in the state now live in and around these second-order trade centers and that a large part of the economic activity takes place there. It is clear, too, that the large urban systems will even more completely dominate the geography of the state in years ahead than they do now. They will be the nerve centers, the brain if you will, and the areas around them will tend more and more to serve them in one way or another or to be served by them. Lines of interconnection can be visualized as having two components: first, interconnection between major urban systems and second, interconnections converging from all points of the trade areas toward the great cities that dominate them. This may be hypothetically diagrammed (see Fig. 116).

Suppose the diagram showing composite nodal regions and lines of interconnection is superimposed on a mosaic of composite uniform regions. There are now three nodal regions and four uniform regions in the diagram (see Fig. 116). Now it makes sense to argue that nodal Region I should be set up as a planning and development region with an administering regional council. Region I, however, has substantial areas in uniform regions A and B and a tiny area in C. It should be subdivided (and perhaps put under separate divisions of the regional council) because different kinds of planning will be needed, different programs set up, and different regional performance expected in subdivisions A and B. It is reasonable that the small area of C should be put under the jurisdiction of Region II. The result of following such a procedure for New York State is shown in Figure 117. There is now discernible, for example, a Rochester Planning and Development Region, the southern part of which has a different resource base, has exhibited a different historical development, now has a different economy, and most certainly does not have the same potential as the northern part. The regional planning and development council, which holds the responsibility for this region, should have two divisions, one for each of the uniform zones, and perhaps a third—directed primarily to the urban landscapes of metropolitan Rochester. All divisions, in any case, have as their responsibility the "highest and best" use of the region.

When several different regional councils have divisions concerned with a single composite uniform region such as the Appalachian Country, intercouncil cooperation is obviously required. In fact, in such a case as this, perhaps an intercouncil Appalachian Association should be established to coordinate the work of the various regional divisions concerned.

Adjusting Boundaries According to Political Reality and Statistical Practicality

There are still, however, two important considerations missing from the regionalizing process. They are in fact most practical considerations. No matter how otherwise sensible a planning or development region may be, unless it fits into some kind of reasonable political framework and unless statistical data are available for it, carrying out plans and analyzing results will be difficult, if not impossible. Because statistical data are ordinarily collected and summarized by political units, both administration and analysis can be facilitated if planning and development regions coincide with political units or groups of political units. Ordinarily counties or groups of counties are used, and Figure 118 shows how the New York State Office for Regional Development, using county boundaries, has regionalized the state for the purpose of preparing regional plans and establishing regional councils. There is not complete dissimilarity between this map and that of composite nodal regions, and much of the dissimilarity that does exist is related to the use of counties only in drawing regional boundaries. These regions, ten in number, are likely to result in unnecessary duplication, and councils will find it difficult to handle the widely diverse regional conditions under their jurisdiction. It would, for example, seem to be inefficient for four regional councils to be concerning themselves with the Adirondack Wilderness uniform region or for seven councils to wrestle with how best to use the Appalachian Country. Although these kinds of problems will not be completely solved no matter what the regional breakdown, if a single development council could handle a region with general internal functional cohesion and as little areal diversity as possible, better results and more efficient programs might be expected. To help solve this problem it is suggested that town boundaries be used when county boundaries depart too far from rational regional boundaries. Many whole counties will remain intact under this procedure. If town boundaries are used when desirable, the internal cohesion of the nodal regions and the subdivisions based on uniformity can be more closely approximated without giving up the advantages of political boundary delineations and ready sources of statistical material. In this day of data-processing, working with town census figures instead of county ones should not be an insurmountable task. State collecting agencies, on the other hand, might find it advisable to modify some of their collecting and summarizing techniques if they were to settle on the regional organization suggested here.

Figure 119, incorporating the concepts of nodality and uniformity, as well as political reality and statistical practicality, is offered as a rational regionalization scheme for planning and development in New York. It will be noted that the northern portions of the unwieldy and large Syracuse and Albany nodal regions (see Fig. 117) have both been severed from those regions and made into a northern New York planning and development region. This seems reasonable, for this whole area has much in common, involving at least some internal cohesion and regional identity. Although a similar argument might be put forth for an Appalachian planning and development region, it is unlikely that the internal cohesion and regional identity here anywhere near equals the divergent orientation to the second-order nodal foci in significance. In any case, the boundaries of the regions are not meant to be inviolable but rather subject to adjustment as new information becomes available or as particular planning and development interests and objectives evolve.

Table 41 summarizes certain characteristics of the planning and development regions and their divisions. Density of population, which is in a way a summary of the extent of economic development, supports the justification for subdividing regions into divisions with special administrative and action focus. For example, in the case of the Buffalo region the Heartland Division has a population density of 640 per square mile while the Appalachian Division has a density barely a tenth of that. Clearly planning and development problems in the two divisions will be different even though they are both within one nodal region. The northern New York region, with 31 per cent of the state's area and less than 3 per cent of its population, has by far the lowest population density of any of the regions, and even for the few people there jobs are difficult to find. The importance of the New York region is reflected

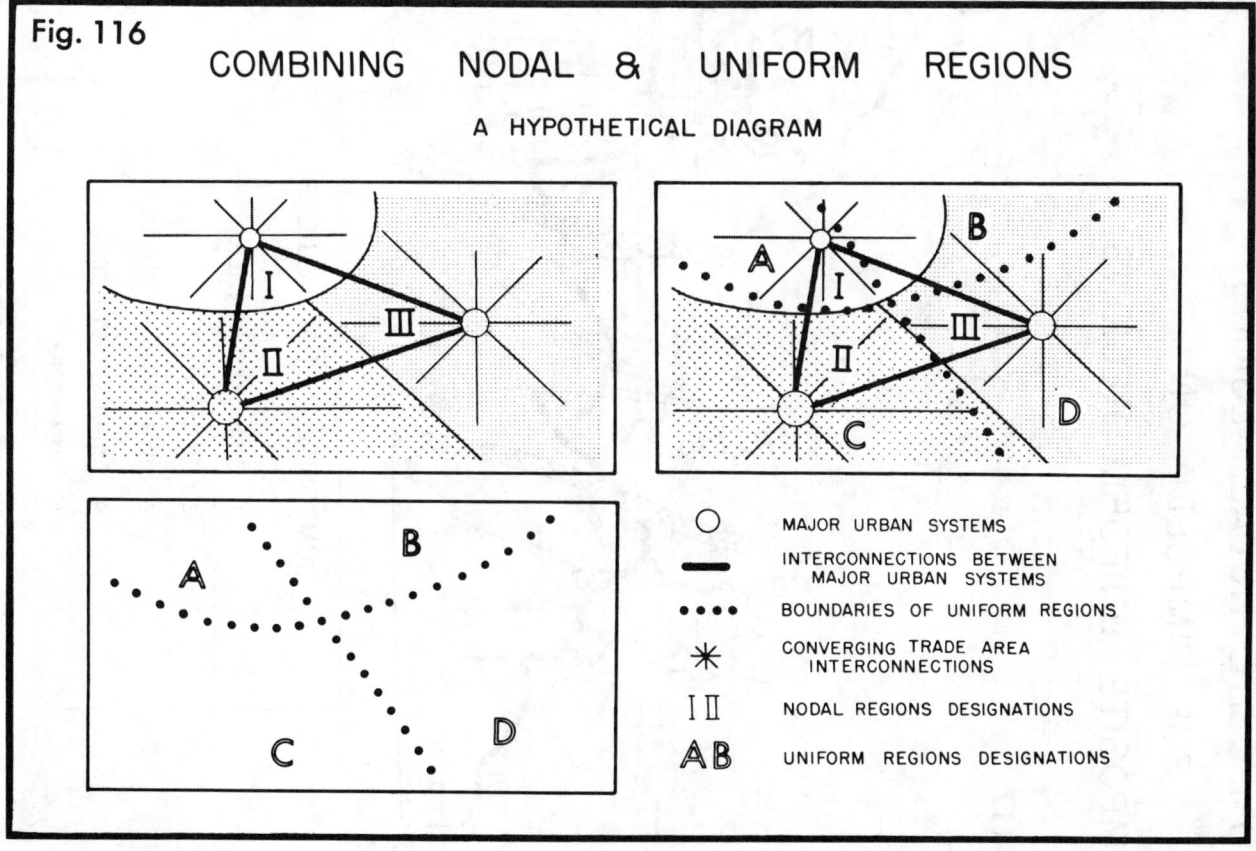

Fig. 116 COMBINING NODAL & UNIFORM REGIONS
A HYPOTHETICAL DIAGRAM

○ MAJOR URBAN SYSTEMS
— INTERCONNECTIONS BETWEEN MAJOR URBAN SYSTEMS
•••• BOUNDARIES OF UNIFORM REGIONS
✳ CONVERGING TRADE AREA INTERCONNECTIONS
I II NODAL REGIONS DESIGNATIONS
A B UNIFORM REGIONS DESIGNATIONS

CHAPTER 16   PLANNING AND DEVELOPMENT REGIONS

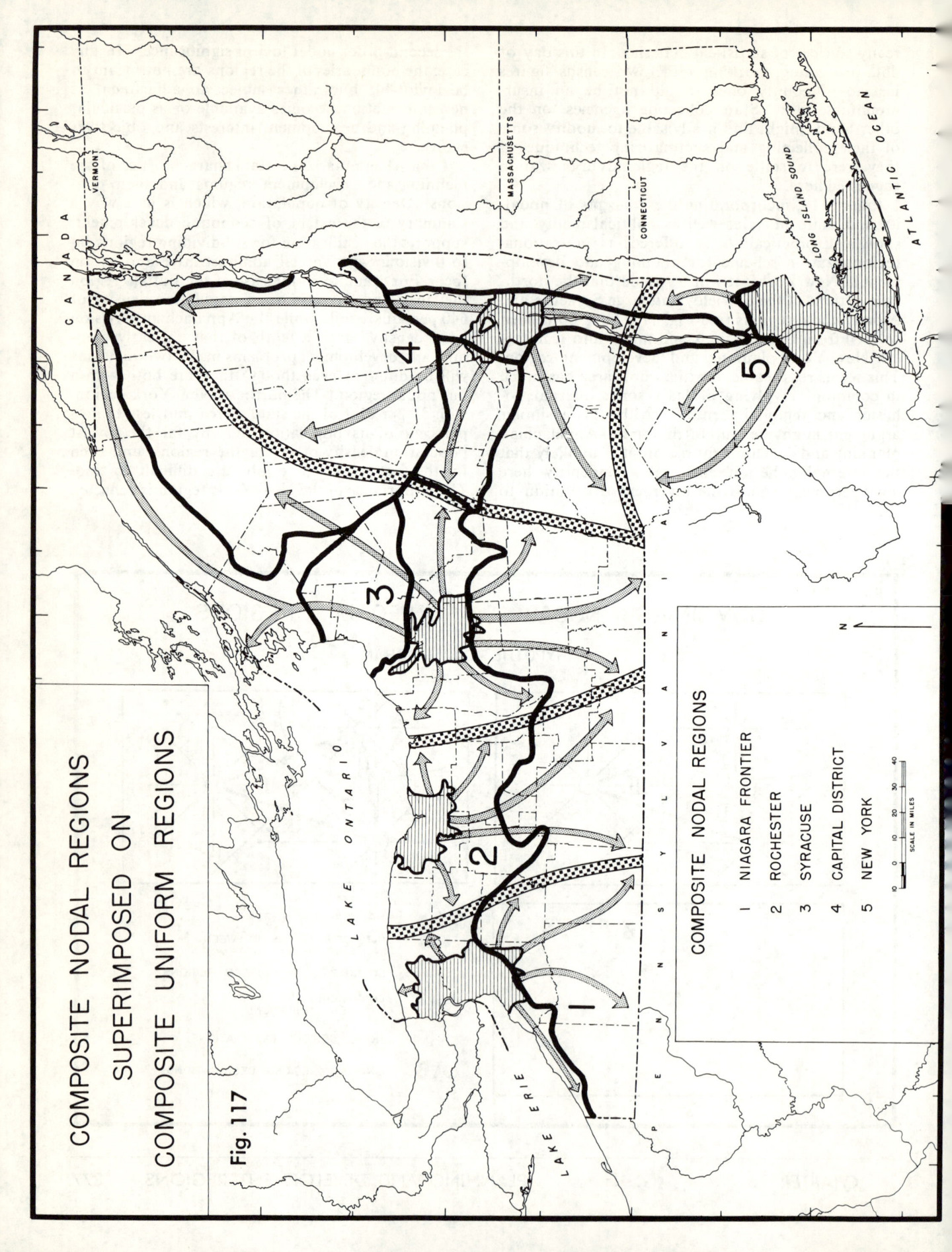

COMPOSITE NODAL REGIONS SUPERIMPOSED ON COMPOSITE UNIFORM REGIONS

Fig. 117

COMPOSITE NODAL REGIONS
1 NIAGARA FRONTIER
2 ROCHESTER
3 SYRACUSE
4 CAPITAL DISTRICT
5 NEW YORK

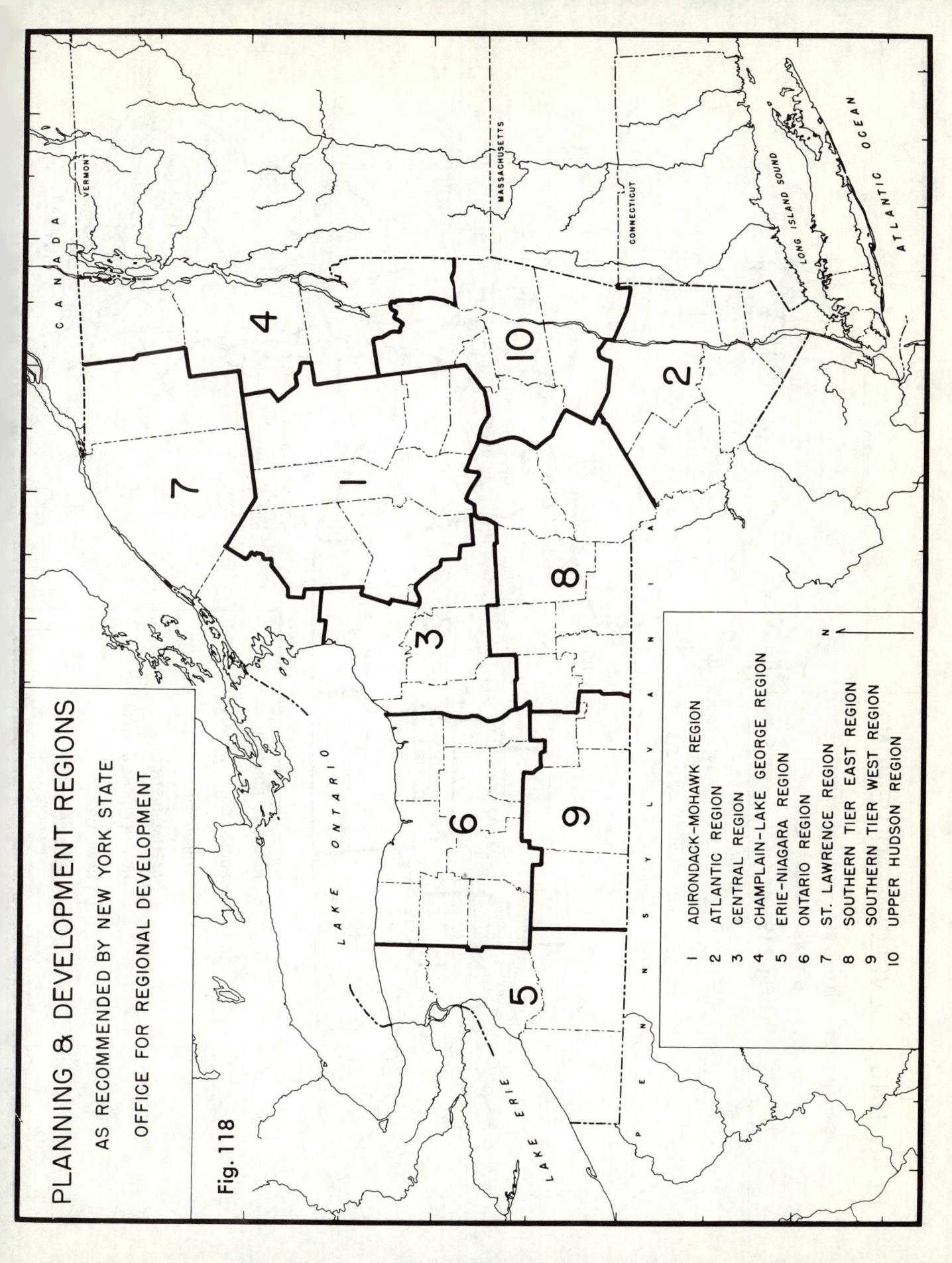

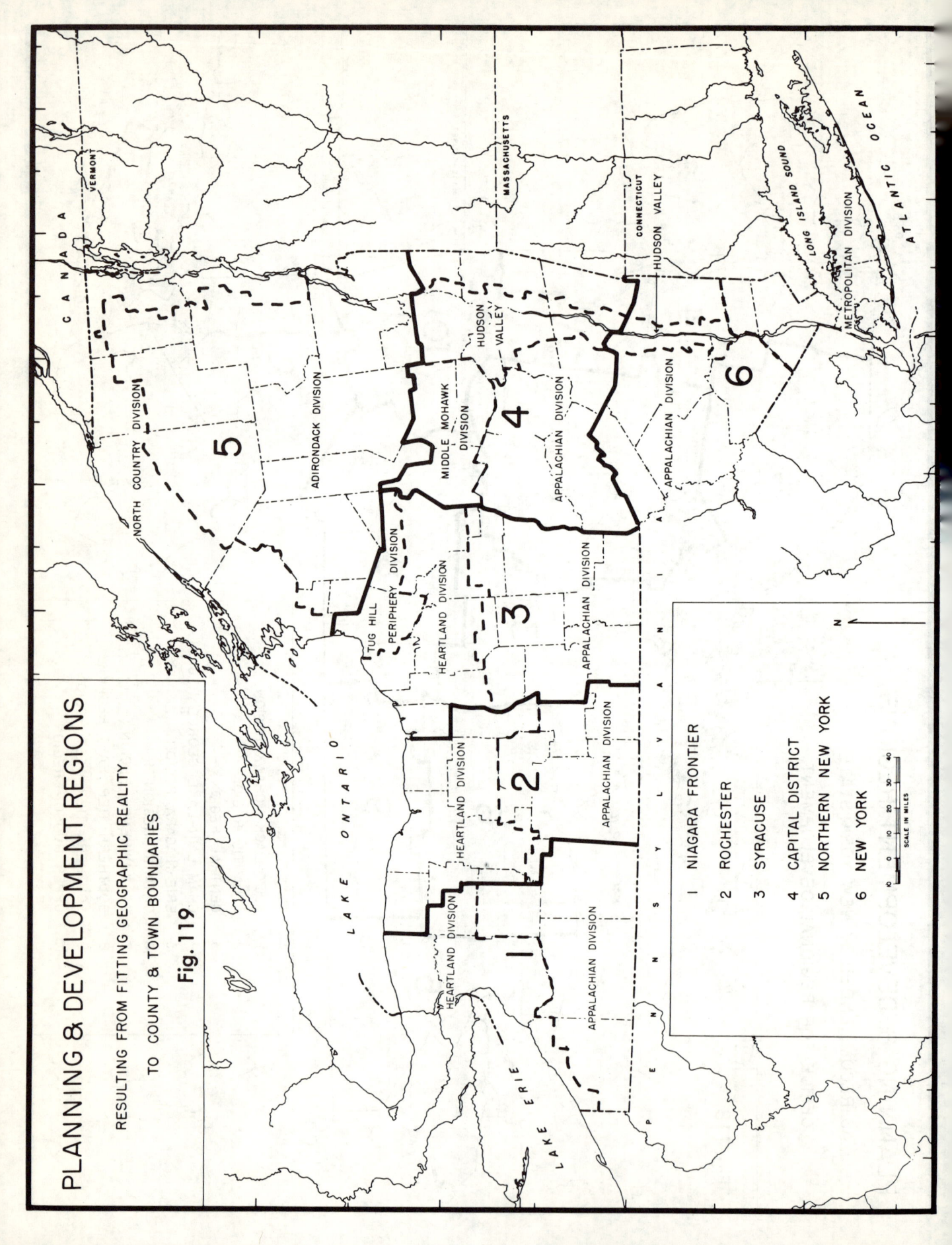

TABLE 41
PLANNING AND DEVELOPMENT REGIONS

	AREA		POPULATION		
	Square Miles	Percentage of State	Total	Percentage of State	Density
Niagara Frontier	5,872	11.8	1,639,731	9.8	279.2
Heartland Division	2,183	4.4	1,397,616	8.3	640.2
Appalachian Division	3,689	7.4	242,115	1.5	65.6
Rochester	6,273	12.7	1,088,252	6.5	173.5
Heartland Division	3,449	7.0	849,846	5.1	246.4
Appalachian Division	2,824	5.7	238,406	1.4	84.4
Syracuse	7,299	14.7	1,296,638	7.7	177.6
Heartland Division	2,323	4.7	818,573	4.9	352.4
Appalachian Division	4,047	8.1	437,635	2.6	108.1
Tug Hill Division	929	1.9	40,430	.2	43.5
Capital District	7,710	15.6	1,012,260	6.0	131.3
Hudson Valley Division	1,757	3.5	659,922	3.9	375.6
Appalachian Division	4,440	9.0	181,017	1.1	40.8
Middle Mohawk Division	1,513	3.1	171,321	1.0	113.2
Northern New York	15,246	30.7	484,839	2.9	31.8
Adirondack Division	10,908	22.0	198,395	1.2	18.2
North Country Division	4,338	8.7	286,444	1.7	66.0
New York	7,176	14.5	11,260,584	67.1	1,569.2
Metropolitan Division	2,807	5.7	10,726,355	64.0	3,821.3
Appalachian Division	3,871	7.8	293,203	1.7	75.7
Hudson Valley Division	498	1.0	241,026	1.4	484.0

by the 67 per cent of the state's population it contains. Even so, within it, the Appalachian Division has not greatly felt the impact of the city as reflected by population density.

Selected References

Change, Challenge, Response: A Development Policy for New York State. Albany: Office for Regional Development, 1964.

Duncan, Otis, W. R. Scott, Stanley Lieberson, Beverly Duncan, and H. Winsborough. *Metropolis and Region.* Baltimore: Johns Hopkins Univ. Press, 1960.

Freeman, T. W. *Geography and Planning.* London: Hutchinson, 1958.

Friedmann, John. "Regional Planning in Post-Industrial Society," *Journal of the American Institute of Planners*, XXX, No. 2 (May, 1964), 84–90.

——— and William Alonso. *Regional Development and Planning: A Reader.* Cambridge: M.I.T. Press, 1964.

James, Preston E., and Clarence F. Jones. *American Geography: Inventory and Prospect.* Syracuse: Syracuse Univ. Press, 1954. Chapter 2, pp. 19–69.

Levine, Lawrence. "Land Conservation in Metropolitan Areas," *Journal of the American Institute of Planners*, XXX, No. 3 (August, 1964), 204–16.

Meyer, John R. "Regional Economics: A Survey," *American Economic Review*, LIII, No. 1 (March, 1963), 19–54.

Perloff, Harvey, *et al. Regions, Resources, and Economic Growth.* Baltimore: Johns Hopkins Univ. Press, 1960.

Pitts, Forrest R. (ed.). *Urban Systems and Economic Development.* Papers and Proceedings of the Conference on Urban Systems Research in Underdeveloped and Advanced Economies, School of Business Administration, University of Oregon, 1962.

PART FIVE

THE GREAT URBAN SYSTEMS

BOTH statistical evidence and development theory lead to recognition of the ever increasing dominance of urbanism in New York, and especially to the recognition of the increasing importance of the largest of the urban systems. Part Five contains one chapter on each of the seven great urban systems of the state: Albany-Schenectady, Binghamton, Buffalo, New York, Rochester, Syracuse, and Utica.

The urban system, as portrayed in Chapter 14, contains concentric zones extending from the CBD to the urban frontier. Obviously, some areas beyond the urban frontier exhibit functional ties with areas within it. Therefore, chapters which follow quite correctly deal with areas at least as broad as the census-designated Standard Metropolitan Statistical Areas (SMSAs). Implied synonymous usage of the terms "SMSA" and "urban system" appears as chapters are concluded with population projections for SMSAs. It should be kept in mind, though, that an urban system can best be defined only in broad functional terms, that its outer boundary is subject to change, and that this boundary does not exactly coincide with any political or census designation.

A local authority prepared each of these chapters, but the editor has taken a free hand in trying to maintain reasonably parallel presentations. Population projections at the end of each chapter were added by the editor and are his responsibility.

Although each urban system had a particular kind of beginning and a somewhat different historical development, the various authorities found much in common among them. All had great transportation advantages and attracted a substantial concentration of industry and people early in their history. Once large, they continued a steady climb to ever greater size, while less fortunate neighboring centers fell behind. Decline or static population trends of central cities and skyrocketing population explosions in outlying districts, congestion and obsolescence in the CBD, frame, and middle city, concentrations of low-skill and low-income populations in slums, need for urban renewal and transport net revitalization, and frustrating internal political fragmentation are problems common to all.

Over 90 per cent of the state's population growth between now and the year 2000 will occur in the seven urban systems. They will provide the best jobs, and the brightest young people will tend to gravitate to them. They are where the specialized goods and services are to be found, the places which will attract most of the industrial expansion, the locations of enlarging sales and service functions, the overwhelmingly significant economic and cultural foci of the future.

CHAPTER 17   HOWARD J. FLIERL

Albany-Schenectady

The Albany-Schenectady Urban System is located in the Hudson Valley, where the Mohawk joins the Hudson. Probably few places in the country have enjoyed a more favorable transportation orientation over the years. The state capital function of Albany itself and certain raw material advantages have contributed much to the urban growth that has taken place.

From the earliest days there have been three dominant central places in the area. Albany, the largest, with a 1960 population of 130,000, is followed in order by Schenectady (82,000) and Troy (68,000). All three cities have lost population in recent years as people and businesses have moved to areas outside their political limits. Albany and Troy on the Hudson River and Schenectady on the Mohawk roughly form the three corners of a triangle. (See Fig. 120.) Growth has occurred principally within the confines of this triangle but recently has extended outward to some degree too. Close economic ties extend as far north as the Glens Falls area in Warren County. Actually, orientation of the urban system to the north is greater than to the west up the Mohawk. In the colonial period Schenectady represented the westernmost extent of Hudson Valley settlement; in the modern period the Middle Mohawk Valley beyond Schenectady in reality forms a separate geographic unit.

The present urban system encompasses numerous smaller central places including Cohoes (20,000), Rotterdam (16,900), Saratoga Springs (16,600), Watervliet (14,000), Rensselaer (10,500), Scotia (7,600), Colonie (7,000), and Mechanicville (6,800). Many of these merge, or nearly merge, with each other or with the three dominant centers.

Figure 120 shows that the urbanized area follows along two sides of the triangle formed by the large cities, but fails to extend along the banks of the Mohawk. Growth, however, is now occurring along N.Y. Route 7 between Schenectady and Troy and subsequent censuses will show the urbanized area as occupying most of the land south of the Mohawk River. The Northway (Interstate 87) and other high-speed highways are now opening up areas north of the Mohawk to urbanization too. In its maximum dimension from northwest to southeast the urbanized area extends for approximately twenty-five miles. It contained a population of 455,000 in 1960, 175,000 of which was outside the limits of the three large cities.

The urban frontier (see Chapter 14 for definition and Fig. 120 for location) encloses an area of about 412 square miles and contains an estimated population of 527,000. From northwest to southeast, its maximum extension is at least 30 miles, and from southwest to northeast it averages about 15 miles. Its shape reflects the location of the major centers, access to these centers, and, to a certain degree, terrain conditions. Of special interest is the large enclave in the bend of the Mohawk River within the general urban frontier. Land values up to 1963 had not risen here. This resulted in part because of the low-lying, badly drained land near the river, but probably more significantly because access from this area to the major cities was not especially good. Now with the Northway and its bridge across the Mohawk it can be assumed that development will take place rapidly at least on the more attractive sites and that the enclave characteristics will largely disappear.

The Albany-Schenectady-Troy SMSA, composed of Albany, Schenectady, Rensselaer, and Saratoga counties, contained a population of 658,000 in 1960, having increased by 127,000, or about 25 per cent, since 1940. This rate of growth is similar to that of the more slowly growing upstate urban systems, Binghamton and Utica, rather than to that of rapidly expanding Syracuse or Buffalo.

The tributary area around Albany, usually including Warren, Washington, and Schoharie counties as well as the Albany-Schenectady-Troy SMSA, is frequently referred to as the Capital District. It is so designated by the New York State Department of Commerce, which issues statistical digests for the various parts of the state.[1]

---

[1] These publications, called *Business Fact Book* and issued every few years by the Department of Commerce in Albany, contain useful data which may prove valuable to the reader who desires additional information or who wants to update the contents of these chapters on the great urban systems.

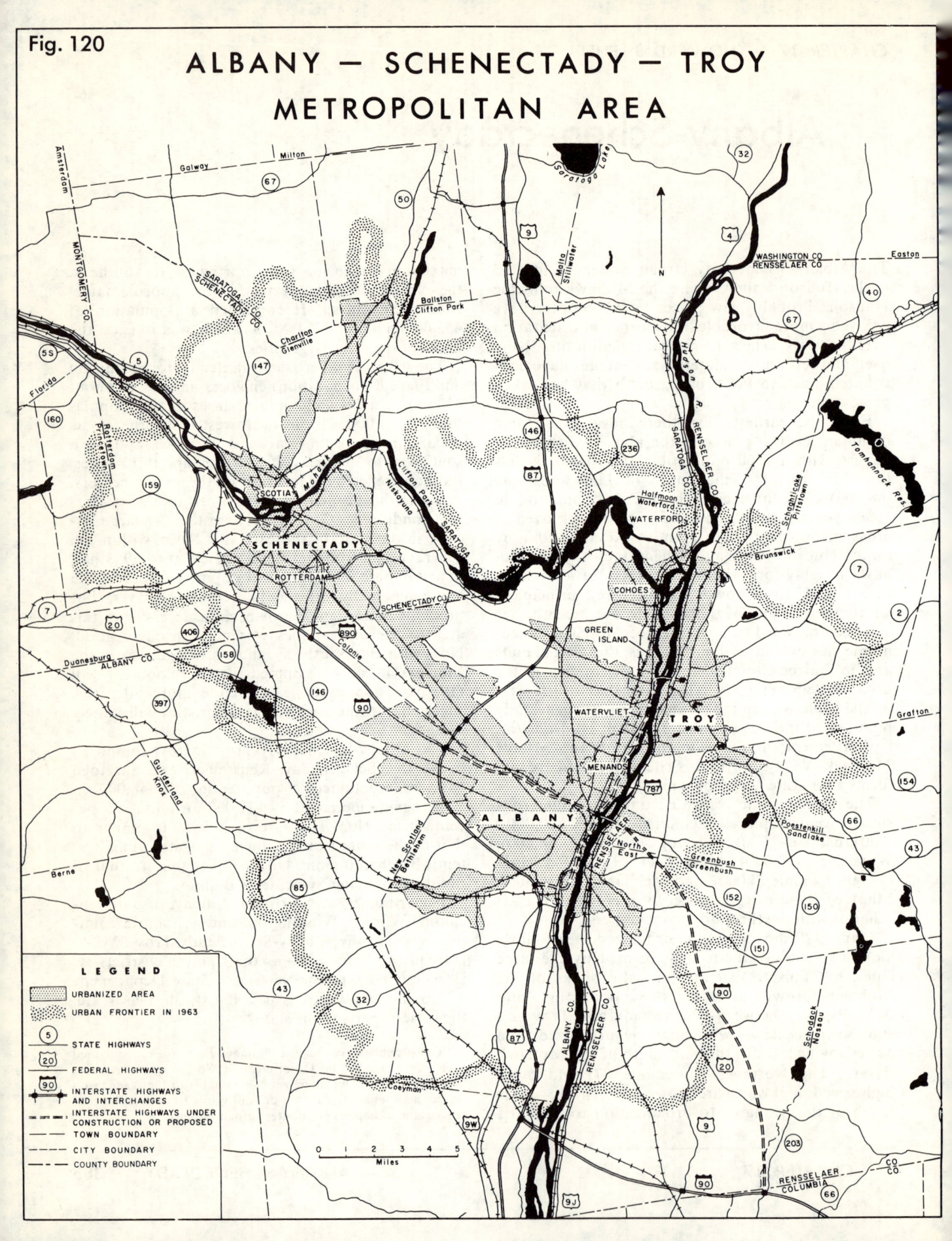
Fig. 120 ALBANY — SCHENECTADY — TROY METROPOLITAN AREA

An Early Beginning

Settlement and economic development at the confluence of the Mohawk and Hudson rivers was well under way when the Syracuse and Rochester areas, for example, were still wilderness. Henry Hudson's voyage and the establishment of Dutch settlements at Manhattan and Albany during the first part of the seventeenth century heralded the development of the Hudson River axis. From the vicinity of Albany and Schenectady trade tentacles spread out to the west and north, and Yankee trails across the Berkshires to the east added to the crossroads function. By 1686 Albany had become the chief fur-trading center for the English colonies and was clearly the westward-facing doorway to the yet unoccupied interior.

Establishment of Albany as the state capital in the 1790's provided a special impetus for growth—one which has contributed substantially to the economic success of the whole urban system. Of great importance, too, was the area's function as the eastern portal of the Mohawk route and Cherry Valley turnpike and later as the Hudson River terminus of the Erie Canal. The canal made possible early eastern movement of potash, Genesee flour, and Onondaga salt and western movement of manufactured products.

There were raw material advantages, too, which definitely helped local development. Nearby supplies of molding sand and limestone for flux and of iron ore from the northeastern part of the state established Troy as the major iron and steel center in the country by the time of the Civil War. This pre-eminence was short-lived, but early production of such things as stoves, armor plates (for the *Monitor*), horseshoes, coaches, and bells contributed to the city's industrial character. The state's most highly developed concentrations of Portland cement manufacture are based on the limestones, shales, and clays along the Hudson south of Albany; in like manner Hudson Valley clays have long supported the manufacture of brick, and the deposits of valuable molding sand—a very fine-textured sand with a high clay content—occur among the glaciolacustrine deposits in the vicinity. During the nineteenth century the large timber resources of the Adirondacks and northern New York enabled the area to become an outstanding processor of such wood products as lumber, paper, and furniture. Albany, and places like Glens Falls and Mechanicville, profited a great deal from the wood-using industries.

Troy also had an early beginning and a lustrous history of clothing manufacture beginning with the invention of the detachable collar in 1825. In 1860 the city contained 39,000 people.

In 1831 the first railroad from Albany (Fig. 121) was built westward to Schenectady and shortly manufacturing began to flourish there. Production of locomotives, which began in 1848, gave rise to the development of the American Locomotive Company, which was to become one of the cornerstones of industrial Schenectady. Certainly one of the greatest single events in the whole urban system's history was Thomas Edison's decision in 1886 to buy two factory buildings in Schenectady and move his small electric machine works from New York City to these buildings. This business later became the General Electric Company, the most significant single element in Schenectady's entry into its modern phase as an industrial center.

Physical Setting

South from Albany the Hudson Valley is confined between the Taconic Mountains on the east and the Helderberg Hills section of the Appalachian Upland on the west. Near the confluence of the Mohawk and Hudson rivers, however, the valley opens westward to form a roughly triangular lowland bounded on the southwest by the Helderberg Escarpment, on the north by the foothills of the Adirondacks, and on the east by the Taconics. The Albany-Schenectady Urban System occupies much of this triangular lowland and will eventually fill it up entirely.

Elevations within the urban system are low; in fact, the entire area may be likened to a vast bowl whose rim is the surrounding mountains and plateaus. Parts of Albany itself are practically at sea level, for the mean elevation of the Hudson at this point is approximately three feet, and some slight tidal action in the river is observable, even as far north as Troy. Its waters, though hardly to be described as brackish, do contain more than the normal quantity of salts of fresh-water streams.

From the river, the land rises rather rapidly up to elevations of some 200 feet. In both Albany and Troy older built-up sections rise as a series of steep steps or terraces from the river below. Schenectady, on the other hand, spreads out far beyond the level plain of black, rich soils which formed its founding site along the Mohawk. The whole area was under the water of a temporary lake during the Pleistocene Epoch, and a large sandy plain extending several miles along the Mohawk Valley and southward nearly to the Helderberg Escarpment resulted. These sands, although used in a variety of ways, provided a poor base for the development of soils and limited local agricultural development. On the other hand, the relative flatness of this old lake bottom makes the spread of urbanization easy.

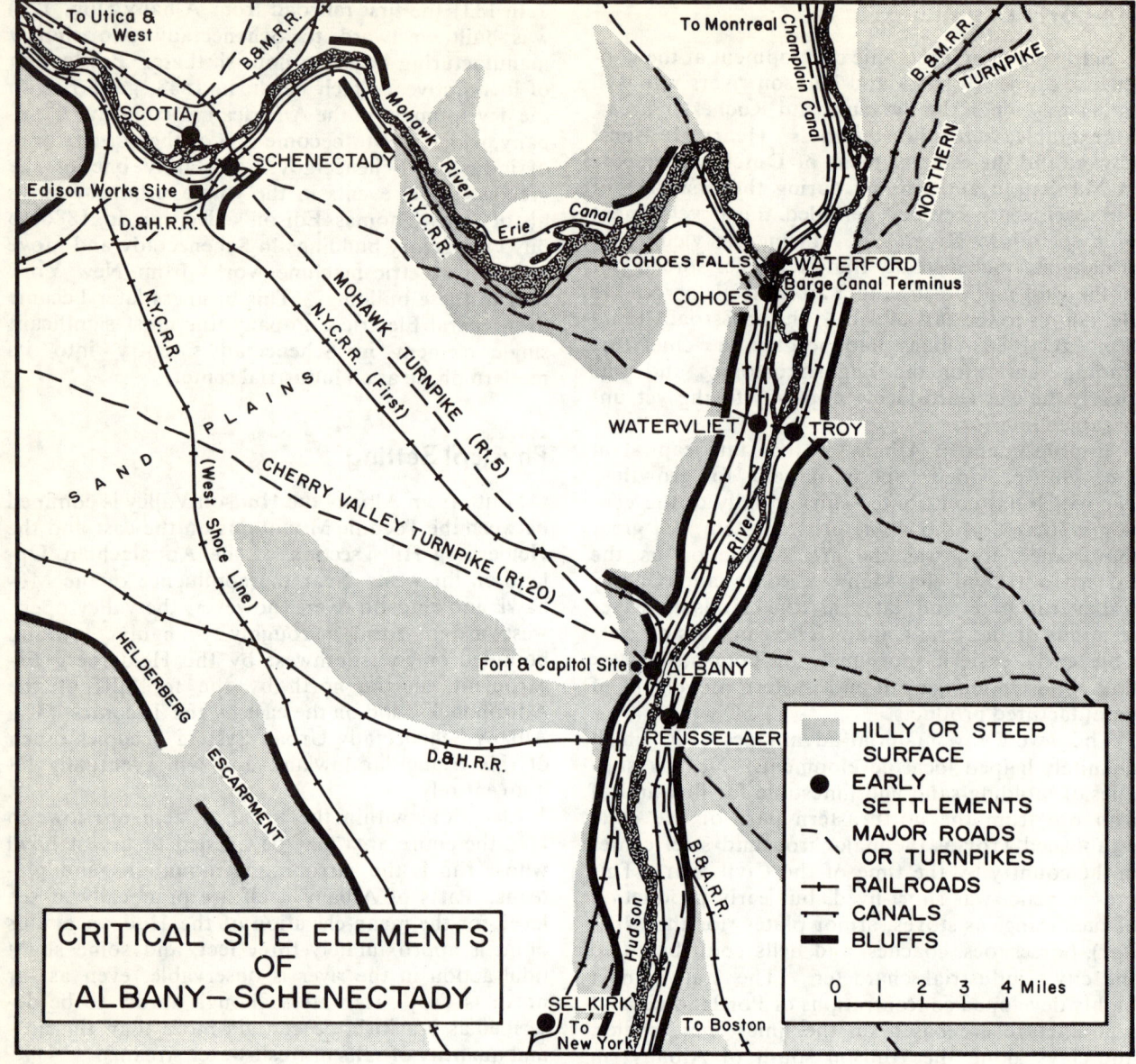

Fig. 121

Transportation Setting

The advantages and importance of the Hudson–Mohawk–Lake Plain axis have been repeatedly extolled in this volume and in practically all other interpretations of New York State's development. The location of the Albany-Schenectady Urban System is one of the finest on that axis. It is today, as it was in colonial times, strategically situated for transportation of all kinds, commanding all north-south and east-west movements.

Northward, the Hudson River and the Champlain Lowland form a corridor that reaches Canada and the St. Lawrence Valley. Montreal, one of Canada's greatest cities and one which promises increasing growth since completion of the St. Lawrence Seaway, lies just west of this corridor and only a little upstream from where the Richelieu Canal reaches the St. Lawrence River.

Land communications eastward are partially blocked, first by the barrier of the Taconics and then by the Berkshire Hills of Massachusetts. Nevertheless, major rail and highway connections have for a long time linked the Capital District with the principal cities of New England.

A bulk breaking point or transshipment point is generally regarded as a likely site for the development of cities. Both Schenectady and Albany have such sites. At Schenectady, the Mohawk River begins a series of meandering loops, finally passing over several falls and rapids, before it enters the Hudson River. Before the advent of the Erie Canal, river cargoes on the Mohawk had to discharge at Schenectady for an overland haul to the Hudson, reaching it at Albany. Albany lay essentially at the end of a nearly straight line across the sand plain, and on tidewater. The route across the plain provided the shortest overland haul between these two rivers; and each city benefited from its position at the end of the route.

After 1825 the Erie Canal provided a bypass around the falls of the Mohawk River below Schenectady. This caused Schenectady to lose importance as a transshipment point, but Albany has continued to function in this capacity between river or ocean traffic on the one hand and canal, railroad, and highway traffic on the other.

Railroad Transportation

Three railroads, with their branches, serve the Albany-Schenectady Urban System. (See Fig. 85.) The main line of the New York Central connects the cities of the Capital District with points south and west. The multiple-track section through the Mohawk corridor west from Albany carries the second largest freight traffic in the United States. Only a portion of the Pennsylvania exceeds it. A New York Central branch, the West Shore, links New York City and Albany and has a side branch to Schenectady. Another branch, the Boston &

ALBANY. This view is northwest over the Hudson River, the City of Albany, and across the broad lowland which stretches to Schenectady and beyond. The vacant land near the city center is the South Mall urban renewal project and will be developed as a state government office complex. *N.Y.S. Dept. of Commerce.*

CHAPTER 17   ALBANY-SCHENECTADY

TRANSPORTATION on the Hudson. Ocean freighters, oil barges, and many other vessels ply the Hudson as far north as Albany. This view just south of Albany shows a variety of oil and grain storage facilities and an electric plant supplied by barge-hauled coal. *N.Y.S. Dept. of Commerce.*

Albany, extends east to Boston from Albany. Thus a traffic "T" is formed, involving through east-west traffic and a southward flow to New York.

The Delaware & Hudson Railroad runs southwest from Albany to Binghamton and the Southern Tier, and north from Albany along the upper Hudson and Champlain valleys to Plattsburgh and beyond. A branch also extends into Vermont along Lake Champlain. The Boston & Maine connects with the Delaware & Hudson and provides competition for the Boston & Albany to New England. In all, six major rail lines converge like the spokes of a wheel on the urban system.

Selkirk, immediately southwest of Albany, is a freight routing center for the whole area. In its freight yards trains are made up to service much of the Northeast. The range of freight carried is diversified. It includes various types of bulk materials such as grains and feed, coal and coke, ores, sand and gravel, oil, and cement, as well as wood and wood pulp, fabricated machinery, food products, and general merchandise. Albany is a dividing point for rail traffic originating west of the city. Freight density is higher toward Boston than along the Hudson to New York City. Thus freight passing through Albany contributes only a small part to the total volume reaching New York City. In fact, much of the freight moving toward Albany represents intrastate movements alone.

Grain products and petroleum products offer interesting contrasts in their direction of movement relative to the area. It is almost axiomatic that grain moves east. Within the state, then, the principal point of origin of grain product movements is the Niagara Frontier, principally Buffalo which has received down-lake grain shipments. That part which continues east by rail declines steadily as major distribution centers are passed—Rochester, Syracuse, and to some extent Utica. About half of the tonnage reaching Albany remains there for processing and distribution; most of the remainder continues toward New York City.

Petroleum products reverse the grain picture. Albany receives large quantities of petroleum products by tankers, via the Hudson River. Most of this moves west by rail toward Buffalo much in the manner in which grain products travel eastward to Albany, though water transportation is more important in the case of petroleum. Some goes north by rail, too.

HIGHWAY TRANSPORTATION

The New York State (Governor Thomas E. Dewey) Thruway (Interstates 87 and 90), the Northway (Interstate 87), major east-west federal and state routes, a diagonal route, and north-south federal highways serve the Albany-Schenectady Urban System. Many subsidiary or alternative roads complement these.

After following northward up the west bank of the Hudson, the Thruway curves gradually toward the Mohawk Valley between Albany and Schenectady. Four interchanges—two at Albany and two on the approaches to Schenectady—serve these cities. The result is essentially the same as that occurring in greater or lesser degree at cities all along the Thruway. Tourist and convention facilities, warehousing and trucking industries, and service and manufacturing industries—all of which depend to a large degree on Thruway passenger and truck traffic—are developing in the vicinity of these interchanges.

N.Y. Route 5, one of the state's oldest east-west roads, links the principal cities and intervening small towns along the Mohawk–Lake Plain axis. Following the north bank of the Mohawk eastward to Schenectady, it crosses there to proceed due southeast over the old sand plain route to Albany. Except for trucks using the Thruway, this is the principal freight route. It avoids the steep, long slopes of U.S. Route 20, which essentially parallels it some miles to the south and which in west-central New York is identical with it. Route 5 is also less prone to fog conditions than is Route 20.

East of Albany, U.S. Route 20, the Berkshire Thruway spur (Interstate 90), and the Massachusetts Turnpike form the principal connections with Boston and eastern New England. N.Y. Route 7 links Schenectady with Troy and Troy with northern New England. Equally important is this highway's link with Binghamton in the Southern Tier. U.S. Routes 9 and 9W extend northward along the east and west banks respectively of the Hudson as far as Albany, where they then merge as one route west of the river extending north. Troy, on the east bank, lies just off the principal north-south highways.

WATER TRANSPORTATION

It is of some interest that what was the head of navigation for Hendrick Hudson in 1609, when he sailed his ship *Half Moon* up the river named for him, remains today approximately the head of navigation for ocean vessels. Directly related to the deep water channel of the Hudson River is the creation of the Albany Port Authority and the development of the Port of Albany. Authorized several years earlier, the port was officially opened to waterborne commerce in 1932. For many years a 27-foot channel permitted all but the largest freighters to ascend the Hudson to Albany, and a 32-foot channel now being developed will permit passage of even larger ships.

Most of the products handled by the Port are bulk cargo. Extensive gasoline and petroleum storage facilities have been developed, and Albany is a principal distribution center. Tanks also store large quantities of molasses from the West Indies, and wood pulp from Sweden, Finland, Canada, and other countries arrives at the Port for distribution to area paper mills. Other imports include lumber and lumber products, wool, and canned goods. Wheat is the principal export from the Port, which handles about seven million tons of cargo annually. Rail connections plus trucking accomodations facilitate the collection and distribution of goods handled by it.

Shallow draft boats and barges ascend the Hudson beyond Albany to where the New York State Barge Canal may be entered at Waterford. A series of five locks involving a vertical lift of 169 feet permits continuation along the canalized Mohawk River to Schenectady and the west. A water route northward follows the Champlain canal from Troy on the Hudson River to Whitehall at the south end of Lake Champlain. Beyond Lake Champlain the Richelieu Canal forms a continuous water link through Quebec to the St. Lawrence River. While these canals have played significant historical roles, business on them has been in a state of decline for many years. There is even serious question as to whether they can be economically maintained as commercial routes at all. Pipe lines and the great expansion of truck freight have been particularly contributory to the near demise of the canals.

The Albany Port Commission, like many eastern interests, opposed the building of the St. Lawrence Seaway as a threat to local development. The Seaway's opening revived interest, consequently, in a long-discussed deep water channel that would link Albany and the Hudson River with the St. Lawrence along the Champlain corridor. Thus far nothing has come of it; nor, in view of the general low demand for canal traffic, is there likely to be any action taken in the foreseeable future.

AIR TRANSPORTATION

Although both Schenectady and Troy maintain municipal airports, only Albany County Airport, situated centrally between the major cities, has regularly scheduled air-lines service. Its newly modernized facilities are used by TWA, American, and Mohawk air lines. Strategically situated approximately halfway between New York City and Plattsburgh in northern New York, and on the main east-west air route between Boston and Buffalo, Albany's airport handles a reasonably large passenger and freight volume.

Current Livelihood Structure

In 1960 the Albany-Schenectady-Troy SMSA had 2.0 per cent of its labor force in the primary sector,

36.5 per cent in the secondary sector, and 61.5 per cent in the tertiary sector. These figures show it to be relatively less oriented toward manufacturing and relatively more oriented toward tertiary activities than any of the other upstate urban systems. This is not surprising, for the long-standing importance of trade and commerce plus the capital function of Albany could hardly be expected to be matched by proportional development of manufacturing. The industrial base is diversified but leans heavily on electrical and nonelectrical machinery. Neither the secondary nor tertiary sector has exhibited a recent growth trend as favorable as other of the state's urban systems. Economic activity is concentrated in and around the major cities of Schenectady, Albany, and Troy. The east and south sides of the triangle formed by these cities are extensively developed, but the north side along the Mohawk has lagged behind. The interior of the triangle is now an area of rapid residential and commercial growth.

Manufacturing

Between 1947 and 1963 manufacturing employed between 55,000 and 80,000 persons; in 1947 the figure was 79,000. This 1947 level was maintained until 1954, when a decline in employment set in. Schenectady, the smallest county within the SMSA, has the largest amount of manufacturing. In 1958 its value added by manufacture was more than half again as much as Albany County's, nearly three times Rensselaer's, and far greater than Saratoga's.

There are several noteworthy specific concentrations of manufacturing. The most important is found in and immediately around the city of Schenectady; the second in importance lines both sides of the Hudson River. There are in addition a number of widely spaced clusters of small industrial centers north of the main urban system.

Manufacturing in Schenectady is so dominated by the great General Electric complex that the city is sometimes described as a one-industry town. The General Electric facilities are world famous for the production of turbines, electric motors, generators, and electronic equipment. The area also, however, is a major producer of diesel locomotives and tanks, as well as electrical insulating material. The principal districts are near the railroad and the Mohawk River, not far from the center of the city. Recently industrial and federal research laboratories have been built at sites peripheral to the city.

The Hudson River manufacturing zone forms a virtually continuous belt nearly fifteen miles long and lines the narrow river terraces on both sides of the Hudson River from a point a few miles south of Albany to north of Troy. Here are not only a variety of heavy and light industries but also numerous warehousing and wholesaling establishments. In Albany itself, food products and related industries—including a sizable meat-packing industry—are of major importance. Associated with pulp and paper production in the general vicinity are a number of concerns manufacturing paper products such as towels, bags, and drinking cups. Albany is also a printing and publishing center; several well-known magazines and directories are printed there. Another long-established type of production, and one also associated with the paper industry, is the manufacture of paper-making felts. Production of these felts occurs both in Albany and Rensselaer. Small machine shops produce a variety of fabricated metal goods. Just north of Albany, in Watervliet, are two metal goods industries worthy of special note. One produces ordnance equipment, and the other manufactures high-grade alloy steels in electric furnaces. In Watervliet and Green Island are companies which produce such products as abrasives, adhesive tapes, chemicals, and automobile parts. On the east side of the Hudson pharmaceutical and dyestuff manufacture as well as paper-felt production is carried on in the city of Rensselaer.

The story of Troy's initial development as an apparel center following a Troy housewife's invention of the detachable collar in the early nineteenth century is familiar. Though no longer the "collar and cuffs" city once described in geography texts, Troy is still dominated by its apparel industry. In fact, a list of its largest companies is made up almost wholly of apparel firms. Men's shirts, sportscoats, jackets, and suits, as well as women's dresses, are principal products.

The historically significant iron and steel industry of Troy is now represented by but a single iron smelter, a marginal producer which comes into production only when economic conditions or heavy demands warrant it. The plant, when it operates, uses sintered (concentrated) Chateaugay ore from the vicinity of Lyon Mountain in the Adirondacks. Transportation costs, even for these sintered ores, are comparatively high, as are the costs of coke. The result is high-cost production.

Directly across the river from Troy, on the falls of the Mohawk, is Cohoes. At one time this city was a major apparel center specializing in knit goods. Although some of its old factories still stand, many have been converted to multiple-use warehousing and light industrial purposes. The apparel industry, once the basis of prosperity, is essentially gone.

The northern industrial clusters, although physically separated from the main urban system, are economically tied to it. Major among these is Mechanicville, some six miles up the Hudson from

Troy, and the cluster of cities forty miles farther north including Glens Falls, South Glens Falls, Hudson Falls, and Fort Edward. The word "falls" in many of these place names suggests the reason for their industrial orientation.

Many of the cities in these northern clusters have in common their early development as centers of the pulp or paper industry. They were close to the basic raw material of the industry, wood, and had substantial excellent supplies of industrial water. The water has been used for processing, power, and transportation. Much of the industry now relies on wood pulp imported from such places as Canada, Sweden, and Finland. Mechanicville is notable in that within comparatively recent times it has introduced paper-making based on hardwoods in some of its operations. This has opened up new local raw material sources.

Manufacturing in these northern centers is not restricted to paper and related production. Apparel and textile industries are important too, supplying almost as much employment as paper. Factories produce woven cotton and rayon underwear fabrics, knitted gloves and hosiery, men's shirts, and women's dresses. Some metal goods and machinery manufacture is carried on too, and various electrical products are turned out. Glens Falls is noteworthy, also, as one of the principal centers in the state for the production of Portland cement.

It is often thought surprising that, except for limited early development near Cohoes, factories did not locate along the Mohawk between Schenectady and the Hudson River. There are reasons for this. For one thing, the Mohawk in this area flows between steep banks with little room for factories on the flood plain or terraces. A branch of the New York Central Railroad which follows the river here clings to the south bank, sometimes appearing to have scarcely enough room to accommodate the tracks. In those places where room does exist along the river, the land is low and swampy, hardly favorable for industrial sites.

Some manufacturing is found along the main transport lines across the sand plain between Albany and Schenectady, but substantial development has not taken place here. A change may be in the offing, however, for both available space and good transport access exist.

Although there has been some relative decline in the importance of both electrical and nonelectrical machinery, these two groups are still the leaders and together employ over one-third of the manufacturing labor force. Production is centered in Schenectady, where the General Electric Company has one of the largest manufacturing complexes in the state, and the world's largest turbine plant. The company, which has been in this location since 1886, is involved in the production of both electrical and nonelectrical machinery.

Paper manufacturing and apparel production, each with about one-tenth of the factory workers, rank third and fourth. The paper plants are widely distributed, but many are in the detached industrial areas to the north. The apparel industry is concentrated in Troy. Ranking next in order of importance are food, textiles, printing and publishing, and chemicals.

The fifteen largest manufacturing concerns, together with principal products and approximate recent employment, are shown in Table 42.

TERTIARY ACTIVITIES

In terms of the percentage of the labor force in the tertiary sector, the area ranks well above all the major upstate urban systems and not substantially below New York. Of the 247,000 employed persons in the SMSA in 1960, about 152,000 were employed in the tertiary sector. In terms of growth of the sector in the 1950–60 decade, however, there is little

TABLE 42
MAJOR MANUFACTURING CONCERNS OF THE ALBANY-SCHENECTADY URBAN SYSTEM

Company	Products	Approx. Employment Early 1965
General Electric Co.	Turbines, motors, generators, electrical equipment	22,500
Behr-Manning Co. (Div. of Norton Co.)	Abrasives, pressure-sensitive tapes	2,800
Alco Products, Inc.	Diesel locomotives	1,900
Cluett Peabody & Co., Inc.	Shirts	1,600
Allegheny Ludlum	Alloy steel forgings and castings	1,500
International Paper Co.	Specialty papers	1,400
Williams Press, Inc.	Magazine-printing	1,200
Tobin Packing Co., Inc.	Meat-packing	1,100
Minnesota Mining & Manufacturing Co Co., Inc.	Electrical insulating materials	1,000
Ford Motor Company	Automotive parts	800
Albany Felt Company	Industrial felts	800
West Virginia Pulp & Paper Co.	Paper and paperboard	800
Van Raalte Co., Inc.	Knit fabrics	700
General Aniline & Film Corp.	Dyestuffs	700
Marshall Ray Corp.	Men's sport coats and jackets	600
Sterling Drug, Inc.	Pharmaceuticals	600
Huyck Corporation	Papermaker's felt	600

room for optimism, for it ranked behind all seven of the major urban systems with a gain of but 13.9 per cent. (See other chapters in Part Five.)

In 1960, 21.8 per cent of the SMSA's total employment was in service industries. This places the area behind only New York among the seven urban systems. Of all workers 18.6 per cent are in wholesale and retail trade, 7.4 per cent in public utilities, 3.8 per cent in finance, insurance, and related fields, and 9.9 per cent in public administration. Because Albany is the state capital, the area ranks first in public administration among the major urban systems.

The total sales and service function of the SMSA in 1960 produced a tertiary potential (see Chapter 12) of 865,000 and a substantial tertiary surplus of 208,000. The surplus is higher than that of any of the upstate urban systems, including much larger Buffalo. It is especially high in Albany County, amounting to 220,000. Industrial Schenectady County has a small surplus of only 12,000 and Rensselaer and Saratoga counties both have deficits. Of key significance to the economic health of the urban system would seem to be Albany County's success at keeping the tertiary sector expanding.

Government and administration provide employment for large numbers of people in state offices. In fact, in 1950, 18 per cent of all employees in Albany County and 20 per cent of all employees in the city of Albany were government workers. The vast number of state employees and offices has gradually developed to the point where the centrally located State Office Building could provide only a fraction of the space required. As a result offices have been dispersed as space has become available. Recently, the state began development of a state campus near the west edge of the city which will concentrate certain governmental functions as well as educational facilities at that location. Also, massive urban renewal near the center of Albany will greatly increase office space.

All large urban systems are experiencing expansion of suburban shopping facilities, but there is one so geographically striking in Albany County—at Latham—that it merits special attention. As pointed out earlier, development of the Albany-Schenectady Urban System historically was focused at three points of a triangle, leaving the interior of the triangle relatively undeveloped. Within this triangle major north-south and east-west highways now intersect at Latham. Its location is such that it is only three or four miles west of Troy, about nine or ten from Schenectady, and some five or six from Albany. Even to a casual observer, the number of businesses is far out of proportion to the population of the immediate vicinity. The fact is that this intersection is within relatively short driving time of the large urban population on its periphery. This, together with the rapidly expanding road system, has helped to create a complex of primarily retail outlets centered on the Latham intersection. Nearly all of the businesses have opened since World War II. The variety of services and property values are continuing to soar. So rapid and sudden has been this development that only in 1953 were a planning board and building commission created to cope with such problems as water supply, sewage disposal, building codes, and zoning.

It is interesting to speculate on whether this area might develop sales and service facilities so extensive as to usurp the CBD function for the urban system as a whole. Certainly it is centrally located and its high degree of accessibility is an asset. It seems likely that for such a new CBD to emerge, however, there would have to be actual help from metropolitan-wide planning. Resistance from existing cities could be counted on to be substantial.

Problems and Prospects

Aging and congestion in the frame and middle city areas are striking problems. Perhaps, however, they have been even more widely evident in this urban system than in some others because so many central places, large and small, have their own problem areas here. Also, of course, these centers are all old by upstate standards and thus have older structures, street patterns, and the like in and near their CBDs and industrial districts. Active urban renewal and rebuilding programs are under way, especially in Albany and Troy, to solve problems of blight and provide space for new acceptable housing and centralized office space.

The tendency for what might be called overspecialization, as in the case of Schenectady's machinery emphasis, or Albany's governmental function, or the paper orientation of the northern clusters, may in some instances turn out to be a handicap. On the other hand, these three emphases, because they all occur in one urban system, actually provide a measure of diversity instead of overspecialization as far as the entire system is concerned. The more critical question is which of these three emphases, or additional ones, can be cultivated to support the growing need for jobs in the area. Evidence would suggest that substantial increases in employment in machinery or paper plants, or other manufacturing lines for that matter, are unlikely. Stability and some growth will undoubtedly accrue from Albany's governmental function, for expansion in government is the order of the modern day. Prospects for additional growth would seem to be tied up with expansion of facilities for higher education and with the establishment of the urban system as a more rapidly growing trade center serving a huge area lying between New York to the south, Syracuse to the west, Montreal to the north, and Springfield to the east. Figure 117 graphically portrays the Albany-Schenectady Urban System's wide reach in New York State. Trade, particularly that related to Port functions, should increase. The concern that the St. Lawrence Seaway would depreciate it as a handler of bulk cargoes has proved largely unfounded, and now the deepening of the river and general improvement of Port facilities may enhance its future as a true seaport.

The favorable transport conditions that have contributed so extensively to development are still pretty much present, and the status of the urban system as a second-order trade center almost assures substantial population growth in the future. Both economic and aesthetic assets are present. Certainly the high degree of access to both New York City and some of the most attractive mountain and lake country in the Northeast must be numbered among these assets. The population of the SMSA, at 658,000 in 1960, should reach 910,000 in 1980 and be roughly 1,270,000 by the year 2000. If these predictions approximate actual trends, the Albany-Schenectady Urban System of the year 2000 will be far larger than Utica and Binghamton but significantly smaller than Buffalo, Syracuse, and Rochester.

Selected References

Beers, S. N., and D. G. Beers. *New Topographical Atlas of the Counties of Albany and Schenectady, N.Y.* Philadelphia: Stone and Stewart, 1866.

Bureau of Municipal Research. *Longterm Financial and Improvement Program for the City of Schenectady, New York.* Schenectady: Prepared by the Capital Budget Committee, 1929.

Carmer, Carl. *The Hudson.* ("The Rivers of America Series.") New York: Rinehart, 1948.

City Planning Commission. *Preliminary Report on Major Streets, Transit, Parks and Playgrounds.* Schenectady: The Commission, 1924.

Hislop, Codman. *The Mohawk.* ("The Rivers of America Series.") New York: Farrar and Rinehart, 1939.

Howell, G. R., and Jonathan Tenney. *Bi-centennial History of Albany; History of the County of Albany, N.Y., from 1609 to 1886.* New York: W. W. Munsell, 1886.

Munsell, Joel. *The Annals of Albany.* 10 vols. Albany: J. Munsell, 1850-59.

Temporary State Commission on the Capital City. *Report, 1961-62.* Albany: The Commission, 1962.

Vosburgh, Frederick. "Drums to Dynamos on the Mohawk," *National Geographic Magazine*, XCII (July, 1947), 67-110.

Weise, A. J. *The History of the City of Albany, New York, from the Discovery of the Great River in 1524, by Verrazano, to the Present Time.* Albany: E. H. Bender, 1884.

——. *History of the City of Troy from the Expulsion of the Mohegan Indians to the Present Centennial Year of the Independence of the United States of America,* 1876. Troy: W. H. Young, 1876.

CHAPTER 18   JOSEPH E. VAN RIPER

Binghamton

The Binghamton Urban System is situated in the southern part of the Appalachian Country and is generally regarded as the principal trade center of the Southern Tier. Its location apart from the Hudson–Mohawk–Lake Plain axis makes it unique among the major urban systems of the state. It is a valley community, and its form, function, and unique developmental problems are intimately related to the characteristics of the local terrain.

There are three relatively major central places: Binghamton (76,000), Endicott (19,000), and Johnson City (19,000). In addition to these, which give the area its local name, the Triple Cities, there are several secondary centers, Vestal, Endwell, Chenango Bridge, and Hillcrest being among the largest.

The urbanized area (Fig. 122) extends along the river valleys and spills up onto the hill country in various directions. In 1960 it contained 158,000 people, 44,000 of whom were outside the incorporated limits of the three major central places. The urban frontier encloses an area of about 178 square miles which had an estimated 1960 population of 193,000. Its limits are roughly twenty miles apart in an east–west direction and ten miles apart north to south. The river valleys have had some effect on its shape, but the forces of urbanization have extended well into the uplands. Outside the urbanized area but inside the urban frontier residential construction is scattered along roads wherever they traverse the hill country. That portion of the SMSA in New

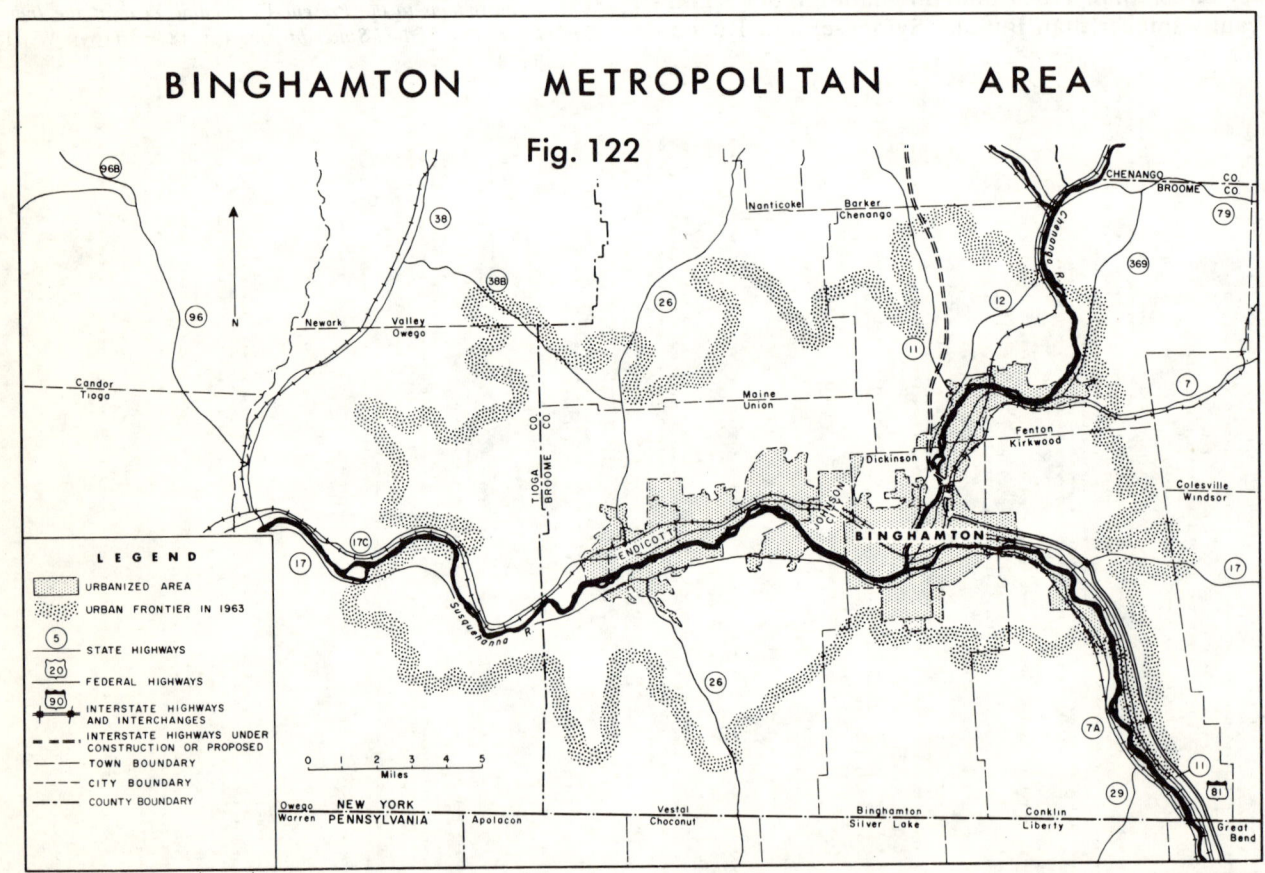

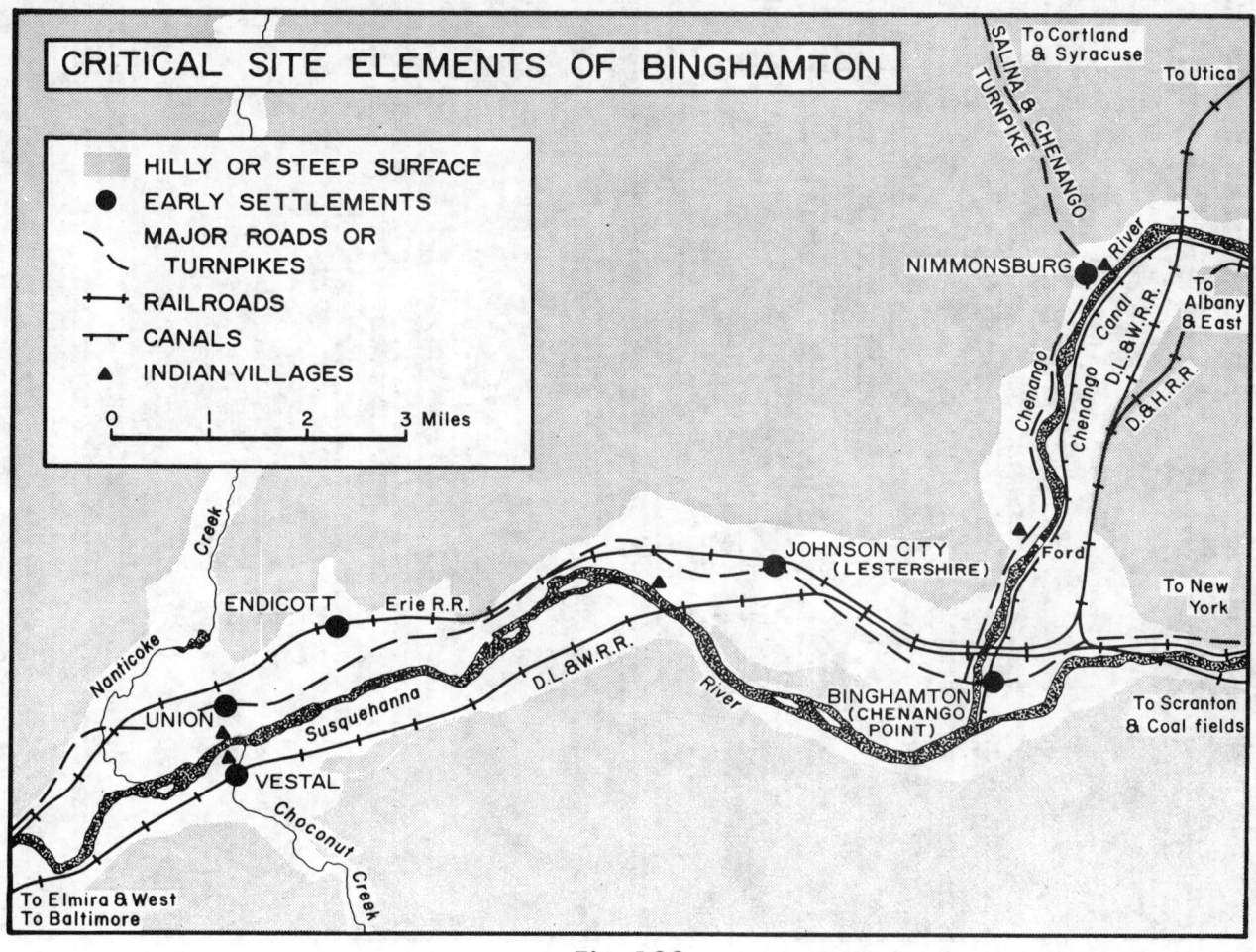

Fig. 123

York State is composed of Broome and Tioga counties. The 1960 population had reached 250,000, having increased by 48,000, or 25 per cent, since 1940. This is less impressive growth than enjoyed by Syracuse, for example.

In general characteristics the urban system is much like the other major systems in the state. It has its CBD and frame, its network of arterial highways and back streets, its sprawling mass of suburban residences, shopping centers, and outdoor movies. It is an important industrial center and as such is a noteworthy unit of the American Manufacturing Belt. Into and out of its factories pour a wide variety of materials and products. Parking lots for automobiles near factories and railroad tracks and a broad swath of factory roofs stand out in aerial photographs. Binghamton differs somewhat from the other urban systems in physical setting, historical development, and specific economic orientation.

Historical Development

Four factors have had important influence on the geography of the Binghamton Urban System. In the first place, terrain has strongly influenced settlement patterns and routes of access from Indian days to the present; second, manufacturing has been unusually outstanding in the economy since early white settlement; third, there have been close social and economic interrelationships with the Scranton-Wilkes Barre area of Pennsylvania; and fourth, amenable labor-management relations within the major industries have produced a consistently favorable labor market.

THE ROLE OF TERRAIN

Terrain in the Binghamton vicinity is typical of the Appalachian Upland. Valley bottom land along the Susquehanna River, containing the major

This Overview of Binghamton (looking northeast) shows the CBD that has developed near the junction of the Chenango River (left) and the Susquehanna (right). In the distance can be seen the rolling upland surface of the Appalachian Plateau. Note how in some sections the residential property is working its way up the side slopes of the valley. *Bob Garvin, Airphoto Associates.*

portion of the present urbanized area, is approximately a mile wide, and lies at an elevation of about 850 feet above sea level, and from 400 to 800 feet below the summits of the adjacent hills (Fig. 123). Several tributary valleys, as well as the main valley, contained important Iroquois trails prior to white settlement, and today are closely followed by state or federal highways.

The focusing of valley routes from so many directions favored the development of nucleated settlements, even during the Indian period. The most favored kind of site for an Indian village was on a gravel terrace, above flood level, near a suitable ford across the river, and near the mouth of one of the tributary valleys. Taking advantage of these kinds of sites, Indians built their villages at the present location of Vestal, western Endicott, Johnson City, and near what is now Nimmonsburg. The latter was the largest, and bore the Indian name Ochenang (bull thistle). It lay near an easy ford across the Chenango River and the main Susquehanna east–west trail, with ready access to the whole Indian trail system. Here was the junction of the routes from the north and northeast with the main east–west valley route.

The first white settlement took place shortly after the Sullivan-Clinton military campaign against the Iroquois (1779). The Indian villages were burned during the campaign, but their sites, represented by grassy glades in the midst of the forest, were selected as the first white settlements. Ochenang became Chenango Village, and the two village sites opposite each other across the Susquehanna later became Vestal and Union (the name Union was later changed to Endicott).

The site of present-day Binghamton, at the junction of the Chenango and Susquehanna rivers, was not used until 1800, or about fifteen years after the first pioneers entered the area. By this time Chenango Village, two to three miles to the north, had become a thriving little community of a couple of dozen families. The story of the origin of Binghamton well illustrates how less than the most obvious reasons can determine the specific site of a major city.

The present location of the Binghamton CBD was originally avoided because the river crossings there (on both the Chenango and Susquehanna) were relatively poor for fording by wagons. At an early time a wealthy Philadelphia land broker and diplomat, William Bingham, after whom Binghamton was named, secured title to a large block of land in southern New York and northern Pennsylvania. The northern boundary of the Bingham Patent was located about a mile or two south of Chenango Village. His local business representative, Joshua Whitney III, an enterprising and loyal young man, prematurely but bravely announced plans for a bridge to be built across the Chenango on his employer's land. He even organized a tree-cutting bee as a mark of his good intentions. Although the bridge was not completed until eight years later (1808), several settlers purchased parcels of land near the bridge site. Once the bridge was completed, Chenango Point (the original name of the settlement, see Fig. 123) soon replaced the older Chenango Village to the north as the commercial nucleus of the area. This was primarily because the main east-west post road in the Susquehanna Valley had to bend well to the north to take advantage of the ford at Chenango Village. The bridge, at what was later to become Binghamton, thus provided a short-cut for east-west travel along the Susquehanna Valley.

The rapid growth of Binghamton itself is illustrated by Table 43. Only ten years after the first parceling of land in Chenango Point, and two years after the construction of the first bridge, the community had two hundred residents.

The influence of surface configuration on the historical development of the area continued long after pioneer settlement. The early popularity and success of the Erie Canal (opened in 1825) and other waterways as routes for bulk transport led to a search for a north–south route connecting the Erie Canal with the Susquehanna River. Several routes were proposed, but the one finally selected for construction followed the valley of the Chenango River from its mouth at Binghamton to its headwaters, thence northward to Utica. The Chenango Canal was completed in 1837 after three years of construction work. At first its main function was to carry lumber from the mills near Binghamton. Coal from the Scranton area of Pennsylvania soon replaced lumber as the major cargo. This was first carried overland between Scranton and Binghamton by wagons. The canal terminus at Binghamton provided an important stimulus to the commercial growth of the community, but its influence waned rapidly after the tapping of the area by three railroads. The first rail line to be built was the Erie, and the first train from New York City entered Binghamton in December of 1848. The Delaware, Lackawanna & Western and the Albany & Susquehanna (later to become the Delaware & Hudson) were built immediately after the Civil War. Today, reflecting mergers and line abandonment, the urban system is served by two railroads, the Erie-Lackawanna and the Delaware & Hudson.

Terrain has also strongly affected directions of expansion and distribution of built-up areas. Except for a few small sections, the contiguously built-up areas are still confined to the valley floor and the lower slopes. The greatest pressure for suburban development on the steeper valley sides is immediately adjacent to the city of Binghamton. Two considerations have tended to discourage expansion of dense development up the valley sides and into the uplands. The first is the difficulty of access during the winter following heavy snow or ice storms. The second is the effect of soil creep on these slopes, and the resulting damage to basements and foundations. Most of the valley sides are heavily plastered with glacial till, which contains large amounts of clay, and an extremely slow, insidious movement downward is almost inevitable. Terracing into the till tends to reduce but not eliminate this problem. While the uplands within the urban frontier now have a large number of residences, few contiguously built-up areas have developed. Instead, homes are scattered along the hard-surfaced upland roads, generally on lots of from one to ten acres.

MANUFACTURING, THE DOMINANT FUNCTION

Binghamton and the surrounding communities developed an important industrial function almost

TABLE 43
POPULATION OF BINGHAMTON
FOR SELECTED YEARS

Year	Population
1810	200
1834	1,500
1855	9,000
1867	11,000
1880	18,000
1900	40,000
1930	76,662
1950	80,674
1960	76,000

WEST END OF THE BINGHAMTON URBAN SYSTEM, the Endicott-Vestal Area. Endicott lies on the far, or north, side of the Susquehanna River. Largely because of terrain conditions, the north side of the valley has been more broadly developed than the south side, where steep slopes 400 to 500 feet high (shown as wooded areas) have made expansion away from the river difficult. *Bob Garvin, Airphoto Associates.*

as soon as they were formed. At first, as was the case in so many communities on the American frontier, milling was the principal industry. Grain had to be milled to provide flour for local use, and lumber mills producing rough lumber turned out the first major product to be shipped from the area. Diversification of industry and the production of specialty products began only ten years or so after white men arrived. By the time of the Civil War, or only fifty years after the original settlement, Binghamton was a full-fledged manufacturing community, producing plows, carriages, sleighs, cigars, flour, shoes, plaster, and lumber. An ironworks was set up in the 1860's.

Shoe manufacture, which later was to become the leading industry, started in 1854 with the Lester Shoe Company. In 1888 this shoe company expanded its production by building a new plant in Lestershire (now Johnson City). Hired as a foreman in the new plant was a New England shoe worker, George F. Johnson, who later persuaded another New Englander, John B. Endicott, to finance a new, modern plant in the western part of the county. The Endicott-Johnson Corporation now is one of the three largest shoe-manufacturing companies in the United States, employing about fifteen thousand workers in twenty-eight plants in the urban system. The early importance of shoemaking resulted in a significant subsidiary industry, the tanning of leather. In early days, hemlock bark was the main source of tannin, and hemlock abounded on the forested slopes throughout the southern part of the state. In the 1860's and 1870's long lines of wagons laden with tanbark could be seen converging from all directions on the local tanneries. The shoe industry today produces its own leather in local tanneries, but uses imported quebracho bark and other tanning materials.

Cigar-making was another industrial specialty of Binghamton that developed in the mid-1800's. In

1880 it was one of the leading cigar-manufacturing cities in the entire nation, and by the end of the century there were thirty factories, employing some five thousand workers locally. Much of the tobacco used was grown along the bottom lands of the Susquehanna Valley from east of Binghamton westward to beyond Owego. The competition of cigarettes and the shift of cigar-leaf production to other areas of the United States was disastrous to Binghamton's cigar industry, and today neither tobacco nor cigars are produced there.

Other large industries began about the turn of the century. The forerunner of the International Business Machines Corporation was the Bundy Time Recorder Company, which originated in 1888 and employed eight men to develop a time-recording device. It moved from Binghamton to Endicott in 1902 and remained a small local specialty industry until Thomas J. Watson, a superb salesman, joined the company and used his persuasive powers to develop its expanding horizons. The photographic industry, represented today by the Ansco and Ozalid companies, began in 1902 with a small partnership, Anthony and Scoville, Inc., which produced film under a new patent. This firm later became the well-known Ansco Corporation.

Relationship with the Scranton-Wilkes Barre Area of Pennsylvania

Throughout much of its history, the Binghamton Urban System has had a close interregional relationship with the Scranton-Wilkes Barre anthracite area in the Wyoming Valley of Pennsylvania. Initially, much of the settlement in the Binghamton area was made by way of Pennsylvania, the settlers moving up the Susquehanna and Delaware drainage basins. Many of these settlers were New Englanders, and owing to the antagonistic Dutch in the Hudson Valley and the rugged terrain of the Catskills to the east, they found it easier to reach newly purchased acres in southern New York by moving north from Philadelphia. The distance from Scranton to Binghamton by way of the Susquehanna River is almost three times as great as the direct route overland, so the latter route became a well-traveled road from the end of the eighteenth century. Mention already has been made of Binghamton as a major distribution point for Wyoming Valley anthracite during the 1800's by way of canal and railroad.

The most significant relationship between the two areas, however, was to come during the present century. This has been the migration of people from the Wyoming Valley to the Triple Cities as employment in the anthracite mines declined. The replacement of coal by oil and gas for domestic heating led to widespread unemployment in the coal fields. Fortunately, the expansion of the new industries to the north in the Binghamton area provided a market for this labor surplus, and Pennsylvanians swarmed across the border during the 1920's and 1930's. Although the rate of migration has slowed, the movement continues, and welfare offices currently report that a considerable proportion of their new welfare cases are recent immigrants from Pennsylvania. The social and cultural alignment with the Pennsylvania anthracite area is illustrated in many ways. For example, some people drive to Scranton to shop, although shopping facilities generally are good in the Triple Cities. The Scranton *Times* has a surprisingly large circulation in Broome County, and Binghamton radio and television stations regularly present Wyoming Valley news.

Favorable Labor-Management Relations

Perhaps one of the most significant features of the Binghamton industrial scene in recent decades has been labor-management harmony. The role played by labor unions has been relatively small. Despite strenuous organizational attempts on the part of the unions, they have not as yet been able to organize the largest companies. Newer industries have been unable on the whole to resist union organization, but in general the unions that are active in the area get along relatively well with management.

Causes for labor-management peace may be traced to the labor policies of Charles F. Johnson, one of the founders of the Endicott-Johnson Corporation, and Thomas J. Watson, who developed I.B.M. into a national industrial giant. Both of these men firmly believed that a satisfied factory worker is an efficient worker; therefore they provided a wide range of fringe benefits, including free medical care, liberal vacation policies, Christmas bonuses, housing developments, and many others. Even more important, however, were carefully developed programs designed to increase morale, achieve close personal relations between management and labor, and reinforce the workers' personal identification with company welfare. These programs have been implemented by numerous mass meetings, company publications, dinners, and gala celebrations.

Such policy made an enormous impression upon an immigrant labor force that had been accustomed to domination by a social or industrial aristocracy. Many of the newcomers were recruited from the shoe factories of Czechoslovakia after World War I. Others were new Italian or Polish immigrants. Swelling the industrial ranks later were the growing numbers of discouraged Pennsylvanians from the labor surplus anthracite area. To all of these newcomers the management policies seemed to warrant support and loyalty.

Some Details of the Current Geography

The densely built-up parts of the Binghamton Urban System today form a long and narrow zone with an enlargement near its eastern end where the junction of the Chenango River Valley adds to the amount of level land for urban development.

The principal central business district of the urban system at the original site of Chenango Point contains the typical department stores, banking and other financial services, and specialty retail stores. The tallest buildings are twelve to fifteen stories high.

Endicott and Johnson City have developed CBDs of their own, although the smaller volume of their central place function is reflected in the lower buildings, smaller stores, the lesser variety of retail specialties, and the general absence of wholesale establishments. These secondary CBDs are simple and tend to be linear, following major through streets.

Binghamton is large enough to have at least the rudiments of internal areal specialization. Although not sharply defined, there are a banking district, a theater district, a wholesale services district, an auto sales district, and others. The residential areas of Binghamton also tend to be arranged in distinct zones of different quality housing, with the lowest quality being in the frame zone or near factories or railroads. The range and zonation of residential quality are much less in Endicott and Johnson City.

Industrial plants in the urban system until recently have been located mainly along the railroads. Trucking now, however, makes it possible for plants that do not require large quantities of bulk materials to locate almost anywhere that is not zoned against them. Decentralization of industry has been pronounced; in fact, there is considerably more industrial employment in Endicott and Johnson City than in Binghamton, the latter's function being more commercial than industrial. Major industries, especially in the electronics field, have abandoned production of minor items used in their line products, preferring to purchase these from small independent manufacturers. Thus, a number of satellite-manufacturing plants have located in the area, and because of their lack of dependence on heavy or bulky commodities, may be found along the major highways outside the older, congested industrial areas. The most common location of these satellite concerns is along the major east-west highway south of the Susquehanna River. Both I.B.M. and Endicott-Johnson are decentralized locally, with plants and installations spread throughout the area.

A number of gaps occur within the built-up area of the valley. The largest of these involves low ground that has been previously susceptible to floods. A system of dikes and flood walls recently built by the United States Corps of Engineers is intended as a solution to this flood problem. The general process of removing till or gravel from terraces and benches along the sides of the valley and depositing it in low, poorly drained depressions in the valley floor provides some new land each year, but space for urban expansion on level valley-bottom land is becoming scarce.

The network of roads and streets clearly illustrates the influence of land surface configuration and the major rivers. On level land the alignment is roughly rectangular, with the main arterial routes extending along the east-west valley axis, and the other roads and streets forming a grid pattern based on them. In the hilly areas, however, valleys strongly influence the location and direction of roads.

The whole urban system is undergoing considerable geographic modification in population, with declines in the older, congested portions and rapid growths in the outlying suburban sections. Population trends for the various political units are shown in Table 44.

Major growth trends may be summarized as follows:

1. The several termini of the built-up area, contained within the gentler slopes of the major valleys, are growing most rapidly. The southwestern sector (Vestal) is experiencing singularly rapid growth. Expansion also is substantial up the Chenango River Valley in the vicinity of Chenango Bridge, southeast of Binghamton in the Conklin area and on the gentler south-facing slopes northeast of Endicott in the Endwell area.

2. Population is beginning to decline appreciably within Johnson City and Endicott and certain parts of the middle city of Binghamton. This is because of the shortage of vacant lots and expansion of commercial and industrial functions at the expense of residences.

3. Population growth is occurring along hard-surfaced roads on the uplands, especially within the urban frontier.

Accessibility to jobs and to the urban service centers continues to be important. This helps to explain why the residences persist in marching up the slopes adjacent to Binghamton and why new bridges across the Susquehanna in the western end of the county produce almost explosive population growth reactions. The new state highway bridge at Johnson City, built in 1955, has been largely responsible for the remarkable recent growth of Vestal. Two new, wide highway bridges are scheduled for early completion in the Endicott area. When these are completed, the broad flats and

TABLE 44
POPULATION TRENDS WITHIN THE BINGHAMTON URBAN SYSTEM
(By Political Units)

	1930	1940	1950	1957	1960
Broome County	147,022	165,749	184,698		211,366
Binghamton (City)	76,662	78,309	80,674		75,085
Binghamton (Town)	1,092	1,576	2,073	2,988	3,407
Chenango (Town)	2,074	3,265	5,747	8,601	9,459
Conklin (Town)	1,332	2,156	2,872	3,751	4,335
Dickinson (Town)	4,255	5,060	5,450		6,398
Union (Town)	42,579	50,195	55,676	62,741	64,265
Endicott (Village)	16,231	17,702	20,050	19,636	18,708
Johnson City (Village)	13,567	15,039	19,249	19,398	19,062
Vestal (Town)	2,848	5,710	8,902	14,107	16,804
Kirkwood (Town)	1,237	2,150	2,997		4,669

Source: U.S. Census Reports

gravel terraces west of Vestal and south of the river will provide ideal areas for further urban expansion. It is likely that the urbanized area, within ten to twenty years, will include the village of Owego, which lies about twelve miles west of Endicott. The erection of a new, huge I.B.M. plant in Owego in 1958 also has tended to direct major expansion westward.

Current Livelihood Structure

In 1960 the Binghamton SMSA had 3 per cent of its labor force in the primary sector, 51 per cent in the secondary sector, and 46 per cent in the tertiary sector. Its orientation toward the secondary sector is greater than that of any of the other major urban systems in the state and its orientation toward the tertiary sector is less than that of any of the other systems. In view of the growing general importance of tertiary activities, it would seem that expansion of this type would be advantageous.

Manufacturing

Between 1947 and 1963 manufacturing employed between 36,000 and 42,000 persons; in 1947 the figure was just under 36,000. Consistent growth in manufacturing has not been evident, although between 1950 and 1960 there was a 10 per cent gain.

Most of the manufacturing can be classified as light, with factories generally using semiprocessed materials. This is in contrast to heavy manufacturing, such as the steel industry of Buffalo, which uses bulky raw materials. Ordinarily at least a quarter of all factory workers are engaged in the production of leather goods. This segment of manufacturing is dominated by one of the nation's largest shoe manufacturers. The second most significant type of manufacturing is high-value machinery, including electrical machinery; this is followed by instruments and related products, including photographic equipment and films. Light transportation equipment, food and kindred products, printing and publishing, paper and allied products, and furniture are important types of production, too.

There were sixty-two establishments in 1958 which employed over one hundred workers each, but a few very large concerns dominated. Leading manufacturers, their principal products, and approximate recent employment are shown in Table 45. The big five—Endicott-Johnson, I.B.M., Ansco, General Electric, and Link Aviation—account for at least three-fourths of manufacturing employment in the SMSA. Economic success of the whole area will be strongly affected by the employment trends of these five companies.

Tertiary Activities

Binghamton does not depend as heavily on tertiary employment as the other major urban systems, but that is not to suggest that the tertiary sector is unimportant. Of the 97,000 employed workers in the SMSA in 1960, roughly 45,000 were employed in the tertiary sector. During the decade 1950–60 gain in tertiary employment was 17.7 per cent. Being a third-order, rather than a second-order, trade center means that neither the geographic reach nor the variety of sales and service functions of Binghamton anywhere near matches that of Syracuse or Rochester. On the other hand it is the major central place of the Southern Tier and will continue to dominate the trade and service function of that area plus a part of northern Pennsylvania.

In 1960, 19.1 per cent of the SMSA's total employed persons worked in service industries, 15.7 per cent in retail and wholesale trade, 5.0 per cent

TABLE 45
MAJOR MANUFACTURING CONCERNS OF THE BINGHAMTON SMSA

Company	Products	Approx. Employment Early 1965
Endicott-Johnson Corp.	Shoes, leather products	14,000–15,000
International Business Machines	Electronic apparatus, etc.	8,000–8,500
Photo & Reproduction Div. of Gen. Aniline & Film Corp. (Ansco)	Films, cameras, etc.	3,700
General Electric Corp.	Aircraft armaments	2,400
Link Aviation	Flight simulators, etc.	1,700
Kroehler Mfg. Co.	Furniture	700
Vail-Ballou Press	Printing	600
Cadre Industries Corp.	Cables, harnesses	500
Crowleys Milk Co.	Milk and dairy products	400
Dunn and McCarthy, Inc.	Shoes	400
Binghamton Press	Printing	300
Endicott Forging & Mfg. Co.	Forgings	300
The Fairbanks Co.	Valves	300
Kason Hardware Corp.	Hardware	300
Kupfrian Mfg. Div.	Shock mounts, copper tubing	200

in public utilities, 3.0 per cent in finance, insurance, and related fields, and 3.2 per cent in public administration. The total sales and service function in 1960 produced a tertiary potential (see Chapter 12) of 261,000 but a tertiary surplus of only about 10,000.

External Functions and Relationships

No large urban system functions entirely as an independent unit serving only the population within its borders or even its immediate tributary trading area. The products of factories in the Binghamton Urban System are distributed throughout the United States, and the railroad cars and trucks that bring raw materials into the area have come from distant and diverse shipping points. Binghamton benefits from its position on major rail facilities leading from New York City (Fig. 85). The Erie-Lackawanna system provides direct access not only to New York City but to the steel centers in Ohio and beyond, to the docks of Hoboken, to the coal mines of Scranton, and to the heavy industries of Buffalo.

The importance of these through rail connections can be seen in the development of industrial communities at intervals of about thirty to fifty miles along them. Elmira, Corning, Olean, Salamanca, and Jamestown follow in order across the southern portion of the state. With the Scranton-Wilkes Barre and Binghamton areas, these communities form an industrial corridor connecting the Pennsylvania and Ohio steel cities with the New York metropolitan area.

Three commercial passenger air lines and a freight airline provide excellent air service. The Broome County Airport is somewhat unique in that it lies at one of the highest elevations in the county (1,600 feet above sea level). When many of the valley landing strips are closed because of lowland radiation fogs, this field lies high and clear.

Part of the growth of the industries and communities in the Binghamton area definitely is related to the continued growth of New York City and its services. As the New York area grows, it stimulates neighboring communities to a degree that increases with nearness and with effectiveness of the transportation and communication system. Binghamton is only five to six hours from Gotham by train or automobile, and only forty minutes by air. If it is easy and desirable for the average reisdent of Binghamton or Endicott to spend a week end in New York shopping or enjoying the sights, it is similarly easy for the New York investment broker to become optimistic over the opportunity for a new electronics plant within easy supervisory reach and control.

Problems and Prospects

Some of the problems with which the Binghamton Urban System must contend are common to most expanding urban areas; others are uniquely local. The problems of deterioration in the older central areas that accompany increased age and congestion are found in Binghamton as they are in nearly all northeastern United States cities. The high costs of demolition, slum clearance, and street-widening are continuing challenges to the city of Binghamton. In recent years they also have begun to be felt in Endicott and Johnson City.

A major problem is providing cohesiveness to an urbanized area that is becoming unusually long, narrow, and politically unwieldy. Access to the CBD from the urban frontier is not easy, and continued improvement of arterial routes is needed. The erection of neighborhood shopping centers is proceeding slowly, but these supplement, and do not replace, the centralized downtown services. The CBD of Binghamton needs to be enlarged by vertical growth, and there is a serious shortage of off-street parking. A new limited-access highway system that will separate all through traffic from the local flow is now under construction and will be completed during the 1960's. This should greatly relieve a serious clogging of the present east–west routes through the valley.

The growing tendency for spread of residential areas up valley slopes and onto the uplands, especially when the spread proceeds without subdivision control, should be carefully watched. Health hazards resulting from inadequate functioning of septic tanks may become locally acute and, as pointed out above, damage to homes on steep clay slopes owing to cracked foundations and basement flooding during spring thaws and heavy thunderstorms is likely to continue.

Most of the upland has clay hardpans a short distance below the surface, and the general imperviousness interferes with septic tank performance. There are problems with wells, too. Shallow surface wells are exceedingly susceptible to pollution unless they are far from sources of contamination and often are not generous producers. Deeper and safer water sources may be found by drilling wells into the underlying strata of siltstones, sandstones, and shales, but such wells, which may have to be several hundred feet deep, often produce water too high in mineral content to be palatable. Poorly drained upland soils, furthermore, are generally poor for agriculture, and former farms, because of their low value, have been commonly divided into relatively large lot sizes for suburban residential use.

Perhaps the most difficult problem to face is the integration of local governmental functions. As the urban complex grows, it engulfs many separate local governmental units. Few of these are willing to give up political sovereignty or presumed noteworthy financial advantages. Thus there is much duplication of services with resultant inefficiency. Separate highway departments, for example, are maintained by the city of Binghamton, the State of New York, Broome County, and each of the towns. A state snowplow after a winter storm raises and lowers its plow over and over again as it passes through areas of different political jurisdiction along the main highways through the metropolitan area.

Endicott and Johnson City used to view gleefully the financial woes of Binghamton, caught in the dilemma of declining assessed valuations and mounting public service costs. Cushioned with the large tax payments of the big industrial corporations and with only moderate local services, they were until recently favored by extremely low tax rates. Now, however, unable to expand industrially within their restricted borders, facing declining residential property valuations, and with the cost of their municipal services increasing, Endicott and Johnson City view the rapidly growing suburban areas beyond their village limits as potential sources of municipal revenue and are becoming interested in annexation. The suburban areas, on the other hand, have problems of their own and do not relish the idea of helping their older neighbors. Localized civic pride may be an advantage, but in a highly fragmented form it is an expensive luxury.

The Broome County government has been unusually progressive. It has financed the local air terminal, entered into a joint venture with the State of New York in building and maintaining a new two-year technical institute, and supports an active and excellent county planning board. A rural-urban division of the County Board of Supervisors, however, in which the balance of voting power still resides in the town supervisors, is a deterrent to more active implementation of urban needs. Representation on the board has in the past been anything but proportional. The town of Union, for example, with a population of about 63,000, has only one supervisor, as does the town of Nanticoke, with only 675 people. This is a nation-wide problem and will presumably be corrected as the result of recent court decisions.

The manufacturing base can be expected to survive, but it will be relatively less significant in the future economy. More manufacturing diversity and less dominance of a few very large companies would seem to be in order. Unionization probably will increase and thereby lessen what has been viewed as a local labor advantage. On the other hand, the general labor population should retain for some years its long-standing tendency to be fairly tolerant of management. Competitive wages and salaries in electronic and other high-value-added industries make for stiff labor competition for the shoe industry. To retain its position both nationally and locally, the shoe industry will have to modernize its plants and management policies and introduce labor-saving devices wherever possible. Such changes are already taking place. Increasing orientation toward high-value-added industrial types will strengthen the manufacturing base.

As pointed out above, the Binghamton Urban System, being less oriented toward the tertiary sector than it might hopefully be, may not experience quite as favorable over-all growth rates as some of the other urban systems in the state. The population of the SMSA, at 251,000 in 1960, should reach 350,000 by 1980 and be roughly 480,000 by the year 2000.

Selected References

Blair Associates. *Central Binghamton 1970: A General Neighborhood Renewal Plan for the Central Area of Binghamton, N.Y. (Project NYR-60 [GN].)* Syracuse: Prepared for the city of Binghamton in cooperation with the Broome County Planning Board and Urban Renewal Administration, 1961.

Broome County Planning Board. *Economic Base of Broome County*. Binghamton: The Board, 1953.

———. *Physical and Historical Background of Broome County*. Binghamton: The Board, 1953.

———. *Use of Land in Broome County*. Binghamton: The Board, 1953.

Merrill, Arch. *Southern Tier*. 2 vols. New York: American Book-Stratford Press, 1952.

Seward, William Foote (ed.). *Binghamton and Broome County, New York: A History*. 3 vols. New York and Chicago: Lewis Historical Publishing Co., 1924.

Smith, Henry P. *History of Broome County*. Syracuse: D. Mason and Co., 1885.

CHAPTER 19   KATHERYNE THOMAS WHITTEMORE

Buffalo

THE Buffalo Urban System is located where the Upstate Heartland meets Lake Erie, and functions as the western anchor of the Hudson–Mohawk–Lake Plain axis. Its historically favorable geographic location relative to transportation and cheap hydroelectric power has made it a major transshipment point and industrial center.

The city of Buffalo, with a 1960 population of 533,000, is by far the largest central place, but Niagara Falls (102,000) is a major secondary core within the urban system. In addition, the 1960 census lists some dozen other incorporated and unincorporated places with over 10,000 people each.

Figure 124 shows that the irregularly shaped urbanized area extends from the vicinity of North Evans on the south roughly 40 miles northward to Lewiston on the Niagara River. Maximum east–west dimension is reached from Buffalo to Lancaster in the central portion. There were slightly more than a million people living in this urbanized area in 1960. The urban frontier encloses a huge area of about 744 square miles and contains a population of at least 1,250,000. It averages about 40 miles north to south and 18 or 19 miles east to west, with irregularities in its outline largely reflecting degrees of highway access. The SMSA, composed of Erie and Niagara counties, contained a population of 1,307,000 in 1960, having increased by 348,000, or 36 per cent, since 1940.

An Overview

An air passenger coming into Buffalo from the west can see clearly the extent and pattern of the urban settlement at the eastern, lower end of Lake Erie. No hills obscure or complicate the scene. The city, its suburbs, and its urban fringe spread across a level plain to an even horizon. The traveler notices first the long breakwaters and the jutting piers of the outer harbor, then the grain elevators and warehouses that mark the winding inner harbor. To his right, at the southern end of the harbor, the tall smoking chimneys indicate the industrial districts of Lackawanna and Woodlawn.

Across the inner harbor rises the elevated bridge called the Buffalo Skyway; broad express highways sweep to the south, east, and north. Beyond the complex curves of these highways rise the tall buildings of the central business district. The main streets of Buffalo radiate like spokes of a half-wheel from the waterfront.

From the lake front to the airport, the air passenger crosses the greatest east–west extent of the urbanized area. As he does so, he sees beyond the CBD the older residential sections of the city spreading out to the north, east, and south.

A chartered flight in a south–north direction would pass over the full forty-mile length of the urbanized area. At the southern end, a narrow stretch of built-up land begins at North Evans and follows the lake shore as far as Woodlawn and Lackawanna, where still other suburbs extend the area to the east. To the north of Buffalo it is difficult to distinguish the city from its suburb of Kenmore and the built-up areas in the towns of Tonawanda and Amherst. Ahead are the twin cities of Tonawanda and North Tonawanda. Beyond these cities, a thin band of continuous settlement along the Niagara River connects the Tonawandas with Niagara Falls. The urbanized area broadens again at the city of Niagara Falls. This city fills in almost completely the triangle formed by the bend in the Niagara River. The built-up area extends north to include the city of Lewiston. This quarter-moon crescent of urbanization lying along Lake Erie and the Niagara River from North Evans to Lewiston is the heart of the larger Buffalo Urban System.

Outside this central zone in eastern Niagara County is Lockport, a busy manufacturing city. In Erie County, just north and west of the Appalachian Upland, are Gowanda, Hamburg, Orchard Park, and East Aurora. Southeast of this circle of cities, but still within the SMSA, are several villages in the valleys of the upland. They contain only a few small manufacturing establishments and serve mainly as rural trade centers. Almost the entire southern third of Erie County is rural, although within recent years some of the residents have been commuting daily to Buffalo and other cities within the urbanized area.

An overview of the Buffalo Urban System, then, discloses that there are features common to all

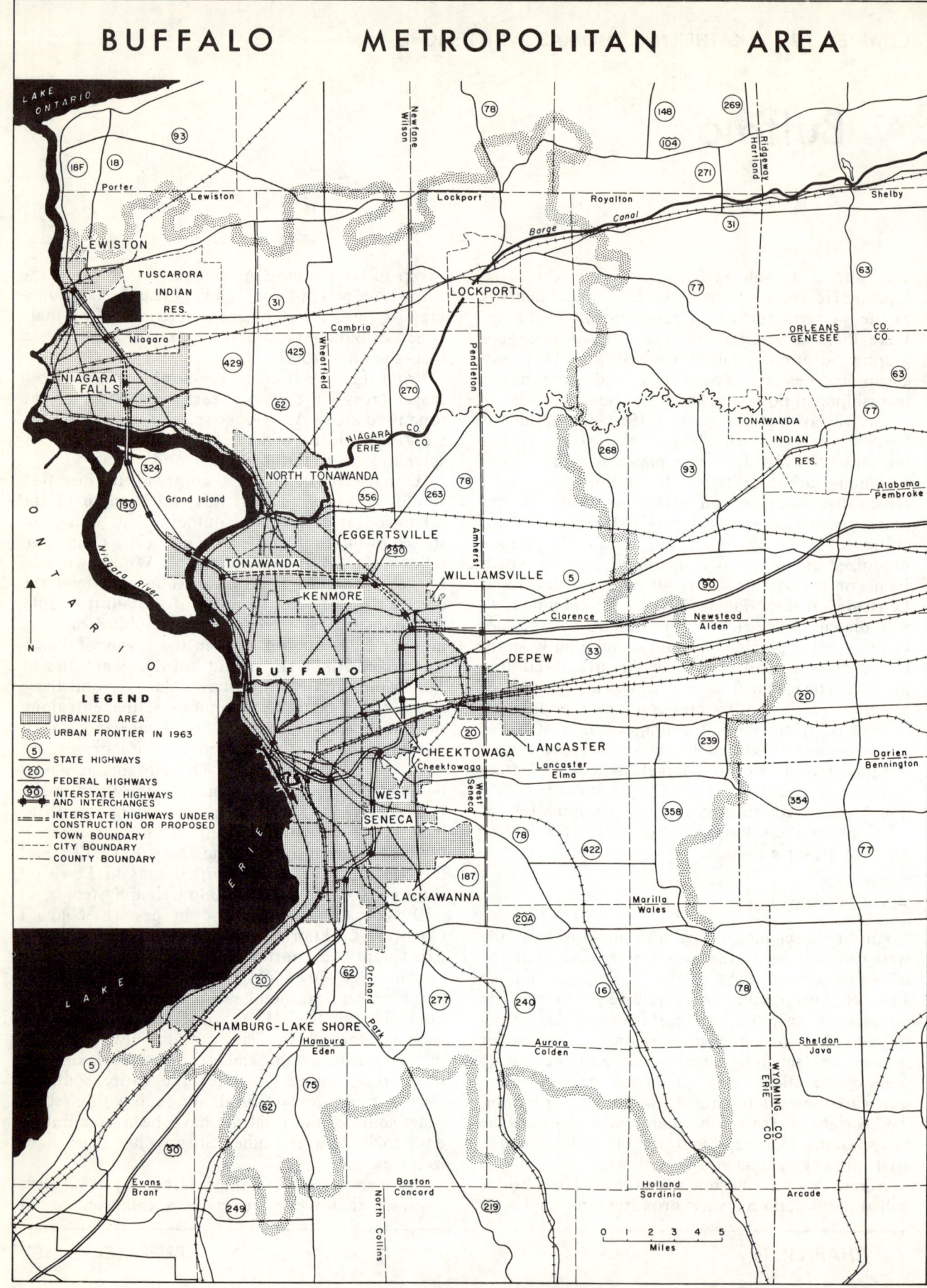

Fig. 124

BUFFALO AND ITS LAKE FRONT. Buffalo has grown and prospered where land and water transportation meet. Dominating this view are highway, rail, and dock facilities. Part of the CBD is seen on the right. High-rise apartments, originally built for use as low-income housing, are utilized by relatively high-income groups. The large building in the foreground is the War Memorial Auditorium. Terminal facilities of the New York State Barge Canal lie behind the breakwater on the left. *Buffalo Area Chamber of Commerce.*

urban systems. These include an urbanized area composed of: a large central city with its distinctly industrial, commercial, and residential zones; the older, well-established suburbs and the newer and still raw suburban developments; the outlying towns and small cities; and the outer fringe with the intermixing of urban and farm land uses.

The overview also shows several unique characteristics that differentiate the Buffalo Urban System from other urban systems in the state. One such characteristic is the crescent shape of the built-up area which results from the position of the lake and the river. The waters of Lake Erie and the Niagara River form a sharp boundary indeed, for urbanization has hardly crossed the east branch of the Niagara River onto Grand Island. Only at Niagara Falls has the river failed to halt the westward spread of urbanism; there twin cities face each other across the Niagara gorge.

Except for the water bodies on the west, there are no physical features that limit or control expansion. The surface of the Erie–Ontario Lowland is everywhere suitable for urban use. The poor drainage condition that exists in some places has been alleviated by fill, drainage ditches, sump pumps, and storm sewers. Although the Appalachian Upland extends into southern Erie County, the area of concentrated urban population has not yet penetrated far enough south to be influenced by the change in topography. The rivers that flow across the lowland are relatively small and no barriers of any consequence to urban growth.

Residents of Buffalo or Lewiston or Orchard Park are not likely to refer to their section of New York

State as the Buffalo Urban System. They are more likely to call it the Niagara Frontier. This name is usually employed to refer to the general area adjacent to the Niagara River and sometimes includes the Canadian side of the river as well as the American side.

Historical Development

Factors that have encouraged urban development through time must be examined if the modern geography of the Buffalo Urban System is to be understood. The present is but one stage in a trend started by factors now obscure. The present urban system did not develop by a spreading out from Buffalo alone but represents a coalescence of a number of central places. Actually, at one time, advantages for city growth existed to about the same degree at a number of sites. This section on historical development will first consider the factors that encouraged urban settlement at these several sites, then examine the special advantages that brought about the ensuing patterns of growth.

Early Trade Centers—Before 1830

Settlement in western New York is recent when compared to that in the eastern part of the state. Albany was nearly two hundred years old when the first houses were built near the mouth of Buffalo Creek. While the settlement of the Hudson Valley was a part of the colonizing of the Atlantic Coast, that of western New York was a part of the post-Revolutionary movement across the Alleghenies and into the Old West.

Although all the concentrations of population—the large cities and the smaller villages as well—share the over-all locational advantages of the area, an individual complex of factors has encouraged growth in each case. Most of the early villages were trade centers that developed where a break occurred in the flow of traffic.

The first important route of travel in western New York lay in a north-south direction along the Niagara River. This route was used by travelers who had reached Lake Ontario either by way of the St. Lawrence River or the Mohawk and Oswego rivers and who were on their way to Lake Erie and the Upper Lake Country. The lower Niagara River was navigable to the northern end of the gorge. At the entrance to the gorge, cargoes were unloaded and with great difficulty carried up the Niagara Escarpment. From the top of the escarpment, portage routes on both the eastern and western sides of the Niagara River led to points above the falls where navigation was again possible. Places along this route that were strategic for military protection and for trade became the sites of small settlements. Forts were built on both sides of the Niagara River at its mouth. The ends of the portages were also natural locations for trading posts and garrisons.

Before 1800 there was little east-west movement overland, and only a thin trickle of travelers made their way on foot or by horseback over the Indian trails, the only paths available. After 1800 a considerable increase in the flow of travelers resulted from the extinguishing of the Indian titles, the surveys of the Holland Land Company, and the cutting of passable roads through the forest. The first woods road for wagons was cut through from the Genesee River to Buffalo Creek (see Fig. 125) in 1798. This road, called the Genesee Road, followed the Indian trail that lay along the well-drained crest of the Onondaga Escarpment, the present location of N.Y. Route 5. About 1804 the Ridge Road came into use. This was a natural highway located on an old barrier beach which was formed near the shore of a former glacial lake. This is the present location of U.S. Route 104 between Rochester and Lewiston.

The Genesee Road and the Ridge Road became two of the most important land routes. Travel by carriage and stagecoach was added to that by horseback and wagon. The low mileage that could be covered in a day's journey resulted in a close spacing of inns. At one time the Ridge Road averaged one inn per mile between the Genesee and Niagara rivers. At crossroads, or where an increasing number of settlers created the need for a trade center, blacksmith shops, general stores, and homes clustered around the inns.

On the Genesee Road, Clarence Hollow, now called Clarence, provides an example of one of these early centers. Encouraged by the Holland Land Company, the first settler established an inn there. Later he utilized two small streams where they dropped down the face of the Onondaga Escarpment for power to operate a gristmill and sawmill. These enterprises formed the nucleus of a settlement which grew, though slowly, with the increasing travel on the Genesee Road and with the coming of pioneer farmers.

When overland travel was light a location at the east end of Lake Erie held few advantages. Black Rock, a small settlement a few miles down the Niagara River, equaled in importance the site on Lake Erie at the mouth of Buffalo Creek (later Buffalo). At Black Rock a ferry carried travelers across to Fort Erie on the Canadian shore. In addition, the Niagara River channel location, which was sheltered by an island, afforded protection and anchorage for the few ships sailing on Lake Erie. The small harbor became the location of an early shipyard.

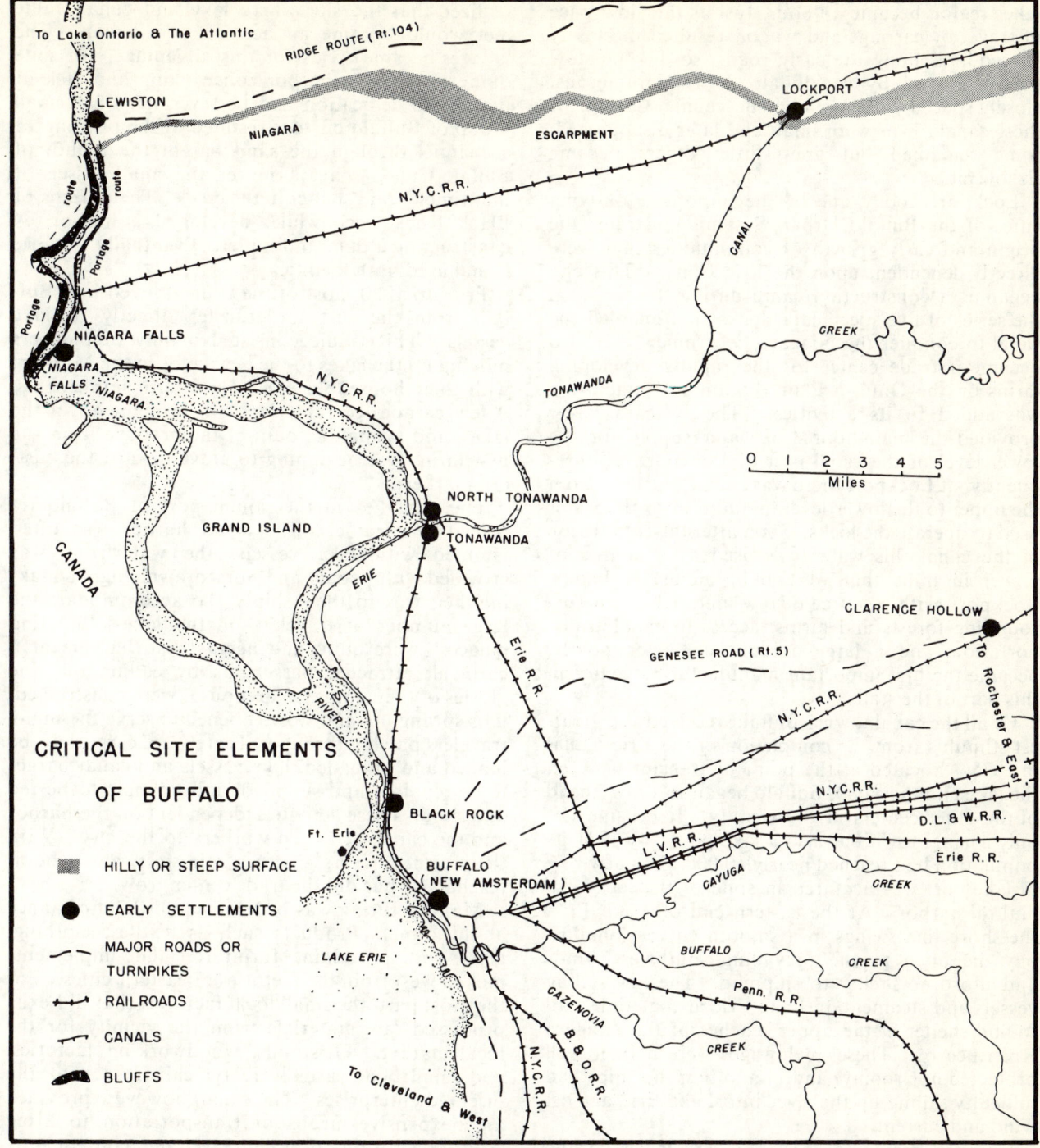

Fig. 125

Only a few houses existed at the mouth of Buffalo Creek before 1800, but 1804 brought the founding of the village of New Amsterdam, soon to be called Buffalo. Land travel and lake commerce developed slowly, and the growth of Buffalo lagged correspondingly. Nevertheless, by 1810 the population numbered 1,508. Burned during the War of 1812, the village recovered from this disaster and by 1820 had 2,095 inhabitants.

The building of the Erie Canal had much greater impact. As a result of the canal, the slow trickle of settlers into western New York and the upper lakes region became a flood. Just as the slow pace of travel by carriage and wagon resulted in closely spaced inns along the early roads, so did the slow pace of travel by canalboat result in numerous, closely spaced villages along the canal. Certain of these canal towns were small and later disappeared; some remained but grew little; others became sizable cities.

Lockport, today one of the important outlying cities of the Buffalo Urban System, illustrates the origin and early growth of communities that were directly dependent upon the Erie Canal. This city began as a construction camp during the work on the series of five spectacular locks that enabled the canal to conquer the Niagara Escarpment. It also became a trade center for the rapidly developing farms on the Ontario Plain. Soon manufacturing was added to its activities. The Niagara River provided the main source of water supply for the lower level of the canal east of Lockport. Consequently, at Lockport there was a constant flow from the upper to the lower level in addition to the water used to operate the locks. Soon after the completion of the canal this water was used as a source of power in mills that were built along its banks. Lockport, with a source of raw materials from surrounding forests and farms, access to canal transportation, and a large amount of water power, became the first important manufacturing center in this part of the state.

Of all the canal towns, Buffalo received the greatest stimulus from the completion of the Erie Canal in 1825. Located at the point of transfer between the lake and the canal, Buffalo began to draw ahead of the other centers in the vicinity. It became not only a canal town but also a lake port. By 1831 its population had reached nearly 10,000. The new role of lake port was acquired in spite of the lack of a natural harbor. At the eastern end of Lake Erie, the shore line swings in a smooth curve. Shallow bays lie open to the prevailing southwest winds and afford no shelter to shipping. The few sailing vessels and steamers on Lake Erie at that time had to find shelter in the upper reaches of the Niagara River nearby. These anchorages were neither well protected nor roomy, and in addition the ships had difficulty sailing up the river into Lake Erie against wind and current.

Near the beginning of the Niagara River lay the mouth of Buffalo Creek, a stream that rose in the hills southeast of Buffalo and meandered across the level country to Lake Erie. A sand bar near the mouth of this creek made the development of a harbor difficult. When the decision was made to extend the Erie Canal across the state to Lake Erie, and not to Lake Ontario as had first been proposed, the alert citizens of both Buffalo and Black Rock realized that the site where lake and canal traffic met would become an active trade center. The villages became rivals for this advantage. In spite of inexperience in harbor construction and lack of funds, the dedication and perseverance of a small group of Buffalo citizens resulted in the opening of a channel through the sand bar at the mouth of Buffalo Creek. This permitted the small vessels of that time to find shelter in the creek. The citizens of Black Rock, meanwhile, developed a harbor by constructing a dam and a pier. Eventually the Erie Canal used both facilities.

Prior to 1830 most of the traffic moved into Buffalo from the east for transfer directly to lake vessels. This traffic consisted mainly of pioneers moving up the lakes to the new lands in the Midwest with their household effects and farm implements. A few cargoes of furs were still brought down the lakes, and a small amount of the produce from the new farms was beginning to move toward the eastern markets.

The activities in the young port of Buffalo reflected this traffic. The small harbor was often jammed with lake vessels; the waterfront was crowded with people and horsedrawn wagons making their way to their ships. To accommodate the large number of travelers on the lakes, "floating palaces" were built, and agents, like barkers at a carnival, attracted passengers by calling out the glories of their ships. Warehouses were constructed, inns sprang up, and stores opened to serve the many travelers passing through. There were cargoes to be loaded and unloaded, lake vessels and canal barges to be repaired and supplied, and animals to be fed and shod. These activities, dependent on the harbor and the canal, attracted workers to the city. With them came lawyers, doctors, ministers, and schoolteachers. Thus did the trade center grow.

Manufacturing was of relatively little importance at this time. Products such as textiles, clothing, glass and porcelain, furniture, and implements moved west from the factories of eastern cities. For the most part, the small local factories that did exist processed raw materials from the vicinity for the local market. Gristmills, woodworking factories, and small boat yards were typical of such locally oriented enterprises. The canal, however, provided an inexpensive means of transportation to bring Adirondack iron from the east, and consequently some small foundries and machine shops were operated in Buffalo almost from the beginning of canal traffic.

The Great Growth—1830–1900

Although a few reversals resulted from panics and depressions, Buffalo grew fairly steadily as the decades of the nineteenth century passed. Transportation and trade were the great inducers of growth. From a small town crowded near its harbor, Buffalo spread north and east until it became a city of tree-shaded streets lined by elegant mansions and comfortable middle-class homes. From 10,000 inhabitants in 1831, the population had increased to 42,361 by 1851. By the turn of the century the city ranked eighth in the nation with a population of 352,387.

During the first period of settlement along the Niagara Frontier, the site of Buffalo was no more favorable than other sites, and Buffalo was but one of a number of small settlements. From 1840 to the end of the century, however, advantages induced largely by Buffalo's location on the Canal and Lake Erie resulted in its far outstripping the other towns and cities, which did not have this "canal-lake" site. In 1880, when Buffalo's population had reached 155,000, Lockport's was 13,522, while Niagara Falls and Tonawanda each had only a few more than 3,000 inhabitants. Black Rock, Buffalo's first competitor, had failed to pass the 1,000 mark.

Wheat was of great importance to Buffalo's growth in the nineteenth century. The decrease in the importance of grain production in the Genesee Country and the opening of fertile lands in Ohio and elsewhere resulted in eastward shipments of wheat which grew quickly to large proportions. The completion of the Ohio canals in 1833 turned the flow of grain and flour from the Ohio River north to Lake Erie. Soon after, grain began to reach Buffalo in appreciable amounts not only from Erie ports but also from Chicago and other cities farther west. In 1840 the lake vessels delivered about one million bushels to Buffalo elevators. In 1855 the amount reached more than 25 million bushels, and Buffalo became the leading grain port on the continent—if not in the world. During two record years toward the end of the century, the port handled more than 76 million bushels. This growth in the grain trade was reflected in the changing appearance of the harbor. By 1855 ten great elevators towered along the water's edge.

The rapidly developing farm lands tributary to the Great Lakes contributed other cargoes to the lake traffic. Pork, bacon, and other meats were important for a time. This type of cargo reached a peak in the 1850's.

Later, when the forests of the upper lakes region became the major source of lumber for the wood-hungry eastern cities, lumber was added to the important items of lake commerce. After 1850 Buffalo, Tonawanda, and North Tonawanda became lumber transfer points. Lumberyards occupied many miles of lake front and riverbank. In 1876 nearly 120 million board feet of lumber were received in the Buffalo area, half of which moved on eastward via the Erie Canal.

Iron ore did not enter lake commerce to any large extent until the completion in 1855 of the "Soo" Canal on the American side, which permitted vessels of appreciable size to leave Lake Superior. Buffalo received few shipments of iron ore until after 1880, and even then the amounts were small when compared to those received by other Lake Erie ports.

The tremendous growth in various kinds of raw-material freight moving down the lakes reversed the up-lake pattern of traffic of the earlier years which had been dominated by pioneers and their possessions. Up-lake passenger business, however, did remain significant for several decades. In 1833 more than 40,000 passengers left Buffalo bound westward by steamboat. In 1845 the number had reached more than 93,000. At about this time, though, railroads began to take passenger business from the lakes, and as a result lake passenger traffic and the number of steamers decreased.

Railroads, while cutting passenger traffic, introduced a new type of up-lake freight traffic. Railroads were built to tap first the anthracite coal fields, then the bituminous fields, and coal began to move to Buffalo for shipment up the lakes. This cargo, which became important in the 1870's, reached one million tons in 1882 and amounted to between two and three million tons almost every year of the last decade of the century.

The economic development of northeastern United States and the increase of commerce which benefited Buffalo would not have been possible without the growth and diversification of transportation facilities. The improvement of the harbor and conditions for navigation on the lakes, the enlargement of canals, and the building of railroads all were important factors. At first the growth of lake commerce brought an increase in the number of vessels sailing the lakes, the size being limited by the shallow channels of rivers and harbors. Later the ships increased in size as a result of improvements in the conditions of navigation, technological developments, and the growing importance of bulk cargoes. To accommodate this lake fleet, improvements in the harbor were almost continually under way. Improvements in depth, size, area, and protection were achieved through the building of breakwaters, the deepening of channels, and the construction of slips and basins.

Transportation facilities between Buffalo and the East were constantly improved and increased. The Erie Canal, which had contributed so much to Buffalo's early growth, was enlarged between 1834 and 1862 and continued to be of great value. Freight on the canal reached 2 million tons nearly every year in the period 1860–1900 with more than 4.5 million tons moved during the peak year of 1880.

Throughout the nineteenth century commerce remained paramount and Buffalo continued to be a typical trade center. The city was, however, not without considerable industrial development. Manufacturing, which had begun on a small scale to satisfy the needs of the local market, increased in amount and diversity, becoming especially important during the last part of the century. However, inasmuch as the character of the industries was closely related to the commerce of the area, they testified to the continued importance of trade.

Flour-milling illustrates clearly this relation between manufacturing and commerce. With the increase in the grain traffic on the lakes, flour-milling became important in Buffalo. Some of this milling supplied the local market, but it was not long before flour was moving to eastern markets. By 1856 there were at least six mills in operation. By 1888 the flour mills in Buffalo had an operating capacity of 7,000 barrels a day.

In addition to flour-milling, the grain traffic gave rise to a number of other industries. Toward the end of the century factories were built to process the flax that moved down the lakes, and Buffalo became the principal center of the linseed oil industry. Malt houses and breweries, too, sprang up in the city, using the barley that was a large item in the grain traffic.

Tanning is another example of an industry closely related to commerce and transportation routes. Hides which were shipped down the lakes from the West were obtained cheaply and in large quantities. Hemlock bark was available from the forests of the vicinity, especially the mixed forests of the Appalachian Upland a short distance away. By the middle 1850's Buffalo boasted a dozen tanneries employing several hundred workmen. The advantages that gave rise to this industry were shared by all of the communities near the lake, and almost every town in Erie County had at least one tannery.

The iron-working establishments increased in number and size. Their products included implements for neighboring farms and marine boilers and engines for lake vessels, as well as machinery for other factories in the area. Shipbuilding, too, was an industry of some importance. During the 1850's, thirty-eight vessels were being built in the shipyards at one time, providing employment for nearly two thousand people. Other manufacturing establishments included a few textile mills and an infant glass industry.

On the eve of the Civil War, the city claimed more than five hundred manufacturing firms, and by 1884 there were more than one thousand.

At the close of the nineteenth century, Buffalo was almost one hundred years old. During this span of time, the pattern of settlement on the Niagara Frontier had changed from one of many small communities scattered over the area to the emergence of Buffalo as the overwhelmingly dominant center. Its role as an outstanding trade center remained clear throughout this period, but a trend toward diversified manufacturing had become discernible. The appearance of iron ore in the commerce of Buffalo heralded the coming of heavy industry, and experiments at Niagara Falls were about to yield results in the development of hydroelectric power.

The Power Era—The Twentieth Century

The twentieth century is set off from earlier times by the development of large power sites and the resulting importance of hydroelectric power. Prior to this the presence of small water-power sites had been an important factor in the development of small industries and small communities in many parts of Niagara and Erie counties. Within the present limits of Buffalo, early mills utilized Niagara River water in the Black Rock Channel and small sites between Big and Little Buffalo creeks, but large sources of hydroelectric power had not been developed.

The tremendous power potential of the Niagara River could not be fully developed until the nineteenth century neared its end. Technology had not advanced to the point where large amounts of water and great heads could be handled, and markets for huge quantities of power were unavailable and unforeseen. Finally, in 1895, three generators went into operation in the first large hydroelectric power plant at Niagara Falls. Power from this plant began to move streetcars on the streets of Buffalo in November, 1896, proving that electricity could be transmitted and marketed many miles from its source.

Markets for the electricity increased rapidly not only near the falls but throughout the Niagara Frontier. Following the success of this first plant and in response to the increasing demand for power, three large hydroelectric power stations were built in quick succession on the Canadian side of the Niagara River and another on the American side. By 1922 a fourth Canadian plant, constructed near the northern end of the Niagara Gorge, was in production. This plant was more efficient than the five older ones as the head available was far greater.

A tremendous increase in the demand for hydroelectric power which began with World War II and continued during the postwar period brought even greater harnessing of Niagara water power. The capacity of the newest Canadian installation was increased, and a gigantic new project was completed on the American side. The major power station of this project is located near the northern end of the Niagara gorge and is one of the largest installations in the world. The treaty between the United States and Canada sets a limit on the amount of water that can be diverted for power, and this limitation on the amount of water available will curtail further hydroelectric power expansion.

The limitation on the diversion of water from the river for power plus the increased demand for power during World War II resulted in the construction of a large steam plant on the bank of the Niagara River in Tonawanda. A continued rise in demand for power resulted in several enlargements of this plant. Other recently built steam plants are tied into this large system of steam and hydroelectric power plants that serve the Niagara Frontier.

Accompanying the growth of power sources in the twentieth century has been continued expansion of transportation facilities. As shown in Chapter 12, Buffalo has a transportation potential of 1.6 million and a transportation surplus of 560,000. This is small by New York City standards but very large compared to any other urban system in the state. Buffalo has become one of the major railroad centers of the country. Furthermore, when the automotive era dawned, the city was well located to derive the benefits from the improved highways. Paved roads, the successors to the early turnpikes and plank roads, converged upon Buffalo from all directions. As was true with canal, turnpike, and railroad, the Mohawk–Lake Plain axis was followed by modern paved highways, the most recent case in point being the New York State Thruway. Completion of the Welland Ship Canal in 1930, continual enlargement of harbor and port facilities in Buffalo, and completion of the St. Lawrence Seaway in 1959 all have contributed to an economic base conducive to industrial development. Recently thoughts have been expressed concerning the feasibility of an all-American canal that would parallel in function the Welland.

After 1900 the pattern of urbanization shifted noticeably. The second half of the nineteenth century had seen the rise of Buffalo as a dominant central city; no other city in Erie or Niagara County had grown proportionately. Although Buffalo retained its position as the central city of an expanding urban system, the twentieth century brought slower growth within the city limits. The 1950 census reported the population as 580,132, but by 1960 it had dropped to 532,759, a decrease of 8.2 per cent. In the same decade the population of Erie and Niagara counties increased from 1,089,230 to 1,306,957. The ring towns nearest Buffalo jumped 75 per cent in population from 1950 to 1960, while towns a bit farther out increased by 50 per cent. The growth of the city of Niagara Falls reflects its rise as both an industrial and tourist center.

Current Livelihood Structure

In 1960 the Buffalo SMSA had 1.3 per cent of its labor force in the primary sector, 45.5 per cent in the secondary sector, and 53.2 per cent in the tertiary sector. Its manufactural base is oriented toward heavy types of *producer goods* (goods to be used by manufacturers rather than by ultimate consumers as are *consumer goods*), although there is considerable diversity. Both the secondary and tertiary sectors have exhibited moderately successful trends when compared to other of the major urban systems in the state.

Manufacturing

Between 1947 and 1963 manufacturing employed between 155,000 and 200,000 persons; in 1947 the figure was 184,000. This makes it the second-greatest manufacturing complex in the state and a high-ranking center nationally.

Buffalo's reputation as an iron and steel center is supported by the fact that roughly 20 per cent of the factory workers are engaged in production of primary metals. Although there is considerable variation from census year to census year, since World War II primary metals have had to share first place with transportation equipment. This is the only urban system in the state in which these heavy types of industries take first or second place. Following in order of importance are machinery, food and kindred products, and chemicals. Printing and publishing, metal fabrication, and the manufacture of paper and stone, clay, and glass products contribute to what must be described as considerable industrial diversity.

The iron-and-steel industry developed rapidly after 1900. Buffalo and Lackawanna shared with Gary, Toledo, Cleveland, and other lake ports the advantages of a location where iron ore and limestone could be assembled by lake transportation, where coal could be obtained by a relatively short rail haul, and where the eastern markets for iron and steel could be easily reached.

In the 1958–61 period about seventy-five establishments in the primary metal category employed about 31,000 workers. Metal-fabricating plants, attracted by the primary metal producers, number several hundred. The large integrated iron-and-steel plants lie along the lake and the Buffalo River, particularly in Lackawanna, but some of the highly specialized metal-fabricating industries, especially

PRINCIPAL POWER-GENERATING STATION of the New York State Power Authority at Lewiston. A hydrostatic head equal to the drop over the falls plus the amount of fall in the rapids both above and below the falls exists here. Water is pumped into a reservoir above the dam at night when flow over the falls is not demanded and then released through turbines as needed to satisfy peak power demand periods. This, together with similar facilities on the Canadian side of the river, constitutes one of the world's largest concentrations of hydroelectric power generation. *Buffalo Area Chamber of Commerce*.

those requiring large amounts of electricity, are located in or near Niagara Falls.

Almost from the beginning, Buffalo craftsmen carried on work related to the transportation facilities. Boats and sails, marine engines, wagons and carriages, and locomotives and railway cars were made or serviced. After the turn of the century the Buffalo area attracted the newly developing automobile industry and, later, aircraft manufacturing. Expansion during the 1950's was tremendous in some of these lines. Since 1960 production has been more oriented to automobile engines, axles, forgings, radiators, and car bodies, as well as rocket engines, space vehicles, and the like. In the 1958-61 period the transporation equipment industries employed about 31,000.

Grain remained a major lake cargo, and the milling of flour and feed increased until Buffalo surpassed the older milling centers near the wheat belts. The comparatively recent popularity of packaged cereals and mixes has contributed to the growth of this industry in Buffalo, as elsewhere. In 1958 the grain-products industry included thirty-two establishments employing more than 5,000 workers. The towering elevators and mammoth mills that line the winding lower course of the Buffalo River give evidence of the importance of Buffalo as a grain-processing center.

One of the most significant chapters in the story of manufacturing in the Buffalo Urban System is the one dealing with the rise of the electrometallurgical and electrochemical industries. That these types should develop near great power sites is now to be expected, but the pioneers of Niagara power, who believed that the transmission of electricity to the city of Buffalo was necessary if large amounts of power were to be marketed, did not foresee the range and quantity of new industries that would be drawn to Niagara Falls by the availability of plentiful, cheap electricity.

One of the first industrial users of electricity at Niagara Falls was a plant built to manufacture acetylene gas. Many uses have since been developed for this gas which burns with a brilliant white light and which has high heating power—lighting homes in rural areas, enriching gas in city mains, welding and cutting metals. In addition, when the synthetic organic chemical industry was born in the late 1920's, acetylene was used in the production of these man-made elements. Currently pipe lines carry the

BUFFALO SKYWAY AND HARBOR AREA. This is the meeting place of highway and railway with the Buffalo River and harbor facilities. At the left new autos are unloaded from lake boats after being transported from Detroit. In the center is one of the world's largest food-processing plants, and just beyond the Buffalo River may be seen major rail freight terminals. *Buffalo Area Chamber of Commerce.*

BUFFALO HARBOR. The foreground is dominated by the extensive docks of Buffalo's steel industry. Beyond, flour-milling and storage facilities line the banks of the Buffalo River. Extensive rail facilities serve the area when lake and river navigation cease in the winter season. *Buffalo Area Chamber of Commerce.*

gas from the plant where it is manufactured to several chemical plants in the vicinity which use it in their manufacturing processes. The acetylene plant thus was not only an important early user of electricity but an attractor of other industries.

The history of this company from its first small operation to a multimillion-dollar industry is similar to that of many such industries in and around Buffalo and Niagara Falls. Today the products of the area's chemical plants are many: aniline dyes, caustic soda, chlorine bleaches, detergents, drugs, silicones, polyethylene, cellophane, fertilizers, hydrogen peroxide, and dozens of others. In recent years 13,000 to 17,000 workers have been employed by these chemical industries.

Establishments that manufacture machinery, both electrical and nonelectrical, number more than 300. Still another 200 fabricate metal products. In addition the products of area factories include gasoline and fuel oil, clothing, paper, rubber items, cement, and many others.

The fifteen largest manufacturing concerns, their principal products, and approximate recent employment are shown in Table 46.

TERTIARY ACTIVITIES

Although the employment figures emphasize the importance of manufacturing in the Buffalo Urban System, many more workers are employed in terti-

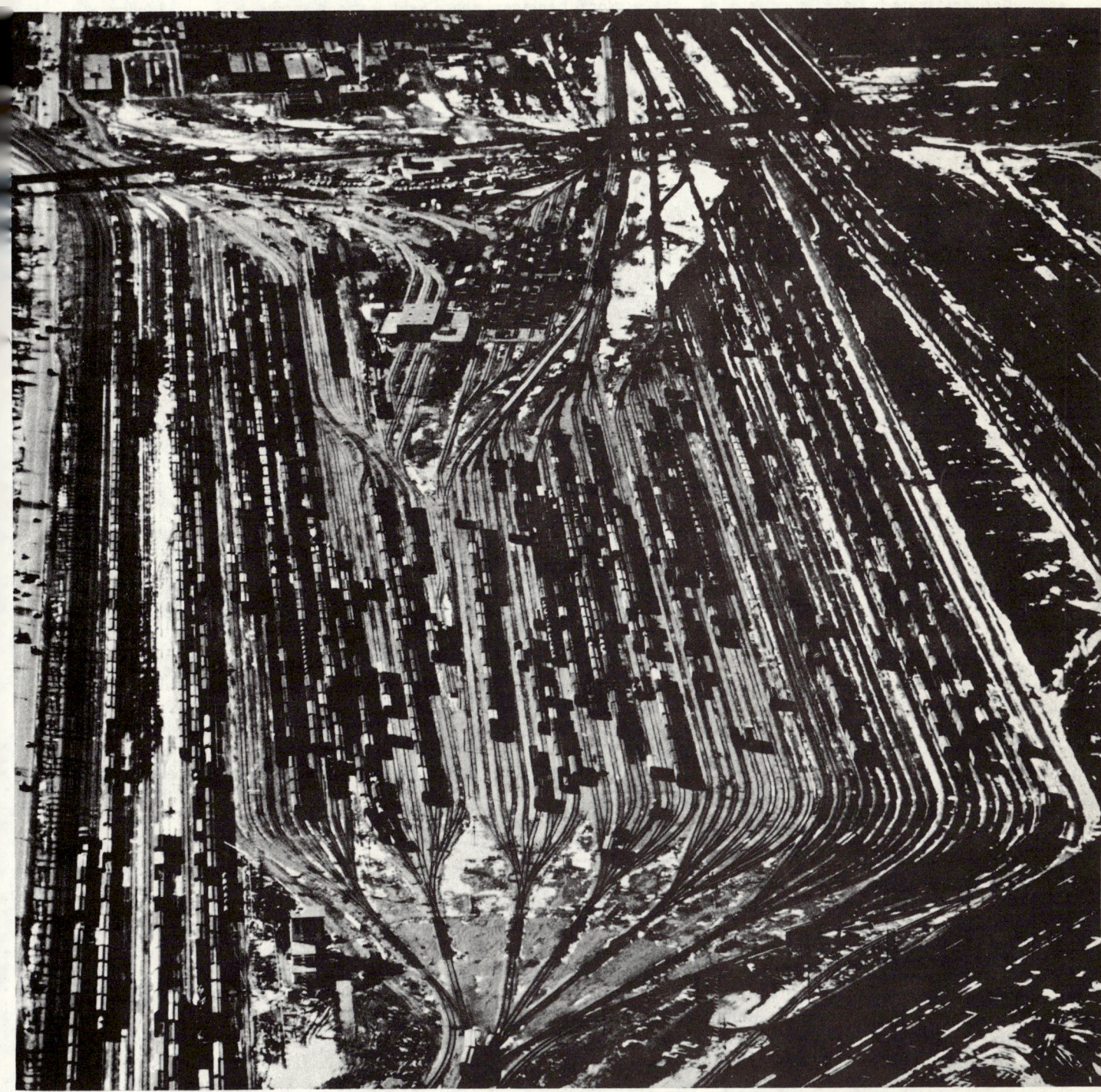

BUFFALO, A TRANSPORTATION FOCUS. Freight cars are assembled here in one of the world's largest distribution yards. *N.Y.S. Dept. of Commerce.*

TABLE 46
MAJOR MANUFACTURING CONCERNS OF THE BUFFALO SMSA

Company	Products	Approx. Employment Early 1965
Bethlehem Steel Co.	Steel sheets, strip, bars, and structural shapes	18,000
General Motors Corp.	Engines, axles, radiators, forgings	17,900
Westinghouse Electric Corp.	Motors, controls, welding equipment, and copper wire	6,900
Bell Aerosystems Co.	Rocket engines, space vehicles, and servo-mechanisms	4,600
Ford Motor Co.	Automobile stampings	4,500
Union Carbide Corp.	Ferro-alloys, carbon and graphite, industrial gases, and automotive products	4,100
Trico Products Corp.	Automotive equipment	4,000
E. I. duPont de Nemours, & Co., Inc.	Cellophane, polyester films, inorganic chemicals	3,600
The Carborundum Co.	Abrasives, grinding wheels, refractories	3,500
Republic Steel Corp.	Steel bars	3,400
Sylvania Electric Products, Inc.	Defense electronics and communications systems	2,400
Dunlop Tire and Rubber Corp.	Tires, tubes, golf and tennis balls	1,900
National Gypsum Co.	Gypsum board and plaster	1,900
National Aniline Div., Allied Chemical Corp.	Dyes and chemicals	1,800
Spaulding Fibre Co., Inc.	Vulcanized fibre	1,500

ary activities. In 1960 these workers numbered 250,000 and constituted 53.2 per cent of the 476,000 employed persons. The number employed in the tertiary sector seems large, but it may be pointed out it is not as significant in proportion to the total population as is that of Syracuse, for example. In terms of the percentage of its labor force in the tertiary sector, Buffalo ranked fifth among the seven major systems of the state. (Compare the comments in the various urban chapters relative to the tertiary sector and the figures in the tables of Chapter 12.)

Of the total number employed in the Buffalo Urban System, 19.1 per cent was in the service industries, 18.6 per cent in wholesale and retail, 7.8 per cent in public utilities, 4.0 per cent in public administration, and 3.7 per cent in finance, insurance, and related fields. Although the employed persons in the service industries amounted to 19.1 per cent of the total labor force, it should be noted that Buffalo ties with Binghamton for the lowest service orientation of the seven urban systems.

The total sales and service functions of the Buffalo Urban System in 1960 produced a tertiary potential of 1,463,000 and a tertiary surplus of 156,000. It is especially interesting to note that Niagara County by itself shows a tertiary deficit of 45,000. This reflects both the manufacturing orientation of Niagara County cities and their dependence on Buffalo for central city sales and service functions.

In the decade 1950–60 a significant amount of growth took place in the tertiary sector with a gain of 17.9 per cent. In terms of growth in this sector the Buffalo Urban System ranked fourth among the urban systems of the state. Groups concerned with the development of the area might carefully analyze this tertiary sector of the economy for possibilities of future expansion.

One tertiary activity, tourism and recreation, began early in the Niagara Frontier and expanded to tremendous proportion in recent years. Among the first travelers who struggled over the paths and woods roads to the Niagara Frontier were adventurous sightseers bound for Niagara Falls. Although historic sites such as Fort Niagara and man-made wonders such as the power developments receive their share of visitors, Niagara Falls remains a principal attraction of world-wide fame. The license plates seen in the parking lots and on the roads, and the variety of languages and dialects overheard at the observation points testify to the fact that visitors come from all parts of the United States and Canada and from many other countries as well. Restaurants, hotels, and motels, which profit from these tourists, are located not only in the city of Niagara Falls itself but also along Niagara Falls Boulevard and other highways approaching the city.

The snowy hills of southern Erie County are another attraction. They are dotted with ski resorts and clubs which attract thousands of urban residents, largely local. Almost unlimited opportunities for boating are provided in the Niagara Frontier by Lake Erie, Lake Ontario, the Niagara River, and the New York State Barge Canal.

Problems and Prospects

Although population in the city of Buffalo is steadily declining, growth of the whole urban system continues. Accompanying these trends are the usual problems. The movement of many affluent residents, as well as businesses and industries, to the suburbs tends to result in dwindling revenues but increasing costs in the older parts of the urban system. In the newer sections new functions are thrust on the people that they are ill organized to carry out. These include elaboration of water mains, sewers, roads, and schools. Slums develop in the middle city or around the industrial areas; traffic congestion makes reaching the CBD tedious and difficult; port facilities struggle to keep up with rising demands; increased mobility is demanded everywhere.

The Buffalo Urban System would be fortunate, indeed unusual, if it were not confronted by these kinds of problems. Solutions are being approached from several directions. Planning boards have been established at local, county, and metropolitan-wide levels and privately sponsored development efforts have been undertaken. A port authority is contributing its efforts. Urban renewal projects are attacking obolescent areas.

The Thruway (now Interstate 90) not only touches Buffalo but also is linked to the heart of the city. Access is provided to the Buffalo Skyway across the harbor and two extensions carry traffic to Niagara Falls. Other high-speed access routes are under construction. Municipal-built, but privately operated, parking garages aid the auto-driving customer or freight carrier to the CBD.

In spite of efforts to solve the problems of deterioration in Buffalo, Niagara Falls, and other of the larger centers, greatest future growth will continue to occur in outlying districts, particularly in the towns to the southeast and east of Buffalo. Access to jobs and access to the CBD provided by expressways will strongly affect the extent of growth in any one direction. Driving time, not distance, will become paramount as the size of the urban system increases.

Forecasting the directions that manufacturing and trade will follow is difficult. One metalworking plant shuts down and another moves away, but a large foundry takes their place. Lake freighters bring less grain into the port, but the number of ocean freighters increases slowly but steadily each year. A grain elevator is closed, but nearby a major flour-milling company erects a multimillion-dollar addition of the most modern and efficient type. Evidence indicates that the Buffalo Urban System will remain a major producer of iron and steel, transportation equipment, machinery, chemicals, flour, and other commodities for which it is now famous.

A significant industrial development in the area is the rising importance of research and engineering. Some of the large, well-established industries have built laboratories and pilot plants which serve their branches in other parts of the country. Other companies conduct research on a contract basis. The clean, modern lines of the new laboratory buildings have brought a change in the industrial picture throughout the Niagara Frontier. Some of these new structures stand among the old factories, but others are scattered in the rural areas or in sections that were once completely residential. This type of development is highly desirable to any community.

The favorable conditions that originally contributed to the growth of commerce and industry are still present. Foremost among these is a location in the industrial Northeast at a point well served by land and water routes. Also of primary importance is the availability of large amounts of power from hydroelectric and steam plants. Nor should it be forgotten that the area offers the amenities of life which make it attractive to employers and employees alike—lakes and rivers, hills and woods, and the cultural advantages of a large urban system. The population of the SMSA, at 1,307,000 in 1960, should reach 1,800,000 by 1980 and be roughly 2,500,000 by the year 2000.

Selected References

Adams, Edward Dean. *Niagara Power: History of the Niagara Falls Power Company, 1836-1918.* Niagara Falls: Niagara Falls Power Company, 1927.

Bingham, Robert W. *The Cradle of the Queen City: A History of Buffalo to the Incorporation of that City.* Buffalo: Buffalo Historical Society, 1931.

Boyd, Walter H. *The Niagara Falls Survey of 1927.* ("Canadian Geological Survey Memoir," 164.) Ottawa: F. A. Acland, printer, 1930.

Buffalo Historical Society. *Niagara Frontier.* Buffalo: The Society, Winter, 1953, to date.

City Planning Associates—East, Inc. *Buffalo Master Plan, First Study Phase.* Buffalo: Buffalo Planning Commission, 1964.

Community Welfare Council of Buffalo and Erie County (comp.). *Tract Facts.* Buffalo: The Council, 1964.

Dow, Charles Mason. *Anthology and Bibliography of Niagara Falls.* 2 vols. Albany: The State of New York, J. B. Lyon Company, printers, 1921.

Glazier, A. E., and E. F. Rundell. *Buffalo, Your City.* Buffalo: Foster and Stewart Publishing Corp., 1947.

Graham, Lloyd. *Niagara Country.* ("American Folkway Series.") New York: Duell, Sloan and Pearce [c. 1949].

——— and Frank H. Severance. *The First Hundred Years of the Buffalo Chamber of Commerce.* Buffalo: Foster and Stewart Publishing Corp. [c. 1945].

Harris, F. R. *Report on the Port of Buffalo.* Buffalo: Prepared for the Division of the Port, City of Buffalo, State of New York, 1955.

Holley, George W. *Niagara: Its History and Geology, Incidents and Poetry.* New York: Sheldon and Company, 1872.

Horton, J. T., E. T. Williams, and H. S. Douglas. *History of Northwestern New York; Erie, Niagara, Wyoming, Genesee, and Orleans Counties.* 3 vols. New York: Lewis Historical Publishing Company, 1947.

Howells, W. D., Mark Twain, and Professor Nathaniel S. Shaler. *The Niagara Book; a complete souvenir of Niagara Falls, containing sketches, stories, and essays.* Buffalo: Underhill and Nichols, 1893.

Johnson, Fred H. *Guide to Niagara Falls and its Scenery, including points of interest both on the American and Canadian sides.* [including] "Geology and Recession of the Falls" by Sir Charles Lyell. Philadelphia: G. W. Childs, 1863.

Larned, J. H. *A History of Buffalo, Delineating the Evolution of the City.* New York: The Progress of the Empire State Company, 1911.

Parkman, Francis. *Historic Handbook of the Northern Tour: Lakes George and Champlain; Niagara; Montreal; Quebec.* Boston: Little, Brown, 1885.

Severance, Frank H. *Old Trails on the Niagara Frontier.* 2nd ed. Cleveland: The Burroughs Brothers Company, 1903.

――――. *The Picture Book of Earlier Buffalo.* Buffalo: The Buffalo Historical Society, 1912.

――――. *Studies of the Niagara Frontier.* Buffalo: The Buffalo Historical Society, 1911.

Smith, Henry Perry. *History of the City of Buffalo and Erie County.* 2 vols. Syracuse: D. Mason and Company, 1884.

Thomas, Robert E. *Salt and Water, Power and People—A Short History of Hooker Electrochemical Company.* Niagara Falls, N.Y.: The Company, 1955.

White, Truman C. (ed.). *Our County and Its People; a descriptive work on Erie County, New York.* Boston: The Boston History Company, 1898.

Whittemore, Katheryne Thomas. "Buffalo and Niagara Frontier," in *Industrial Cities Excursion Guidebook, 1952*, ed. H. M. Mayer. Publication No. 2 of the Seventeenth International Geographical Congress. Washington, D.C.: The Congress, 1952.

CHAPTER 20   ROBERT McNEE

New York

THE massing of population and urban functions around the lower Hudson Estuary and associated bays and inlets is one of the wonders of the modern world. Nowhere else is there anything comparable to this "city of cities." The urban web is so vast and complex that it sometimes seems more like a nation or an empire than a single urban system. In this chapter the unity of the whole complex is emphasized, despite the somewhat arbitrary political boundaries which divide the area into separate and sometimes quarreling states and counties. This book is about New York State, but in treating the New York Urban System it would be unrealistic to exclude completely those parts of it which are in New Jersey and Connecticut.

Few authorities agree in all details what constitutes the system; the definition used here will be that of the New York Metropolitan Region accepted by the Regional Plan Association and the Metropolitan Regional Council. Included are the twenty-two counties that surround the Hudson Estuary; twelve are in New York State, nine in New Jersey, and one

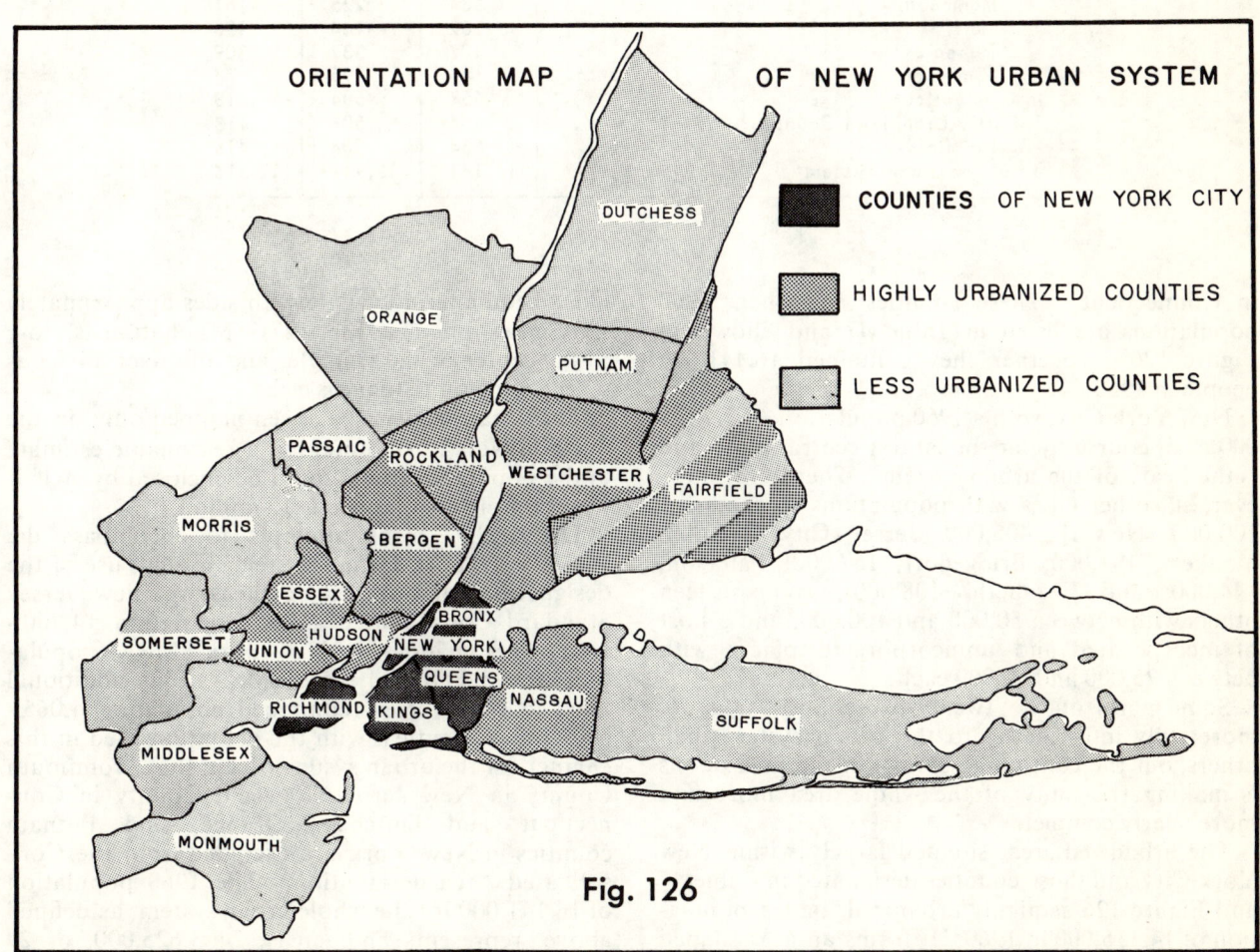

Fig. 126

TABLE 47
POPULATION OF COUNTIES INCLUDED IN THE NEW YORK URBAN SYSTEM

	Population In Thousands		
	1960	1950	1940
In New York State	11,087	9,865	8,984
Boroughs of New York City	7,782	7,892	7,455
Bronx (Bronx County)	1,425	1,451	1,395
Brooklyn (Kings County)	2,627	2,738	2,698
Manhattan (New York County)	1,698	1,960	1,890
Queens (Queens County)	1,810	1,551	1,298
Richmond (Richmond County)	222	192	174
Highly Urbanized Counties	2,246	1,388	1,054
Nassau	1,300	673	407
Rockland	137	89	74
Westchester	809	626	573
Less Urbanized Counties	1,059	585	475
Dutchess	176	137	121
Orange	184	152	140
Putnam	32	20	17
Suffolk	667	276	197
In New Jersey	4,400	3,580	3,114
Highly Urbanized Counties	2,819	2,490	2,227
Bergen	780	539	410
Essex	924	906	837
Hudson	611	647	652
Union	504	398	328
Less Urbanized Counties	1,581	1,090	887
Middlesex	434	265	217
Monmouth	334	225	161
Morris	262	164	126
Passaic	407	337	309
Somerset	144	99	74
In Connecticut	654	504	418
Partly Urbanized County	654	504	418
Fairfield	654	504	418
Total for Urban System	16,141	13,949	12,516

in Connecticut. These counties and their 1960 populations are listed in Table 47 and shown in Figure 126. Together they contained 16,141,000 people in 1960.

New York City, with a 1960 population of 7,782,-000, is of course by far the largest central place and is the heart of the urban system. There are, however, six other cities with populations in excess of 100,000 (Newark, 405,000; Jersey City, 276,000; Yonkers, 191,000; Bridgeport, 157,000; Paterson, 144,000; and Elizabeth, 108,000), some fifteen others with between 50,000 and 100,000, and a host of incorporated and unincorporated places with between 25,000 and 50,000 each.

Some parts of the twenty-two-county area are more truly integrated into the urban system than others, but the continuing sprawl of the macropolis is making the unity of the whole area more and more nearly complete.

The urbanized area, situated largely within New York City and those counties designated in Table 47 and Figure 126 as highly urbanized, had a population of 14,115,000 in 1960. It forms an area shaped like an equilateral triangle with sides approximately sixty miles long (see Fig. 127). Manhattan is close to the center of the triangle, and the over-all focus on the Hudson Estuary is clear.

The urban frontier has been mapped only on the north in New York State, but a reasonable estimate of the population that would be enclosed by it if it were completely mapped is 15 million.

The U.S. Census attempts to encompass the whole New York Urban System through use of the designation New York–Northeastern New Jersey Standard Consolidated Area (population 14,760,-000). This includes the New York SMSA (population 10,695,000) and three SMSAs plus additional areas in New Jersey (combined population 4,065,-000) and is identical with the definition used in this chapter for the urban system except that Monmouth County in New Jersey, Fairfield County in Connecticut, and Dutchess, Orange, and Putnam counties in New York are excluded from the Consolidated Area designation. The 1960 population of 16,141,000 for the whole urban system, as defined above, represents an increase of 3,625,000, or 29 per cent, since 1940. This is an intermediate growth rate when compared to upstate urban systems, being larger than Binghamton's or Utica's, but less than Syracuse's or Buffalo's. During this twenty-year period the New York State part grew by 23.4 per cent, the New Jersey part by 41.3 per cent, and the one county in Connecticut by 56.5 per cent. Table 47 illustrates how slow growth has been in New York City as compared to other areas.

This treatment of the New York Urban System is organized in four parts: (1) development of the system, (2) livelihood structure, (3) internal form, and (4) problems and prospects. Development is particularly stressed because the most important single aspect of the system is its dynamism—its ability to grow and prosper despite many limitations and handicaps.

Development of the Urban System

Since the New York Urban System is world-wide in its relationships, a full explanation of its development would require a detailed examination of the economic and urban geography of most of the earth over several centuries. To express this idea in another way, the New York Urban System of today is the product of many forces operating from many places over great distances. Only the most basic aspects of these forces and their results can be examined here. Three major phases in the development of the urban system are particularly pertinent to this analysis: (1) emergence as a regional trade center, 1624–1815, (2) emergence as the primate city of the United States, 1815–1914, and (3) emergence as the primate city of the occidental world, after about 1914.

A REGIONAL TRADE CENTER

New York grew slowly after its founding in the early seventeenth century; nearly two centuries after settlement it still dominated only a limited hinterland along the East Coast and had attained no more stature than that of a regional or second-order trade center. As such it competed more or less on a par with other similar trade centers along the Atlantic Seaboard (see Fig. 128). The *site* (precise location) of New Amsterdam, though well chosen in terms of defense against land-based attack, and in terms of its excellent harbor, was not markedly superior to the sites of the rival regional trade centers on the East Coast such as Boston, Newport, Philadelphia, and Charleston. The hard-rock base of Manhattan, later an advantage when erecting skyscrapers, was no particular asset to colonial New York. The ships of the day were small, usually of 100 tons in capacity or less, and therefore the great size of the harbor was largely superfluous. Nor was the *situation* (general geographic location) of New York outstanding. Its situation relative to the triangular trade between Britain, the Atlantic Coast, and the West Indies was advantageous, but no more so than that of Boston or Philadelphia. Its hinterland, primarily the area described here as the urban system plus the Hudson Valley, was less populated and less productive than the rich hinterland of Philadelphia. New York, like the other regional trade centers of the time in America, was in the orbit of London. Its relationship to London was somewhat analogous to Syracuse's relationship to New York today. Consequently, although some of the urban functions that later made the city famous had an early start, their development was slow. In fact, though colonial Philadelphia was not clearly dominant over New York, it had a larger population and a greater development of banking, insurance, publishing, and manufacturing than New York in 1790.

The form, or physical structure, of old New York is still reflected in many aspects of Manhattan south of Fourteenth Street. The city huddled near its principal economic base, the East River docks. Northward expansion was retarded by a succession of walls, in much the same fashion as for the historic cities of Europe. The present Wall Street marks the most famous wall, which stood from 1653 to 1699. There were several other walls, too. The last, in the general area of Chambers Street, was built and abandoned in the mid-eighteenth century.

PRIMATE CITY OF THE UNITED STATES, 1815–1914

The century from 1815 to about 1914 was a golden period, for New York was not only transformed from a regional or second-order trade center to a first-order trade center, but it became the primate city of the emerging economic colossus of the earth, the United States.[1] The basic cause of this transformation was a series of changes in New York's situation. These changes in turn resulted in a new meaning for its site, a clustering and pyramiding of urban functions, and a radically altered form.

Changes in Situation

The changes in New York's situation were not isolated phenomena; rather, they were a part of

---

[1] For definitions of trade-center orders, see Chapter 12. A *primate city*, by concept, is one that stands out all by itself in terms of magnitude and significance. It has the finest wares, the rarest articles, the greatest human talents, and the most skilled workers. To it flows an unending stream of the most able people. It is the market for all that is superlative in intellectual and material productions. Its supereminence as a market runs parallel to its supereminence in size. (The definition of a primate city as paraphrased from: Mark Jefferson, "The Law of the Primate City," *Geographical Review*, XXIX (April, 1939), 226–32.)

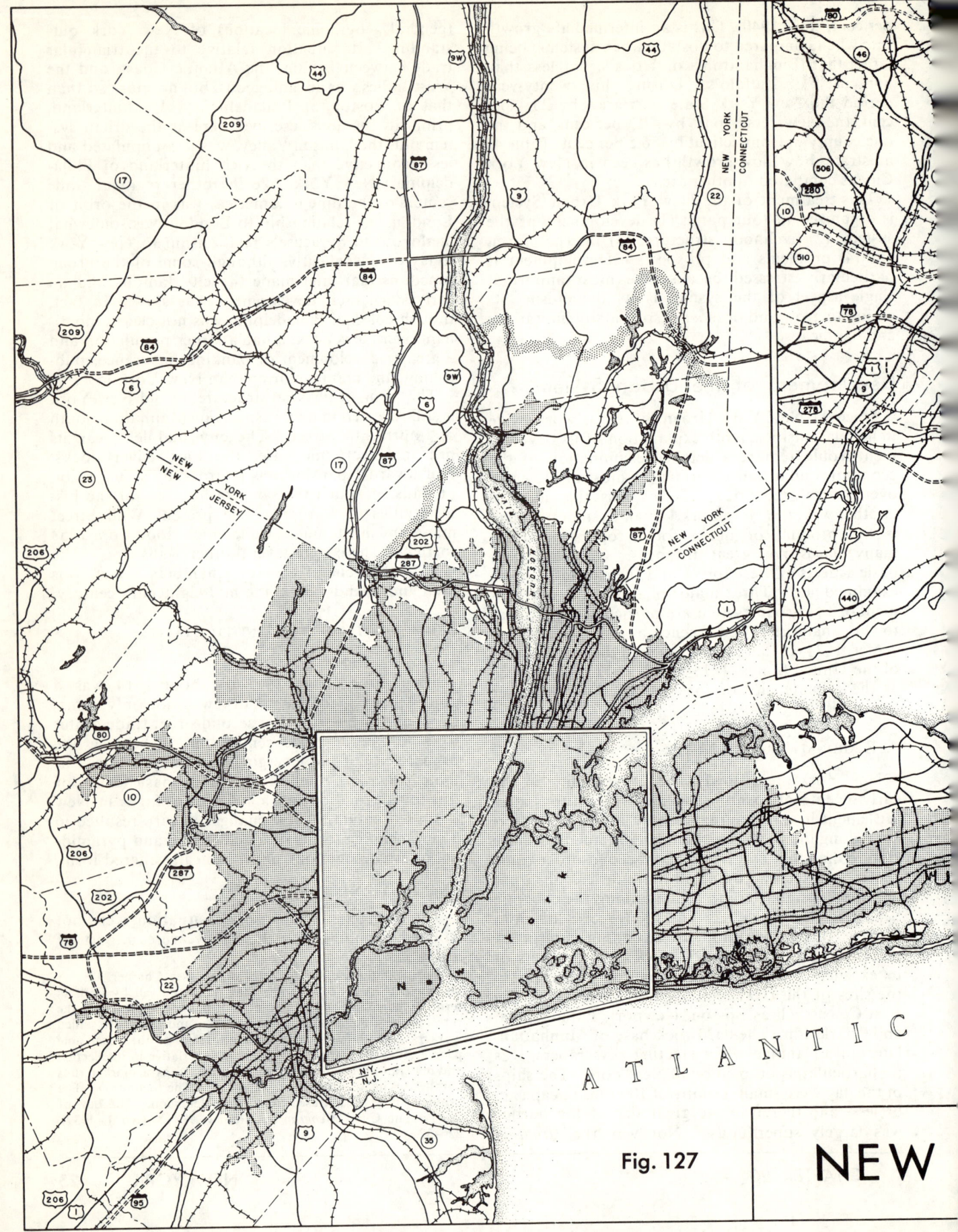

Fig. 127

NEW

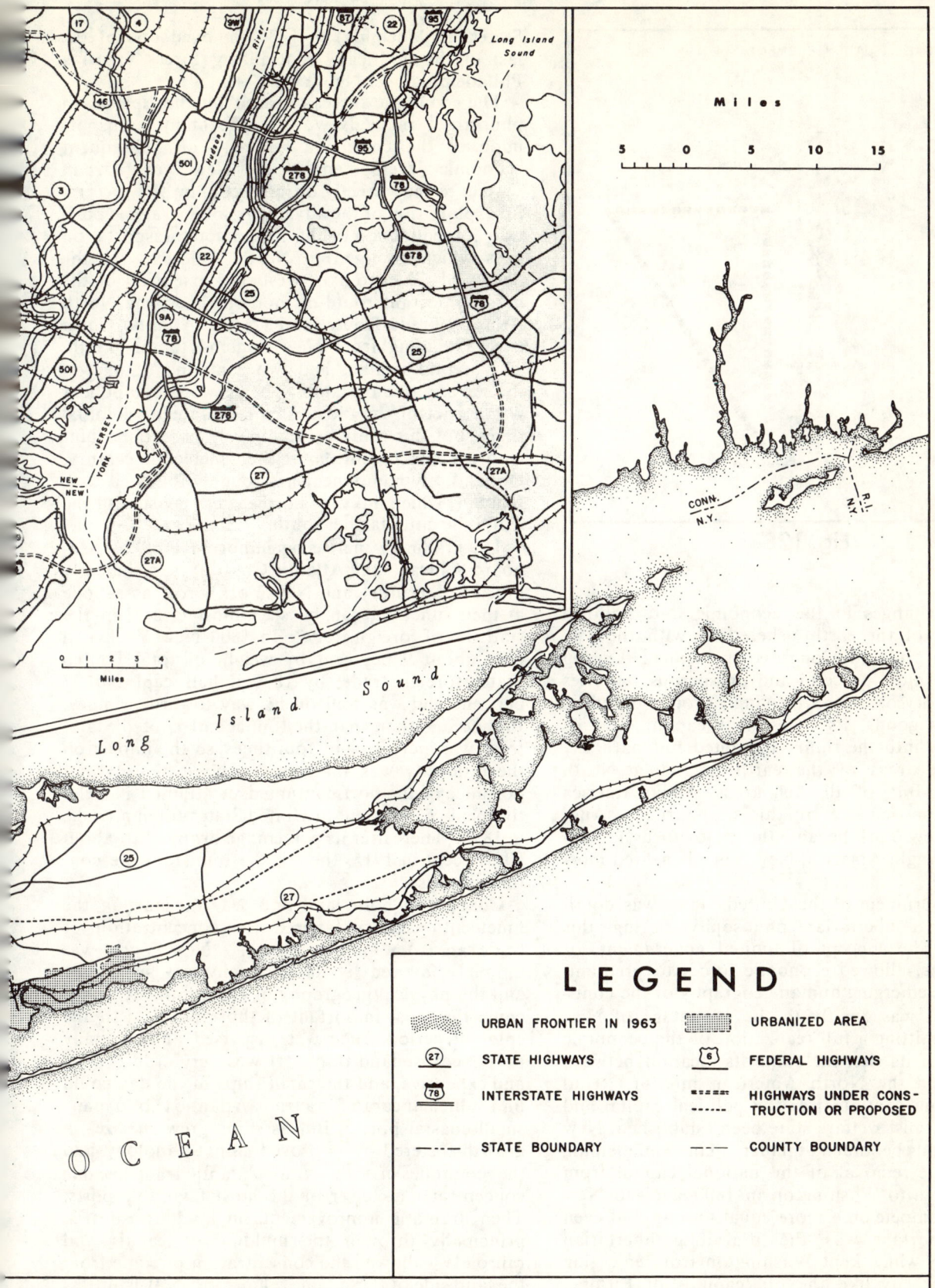

YORK METROPOLITAN AREA

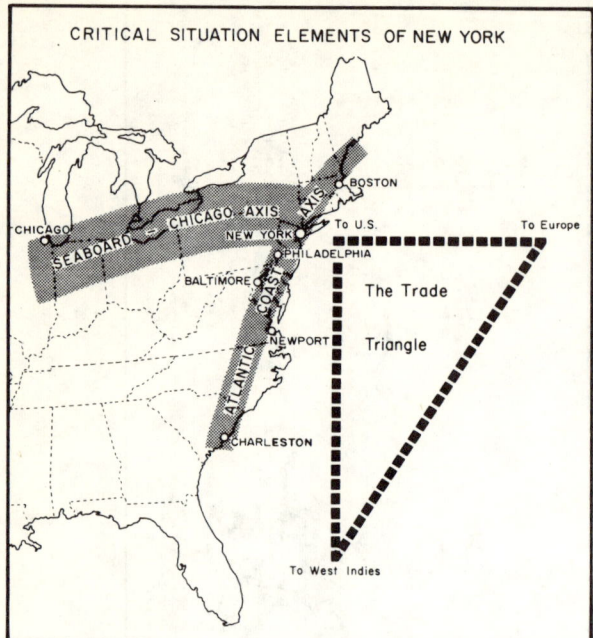

Fig. 128

dramatic changes in the economic and political geography of the earth ushered in with the Pax Britannica after the Congress of Vienna. During one of the most peaceful and prosperous centuries in the history of man, a thriving Europe poured its people, its goods, its investment capital, and its technology into the thinly inhabited but naturally productive areas of the earth, including North America. But, of all such areas, Anglo-America benefited particularly by this outpouring. And, because New York became the major focus for developing Anglo-America, New York benefited most of all.

The government of the United States was dominated by a libertarian philosophy during this century. The concept of limited government, of *laissez faire*, allowed economic forces to predominate in the emerging human geography of the continent. This was greatly to the advantage of New York, permitting a full realization of the economic potential of its site, including its location between Europe and the North American interior. Had Philadelphia continued as the political capital and had a centralist welfare state been established, New York might be much smaller than Philadelphia today. The removal of the national capital from Philadelphia to Washington in 1800 allowed New York to compete on a more equal footing, but even more important was the prevailing libertarian philosophy which kept Washington from emerging as a major city like the old capitals of Europe. Jackson's destruction of the Bank of the United States (Philadelphia) in 1836 removed one of the last of the politically centralizing forces favoring Philadelphia over New York.

The economic forces released by the libertarian philosophy made New York the major transport node of the continent. Nearly all subsequent economic development of the New York Urban System rested, directly or indirectly, on the successful exploitation of the advantages of a strategic position at the junction of the two major transportation axes which emerged in North America: (1) the Seaboard–Chicago axis, westward extension of the most important route of modern times, the North Atlantic Trade Route, and (2) the Atlantic Coast axis, extending from New England to the South. This second axis has had two components, one on the sea and one on land. The seaward component was a part of the old triangular route of colonial times, but this route became much more important with the rise of cotton in the nineteenth century. On land, the component has roughly followed U.S. Route 1, which links the northeastern coast with the Fall Line cities farther south.

The gradually increasing importance of the Seaboard–Chicago and Atlantic Coast axes in the early nineteenth century and New York's role as a node at their junction, can be seen most clearly in the statistics of foreign trade. In 1800 New York had only about 9 per cent by weight of total United States foreign trade; by 1830 it had captured 37 per cent and was well on its way toward primacy. In the same time span the tonnage of United States trade was increasing by four times, so the impact on the little city was immense. The increasing importance of the port continued; it attained its relative peak, 57 per cent of United States foreign trade, in 1870. Thereafter its percentage dropped to about 50 per cent in 1915, though the total tonnage continued to rise.

Changes in both land and sea transport in the nineteenth century favored the concentration of trade at a few major points. New York was uniquely favored by its local physical geography and the physical geography of the continent to become the most important of those points. In the colonial period trade was dispersed among many ports because land transport was very crude, slow, and expensive, and the small ships of the day could find sufficient cargo for a trans-Atlantic trip to many small coastal ports. But as ships grew in size—a growth fostered by improved ship technology and the economies of scale—trans-Atlantic trade became concentrated at fewer of the most favorable ports. Then dramatic improvements in land transport—principally through the building of canals and railroads—allowed the concentration of trade from the land side of these same major ports. It is in this sense, as a part of a general pattern of concentrating

AN OVERVIEW OF THE NEW YORK SETTING. Within the confines of this photo are the greatest concentration of people and business and the best and most extensive port facilities on the face of the earth. Clockwise from the tip of lower Manhattan near the center of the picture may be seen (1) Brooklyn, including Brooklyn piers; (2) Staten Island and its Free Trade Piers at lower left. A new bridge connecting Brooklyn and Staten Island has just been opened across the narrows at the bottom of the photograph; (3) Bayonne, Middle left; and (4) Hackensack River and Jersey Meadows, upper middle left. *Port of New York Authority.*

trade, that the Erie Canal was so important. The Canal, opened in 1825, allowed the Port of New York to break out of its colonial hinterland, the circumscribed Hudson Valley, and to tap the Great Lakes basin. Philadelphia's attempts to compete with New York by building a route over the mountains were relatively unsuccessful. The long-range significance of the Erie Canal route was not fully realized at the time, for in the early nineteenth century the Ohio Valley was more developed than the Great Lakes littoral. Chicago was a minor settlement, unable to obtain a loan from the Ohio River bankers because it was considered to have no future! But in the succeeding decades of the nineteenth century, the Ohio Valley area grew relatively slowly, while the Great Lakes littoral was one of the most rapidly developing areas of the United States.

Nor did the rails reduce the initial advantage the canals had given to New York; in fact, New York became even more important as a transport node in the railroad era. Railroads, following the same water-level Erie Canal route, penetrated to Chicago and the vast plains beyond. Other rail lines, extending inland from Philadelphia and Baltimore, enabled these ports to compete with New York in the midwestern hinterland. Yet New York already had a commanding lead over other ports and the rail lines were built primarily to serve the ports rather than the ports being built to serve the rails. Consequently, the lines extending inland from some of these other ports simply continued to New York and many goods followed these competing lines to the port with the largest volume and the greatest economies: the Port of New York. The expenses of attempting to bridge the Hudson Estuary were formidable; hence several of the western rail lines terminated on the Jersey shore opposite Manhattan.

The flow of traffic on the Atlantic Coast axis, both water and rail, worked to the advantage of the most central port on the route: New York. The more trade tended to concentrate at New York, the more this trade nourished the growth of related services, such as wholesaling, banking, insurance, and trade specialties of all kinds. In turn, the great variety of such services made the port even more attractive, creating an accelerating spiral of trade growth.

A closely related aspect of New York's change in situation was its development as a great wholesaling center for the United States. Initially, wholesaling was closely related to port function. New York early acquired a near-monopoly of textile imports, and in the mid-nineteenth century about one-third of all imports to the United States were textiles. The city's dominance in textiles was a factor of some importance later in the growth of the garment industry. New York also became the major import wholesaling center for many other products, such as jewelry, chinaware, furniture, and delicacies.

But gradually the internal circulation on the transportation system of the United States became even more important than the export-import trade. With a rapidly growing population and the availability of the resources of a continent, the American economy gradually became more and more self-contained and self-sufficient. The great American Manufacturing Belt emerged, stretching from the Atlantic nearly to the Mississippi, north of the Ohio. Within this broad belt there were many specialized manufacturing regions, such as the textile region of New England or the steel region of Pittsburgh. Agriculture, too, developed regional specializations such as the Corn Belt and Dairy Belt, producing for the rising American population as well as for export. All of this regional specialization required exchange services, including wholesaling. New York was poorly located to serve as the ideal focus for such internal circulation because its location was peripheral rather than central. Hence other centers, such as Chicago, captured part of the wholesaling function. Yet New York's established position at the junction of the nation's two major transport axes, its location in the middle of the most densely populated area of the nation—the Atlantic Seaboard, its great variety of business services, and its major development of import wholesaling, made it a major center in domestic wholesaling as well.

New Meaning for the Site

As New York's situation changed, the hitherto latent advantages of its site were realized. The deep, commodious, and largely ice-free harbor, and the hundreds of miles of easily developed shore line were important factors in allowing the concentration of over half of America's foreign trade in this one small area. No other natural harbor, from Nova Scotia to Florida, was so well suited for such concentration. Likewise, the complex of islands and peninsulas at the junction of the Hudson Estuary and Long Island Sound were easily linked by local water transportation.

Yet the site had disadvantages, too, and these became more and more apparent with the growth of land transport. The same water bodies that were avenues for ships were barriers to land vehicles. It was only through a series of heroic and expensive engineering efforts, beginning with the Brooklyn Bridge in 1883, that these physical barriers were partially overcome. The Harlem and East rivers were more easily dealt with than the Hudson Estuary. Hence, by World War I there were several links between Manhattan and areas to the north and Long Island, but there were still no bridges across the Hudson. Consequently, Brooklyn and the Bronx were developed much more rapidly than either Staten Island or the Jersey shore. The major rail lines coming from the west developed their main freight terminals on the Jersey shore opposite lower Manhattan, relying on car ferries to link the shore with other parts of the region rather than attempting to bridge the Hudson. This had important effects on the distribution of manufacturing within the urban system, leading to concentration of heavy and transport-sensitive industries on the Jersey side of the Hudson and manufacturing for local or New England consumption or for European export on the east side of the Hudson. The rail terminal network of Jersey City and Hoboken was well established by 1870; it remains basically unchanged as an element of the urban system's structure today.

The costs of internal circulation were high, among the highest of any major city in the world. Part of these costs represented the congestion resulting from success, but most of them represented long-range disadvantages of the site for modern land transport. The site also had disadvantages for constructing buildings. Though the hard-rock base of Manhattan had advantages for vertical construction, large areas of Brooklyn, Queens, the Bronx, and even parts of Manhattan were low and swampy.

Land-fill began early in the nineteenth century and has continued to today, particularly along shore lines. The present-day tip of Manhattan is man-made land.

Clustering and Pyramiding of Functions

New York's transport role and wholesaling activities proved to be magnets attracting other functions to the same area; a whole cluster of urban functions grew out of these initial advantages. New York became the leading manufacturing city and the leading service city of the United States.

Manufacturers in New York began to stress those products for which the ease and frequency of shipment were more crucial than actual transport costs; others stressed products having high value in relation to their weight: clothing, cigars, toys, tools, and the like.

As the leading transport node and as a major job market, New York was the major center for immigration. Immigration created a steady flow of cheap labor for factories. This flow offset the general tendency for wages to rise in the thriving macropolis and allowed low-wage industries such as the needle trades to flourish. The large skilled labor pool of both natives and immigrants encouraged the creation of new industries and the expansion of old ones. The steadily rising population of the New York Urban System created a very large local market, thus stimulating a phenomenal development of such ubiquitous industries as breadmaking, brewing, candy-making, and brick-making.

The interrelationships of the many attractions of the city can be seen particularly in the case of the garment industry. New York had been a congregating point for buyers since its development as the major textile import wholesaling center early in the nineteenth century. Then, in the late nineteenth century, a flood of Jewish tailors began to arrive from Eastern Europe at just the time when the invention of the sewing machine and other technological advances had made the ready-to-wear industry possible. The abundance of loft space in old factories near the slums of the Lower East Side was ideally suited to the space requirements of the new industry. New York's unparalleled role as a transport node made the rapid shipment of clothing to all parts of the country relatively easy. Finally, the general prestige of the great city helped to establish it as a center of style. The garment industry was one of the outstanding growth industries in America from about 1870 to 1910, and New York was its center.

A somewhat similar cluster of functions developed in the service industries such as banking, insurance, securities and exchange markets, and corporate management. From the beginning of the nineteenth century, these industries indirectly helped the city to grow by adding prestige and by attracting certain types of related manufacturing such as printing. Yet at first the direct effects of such functions on local employment and population growth were relatively small, for even though New York gained the lion's share of such national service functions, the total number of workers in the United States employed in such services was relatively small until the late nineteenth century. But from that time onward, the importance of these service functions and the employment in them increased much more rapidly than in either the primary or secondary sector. From the nucleus of such functions in the early nineteenth century emerged the primate banking, security, and central office management character of twentieth-century New York.

Growth of these functions was initially closely related to New York's role as a transport node and the generally thriving economy of the city. At first the bankers were concerned primarily with the financing of foreign trade, but gradually they generalized their activities to meet the needs of the developing hinterland. New York was already a leading banking city by 1836, but the elimination of its most serious competitor, the Bank of the United States (Philadelphia), in that year confirmed New York's dominance, a position it never relinquished. Thereafter the banking function in New York grew with the developing interior. Its lustrous position as a banking "capital" was cemented and institutionalized with the creation of the headquarters of the Federal Reserve System there early in the twentieth century.

The marine and casualty insurance business also developed in relation to the port at first, with the initial marine emphasis gradually broadening to include related insurance activities all over the continent. New York and Hartford, Connecticut, became the chief centers for such insurance in America.

The securities and exchange markets developed slowly at first, abetted by the stock issued by the new banks and insurance firms. The growth of the railroads and the general industrialization of the late nineteenth century made such activity much more important. Although stock exchanges were established in many other cities, New York's position as banking capital and especially as the center of investment banking enabled it to attract a preponderant share of the total volume of securities transactions.

Life insurance firms began to flourish in New York not only because of the large size of the city and its general national importance, but also because the early life insurance firms were active traders on the stock exchanges. In the late nineteenth century, New York became the life insurance capital of the United States, particularly the headquarters of the largest firms.

With the growth of great industrial corporations in the late nineteenth century, New York became the leading center for corporate central offices. This was a logical development, in view of New York's importance as a manufacturing center, the role played by the banks in corporate mergers, and the great variety of services available which were appropriate to corporate decision-making.

In addition to banking, securities, insurance, and corporate headquaters, all kinds of other lesser but related services grew and nourished one another. Included were law offices, patent firms, and trade organizations. And, of course, religious, educational, and cultural services expanded to meet both local needs and the needs of the developing hinterland. This cluster of services rested ultimately on the advantages of concentrating certain urban services rather than dispersing them among many cities. In short, as long as the economic and social forces tending toward centralization in the United States outweighed the forces tending toward decentralization and dispersal, New York grew rapidly.

The transport system, the wholesaling function, specialized manufacturing, and a great clustering of highly specialized services all combined to make New York the primate city of the United States. It had no competition in size and no competition in the nation-wide orientation of its functions. This primacy became widely recognized and with such recognition came still further stimuli to growth. The city acquired a certain magic aura, the "sweet smell of success." It became a desirable address for business firms, a favored residence of the wealthy, a center of fashion, a magnet for art galleries, a leader in music and the theater, and a favored locale for American fiction. It became, for millions of Americans, a magic place to be ambiguously admired and feared from afar, in tiny hamlets and big cities alike. In short, New York became more truly the focus of American life than Washington, the mere political capital. Except in political function, it became more like a London or Paris than any other city in America.

A Radically Altered Form

The golden age of growth for New York, 1815–1914, was a period of radical change in the form or internal structure, of the city. *Form* refers to the physical distribution and appearance of the various functional parts of the city. The changes of this period were even more striking and far-reaching than those of either the preceding centuries or the post–World War I period. In fact, the basic pattern of the internal geography of the New York Urban System and even the main directions of change were fairly clearly established by 1914. Major among the changes were: (1) upward growth and outward sprawl, (2) development of specialized functional areas, and (3) a geographic shift in the location of the chief point of focus within the urban system.

Upward Growth and Outward Sprawl.—Upward growth and outward sprawl were equally important and the product of technological advances in both communication and transportation which permitted the concentration of jobs and services at the center concurrently with a dispersal of residences. More than any other city in America, New York took full advantage of the new technological advances: steel-framed buildings, elevators, centralized electrical power production, subways, commuter trains, and the telephone.

Upward growth took the form of taller buildings, more closely spaced and containing more square footage per acre of land than found anywhere else on earth. True, much higher skyscrapers were to come in later years, but well prior to World War I New York was the "skyscraper city." Commuters came into town to work in them; tourists came in to crane their necks and look at them.

Outward growth was perhaps less impressive but equally important. In 1820 the city's half-million people lived within 10 square miles of built-up land; by 1900, 140 square miles of built-up land were needed by the area's 5.5 million people. Thus, while the population was increasing by eleven times, the built-up area was increasing by fourteen times, and the average population density was declining. Outward expansion, marked physically by improved transportation lines, was marked politically by the creation of Greater New York in 1898. Herein lay a continuing contrast in the structure of the urban system, the contrast between: (1) the socioeconomic area unified by commuting patterns, the movement of goods, and the functional interrelationships among the various parts, and (2) the political area, the area unified under a single governmental structure. This discrepancy has grown and become magnified in the twentieth century. The inclusion of the Bronx, Brooklyn, and Queens in Greater New York was justified no doubt by the functional links between these areas and Manhattan in 1898, even though some boroughs such as Brooklyn continued to have unique characteristics based on their earlier, separate origins. But the inclusion of Staten Island (Richmond) in Greater New York was of more doubtful value, both for New York and for Staten Island. The island was linked with Manhattan by ferry but not by subway, and hence it was not highly urbanized. For New York this meant the relatively minor cost of subsidizing the ferry service as a sop to Staten Island; for Staten Island it meant having nearly all the disadvantages of belonging to a centralized urban government with almost none of the advantages. From a functional standpoint, it would have been far more logical to add Jersey City and Bayonne to New York, for these areas were more closely tied to Manhattan in both physical and functional terms. But they were in New Jersey and state sovereignty did not permit such cross-state merging of city administration. The New Jersey–New York boundary line was, and is, a barrier within the urban system and is highly important in determining its structure.

Development of Specialized Functional Areas—The conquest of time and space by new means of transportation permitted the growth of large, specialized, functional areas within the city. Subways and commuter railways allowed a separation of place of work and place of residence. This permitted choice of residential areas, with the choice determined largely by the gradual obsolescence of the older residential areas and by the income level and ethnic affiliation of the residents.

Residential succession began and specialized residential areas emerged. The oldest specialized residential area was the Lower East Side, near the docks, north of what was then the city's core (the Wall Street area), and east of the new factory district. The East Side became a slum filled with successive waves of poor immigrants, first the Irish and the Germans, later the Italians and East European Jews. With rising income, the Germans moved northward to establish an ethnic colony in Yorkville on the Upper East Side, just as the Jews later moved out to the Bronx and beyond. Other groups followed similar patterns. Greenwich Village began as a suburb for the wealthy families of Manhattan; later it became a center for newly arrived Italian immigrants (it was not until after World War I that it became a center for "bohemianism" and "beatnikism"). The "Old Americans" (pre-1815 vintage) and others of higher income moved outward, providing the cutting edge of the urban frontier, just as they do today. The wealthiest commuter groups settled in commuter towns beyond the city, along the rail lines radiating outward. These towns developed particularly in the more isolated and more scenic areas such as the shores of Long Island Sound and along the east shore of the Hudson Estuary.

Meanwhile other specialized functional areas developed, too, such as transport areas, manufacturing areas, and commercial zones. The main outline of the railroad and dock areas was already established by the late nineteenth century. The chief rail terminals and associated docks were developed in Jersey City. The general cargo piers spread from the original nucleus on the East River side of Manhattan's tip to the west shore of Manhattan northward to Forty-second Street and southward to the shores of Brooklyn, facing Upper New York Bay. Bulk cargo piers developed in Brooklyn, along the Kill van Kull, and around Newark Bay. Wholesaling, originally concentrated near the East River piers, shifted to the lower West Side, adjacent to the Hudson River piers. Water transportation was highly important within the region, particularly because of the use of barges to make connections with the rail terminals in New Jersey. Consequently, the chief manufacturing areas were developed along the waterfronts, generally adjacent to the dock areas. To a much lesser degree, manufacturing spread out along the rail lines back from the shores. Several areas became noted for "nuisance" types of manufacturing, those characterized by noise, smells, smoke, and heavy machinery. The most notable of such areas were Hunt's Point in the Bronx, Newtown Creek in Brooklyn, and the shores of Newark Bay. The Jersey shore, with the most direct rail connections to the interior, tended to stress heavy manufacturing and manufactured products for the national market. Brooklyn, with better access to the consumers of the metropolitan area, stressed manufacturing for the local market. Manhattan, with the greatest variety of services, stressed light consumer manufacturing for the national market. The CBD of Manhattan, south of Central Park, was the manufacturing core of the whole region, particularly for those types of manufacturing for which the region was nationally famous, such as the garment industry. Although the other boroughs of the city had some manufacturing, they served primarily as commuter zones for the manufacturing focus, Manhattan. Manhattan also served as the chief concentrating point for the cluster of service industries for which the city was nationally famous: wholesaling, banking, insurance, securities, corporate management, and high-quality retailing. A few outer areas, such as Newark and downtown Brooklyn, developed as major service centers, but most of the commercial areas developed outside Manhattan were of a decidedly secondary and local nature.

Shift in the Chief Point of Focus—One of the most dramatic changes in the form of the city was the shift of the urban system's major focus from Downtown (the Wall Street area) to Midtown (the Forty-second Street area). The gradual outward growth of settlement from the old historic core in lower Manhattan had a strong northerly component because of the physical geography of the area, the peculiar pattern of land and water areas. Thus, by the late nineteenth century, the most central and most easily accessible part of the city was in mid-

Manhattan, several miles north of the original core. This was a major reason for the selection of midtown for the site of the Grand Central Terminal (1871), Pennsylvania Station (1910), and the major foci of subway lines. The slow northward migration of business in Manhattan was accelerated by the creation of this new transport node in Midtown. A process of detaching service clusters from the old Wall Street focus and regrouping of these services in Midtown began. This process is still not complete, but by 1914 the dominance of Midtown over Downtown was already established. Corporate offices, hotels, the entertainment industry, department stores, and specialized shops were attracted rapidly to the new focus, Midtown. The great commercial banks and associated services such as the clearing house were less attracted to the new focus. Few banks had even established Midtown branches by 1914. Some insurance firms moved northward, while others clung to the Downtown focus or settled in the broad zone between the two foci. This middle ground, which later became known as the "valley" because most of its buildings were much lower than those of Midtown to the north or Downtown to the south, became the most important manufacturing area of Manhattan, "loftland."

Primate City of the World, After 1914

In the years after 1914 New York increased greatly in world-wide importance. It became not only the world's largest urban system but the one exerting the broadest and farthest-reaching influences. If any urban system had the right to be called a world primate city, New York did. It rose as America rose and changed as America changed. The United States became an economic giant, producing about half of all the world's goods and services. World economic leadership passed from Western Europe to the United States, i.e., from Paris and London to New York. New York's role as the principal center of the new Federal Reserve System powerfully aided it in becoming a major international financial center; its great banks established branches around the earth. Industrial corporations with New York head offices spread their activities to other continents. This was particularly true of companies interested in minerals, such as the petroleum, steel, and copper firms. New York became an important importing point for these and other raw materials. The zone of influence of New York's music, arts, theater, and publishing extended to the ends of the earth. It became a major node, if not *the* major node, in a developing world-wide system of airways. In spite of this latter point, however, it is unlikely that New York can ever be the air gateway to North America to the same degree that it was the sea gateway in the nineteenth century; the air-ocean has no boundaries, and hence New York can be bypassed. New York's global primacy was recognized, at least symbolically, by its selection as the site for the United Nations after World War II.

A Declining Status within the United States, After 1914

In spite of the improvement in New York's world-wide position, its situation within the United States exhibited post-World War I deterioration. This deterioration first took the form of a slowing down of population growth. In 1830 the area here referred to as the New York Urban System had less than 5 per cent of the population of the United States; in the golden century thereafter, the area consistently grew more rapidly than the nation as a whole, even during the period of rapid western expansion. Thus by 1930 it had over 9 per cent of the population of the United States. But after 1930 growth was slower, approximating the same rate as the nation or even a little less. Meanwhile, other centers, such as the nation's capital and the cities of the South and Far West, grew much more rapidly than the national average. Broadly speaking, certain aspects of both decentralization and centralization worked against New York. Noteworthy among these aspects were decentralization of the economy and population throughout the United States and centralization of political power and administrative function in Washington, D.C. Changes in the distribution of consumers, in the location of manufacturing, in the direction of trade, and in the methods and costs of transportation—factors that had long worked to increase the importance of New York—now began to work against the city. Relative declines in trade, wholesaling, and to a lesser degree manufacturing, but success in service expansion, characterized New York's new position in America. More and more, it became the giant in a country of giants rather than a giant in a country of pygmies, as it had tended to be earlier.

Trade Status

The long, slow decline in New York's percentage share of total United States foreign trade, which began in 1871, accelerated sharply after World War II, falling to 32 per cent of the total foreign trade and 38 per cent of the ocean-borne foreign trade in 1957. Though no other port had more than 10 per cent, giving New York a comfortable superiority, the gap between it and other ports had narrowed greatly. As a result, New York had lost much of its uniqueness; now other ports could claim frequent sailings to widely dispersed destina-

AIR GATEWAY TO A CONTINENT. The vast complex of Kennedy International Airport is unequaled, and fortunate it is that it could be built so close to the center of the New York Urban System by filling in swampland. Large and versatile as this facility is, it is already becoming congested. *Port of New York Authority.*

tions and could provide many of the specialized services formerly available only at the Port of New York. The greatest declines were in bulk cargo, but even in New York's specialty, general cargo, trade was lost to other ports. Similarly, although coastal shipping between New York and other American ports such as those of the Gulf and the Pacific Coast remained important, New York lost much of the local coastal transshipment trade it had formerly enjoyed between New England and the South. There will probably be a continued but more gradual decline in New York's percentage of all trade but a modest increase in volume because of the expected over-all increase in United States foreign trade as a whole.

Wholesaling Status

A relative decline occurred in wholesaling, too, though the inertia in the location of the offices of wholesalers tended to blunt the decline. Even here, however, a separation of functions occurred to some degree; sometimes, while the paper work continued to flow through New York offices, the goods themselves bypassed the city. Similarly, the role of the wholesaler in the economy declined to some degree because of the rise of vertically integrated firms. When the whole production, storage, transportation, and retailing sequence takes place within a single firm, there is much less need to route trade through the larger cities with their highly developed wholesaling and related services.

Manufacturing Status

In manufacturing the urban system fared better, though here, too, there were losses. The port became somewhat more important as a source of raw materials. However, the most rapid development of the refining industries was in the nearby Delaware Valley, rather than in the New York Urban System. New York was at a disadvantage in national marketing because of high transport costs due to its off-center location even in the latter part of the nineteenth century. But now this factor became much more important through continued population dispersal westward, the reduction in regional disparities in purchasing power, and the rising ton-mile costs of transportation. Hence New York was more and more at a disadvantage in manufacturing activities for which transportation costs were a significant factor. A major New York industry, printing for national markets, began to leave the area. Improved transportation and communication allowed the separation of printing and publishing functions; New York remained the publishing capital though declining as a center for printing.

New York's skilled labor pool, once a major advantage, began to decline in importance as the nation urbanized and skills became widespread. Though New York remained the principal immigrant gateway after 1914, the former flood became a trickle with the immigration restrictions of the twenties. Further, many of the newer immigrants who did come, particularly the refugees, were often professional people unwilling to enter low-wage industries. A new flow of cheap labor from Puerto Rico and the South began. However, although the Puerto Ricans first concentrated almost exclusively in New York they soon established colonies elsewhere and began to fan out over most of the old manufacturing belt. Puerto Rican and Negro women entered the needle trades, but the men did not because they considered them a woman's field. Furthermore, the new mobility of industry tended to reverse the nineteenth-century pattern, i.e., factories began to move to the sources of cheap labor in Puerto Rico and the South rather than requiring the cheap labor to move to the factories as in the past.

The garment industry illustrates both the process of industrial dispersal and the holding power of a primate city. In the nineteenth century it was concentrated almost entirely in Manhattan. Gradually, however, the industry spread outward, first to outer parts of the New York Urban System, then to cheap labor areas in continental United States such as the anthracite coal region of Pennsylvania, the declining industrial towns of southern New England, and the South, and finally to more distant areas such as Puerto Rico and even Hong Kong and Japan. The bulk of the garment industry remained in New York, but the geographic dispersal had a major effect. At first the movement to cheap labor areas occurred largely in the more standardized types of production, such as women's house dresses and men's work clothing, with New York easily retaining the production of items more subject to rapid changes in style. But then with new ease of transportation, particularly induced by trucks, separation of functions within the industry became possible. Dresses were designed and cut in New York, shipped to low-wage areas for sewing, then returned to New York for final finishing and selling. Thus, even though geographic dispersal occurred, New York remained the focus, retaining the selling function, much of the managerial function, the designing, the cutting, the final finishing, and, in some cases, the entire operation. The garment district, rather than being *the* garment district, became the center of a far-flung garment industry. It was able to retain this role not only because of historical inertia but also because of the great advantages of New York as a selling and style center.

In spite of the slow growth of many types of manufacturing in the New York Urban System after 1914 and actual declines in some fields, New York remained a major national manufacturing center. A primary reason for this was its role as a center for the origin and dispersal of new industries. Industries, such as plastics and electronics, kept developing to take the place of industries moving away. Infant industries, still unstandardized and experimental, thrive in New York, with its great clusters of services and suppliers and resultant flexibility.

Service Status

There is one area of activity within the United States in which New York did very well after World War I, and that is services. Physical expression of this success took the form of the great office-building boom in Manhattan. The prestige of the primate city and the great cluster of existing services were strong attractions for new services. Two of the most important of these were advertising and public relations. A large percentage of the newly important foundations, such as the Ford Foundation, settled in Manhattan. National organizations of all kinds, including religious ones, flocked to the city. The new Interchurch Center near Columbia University is one example. This building, built to house national and international Protestant activities, was located in New York because of the city's transportation and numerous services, even though some site much farther west and south would have been nearer the Protestant center of gravity in the United States. The entertainment industry boomed, too, as did the tourist industry. Yet of the major components of the old service cluster (commercial banking, insurance, corporate management, and securities exchange), only the latter two grew rapidly enough to increase New York's stature as a national center of such activities. Domestic commercial banking and insurance grew somewhat more slowly than the national average. New York was still the great national center for both activities, but its dominance was somewhat tarnished by new competition from other centers. In balance, however, New York increased in importance as a national service center; this growth greatly enhanced the city's prestige and to some extent offset its declining fortunes as a transport node and as a wholesaling and manufacturing center.

Changes in Internal Form, After 1914

In the last half-century, there have been great changes in the form of the New York Urban System and closely related changes in the meaning of its site. For the first time, there were widespread attempts to control form through public and quasi-public planning; however, these attempts have been more successful in modifying details of the form than in directing the over-all growth pattern. The region expanded rapidly in area through suburban sprawl; yet, contrary to the experience of most other American cities, there was also great upward growth at the core, the Manhattan central business district. Between the two rapidly developing areas, the expanding suburbs and the dynamic core, there emerged a vast area of slow decay, the "gray zone." Despite the vitality of the old Manhattan core, the region became less unifocal and more multifocal.

Impact of Planning, Zoning, and Regulating

Public and quasi-public planning became important for the first time. While the region had been growing most rapidly in population and changing radically in form from 1815 to 1914, a philosophy of limited government prevailed; hence public attempts to direct the pattern of growth were few. A few regulatory ordinances were enacted, such as those which banned certain "undesirable" types of manufacturing from New York City and thereby relegated them to outlying areas such as Newtown Creek or the New Jersey meadows. But in the main, like Topsy, the system just grew. Then in the twentieth century a subtle change in thinking took place; planning by private corporations such as banks and insurance firms and by various public and quasi-public agencies became fashionable and increasingly gained momentum. New York City adopted its first comprehensive zoning law in 1916; many outlying towns and suburbs quickly emulated the city. The Port of New York Authority, a bistate, quasi-public agency, was created in 1921 to deal with the long-festering transportation problem. The incorporating law limited the authority to a circle 25 miles in radius, focusing on the southern tip of Manhattan. A new private study and planning group, the Regional Plan Association (1929), became the major source of comprehensive information and planning for the urban system. A wide variety of special consulting, planning, and regulating agencies blossomed, most of them bistate or tristate in composition: The Interstate Sanitary Commission (the 1930's), the Civil Defense Committee, the Waterfront Commission (1953), and the Metropolitan Regional Council (1956). Housing authorities attacked the slum problem; urban renewal became the order of the day. Private groups, such as banks, insurance firms, utilities firms, welfare agencies, and church groups began to plan on a regional and comprehensive basis.

How effective was all of this consultation, planning, and regulation in directing the region's growth, in modifying or determining the form it was taking? To what degree did planning take the place of the simple, unregulated, organic growth of the nineteenth century? Without minimizing the very real accomplishments of these groups, it must be recognized that the forces creating changes in the form of the urban system largely were beyond their powers to forecast or control. Thus, in the main, the system just grew, as in the past. Much of the zoning of Manhattan tended to "freeze" the location of the garment district, increasing the congestion there and not allowing a continuation of its traditional northward migration. The congestion of the garment district and the lack of new loft space have been major factors driving the industry from Manhattan. Tight zoning in some suburban areas, particularly those zoning regulations aimed at controlling population density, tended to slow outward growth, somewhat limiting but by no means overcoming the pressures for decentralization. That is, to some degree, the new zoning laws had the same effect as the walls of the medieval city: "containing" the city.

But the suburban zoning laws were not uniform. Always there were suburban areas without zoning laws or with very flexible zoning regulations. These became the "holes in the dike" through which the outward-flowing urban tide could burst. Thus suburban zoning affected the pattern locally but not generally, producing "rivers" and "lakes" of concentrated and very rapid residential, commercial, and industrial development in some areas while leaving "islands" of slower and more ordered development elsewhere. There were many slum clearance projects in the older urban areas, but new slums grew in other areas; overall, the growth of slums and the spread of blight were much more rapid than slum clearance. The Port Authority, because of its own predilections and because of certain legal restrictions in its charter, concentrated on improvements in the port, the development of a regional system of airports and jetports, and the building of tunnels and bridges, while doing relatively little to improve the archaic rail and subway commuting systems. The Port Authority and other groups did affect the form of the region, but primarily in the field of transportation and primarily within the area already highly urbanized in 1914. Meanwhile, the greatest growth of population and urbanization occurred in the outer areas, areas beyond the reach of any single governmental authority or regulating body. The Metropolitan Regional Council, which initially included representatives from most of the governing bodies of the twenty-two-county region, was of an advisory nature, without enforcement powers. The various groups set up after 1916 to regulate and control found themselves reacting to changes in form more often than controlling the form.

Suburban Explosion

The filling up of the region, the pronounced outward sprawl of the urbanized area, was evident to alert observers even before World War I. Yet few, except perhaps the more canny land speculators, were prepared for the revolution in land use set off by the widespread use of automotive transport. The American Dream of a house in the country with space for "the kids" was old and persistent; indeed, it had produced the sprawl of the late nineteenth century. But few realized the strength of this dream until the automobile suddenly made the dream realizable for the middle classes as well as the rich. Though the 5.5 million people of the region used only 140 square miles of urbanized land in 1900, the less than 15 million of 1954 used about 1,100 square miles of built-up area. Thus, while the population was increasing by less than three times, the urbanized area was increasing by eight times. The areal expansion continued without signs of abatement after 1954, awaiting only another detailed land use survey to set new records in the profligate use of the area's limited space. The old commuter towns of the railway age had been like beads of a string, well separated by open country. Outlying cities, such as White Plains, were separated from New York City by wide belts of open land, even though they were to some degree functionally tied to Manhattan. Now all of these open spaces, the interstices between developed areas, were accessible by automobile. Residential development filled these spaces where terrain permitted and went still farther, extending hundreds of tentacles of urban development along the major highways and expressways in contrast with the dozen or more urban tentacles of the railroad era. The urbanized area has increased by an average of more than 30 square miles every year since the end of World War II.

Nor has the suburban explosion been confined to residences. Manufacturers, too, sought the open country, not so much because of dreams of the good life as because of changes in manufacturing technology. The new assembly line processes of industry had made the old multistory factory buildings of the past obsolete while the new accessibility of the outlying areas for supplies and workers freed the manufacturer of the tyranny of a central location. This suburbanization of employment opportunities in turn accelerated suburban residential growth, though in some instances it also created a reverse commuting pattern, i.e., the movement of workers from residences in the decaying core cities to the jobs in the new plants of the suburbs. Retailing, except that of the most specialized sort, perforce followed the customers to the suburbs too. Many suburban shopping centers became far more than mere neighborhood shopping areas; instead, they became regional centers with great drawing power. The shopping complex in Paramus, New Jersey, composed of what might be defined as several regional shopping centers, drew trade not only from the surrounding suburban zone but also from the older commercial zones of northern Manhattan to the east and Paterson-Passaic to the west. Wholesalers, too, felt the pull to the open spaces, partly because their customers, the retailers, were moving outward. But more important, the flexibility of the truck permitted local wholesalers to leave the core. Other service industries, such as local commercial banks and local insurance businesses, moved outward, too, primarily through the establishment of branches.

In many respects this pattern of rapid outward growth was not unique to New York; cities all over America were experiencing similar areal expansion. Yet New York's experience was distinctive; it had been much more densely populated than most American urban systems and had taken much greater advantage of vertical construction and subway transportation. Therefore, the trend toward automotive transport and single-story construction was much more of a revolution. Secondly, the very massiveness of the New York system created differences of kind as well as differences of degree in suburbanization. For example, each of the great regional shopping centers of Paramus, New Jersey, is larger than most such suburban shopping centers around the country. Within a circle of a mile radius, there are several such mammoth regional shopping centers plus many large stores not directly part of the planned shopping centers. The cluster as a whole has more downtown services and a greater volume of business than the central business districts of cities the size of Cleveland. A nodal highway location in a rapidly growing suburban zone plus imaginative planning have created a new urban phenomenon. Paramus has passed a certain threshold of growth and is now a major urban focus within the New York Urban System.

Continuing Upward Expansion

Even with all of the suburban expansion, there has also been upward expansion at the core in Manhattan. For New York this has not been historically unique. As previously outlined, the pattern of growth in the late nineteenth century had been of this type; upward growth at the core had gone hand in hand with outward sprawl. But among great American cities of the twentieth century, this experience was highly unusual if not unique. While most American cities were complaining of dying cores and seeking by heroic urban renewal programs to stave off decay, Manhattan gloried in the greatest building boom of its history. The office-building boom of the twenties staggered the imagination, creating the Manhattan skyline of song and story. But this was child's play compared to the boom that began after World War II, a boom of more than twice the dimensions of that of the twenties. Between World War II and 1962, more than 50 million square feet of office space were added to the 100 million square feet available in Manhattan before World War II. This was more new office space than was being added in nearly all other American cities combined. At the same time much of the older office space was being thoroughly reconditioned and modernized. The major center for such new office construction was in the Grand Central–Park Avenue area, although some new structures were erected around the old Wall Street focus too. This new office construction increased office employment in Manhattan but not as much as it might seem, since the new affluence of American life was partially expressed in an increase in office space per worker. Some of the new office construction was related to local services within the region such as local commercial banking or newspaper publishing, but most of the new office construction related to Manhattan's role as a national service center. Similarly, the growth of hotels in Manhattan related primarily to New York's national service role. This buoyancy of the core of the New York Urban System was particularly striking because it occurred as a natural organic development and not as a result of the activities of the many planning agencies.

The Developing Gray Belt

Between the rapidly expanding suburbs and the dynamic core in Manhattan was a broad zone of slow decay. This area, termed the *gray belt*, included nearly all areas developed before 1914 except the central business district of Manhattan. That is, it included upper Manhattan, most of the Bronx and Brooklyn, part of Queens, Bayonne, Jersey City, Hoboken, Paterson–Passaic, much of Newark, part of Yonkers, and the cores of hundreds of old satellite towns and commuting centers. This is pretty much the middle city described in Chapter 14. These gray belt areas acquired some of the manufacturing and services formerly concentrated in Manhattan, so that they became less commuter areas for Manhattan and more commercialized communities. Housing aged and became available to the low-income families from the core. Though there were many urban renewal projects, these hardly touched the problem of hundreds of square miles of aging, deteriorating urban land.

Multifoci

Despite the building boom of Manhattan and the many other evidences of Manhattan's vitality as a national service center, the region was gradually becoming less unifocal and more multifocal. That is, this building boom masked the fact that while the central business district was becoming more and more important as the "Main Street" of the nation, it was becoming less important as the "Main Street" of the urban system. Manhattan did remain the principal focus of the urban system, but many lesser foci developed too. The central business district still had about one-third of all the jobs in the region in 1956, demonstrating its focal dominance. And in wholesale trade, finance, the theater, garment-making, printing, and publishing, it had much more than half of the urban system's jobs. Yet old New Yorkers could remember the time when the CBD had two-thirds of all jobs. The decentralization of jobs was not a simple dispersion outward in all directions; rather, lesser foci emerged such as Bayside, Jamaica, Garden City, Hempstead, White Plains, Paramus, and Yonkers. Many older outlying foci such as Newark retained or acquired important service functions. The multiplication of foci and the development of expressways between these foci as well as with Manhattan created an amorphous, fluid structure for the region coexistent with the remnants of the old rail-subway unifocal pattern of the past.

A New Meaning of Site

These radical changes in form changed the meaning of the site. A large part of the suburban residential, commercial, and industrial development was carried out by the new large-scale developers. These developers, planning vast projects on the drawing boards far from the sites, needed large tracts of flat, well-drained land; they found small, irregular, steep, or swampy sites expensive not only in construction but also in planning. Hence the most rapid urban growth was on the flat, well-drained plains of Long Island and north-central New Jersey. The development of the hilly land to the north and west was much slower. Little by little, the urbanized area lost its simple radial pattern and assumed a more linear east-west orientation in conformity with the terrain and the direction of the coastline. Large-scale land-fill projects, directed by the Port Authority, transformed formerly unusable swamps, as in the case of La Guardia, Kennedy, Teterboro, and Newark airports, and Port Newark. Although talk of draining the Hackensack Meadows of New Jersey had persisted since colonial days, for the first time the project seemed economically sound and technologically feasible.

RETROSPECT ON THREE AND A HALF
CENTURIES OF DEVELOPMENT

For nearly three and a half centuries, the only constant ingredients of the geography of the region have been those aspects of physical geography highly resistant to the transforming power of man, such as the climate, the tides, and such terrain barriers as the Palisades. The human geography of the area, on the other hand, has been highly dynamic, perhaps more dynamic than that of any other area of similar size on earth in all the history of man. The situation changed repeatedly in response to changes in places both near and far, as New York became in succession a regional capital, the primate city of the United States, and, at last, the leading world macropolis. The site had one meaning for the colonists clinging to the edge of a continent with their small ships, ineffective land transport, and small-scale economy. It had quite another meaning in the age of large ships, continental railroads, a developing continent, and subways, and still another in our age of nation-wide economic decentralization and of suburban expansion and expressways. Similarly, though New York has been multifunctional for over a century and a half, the precise composition of functions has been constantly changing. Though the area over time became less and less attractive for certain functions, it always succeeded in attracting other, newer functions. In terms of form there have been many New Yorks: the tight colonial cluster of lower Manhattan; the sprawling, subway-railway city of many specialized functional areas but with a single great heart, Manhattan; and the amorphous, multifocal, expansive city with bright new suburbs, bright and shining towers in Manhattan, and miles of aging slums in between.

Livelihood Structure

In 1960 the New York Urban System had only 0.7 per cent of its labor force in the primary sector, a low 34.1 per cent in the secondary sector, and an outstandingly high 65.2 per cent in the tertiary sector. All of the upstate urban systems had relatively greater orientation toward manufacturing and relatively less orientation toward tertiary activities.

During the decade 1954–64 manufacturing employed between 1.9 million and 1.8 million persons; in 1947 the figure was only slightly below this at 1.75 million. Although the highest level was reached in 1954, the figures exhibit a rather static situation in employment. This is encouraging in view of congestion problems in the city. Growth in suburban counties has roughly kept pace with losses in central areas.

Manufacturing tends to be dominated by small-scale, light, specialized types. Ranking first with nearly a fifth of the factory workers is the production of apparel and related products; next is machinery (electrical and nonelectrical), followed by printing and publishing. Of considerable importance are a large number of other types including food, chemicals, fabricated metals, transportation equipment, textiles, and scientific instruments.

Actually, the significance of manufacturing is not widely appreciated, even by many of the residents. Though a few large plants exist in peripheral parts of the urban system, the area generally lacks such obvious evidence of manufacturing as the steel mills of Buffalo or the electrical plants of Syracuse and Schenectady. Far more characteristic are small plants tucked away in tall loft buildings resembling offices or tenements. Indeed, many Manhattan buildings house both offices and factories, and hardy artists have been known to coexist with loft-quartered factories. But, while manufacturing is generally unobtrusive, the office towers of Midtown are anything but unobtrusive. Rather, they are monuments built to impress the observer, like clusters of arresting exclamation points. Small wonder that New York is more famed for its eye-catching office functions than its less evident but significant manufacturing. Similarly, though the

New York Urban System is the leading center for manufacturing in the United States in both employment and value added, its share of total American manufacturing (12 per cent) is much less than its share of the nation's business and professional services (37 per cent), wholesaling (42 per cent), and finance (49 per cent).

Of the 6.5 million employed workers in the twenty-two county urban system in 1960, roughly 4.25 million were employed in the tertiary sector. An 18.1 per cent gain in tertiary employment between 1950 and 1960 meant an addition of 650,000 jobs. In 1960, 23.1 per cent of all employed persons worked in service industries, giving the New York Urban System not only a much larger service function but a higher relative orientation toward service than any of the upstate urban systems. Unusual concentrations of specialized national services such as banking, insurance, wholesaling, and management exist. Retail and wholesale trade accounted for 19.3 per cent of all workers, public utilities 8.6 per cent, finance, insurance and real estate a very high 7.3 per cent, and public administration 4.8 per cent. The total sales and service function of the twenty-two-county area in 1960 produced a tertiary potential (see Chapter 12) of 24.5 million, and a tertiary surplus of 8.4 million.

The economic activities of any city may be placed into two categories: (1) city-serving or nonbasic, and (2) city-forming or basic. The first category includes those activities which are oriented toward local needs and do not ordinarily involve an inflow of wealth to the city; the second category includes those activities which are oriented to customers outside the city and so presumably result in an inflow of capital and general growth. The distinction between city-serving and city-forming activities often is clear-cut. Obviously the many new hotels being built in Manhattan are intended to serve primarily the millions of annual visitors and are therefore city-forming or basic; quite as obviously the recent growth of metal-working industries in the region results from the attraction of the large local market rather than any new-found ability of the region to compete nationally with the steel centers of the Midwest. But the great complexity of the urban system precludes easy distinctions between city-forming and city-serving activities. For example, the national reputation of the Broadway theater might seem to make this a city-forming activity. Yet the best available estimates indicate that only about 20 per cent of the tickets sold are used by visitors. Similar estimates show that only about 10 per cent of the restaurant and night club revenue is derived from visitors, despite the national fame of a few clubs. The central offices of giant corporations are clearly city-forming just as local or branch offices are city-serving, but the available statistical evidence on these types of activities is often ambiguous. Despite such analytical problems, broadly accurate estimates of the sources of the region's livelihood can be made. In 1956 about 40 per cent of the region's jobs were city-forming and about 60 per cent were city-serving, a ratio not unusual for large urban systems in the United States.

CITY-SERVING ACTIVITIES

Locally oriented or city-serving activities of New York are broadly similar to those of other American cities in terms of major categories. However, within each of these categories the degree of refinement and specialization is greater than is common in smaller cities. As indicated, the theater is more important in the lives of New Yorkers than elsewhere, though only a minority of the population ever attends. Similarly, there are many more restaurants specializing in a particular and often exotic cuisine than in most other cities. Retail trade and consumer services are numerically much the most important city-forming activities, employing nearly one-fifth of all workers and almost one-third of all city-serving workers. Other important city-serving activities in declining order of importance include government, local business and professional services, public utilities, wholesaling, manufacturing, construction, and finance.

Wholesaling is much more important as a city-serving activity than might be assumed. About 7 per cent of the workers of the region are employed in wholesaling, according to the census. However, the census classifies as manufacturing workers many employees who devote secondary but important parts of their time to wholesaling activity. In any case, about 60 per cent of the census-classified wholesaling is city-serving.

Numerically, manufacturing is one of the least important city-serving activities; only about one-eighth of all manufacturing employment is in the city-serving category. Yet it is this small sector which does the most to create the common image of diversity in the area's manufacturing. The ubiquitous industries, such as brewing, baking, newsprinting, and upholstering, are highly developed and are city-serving. But more significant are the many industries which are often city-forming on a national scale but which are city-serving in New York, such as auto assembly, metal working, cigarette-making, and boatbuilding. Until recently, few other local markets in the United States were large enough to permit development of such diversified city-serving manufacturing.

CITY-FORMING ACTIVITIES

Many of the unique aspects of the urban system's livelihood structure are found among activities in the city-forming category. Three activities of but minor city-serving importance are outstanding in the city-forming category: (1) manufacturing, (2) finance, and (3) business and professional services. In 1956 these three provided only about 25 per cent of all city-serving employment but almost 90 per cent of all city-forming employment. Hence, it is to these three that our attention is particularly directed.

Manufacturing Activities

The real core of the city-forming manufacturing is a group of unstandardized, small-scale, semihandicraft industries characteristic of primate city manufacturing. New York, like other primate cities, has many disadvantages for manufacturing, such as high rents, high taxes, congestion, architectural obsolescence, and high labor costs. On the other hand, it has unique advantages for unstandardized production. There is no better place to keep in close touch with changes in demand or in the development of new products and styles. There is no better place to obtain quickly almost any kind of raw materials. No place has a greater variety of services for hire, permitting the producer to concentrate on only part of the productive process. Shipment is fast, frequent, and adapted to the less-than-carload lot needs of small producers. The labor force is extraordinarily diversified, including nearly every skill and every wage range. No city on earth has a greater variety of manufacturing space for rent.

Industries that might be considered typical include women's and misses' outerwear, children's outerwear, millinery, handbags and small leather goods, furs, periodicals, scientific instruments, costume jewelry, and notions. These are all definitely city-forming industries, the urban system having at least 40 per cent of national employment in all and considerably more in several. These industries employ only one-twentieth of the manufacturing workers nationally but about one-fifth of those of the urban system and about one-fourth of all city-forming manufacturing workers. Furthermore, these eight industries are typical of the many other city-forming industries of the urban system. Most products are nondurable consumer goods. The nondurability may be based on perishability, as in periodicals, on style vagary, as in clothing, or on rapid technological change, as in scientific instruments.

The New York Urban System is unique in America in having such a large proportion of its total manufacturing employment in small-scale, unstandardized, semihandicraft industries. Although it has less than one-eighth of all manufacturing employment in the United States, it has over one-fifth of all factories. Small single-plant firms predominate rather than the multiplant industrial goliaths so characteristic of America. This industrial complex seems an anomaly in twentieth-century America, with its strong emphasis on standardized items and mass production. But the superficial resemblances to the manufacturing of certain European regions or to early nineteenth-century manufacturing in the United States are misleading too. This manufacturing complex is the result of a long and still active selection process in which the more easily standardized types of production gravitated to various parts of the United States favorable to mass production and mass distribution while the less easily standardized types of production gravitated to the macropolis. The unstandardized industries have achieved some of the advantages of specialization and division of labor through geographic clustering that other industries have achieved through standardization, mass production, and vast plants. Though individual plants are small, the production of each plant is normally functionally integrated with that of many other smaller plants, almost as in the steps or stages of an assembly line but with more flexibility. Thus, though the garment district is often considered simply a collection of many small establishments, there is actually a very complex system of production flow from establishment to establishment. In a sense, the area operates as one giant factory with the same sort of specialization by plant or firm that would be specialization by departments or divisions within a single large plant in some industries. Other New York specialties, such as furs, buttons, or foundation garments, cluster together too. They nurture each other through geographic proximity and functional interrelationships, creating together an economic complex which can not only survive but also grow, even in a world of standardization and mass production. Thus the characteristic factory of the New York Urban System is a flexible, small unit producing a highly variable product on short notice and on short production runs. This contrasts sharply with the characteristic huge plant of the Buffalo Urban System, which produces a standardized product in endless quantities on a long production schedule.

However, the emphasis on unstandardized production has had one unfortunate effect on the livelihood structure of New York. This is the general tendency for the average manufacturing wage to rise more slowly than in the United States as a whole. New York is a high-wage town only in the sense that the average wage in a given industry is generally at or above the wage prevailing nationally for that industry. The generally rising wage rate in the United States is perforce tied ultimately to increases in productivity. Spectacular increases in productivity have occurred primarily in the mass production industries. Productivity increases have been much slower in the unstandardized types of manufacturing characteristic of New York.

Tertiary Activities

The New York Urban System is world famous as a financial center, the area term, Wall Street, being symbolic of power and importance for nearly a century. This image is so strongly developed that most people tend to overemphasize the importance of finance as a city-forming activity in the region. Finance does provide about 5 per cent of all employment in the region, but about three-fifths of this is actually city-serving in nature. Finance provides less than 6 per cent of all city-forming jobs in the region. As city-forming activities, manufacturing is twelve times as important, and business and professional services are three times as important. In fact, retail trade is almost as important as a city-forming activity as finance. Finance, however, has been of extraordinary importance historically in attracting other city-forming activities, such as corporate central offices, to the region.

Image-building or iconograpy is a slow historical process. The image of the New York Urban System as a center for business and professional services other than finance is only slowly emerging. Since World War II, a new area term, Madison Avenue, has been coined to reflect the new image. This term is not very accurate either geographically or functionally, but its vogue indicates the need for a new term to describe the new New York. The Managerial Revolution, potentially as significant as the earlier Industrial Revolution is creating a new livelihood base for the urban system. Today business and professional services provide about one-sixth of all city-forming employment, nearly one-fourth as much as manufacturing. The most important group, though not the numerical majority of such workers, is in corporate central offices. Of the five hundred largest industrial firms in the country, more than 30 per cent have their headquarters in New York while nearly all others maintain major branch offices there. The very large firms are particularly prominent; of the top fifty industrials, nearly half have New York headquarters. Many of the office jobs are relatively routine, but a growing percentage are highly skilled, highly specialized jobs in various aspects of economic planning. In terms of planning policy for these industrials, head offices resemble in a sense the central bureaucracies of totalitarian states. That is, much of the economic planning that is concentrated in the political capital in a totalitarian state is concentrated in New York, the economic capital of the United States. Head offices are functional magnets attracting other activities just as finance and wholesaling have long done. Such related enterprises in-include business associations, labor union offices, public relations, advertising, law, accounting, engineering, tax consulting, location planning, and so on. Large head offices have their own departments of law, accounting, public relations, and the like. However, even the largest firms are not large enough to be completely self-servicing. The various specialized departments within a head office function partly to meet company needs directly from within and partly as liaison agencies to provide appropriate links with outside servicing groups such as advertising agencies, law offices, and engineering firms. In totality, this managerial complex provides some of the same advantages of geographic clustering noted in unstandardized manufacturing.

Though corporate head offices may be considered the core of this complex, there are many managerial activities not directly related to the activities of such offices. For example, New York is a preeminent center for the great foundations, such as the Ford Foundation. The principal activity of such foundations is much like that of the corporate central offices, i.e., the managing and planning of vast investment funds. Though the Ford Foundation specializes in financing education and cultural pursuits, its managerial activities are in some ways quite similar to those of an industrial corporation.

The emergence of this tremendous managerial function in New York has both positive and negative aspects. On the bright side, managerial employment is increasing rapidly, maintaining the urban system as a primate center. Similarly, since the wages and salaries paid are relatively high, the impact on city-serving activity is relatively great. But there is a dark side to the picture, too. The growth of the managerial function is a major element in the explosive social geography of the region. The well-educated and relatively well paid managerial elite concentrates in the more affluent suburbs and in inner areas such as the East Sixties of Manhattan, isolating itself geographically from the social, political, and economic problems of the masses. Wage rates in the unstandardized manufacturing industries characteristic of the area not only are lower, but they are not increasing very rapidly for reasons already cited. The masses, and particularly the Negroes and Puerto Ricans at the bottom of the socioeconomic scale, understandably feel frustrated by the contrast between their lives and those of the elite. This problem is common to all urban systems, but it is acute in New York because of the proportionally greater percentage of such elite jobs and the relatively small importance of that great leveler of manufacturing wages, mass production. Nor is this purely a problem of the city versus the suburbs. The problem is also found within many older suburbs.

Internal Form

This interpretation of the current internal form of the New York Urban System is presented below in three parts: (1) sources of New York's distinctive form, (2) concentric growth rings, and (3) commuting patterns.

Sources of New York's Distinctive Form

Large cities tend to become more and more alike in the twentieth century. Thus there are many ways in which the structure of the New York Urban System resembles that of other large American cities and other primate cities in other continents. And yet New York is distinctive, too. In many respects, there is really only one New York.

Perhaps the most obvious source of New York's distinctiveness is its great population and physical size. An urban area nearly fifty miles in radius with almost 16 million people is akin to a nation, a phenomenon of a different order than most urban systems. The large population is at least partly responsible for the great pressure for space near the center, the emphasis on tall buildings and the unusually high population densities in the middle city near the core. The great population has also allowed a higher degree of specialization among geographic areas within the urban system than is common. There is more differentiation among types of shopping areas, more specialization in service areas, more diversity in kinds of manufacturing areas, and a greater range of kinds of residential areas than is possible in smaller, less populous urban systems.

Another source of distinctiveness in form is the highly diversified livelihood base. Many specialized areas are needed for such a diversified kind of manufacturing and such a diversity of special services. Similarly, the great range of income of the inhabitants leads to great diversity in residential areas.

Few native New Yorkers fully recognize the importance of the site as an enduring source of New York's distinctiveness. Yet it is one of the most important features of the system's form. The area is highly fragmented, a complex of bays, inlets, peninsulas, headlands, and islands. The mixture of cliffs such as the Palisades, swamps and tidal marshes such as those of the south shore of Long Island or the Jersey Meadows, morainic hills such as those of the north shore of Long Island, the sandy outwash plains of central Long Island and Staten Island, and the rugged hills to the north in Westchester provide a site with great internal diversity. Much of the beauty of the region results from these nature-induced vistas, though there are many ways in which the full potential of the site has not been used effectively. Among major world macropoli only a few such as Rio de Janeiro have similar site complexity.

Perhaps the most important source of New York's distinctiveness is its peculiar blend of historical remnants and dynamic modernism. Like other cities of the Atlantic Seaboard, New York's form reflects its early beginnings in the seventeenth and eighteenth centuries. This is readily apparent in certain sections, including not only parts of lower Manhattan such as Greenwich Village but also parts of Brooklyn and parts of suburban towns such as Hackensack, New Jersey. Narrow streets are not confined to lower Manhattan, nor traffic jams to the core and middle city. These remnants from the preindustrial era set New York apart from the American cities of the interior and link it with older cities in Europe and on the Atlantic Seaboard.

On the other hand, New York is quite different from most other seaboard cities such as Philadelphia or Boston because it has been so dynamic for the last century and a half. The old is constantly giving way to the new. The mixture of the new and the old gives New York a pattern of great distinctiveness.

Concentric Growth Rings

As suggested in Chapter 14, every urban system has an identifiable CBD or downtown area with a high degree of transport nodality. Around this core there are successive rings of urban development, decreasing in the intensity of land use and in age as one proceeds outward to the urban frontier and the strictly rural land beyond. Some critics have pointed out that this concentric ring model is so simple that it distorts the true structural pattern of the New York Urban System. For example, although Manhattan forms the major core, there are also lesser cores at Yonkers, White Plains, Garden City, Newark, Paterson-Passaic, and so on. Each of these cores has transport nodality and is surrounded by successive rings of urban development. In reality the system consists of many cores and the various rings around the numerous cores overlap, producing a very complex urban pattern. On the other hand, Manhattan contains without doubt the principal core of the urban system, thus exerting great influence on the structure of the entire area. Therefore, for the purposes of this broad overview of the urban system's structure, a concentric ring analysis is of substantial value. (See Fig. 129.)

The principal core or CBD is primarily the part of Manhattan that lies south of Central Park. Manhattan may also be thought of as containing much

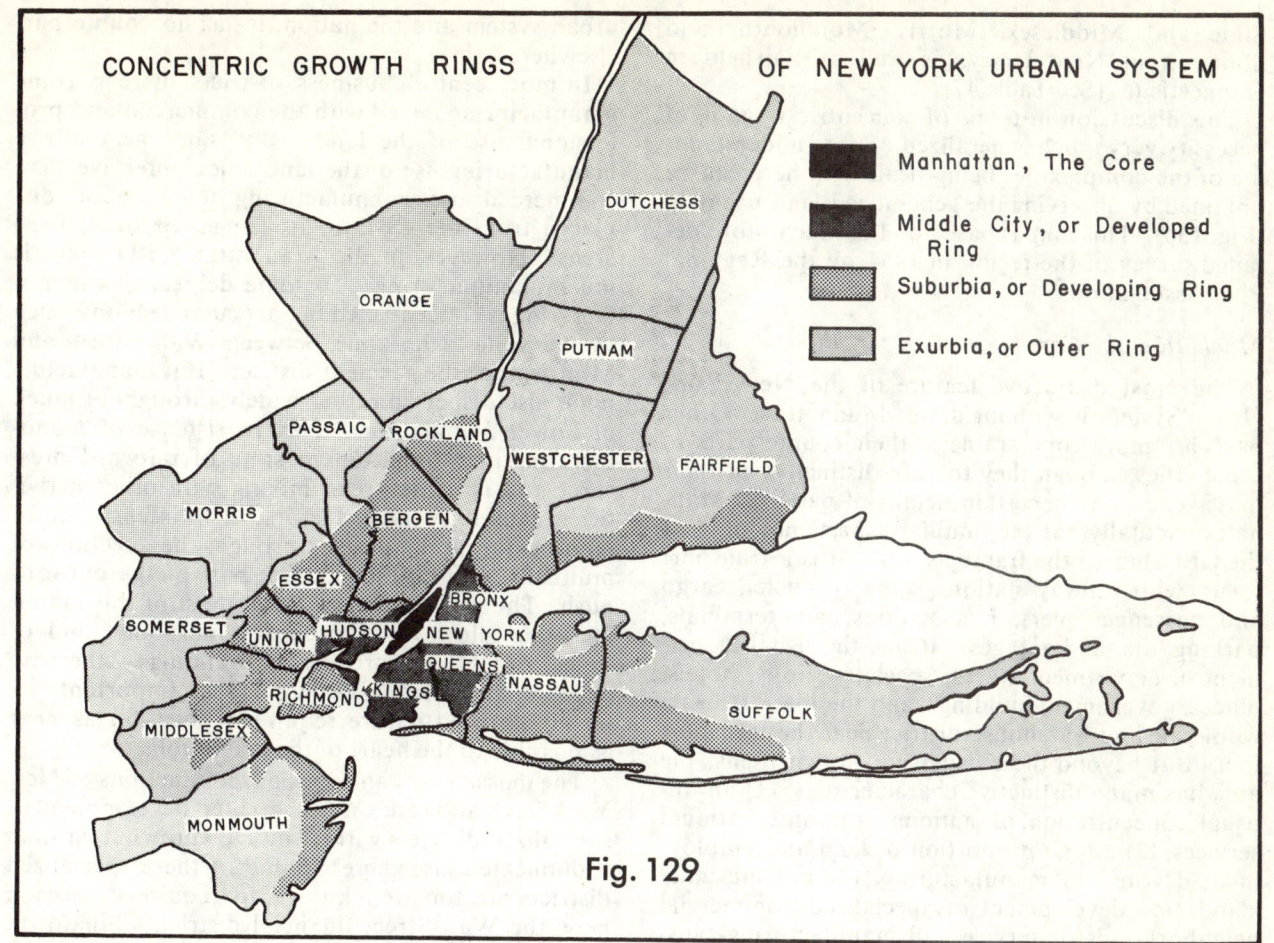

Fig. 129

of the frame zone (as defined in Chapter 14). Surrounding this core is a *developed ring* (largely the middle city of Chapter 14 but including part of the frame), which includes the rest of the Port District and all of the areas linked with the core by the subway system and the tubes. Though historically this zone has been principally a residential commuter zone for the core, today it has considerable commercial and manufacturing development. Much manufacturing and warehousing formerly in the core have recently shifted outward to this belt. Beyond this ring is an area of markedly more recent and less intensive urban development, the *developing ring* (approximates the fringe zone or suburbia of Chapter 14). Though much of this belt has been linked with the core by commuter railroads for half a century or more and by automotive transport for three decades or more, its major development has occurred since World War II. Rapid increases in population density and profound changes in land use are currently in progress. Beyond suburbia, or the developing ring, is exurbia, sometimes called simply the *outer ring*. It is just now beginning to feel the impact of the growth of the macropolis. Much of exurbia is still open, and urban development is sporadic rather than general.

As can be seen from Figure 129, the rings described correspond only very roughly to county units. It is necessary, however, to refer to county statistical data from time to time even though such references to some degree distort the actual geography of the rings. Each county has been categorized according to its predominant ring orientation. Thus Staten Island or Richmond County is grouped with the developing counties even though a small portion of it facing Upper New York Bay has been developed for many decades. Manhattan (New York County) as a whole is the core. The middle city or developed counties are Kings (Brooklyn), Queens, Bronx, and Hudson in New Jersey. The suburban or developing counties are Richmond, Westchester, and Nassau in New York State and Essex, Bergen, Passaic, and Union in New Jersey. The counties of exurbia which are experiencing spatially limited development are Suffolk, Putnam, Orange, Dutchess, and Rockland in New York

State and Middlesex, Morris, Monmouth, and Somerset in New Jersey, as well as Fairfield in Connecticut. (See Table 47).

This discussion in terms of concentric rings is of necessity very much generalized. Some understanding of the complexities being dealt with here can be obtained by observing the generalized land-use map (Fig. 130). This map is based on the much more detailed survey of the region in 1954 by the Regional Plan Association.[2]

Manhattan, the Core

The most distinctive feature of the New York Urban System is without doubt Manhattan. Other parts are more comparable to their counterparts in other cities, though they too are distinctive in their massiveness and in certain details of pattern. Manhattan actually has the familiar aspects not only of the CBD but of the frame as well. It is a transport focus, with railway stations, subway nuclei, cargo and passenger piers, bus stations, air terminals, parking lots, and garages. It has the usual assortment of department stores, specialty shops, hotels, offices, government buildings, and the like. It has a wholesale and warehouse district near the transport foci. But beyond these usual downtown items, the area has many distinctive characteristics: (1) an unusual concentration of national and international services, (2) a high proportion of land and employment devoted to manufacturing, (3) an unusually elaborated development of specialized commercial neighborhoods or service and manufacturing clusterings, (4) a large number of permanent residents for a central business district, and (5) a very rapid rate of redevelopment and change.

Most of those activities which set New York off as a primate city and make it unique among cities are concentrated in Manhattan: shipping, banking, insurance, corporate management, publishing, wholesaling, entertainment, and the like. Although these activities are found to some degree in the central business districts of all large cities, their concentration here is so great as to change the balance or mix of activities normally associated with a central business district. Since the CBD of Manhattan is in a sense the "downtown" for both the urban system and the nation, it has no counterpart elsewhere.

In most central business districts there is some manufacturing mixed with the commercial and professional use of the land. But since generally a manufacturing use of the land is less intensive than commercial use, manufacturing tends to be displaced from the CBD to its fringes or to outlying areas. However, in the Manhattan CBD there is much manufacturing. To some degree, this manufacturing is concentrated in particular sections, such as the valley (the zone between Wall Street and Midtown) or the garment district. But manufacturing is also dispersed rather widely throughout much of Lower Manhattan. This persistence of manufacturing in Manhattan in spite of outward pressures and its widespread mixing with other activities is caused by many factors. The intensifying of manufacturing land use through the erection of multistory loft buildings offset part of the outward push. The small-scale nature of most of this manufacturing allows it to occupy almost any kind of space, including portions of buildings otherwise devoted to offices. Perhaps most important, its unstandardized nature requires a location as near as possible to the heart of the macropolis.

The massiveness and specialized functions of New York have allowed a much greater development of specialized districts within the Downtown area than is duplicated elsewhere. Some of these specialized districts are too well known to require discussion here: the Wall Street financial district, Chinatown, the garment district, the fur district, the wholesale flower district, the Broadway theater district, and Greenwich Village. What is less generally realized is that there are many, many such specialized areas in Manhattan and that new specialized zones are constantly developing. One example of such a specialized district is the zone of automobile retailing which has been well established along Broadway south of Columbus Circle for nearly half a century. New urban redevelopment is dislocating this cluster; the Columbus Circle area is becoming a great center for culture, education, entertainment, and office work. Perhaps the most important of the newer specialized districts is the zone of corporate central offices in the Grand Central–Park Avenue area: trade organizations, engineering consultants, advertising, public relations, and so on. Although relatively few advertising agencies are actually located on Madison Avenue, the term Madison Avenue has become synonymous with this managerial cluster. Other clusters include the antique shops on upper Third Avenue, the art gallery area on Madison Avenue north of Fifty-seventh Street, the men's sportswear wholesaling zone of lower Fifth Avenue, the women's millinery wholesaling of

---

[2] Readers wishing more detailed statistics should consult the *New York Metropolitan Region Study*, directed by Raymond Vernon for the Regional Plan Association, Inc. This is a nine-volume study published by Harvard University Press. The most useful volumes for interpreting the form of the New York Urban System are the introductory volume, *Anatomy of a Metropolis*, by Edgar M. Hoover and Raymond Vernon, and the concluding volume, which projects present trends into the future, *Metropolis 1985*, by Raymond Vernon. These volumes are organized in terms of the concentric ring model, though the terminology and the groupings vary somewhat from those presented here.

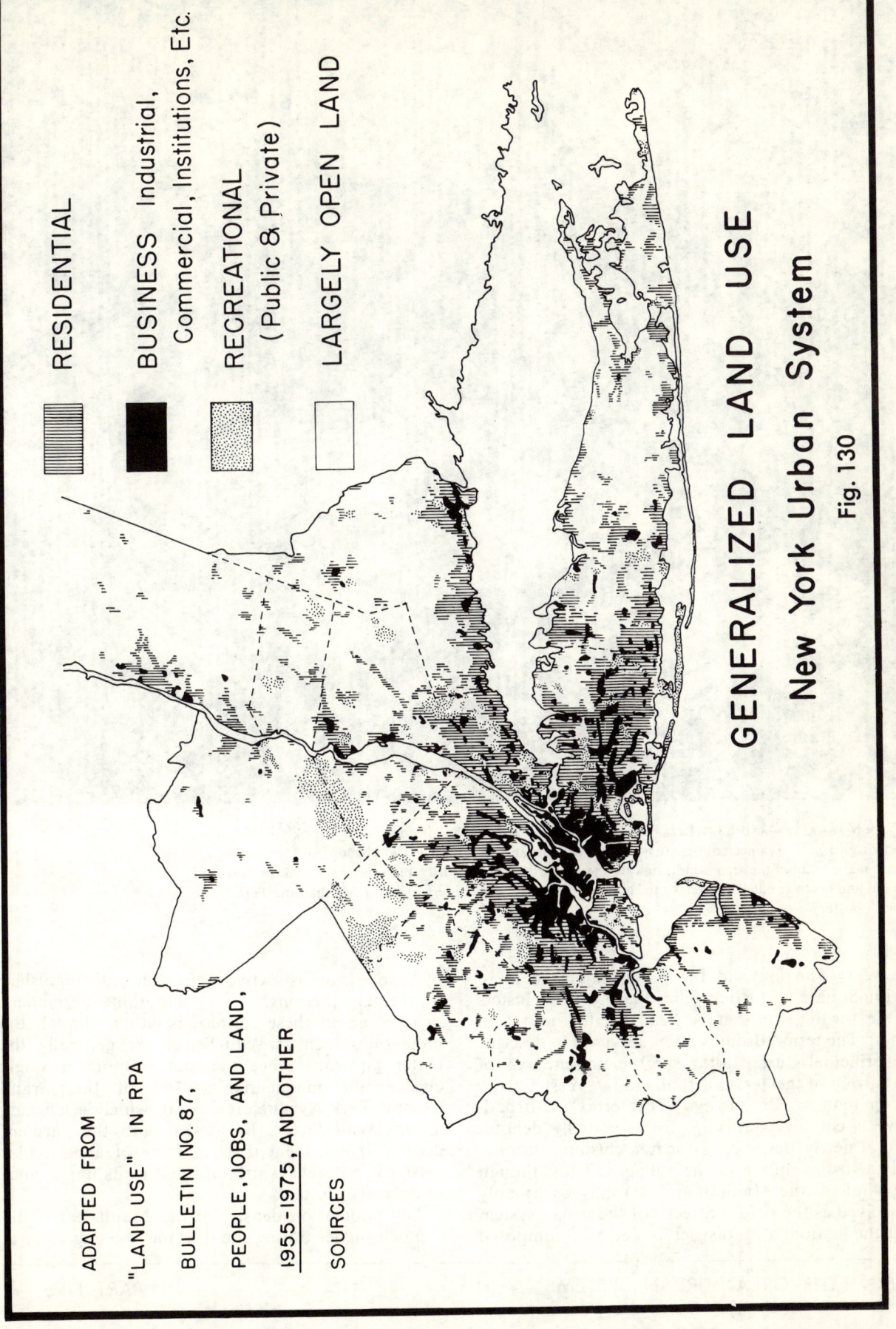

GENERALIZED LAND USE
New York Urban System
Fig. 130

Lower Manhattan and East River Docks. The area around the approaches to Brooklyn Bridge (upper right) reflects nineteenth-century New York, as do the ferry slips at the tip of Manhattan. Continued vitality of the area is indicated by the modern Port Authority piers in the Brooklyn foreground and postwar office buildings mixed among the older skyscrapers of lower Manhattan. *Port of New York Authority.*

the West Thirties, and the publishing clusters of Union Square and Rockefeller Plaza. New clusters in the theater world have been created in recent years. The term "Broadway" is still used to describe the principal cluster, although the present axis of that group of theaters is actually in the West Forties rather than on Broadway. The term "off-Broadway" is equally confusing, for it actually denotes several newly developing theater clusters, particularly a strong cluster in the Village. Thus, though the whole of the Manhattan CBD may be broadly conceived as the "Main Street" of the urban system and the nation, this district is actually composed of hundreds of distinctive main streets distinguished by special functions. There is nothing permanent or fixed about these clusterings either, though the older ones such as Wall Street are generally the better known. Since the most important single cluster, the managerial section of the Grand Central–Park Avenue area, is not widely celebrated as are Wall Street, Broadway, and the garment district, it is evident that the fame of a particular cluster is not always proportional to its importance to the urban system.

The population density of the Manhattan CBD is much higher during the day and evening than at

MANHATTAN (looking south). The Empire State Building stands out near the center of the photograph. Midtown office complex is at lower right. Note the variety of construction in progress here, including the new skyscraper being erected over Grand Central Station just to the left and below the Empire State Building. At the left is the United Nations area, and beyond, the Lower East Side and Brooklyn. The older business and financial center stands out at the tip of Manhattan. *Port of New York Authority.*

night. Yet the nighttime population is unusually high for a CBD. The population of Manhattan has been declining for about half a century, and a preoccupation with this decline by many authors has obscured the fact that in spite of the decline, there is still a very large round-the-clock population. At no time is the CBD deserted in the way that the CBD's of many other cities are late at night. As with manufacturing, the pressure for space near the core has not resulted in as much outward dispersal of residences as might be expected. The question is not why Manhattan's nighttime population is declining but rather why it has not declined more rapidly. There are many possible reasons for the persistence of a large core population. One is the attraction of the core for young, unmarried adults and young marrieds without children. If one seeks a job in Manhattan because of the lure of its glamour and sophistication, why settle in some suburb remote from this glamour and sophistication? Small apartments are in great demand in Manhattan. Somewhat similar factors are involved in the decision of many high-income residents to find luxury housing in the core rather than to seek the affluent suburbs. The sprouting of luxury housing in Manhattan, especially in the Village and the East Sixties, has been a marked feature of postwar years. Manhattan is also becoming the home of the very poor. A common solution to the high-rent problem is the parcelization of formerly spacious apartments and the crowding of more than one family into the newly created tiny apartments. The strict enforcement of the health and housing laws would eliminate much of this crowding, but the city has found strict enforcement most difficult. The various slum clearance and redevelopment projects in the core have not reduced population density as much as might be expected because the new apartment houses generally have much greater vertical development than the previous slums.

Perhaps the most distinctive feature of the Manhattan CBD is the very rapid rate of redevelopment and change. Often relatively new buildings are torn down to make way for still newer and taller buildings. Third Avenue, long a zone for the deteriorating El (elevated railway) and decaying buildings, has changed very rapidly in recent years to become one of Manhattan's leading streets. The nostalgic air of the Village is fast disappearing under an apartment-building boom. Most new construction tends to follow the very latest architectural trends rather than local or neighborhood styles. Rockefeller Center was a new development for its time. The United Nations attempted certain innovations in style in order to set the area off from the rest of the core. The head offices of the giant corporations are monuments to the power and prestige of the corporation as well as functional places for decision-making; hence many of these buildings strain for modernity and outward impressiveness. Because of this rapid rate of development and change, Manhattan lacks the stately, calm, and ordered air of the core areas of many foreign capitals or such American cities as Philadelphia and Boston.

Middle City or Developed Ring

The Middle City or developed ring includes most of the City of New York, i.e., upper Manhattan, the Bronx, Brooklyn, most of Queens, and parts of Staten Island. In addition, it includes a portion of the Jersey side of the Hudson facing Manhattan, principally Hudson County or the ridge between the Hudson and the Jersey Meadows. Nearly all of this zone is already highly developed and has been for some time. It is used principally for residences but also for commercial and industrial purposes. Over 90 per cent of all usable land is now in urban use; most of it is very intensively used. Population density is high compared with most middle city zones around the nation because this is the land of subways and multistory apartment houses. Until after World War II there were still many areas of single-family homes in the outer parts, but these are fast giving way to apartment houses. Together, Manhattan and the developed counties have about 53 per cent of the urban system's total population. The developed ring as a whole has probably reached or passed its population peak, for the population decline which began in Manhattan half a century ago has slowly spread into it. Thus it can be assumed that all of New York City proper has passed its population peak. This is supported by the decline noted in the 1960 census.

There are many fine residential areas within the middle city, but the zone as a whole is declining as a residential area. The great majority of the dwellings are relatively old, between thirty and eighty years. With the aging of residential buildings has come a general aging of the population and a reduction of the younger middle-income population. The homes of the middle-income population who have moved to the suburbs characteristically are being occupied by the poor. Because Negroes and Puerto Ricans are generally near the bottom of the income ladder, this means a rapid increase in Negro and Puerto Rican population within the middle city. Today about one-fifth of the population of Manhattan is Negro; for Manhattan and the developed counties combined, about one-tenth is Negro. About 80 per cent of all Puerto Ricans in continental United States live in the core and middle city.

Unfortunately, the general problem of aging and obsolescence of the middle city is not confined to residential buildings per se. If it were, the problem could be solved relatively simply by public redevelopment of housing or tax incentives to private builders to clear slums. Rather, the problem of obsolescence is a general problem affecting nearly all buildings and streets. This zone has the oldest land use pattern of the region, the pattern most out-of-joint with current needs. It reflects the requirements and advantages of waterfront locations and rail-subway transportation. It is generally poorly laid out for truck and automobile transportation, and yet it has had these modern means of transportation thrust upon it. The resultant congestion and decay are widespread and difficult to allay. The land use pattern similarly reflects the former status of this zone as a developing or suburban ring of the late nineteenth and early twentieth centuries, the commuting zone for Manhattan. The industrial activities considered undesirable for central areas were developed here: the Newtown Creek area of Brooklyn and Queens, Hunt's Point in the Bronx, and the oil-refining area of the Jersey shore. A special aspect of the former role of the developed ring or middle city as an outlying area is the great cemetery area of Queens.

Lastly, the land use pattern of the middle city reflects the lack of planning and public regulation of land development prior to 1916. By that time, the basic pattern of urban development for most of the zone was already well established. Thus heroic efforts at slum clearance and residential redevelopment in recent decades have changed local areas without resolving the more general and pervasive problem of urban decay throughout. In the period since World War II, residential redevelopment projects within the City of New York have required the moving or relocation of about half a million people. This is action on a truly grand scale. However, emphasis is now shifting to the more general problem of urban decay, including efforts at planning on the neighborhood level for the conservation and upgrading of existing structures. The future hinges on the success of such efforts, since experience has indicated already that slum clearance alone cannot cope with the problems of a middle city so vast and complex.

Suburbia or Developing Ring

Suburbia, or the developing ring, is found principally in three counties of New York State—Nassau, Westchester, and Richmond, and in four counties in New Jersey—Union, Essex, Bergen, and Passaic. As a zone, it is generally referred to as the suburbs by New Yorkers. It is very different from the above-described developed ring or middle city in certain respects while closely resembling it in others.

A major difference between the middle city and suburbia is the rate of population growth and new urban development. As would be expected, the population is increasing faster in suburbia than in any other part of the New York Urban System. The population will continue to grow for some time, probably not reaching its peak for several decades. There are substantial areas still open for housing development. On the other hand, the inevitable second wave of urban development has already begun in the inner fringes. This takes the form of substitution of high-rise apartment houses for single-family homes. Thus some parts of the inner portion of suburbia do not differ so much from the middle city as might be expected, except in their newness. Today, suburbia as a whole has about 5 million residents, or about 30 per cent of the population of the region. By 1985 it probably will have over 8 million people, or about one-third of the population of the entire urban system. Population densities vary widely because of great differences in zoning laws and their enforcement, but about half of the present population is found within strips four miles wide extending along the major transport arteries.

Unlike the middle city, suburbia still has large areas of open space remaining. Yet, except at the outer fringes of the belt, this open space is principally land difficult to develop for urban use. New York State has its share of such areas: the swampy south shore of Long Island, small hilly zones on the north shore of Long Island, many hilly areas in Westchester, and the swampy shores of the Kill van Kull in Staten Island. However, the largest areas of such difficult terrain are in New Jersey. The outstanding example is the Hackensack Meadows. Land fill of the Meadows is now in progress. If present plans for vast industrial zones in the Meadows materialize, New Jersey may attract more industry than New York, shifting the industrial balance westward with profound effects on the future geography of the whole urban system. A smaller but important open zone is the wooded back slope of the Palisades. Even after the opening of the George Washington Bridge this area developed slowly, primarily because of the steepness of the land and generally poor drainage. Poor drainage seems illogical on such steep slopes, but the impermeability of the underlying rocks produces the problem. Gradually this zone is yielding to luxury housing and parts of it may be saved for park development under present programs. The gradual disappearance of open space will in time make suburbia like the middle city.

One of the sharpest differences between the middle city and suburbia is the strong orientation of the latter toward the automobile rather than railroads and subways. Housing follows the highways and most housing is single-family residences. A few of the older industrial plants are located along railroads or rivers, but most of the new industrial development is highway-oriented and widely dispersed. The New York Urban System has many industrial parks, most of which are located in suburbia. Retailing zones tend to be highway oriented and linear. The highway orientation of the ring creates an amorphous settlement pattern and is leading to great changes in commuting. At present there is much commuting within the ring, as distinguished from commuting to the core or to the middle city. In the future, this tendency toward the self-sufficiency of suburbia is expected to increase. Since it now seems unlikely that the subways will be extended outward, major differences between suburbia and the middle city will tend to persist.

Newness is now a dominant characteristic in suburbia, yet it has sporadic areas of incipient decay and obsolescence resembling those of the middle city but smaller in areal dimensions. Suburbia was already urbanized in places and linked with Manhattan by commuter railroads half a century ago. Thus there are many old urban cores interrupting the general pattern of recent development. Some of these large cores, such as Yonkers or White Plains, are large cities. Unless checked by appropriate public action, suburbia will gradually come to resemble the middle city with its pervasive problems of blight and decay.

Exurbia or Outer Ring

Manhattan's influence extends for at least 50 miles, well beyond the limits of suburbia. However, on the outer fringes of the New York Urban System, urban development is very limited, often extending great distances along major transport lines but leaving large open spaces between. The new limited-access highways such as the Thruway are greatly extending the range of influence by sharply reducing transport and commuting time. The varied development of exurbia is tied to the access points of these highways as well as to conventional highways and the railroads.

Of the twenty-two counties considered here as part of the New York Urban System, the ten on the outer edge are still sufficiently thinly populated to be designated as the counties of exurbia. In 1960 about 3 million people lived in these ten counties, less than one-fifth of the total population of the urban system despite the great area involved. In recent years, three of the outer counties (Suffolk, Fairfield, and Middlesex) have developed much more rapidly than the others. Together these three counties have more people than the other seven counties of exurbia combined.

Viewed in terms of continuity of urban development, much of the area of the outer ten counties would have to be excluded from the New York Urban System. But viewed dynamically, in terms of the anticipated rapid development of these counties in the next two decades, it is wholly realistic to include them. Also, all of the counties, with the possible exception of parts of Dutchess County, are already functionally related to Manhattan and the New York Urban System in general. Exurbia probably will increase dramatically in population, from the present 3 million to nearly 8 million in 1985. If this forecast is correct, the population of the New York Urban System by that time will be almost equally balanced between: (1) the Manhattan core plus the middle city, (2) suburbia, by that time not unlike the present middle city in many respects, and (3) exurbia, by that time not unlike present suburbia in many respects. The total population of the region will be nearly 24 million.

COMMUTING PATTERNS

The New York region is functionally unified by internal flows of people, goods, and ideas. In the last analysis, these movements contribute much to internal form. However, the very dynamism of this aspect of the urban system makes interpretation difficult. Our comments here are confined to one type of flow, the commuting pattern.

According to the popular image, the commuting pattern is a simple flow between the core and suburbs. Actually, a thorough study of the incredibly complex commuting pattern has never been made, although there is much evidence to suggest the complexity of the pattern. For example, an analysis of the home locations of the workers for each of the industrial sections of Paterson-Passaic shows a very wide dispersion of residences. The majority of the workers commuted from nearby communities, but some commuted from Rockland County, New York City, and various New Jersey counties. The same pattern is observable throughout the New York Urban System. Each job center, whether a university, office building, factory, or department store, draws a variety of employees from a great many differing places within the over-all urban system. Conversely, each residential area, whether a particular ethnic cluster or a mixed neighborhood and whether predominantly of a certain income level or a mixed economic grouping, supplies workers to widely varied types of employment and sends them to widely scattered parts of the urban system. Because New York is larger than any other urban system and because it has such a great diversity of employment opportunities and types of residential areas, the commuting pattern is probably very much more complex than encountered anywhere else.

Some understanding of this complexity can be gained from a generalized journey-to-work survey of the Regional Plan Association in 1956. According to the survey, the most important commuter counties are still the old commuter counties of New York City: Queens and Bronx. Approximately three-fourths of the workers in these counties commute to work in other counties. Not surprisingly, about half of the Queens workers and nearly three-fifths of the Bronx workers commuted to Manhattan. But, contrary to the popular image, these counties also supplied significant numbers of workers to much of the rest of the urban system: to the middle city, to all suburban counties, and to four counties in exurbia—Morris, Fairfield, Rockland, and Suffolk. With the continuing growth of employment in suburbia and exurbia, this dispersal of commuting will no doubt increase. Other counties with relatively high rates of commuting were Richmond, Bergen, Kings, and Putnam. The percentage commuting to jobs outside their county of residence varied from about 63 to 67. The workers in all four counties showed a preference for Manhattan jobs, though they also commuted to most of the counties of the middle city and suburbia. Many workers from Putnam County worked in adjacent Dutchess

County. At the other extreme, less than 10 per cent of the workers of Orange and Dutchess counties commuted to jobs in other counties of the urban system. In terms of commuting, these two counties are marginal to the New York Urban System. The percentage of out-of-county commuters for Fairfield is only slightly higher, and yet there is a marked commuting tie to the core, New York County. Most of the workers in Manhattan find jobs within the county, but the 30 per cent who commute outside went to every county of the middle city, every county of suburbia, and even to Orange, Fairfield, and Suffolk. This "reverse" commuting *from* the core is a marked development of recent years.

It is thus clear that the relatively simple commuting pattern of the railway and subway era has given way to a complex pattern based upon remnants of the railway-subway pattern plus the much more amorphous bus and automobile pattern. These patterns tie the region together quite as firmly as before but in much less easily predictable ways. This truism applies with equal force to most of the other flows within the urban area. The movement of people, ideas, and goods is taking new paths and thereby radically changing the meaning of the urban system's form. The problem for plain citizens and planners alike involves much more than slum clearance and attention to the more obvious social and economic problems. Above all, it requires an expanded vision to understand the revolution in land use now in progress and hence the probable form of the urban system in the future.

Problems and Prospects

Forecasting is difficult for any urban system, however small. The difficulty increases in more or less direct proportion to the size and complexity of the tributary area of the system and the range of influences focusing on it. Thus, forecasting for New York is more difficult than for any other urban system. However, certain outstanding assets and problems can be pinpointed and certain prognostications probably should be attempted.

Assets

Far and away the greatest asset for the region is its established position, its significance as a very large "going concern." All other assets, such as the truly magnificent harbor, are conditioned in their importance by variables such as transportation changes, changes in the direction of trade, and so on. Closely related to the system's established position is its historic dynamism. Time after time there have been imaginative solutions by the city's leaders to the problems of growth. Yet these characteristics are themselves partly, and perhaps largely, a product of established pre-eminence. A primate city tends to attract the best of a nation's talent. So we are still faced with a conundrum. New York may remain a primate city for a thousand years or more, like Byzantium. Or, if it loses its dynamism, it may decline to a tourist curiosity like Venice. In a highly mobile world, such a transition could occur too rapidly for the inhabitants to adjust and to take action in time to restore the city to prominence.

Problems Related to Situation and Function

Several threats to the situation and functions of the system seem currently of long-range significance. Perhaps the most important threat is the current tendency toward the economic fragmentation of the earth. Other significant threats are the general tendency toward economic decentralization in the United States, changes in the nature of manufacturing innovation, and the rise of Washington, D.C., as a significant competitor in key activities.

The economic one-worldism created by the British in the nineteenth century seems to be giving way to great economic blocks: Communist Eurasia (Poland to China), Western Europe plus parts of Africa and Asia, the remnants of the British Commonwealth, and the American system. Thus far, the chief restriction on New York's zone of operations has been the self-contained nature of the separatist, totalitarian block of Eurasia. But new threats now arise. The increasing economic integration of Western Europe and the attendant rapid growth of a self-contained Continental economy may pose a new threat of even greater importance. Possibly this economic integration may be gradually extended to Great Britain, other European countries, and across the Atlantic to include Canada and the United States. Should this occur, New York would greatly prosper as the principal urban center of an expanding free-trading area. Significantly, New York's prosperity has been based on its role as a link between North America and Europe. With the rise of a great free-trading area in North America, i.e., the westward expansion and industrialization of the United States, New York parlayed this linkage role to dazzling proportions. The emergence of a North Atlantic free trade zone would create opportunities for New York of incalculable dimensions. If, instead, the Atlantic Basin splits into rival trade areas, New York may soon lose much of its dynamism. More than any other city on earth, New York has a stake in the easing of world tensions and the growth of a true world economy rather than separatist continental economic blocs. If one could confidently predict such an era of world peace and prosperity, a modern Pax Romana, there is little doubt that New York's established position as the leading world macropolis would carry it forward to new levels of power and prestige.

Another broad threat to New York is the current general tendency toward economic decentralization in North America, not toward rural areas and small cities but toward other large urban systems. Apparently, this trend has not yet run its course. For example, the ease and speed of modern trucking and the attraction of low-wage areas outside the region have now begun to affect such historically local industries as bread-baking. Recently "foreign" baked goods from the depressed coal-mining regions of eastern Pennsylvania have begun to invade New York's markets. As long as economic decentralization continues, the urban system should theoretically expect but modest growth rates, at or below the national average. On the other hand, such locational trends are always potentially reversible. A new trend toward economic centralization could develop at some unforeseen time in the future. This could lead to trends comparable to New York's phenomenal growth in the nineteenth century. Currently, the major possibility of this sort seems to lie in certain services, such as corporate management. It is clear that the Managerial Revolution has just begun. Ultimately, this revolution in employment may become as important as the earlier Industrial Revolution. New techniques such as those implied in the term automation most certainly will cause continuing declines in the proportion of the nation's workers employed in primary and secondary activities while proportionally increasing the number of workers in the tertiary sector. Decision-making, and tertiary-oriented centers such as New York, may become more and more important.

Growth and dominance of great industrial corporations is not an unmixed blessing for New York. The concentration of the head offices of so many of the larger firms in the region buoys certain types of employment. Yet this concentration of industrial power also threatens the region in another sector of employment: manufacturing. For several decades the principal factor maintaining the region as the chief manufacturing center of the country has been its attractiveness for small firms producing new, unstandardized items. As industries have matured and standardized they have left the region, but new industries, still unstandardized, have always emerged to take their place. Whether this historical process will continue unchanged is open to question. Today, more and more new products are the result of long and elaborate laboratory research; fewer and fewer are the result of common-sense variations on ideas or products in general use. Laboratories for industrial research are controlled and financed largely by the great corporations, whether they have their head offices in New York or elsewhere. Laboratory work, and employment related to work in laboratories, is a small but growing source of employment in New York. But once a large firm has designed and tested a new product in its laboratories, there may be no particular reason why it should choose New York for manufacturing the product. Unlike the small, struggling firm, the giant corporation will likely not seek the safety of the core of a large primate city for its plant location. It can locate the new production almost anywhere, either at one of its dozens of existing plants spread across the country or at some carefully selected new site. Today, much more than in the past, location decision-making in the large firm is the product of careful research and evaluation of alternative sites. Hence inertia, the attraction of old centers for new industry, is less important than formerly. Unless new industries continue to emerge independently of the existing industrial giants, the New York Urban System is likely to decline somewhat as a manufacturing center.

New York is not without competition in the service sector of its economy either. For the long range its most important rival appears to be the nation's capital, Washington, D.C. An apparent trend toward the growth of a centralized national bureaucracy seems to have been working against New York for several decades. This kind of bureaucracy benefits the political capital, Washington, far more than New York, with its historical stake in the private bureaucracies of the great banks, insurance firms, and industrial corporations. Ultimately Washington may advance far enough to attract many of the national services now clustered in New York such as banking, insurance, and the head offices of firms and organizations of all kinds. This kind of accruement has taken place in the capital city of Japan, Tokyo.

In any event, the detaching of services from New York and the adding of them to Washington would be a slow process, perhaps much like the slow transition in Manhattan from the dominance of the Wall Street focus to the Midtown focus. The future of every primate city is ultimately based on such imponderables.

PROBLEMS RELATED TO SITE AND FORM

The current problems related to site and form are more immediate and obvious than those related to situation and function. Traffic congestion grows. Slums and crime spread. Suburban sprawl increases. Open space which should be retained for recreational uses is instead gobbled up for homes or industry. The zoning regulations of the 1,400 separate governments of the urban system reflect a lack of coordinated, comprehensive planning.

Perhaps the most important problem of all is the lack of regional consciousness. Many—perhaps most—of the inhabitants of the New York Urban System do not feel part of the system as a whole. They do not identify with it. Rather, they identify with a part, usually a small part, of it. In a democracy the governments cannot proceed very far toward integrated regional planning until the people feel the need for such a practical expression of regional unity. With the merging of the New York Urban System, in time, with the larger Megalopolis which extends along the East Coast from north of Boston to south of Washington, the problem will be greatly magnified. For identification with this huge area will require even greater stretching of local loyalties.

The changing make-up and distribution of population is another major problem. In the last decade and a half nearly one million middle-class white people have moved from the core and middle city to suburbia and exurbia. These people, substantial earners and taxpayers, and often employers too, have been replaced as city residents by Negroes and Puerto Ricans, mostly unskilled, many unable to find work or carry much of the tax burden. Staggering numbers of persons are receiving some sort of welfare aid at an average cost to society of close to $1,000 each per year. The 1965 welfare budget submitted was 81 million dollars higher than that for 1964. The human cost of such poverty is incalculable.

Inadequate housing is a third problem. It has been estimated that at the end of 1964, 1.25 million residents of New York City alone were in substandard housing, dilapidated, uncomfortable, often unhealthful. Public housing and urban renewal programs, although massive by all known standards, come nowhere near solving housing problems.

Traffic is a fourth problem. Billions have been spent on highway construction and most experts feel that traffic problems, especially in the core and middle city, are getting worse by the day. It is not only a matter of the tremendous number of local cars moving about, but also of the hundreds of thousands of others which move into the city area each weekday; all add to the mounting air-pollution problem. Nor are commuter railroads and subways without troubles; in spite of rising fares they must look more and more to government subsidy.

A fifth problem is racial violence, vice, and a general rise in crime, especially in heavily populated Negro ghettoes such as Harlem and the Bedford-Stuyvesant section of Brooklyn.

A major site problem, the draining and filling of many marsh areas, proceeds piecemeal.

All of these problems are serious ones, the kind that other urban systems already are experiencing or may experience in time. Solutions will require the attention of the most able people of the community; if the most able and articulate all move to the suburbs and remain insulated from the problems, locally induced solutions most certainly will be few and far between.

A new kind of urban system is in the making. As the form of the new system crystallizes, many of the most urgent current problems may prove transitory. Others, it is hoped, will be solved more easily than now appears possible. If the people develop sufficient understanding, and particularly if they develop the system-wide regional consciousness needed for a broad, democratic attack on the problems, solutions can evolve democratically. If this broadness of vision and unity of purpose fails to develop, it is almost certain that the state or federal government will find it necessary to impose solutions.

In any case problems of site and form may in the long run appear much less important than the more elusive problems of situation and function. Site and form may be altered by direct action of the inhabitants. The situation of New York, on the other hand, is deeply influenced by forces far beyond the power of New Yorkers to alter except on a limited scale.

PROSPECTS IN SUMMARY

The greatest asset of the urban system, its established position, will probably enable it to grow and prosper for many decades. Quite probably it will continue to increase in international importance, though perhaps only holding its own nationally. The many current problems within the region will have to be resolved through a combination of land use evolution and planning. If sufficient regional consciousness develops, the planning will be locally controlled and democratic. If not, planning will be imposed by state governments or by Washington, D.C.

The population of the urban system (the twenty-two counties), at 16,141,000 in 1960, should reach nearly 22,000,000 by 1980 and be roughly 30,000,000 by the year 2000. The New York SMSA (the city plus Nassau, Suffolk, Westchester, and Rockland counties), with a 1960 population of 10,695,000, can be expected to contain 12,380,000 people by 1980 and about 14,750,000 by 2000. New York City proper (the five boroughs) very likely will remain at approximately its present population level in the decades ahead; tremendous increases in suburban and exurban counties will account for the system-wide growth. (See Table 48.)

TABLE 48
POPULATION PROJECTIONS FOR COUNTIES INCLUDED IN THE NEW YORK URBAN SYSTEM

	Population (In Thousands)				
	1960	1970	1980	1990	2000
In New York State	11,087	12,150	13,325	14,875	16,945
Boroughs of New York City	7,782	7,800	7,800	7,800	7,800
Bronx (Bronx County)	1,425	1,400	1,350	1,325	1,300
Brooklyn (Kings County)	2,627	2,550	2,450	2,325	2,300
Manhattan (New York County)	1,698	1,675	1,650	1,600	1,600
Queens (Queens County)	1,810	1,860	1,900	1,950	2,000
Richmond (Richmond County)	222	315	450	600	600
Highly Urbanized Counties	2,246	2,690	3,170	3,575	4,050
Nassau	1,300	1,400	1,500	1,575	1,650
Rockland	137	230	370	530	700
Westchester	809	1,060	1,300	1,470	1,700
Less Urbanized Counties	1,059	1,660	2,355	3,500	5,095
Dutchess	176	290	410	620	920
Orange	184	300	440	720	1,100
Putnam	32	50	95	130	175
Suffolk	667	1,020	1,410	2,030	2,900
In New Jersey	4,400	5,850	7,475	9,300	11,355
Highly Urbanized Counties	2,819	3,450	4,010	4,600	4,970
Bergen	780	1,070	1,280	1,530	1,700
Essex	924	1,130	1,340	1,530	1,640
Hudson	611	610	600	600	600
Union	504	640	790	940	1,030
Less Urbanized Counties	1,581	2,400	3,465	4,700	6,385
Middlesex	434	660	950	1,300	1,750
Monmouth	334	550	890	1,260	1,735
Morris	262	400	550	710	950
Passaic	407	530	645	760	900
Somerset	144	260	430	670	1,050
In Connecticut	654	900	1,200	1,430	1,700
Partly Urbanized County	654	900	1,200	1,430	1,700
Fairfield	654	900	1,200	1,430	1,700
Total for Urban System	16,141	18,900	22,000	26,300	30,000

Selected References

Albion, Robert G. *The Rise of New York Port: 1815–1860.* New York: Scribner's, 1939.

Brush, John E. *The Population of New Jersey.* New Brunswick, N. J.: Rutgers Univ. Press, 1956.

Feininger, Andreas. *The Face of New York.* Rev. ed. New York: Crown, 1964.

Friis, Herman R. *A Series of Population Maps of the Colonies and the United States, 1625–1790.* New York: American Geographical Society, 1940.

Gottmann, Jean. *Megalopolis.* Cambridge: M.I.T. Press, 1961.

Green, Howard L. "Hinterland Boundaries of New York City and Boston in Southern New England," *Economic Geography,* XXXI (October, 1955), 283–300.

Griffin, John. *Industrial Location in the New York Area.* New York: City College Press, 1956.

———. *The Port of New York.* New York: Arco, 1959.

Jefferson, Mark. "The Law of the Primate City," *Geographical Review,* XXIX (April, 1939), 226–32.

Journey to Work. New York–New Jersey Transportation Agency, 1963.

Kenyon, James B. *Industrial Localization and Metropolitan Growth: The Paterson–Passaic District.* ("Department of Geography Research Paper," #67. Chicago: University of Chicago, 1960.

Kieran, John. *A Natural History of New York City.* Boston: Houghton Mifflin, 1959.

Lyman, Susan E. *The Story of New York.* New York: Crown, 1964.

Neft, David. "Some Aspects of Rail Commuting: New York, London, and Paris," *The Geographical Review,* XLIX (April, 1959), 151–163.

Nelson, Howard J. "Walled Cities of the United States," *Annals of the Association of American Geographers,* LI (March, 1961), 1–22.

The New York Sun, special issue, "The Shopping Place of Millions," 1925. Map shows retail clusters in Manhattan of that date. Copy in American Geographical Society Library.

Patton, Donald J. "General Cargo Hinterlands of New York, Philadelphia, Baltimore, and New Orleans," *Annals of the Association of American Geographers,* XLVIII (December, 1958), 436–455.

Sayre, W. S., and H. Kaufman. *Governing New York City: Politics in the Metropolis.* New York: Russell Sage Foundation, 1960.

Stern, Peter. "New York City," *Focus,* II, No. 19 (June 15, 1962).

Torrey, Raymond H., Frank Place, Jr., and Robert L. Dickinson. *New York Walk Book*. 3rd ed., rev. New York: The American Geographical Society, 1951.

U.S. Department of Commerce. *The Port of New York, New York and New Jersey* (No. 5, rev. 1952), 3 vols. Washington, D.C.: Govt. Printing Office, 1955.

Van Burkalow, Anastasia. "The Geography of New York City's Water Supply: A Study of Interaction," *The Geographical Review*, XLIX (July, 1959), 369–86.

The New York Metropolitan Region Study series (Cambridge: Harvard Univ. Press, 1956–59) contains the results of a detailed survey of the region. The series includes the following books:

Chinitz, Benjamin. *Freight and the Metropolis*. 1960.

Hall, Max (ed.). *Made in New York*. Studies by Roy B. Helfgott, W. Eric Gustafson, and James M. Hund. 1959.

Handlin, Oscar. *The Newcomers*. 1959.

Hoover, Edgar M., and Raymond Vernon. *Anatomy of a Metropolis*. 1959.

Lichtenberg, Robert M. *One-tenth of a Nation*. 1960.

Robbins, Sidney, and Nestor Terleckyi. *Money Metropolis*. 1960.

Segal, Martin. *Wages in the Metropolis*. 1960.

Vernon, Raymond. *Metropolis 1985*. 1960. This is the summary volume of the series, with detailed forecasts of the region.

Wood, Robert C. *1400 Governments*. 1960.

OTHER SOURCES

Maps and Air Photos: U.S. Geological Survey for topographic maps and Coast and Geodetic Survey for harbor charts. Humble Oil Company, Inc., and other major oil companies publish metropolitan road maps with detailed insets of the Manhattan Central Business District.

Newspapers and Magazines: The *New York Times* is particularly helpful because it includes detailed maps of areas under discussion. Magazines of value include *Holiday*, special issue (October, 1959), with maps, and *Fortune*, special issue, LXI, (February, 1960), 47–272.

Port of New York Authority: The Port Authority supplies popular material suitable for use in teaching such as "Crossroads of World Trade," a packet of general materials on the port and region. The Port Authority has published many more detailed studies, including *The Port of New York* in 1941. Various staff reports from the research division are available for study by qualified persons.

Public Planning Agencies: The various cities and counties of the region publish a variety of detailed and useful materials. Recently the Department of City Planning of New York City completed a new series of detailed land use maps.

Regional Plan Association: This association has published a great variety of studies on different aspects of the region since the publication of its first survey of the region in 1928. Two excellent studies are *People, Jobs, and Land 1955–1975*, RPA Bulletin #87, 1957, with maps, and *Hub-based Travel in the Tri-state Metropolitan Region*, RPA Bulletin #91, 1959.

State Agencies: Particularly the New York State Department of Commerce and the New Jersey Department of Conservation.

CHAPTER 21   ROBERT B. HALL, JR.

Rochester

The Rochester Urban System, located on the south shore of Lake Ontario along the banks of the Genesee River, is in the Upstate Heartland halfway between Syracuse and Buffalo. Its excellent location provides easy access to rail, highway, and water transport facilities of the Mohawk–Lake Plain axis, as well as to the St. Lawrence Seaway.

The city of Rochester, with 319,000 people in 1960, stands alone as the only large central place, although some suburban towns have sizable populations. Irondequoit, with 55,000 people, has the most inhabitants of these towns. The degree of dominance of Rochester over the whole urban system is greater than that of any one city in any of the other urban systems.

The urbanized area (Fig. 131) is reasonably compact, extending farthest outward from Rochester in a southeastward direction along the access routes to the Thruway. In 1960 it contained 493,000 people, 174,000 of whom were outside the limits of the city of Rochester. The urban frontier encloses an area of about 473 square miles and contains an estimated 1960 population of 572,000. Its greatest dimension of about 35 miles is in a northwest–southeast direction. The positions of Rochester and Lake Ontario have influenced its shape. Composed of Monroe, Wayne, Orleans, and Livingston counties, the Rochester SMSA includes not only the large urbanized area but a considerable amount of excellent farm land as well. The 1960 population of the SMSA was 733,000, having increased by 175,000 or 31 per cent since 1940.

The Rochester Urban System is outstanding for manufacturing. World famous as a center of the photography industry, particularly for the Eastman Kodak Company, it is also a major producer of scientific instruments, optical goods, electrical equipment, men's clothing, and many other items. These light industries require much technical know-how and have resulted in one of the highest percentages of professional, technical, and skilled workers in the country. They also have resulted in the highest per capita income upstate.

General prosperity and low unemployment rates are characteristic. Prosperity is reflected in attractive, well-kept residential areas, many beautiful parks, and a wide array of cultural facilities, including educational institutions and museums, recreation centers, and fine offerings in music.

A major urban renewal program, one of the more comprehensive in the country, has been undertaken and holds great promise for the future. Ambitious plans, already well under way, call for modernization of the CBD, a major overhauling of approach highways, and a rehabilitation of the blighted areas.

The country surrounding Rochester is as good an agricultural area as exists in the state and contains some minerals as well. Dairy products, fruits, field crops, potatoes, and vegetables are all important in the prosperous rural economy. Important food-processing industries occur within the SMSA and in nearby counties. While some mining is carried on in the vicinity, notably salt, gypsum, and sand and gravel, it is not very significant to the general economic development of the urban system. Rochester industries obtain the bulk of their materials from distant sources, both in the United States and overseas.

Historical Development

The site of Rochester was originally determined chiefly by the Genesee River (Fig. 132), which descends over a series of three falls within the city for a total drop of about 235 feet. This of course meant water power, which was so important to early settlement. As might be expected, a sawmill and a gristmill were the first business ventures. Established about 1789 on the west bank near present-day Main Street, they used the Genesee's power. Subsequently, flour-milling expanded and by 1834 Rochester had become perhaps the greatest flour center in the world, getting wheat from the rich grainfields of western New York State and newly opened producing areas still farther west. Excellent transportation facilities, development of the right kinds of manufacturing, and particularly forward-looking industrial leaders all have contributed to successful growth of the Rochester Urban System.

Role of Transportation

A major contributing factor to Rochester's economic development has been its location along the

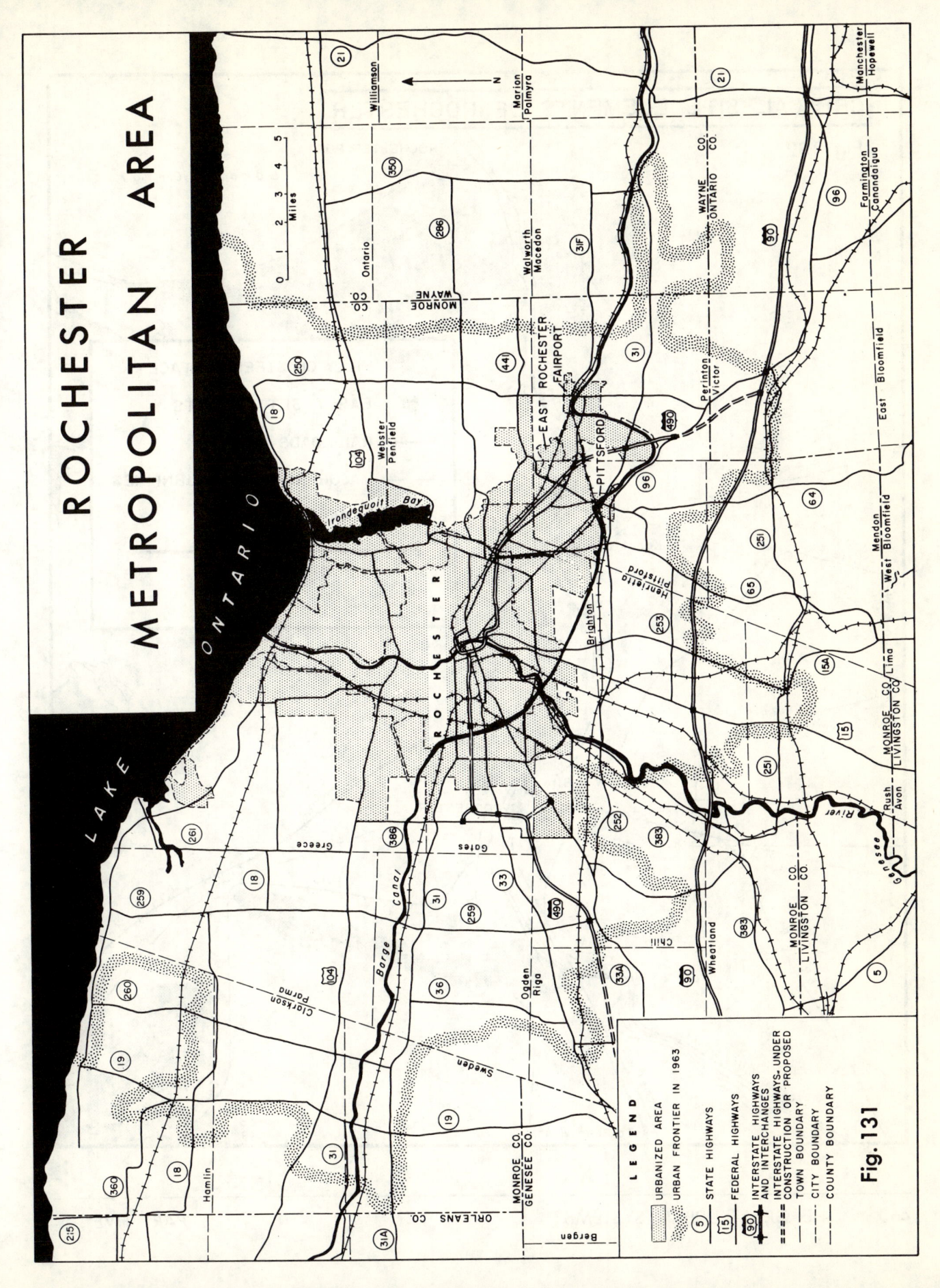
Fig. 131 — ROCHESTER METROPOLITAN AREA

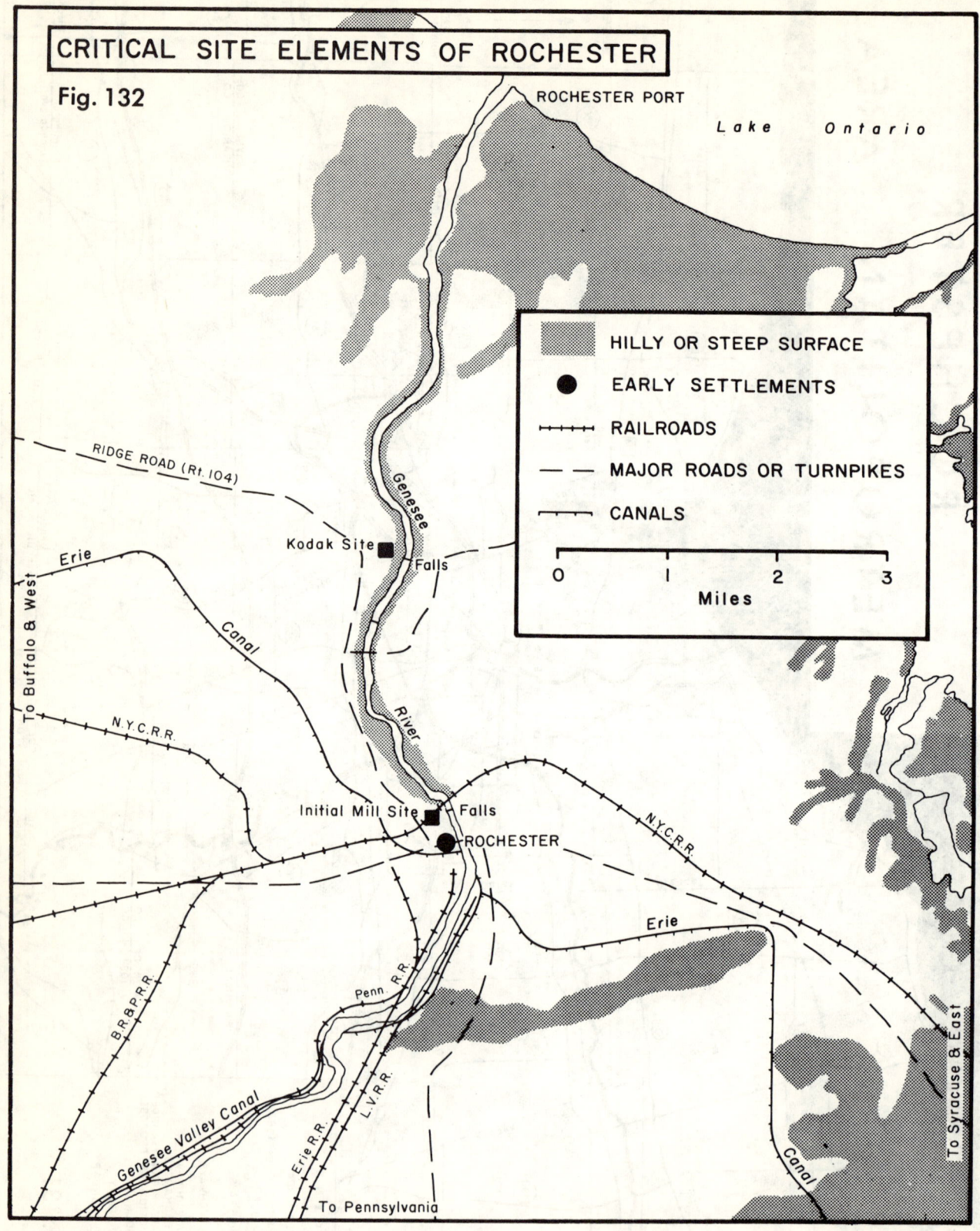
Fig. 132 CRITICAL SITE ELEMENTS OF ROCHESTER

important transportation lines extending west out of the Upper Mohawk Valley and along the plain south of Lake Ontario. The funneling of transportation south of Lake Ontario and north of the Finger Lakes along the Mohawk–Lake Plain axis meant Rochester was on the main path of commerce from New York City to Buffalo and points west. The opening of the Erie Canal in 1823–25 brought new settlers to Rochester and made available large new markets for Rochester flour in the East. Growth at this time was rapid and Rochester became known as the "Flour City." Grain-milling remained important up to about the middle of the nineteenth century, but the opening of extensive new grain lands to the west and milling centers in Buffalo and elsewhere led to the decline and eventual disappearance of this activity.

The first half of the nineteenth century also saw the development of railroads in the Rochester area, with the eventual linking of east–west lines in 1839 and the later merger of these into the New York Central. In addition to the New York Central the city is now serviced by four other major railroads and over one hundred truck lines, the latter making increased use of the recently built Thruway. A system of new approaches from the Thruway to the city is being constructed. These approaches will give faster access to and from the city and help to create sites for future industrial, commercial, and residential expansion.

Monroe County operates an airport southwest of the city, where daily flights are scheduled by three air lines. This is particularly important to businessmen who need quick access to New York City and other urban centers. Medium-range jet service is available.

Water transportation is now available to Rochester via the New York State Barge Canal and the Port of Rochester. While both this canal and its predecessor, the Erie Canal, were of importance in the past, canal traffic is insignificant today. Traffic on the canal now amounts to but a minor flow of bulky materials, mostly petroleum products. Its future as a commercial transportation route for Rochester is not bright, and competing methods of transportation may eventually absorb what little business it now has. It has even been suggested that its only future is as a scenic route for pleasure boats or as a source of water for agriculture.

The Port of Rochester is located on the St. Lawrence Seaway at the mouth of the Genesee River. The initial development of port facilities has been completed and includes a 23-foot-deep lower harbor and a 600-foot turning basin. The port also has docks for general cargo, cement, and coal, as well as warehouses, a coal-loader, and a spur rail line.

The location as the first port east of the Welland Canal, its past importance as an exporter of coal to Canadian markets, and its excellent access to land transportation facilities for the transshipment of general cargoes at one time gave rise to considerable optimism for an expanding business. However, as things now appear, even though some estimates of the port's potential remain optimistic, its future is not at all clear. Predictions for greatly expanded business have not materialized, and the port has been operating at a deficit. The value of general cargo handled has recently been declining, and failure to modernize coal-loading facilities may lead to the closing down of coal-shipping.

Manufacturing Orientation and Industrial Leadership

Some types of manufacturing grow and prosper as a nation grows and prospers; also, some industrial management contributes to general development of the community as well as to industrial success. Rochester was fortunate to evolve an industrial base with both the types of manufacturing which have done well over the years and unusually contributory management.

The period of flour-milling and railroad-building during the first half of the nineteenth century was a period of diversification of manufacturing. Among the industries developed were leather tanneries and shoemaking concerns, carriage shops, machine shops, foundries, and, most important to the continuing economy of the city, the ready-made clothing industry. With the development of railroads and expansion of the West, flour-milling declined and was succeeded in importance by the nursery industry. By 1850 orchards, nurseries, and seed houses had become so significant in the local economy that Rochester became known as the "Flower City," rather than the "Flour City." But, except for clothing, these were not the things which produced the Rochester of today.

It was such men as John Bausch, George Taylor, Henry Lomb, William Gleason, Frank Ritter, and George Eastman that gave Rochester a vigorous, new, and lasting industrial orientation. During the latter part of the nineteenth century, these men and others like them established the technical and scientific industries—optical equipment, scientific instruments, and the photographic industry. These light, high-skill types, plus the earlier established clothing industry, came to form the principal base of Rochester's economy. They are all the kinds of manufacturing which have grown vigorously with the American economy, and which do relatively well in industrial areas of the older Northeast. Textiles, gloves, and knitwear, for example, would have produced a less successful and less enduring economic base.

Many of the products now manufactured are the result of local inventions or innovations, and the bulk of the industrial production comes from firms that originated and stayed in Rochester. Because of this indigenous character, entrepreneurs have always had a vital interest in the over-all development of the community. By instigating commercial and institutional training programs to develop the skills of the labor force and encouraging business participation in all forms of community life and service, industrial leaders have done much to shape not only the economic but also the social and cultural aspects of the urban system. Moreover, many of these men, George Eastman of Eastman Kodak in particular, gave large gifts of land and money for parks, educational and cultural facilities, hospitals, and other important landmarks in the city. The Eastman Dental Dispensary, Eastman School of Music, and Durand-Eastman Park are just a few examples.

Industrial leaders also seem to have been particularly sensitive to the value of good labor-management relations. George Eastman, again, is often given credit for pioneering in the field of labor relations, including high wages, wage dividends, various forms of insurance, and welfare funds. His was one of the nation's first efforts in the field and has, along with the policies of other entrepreneurs in Rochester, led to a generally prosperous and satisfied labor force. As a result of Kodak's and other companies' progressive and at times perhaps paternalistic policies, the urban system has a relatively stable labor force, a minimum of labor strife and strikes, and relatively little unionization.

A Face-Lifting

Relative business losses of the CBD to outlying shopping areas are common, as is the blighting of areas immediately surrounding the CBD. In and around the CBD traffic is heavy, parking space scarce, and the general appearance of things unattractive. Such statements apply not only to Rochester but also to most other large urban systems. Few cities, however, have been more active in trying to meet these problems than Rochester. Specifically, what Rochester is doing to solve these problems may not be of major interest to the ordinary reader of this volume, but the kinds of things being done should be, for most cities are already embarking, or will embark, on similar programs if they are to keep up with the times.

CBD

Rochester's CBD occupies both banks of the Genesee River in the south-central part of the city. That this is the same general locality occupied by the earliest gristmills and sawmills suggests the long and continued influence of the river. Although strips or ribbons of commercial development extend in various directions along major streets from the CBD, the major concentration of business activity is in one small area. Within the CBD are seven of the eight major department and specialty stores and sufficient other commercial facilities to make it by far the most important shopping district between Buffalo and Syracuse. This high concentration of shopping facilities has strongly affected the layout of the new access routes and the general plans for urban renewal.

Rochester is now (early 1960's) embarking on an unusually extensive building program aimed at beautifying and modernizing its CBD and thereby hoping to make it an attractive commercial center for a wide area around it. Old buildings and landmarks are coming down, and new ones are going up. Underground parking is being provided, streets are being widened and in some cases relocated, and limited access arterial routes now lead into a maze of new patterns of traffic flow. An Inner Loop is being completed to relieve the congestion caused by through traffic. This Loop rings the CBD, providing the possibility of high-speed travel all around it. Several thousand new parking spaces have already been added and more are scheduled as part of the continuing program of redevelopment. The expansion of parking facilities has been one of the most successful aspects of the program.

A War Memorial has been completed on the west side of the river in the southwestern section of the core area within the Loop. It has a large auditorium which seats 10,000 and an exhibit hall and other features for conventions and trade shows. Hockey games and other athletic events are held in the building, too.

A Civic Center complex of buildings to the west of the War Memorial is being constructed to house both city and county offices.

The project which has received the greatest publicity and attention is the Midtown Plaza, located to the east of the river about three blocks from the banks of the Genesee. Here, on about 7½ acres of land, private enterprise and the city have joined forces to build a downtown shopping center with an underground municipal garage that has two thousand parking spaces. Within this site are two of Rochester's leading department stores as well as a major hotel. The center has, in addition to the three original buildings, an enclosed two-story pedestrian shopping mall with a skylight, heating, and air-conditioning. The mall area includes shops, offices, exhibition areas, plantings, and an open café. Finally, there is an eighteen-story office and hotel building which rises above the complex. A bus terminal at the southeastern corner of the plaza increases access.

In addition to Midtown Plaza, several new major projects are under way and planned for the near future. Ground has been cleared next to Midtown Plaza for the thirty-story Xerox Square building scheduled for completion in 1967. The new building will include a public skating rink and restaurant area.

Demolition of old buildings in a blighted area along both sides of the river to the north of Midtown Plaza has begun for the development of the Genesee Crossroads Urban Renewal Project. This area covers more than thirty acres and is scheduled to be rebuilt with a large new Federal Building, high-rise apartments, a park, and other developments, private and public. Planners hope for at least partial completion by 1970. The Crossroads Project will open the river to view along Main Street again, and should add greatly to the attractiveness of the CBD.

In addition to the above developments, several new individual buildings have been built or are being built in what promises to be a highly succesful face-lifting job for downtown Rochester.

While few statistics are available at this time on the results of this general program to attract shoppers back to the city, general impressions and a few studies suggest considerable progress. Figures compiled by the Rochester Chamber of Commerce show a significant upturn in business following the completion of the Midtown Plaza project and indicate considerable improvement in the competitive position of the downtown stores with suburban shopping areas.

OUTSIDE THE CBD

The question of access to the CBD is important. No matter how functional and attractive the CBD, if access is unduly difficult it cannot serve the urban system efficiently. Road-building programs in which the city, county, and state are now engaged are bringing great changes to the face of the whole area. When the program is eventually completed, there will be an Outer Loop of new and improved roads which will essentially encircle Rochester at about the city limits. This Loop will provide rapid transportation to all parts of the urbanized area and will lessen the pressure on radial arterial routes. The Outer Loop will eventually tie into the Thruway to the southeast and southwest, and major arteries will extend along the lake to the northeast and northwest, bringing the major highways from all areas surrounding the city into the circling flow. Arterials will lead from the Outer Loop inward to the Inner Loop and there provide access to the CBD. Such a system will not only afford a more orderly flow of traffic and relief of congestion, but will also have a profound effect on the future development of all parts of the urban system as outward expansion occurs.

Current Livelihood Structure

In 1960 the Rochester SMSA had 3.7 per cent of its labor force in the primary sector, 47.3 per cent in the secondary sector, and 49 per cent in the tertiary sector. Among the urban systems of the state, only Binghamton was more oriented toward manufacturing and less oriented toward tertiary activities. This probably reflects the very successful industrial history of Rochester and the general preoccupation of civic leaders with manufacturing rather than with the tertiary part of the economy.

MANUFACTURING

In the years 1947 to 1963 manufacturing employed between about 100,000 and 122,000 workers, with high points being reached in the mid-1950's and another upswing occurring by 1963. Reflecting increased worker efficiency, statistics show a substantial and generally continual increase in the amount of value added by manufacture. All in all, it is probable that Rochester's manufacturing base has been more vigorous and has exhibited better relative growth potential than that of any of the other urban systems in the state. Manufacturing is more concentrated in the central part of Rochester, too, than in the other urban systems, the big pull to the suburbs not yet having taken place. Within Rochester itself about 45 per cent of the labor force is engaged in factory work.

Manufacturing is largely of high-value-added types, dominated by the above-mentioned photographic equipment, scientific instruments, and optical goods. The urban system also produces a wide variety of men's clothing, machinery of various kinds including business machines, electronic equipment, and automotive equipment. In 1961 instruments and related products accounted for more than 37 per cent of the manufacturing employment, electrical machinery for 15 per cent, nonelectrical machinery for 9 per cent, food for 8 per cent, and apparel for 7 per cent.

By far the largest employer in the urban system is Eastman Kodak. In addition, though, there are a surprising number of nationally and internationally known firms, most of which are listed in Table 49.

The great importance of Kodak in the economy and cultural structure of Rochester warrants a few special remarks. This company operates several plants in Rochester as well as elsewhere in the United States and overseas. The largest unit, the

CENTRAL ROCHESTER. The newly constructed Inner Loop surrounds the CBD and part of the frame. The original Genesee crossroads where Rochester originated is just to the left of where the loop and railroad cross the river in the center of the photograph. The cloverleaf on the left provides connection with the Thruway. Just above the cloverleaf new buildings of the civic center complex are visible. Rochester has been a leader in city "face lifting." *Martin R. Wahl.*

TABLE 49
MAJOR MANUFACTURING CONCERNS OF THE ROCHESTER SMSA

Company	Products	Approx. Employment Early 1965
Eastman Kodak Co.	Film, cameras, projectors, and industrial chemicals	32,000
Xerox Corp.	Xerography machines and supplies, photocopy machines	6,000
Bausch and Lomb, Inc.	Scientific optical and electronic instruments, ophthalmic instruments	4,800
Delco Appliance Div., General Motors Corp.	Air conditioning, humidifiers, windshield-wiper mechanisms, small motors	4,300
General Dynamics/Electronics Div., General Dynamics	Military and industrial communications and electronic equipment	4,200
Stromberg-Carlson Div., General Dynamics	Telecommunications equipment	3,700
Rochester Products Div., General Motors Corp.	Carburetors, fuel pumps, fuel injection systems, locks and keys	3,200
Garlock, Inc.	Industrial rubber products, gaskets, oil seals, and electronic products	2,700
Bond Stores, Inc., (includes retail stores)	Men's clothing	2,700
Gleason Works	Machinery, gears, couplings, and engineering services	2,400
Taylor Instrument Cos.	Thermometers, barometers, controlling instruments for industrial processing	2,100
Commercial Controls Corp. (parent firm: Friden, Inc.) Subdiv., Singer Manufacturing Co.	Data-processing equipment	1,700
Hickey Freeman Co.	Men's suits, coats, and sportswear	1,700
Ritter Company, Inc.	Medical and dental equipment, chairs, tables, and sterilizing and X-ray equipment	1,500
Hickock Manufacturing Co., Inc.	Men's clothing, accessories, leather goods, seat belts	1,300
General Railroad Signal Co. Div., General Signal Corp.	Transportation control systems	1,200

Courtesy Rochester Chamber of Commerce.

Kodak Park Works, produces film, paper, chemicals, and other products, covers an area of 1,000 acres, and provides about 20,000 jobs. In total the company employs more than 30,000 people in Rochester, nearly 30 per cent of the entire manufacturing labor force. It is more than just the largest and best-known company in the urban system. As pointed out above, George Eastman's gifts to the city are evident everywhere—parks, theaters, dental dispensary, and the like. The company has been a leader locally in the development of social and civic responsibility and in support to education, and has on the national level pioneered in bettering labor-management relations. High wages, fringe benefits, and fewer work stoppages have all helped to create a strong feeling of loyalty for the company in employees and in other people in the city as well. Such heavy dependence on one company and industry can often create difficult economic problems for a community, but fortunately this company has been very successful; it has grown rapidly, diversified and expanded facilities, and introduced new lines of products including chemicals, fibers, and plastics. It is one of the largest and most prosperous manufacturing concerns in the nation. Among the many other outstanding and expanding firms in Rochester is the Xerox Corporation. It has attracted special attention through development of an electrostatic copying process, Xerography, and through its excellent performance on the stock market in recent years. The Xerox building in downtown Rochester will further strengthen its image in the community.

Tertiary Activities

Of the 284,000 employed workers in the Rochester SMSA in 1960, approximately 139,000 were employed in the tertiary sector. This is a large number, but reflects a proportionately light orientation toward tertiary activities when compared to most other major urban systems in the state. On the brighter side, during the decade 1950–60 Rochester ranked third among the seven urban systems in per cent gain in tertiary employment. This, added to a strong secondary sector, suggests good growth potential.

In 1960, 20.2 per cent of the SMSA's total employed persons worked in service industries, 16.8

per cent in retail and wholesale trade, 5 per cent in public utilities, 3.4 per cent in finance, insurance, and related fields, and 3.4 per cent in public administration. Within the tertiary structure, and not shown by the above figures, is a strong orientation toward education. Rochester has several institutions of higher learning including the University of Rochester with its well-known schools of medicine and music, Rochester Institute of Technology, Colgate-Rochester Divinity School, and St. John Fisher College. A community college also has been recently established.

A total sales and service function of the SMSA in 1960 produced a tertiary potential (see Chapter 12) of 810,000 and a tertiary surplus of 78,000.

External Relationships

As a second-order trade center Rochester serves a broad area of west-central upstate extending from the shores of Lake Ontario to the Pennsylvania border (see Fig. 117). From the northern portion of this area, particularly, it buys dairy products, fruit, and vegetables to supply its food-processing industries; to the surrounding counties it sells specialized services and manufactured products of many kinds.

The impact of this urban system, though, is felt far beyond the limits of its upstate trade area. Most of its industries secure materials from more distant sources in the state, nation, and world. Chemicals come from Syracuse and Niagara Falls, silver from Canada and Mexico, and monazite (used in the optical industry) from India and Brazil, to mention a few examples. Eastman Kodak is said to be the largest importer of silver in the country besides the United States Mint.

A much larger volume of Rochester's products are sold in national and international markets than regionally, too. Some firms ship well over one-fourth of their output outside the United States. Cameras and photographic film and paper, optical goods, instruments, dental equipment, electronic and communication equipment, gear-cutting machinery, and many other products are sold all over the world as well as in nation-wide markets. Men's clothing, baby foods, heating equipment, steel tanks, automotive equipment, and buttons all have national markets. Rochester firms have branch operations in other localities, too, ranging from Tennessee to Japan. Perhaps color film alone has made Rochester one of the most widely known names among cities of its size.

Problems and Prospects

Although one of the most prosperous urban systems in New York, Rochester, like the others, is faced with problems and challenges. The past ten years have brought a great increase in population of the suburbs and exurbia, while the population of the city of Rochester has declined. This pattern of a static core and an ever-growing and -widening ring of population around it poses serious problems of adjustment. The increasing population of the urban system puts a greater demand on the facilities of the central city, and at the same time newly developing commercial areas in the surrounding area offer increased competition to the central city. In order to keep as large a share of the business as possible, the CBD must be made attractive and convenient to potential users. Cultural and educational facilities should be expanded to keep pace with the growth of the urban system. Considering the kind of industry in the area and the importance of technology, it is urgent that local government and industry continue to support the expansion of educational facilities to meet the growing need for educated and skilled labor and management. Like other northern cities, Rochester is experiencing a large influx of Negroes from the South, and serious racial and economic problems have resulted. Many of the newcomers lack the necessary training and skills required by Rochester industries, and serious housing problems have resulted from the concentration of large numbers in the overcrowded slum and blighted areas of the middle city. While there are no simple solutions, there is clearly an urgent need for more and better housing, and a need for expanded job training programs.

The city has made a good start on its urban renewal plan, but it is imperative that the renewal be carried out as rapidly as possible and in an orderly fashion, rather than on a piecemeal basis. There is a real need, too, for realistic comprehensive long-range plans to meet the needs of the urban system, not just of the city of Rochester. The lake front and river provide an opportunity for imaginative development and should be made the most of.

A promising recent trend has been the amalgamation of several city and Monroe County departments and offices and the growing recognition that the city and county are really parts of a greater whole, sharing many problems and dependent on each other. This understanding probably should extend beyond the borders of Monroe County. The city of Rochester is now part of the greater urban system and, as a matter of fact, will become a less dominant part as the system grows.

Future economic prosperity depends on continued expansion of manufacturing and healthy growth of the tertiary sector. Manufacturing must be innovative, diversifying its products to enter new markets and meet new sources of competition. Recent growth in the number of manufacturing firms with operations in the urban system, the upward trend in the number of manufacturing jobs in the last year or two, and the relatively large capital investments that are being made to expand and modernize industrial plants are encouraging signs for the secondary sector. The population of the SMSA, at 733,000 in 1960, should reach 1,155,000 by 1980, and be roughly 1,840,000 by the year 2000.

Selected References

Committee on Economic Research of the Rochester Chamber of Commerce. *Basic Economic Information, Rochester, New York.* Rochester: The Committee, 1963.

Federal Writers' Project, Works Progress Administration. *Rochester and Monroe County.* Rochester: Scrantom's 1937.

Foreman, E. R. (ed). *Centennial History of Rochester, New York.* 4 vols. Rochester: Under Direction of the Rochester Public Library, 1931-34.

McKelvey, Blake. *Rochester: an Emerging Metropolis, 1925-1961.* Rochester: Christopher Press, 1961.

——. *Rochester the Flower City 1855-1890.* Cambridge: Harvard Univ. Press, 1949.

——. *Rochester, the Quest for Quality 1890-1925.* Cambridge: Harvard Univ. Press, 1956.

——. *Rochester the Water-Power City 1812-1854.* Cambridge: Harvard Univ. Press, 1945.

Merrill, Arch. *Rochester Sketchbook.* Reprinted from the *Rochester Democrat and Chronicle.* Rochester: The Gannett Company, 1946.

Monroe County Planning Council. *Design for the Future, Rochester-Monroe County Metropolitan Area.* Rochester: The Council, March, 1964.

——. *Economic Study, Rochester-Monroe County Metropolitan Area.* Rochester: The Council, March, 1963.

——. *Population Study, Rochester-Monroe County Metropolitan Area.* Rochester: The Council, June, 1962.

New York State Department of Commerce. *Business Fact Book, Rochester Area.* Albany: The Department, 1963.

——. "The Photographic Industry of New York," *New York State Commerce Review,* XI, No. 8 (August, 1957), 1-12.

——. "The Port of Rochester," *New York State Commerce Review,* XII, No. 6 (June, 1958), 9-12.

New York State Department of Public Works. "Report on Arterial Routes in The Rochester Urban Area," *Urban Area Report, Rochester 1947.* Albany: The Department, 1947.

"Rochester Rebuilds Downtown," *Engineering News-Record* (February 18, 1961).

Tanner, Ogden. "Renaissance on the Genesee," *Architectural Forum,* CXI, No. 1 (July, 1959), 104-9.

U.S. Department of Labor. *Area Manpower Guidebook, 174.* Metropolitan Labor Market Guides (1957), 199-200.

CHAPTER 22   DAVID J. DE LAUBENFELS

Syracuse

THE Syracuse Urban System is situated in the Upstate Heartland on major rail, water, highway, and air lines. The city of Syracuse, with a 1960 population of 216,000, is by far the largest central place. Figure 133 shows how the urbanized area extends substantially beyond the Syracuse boundary to the east, north, and west but has little penetration to the south. It contains 333,000 people, with 117,000 living beyond the Syracuse city limits. The urban frontier encloses an area of about 482 square miles with an estimated population of 420,000. In its longest dimension this area extends about 43 air line miles from Fulton in the northwest to Cazenovia in the southeast. For many years the SMSA was composed of only Onondaga County, but recently economic ties between Syracuse on one hand and Oswego and Madison counties on the other have become so strong that the three-county SMSA was established. The 1960 SMSA population was 564,000, having increased by 159,000, or 39 per cent, since 1940.

Geographical Advantages

For some time the Syracuse Urban System has been characterized by favorable growth trends and better than average economic health. Several important factors have been responsible. Its focal position on transport routes in Upstate New York has been critical to its growth throughout history. In the earliest period the production of salt did much to explain the existence of Syracuse, and for many years the locality held a virtual monopoly on this vital commodity. Now its substantial size plus a healthy array of secondary and tertiary enterprises contribute heavily to economic strength. These four factors—focal position, salt, large size, and the right kind of economic activities—have been and are Syracuse's advantages.

Syracuse lies at the center of Upstate New York in the eastern portion of the Upstate Heartland. Surrounding the Upstate Heartland are barriers, including the Appalachian Upland, the Adirondacks, and Lakes Erie and Ontario. Four major gateways pierce the terrain barriers and lead to neighboring regions. To the east lies the Mohawk Corridor; to the south Onondaga Valley permits easy contact with the Susquehanna tributaries, which in turn form a gap of sorts in hilly terrain less rugged than that farther east or west; to the west across the Erie-Ontario Lowland and through Buffalo is the very important route leading to the interior of the continent; to the north a strip of plain between Lake Ontario and the Tug Hill Upland provides access to the North Country and Canada. Lines drawn from east to west and from north to south to connect these four gateways would cross at Syracuse. It should therefore be no surprise that Syracuse has always been a hub of transportation. Major Indian trails, turnpikes, canals, railroads, highways, and now expressways have intersected here.

The site of Syracuse at the foot of the Appalachian Upland was also originally developed because of the nearby natural salt springs. The importance of salt in the Syracuse economy has declined over the years, although it still supports one major local manufacturer. The history of salt has, however, given distinctiveness to Syracuse reflected in its nickname, the Salt City. The name of the main street, Salina, also commemorates the early dominance of salt in the economy of the young city.

Large size is a great attribute to an urban system in the modern world. Syracuse clearly is above the minimum threshold size necessary for vigorous modern urban growth. Ever since the depression of the 1930's, Syracuse has been the fastest growing of New York's large urban systems, and it will likely continue to hold that distinction in the decades ahead.

The favorable growth trend seems attributable to two things. In the first place, Syracuse is becoming more and more the regional trade and service center for Upstate New York. In the second place, Syracuse boasts a good group of diversified "growth" industries, such as electronics, appliances, and electrical machinery, for which there has been a rising demand in the national market. Because of this orientation toward tertiary activities and growth industries, in times of general unemployment distress Syracuse has been less severely affected than most other cities in the northeastern

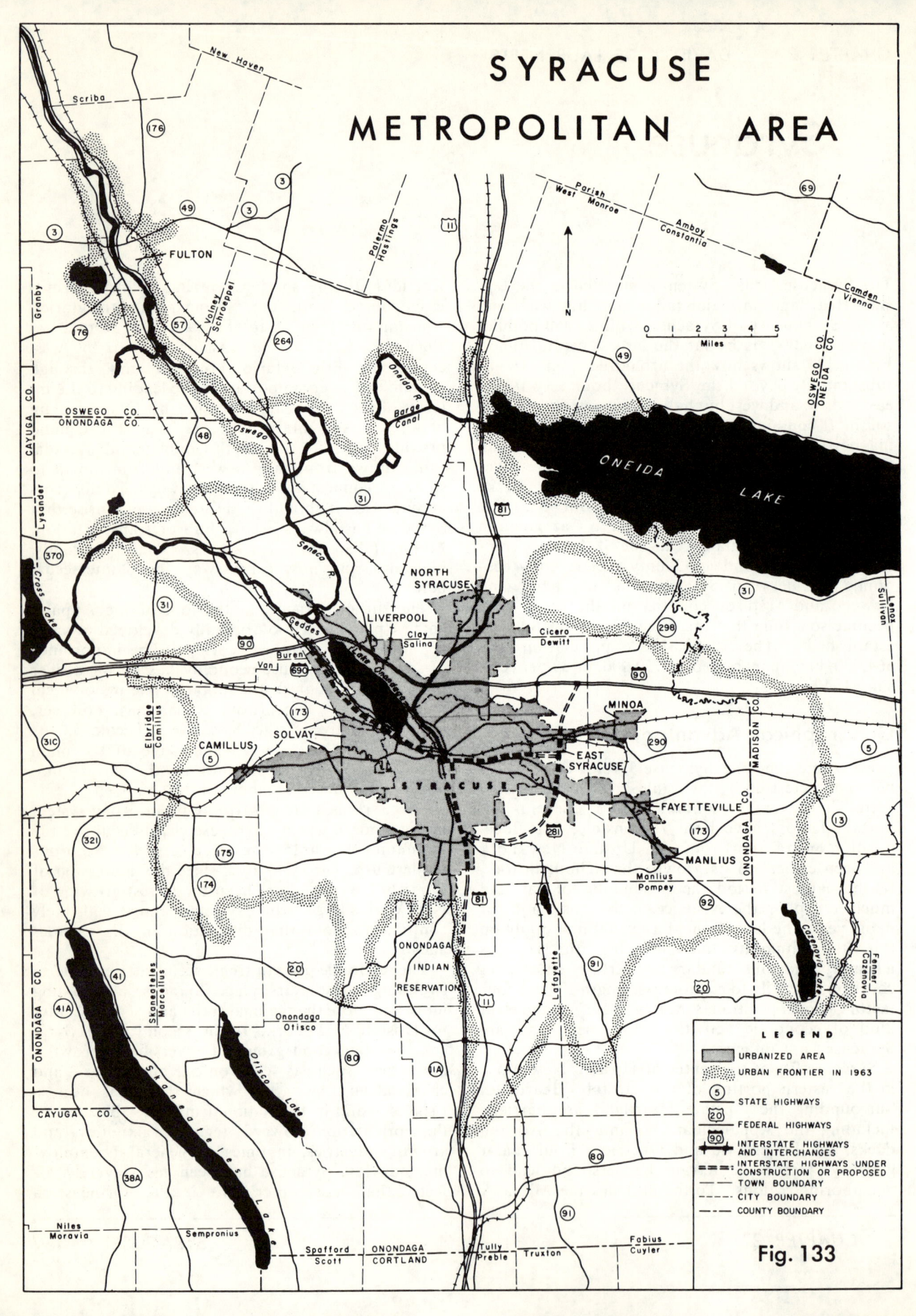

Fig. 133

TABLE 50
RELATIVE GROWTH RATES OF THE LARGE URBAN SYSTEMS

Metropolitan Area	Per Cent Increase of Population	
	1940–50	1950–60
Syracuse	15.5	21.2
Rochester	10.6	20.3
Buffalo	13.3	20.0
Utica	7.8	16.4
Binghamton	11.4	15.1
New York	10.0	11.9
Albany-Schenectady	10.1	11.6

part of the United States, and in good times general growth of the economy proceeds at an above average rate.

Historical Development

Highlights in the evolution of the Syracuse Urban System include: (1) early Indian usage and the attraction of salt, (2) coming of the Erie Canal, (3) impact of the railroads, and (4) appearance of the automobile-induced urban sprawl. Transportation played an important role in all of the above.

INDIANS AND SALT

The general area around Syracuse was very important to the powerful Iroquois Confederacy. In the southern part of the city their great central trail from the Hudson to Lake Erie, which passed through the principal villages of each member nation, crossed the north-south trail leading from the headwaters of the Susquehanna north to the shore of Lake Ontario and on to Canada. Near this junction the Onondagas had their village where, because of its central position, the Iroquois kept their Grand Council Fire. Figure 134 shows the critical early elements of the Syracuse site. Not far from the trail junction were a series of salt springs, important to Indian and white man alike. The village of the Onondagas, headquarters of the Iroquois Confederacy, attracted ambassadors, traders, and military expeditions; but it was primarily the salt that brought white settlement into the area that is now Syracuse.

The actual site of the village of the Onondagas changed from time to time. It normally occupied high ground along one of the creeks that flowed northward out of the Appalachian Upland. The great central trail, too, was just south of the plain traversing the northern edge of the plateau near the village sites. The reservation set aside for the Onondagas today lies more than five miles south of the center of Syracuse.

The salt springs were located all around the southern half of Onondaga Lake, which joins Syracuse on the northwestern side. At the Treaty of Fort Stanwix in 1778, a salt reservation was set aside extending around the shores of the lake. Within a dozen years many people began settling as squatters on the salt lands in order to engage in the manufacture of salt. At this time, improvements began to be made in the old Indian trails in order to convert them into roads. Getting salt to market from the salt springs was as much of a motive for this road development as any other factor.

By the beginning of the nineteenth century a measure of organization came to the Salt Springs Reservation and the roads giving access to it. It began in 1797 when the state took charge of the land and issued leases to fifteen-acre lots. The Seneca Road Company was chartered in 1800 to build a turnpike over what had been the great central trail of the Iroquois, a route known in the Syracuse area as the Seneca Turnpike. Within a few years they had also constructed a north branch through the Salt Springs Reservation, which became the main line of the Genesee Turnpike, now Genesee Street. At the same time the Salina and Chenango Turnpike was built from the salt springs into southern New York and another road leading north to Oswego appeared. Clearly, the salt springs became the local focus as major activity developed near the junction of the Genesee Turnpike and the north-south routes rather than at the junction of the latter with the Seneca Turnpike. A vigorous community called Salina, for its product, grew up next to the salt springs after the turn of the century. The actual crossroads, where the center of Syracuse lies today, was on the edge of a swamp. Salina, a mile and a half to the northwest on a bluff, was organized in 1809.

COMING OF THE CANALS

The real nucleus of the city of Syracuse itself did not form until the Erie Canal was built. Inevitably older Salina became a great rival to the new center but it was not long before it was absorbed by growing Syracuse.

Before the construction of the Erie Canal, there had been only half a dozen log cabins, a tavern, and several mills at the crossroads in the swamp. The mills were there because a small dam had been built where the Genesee Turnpike crossed Onondaga Creek. The middle section of the canal from Utica to Montezuma near the port of Cayuga Lake was started in 1817, and a camp of laborers swelled the population. Six different names were given the crossroads settlement at various times, with the name Syracuse finally being settled upon in 1820, at which time the population was approximately 250.

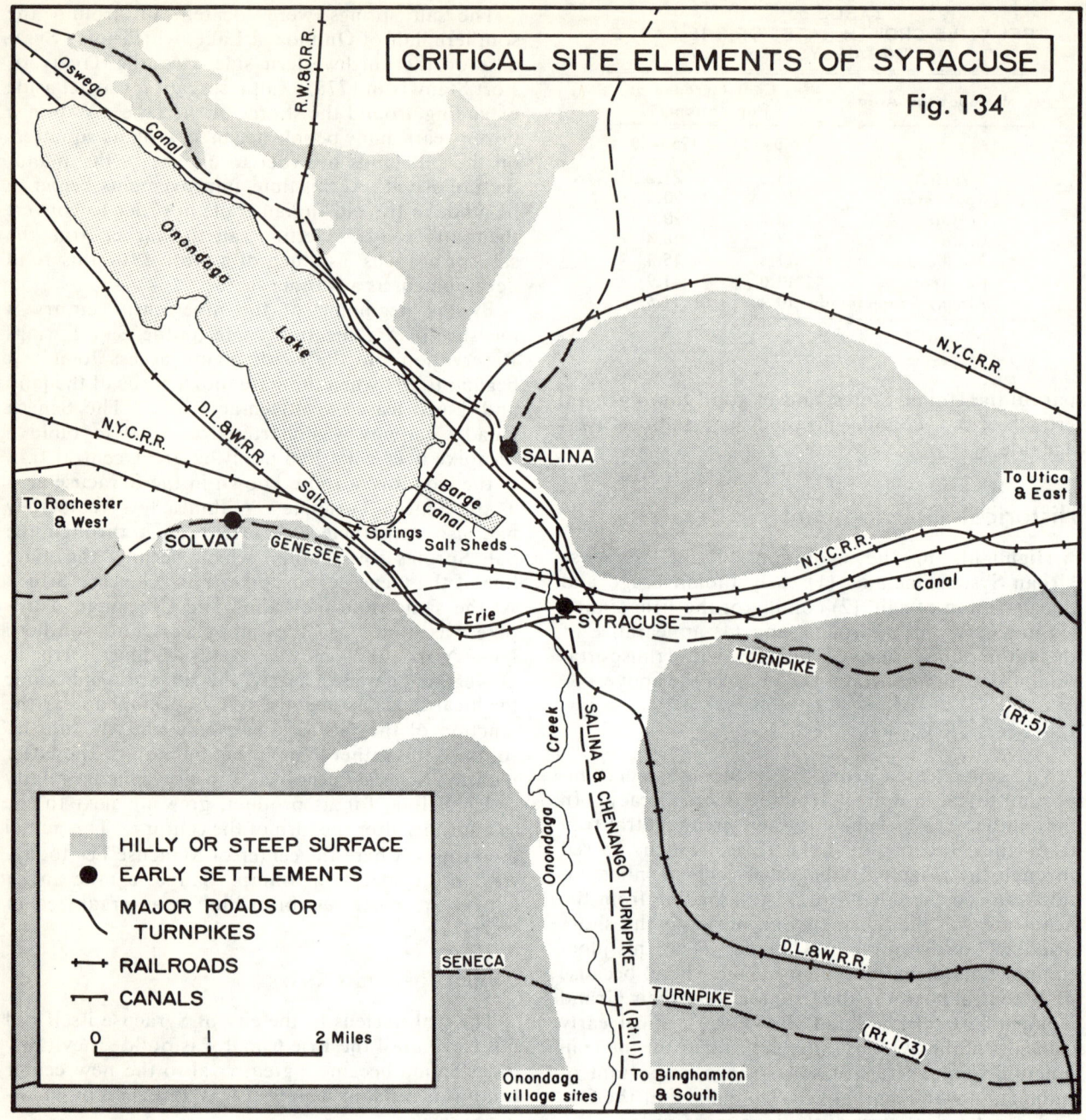

CRITICAL SITE ELEMENTS OF SYRACUSE
Fig. 134

The first packet boat reached Syracuse that same year. The Erie Canal passed directly through the crossroads and a branch canal to Salina left the main canal two blocks to the east. This branch canal was extended to Onondaga Lake, thence to the Oswego River, and finally to Lake Ontario at Oswego, reaching there in 1828. Meanwhile the Erie Canal itself had been completed to Buffalo in 1825, the same year Syracuse was incorporated. Because the swamp just south of Onondaga Lake had proved unhealthful and a hindrance to travel, the outlet to Onondaga Lake was lowered and the swamp was largely drained.

Salt manufacturing increased with a change in technique. At first the brine had been boiled down into fine salt in cauldrons. With the development of new salt-producing techniques employing solar evaporation, recently reclaimed swampland just to the west of Syracuse became the center of the salt industry.

Syracuse's more favorable location with respect to the canals and the expanded salt industry caused it to outdistance Salina in size. By 1830 it had twenty-five hundred people; by 1840, eleven thousand. Finally, in 1847, Syracuse and Salina joined to form the city of Syracuse with a combined population of twenty thousand. The assemblage of so many people and the buildings they built into an area that was a virtual wilderness half a century before is a typical story of Upstate New York.

IMPACT OF THE RAILROADS

Important as were canals to Syracuse and central New York, they soon had serious competition from railroads. For a long period, between 1860 and the 1920's, the railroad was pre-eminent. During this period, when population was increasing at a rate of about one-quarter each decade, significant changes were taking place in the complexion of Syracuse. The importance of canals and salt faded away while excellent railroad access attracted a diversified manufacturing economy. It would be difficult to weigh the relative importance of the two factors at this time, but it is clear that Syracuse grew both because of its focal position on the railroads and because of the momentum provided by the pre-existing population mass.

The first railroads reached Syracuse in 1838 and 1839, connecting east to Utica and west to Auburn, but it took until the time of the Civil War for anything like a complete railroad network to develop Upstate. The main line of the New York Central, after it was organized, ran east-west through Syracuse, paralleling the Erie Canal. These two mediums were only a block and a half apart through the heart of the city. Lines were built to the north and the south as well as to the east and the west, and all of these also passed near the center of the city. The competition between canals and railroads was good for Syracuse, particularly as both mediums focused traffic on the city. Early in the twentieth century the Erie Canal through Syracuse was abandoned; its successor, the Barge Canal, has a branch through Lake Onondaga, which at the time of this writing still permits commercial contact with Syracuse.

The salt industry reached its peak during the Civil War and then declined. By the time the springs began to be deficient in salt, Syracuse's monopoly on salt production in North America had been broken as other sources were discovered. Local production of salt dwindled in the last decades of the nineteenth century, and by 1908 the government had sold all the salt lands for other uses. Before the salt springs were abandoned, however, Solvay Process Company brought in brine wells south of Syracuse and engaged in the manufacture of various chemicals including caustic soda and bicarbonate of soda. For many years after the demise of the original salt industry, Solvay was the leading industry of the area and today it still ranks as the third largest manufactural employer. The abandoned salt lands fortunately were located near the center of growing Syracuse and so became valuable for industrial and other purposes.

Many different kinds of manufacturing followed salt-making, with the production of metal products being most prominent. Among the earlier products were ploughs, bicycles, and guns. Some of the abandoned canal lands were taken up by such companies as the L. C. Smith Typewriter Company and the Continental Can Company. The abandoned salt lands were the locale for a major development of automotive industries; for example, the Franklin Motor cars were made here. In addition Syracuse became a center for pottery-making, candles, and various kinds of small machinery. During the early part of the twentieth century there was generally an increasing demand across the nation for the kinds of products made in Syracuse. It is interesting to note that the city never did turn in a major way to the production of textiles or other soft goods as did so many other northeastern cities.

By growing at a steady rate, the population reached 200,000 during the 1920's and Syracuse had become a major center within the American Manufacturing Belt. The city had spread out in the shape of a diamond with the points oriented in the directions along the major east-west and north-south axes. The continuously built-up area extended outward roughly four miles from the center, and several small suburban communities were taking form. The village of Solvay together with the factory for which it was named was located on rail lines beyond the city limits to the northwest. A bypass rail line was built around the north side of the city. Where this bypass joined the main line east of the city, the New York Central freight yards, which contributed strongly to the support of the communities of East Syracuse and Minoa, evolved. Later Mattydale grew on the north side of Syracuse near the railroad bypass. Other villages, such as Fairmount, Liverpool, North Syracuse, and Fayetteville, were nearby but separated from Syracuse by open stretches of rural land. None of the communities outside the city limits of Syracuse had more than a few thousand inhabitants in the twenties.

THE HIGHWAY REVOLUTION

The automobile and the highway began to take on major significance in the 1920's. By this time the growing use of the automobile had encouraged spread of urban dwellings well beyond the city boundaries and trucks had begun to carry some interurban freight. It was not until after World War II that the great suburban explosion and extensive take-over of freight by trucks really occurred. People and factories both literally fled the city, producing a peripheral sprawl extending outward several miles beyond the previously built-up area. Left behind were a group of different problems including neighborhood population and tax-base declines.

Accelerated peripheral growth occurred largely along major-access highways, so that arms of urban land use began to probe deeply into the surrounding countryside; these projections have now reached ten miles and more from the center of Syracuse. Communities still farther out have experienced a quickening growth and, in addition, city-employed people are establishing their homes in the surrounding rural landscape out to about a forty-minute driving time—roughly twenty miles. In any given recent decade, there has been more than doubling of population in each whole town centered about eight miles out, with the exception of those in the hill country to the south of Syracuse. Even these are now joining the growth trend. The time is fast approaching when the population in the built-up areas outside the Syracuse city limits will more than equal that of the city itself.

Outlying villages such as Fayetteville, North Syracuse, Liverpool, and Fairmount, located along major highways, have expanded their built-up areas both toward and away from Syracuse. Growth of older suburbs has reached the point where now the typical one has a population of eight to ten thousand. Some have a restricted incorporated area; others like Dewitt and Westvale are simply filling the space that once separated them from Syracuse and surrounding villages. Whereas in 1930 only one of nine urban dwellers lived beyond the political boundaries of the city of Syracuse, by 1960 at least four out of nine lived beyond these boundaries. Suburban population increased by eight times during that thirty-year period. Table 51 shows the 1960 level of population in the suburban sectors.

The highway revolution has been felt not only in the suburbs but in the central city as well. As growing numbers of residents flee the city to the suburbs and as their former residences are occupied by people of less affluence, residential blight sets in and substandard housing inexorably expands. The problems of an increasing proportion of substandard housing with its attendant social erosion are obvious. Suburbanites, trying to get to work in the city or secure goods and services there, have for a long time flooded narrow streets with traffic volume these streets were never meant to handle. Serious congestion has resulted. Population stagnation in some neighborhoods has caused considerable concern, too. Actually, a distinct population decline has occurred near the center of Syracuse, barely balanced by a slight growth just inside the city limits. The result was a net growth of only 4 per cent in the thirty years between 1930 and 1960 for the political city of Syracuse.

Even growth of the central business district has been small. Suburban trade centers have sprung up closer to the new centers of population growth and drained potential business away from the CBD. Among these, three are regional shopping districts (see chapter 14). In spite of the obvious effects of such shopping centers on the CBD, it is interesting to note that not even the largest does as much as one-tenth the business of the CBD.

Factories have moved out of the city, too, to escape congestion and to have better access to highway-oriented workers, suppliers, and markets. The effect on the tax base and on local employment opportunities has been substantial. Suburban manufacturing development occurred primarily along the northern edge of the urbanized area. The location of new industries to the north of the city is conditioned by ample level space in this direction and proximity to highway—particularly Thruway—service. The fact that they are also located close to the east-west railroad bypass is now of secondary importance. Some do use the railroad, to be sure, but to a limited degree, and others do not use it at all. The factories, the Thruway, and the railroad are together because each requires space and level terrain, a combination not found south of Syracuse. The most important expansion has been in the electronics lines, but new factories and expansion of old ones in metal goods, auto parts, and air-conditioning equipment have been significant, too. Some migration of factories from the general area has taken place over the years, but by and large it has been small compared to the trend for northeastern cities.

The highway revolution has been particularly significant in shaping the modern geography of Syracuse, in the older parts of the urban system as well as in the peripheral parts, in the manufacturing and commercial zones as well as in the residential areas. Precise figures are not available, but estimates suggest that as much as 80 per cent of the freight originating or terminating in Syracuse is now carried by truck. The decline of rail passenger traffic, and therefore service, is familiar. Even the

TABLE 51
ESTIMATED POPULATION IN EACH OF THE
SUBURBAN SECTORS OF SYRACUSE, 1960

East (Dewitt, Fayetteville, etc.)	25,000
Northeast (East Syracuse, Minoa)	15,000
North (Mattydale, North Syracuse, Cicero)	45,000
Northwest (Liverpool, Solvay)	25,000
West (Westvale, Fairmount, etc.)	25,000
South (Nedrow, Lafayette, etc.)	15,000

principal railroad passenger terminal in the city has been closed! Resources are being directed heavily toward improving highway service in and through the city. Years ago the abandoned Erie Canal bed was converted into a boulevard; now the main east-west rail right-of-way, which is elevated through the center of the city, is being converted into an express highway.

Some Details of the Current Geography

Geographical persistence in Syracuse is everywhere evident as modern geography reflects past geography. Indian trails and salt springs, canals, railroads, and the highway revolution have all in turn had their effect on the positioning and growth of the various parts of the urban system. Terrain, too, has played a major role.

Near the Center

The earliest CBD was situated just north of present Clinton Square where turnpikes intersected the canal; subsequent development has occurred in a four-block area just to the south along South Salina Street where room for growth existed. Rail yards, warehouses, and factories to the west and institutional buildings to the east (churches, city, county, and state offices, etc.), tightly restrict the CBD, permitting only limited spill-over onto adjacent parallel streets on either side. Numerous parking lots and parking garages, now located in the frame near the CBD, serve the increasing number of automobile-driving customers. In the frame to the east of the CBD, urban renewal is making way for a large plaza, with construction of high-rise apartment buildings to follow. The apartment buildings represent an effort to "bring people back to the city center" by providing suitable nearby housing. North and south of the CBD businesses of one kind or another continue for several miles, especially along Salina Street. Genesee Street, which extends east and west from the CBD, presents two specialized areas, "automobile row" to the west and a concentration of medical offices to the east.

In the Middle City

The middle city, the solidly built-up part of the urban system approximately coinciding with the incorporated area, extends outward three or four miles from the CBD. Scattered through it are neighborhood and community shopping districts. Many of these are not doing especially well, depending as they do on nearby static populations with declining buying power and being badly located with respect to the faster-growing, newer suburbs. In essence they are feeling the competitive pinch between the CBD on one hand and the vigorous regional shopping districts of the suburbs on the other.

By 1865 most of the level land within two miles of the CBD had been taken up by closely spaced housing or other uses. This century-old inner part of the middle city contains nearly all of the present-day substandard housing and many vacated factories and warehouses, and is undergoing urban renewal or other intensive redevelopment planning. Both Syracuse University and the nearby extensive concentration of hospitals and medical facilities of the Upstate New York Medical Center are situated to the southeast just beyond this deteriorating zone. Farther out, on the drumlins and hills to the southeast, southwest, and northeast, newer and more expensive homes remain in good condition and form good to excellent residential districts.

On the Periphery

Although some open land can be found within a radius of four miles of the CBD, suburban agglomerations reach out ten miles and more. Growth has not been equal in all directions, there being most to the north and least to the south. In spite of the airport and poor drainage to the north, the most rapidly growing suburban sector is in this direction. Level terrain here has been attractive to transportation, manufacturing, and the spread of utilities. Although characterized by attractive new property, the area south of the city has experienced only limited growth. The Onondaga Indian Reservation is located there, and the high, hilly country is difficult and expensive to supply with municipal water. Growth along major transport arteries to the west and east has been rapid and extensive but not equal to that to the north.

Older villages and shopping districts have been "imbedded" in the suburban growth, and attractive new regional shopping districts with ample parking space have been built. Three of these, Shoppingtown in Dewitt, Northern Lights in North Syracuse, and Fairmount Fair in Fairmount are of major proportions. Each is located along a principal transport artery and serves the eastern, northern, and western suburban sectors respectively. All draw business from the suburbs, from nearby parts of the city itself, and even from areas beyond the urban frontier.

Figure 133 shows the urban frontier to be from 10 to 25 miles from the CBD. Generally all areas within it have nonfarm populations amounting to 90 per cent of the total. Greatest outward penetrations of the urban frontier occur at Fulton and Cazenovia. In the first instance there is a response to the Oswego River and an industrial satellite; in the second, to an attractive residential node. In general the frontier is very irregular, being affected

by such things as the Onondaga Indian Reservation in the south, the Cicero swamp south of Oneida Lake, and, of course, the major transport arteries in various directions. Some degree of regularizing of the frontier at a radius of 15 to 18 miles from the CBD is likely to occur, but terrain, swamps, water bodies, and degree of access will never permit an approach to a perfect circle.

Current Livelihood Structure

In 1960 the Syracuse SMSA had 3.6 per cent of its labor force in the primary sector, 41 per cent in the secondary sector, and 55.4 per cent in the tertiary sector. Its manufactural base is well oriented toward the types of industries that are doing relatively well in the northeast and its tertiary sector is diversified and vigorous.

Manufacturing

In the years 1947 to 1963 manufacturing employment remained quite steadily at between 65,000 and 68,000. If 65,000 is taken as a median current figure and if it is assumed that 2.8 people[1] are supported by each factory job, then about 180,000 people in the Syracuse Urban System can be said to be gaining their livelihood from manufacturing.

Syracuse is a hard-goods center, emphasizing machinery, metal products, and other heavy goods. One-third of all factory workers are engaged in electrical machinery production, and another 15 per cent are in nonelectrical machinery production. Following in order are food and kindred products, paper and allied products, primary metals, and chemicals. The greatest growth in the last twenty years has been in electrical machinery, with one company, General Electric, assuming a pre-eminent position in the industrial scene. Losses have been particularly noteworthy in nonelectrical machinery.

Changes in manufacturing have been related to the elimination of small and obsolete factories, often on crowded or otherwise inadequate sites. Numerous small plants have failed and many a large and venerable factory stands empty. Some companies such as Carrier Corporation have built new plants in the suburbs; others such as Smith Corona have moved away. The net result has been to develop an emphasis on newer, larger factories in locations peripheral to the urban system. That most of these new factories are manufacturing products in growing demand in the nation is Syracuse's good fortune. There are sufficient diversity and enough large industrial concerns to give the Syracuse Urban System a reasonably secure industrial future. This diversity expresses itself among the fifteen largest concerns. These are listed in Table 52 together with principal products and approximate employment for January, 1965.

The relative importance of manufacturing in the over-all employment picture is less than in years past and will probably decline still further as the population of the urban system grows. Although total factory output should rise in the future, perhaps at 2 or 3 per cent per year, it appears now that the tertiary sector will have to take care of most of the demand for new jobs.

Tertiary Activities

That Syracuse is depending more and more on service and trade should be regarded as a healthy trend and demonstrates its rising importance as a trade center. Of the 210,000 employed workers in the SMSA in 1960, roughly 117,000 were employed in the tertiary sector, and (using the 2.8 multiplier) 330,000 were supported by it. In the next twenty years alone over 100,000 new jobs will have to be developed in the tertiary sector. Such an addition may seem unattainable on first thought, but, in view of the fact that the decade 1950–60 saw a tertiary employment gain of approximately 23 per cent, it does not appear impossible. Not only is this 23 per cent gain larger than for most other urban systems in the state, but Syracuse has a very large trade area including Binghamton and Utica and extending far

[1] This is only an approximate figure and was arrived at by dividing the number of employed persons in the SMSA into the population and assuming a small drive in to work from beyond the SMSA boundaries.

LARGE-SCALE MANUFACTURING at the Syracuse periphery—Electronics Park (upper), Vicinity of Thruway Interchange 35 (lower). The Syracuse Urban System, located where the Ontario Lowland meets the Appalachian Upland, has plenty of level land around its northern periphery. With good railroad and highway access as additional assets, this northern periphery has experienced considerable manufacturing expansion during the last quarter-century. The upper photograph shows the single largest industrial unit in the Syracuse Urban System, General Electric Company's Electronics Park north of the city. Electronic equipment and electrical machinery are the chief types of products manufactured there. The Thruway is seen on the left and an interchange is nearby. This type of industrial operation requires large uncongested areas of relatively level land for buildings, parking space, and access. The immediate vicinity is still not congested. *Courtesy General Electric Company.* In the lower photograph the Thruway extends from left to right in the background. Interchange 35 is just off the photo to the left. Here, close enough to Syracuse to benefit from the locational advantages it offers but far enough out on the northeastern periphery to find adequate space, is the area's second largest manufacturing establishment, Carrier Corporation. It produces air-conditioning and heating equipment, industrial compressors, and other products. In the foreground are a number of warehousing and servicing establishments. These kinds of tertiary activities also find interchange locations near large urban systems particularly desirable. *Courtesy Carrier Corporation.*

TABLE 52
MAJOR MANUFACTURING CONCERNS OF THE SYRACUSE SMSA

Company	Products	Approx. Employment Early 1965
General Electric Co.	Electronic equipment and electrical machinery	12,900
Carrier Corp.	Air-conditioning and heating equipment, industrial compressors	5,300
Solvay Process Div., Allied Chemical Corp.	Soda ash, caustic soda, chlorines, potassium carbonate	2,800
Crucible Steel Co. of America	High-speed-tool stainless, alloy and specialty steels	2,500
New Process Gear Div., Chrysler Corp.	Automotive gears and transmissions, sewing machines	2,400
Sealright Corp.	Food containers	2,200
Crouse-Hinds Co.	Conduit fittings, floodlights, traffic signals	2,000
Syracuse China Corp. Div., Onondaga Pottery	Vitrified china	1,600
Ternstedt Div., General Motors Corp.	Plated hub caps, radiator grills, auto trim	1,600
Nestle Co.	Chocolate products	1,400
Bristol Lab. Div., Bristol-Myers Co.	Antibiotics, drugs	1,200
Prestolite Div., Elta Corp.	Starting motors and generators	800
Lamson Corp.	Pneumatic tube systems, conveyors, exhausters, air vacuum systems	650
Porter-Cable Div., Rockwell Mfg. Co.	Portable electric tools, power garden equipment	600
Rollway Bearing Co.	Precision roller bearings	575

northward through the western Adirondacks and on into the North Country. In 1960, 21.5 per cent of the SMSA's total employed persons worked in service industries, 18.4 per cent in retail and wholesale trade, 7.0 per cent in public utilities, 4.5 per cent in finance, insurance, and related fields and 4.1 per cent in public administration. The total sales and service function in 1960 produced a tertiary potential (see Chapter 12) of 674,000 and a tertiary surplus of 110,000. Thus size, growth, and surplus of the tertiary sector all place Syracuse in a strong economic position, both now and for the future.

The level of construction has advanced at a rapid rate in recent years, another sign of prosperity. Extensive highway-building and urban renewal activities should help to keep the level of construction high.

Problems and Prospects

The future promises to bring to Syracuse increasing problems associated with increasing complexity, but it should also bring advantages and sophistication that are the by-products of a successful regional urban system. Problems and prospects are considered below for the central city, the suburbs, and finally the urban system as a whole.

CENTRAL CITY

Incorporated Syracuse—the city—suffers from the typical problems of most central cities: rising costs of crime, welfare, and sanitation, and a declining tax base. Obsolete factories close; the railroad sells its right of way for highway construction. The highways and parking garages built by the city and the state to stimulate business at the center take land off the tax rolls. Although it is not supposed to, even urban renewal may decrease tax revenues. The biggest problem of all is the spreading blight of residential deterioration which takes hold of a house too old to repair but too valuable to destroy. The advent of urban renewal is an indication that a considerable area has become intolerable and remedial measures have been undertaken. But even with remedial measures there will always be an adjacent zone where conditions are not yet intolerable enough to do something. What is needed is some measure either to maintain or to replace those dwellings which are just now being neglected. Many of the problems and their solutions are sociological; only some can be solved through more intensive use of the land—expansion of the business area or construction of apartment houses.

Reduction in growth of CBD business activity caused by the new suburban competition should be temporary. In spite of increased suburban competition, the CBD is not withering away. There continues to be a growth in actual business volume because the center is and will always be the most accessible location for the specialization and comparison in merchandising that are increasingly possible in larger urban places. New construction reveals a major interest in office space, however. This is a reflection of the growth of tertiary activities in the economy in general and of the changes to be expected in the physical make-up of the CBD. Also, many new parking garages are being added with an eye to increasing accessibility and service of the automobile-traveling customer. The central business district of Syracuse, it is predicted, will maintain its position as the overwhelmingly dominant node of a growing metropolitan area.

The depressed residential zone around the commercial core is not being left in a continually deteriorating state. Both government-supported urban renewal and private enterprise are at work to produce more rational land use patterns than this area has known for some time. Multifamily housing projects are widespread—either already constructed or on the planning boards. This wave of renewal near the center is unfortunately moving behind an expanding wave of deterioration farther out. How to solve this problem does not seem to be a part of the understanding of planners and civic leaders. Perhaps at this time there is no solution.

Suburbs

Growth on the margins of metropolitan Syracuse continues to be spectacular. This growth will be further stimulated by the system of expressways now under construction, which will circle the city and feed into a central ring. Both residential zones and industries will find their spread facilitated by the new highway system. Limited water supplies, now somewhat of a deterrent in some instances, will be at least temporarily solved when the already adopted scheme for bringing Lake Ontario water into the urban area is completed.

The suburbs have their own problems, caused mostly by rapid growth. New schools, roads, sewers, water lines, and other utilities must be constructed. What has happened is that metropolitan problems have been divided into city problems and suburban problems, and it is not always clear whether the city or the suburbs got the better deal. One thing about suburban problems is that they vary tremendously. Thus some old suburban cores, like East Syracuse and Solvay, have problems of obsolescence. Some have a fat tax base of industry, while others may have only the workers, their modest value homes, and veterans' tax exemption on each. Some may have ready access to water, while others have almost no water supply of their own. Most are too small to have efficient local government for what amounts to a piece of a city rather than an isolated village. All of these problems must be dealt with in an environment of rapid growth.

Over-all Urban System

The problems of the central city and of the suburbs are parts of the over-all metropolitan problem. As the urban system grows it becomes more complex. Many special jurisdictions appear—water districts, school districts, fire districts—which do not correspond to one another or to any other political area. Coordination is needed, but the fragmentation should not be destroyed because it fills in part very real local and special needs. Syracuse recently adopted a new county charter creating a county executive. This kind of action may bring some greater measure of order even though the county does not include all of the urban system.

To summarize, there are three favorable factors affecting the growth of the Syracuse Urban System. First, the system is already large, thus containing the necessary ingredients of high-level commercial, service, institutional, and market potential that make it more attractive than any smaller center. Second, a diversified manufacturing base with relatively good growth potential gives it advantages over less well endowed urban systems of all sizes. Finally, a central location in the Upstate Heartland at a point where major transporation routes focus is contributory to a bright tertiary future. Regional offices of many kinds find Syracuse a good location. More are moving in each year. As long as these factors continue to be applicable, Syracuse will likely remain the fastest-growing urban system in New York State. The population of the SMSA, at 565,000 in 1960, should reach 1,000,000 by 1980 and roughly 1,850,000 by the year 2000. If these predictions are correct, the Syracuse Urban System presumably will catch up to, and pass, the Rochester and Albany urban systems in size.

Selected References

Blair & Stein Associates prepared the following reports on the Onondaga–Syracuse metropolitan area for the Onondaga County Department of Planning and the New York State Department of Commerce in 1961 and 1962:

Control of Land Subdivision
County Plan
Economic Base (Bureau of Economic Research, LeMoyne College, assisted in preparation.)
Patterns of Land Use
Population: Trends, Outlook, Implications (Bureau of Economic Research, LeMoyne College, assisted in preparation.)
Public Facilities
Recreation & Open Spaces
Transportation
Utilities
Zoning

Department of City Planning, Syracuse, N.Y. *A General Plan.* Syracuse: The Department, 1959.

Faigle, Eric H. "Syracuse: A Study in Urban Geography." Unpublished Ph.D. dissertation, Department of Geography, University of Michigan, 1935.

Martin, Roscoe, Frank J. Munger, *et al. Decisions in Syracuse.* Bloomington: Indiana Univ. Press, 1961.

Metropolitan Development Association of Syracuse and Onondaga County. *A Profile of Onondaga County.* Syracuse: The Association, 1961.

CHAPTER 23   SIDLEY K. MACFARLANE

Utica

The Utica Urban System is located at the eastern end of the Upstate Heartland and overlaps slightly into the Middle Mohawk. It is very near the geographic center of the state and on the Mohawk–Lake Plain axis at the point where the confining Mohawk Corridor opens onto the more level country to the west.

Utica, with a population of 101,000, is by far the largest central place in the urban system, and Rome, population 52,000, ranks second. The dominance of these two cities has resulted in a census designation, the Utica-Rome SMSA. Smaller central places are spread bead-like for forty miles along the Mohawk River from Rome in the west to Little Falls in the east. These include Oriskany (1,600), Whitesboro (4,800), and Yorkville (3,800) in the more open country between Rome and Utica. East of Utica the smaller places such as Frankfort (3,800), Ilion (10,000), Mohawk (3,500), Herkimer (9,400), and Little Falls (9,000) are bottled in by the steep slopes of the narrow valley walls on either side of the Mohawk River.

The urbanized area (Fig. 135) has an unusual shape. It includes a relatively small but irregular area around Utica. A long arm extends south along Sauquoit Creek and another connects Utica with Rome. The large area bounded by the city limits of Rome is by census definition within the urbanized area. In no way, however, does it delineate the true built-up area. The urbanized area contained 188,000 people in 1960, 35,000 of whom were outside the incorporated limits of the two major central places. The urban frontier encloses about 195 square miles and an estimated 1960 population of 215,000. Its greatest dimension—about 25 miles—is in a northwest-southeast direction. Clearly the position of the cities of Utica and Rome has had a major impact on its shape. In the vicinity of Rome the location of the urban frontier illustrates the above-mentioned inadequacy of the census designation of the urbanized area. Composed of Oneida and Herkimer counties, the Utica-Rome SMSA includes a great deal of sparsely occupied, in fact almost uninhabited, country in the northern portions of those counties. The 1960 population of the SMSA had reached 331,000, having increased by 68,000 or 26 per cent since 1940.

The physical setting of the urban system is attractive. The Appalachian Upland with its generally aligned north-south valleys lies to the south. Much of the land on this upland is farmed, but the higher and steeper areas are largely forested. To the immediate north lies the southeast extension of the Tug Hill Upland. This area is farmed but only in a very spotty fashion because of poor drainage, low-quality soils, and steep slopes. Beyond the Tug Hill the Adirondack foothills and the mountains themselves make attractive country. Both summer and winter recreational amenities are within easy reach.

During the last few years drastic changes in the economy have taken place. This is particularly true in Utica and its fringe communities. Two decades ago Utica and nearby towns were famous as soft-goods (nondurable goods) centers, particularly for textiles and knit goods. Today this kind of production has almost vanished, and manufacture of hard goods (durable goods) has replaced it. This metamorphosis has brought about interesting changes in the personality of the urban system.

In recent years Rome has experienced considerable population growth as a result of expansion of facilities at the Griffiss Air Force Base. However, its industrial complex has changed little in size or type of production; it still is known as the "copper city." The favorable trend associated with the air base is in danger of being reversed, if the base should curtail its operation or be closed.

Down the Mohawk Valley to the east are examples of what happens to communities whose internal economic structure is not sufficiently flexible to change with the times and whose history does not include the acquisition of something like a military base. These smaller towns have for some years been losing their old soft-goods industries, with a consequent loss of population. This, in turn, has led to serious fiscal and economic troubles.

Thus the contemporary Utica Urban System reveals three distinct economic personalities. In the west Rome is too dependent on military vagaries, in the center Utica is digesting a major industrial

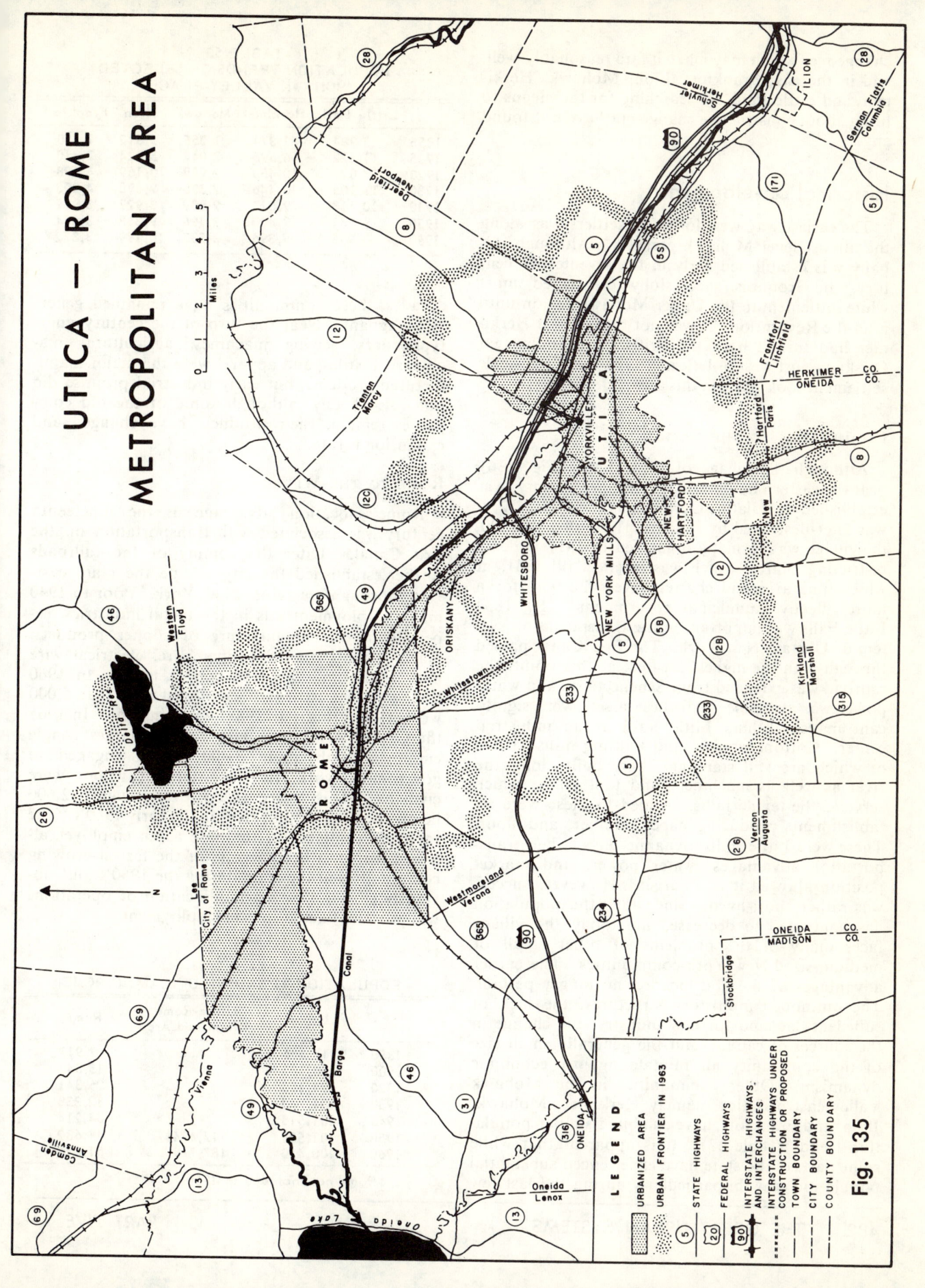

Fig. 135

changeover which may make it fare reasonably well, and in the east Frankfort, Ilion, Mohawk, Herkimer, and Little Falls are searching for the means to bring about constructive change but have not found any.

Historical Development

The easiest way west for early settlers was along the almost level Mohawk corridor. Although Albany was established early in the seventeenth century, the Iroquois successfully obstructed most white settlement in the Upper Mohawk region until after the Revolution. It was not till General Herkimer had turned back the British and Indians at Oriskany, just west of Utica, in 1777 that large-scale settlement took place in this area.

COMMUNITIES TO THE EAST

Although the village of Little Falls (Fig. 136), near the eastern boundary of Herkimer County, was established by Palatine Germans as early as 1752, it was later destroyed by Indians. Resettled in 1790, the village was formally organized in 1811 under a restricting charter which was in effect till 1831, at which time a second charter paved the way for a more effective municipal government. By 1834 Little Falls was important in the transportation system of Upstate New York. The Erie Canal passed through with the highest lift lock at this point, the railroad was extended from Schenectady, and water power was available. All these assets were significant in establishing Little Falls as an industrial center. Cotton, woolen, and knitting mills—some of which are still standing—were built along the river as well as machinery and tool plants which serviced the textile mills. Added to these were establishments producing paper, lumber, and flour. These were Little Falls' dynamic days, when transportation advantages, water power, and market position allowed it to flourish. However, success was rather short-lived. Since 1890 the population has continued to decrease, and today the village faces many of the problems of other small or medium-sized New York communities. The earlier advantages of site and location no longer pertain. The outmoded multistory structures, the lack of sufficient flat land for new industry, the change in the sources of capital, and the generally small size of the community all militate against economic dynamism. Other communities in the Mohawk Valley east of Utica, namely Herkimer, Mohawk, Ilion, and Frankfort, have exhibited similar population trends (Table 53). It is apparent that while some parts of the state have experienced substantial population growth, particularly during the last few decades, these communities have remained generally stagnant. Near the turn of the century guns, typewriters, sewing machines, agricultural machinery, textiles, and apparel were the major manufactured products, but early industrial promise did not last. Today, although some of the old companies remain, their products have changed and expansion is rare.

TABLE 53
POPULATION TRENDS OF SELECTED MOHAWK VALLEY PLACES

	Little Falls	Herkimer	Mohawk	Ilion	Frankfort
1855	3,984	1,371	1,355	812	1,150
1905	11,122	6,596	2,044	5,924	2,870
1920	13,029	10,453	2,919	10,169	4,198
1930	11,105	10,446	2,835	9,890	4,203
1940	10,163	9,617	2,882	8,927	3,859
1950	9,541	9,400	3,196	9,363	3,844
1960	8,935	9,396	3,533	10,199	3,872

ROME TO THE WEST

Rome's location advantage in the nineteenth century was associated with transportation on the Erie Canal. Later the coming of the railroads further established the city astride the main east-west thoroughfare across New York. Prior to 1940 major employment was in the metal industries, especially in the manufacture of copper products, machine tools, tinned copper wire, electrical wire and cable, and brass and copper items. In 1960 these same industries still employed more than 5,000 workers, but this is a 2,300 loss from 1956. In 1965 the Griffiss Air Force Base was the largest single employer, with about 8,400 people engaged in government work, three-quarters of whom were civilians. This figure is however roughly 2,600 below the peak employment in the early 1960's. The rapid development of the base as an employer allowed Rome to become one of the fastest-growing communities in the state during the 1950's, but obviously closing or even modification of operations will produce fluctuations in employment.

TABLE 54
POPULATION TRENDS IN UTICA AND ROME

	Utica	Utica-Rome Urbanized Area	Rome
1850	17,556		7,918
1900	56,383		15,343
1920	94,156		26,341
1930	101,740		32,338
1940	101,518		34,214
1950	101,531	117,424*	41,682
1960	100,410	187,779	51,646

*Rome not included in 1950.

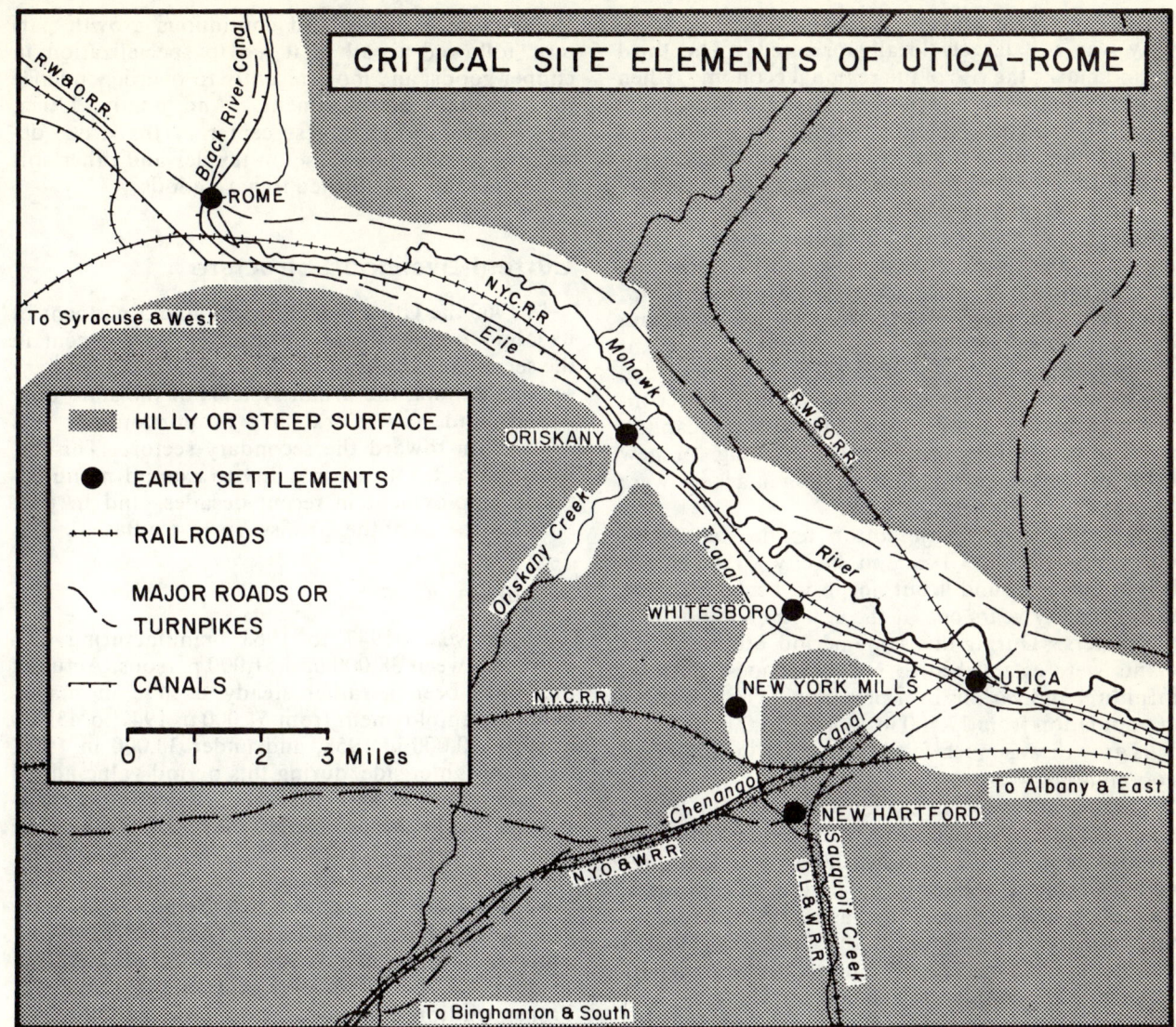

Fig. 136

UTICA, THE CENTER

Utica's history differs from that of the communities to the east in several important respects. It grew faster, first solely as a commercial center, later as a manufacturing city. By the early 1900's Utica had become the knit-goods capital of the world, and there was an air of prosperity and economic vitality that lasted till the mid-1920's. From that time on no population growth has taken place in the city proper. In fact, deterioration of the textile industry has been barely matched by growth in the metal and electrical industries.

Today the urbanized area of Utica extends for seven miles along the south side of the Mohawk River and stretches southward for an equal distance over the terraces and valleys of the Appalachian Upland. It consists of the city of Utica on the east, New Hartford on the south, Yorkville, New York Mills, and Whitesboro on the west. In the early nineteenth century these were all separate villages. New Hartford, New York Mills, and Whitesboro had the advantage of water power from Sauquoit Creek and were the sites of early manufacturing.

By 1830 manufacturing had passed through three distinct developmental periods. First was the frontier period when items processed found their market in the immediate locale. Gristmills and sawmills appeared, as did some fulling mills. The second period essentially grew out of the first. As the population increased, mostly by the moving in of new settlers, iron mills, glassworks, and the first small textile mills were added to the above list. The economy was still largely locally oriented. The third period marks the rise of the regional economy, when both raw materials and markets lay outside the local area. Then textiles and other soft-goods production began to dominate the industrial scene. Utica itself, at that time, although almost twice as large as any of the adjacent communities, had no water power and therefore no important industry. It was a trading town, benefiting from its position at a fording, and later bridging, place across the Mohawk River. However, the stage was set for the use of steam power. Adequate labor was available; excellent cheap transportation for that period was provided by the Erie and Chenango canals, the latter connecting with the coal fields in Pennsylvania; capital, made in commercial ventures, was looking for new fields of investment; and expanding markets were opening up to the east and west.

By 1850 the first large cotton textile steam mill was in operation in Utica, and the city, which earlier had industry around it but not in it, became during the next sixty years one of the country's major textile centers. During this same period other smaller plants were established to produce shoes, pottery, furniture, and various iron products. The city grew in population from 17,500 in 1850 to 94,000 in 1920. This growth was much faster than that in the fringe communities.

The period immediately following World War I marked the end of a century of constant and substantial population and industrial growth for Utica and introduced a thirty-year period of industrial stagnation and decline. Whereas in 1910 forty knitting mills were operating in the area, only half a dozen remained in 1922. The population of the city of Utica has remained at 100,000 for the last three decades, but this does not mean that it is stagnating. Growth has taken place in fringe areas, a pattern repeated in most of the older and larger urban centers of the state. Furthermore, while the population has remained constant in the city, the industrial complex has undergone an almost complete metamorphosis. As late as 1950, 9,500 workers were employed in the declining textile industry, yet by 1957 this number had decreased to 1,100. Fortunately, while this wasting away was taking place in soft or nondurable goods, a balancing growth was taking place in hard or durable goods. The result was a relatively stable labor force as far as size is concerned but an entirely new one in terms of type of production.

To repeat, then, the Utica Urban System has three distinct parts. On the east are the small towns that started with high promise, developed considerable industrial diversification, but have remained stagnant for the last thirty years. On the west Rome, with a steady and continuous growth pattern, followed a path that led to specialization in copper goods, and more recently to overdependence on a military establishment. And finally, Utica, which became the largest center in the area, developed its economic base on textiles and other soft goods but now has turned to hard goods.

Current Livelihood Structure

In 1960 the Utica-Rome SMSA had 5 per cent of its labor force in the primary sector, 40 per cent in the secondary sector, and 55 per cent in the tertiary sector. Among the urban systems of the state only Albany and New York have proportionately less orientation toward the secondary sector. This reflects, in part, Utica's inability to expand manufacturing employment in recent decades, and in part the importance of the Griffiss Air Force Base.

Manufacturing

In the years 1947 to 1963 manufacturing employed between 38,000 and 51,000 persons. Actually there has been a rather steady decline in manufacturing employment from 51,000 in 1947 to 43,000 in 1954, 40,000 in 1958, and under 38,000 in 1963. On the brighter side, during this period value added by manufacture has increased, reflecting a substantial rise in the productivity level of the average factory worker.

Most of the manufacturing is now oriented toward durable goods. Four of the major industry groups—electrical machinery, nonelectrical machinery, primary metals, and fabricated metals—account for over 50 per cent of employment. Transportation equipment, and food and kindred products follow in order. This domination of the industrial structure by a few durable-goods categories is a far cry from the time when Utica was a textile and knitwear center. The major manufacturing concerns of the SMSA, as listed in Table 55, reflect this modern emphasis on durable goods.

The current manufacturing geography of Utica and its immediate surroundings illustrates many of the causes and effects of the changeover to hard goods. Key elements of this geography are shown in Figure 55.

One industrial district (District 1, Fig. 137) follows the east-west edge of the Mohawk flood plain. It is not on the flood plain because of the frequent spring inundations. The old Erie Canal passed through this part of the city before 1911, at which time it was renamed the Barge Canal and moved onto the river flats. Eighteen of the twenty-eight structures built in Utica for textile operations, and

TABLE 55
MAJOR MANUFACTURING CONCERNS OF THE UTICA-ROME SMSA

Company	Products	Approx. Employment Early 1965
General Electric Co.	Electronics and communications products	6,500
Univac Div., Sperry Rand Corp.	Computers	3,000
Oneida, Ltd.	Silver products	2,300
Chicago Pneumatic Tool Co.	Pneumatic tools	1,900
Revere Copper & Brass Co., Inc.	Brass and copper products	1,800
Utica Div., Kelsey-Hayes Co.	Jet aircraft parts	1,100
Rome Cable Div., Alcoa	Wire and cable products	1,100
Beaunit Fibers Div., Beaunit Corp.	Textiles (rayon)	1,000
Utica Div., Bendix Corp.	Starters	1,000
Remington Arms, Inc.	Commercial firearms	800
Camden Wire Co., Inc.	Wire products	500
Bossert Manufacturing Corp.	Stampings	300
Duofold, Inc.	Textiles (sports apparel)	250
Rome Strip Steel Co.	Steel products	250

still standing, are found in this district. Many other old plants are located here, too. In fact twenty-three are over sixty years old, and another sixteen are over forty years old. Accompanying the handicap of age is the fact that many of the buildings are multistory. This part of the city clearly points up such modern site disadvantages as narrow streets, lack of adequate loading and unloading facilities, and insufficient parking space. Fifty years ago a thriving industrial complex functioned satisfactorily in the same location. Workers walked to work in those days and their homes were close to the plants. Today residences bordering the district are crowded closely together and are among the least desirable in the city.

District II runs south-west from the middle of District I along the route followed by the old Chenango Canal. The plants are small and commonly about forty years old. The basic problem here is lack of space, for the plants are squeezed between a city street and a new north-south arterial at their rear. In addition the same handicaps faced in District I exist. Surrounding residences improve somewhat in quality from north to south.

District III forms an arc to the southwest and it is here and in outlying areas beyond that most of the post-World War II plants have been established. Modern structures on sites of adequate size, with parking space, loading facilities, and access roads scaled for efficient use, contrast strongly with the industrial establishments found in the other districts.

Viewing manufacture as a whole in all three districts, one may note several pertinent facts: 73 per cent of the plants were erected before 1920 and 67 per cent of the industrial workers are employed in these older buildings. Less than 30 per cent work in plants built since World War II. In spite of the changeover from soft goods to hard goods in recent years, the majority of the industrial labor force still works in the older buildings. This may be explained by the fact that incoming new industries often are attracted by available space within the old multistory textile buildings. Many of these old mills have been completely refurbished inside and out and look better today with their new coats of paint than they have for a long time. However, the disadvantages accruing from narrow streets, inadequate loading facilities, and the multistory nature of the buildings still remain. It is no wonder that plans are under way for the establishment of industrial parks in the area, but in the meantime new industry has moved into the fringe areas where space is available and today's industrial requirements can be better met.

Utica is like many cities of the Northeast in that it imports all of its industrial materials and exports most of its finished products. Today over 70 per cent (by value) of the industrial materials come from sources outside the state and almost 80 per cent (by value) of the finished products find markets outside the state. For the newer, larger national concerns (General Electric, Sperry Rand, Kelsey-Hayes, Chicago Pneumatic, and Bendix Aviation), the distance from raw materials and markets is not a serious problem. Their finished products in most instances are parts of larger units and are shipped to other branches of these corporations which are generally located elsewhere in the American Manufacturing Belt. Therefore, although the market is outside New York State, a considerable portion of it lies within a relatively short distance and is easily accessible.

Many of the smaller and older industries find that long distances to materials and markets place them in a poor competitive position today. Their old structures, often obsolete operating techniques, and the unchanging attitudes of management make competition from the newer industrial areas insuperable.

One other facet of the industrial change from soft goods to hard goods is revealed in the difference in

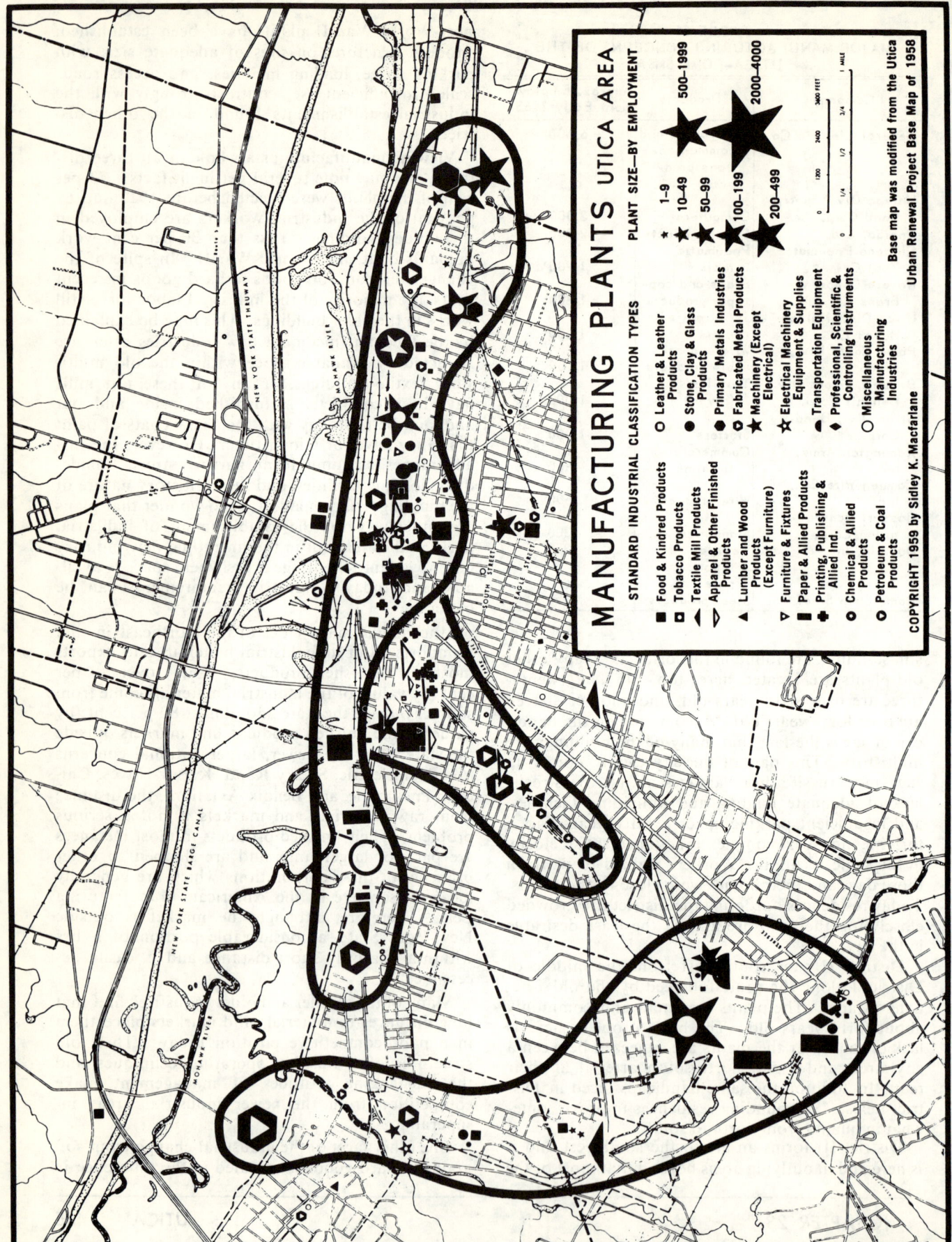

wages paid in the two groups. In 1929 the average weekly salary per worker in textiles was $19, in metals and machinery $27; in 1945, $38 and $58; in 1959, $63 and $95. The switch to hard goods has without doubt raised the economic well-being of the industrial workers. In spite of this switch, however, per capita income levels in 1960 averaged lower than in any other SMSA in the state.

TERTIARY ACTIVITIES

Utica, like Binghamton, is but a third-order trade center (see Chapter 12). Nevertheless, tertiary activities there are important and becoming more so. Of the 118,000 employed workers in the SMSA in 1960, roughly 65,000 were employed in the tertiary sector. During the decade 1950–60 a 27.4 per cent gain in tertiary employment exceeded that of any of the other urban systems in the state. This may be less encouraging than it seems, however, for a significant part of that gain can be attributed to what may turn out to be impermanent employment at the air base near Rome.

In 1960, 20.8 per cent of the SMSA's total employed persons worked in service industries, 16.2 per cent in retail and wholesale trade, 5.5 per cent in public utilities, 3.6 per cent in finance, insurance, and related fields, and 9.3 per cent in public administration. The total sales and service function in 1960 producted a tertiary potential (see Chapter 12) of 364,000 and a tertiary surplus of 33,000.

Problems and Prospects

The Utica Urban System, including as it does a range of situations from those at Little Falls, to those at Rome, to those at Utica and its immediate periphery, makes generalizing about problems and prospects difficult. For example, the lack of industrial space and small size of the urban units are a problem in the Mohawk Valley east of Utica, but not in Utica or Rome. Substantial closing of the Griffiss Air Force Base would be detrimental to Rome's economy but is much less of a problem in eastern portions of the urban system. In general Utica and Rome have been faring better than communities to the east; Utica's immediate periphery is likely to do best in the foreseeable future.

The whole urban system has experienced difficulties inherent in the need for change to a different functional orientation. Fortunately, the change of the industrial structure from nondurable, low-value-added goods to durable, high-value-added goods has been in the right direction. The electronics industry, one of the most dynamic in the country today, has become very important and several major companies with nation-wide markets are manufacturing these kinds of products. The change in the geography of industrial ownership may not be encouraging, although that point is not completely clear. Originally most industrial enterprises were owned and controlled by local entrepreneurs, but today a very high proportion are units of national organizations, and company policy is determined in head offices outside the area. The same trend has been taking place in certain tertiary establishments too.

Other general problems include the overdependence on government spending and the consistently fairly high rate of unemployment. Obsolescence and congestion have plagued the older parts of all the central places. Some dissatisfaction exists too as far as over-all growth is concerned.

There is on the other hand a brighter side to the picture. During the years since World War II significant steps have been taken locally to solve some of the difficulties confronting the urban system. These steps may take time to make themselves felt, but most are highly contributory to future satisfactory development. They include: (1) construction of a new network of superior arterial highways, (2) extension of water pipe lines, (3) expansion of the power grid and its connection with the Niagara and St. Lawrence power sources, (4) completion of the natural gas and oil pipe lines from Texas, (5) carrying out of large urban renewal projects, and (6) establishment of the Oneida County airport between Rome and Utica.

Important as these kinds of changes are, certain forms of cultural transformation might possibly be of even greater significance in the long run. It was recognized even by loyal local citizens that the Utica of 1945–50 was a dismal place in which to live. All around were the signs of urban aging and blight, of physical stagnation. Civic leadership of that time was lacking or uninspiring, and citizens showed little interest or pride in their community. There was no institution of higher learning and no strong cultural program. Potential young leaders left the area for college or careers in more attractive places. But in the last two decades the scene has drastically changed. Two colleges, Utica College of Syracuse University and Mohawk Valley Community College, now offer technical, scientific, and liberal arts courses for more than 5,000 students; high schools have instituted a wide variety of adult training programs; Munson-Williams-Proctor Institute has completed a distinguished art gallery and museum and expanded its music and art activities; the civic symphony orchestra has been revitalized; and there is renewed active citizen interest in local government which has resulted in the redistricting of city wards and modernizing of county government. Fifteen years ago the Utica

Urban System was depressed and depressing. Today problems exist, but the roots for revitalizing economic development are in richer ground and growth seems assured. Although this growth will be substantial it will lag behind that of the other major urban systems upstate. The population of the SMSA, at 331,000 in 1960, should reach 410,000 by 1980 and be roughly 520,000 by the year 2000.

Selected References

Bagg, Moses M. *Memorial History of Utica, New York.* Syracuse: D. Mason & Co., 1892.

Clarke, T. Wood. *Utica, For A Century And A Half.* Utica: Widtman Press, 1952.

Crisafulli, Virgil C. *An Economic Analysis of the Utica-Rome Area.* Utica: The Utica College Research Center, 1960.

Dale, Nelson C. *Geology and Mineral Resources of the Oriskany Quadrangle.* ("New York State Museum Bulletin," 345.) Albany: New York State Museum, 1953.

Durant, S. *History of Oneida County, New York.* Philadelphia: Evarts & Fariss, 1878.

Greene, Nelson. *History of the Mohawk Valley, Gateway to the West, 1614–1925.* Chicago: J. S. Clarke Publishing Company, 1925.

Hardin, George A. *History of Herkimer County.* Syracuse: D. Mason & Co., 1893.

Jones, Pomeroy. *Annals and Recollections of Oneida County.* Rome: A. J. Rowley, 1851.

Macfarlane, Sidley K. "The Characteristics, Problems and Potential of Manufacturing in the Utica Area." Unpublished Ph.D. dissertation, Department of Geography, Syracuse University, 1960.

Neuhoff, M. C. "Trends in Industrial Location." Princeton: Tax Institute Incorporated, 1955.

New York State Department of Commerce. *Industrial Directory of New York State.* Albany: The Department, 1958.

———. "New York's Electric Power Industry," *New York State Commerce Review* (March, 1959), pp. 1–8.

Oneida Historical Society. *Transactions.* Nos. 1, 2, and 3. Utica: E. H. Roberts and Company, 1881.

Outline History of Utica and Vicinity. Utica: New Century Club, L. C. Childs, 1900.

Rapkin, Chester, *et al. Industrial Renewal.* New York: Division of Housing and Community Renewal, 1963.

Rubin, Harold. "Utica, New York: A Case Study in Community Effort and Industrial Change." A Syracuse University Research Report in Public Administration. Syracuse: Syracuse University, 1956.

Thompson, John H., and J. M. Jennings. *Manufacturing in the St. Lawrence Area of New York State.* ("Maxwell School Series.") Syracuse: Syracuse Univ. Press, 1958.

Wager, Daniel E. *Our County and Its People.* Boston: Boston History Company, 1896.

A LOOK FORWARD

JOHN H. THOMPSON

A Look Forward

CERTAIN geographic forecasts concerning the future of the state are in order. Any such look forward must be based on answers to at least two initial questions: How far into the future should the look extend? What aspects of the geography should be considered?

Most demographers, economists, and statisticians who are engaged in prediction admit there is so little likelihood of precise continuance of past trends into the future, and so many variables which enter the picture at any given time or place, that forecasting more than a few years ahead is hazardous at best. It would seem the better part of valor, therefore, to look ahead to perhaps 1970, or 1980 at the outside, and be satisfied. Certain conditions of five or ten years hence probably could be predicted with a fairly high level of accuracy, yet curiosity about the New York State of a later date is so great that there is temptation to forego the relative safety of short-range forecasting and engage in a much less safe prognostication to, say, the year 2000. That temptation has been given in to here, and this look forward extends to the turn of the century. If geographic conditions for New York in the year 2000 turned out exactly as forecast, it would be surprising indeed. Rather, these long-range forecasts should be thought of simply as reasonable projections of past trends into the future against which emerging reality through the years can be compared. They will provide a sort of window to the future even though smudge on the window prohibits an absolutely clear view.

Based on information available from preceding chapters in this volume and elsewhere, population and land use are the elements of the geography which offer the best possibilities for reasonable prognostication. Predictions about a third element, employment, will also be attempted.

Forecasting geographically for an area like New York State means not only arriving at totals of things for the state as a whole, but distributing portions of these totals to appropriate parts of the state. Thus, the forecast should suggest not only how many people there will be in the state by 1980 or the year 2000 but what the relative growth trends and population levels of the various parts of the state will be; not only that urbanization will spread but where it will spread; not only that agriculture will occupy much less land but what agricultural areas are most likely to do best; not only that the tertiary sector will be much larger than now, but how fast it is likely to grow in one locality as compared to others.

Population

Many things affect the growth performance of any given population, and, although no complete discourse on population forecasting is attempted here, some of the obvious things that influence growth should be cited at the outset. Included are: (1) fertility rates (number of births per 1,000 women of childbearing age), (2) mortality rates, (3) migration (in this case intrastate, as well as interstate and international), (4) population distribution among age groups, and (5) economic conditions (this is perhaps most important of all in forecasting what will happen at various localities within the state).

Population projections have been made for each of the great urban systems in the last seven chapters. These are incorporated in this chapter along with predictions for the state as a whole, each county, and selected urban units outside the great urban systems. Of aid has been a recent population study by the New York State Office for Regional Development, which carried forecasts through to the year 1980.[1] Although projections presented in this state study are not followed exactly, they served as useful guides. Other projections published by the Bureau of the Census and certain private research groups have been helpful, too.

STATE TOTAL

A consensus of experts regarding population growth, if no unforeseen economic impact or attitude toward family size were to occur, places the

---

[1] George A. Dudley, unpublished Statement to the New York State Temporary Commission on Per Capita Aid, New York Office for Regional Development.

TABLE 56
POPULATION OF NEW YOUR STATE, 1930–60; PROJECTIONS, 1970–2000
(In Thousands)

Census Figures				Forecasts*			
1930	1940	1950	1960	1970	1980	1990	2000
12,588	13,479	14,830	16,782				
Apparent Possible Forecast Range							
High				19,700	23,000	27,200	31,600
Medium-high				19,300	22,300	25,900	29,600
Medium-low				18,830	21,200	24,200	28,000
Low				18,500	20,300	22,500	25,000

*Falling trends in birth and fertility rates in the 1958–65 period, if continued, should result in slower growth. Up to the time for publication, however, these trends were not of long enough duration to encourage modification of the projections.

1980 population of New York State at between 20 million and 23 million. Of numerous projections examined none was outside this range. An averaging of six different projections, including four made by the Bureau of the Census and two by the New York State Office for Regional Development, resulted in a 1980 state population figure of 21.2 million.

Prognostication about fertility rates, age distributions, migration, and other characteristics of the population beyond about 1980 are not attempted here. In fact, in view of the complex nature of the problem, it was reasoned that continuing the growth beyond 1980 to 2000 at the same rate as projected for the 1960 and 1980 period, with some adjustments, is as justifiable a procedure as any. If the 1980 figure of 21.2 million is used as a point through which to project the population from 1960 to 2000, assuming, as stated, the same rate of growth between 1980 and 2000 as between 1960 and 1980, a forecast of just under 27 million for 2000 results. Carefully prepared population estimates by the Bureau of the Census for the 1960–65 period seem to be running slightly ahead of the above projection schedule and suggest, with continued similarity in trend, a 1980 figure of 22.3 million instead of 21.2 million and a 2000 forecast of roughly 30 million. On the other hand, both crude birth rates and fertility rates in the United States have been falling in the 1958–65 period. Were this kind of trend to be dominant in New York State for the next decade or two, the figure of 21.2 million for 1980 and 27 million for 2000 would turn out to be high.

To illustrate further the effect of variance of the 1980 figure on a constant rate projection to 2000, if a 1980 population of 20.3 million were used the figure for 2000 would be only 25 million; 23 million in 1980, on the other hand, would indicate a 2000 population of nearly 32 million. It is obvious that when constant rate projections are used, even a fairly small departure from what turns out to be the population of 1980 could result in a large error by 2000. In any case it seems reasonable to assume that the state's population in 2000 will be somewhere between 25 million and 32 million.

Table 56 shows figures for the range of conditions considered possible. In order not to complicate matters unduly when forecasting for parts of the state, a single figure between 25 million and 32 million needs to be accepted. Twenty-eight million is used because it represents only a slight adjustment upward from a constant rate projection through the 1980 figure of 21.2 million which was most closely related to the average of other forecasts and because it is most closely in line with the previous twenty-year growth performance of the state. Any degree to which 28 million proves to be in error will be reflected in over- or under-forecasts for parts of the state because all parts are adjusted to fit into that total. If a population forecast for the year 2000 of 28 million is accepted, then the geographic question of how these 28 million people will be distributed within the state can be asked.

Great Urban Systems

Table 57 shows that of the 21.2 million projected for the state in 1980, nearly 19 million will be in the seven great urban systems. Together they should grow roughly 4 million in population during the twenty-year period. Syracuse and Rochester stand out with highest per cent gains, while the New York State portion of New York and Utica rank low, even below the state average.

Looking at the probably less reliable forecasts for the year 2000, one can estimate that 25.4 million of the state's 28 million will be concentrated in the seven urban systems. This involves an increase of 10.5 million or 70 per cent between 1960 and 2000 and represents 93 per cent of all growth likely to occur in the state during the forty-year period. It has been pointed out many times in this volume that the great urban systems are the economic nerve centers of the state. These figures support that contention.

Striking population growth will occur around the margins of the old cities, being most spectacular in the zones designated as suburbia and exurbia today. And it might be repeated too that the suburbs of tomorrow will move into the exurbia of today. Before the year 2000 there may be a renewal of city center growth and a return to a trend of increasing density in older parts of cities as well, but this does not appear evident enough to incorporate in the forecasts for most individual counties. The contrast between the expected growth performance of city center and periphery shows up clearly in county forecasts, especially those for the New York Urban System (see Table 58). For example, of the four counties—Bronx, Kings, New York, and Queens—that make up the present CBD, frame and middle city, changes of from + 11 to – 12 per cent are forecast for between 1960 and 2000; of the highly urbanized counties making up pretty much the suburbia of today—Nassau, Westchester, and Richmond—projected growths ranging from 27 to 170 per cent are expected; and for the outer suburban and exurban counties—Suffolk, Rockland, Putnam, Dutchess, and Orange—forecasts of from 335 to 498 per cent growth are made.

Of the seven great urban systems, forecasts suggest that Syracuse will enjoy the most favorable growth trend between 1960 and 2000. It will be followed in order by Rochester, Albany-Schenectady, Buffalo, Binghamton, New York, and Utica. If only the New York State portion of the New York Urban System is included, it ranks last behind Utica in the growth forecast.

COUNTIES OUTSIDE THE GREAT URBAN SYSTEMS

All of the state outside the seven urban systems contained only 1,851,000 people in 1960 and is expected to experience only a 21 per cent gain by 1980. This is as compared to a 27 per cent gain for the urban systems collectively and a 26 per cent gain for the state. Even by the year 2000 the population of the counties outside the seven urban systems is likely to contain only slightly more that 2.5 million of the state's 28 million.

Growth comes slowly to areas without a major city. Table 58 shows that in the next forty years only three of the thirty-three counties outside the great urban systems are scheduled to experience more than a 100 per cent gain in population. Two of these, Sullivan and Ulster, are just beyond the margins of the present New York Urban System and in the recreation-oriented Catskills. The third, Cortland, lies adjacent to Onondaga, the core county of the Syracuse Urban System, and astride Interstate Route 81 south from Syracuse.

Over half of the thirty-three counties can expect, according to the forecasts, less than a 20 per cent gain by the year 2000. Several are not likely to grow at all. This is indeed poor performance in an age when the national population is likely to rise at a rate close to 1.5 per cent per year, and New York State is forecast to have 67 per cent growth during the forty years. The causes for slow growth are many, but generally they are related to the lack of local potential for urbanization. If nearly 80 per cent of all new jobs are to be in the tertiary sector and the rest in the secondary sector, it is obvious that strictly agricultural counties cannot expect to grow. Schoharie County might be cited as a case in point. Also, some counties, which perhaps at one

TABLE 57
POPULATION OF THE GREAT URBAN SYSTEMS (SMSAs), 1930-60; PROJECTIONS, 1970-2000
(In Thousands)

	1930	1940	1950	1960	1970	1980	Percentage Increase 1960-80	1990	2000	Percentage Increase 1960-2000
Albany-Schenectady	520	531	589	658	770	910	38.3	1,070	1,270	93
Binghamton	173	193	215	251	295	350	39.4	410	480	91
Buffalo	912	958	1,089	1,307	1,530	1,800	37.7	2,130	2,500	91
New York	11,643	12,619	14,951	16,141	18,900	22,000	36.3	25,500	30,000	86
New York (New York State Portion)	8,225	9,085	9,865	11,087	12,150	13,325	20.2	14,875	16,945	53
Rochester	540	557	615	733	940	1,155	57.8	1,500	1,840	151
Syracuse	401	405	465	565	750	1,000	77.0	1,370	1,850	227
Utica	263	263	284	331	370	410	23.9	460	520	57
Total Urban Systems (New York State Portion)	11,034	11,992	13,122	14,931	16,805	18,950	26.9	21,815	25,405	70
All Counties Outside Urban Systems	1,554	1,487	1,718	1,851	2,025	2,250	21.1	2,385	2,595	40
Total for New York State	12,588	13,479	14,830	16,782	18,830	21,200	26.3	24,200	28,000	67

TABLE 58
POPULATION OF NEW YORK COUNTIES, 1930-60; PROJECTIONS, 1970-2000
(In Thousands)

Counties	1930	1940	1950	1960	1970	1980	Percentage Change 1960-80	1990	2000	Percentage Change 1960-2000
Albany*	212	221	239	273	328	400	47	485	590	116
Allegany	38	40	44	44	42	41	-7	41	40	-9
Bronx*	1,265	1,395	1,451	1,425	1,400	1,350	-5	1,325	1,300	-9
Broome*	147	166	185	213	250	297	39	348	405	90
Cattaraugus	72	73	78	80	83	87	9	88	89	11
Cayuga	65	66	70	74	83	95	28	102	110	47
Chautauqua	126	124	135	145	152	160	10	169	178	23
Chemung	75	74	87	99	113	130	31	139	150	52
Chenango	35	36	39	43	46	50	16	52	54	26
Clinton	47	54	54	73	76	80	10	82	85	17
Columbia	42	41	43	47	53	60	28	62	65	38
Cortland	32	34	37	41	50	60	46	72	88	115
Delaware	41	41	44	44	45	47	7	49	51	17
Dutchess*	105	121	137	176	290	410	133	620	920	423
Erie*	762	798	899	1,065	1,230	1,420	33	1,660	1,900	78
Essex	34	34	35	35	36	37	6	37	37	5
Franklin	46	44	45	45	45	45	0	45	45	0
Fulton	47	49	51	51	50	50	-2	50	50	-3
Genesee	44	44	48	54	61	68	26	75	80	48
Greene	26	28	29	31	34	37	19	39	42	35
Hamilton	4	4	4	4	4	5	25	5	6	40
Herkimer*	64	60	61	66	70	75	14	80	85	29
Jefferson	84	84	86	88	93	100	14	103	107	33
Kings*	2,560	2,698	2,738	2,627	2,550	2,450	-7	2,325	2,300	-12
Lewis	23	23	23	23	24	25	9	25	25	8
Livingston*	38	39	40	44	50	57	30	65	80	82
Madison*	40	40	46	55	65	80	45	100	130	136
Monroe*	424	438	488	586	765	950	62	1,240	1,510	158
Montgomery	60	59	60	57	59	66	16	66	66	16
Nassau*	303	407	673	1,300	1,400	1,500	15	1,575	1,650	27
New York*	1,867	1,890	1,960	1,698	1,675	1,650	-3	1,600	1,600	-6
Niagara*	149	160	190	242	300	380	57	470	600	148
Oneida*	199	204	223	264	295	335	27	380	435	65
Onondaga*	292	295	342	423	580	790	87	1,100	1,500	255
Ontario	54	55	60	68	78	90	32	95	100	47
Orange*	130	140	152	184	300	440	139	720	1,100	498
Orleans*	29	28	30	34	40	48	41	57	70	106
Oswego*	70	71	77	86	105	130	51	170	220	156
Otsego	47	46	51	52	53	55	6	57	60	15
Putnam*	14	17	20	32	50	95	203	130	175	447
Queens*	1,079	1,298	1,551	1,810	1,860	1,900	6	1,950	2,000	11
Rensselaer*	120	122	133	143	160	180	26	200	225	54
Richmond*	158	174	192	222	315	450	103	600	600	170
Rockland*	60	74	89	137	230	370	170	530	700	411
St. Lawrence	91	91	99	111	112	112	1	113	115	4
Saratoga*	63	66	75	89	107	130	46	155	190	113
Schenectady*	125	122	142	153	175	200	31	230	265	73
Schoharie	20	21	23	23	22	22	-4	21	20	-6
Schuyler	13	13	14	15	15	16	7	16	16	7
Seneca	25	26	29	32	37	42	31	46	50	56
Steuben	83	85	91	98	102	105	7	110	115	18
Suffolk*	161	197	276	667	1,020	1,410	111	2,030	2,900	335
Sullivan	35	38	41	45	58	75	67	98	125	177
Tioga*	25	27	30	38	45	53	39	62	75	97
Tompkins	41	42	59	66	77	90	36	105	124	88
Ulster	80	87	93	119	165	230	93	247	320	169
Warren	34	36	39	44	51	60	36	65	70	36
Washington	46	47	47	48	49	50	4	50	51	5
Wayne*	50	53	57	68	85	100	47	138	180	165
Westchester*	521	574	626	809	1,060	1,300	61	1,470	1,700	110
Wyoming	29	31	33	35	37	40	14	40	40	14
Yates	17	16	18	19	20	21	11	21	21	11
New York State Total	12,588	13,479	14,830	16,782	18,830	21,200	26.3	24,200	28,000	67

*Counties in the seven great urban systems (SMSAs).

time had a reasonably active manufacturing base but never developed a major city, now find it extremely difficult to attract job-producing enterprises, either secondary or tertiary. Fulton County would be an example of this situation. Still other counties are badly situated with respect to the main stream of American commerce and business. Usually they, too, have no large urban systems within their borders or are not near enough to such an urban system for major benefits. They continue to be agricultural or perhaps try to emphasize recreation, but large population growth does not come from such emphasis. An example would be Franklin County.

In a discussion of this kind the implications always seems to be that slow growth is unsatisfactory, that areas not deemed to have great growth potentials are doomed to disaster. From certain points of view this is more or less true. But there is another side: perhaps it is not only inevitable that some areas will not grow, but perfectly satisfactory from the standpoint of society that they do not. If this notion were generally accepted, then people could settle down to the serious task of optimizing the use of slow-growing areas. Perhaps efforts should go into making these areas attractive places for smaller existing populations rather than hoped-for foci for population growth. In such instances, of course, surplus population, often including the most able people, migrate to faster-growing areas where jobs are more plentiful and varied. However, the higher the success in optimizing the use of the slow-growing areas, the better the chance of keeping at least a reasonable percentage of the favored sons at home. With the pre-eminent development of the tertiary sector, new opportunities there (i.e., other than attracting a new factory) may be adequate for slow growth areas.

LARGER URBAN PLACES OUTSIDE THE GREAT URBAN SYSTEMS

There are a dozen or so cities in the state with 1960 populations between about 20,000 and 50,000. Most have not shown spectacular growth in recent years; in fact, some exhibit distinct signs of distress. Generally, though, these places are now experiencing, and are likely to continue to experience, growth in areas immediately surrounding them. This growth is evidenced by numerous new residences. In some instances the motivating force is new manufacturing, in others it is educational institutions, in some it is migration from the countryside, and in still others it may be that distance to one of the great urban systems is short enough, but at the same time long enough, to be pleasing to many commuters and retirees.

Table 59 presents likely prospects for population growth in and near the larger urban places outside the great urban systems. That Elmira and Corning may coalesce in the future seems likely, and when they do they will form the largest urban unit outside the great urban systems. This area is already well on its way to becoming a third-order trade focus (see Chapter 12). Kingston, halfway between Albany and New York and on major transport arteries, also stands out as a sure prospect for attaining large size and third-order trade center status. Some forecasts for the Kingston area suggest even greater growth than projected here.

TABLE 59
POPULATION OF LARGER URBAN PLACES OUTSIDE
THE GREAT URBAN SYSTEMS (SMSAs),
1960; PROJECTIONS, 1980 and 2000
(In Thousands)

	City 1960	Urban Area* 1960	Urban Area 1980	% Change 1960-80	Urban Area 2000	% Change 1960-2000
Elmira-Corning	64	126	170	35	215	71
Kingston	29	77	140	82	250	224
Jamestown	42	60	75	25	92	53
Ithaca	29	42	65	55	100	138
Geneva	17	36	48	33	66	83
Dunkirk	18	40	48	20	58	45
Plattsburgh	20	34	48	41	70	105
Cortland	19	31	48	55	75	142
Auburn	35	39	48	23	58	49
Watertown	33	33	43	30	56	66
Amsterdam	28	34	37	9	41	21
Gloversville	22	32	33	3	34	6
Olean	22	22	24	9	26	18

*1960 figures for urban areas are from George A. Dudley, *Statement to the N.Y. Temporary Commission on Per Capita Aid* (1962).

The rate of growth in each of these urban areas is somewhat greater than in the counties in which they are situated; in some the growth rates even compare favorably with the great urban systems.

Employment

In this forecast of employment conditions, answers to three questions are sought: How much growth will there be in the number of New Yorkers who work? What changes will there be in demand for employment in the three sectors of the economy? What spatial variation can be expected?

Increase in Numbers of People Working

Because of future growth in population, there is bound to be substantial increase in the number of people seeking employment. To provide new jobs for these people may be one of the most difficult tasks ahead.

In 1960 roughly 39 per cent of New York's people were employed, most of them in the state. If in the year 2000 the same per cent of the population works and the population forecast figure of 28 million is reached, then 10.9 million will have jobs. Compared with the 1960 employment level of 6.6 million, this represents an average annual gross increase in jobs between 1960 and 2000 of about 107,000. This means an average of over 2,000 new employment opportunities per week would have to open and be filled by New Yorkers just to maintain present rates of population participation in the labor market. Of course the figure would be lower in earlier decades than in later ones.

The increase in employment, however, cannot be so easily approximated. What, for example, will be the impact of automation and the increased efficiency of labor? Some experts think automation and increased efficiency will increase the need for new jobs at an even faster rate than growth of the labor force. One estimate for the decade of the 1960's for the United States as a whole shows 55 per cent of all new jobs needed arising out of automation and increased efficiency as compared to 45 per cent due to a growing work force.[1] Translating this into the New York State situation and utilizing the population projection to 1970, roughly 740,000 new jobs will be needed in the decade of the 1960's because of growth in population and about 900,000 because of automation and efficiency.

A related and complicating question is whether or not as high a percentage of the population will be working in the future as now. An examination of United States census data from 1920 to 1965 shows little in the way of a clear-cut trend in the work participation of the total population (i.e., total population divided by total employment). There was a drop from 39.4 per cent in 1950 to 37.2 per cent in 1960, but since 1960 a slight rise has occurred. In any case it seems likely that employed persons will make up a somewhat smaller portion of the total population of the future. Some of the reasons for this are tied up with automation and the higher efficiency of labor; others are not.

Relative to the automation and efficiency issue, society is becoming so much more complex that higher skills are being required for most jobs. Those people who cannot, or will not, meet these skill requirements are destined to join the ranks of the technologically unemployed. Some, at least, will become permanent "unemployables." It is likely that the numbers of the technologically unemployed and the per cent of the total population they represent will rise unless our educational system turns out to be considerably more successful in the future than it has been in the past.

It is obvious, too, that higher portions of the future population will be composed of old people beyond the age of retirement. Perhaps, too, the age of retirement will be lowered. Certainly it is likely that young people will remain in school longer, thus cutting the per cent of the younger population that is working. This is not to say that the labor force participation rate in the normal working years, of say twenty to sixty-five, will drop. As a matter of fact, it may even rise if jobs remain available, for as affluence increases people seem more interested in income than leisure. This attitude may keep the per cent of two-job families high in the future.

A possible way to keep the participation rate of the middle-aged population group at a high level is general acceptance of a shorter work week or longer vacations. This is an attractive idea to labor, providing pay is not proportionately reduced. Reduction in hours without reduction in pay and without increase in efficiency would be inflationary, however. If efficiency rises with reduced hours, the reduction of hours will not have the net effect of increasing the number of jobs. So, the dilemma!

All in all, it is obvious that predicting employment levels for several decades ahead is fraught with even more problems than predicting population growth. In making the predictions shown in Table 60, it is assumed that only 35 per cent of the New York State population in 2000 will be working as compared to 39 per cent in 1960. Admittedly this is a guess, merely reflecting what seems to be a rational assumption that some drop in the per cent of the population that works will occur. The 39 per cent total participation rate is arbitrarily maintained for 1970; 38 per cent is assigned to 1980 and 37

---

[1] *U. S. News and World Report* (April 16, 1962), pp. 44–49.

TABLE 60
NUMBER OF NEW YORKERS WORKING IN THE STATE AND IN THE GREAT URBAN SYSTEMS, 1960; PROJECTIONS, 1970–2000
(In Thousands)

	1960		1970		1980		1990		2000	
	Number	% Working	Number	% Working	Number	% Working	Number	% Working	Number	% Working
New York State	6,600	39	7,350	39	8,050	38	8,950	37	9,800	35
Urban Systems (SMSA Data)	5,942	40	6,716	40	7,370	39	8,276	38	9,183	36
Albany-Schenectady	247	38	293	38	337	37	385	36	445	35
Binghamton	96	38	112	38	130	37	148	36	168	35
Buffalo	476	36	551	36	630	35	746	35	850	34
New York (in N.Y.S.)	4,511	41	4,982	41	5,330	40	5,801	39	6,270	37
Rochester	284	39	367	39	439	38	555	37	644	35
Syracuse	210	37	278	37	360	36	480	35	629	34
Utica	118	36	133	36	144	35	161	35	177	34

per cent to 1990, for state calculations. Similar adjustments are made in projecting employment levels for the great urban systems. Any inaccuracies in the projected employment which result from this procedure are likely to be less than those inherent in the population projections being used as a base.

CHANGE IN EMPLOYMENT DEMAND BY ECONOMIC SECTORS

Changes in livelihood structure which accompany economic development were dealt with in previous chapters, particularly 10, 11, and 12. The trend for the primary sector to decline and the tertiary sector to exhibit a rise was elaborated upon in detail. In 1960, 2.1 per cent of the workers in the state were engaged in primary activities, 35.4 per cent in the secondary sector, and 62.5 per cent in the tertiary sector. It is impossible to be sure exactly what will happen to the employment mix in the decades ahead, but past trends do permit some reasonable assumptions, especially at the generalization level of the state as a whole. State-wide projections in Table 61 presume a drop in the primary sector from 2.1 per cent in 1960 to 1 per cent in the year 2000, a drop in the secondary sector from 35.4 to 32 per cent, and a rise in the tertiary sector from 62.5 to 67 per cent during the same period. Straight line projections are used for intervening years. Thus, of the 9.8 million New Yorkers employed in the year 2000, nearly 6.6 million will be in the tertiary sector, about 3.1 million in the secondary sector, and less than 100,000 in the primary sector. According to these projections, of the 3.2 million new jobs to be formed during this period, 76 per cent will be in the tertiary sector.

That the tertiary sector should so completely dominate new job openings is consistent with the development theory presented earlier. Within the tertiary sector it is likely that such things as professional services (educational, medical, legal, etc.), government, and finance and insurance will grow most rapidly. Most other service categories will do well, too, especially those related to recreation. Automation will strike heavier blows at manufacturing than at trade and service jobs. Half the present number of farmers will be providing increased amounts of food by the year 2000.

SPATIAL VARIATION IN EMPLOYMENT

If it is assumed that roughly the same percentage of the population works in all parts of the state at any given time, the the geography of total employment will follow the projections offered above for the geography of population. There will, of course, continue to be important spatial variations in the employment mix.

To arrive at forecasts for the employment mix of the great urban systems involves many variables and results in more estimations. The figures for the year 2000 in Table 61 are derived by first applying the same percentage changes to each urban system as were applied to the state and then making adjustments based upon recent performance of the sectors in the various systems. Figures for intervening years are from straight line projections. Tertiary orientations of the New York, Syracuse, and Albany-Schenectady urban systems are emphasized by these projections, as are manufacturing orientations of Binghamton, Rochester, and Buffalo.

No attempt is made here to estimate the employment changes by sector for areas outside the great urban systems, but general trends of the state would be present in most places. Even in the most agricultural county there is likely to be nearly a halving

TABLE 61
NUMBER OF NEW YORKERS WORKING IN THE STATE AND IN THE GREAT URBAN SYSTEMS (BY SECTOR), 1960
PROJECTIONS, 1970–2000
(In Thousands)

	1960		1970		1980		1990		2000	
	Total	%	Total	%	Total	%	Total	%	Total	%
State Total	6,600		7,350		8,050		8,950		9,800	
Primary	139	2.1	132	1.8	129	1.6	116	1.3	98	1.0
Secondary	2,336	35.4	2,543	34.6	2,713	33.7	2,945	32.9	3,136	32.0
Tertiary	4,125	62.5	4,675	63.6	5,208	64.7	5,889	65.8	6,566	67.0
The Great Urban Systems (SMSA Data)										
Albany-Schenectady	247		293		337		385		445	
Primary	5	2.0	5	1.7	5	1.5	5	1.3	4	1.0
Secondary	90	36.5	104	35.4	115	34.3	127	33.1	143	32.0
Tertiary	152	61.5	184	62.9	217	64.2	253	65.6	298	67.0
Binghamton	96		112		130		148		168	
Primary	3	3.0	3	2.5	2	2.0	2	1.5	2	1.0
Secondary	49	51.0	56	50.0	64	49.0	71	48.0	79	47.0
Tertiary	44	46.0	53	47.5	64	49.0	75	50.5	87	52.0
Buffalo	476		551		630		746		850	
Primary	6	1.3	6	1.1	6	1.0	6	.8	5	.6
Secondary	217	45.5	244	44.2	271	43.0	311	41.7	343	40.4
Tertiary	253	53.2	301	54.7	353	56.0	429	57.5	502	59.0
New York (in N.Y.S.)	4,511		4,982		5,330		5,801		6,270	
Primary	23	.5	20	.4	11	.2	6	.1	6	.1
Secondary	1,452	32.2	1,554	31.2	1,604	30.1	1,688	29.1	1,750	28.0
Tertiary	3,036	67.3	3,408	68.4	3,715	69.7	4,107	70.8	4,514	71.9
Rochester	284		367		439		555		644	
Primary	11	3.7	11	3.3	11	2.8	11	2.4	11	2.0
Secondary	134	47.3	171	46.2	200	45.2	247	44.1	279	43.0
Tertiary	139	49.0	185	50.5	228	52.0	297	53.5	354	55.0
Syracuse	210		278		360		480		629	
Primary	8	3.6	8	3.0	8	2.3	8	1.7	6	1.0
Secondary	86	41.0	108	38.7	131	36.5	164	34.2	201	32.0
Tertiary	116	55.4	162	58.3	221	61.2	308	64.1	422	67.0
Utica-Rome	118		133		144		161		177	
Primary	6	5.0	6	4.5	6	4.0	6	3.5	5	3.0
Secondary	47	39.6	51	38.2	53	36.8	57	35.4	60	34.0
Tertiary	65	55.4	76	57.3	85	59.2	98	61.1	112	63.0

of the relative importance of the primary sector, with the slack being picked up largely by the tertiary sector. Only in isolated cases where some major manufacturing development occurs—such as the moving in of a large plant or the expansion of an existing one—would there be much likelihood of a relative rise in the importance of the secondary sector.

The extent to which employment opportunities will actually expand in the various localities of the state during the decades ahead will depend in part upon the condition of the national economy and in part upon each locality's ability to keep up with the rest of the country. Wise state and local action may be critical in pairing job demand and job availability.

Land Use

A fitting capstone to a comprehensive treatment of New York State's geography would seem to be an attempt at projected land use. How will man use the state in the future? What will a map of land use for the year 2000 be like? Figure 138 (in back-cover pocket) suggests answers to these kinds of questions; in a sense, it is the culmination of the long series of maps and many pages of text in this volume. It portrays a high degree of selectivity by man, in that, even with a population of 28 million in the year 2000, large portions of the state will remain in only recreation and forest uses.

URBANIZATION

Several things stand out in Figure 138, but the significance of urbanism is perhaps most noteworthy. By using the 1963 urban frontiers of the five largest urban systems and circles approximating the size of lesser urban units, the true extent of urban New York becomes apparent. No longer is there portrayed a large state with a few cities shown by small dots, but rather broad, urbanized zones clearly functioning as dominants in an otherwise less intensively used environment. The major lines of interconnection reflect an adjustment to serving an urban, rather than a rural, society.

Agriculture in Perspective

Agriculture appears appropriately insignificant. The year 2000 may see more agricultural production than 1966, but it will be concentrated in the lowlands of upstate and in favorable valleys. As stated in earlier chapters, much less land and many fewer farm units will be involved. The urban systems will tend to encroach upon some of the best farm land.

The Green Belt Idea

In Figure 138 green belts, supposedly composed of open rural land, encircle the largest urban systems and could have been drawn around Utica and Binghamton and other centers too. They are shown as a zone extending from 5 to 20 miles beyond the 1963 urban frontier. As it should be recalled from Chapter 14, this is an area where the value of open land is still determined by agricultural or lower uses. If something is not done soon to keep these areas open, the urban frontier will sweep over them, and eventually open land at reasonable distances from the centers of the great urban systems will be gone forever. In fact, without establishment of green belts, those areas so shown will be largely engulfed by outward urban sprawl and be properly designated as urban by the year 2000. The notion that open areas are desirable because present and future urban populations will profit psychologically and physically by having ready access to them is the major justification for green belts.

Actually, although it is doubtful that the green belts shown will have been established by the year 2000, it seems that attention should be drawn to the possibility of their establishment by including them in such a map as Figure 138.

Governments at all levels are becoming increasingly interested in the idea; but state-level legislation and action are likely to determine whether or not the belts are to materialize.

A green belt needs only to be kept as open land; it does not have to be—in fact it probably should not be—unused in the usual sense of the word. Some of the land may be in farms, some in golf courses, some in parks, the only requirement being that it not become congested. If the state government is to establish a green belt program there are two possible ways of proceeding. First, land can be purchased outright and thereby become public property administered by public agencies. There are three drawbacks here: (1) it is initially expensive to buy such large acreages, (2) it takes the land off the tax rolls, and (3) it will not maintain the mosaic of farms, woodland, and the like, which Americans have become accustomed to know and enjoy as the open rural environment. Secondly, the government may purchase what has been referred to as scenic, public, or green belt easements, agreements with landowners that enable public agencies to preserve beauty spots and make them accessible to certain public use. In return the landowner gets a payment in cash or tax reductions through guaranteed lower assessments. Where this procedure has been followed, especially in Wisconsin, initial costs have been kept down to about one-fourth what purchase would have been, land has remained in a tax-paying status, and rural environments have remained more or less intact. New York State passed a law authorizing scenic easements in 1960, but Wisconsin has been the first state to apply the principle on a large scale. Many feel the idea of easements may be the biggest contribution to preserving America's beauty in many decades. Certainly it seems the most practical way to establish and maintain green belts.

If green belts are established and urban systems can no longer expand outward, what will happen? The first thing to happen will be the filling in of unused areas within the urban frontier. With rational planning and zoning, the areas within the urban frontier could probably accommodate a doubling of present urban populations without excessive crowding. The cost of providing services might be substantially cut too if more efficient, denser use of suburban and exurban space were accomplished through a filling in process. Looking ahead fifty or a hundred years, one might assume that the larger and faster-growing urban systems might have to jump their green belts in search of space, leaving the green belts surrounded by urban outliers. At such a time the green belts might be looked upon as even a better idea than their founders had thought. After all, New Yorkers generally now agree that Central Park is wonderful to have, but it was William Cullen Bryant who took a walk into the countryside north of the city in 1844 and first thought a large tract of land should be bought for a reservation while land was still cheap. It was so purchased and set aside twelve years later. Westchester County parks form another invaluable heritage of past recognition of the value of open space in or near urban systems.

Forests and Recreation

Over two-thirds of the state, for the foreseeable future, will be largely relegated to forest and used perhaps more for recreation than anything else. The map does indicate that favorable locations in the Appalachian Upland and elsewhere will support some agriculture but certainly at even a lower level than now. If green belts fail to be established near cities, at least New Yorkers, by traveling farther, will be able to reach open land for many years to come. Even here, though, attention must be directed toward keeping these more distant open lands from being defaced by human misuse. Studies should be continually carried on to determine how this two-thirds of the state can best serve the recreation, forest products, and other needs of the people.

All in all, Figure 138 represents a look to the future of a state of fascinating geographic diversity; a state of great economic potential based on urbanization; a state of continual interest in the variety of its landscapes; a state unusually attractive as a place in which to live. A widely traveled man was once heard to say, "Of all the places in the East, I'd prefer to be in New York State. It has everything." This volume, it is hoped, presents a fair assessment of what New York State has, of the past from which it developed, and of the way it may go in the decades ahead.

Selected References

Jones, Barclay, and Burnham Kelly. *Long Range Needs and Opportunities in New York State. A Series of Working Papers.* Ithaca: College of Architecture, Cornell University, 1962.

New York State Department of Labor. *Jobs 1960–1970, The Changing Pattern.* Albany: The Department, 1960.

———. *Projecting New York State Job Patterns.* (Pub. No. B-118.) Albany: The Department, 1961.

New York State Office for Regional Development. *Change, Challenge, Response.* Albany: The Office, 1964.

Steps Toward Economic Expansion in New York State. Report of the Temporary State Commission on Economic Expansion. Albany: The Commission, 1960.

Sufrin, Sidney C. "The Crux: Demand and Employment," *Challenge,* XII, No. 5 (June, 1964), 6–9.

U.S. Bureau of the Census. *Illustrative Projections of the Population of the United States by States: 1970 to 1985.* ("Current Population Reports," Series P-25, No. 301.) Washington, D.C.: Govt. Printing Office, 1964.

———. *Projections of the Population of the United States by Age and Sex: 1964 to 1985 (with Extensions to 2010).* ("Current Population Reports," Series P-25, No. 286.) Washington, D.C.: Govt. Printing Office, 1964.

U.S. Department of Labor. *Manpower Report of the President.* Washington, D.C.: Govt. Printing Office, 1964.

"Where to Find 60,000 New Jobs Every Week," *U.S. News & World Report* (April 16, 1962), pp. 44–49.

APPENDIXES

APPENDIX A
SELECTED SUMMIT ELEVATIONS

Summit	Elevation (In Feet)	County	Summit	Elevation (In Feet)	County
Adirondack Highlands			**Adirondack Highlands (cont.)**		
Mount Marcy	5,344	Essex	Nye	3,895	Essex
Algonquin	5,114	Essex	Kilburn	3,893	Essex
Haystack	4,960	Essex	North River	3,880	Essex
Skylight	4,926	Essex	Blue Ridge (Cloud Cap)	3,868	Hamilton
Whiteface	4,867	Essex	Panther	3,865	Hamilton
Dix	4,857	Essex	McKenzie	3,861	Essex
Boundary	4,840	Essex	Witchopple	3,842	Essex
Gray	4,840	Essex	Bartlett Ridge	3,840	Essex
Iroquois	4,840	Essex	Gooseberry	3,840	Essex
Basin	4,827	Essex	Sentinel	3,838	Essex
Gothics	4,736	Essex	Lyon	3,830	Clinton
Colden	4,714	Essex	Couchsachraga	3,820	Essex
Giant	4,627	Essex	Averil Peak	3,810	Clinton
Santanoni	4,607	Essex	Avalanche	3,800	Essex
Redfield	4,606	Essex	Buell	3,786	Hamilton
Nippletop	4,600	Essex	Boreas	3,776	Essex
Wright	4,580	Essex	Little Nipple Top	3,760	Essex
Saddleback	4,515	Essex	Wakely	3,760	Hamilton
Panther Peak	4,442	Essex	Blue	3,759	Hamilton
Table Top	4,427	Essex	Henderson	3,752	Essex
Rocky Peak	4,420	Essex	Lewey	3,742	Hamilton
McComb	4,405	Essex	The Brothers (Twin Mtns.)	3,721	Essex
Armstrong	4,400	Essex	Wallface	3,700	Essex
Hough	4,400	Essex	Hurricane	3,694	Essex
Seward	4,361	Franklin	Hoffman (Blue Ridge)	3,693	Hamilton
Clinton	4,360	Essex	Cheney Cobble	3,683	Essex
Allen	4,340	Essex	Little Moose	3,630	Hamilton
Esther	4,220	Essex	Calamity	3,620	Essex
Big Slide	4,220	Essex	Sunrise	3,614	Essex
Upper Wolf Jaw	4,185	Essex	Stewart	3,602	Essex
Lower Wolf Jaw	4,175	Essex	Slide	3,584	Essex
Street	4,166	Essex	Gore	3,583	Warren
Phelps	4,161	Essex	Dun Brook	3,580	Hamilton
Donaldson	4,140	Franklin	Noonmark	3,556	Essex
Seymour	4,120	Franklin	Fishing Brook	3,550	Essex
Sawtooth	4,100	Essex	Adams	3,535	Essex
Lookout	4,100	Essex	Little Santanoni	3,500	Essex
Cascade	4,098	Essex	Jay	3,340	Essex
South Dix	4,060	Essex	**Catskills**		
Porter	4,059	Essex			
Colvin	4,057	Essex	Slide	4,204	Ulster
Emmons	4,040	Franklin	Hunter	4,025	Greene
Dial	4,020	Essex	Black Dome	3,990	Greene
Yard	4,018	Essex	Blackhead Peak	3,937	Greene
East Dix	4,012	Essex	Thomas Cole	3,935	Greene
McNaughton	4,000	Essex	Cornell	3,906	Ulster
Green	3,980	Essex	Graham	3,890	Ulster
Blake	3,980	Essex	Peekamoose	3,863	Ulster
McDonnel	3,960	Essex	Table	3,856	Ulster
Cliff	3,960	Essex	Plateau	3,855	Greene
Moose	3,899	Essex	Wittenberg	3,802	Ulster
Snowy	3,899	Hamilton	Sugarloaf	3,782	Greene

Summit	Elevation (In Feet)	County
Catskills (cont.)		
West Kill	3,777	Greene
Panther	3,760	Ulster
Lone	3,740	Ulster
Big Round Top	3,723	Ulster
Big Indian	3,721	Ulster
Balsam Cap	3,700	Ulster
Rusk	3,680	Greene
High Peak	3,660	Greene
Twin	3,647	Greene
Rocky	3,620	Ulster
Fir	3,619	Ulster
North Dome	3,593	Greene
Balsam	3,590	Ulster
Indian Head	3,585	Greene
Balsam	3,565	Greene
Halcott	3,537	Greene
Evergreen	3,531	Greene
Vly	3,529	Greene
Spruce Top	3,520	Ulster
Dry Brook Ridge	3,510	Delaware-Ulster
Windham High Peak	3,505	Greene
Bearpen	3,500	Delaware-Ulster
Round Top	3,470	Greene
Huntersfield	3,450	Greene
Bloomberg	3,448	Delaware
Stoppel Point	3,425	Greene
Belle Ayr	3,406	Ulster
Spruce	3,380	Ulster
Mill Brook Ridge	3,380	Delaware-Ulster
Pisgam	3,365	Delaware
Wildcat	3,268	Ulster
Hemlock	3,264	Ulster
Van Wyck	3,260	Ulster
McGregor	3,253	Delaware
Morrisville Range	3,253	Delaware
Plattekill	3,250	Delaware
Giant Ledge	3,218	Ulster
Catskills (cont.)		
Richmond	3,213	Greene
Utsayanthe	3,213	Delaware
Mongaup	3,150	Ulster
Overlook	3,150	Ulster
Rose	3,123	Ulster
Denman	3,051	Sullivan
Mombaccus	3,000	Ulster
Taconic Hills		
Bald Mountain	2,693	Rensselaer
Harvey Mountain	2,065	Columbia
Perry Peak	2,060	Columbia
Hudson Hills		
North Beacon Mountain	1,602	Dutchess
Spy Rock	1,463	Orange
Bear Mountain	1,210	Orange
Shawangunk Mountains		
Near Sam's Point	2,289	Ulster
Susquehanna Hills		
Hooker Hill	2,325	Otsego
Finger Lakes Hills		
Connecticut Hill	2,095	Tompkins
Cattaraugus Hills		
Alma Hill	2,548	Allegany
Call Hill	2,401	Steuben
Allegheny Hills		
Flatiron Rock	2,387	Cattaraugus
Tug Hill Upland		
Gomer Hill	2,080	Lewis

Sources: *The New York State Conservationist*, various issues; *The World Almanac and Book of Facts* (1961); various U.S.G.S. topographic sheets.

Appendix B

PRINCIPAL WATER BODIES

Name	Area (In Sq.Mi.)	Max. Depth (In Ft.)	Max. Length (In Mi.)	Max. Width (In Mi.)	Elevation (In Ft.)	Approx. Precip. (In In.)	Approx. Snowfall (In In.)	County	Land Form Region
Alcove Reservoir	1.8	NA	2.75	0.75	618	36	60	Albany	Helderberg Hills
Ashokan Reservoir	12.8	98	12	3	588	44	50	Ulster	Hudson Valley and Catskill Mtns.
Beaver River Flow	4.2	60	9.25	1.5	1,660	46	155	Herkimer	Adirondack Low Mtns.
Big Moose Lake	2	70	4	.75	1,824	47	152	Hamilton & Herkimer	Adirondack Low Mtns.
Big Tupper Lake	6.1	90	6.25	1.5	1,542	41	89	St. Lawrence & Franklin	Adirondack Low Mtns.
Black Lake	17.2	30	17.25	2.5	272	31	70	St. Lawrence	St. Lawrence Hills
Blake Reservoir	1	49+	3	1	1,250	41	70	St. Lawrence	Western Adirondack Hills
Blue Mountain Lake	2.1	102	2.5	2.25	1,789	43	110	Hamilton	Adirondack Low Mtns.
Bonaparte Lake	2	69	3	1.75	768	41	85	Lewis	Western Adirondack Hills
Brandreth Lake	1.3	NA	2.5	.5	1,878	45	160	Hamilton	Adirondack Low Mtns.
Brant Lake	2.1	60	5.25	1	801	37	70	Warren	Adirondack Low Mtns.
Canadarago Lake	3.1	42	3.75	1.25	1,276	41	90	Otsego	Susquehanna Hills
Canadice Lake	1	84	3	.3	1,096	31	55	Ontario	Finger Lakes Hills
Canandaigua Lake	16.6	274	15.75	1.5	686	31	48	Ontario & Yates	Finger Lakes Hills
Carry Falls Reservoir	10.1	50	6.5	2.75	1,385	41	90	St. Lawrence	Western Adirondack Hills
Cayuga Lake	66.4	435	37.75	3.5	381	34	56	Cayuga, Seneca & Tompkins	Finger Lakes Hills
Cazenovia Lake	1.7	48	3.75	.75	1,190	39	90	Madison	Susquehanna Hills
Lake Champlain	439	399	125	11	95	28-38	55	Essex & Clinton	Champlain Lake Plain & Adirondack Low Mtns.
Chautauqua Lake	20.9	77	16.25	2	1,308	42	100	Chautauqua	Cattaraugus Hills
Chazy Lake	2.3	70	3.5	1	1,531	37	96	Clinton	Adirondack Low Mtns. & St. Lawrence Hills
Clear Lake	1.6	45	2	1	1,610	41	90	Franklin	Adirondack Low Mtns.
Conesus Lake	5.1	66	7.75	.75	818	31	55	Livingston	Finger Lakes Hills
Cossayuna Lake	1.2	25	3	.5	495	41	50	Washington	Taconic Mtns.
Cranberry Lake	10.7	38	8.5	3.75	1,486	43	132	St. Lawrence	Adirondack Low Mtns.
Cross Lake	3	65	4.75	1.25	380	37	80	Onondaga & Cayuga	Ontario Drumlins

NA: Not Available.

PRINCIPAL WATER BODIES (Continued)

Name	Area (In Sq.Mi.)	Max. Depth (In Ft.)	Max. Length (In Mi.)	Max. Width (In Mi.)	Elevation (In Ft.)	Approx. Precip. (In In.)	Approx. Snowfall (In In.)	County	Land Form Region
Delta Reservoir	4	60	4.25	1.75	550	46	80	Oneida	Tug Hill Upland
Lake Erie	9,900	210	241	57	572				
Follensby Pond	1.4	100	3	.5	1,548	40	90	Franklin	Adirondack Mtn. Peaks
Forked Lake	1.9	74	4.5	.5	1,743	44	135	Hamilton	Adirondack Low Mtns.
Fourth Lake	3.3	85	5.5	1	1,707	48	140	Herkimer & Hamilton	Adirondack Low Mtns.
Lake George	44.4	187	32	2.5	317	35-39	76	Essex & Warren	Adirondack Low Mtns.
Greenwood Lake (N.Y.S. portion only)	3	57	6	.75	621	47	47	Orange	Hudson Hills
Hemlock Lake	2.9	90	7.5	.5	896	31	52	Livingston	Finger Lakes Hills
Hinckley Reservoir	5.1	75	6.75	2.25	1,225	50	90	Herkimer & Oneida	Western Adirondack Hills
Honeoye Lake	2.6	30	4.5	.75	803	32	55	Livingston & Ontario	Finger Lakes Hills
Honnedaga Lake	1.4	NA	4	.75	2,187	52	165	Herkimer	Adirondack Low Mtns.
Indian Lake	6.9	80	13.5	1	1,650	45-51	90	Hamilton	Adirondack Mtn. Peaks
Kensico Reservoir	3.5	155	3.75	1	355	48	28	Westchester	Manhattan Hills
Keuka Lake	17.4	186	19.75	2	709	33	45	Steuben & Yates	Finger Lakes Hills
Lila Lake	2.3	NA	2.75	1.25	1,714	45	130	Hamilton	Adirondack Low Mtns.
Little Moose Lake	1	129	2	.75	1,788	50	160	Herkimer	Adirondack Low Mtns.
Little Tupper Lake	3.8	42	5.5	1	1,718	43	115	Hamilton	Adirondack Low Mtns.
Long Lake	6	45	13.75	1	1,630	40	115	Hamilton	Adirondack Mtn. Peaks & Adirondack Low Mtns.
Lower Saranac Lake	3.5	65	5	1.25	1,534	40	98	Franklin	Adirondack Low Mtns.
Meacham Lake	1.9	62	2	1	1,551	44	94	Franklin	Adirondack Low Mtns.
Middle Saranac Lake	2.1	17	2.25	1.25	1,536	40	95	Franklin	Adirondack Low Mtns.
Nehasane Lake	1	NA	3.8	.5	1,695	45	130	Hamilton	Adirondack Low Mtns.
Neversink Reservoir	2.3	175	5	.5	1,440	44	50	Sullivan	Delaware Hills
Oneida Lake	79.8	55	21	5.5	369	39	83-110	Oneida & Oswego	Oneida Lake Plain, Ontario Ridge & Swampland
Onondaga Lake	4.8	73	4.5	1.25	363	37	83	Onondaga	Oneida Lake Plain
Lake Ontario	7,600	802	193	53	246				
Otisco Lake	3.5	66	5.75	.75	784	38	90	Onondaga	Finger Lakes Hills

NA: Not Available.

PRINCIPAL WATER BODIES (Continued)

Name	Area (In Sq.Mi.)	Max. Depth (In Ft.)	Max. Length (In Mi.)	Max. Width (In Mi.)	Elevation (In Ft.)	Approx. Precip. (In In.)	Approx. Snowfall (In In.)	County	Land Form Region
Otsego Lake	6.2	168	8	1.5	1,194	42	70	Otsego	Susquehanna Hills
Owasco Lake	10.3	177	10.75	1.25	770	35	75	Cayuga	Finger Lakes Hills
Paradox Lake	1.4	52	4	.75	820	38	70	Essex	Adirondack Low Mtns.
Peck Lake	1.1	60	2.75	1	1,360	47	85	Fulton	Adirondack Low Mtns.
Pepacton Reservoir	8.9	180	18.5	5	1,280	44	55	Delaware	Catskill Mtns.
Piseco Lake	4	129	5	1.25	1,661	53	95	Hamilton	Adirondack Mtn. Peaks
Lake Placid	4.4	150	4.25	1.5	1,864	38	130	Essex	Adirondack Mtn. Peaks
Pleasant Lake	2.4	65	3.5	1.25	1,724	50	92	Hamilton	Adirondack Mtn. Peaks
Rainbow Falls Reservoir	1.1	55+	3.5	.5	1,174	40	70	St. Lawrence	Western Adirondack Hills
Raquette Lake	8.5	96	6.5	1.5	1,762	46	131	Hamilton	Adirondack Low Mtns.
Raquette Pond	1.7	12	2.5	1.25	1,542	41	89	Franklin	Adirondack Low Mtns.
Redfield Reservoir	5	50	7	1	937	47	182	Oswego	Tug Hill Upland
Rondout Reservoir	3.3	175	7.5	.5	840	46	55	Ulster & Sullivan	Catskill Mtns.
Round Lake	1.2	NA	2	.75	1,718	40	100	Hamilton	Adirondack Low Mtns.
Sacandaga Lake	2.5	60	2	2	1,724	52	92	Hamilton	Adirondack Mtn. Peaks
Sacandaga Reservoir	41.7	70	27	4	771	44-47	65-77	Fulton & Saratoga	Adirondack Low Mtns.
Saratoga Lake	6.8	96	7	1.5	204	36	65	Saratoga	Hudson Valley
Schoharie Reservoir	1.8	150	5.8	.75	1,130	42	55	Delaware & Schoharie	Catskill Mtns.
Schroon Lake	6.6	121	9	1.5	807	39	70	Essex & Warren	Adirondack Low Mtns.
Seneca Lake	66.7	618	35	3	444	32	50	Seneca, Schuyler, Ontario & Yates	Finger Lakes Hills
Silver Lake	1.3	50	2.5	.75	1,393	36	90	Clinton	Adirondack Low Mtns.
Silver Lake	1.2	37	2.75	.5	1,356	32	76	Wyoming	Cattaraugus Hills
Skaneateles Lake	13.8	300	15	1.5	867	38	84	Cortland, Cayuga & Onondaga	Finger Lakes Hills
Tomhannock Reservoir	2.7	NA	5.25	.75	400	40	60	Rensselaer	Taconic Mtns.
Upper Chateaugay Lake	3.8	78	3.5	1.5	1,310	36	120	Clinton	Adirondack Low Mtns.
Upper Saranac Lake	8.2	100	7.75	2	1,571	40	90	Franklin	Adirondack Low Mtns.

NA: Not Available.

PRINCIPAL WATER BODIES (Continued)

Name	Area (In Sq. Mi.)	Max. Depth (In Ft.)	Max. Length (In Mi.)	Max. Width (In Mi.)	Elevation (In Ft.)	Approx. Precip. (In In.)	Approx. Snowfall (In In.)	County	Land Form Region
Upper St. Regis Lake	1.1	90	2.5	.75	1,617	41	90	Franklin	Adirondack Low Mtns.
Waneta Lake	1.3	29	3	.5	1,098	34	55	Schuyler	Finger Lakes Hills
Wolf Pond	1.3	76	2.5	1	1,563	41	90	Franklin	Adirondack Low Mtns.
Woodhull Lake	1.7	197	4	.5	1,880	49	155	Herkimer	Adirondack Low Mtns.

Sources: *The New York State Conservationist*, various issues; *The World Almanac and Book of Facts* (1961): U.S.G.S. topographic sheets (1:250,000); personal communications with the Board of Water Supply, New York City, and with Donald G. Pasko, Supervisor of Fish Management, N.Y.S. Dept. of Conservation.

Appendix C

PRINCIPAL RIVERS

	Drainage Area (In Sq.Mi.)	Flow (In Cu.Ft. Per Sec.)	Location of Gauge	Length of Main Stream (In Mi.)
I. Great Lakes–St. Lawrence System	≅22,000 (in N.Y.)	239,000	Ogdensburg	
A. Cattaraugus Creek	560	704	Gowanda	64
B. Buffalo River	420			8
1. Cazenovia Creek	141	213	Ebenezer	37
2. Buffalo Creek	145	186	Gardenville	40
3. Cayuga Creek	127	123	Lancaster	34
C. Tonawanda-Ellicott Creeks	610	212	Rapids	84
D. Genesee River	2,446	2,743	Rochester	144
1. Black Creek	212	103	Churchville	46
2. Oatka Creek	208	200	Garbutt	51
3. Caneadea Creek	63	83	Mouth	14
4. Canaseraga Creek	341	147	Dansville	36
5. Honeoye Creek	263	168	Honeoye Falls	36
E. Oswego River	5,002	6,420	Oswego	24*
1. Seneca River	3,433	3,141	Baldwinsville	60†
(a) Onondaga Creek (to junction with Seneca River)	233	109	Syracuse	30
(b) Owasco Outlet	230	282	Auburn	16‡
(c) Clyde River	847			73
1. Canandaigua Outlet	431	145	Chapin	33§
2. Mud Creek–Ganargua Creek	300			50
(d) Cayuga Lake Drainage	813			
1. Fall Creek	152	183	Ithaca	23
2. Cayuga Inlet	173	38	Ithaca	16
(e) Seneca Lake Drainage	707			
1. Keuka Lake Outlet	213			7‖
2. Oneida River	1,404	2,396	Caughdenoy	18
(a) Fish Creek	406			55
1. East Branch Fish Creek		546	Taberg	44
(b) Oneida Creek	149	153	Oneida	26
(c) Chittenango Creek	309	109	Chittenango	40
1. Limestone Creek		139	Fayetteville	26

*To junction of Oneida and Seneca rivers. †To Seneca Lake outlet. ‡43 miles to headwaters. §56 miles to headwaters. ‖33 miles to headwaters.

PRINCIPAL RIVERS (Continued)

	Drainage Area (In Sq.Mi.)	Flow (In Cu.Ft. Per Sec.)	Location of Gauge	Length of Main Stream (In Mi.)
F. Salmon River	285			44
G. Black River	1,930	3,937	Watertown	112
1. Deer River	102	222	Copenhagen	26
2. Moose River	416	839	McKeever	53
(a) Middle Branch Moose River	148	329	Mouth	35
3. Independence River	99	200	Donnattsburg	31
4. Beaver River	338	580	Croghan	59
H. Oswegatchie River	1,609	2,660	Ogdensburg	133
1. Indian River	544			97
2. West Branch Oswegatchie River	272	514	Harrisville	35
I. Grass River	637	607	Pyrites	112
J. Raquette River	1,240	2,089	Raymondville	163
K. St. Regis River	910	1,064	Brasher Center	77#
1. Deer River	212	229	Brasher Iron Works	46
2. West Branch, St. Regis River	280			59
3. East Branch, St. Regis River	347			32
L. Salmon River	452 (in N.Y.)	229	Chasm Falls	46#
M. Chateaugay River	199 (in N.Y.)	179	Chateaugay	30#
N. Lake Champlain Drainage	2,950 (in N.Y.)			
1. Great Chazy River	300	275	Perry Mills	48
2. Saranac River	628	841	Plattsburgh	86
3. Ausable River	519	688	Ausable Forks	57
(a) West Branch, Ausable River		222	Lake Placid	34
(b) East Branch, Ausable River		314	Ausable Forks	40
4. Bouquet River		307	Willsboro	43
5. Lake George Outlet	229	329	Ticonderoga	2
II. Hudson River System	12,200 (in N.Y.)			306
A. Croton River	378	288	Croton	57
B. Fishkill	204	304	Beacon	32
C. Wappinger Creek	182	242	Wappingers Falls	37
D. Rondout Creek	1,148	689	Rosendale	50
1. Wallkill River	779 (567 in N.Y.)	1,062	Gardiner	53#
(a) Shawangunk Kill	149			33
E. Esopus Creek	417	453	Coldbrook	48
F. Catskill Creek	394	123	Oak Hill	37
G. Kinderhook Creek	334 (305 in N.Y.)	476	Rossman	42
H. Normans Kill	168			30
I. Poesten Kill	89	139	Troy	21
J. Mohawk River	3,400	5,693	Cohoes	148
1. Schoharie Creek	947	910	Burtonsville	83
2. Otsquago Creek	60	89	Fort Plain	15
3. East Canada Creek	286	709	East Creek	34
4. West Canada Creek	569	1,337	Kast Bridge	75
5. Oriskany Creek	146			27
K. Hoosic River	730 (partly in Vt.)	920	Eagle Bridge	57
1. Walloomsac River		228	N. Bennington	21
2. Little Hoosic River		105	Petersburg	18
L. Batten Kill	460 (partly in Vt.)	726	Battenville	53
M. Fish Creek (to Saratoga Lake)	253			8
1. Kayaderosseras Creek		139	West Milton	32
N. Sacandaga River	1,055	2,129	Hadley	84
1. East Stony Creek	212			23

#Length of stream in New York State.

PRINCIPAL RIVERS (Continued)

	Drainage Area (In Sq.Mi.)	Flow (In Cu.Ft. Per Sec.)	Location of Gauge	Length of Main Stream (In Mi.)
2. East Branch, Sacandaga River	124	223	Griffin	31
3. West Branch, Sacandaga River	240			29
4. Middle Branch, Sacandaga River	115			23
O. Schroon River	550	819	Riverbank	45
P. Indian River	160+	293	Indian Lake	35
Q. Cedar River		318	Mouth	35
III. Allegheny River System	2,100 (in N.Y.)			325
A. Allegheny River	1,330 (in N.Y.)	2,772	Red House	47#
1. Great Valley Creek	145	221	Salamanca	23
B. Conewango Creek	770 (partly in Pa.)	501	Waterboro	36#
1. Chadakoin River	343	336	Falconer	6**
(a) Chautauqua Lake Outlet	178			7
(b) Cassadaga Creek	165			22
IV. Susquehanna River System	6,267 (in N.Y.)			444#
A. Chemung River	2,518	2,470	Chemung	41††
1. Cohocton River	472	440	Campbell	50
(a) Mud Creek	76	42	Savona	19
(b) Fivemile Creek	68	70	Kanona	21
2. Tioga River	1,370 (partly in Pa.)	1,340	Erwins	12#
(a) Canisteo River	780	341	West Cameron	53
1. Canacadea Creek	59	63	Hornell	11
2. Tuscarora Creek	120	98	South Addison	19
3. Newtown Creek		87	Elmira	16
B. Susquehanna River	3,749 (in N.Y.)	7,483	Waverly	158#
1. Cayuta Creek	148			33
2. Owego Creek	391	275	Owego	30
3. Chenango River	1,582	2,474	Chenango Forks	85
(a) Genegantslet Creek		140	Smithville Flats	26
(b) Canasawacta Creek	80 (approx.)	104	South Plymouth	18
(c) Tioughnioga River	753	1,232	Itaska	70
1. Otselic River	259	390	Upper Lisle	48
4. Unadilla River	561	829	Rockdale	55
(a) Butternut Creek		96	Morris	35
5. Otego Creek	110	174	Oneonta	26
6. Oaks Creek	103	172	Index	24
7. Schenevus Creek	127			26
8. Charlotte Creek	178	255	Davenport Center	31
9. Ouleout Creek	115	171	East Sidney	26
V. Delaware River System	2,580	4,690	Port Jervis	140#
A. Neversink River	346	339	Godeffroy	61
B. Mongaup River	204	340	Mongaup	30
C. Ten Mile River	46	71	Tusten	10
D. Callicoon Creek	112	181	Callicoon	19
E. West Branch, Delaware River	685	1,061	Hale Eddy	73
(1) Oquaga Creek	66	111	Deposit	12
(2) Trout Creek	50	86	Cannonsville	14
(3) Little Delaware River	53	93	Delhi	15
F. East Branch, Delaware River	919	1,252	Fishs Eddy	62
(1) Beaver Kill	322	564	Cooks Falls	37
(2) Tremper Kill	35	61	Shavertown	11
(3) Platte Kill	35	65	Dunraven	11
VI. Long Island Streams				
A. Connetquot Brook	24	39	Oakdale	9

#Length of stream in New York State.   **Between Conewango Creek and Levant.   ††Between mouth and Painted Post.

PRINCIPAL RIVERS (Continued)

	Drainage Area (In Sq.Mi.)	Flow (In Cu.Ft. Per Sec.)	Location of Gauge	Length of Main Stream (In Mi.)
B. Peconic River	75	32	Riverhead	18
C. Nissequoque River	30	4	Smithtown	10
D. Carman's River	71	22	Yaphank	11
E. Swan River	10	12	East Patchogue	4
F. Patchogue Creek	15	20	Patchogue	4
G. Sampawams Creek	24	9	Babylon	4
H. Carll's River	35	28	Babylon	7
I. Massapequa Creek	38	12	Massapequa	6
J. East Meadow Brook	31	17	Freeport	8
VII. Long Island Sound—East River Drainage				
A. Mamaroneck River	24	31	Mamaroneck	10
B. Tenmile River	204	296	Gaylordsville, Conn.	24
C. Blind Brook	10	16	Rye	9

Sources: George Rafter, *Hydrology of the State of New York* (Albany: The State Museum, 1905); various "Water Supply Papers" of the U.S.G.S.; various reports of the Water Pollution Control Board, N.Y.S. Department of Health; U.S.G.S. topographic sheets (1:250,000).

Appendix D
COUNTY DATA

County	Area (In Sq. Mi.) 1960	Population			Change 1940-60 (%)	Distribution 1960			Median Family Income	
		1900	1940	1960		Urban (%)	Rural Nonfarm (%)	Rural Farm (%)	1959 (In Dollars)	Increase 1949-59 (In Dollars)
Albany	531	165,571	221,315	272,926	23.3	83.5	15.2	1.3	$6,199	$2,664
Allegany	1,048	41,501	39,681	43,978	10.8	20.0	64.7	15.4	4,828	2,065
Broome	710	69,149	165,749	212,661	28.3	74.4	23.3	2.3	6,409	2,787
Cattaraugus	1,335	65,643	72,652	80,187	10.4	40.7	48.9	10.4	5,315	2,352
Cayuga	699	66,234	65,508	73,942	12.9	47.7	40.1	12.2	5,384	2,311
Chautauqua	1,080	88,314	123,580	145,377	17.6	57.1	34.2	8.7	5,626	2,484
Chemung	412	54,063	73,718	98,706	33.9	74.8	22.3	2.8	4,758	2,488
Chenango	908	36,568	36,454	43,243	18.6	21.2	59.7	19.1	5,308	2,534
Clinton	1,059	47,430	54,006	72,722	34.7	34.4	56.2	9.5	5,165	2,336
Columbia	643	43,211	41,464	47,322	14.1	23.4	66.4	10.2	5,331	2,549
Cortland	502	27,576	33,668	41,113	22.1	55.5	32.7	11.9	5,505	2,700
Delaware	1,470	46,413	40,989	43,540	6.2	20.7	56.9	22.4	4,856	2,300
Dutchess	816	81,670	120,542	176,008	46.0	41.8	55.0	3.2	6,481	3,038
Erie	1,054	433,686	798,377	1,064,688	33.4	87.6	11.4	1.0	6,395	2,905
Essex	1,826	30,707	34,178	35,300	3.3	23.6	70.3	6.1	4,969	2,476
Franklin	1,685	42,853	44,286	44,742	1.0	41.5	45.4	13.1	4,639	2,165
Fulton	497	42,842	48,597	51,304	5.6	63.0	32.7	4.3	5,379	2,404
Genesee	501	34,561	44,481	53,994	21.4	42.4	46.4	11.3	5,898	2,662
Greene	653	31,478	27,926	31,372	12.3	27.6	64.5	7.9	5,056	2,700
Hamilton	1,747	4,947	4,188	4,267	1.9	—	98.5	1.5	4,511	2,149
Herkimer	1,442	51,049	59,527	66,370	11.5	59.2	32.2	8.6	5,519	2,523
Jefferson	1,293	76,748	84,003	87,835	4.6	42.7	45.8	11.5	5,261	2,402
Lewis	1,293	27,427	22,815	23,249	1.9	15.6	58.4	26.0	4,760	2,199
Livingston	638	37,059	38,510	44,053	14.4	33.5	54.0	12.5	5,607	2,693
Madison	661	40,545	39,598	54,635	38.0	47.0	39.9	13.1	5,451	2,461
Monroe	673	217,854	438,230	586,387	33.8	86.7	12.0	1.4	7,147	3,425
Montgomery	409	47,488	59,142	57,240	3.2	59.9	32.3	7.8	5,411	1,916
Nassau	300	55,448	406,748	1,300,171	219.7	99.7	.3	—	8,515	3,991
New York City (as one statistical unit)	314	3,437,202	7,454,995	7,781,984	4.4	100.0	—	—	6,091	2,565
Niagara	533	74,961	160,110	242,269	51.3	74.0	22.7	3.3	6,692	3,177
Oneida	1,227	132,800	203,636	264,401	29.8	73.0	23.4	3.6	6,180	2,917
Onondaga	792	168,735	295,108	423,028	43.3	80.9	17.2	1.9	6,691	3,232
Ontario	649	49,605	55,307	68,070	23.1	39.2	49.4	11.5	5,734	2,787
Orange	829	103,859	140,113	183,734	31.1	50.7	45.1	4.2	5,721	2,548
Orleans	396	30,164	27,760	34,159	23.1	34.7	50.2	15.1	5,608	2,757
Oswego	968	70,881	71,275	86,118	20.8	42.3	50.0	7.7	5,580	2,808
Otsego	1,013	48,939	46,082	51,942	12.7	30.7	52.0	17.3	4,891	2,137
Putnam	235	13,787	16,555	31,722	91.6	9.7	88.3	2.0	6,539	3,200
Rensselaer	665	121,697	121,834	142,585	17.0	64.6	32.5	2.8	5,747	2,449
Rockland	178	38,298	74,261	136,803	84.2	76.0	23.7	0.3	7,472	3,918
St. Lawrence	2,767	89,083	91,098	111,239	22.1	44.4	43.0	12.6	5,421	2,652
Saratoga	814	61,089	65,606	89,096	35.8	48.2	47.6	4.3	5,676	2,431
Schenectady	209	46,852	122,494	152,896	24.8	86.4	12.8	0.7	6,541	2,723
Schoharie	625	26,854	20,812	22,616	8.7	15.3	60.2	24.4	4,592	2,043
Schuyler	331	15,811	12,979	15,044	15.9	18.7	62.4	18.9	5,041	2,573
Seneca	330	28,114	25,732	31,984	24.3	39.2	49.8	11.0	5,790	2,668
Steuben	1,408	82,822	84,927	97,691	15.0	43.5	46.7	9.8	5,607	2,616
Suffolk	922	77,582	197,355	666,784	237.9	72.3	26.9	0.8	6,795	3,384
Sullivan	986	32,306	37,901	45,272	19.4	21.9	70.2	7.9	5,198	2,576
Tioga	525	27,951	27,072	37,802	39.6	30.1	58.2	11.8	5,626	2,919
Tompkins	491	33,830	42,340	66,164	56.3	47.7	45.3	7.0	6,233	3,017
Ulster	1,143	88,422	87,017	118,804	36.5	42.0	53.5	4.5	5,746	2,921
Warren	883	29,943	36,035	44,002	22.1	48.4	49.8	1.8	5,356	2,530
Washington	837	45,624	46,726	48,476	3.7	37.6	49.3	13.1	5,106	2,323
Wayne	607	48,660	52,747	67,989	28.9	34.9	50.7	14.4	5,667	2,861
Westchester	435	184,257	573,558	808,891	41.0	92.7	7.2	0.1	8,052	3,699
Wyoming	598	30,413	31,394	34,793	10.8	31.7	49.0	19.3	5,428	2,674
Yates	344	20,318	16,381	18,614	13.6	31.0	46.7	22.3	4,799	2,282
NEW YORK STATE	47,939+	7,268,894	13,479,142	16,782,304	24.5	85.4	12.7	1.9	6,371	2,884

Sources: Various U.S. Bureau of the Census publications.

Employment						Cropland Harvested			County
Distribution by Sectors						Total 1959 (Acres)	Percentage of Total Area 1959	Decline 1925-59 (%)	
Primary		Secondary		Tertiary		Worked Outside County of Residence, 1960 (%)			
1840* (%)	1960	1840* (%)	1960	1840* (%)	1960				
62.7	1.4	28.7	27.2	8.6	71.4	12.1	50,693	15	57	Albany
76.6	15.3	21.1	32.7	2.3	52.0	16.7	98,265	15	53	Allegany
76.5	1.9	17.7	50.0	5.8	48.1	3.9	72,347	16	48	Broome
83.0	9.3	13.6	37.2	3.4	53.5	12.7	110,947	13	45	Cattaraugus
73.8	8.6	19.7	40.0	6.5	51.4	16.4	165,347	37	26	Cayuga
80.9	6.9	13.8	45.0	5.3	48.1	4.4	140,573	20	40	Chautauqua
70.2	2.2	23.0	44.2	6.8	53.6	6.4	35,882	14	56	Chemung
81.5	15.6	14.5	38.1	4.0	53.7	14.9	116,373	20	32	Chenango
74.9	11.1	19.8	24.5	5.3	64.4	2.7	92,451	14	38	Clinton
68.2	11.0	22.1	35.1	9.7	53.9	16.8	92,693	23	—	Columbia
81.2	10.1	16.7	41.3	2.1	48.6	10.5	75,717	24	27	Cortland
81.1	23.4	15.4	31.8	3.5	44.8	7.6	142,895	15	26	Delaware
73.6	4.3	22.0	21.7	4.4	74.0	5.7	83,027	16	32	Dutchess
67.5	1.0	22.2	41.6	10.3	57.4	3.9	132,488	20	53	Erie
70.4	14.5	24.0	20.0	5.6	65.5	9.4	32,789	3	57	Essex
83.6	12.6	13.4	24.0	3.0	63.4	12.3	83,733	8	37	Franklin
84.3	3.1	13.4	54.0	2.3	42.9	13.8	27,224	9	60	Fulton
79.8	10.4	16.6	38.2	3.6	51.4	15.6	125,141	39	17	Genesee
70.2	9.1	25.8	32.2	4.0	58.7	17.9	37,478	9	57	Greene
87.3	4.7	12.3	31.8	.4	63.5	14.1	398	—	—	Hamilton
77.4	8.1	17.6	50.0	5.0	41.9	17.4	98,774	11	27	Herkimer
79.9	11.2	15.0	30.0	5.1	58.8	2.8	198,057	24	41	Jefferson
80.9	26.2	16.1	31.9	3.0	41.9	12.6	90,353	11	35	Lewis
74.2	14.2	22.6	33.2	3.2	52.8	22.9	134,308	33	21	Livingston
77.1	12.9	19.1	31.8	3.8	55.3	33.1	119,131	28	24	Madison
59.9	1.4	30.8	47.3	9.3	51.3	1.5	122,026	28	47	Monroe
56.2	7.9	35.7	48.3	8.1	43.8	13.8	90,024	34	23	Montgomery
*	.9		31.2		67.9	43.4	4,268	2	81	Nassau
										New York City (as one statistical unit)
14.8	.2	59.0	30.1	26.2	69.7	39.6	712	—	—	
81.9	2.2	13.5	53.2	4.6	44.6	8.9	113,165	33	45	Niagara
67.2	3.9	25.7	34.8	7.1	61.3	4.0	150,045	19	33	Oneida
64.4	1.7	27.0	39.1	8.6	59.2	2.4	115,170	23	49	Onondaga
74.9	9.9	18.5	31.7	6.6	58.4	23.7	146,698	35	25	Ontario
82.0	5.3	15.4	33.7	2.6	61.0	8.9	105,794	20	24	Orange
82.3	12.2	12.6	43.0	5.1	44.8	28.1	91,702	36	37	Orleans
78.8	6.5	16.8	45.4	4.4	48.1	20.5	73,315	12	56	Oswego
78.5	16.5	18.8	25.0	2.7	58.5	17.4	143,689	22	33	Otsego
73.9	3.0	21.6	30.9	4.5	66.1	44.5	6,493	4	60	Putnam
54.6	2.5	33.4	36.4	12.0	61.1	37.7	62,416	15	54	Rensselaer
68.9	2.0	24.1	34.0	7.0	64.0	26.5	2,412	2	84	Rockland
82.1	15.8	14.3	27.7	3.6	56.5	2.1	226,849	13	40	St. Lawrence
75.4	4.1	19.7	43.8	4.9	52.1	33.5	58,704	11	50	Saratoga
62.6	.8	24.0	42.2	13.4	57.0	11.7	17,900	13	66	Schenectady
85.9	23.5	12.6	23.1	1.5	53.4	18.4	82,441	21	41	Schoharie
*	12.2	—	42.4	—	45.4	30.1	37,912	18	53	Schuyler
74.8	9.1	19.1	34.2	6.1	56.7	16.9	74,920	36	36	Seneca
81.5	8.0	15.3	41.7	3.2	50.3	5.5	194,951	22	40	Steuben
66.3	2.9	14.4	34.5	19.3	62.6	30.6	61,621	10	—	Suffolk
72.5	8.1	20.8	20.2	6.7	71.7	7.7	37,499	6	57	Sullivan
80.2	9.6	15.8	45.8	4.0	44.6	34.0	63,568	19	35	Tioga
74.3	6.5	19.6	21.0	6.1	72.5	3.9	68,089	22	39	Tompkins
64.4	5.2	26.7	40.8	8.9	54.0	13.2	57,084	8	48	Ulster
71.7	2.4	19.6	35.2	8.7	62.4	15.3	4,917	1	86	Warren
79.0	13.2	14.6	40.2	6.4	46.6	24.3	121,722	23	31	Washington
73.4	14.0	21.4	38.5	5.2	47.5	19.3	139,419	36	32	Wayne
66.5	1.0	27.1	28.1	6.4	70.8	29.8	7,169	3	73	Westchester
*	16.5	—	36.1	—	47.4	19.7	123,320	32	26	Wyoming
67.3	16.6	26.9	29.8	5.8	53.6	17.1	69,003	31	37	Yates
66.4	2.0	25.1	33.5	8.5	64.5	28.8	5,032,671	16.4	—	NEW YORK STATE

Nassau, Schuyler and Wyoming counties were not formed in 1840.

APPENDIX D COUNTY DATA

Appendix E
URBAN PLACE DATA
(Incorporated Places of 500 or more; Unincorporated Places of 1,000 or more)

Urban Place	County	Population 1900 (No.)	Population 1940 (No.)	Population 1960 (No.)	Population Growth 1940-60 (Percentage)	Employment Mfg. 1958 (No.)	Tertiary 1960 (No.)	Tertiary Surplus 1960 (No.)	Median Family Income (1959) (In Dollars)
Adams	Jefferson	1,292	1,594	1,944	20.1	NA	NA	NA	NA
Addison	Steuben	2,080	1,617	2,185	35.1	NA	NA	NA	NA
Afton	Chenango	722	806	956	18.6	NA	NA	NA	NA
Akron	Erie	1,585	2,263	2,841	25.5	717	365	1,484	$6,453
ALBANY-SCHENECTADY URBAN SYSTEM (SMSA)		395,209	531,249	657,503	23.8	63,275	173,054	207,767	6,095
Urbanized Area		NA	NA	455,447	NA	NA	133,051	209,808	6,183
Albany	Albany	94,151	130,577	129,726	.7	7,140	65,032	195,434	5,778
Albion	Orleans	4,477	4,660	5,182	11.2	536	1,763	3,633	6,338
Alden	Erie	607	954	2,042	114.0	NA	NA	NA	NA
Alexandria Bay	Jefferson	1,511	1,748	1,583	-9.4	NA	NA	NA	NA
Alfred	Allegany	756	694	2,807	304.5	NA	856	1,473	5,800
Allegany	Cattaraugus	NA	1,436	2,064	43.7	NA	NA	NA	NA
Almond	Allegany-Steuben	NA	533	696	30.6	NA	NA	NA	NA
Altamont	Albany	689	890	1,365	53.4	NA	NA	NA	NA
Amagansett[U]	Suffolk	NA	NA	1,095	NA	NA	NA	NA	NA
Amityville	Suffolk	2,038	5,058	8,318	64.5	617	4,412	13,742	8,385
Amsterdam	Montgomery	20,929	33,329	28,772	-13.7	5,476	6,448	3,468	5,501
Andover	Allegany	954	1,290	1,247	-3.3	NA	NA	NA	NA
Angelica	Allegany	978	928	898	-3.2	NA	NA	NA	NA
Angola	Erie	712	1,663	2,499	50.3	NA	NA	NA	NA
Antwerp	Jefferson	929	817	881	7.8	NA	NA	NA	NA
Arcade	Wyoming	887	1,683	1,930	14.7	NA	NA	NA	NA
Ardsley	Westchester	404	1,423	3,991	180.5	NA	630	-841	10,000+
Arkport	Steuben	NA	618	837	35.4	NA	NA	NA	NA
Arlington[U]	Dutchess	NA	NA	8,317	NA	NA	NA	NA	6,784
Athens	Greene	2,171	1,655	1,754	6.0	NA	NA	NA	NA
Attica	Genesee-Wyoming	1,785	2,379	2,758	15.9	293	977	2,127	5,926
Auburn	Cayuga	30,345	35,753	35,249	-1.4	5,240	8,752	8,511	5,518
Aurora	Cayuga	499	372	834	124.2	NA	NA	NA	NA
Ausable Forks[U]	Clinton-Essex	NA	NA	2,026	NA	NA	NA	NA	NA
Avoca	Steuben	1,006	1,006	1,086	8.0	NA	NA	NA	NA
Avon	Livingston	1,601	2,339	2,772	18.5	NA	859	1,523	6,069
Babylon	Suffolk	2,157	4,742	11,062	133.3	2,809	6,153	19,703	$7,642
Bainbridge	Chenango	1,092	1,450	1,712	18.1	NA	NA	NA	NA
Baldwin[U]	Nassau	NA	NA	30,204	NA	NA	NA	NA	8,728
Baldwinsville	Onondaga	2,992	3,840	5,985	55.9	363	1,063	-670	6,807
Ballston Spa	Saratoga	3,923	4,443	4,991	12.3	660	1,776	3,889	5,378
Balmville	Orange	NA	NA	1,538	NA	NA	NA	NA	NA
Barker	Niagara	NA	452	528	16.8	NA	NA	NA	NA
Batavia	Genesee	9,180	17,267	18,210	5.5	3,088	5,811	10,845	6,100
Bath	Steuben	4,994	4,696	6,166	31.3	174	2,332	5,494	5,746
Baxter Estates	Nassau	NA	760	932	22.6	NA	NA	NA	NA
Bayville	Nassau	NA	1,516	3,962	161.3	0	411	-1,907	7,109
Beacon	Dutchess	9,480	12,572	13,922	10.7	2,662	3,698	4,568	6,497
Bellerose	Nassau	NA	1,317	1,083	17.8	NA	NA	NA	NA
Bellmore[U]	Nassau	NA	NA	12,784	NA	NA	NA	NA	7,736
Bellport	Suffolk	NA	650	2,461	278.6	NA	NA	NA	NA
Belmont	Allegany	1,190	1,146	1,146	0	NA	NA	NA	NA
Bergen	Genesee	624	658	964	46.5	NA	NA	NA	NA
Bethpage-Old Bethpage	Nassau	NA	NA	20,515	NA	NA	NA	NA	7,772
BINGHAMTON URBAN SYSTEM (SMSA)		69,149	165,749	212,661	28.3	38,057	46,701	20,844	6,409
Urbanized Area		NA	NA	158,141	NA	NA	42,468	54,199	6,537
Binghamton	Broome	39,647	78,309	75,941	-3.0	10,896	28,372	65,909	6,251
Black River	Jefferson	949	897	1,237	37.9	NA	NA	NA	NA
Blasdell	Erie	415	2,322	3,909	68.3	256	460	-1,609	6,343
Bolivar	Allegany	1,208	1,344	1,405	4.5	NA	NA	NA	NA

NA: Not Available. [U]Indicates unincorporated urban places.

URBAN PLACE DATA (Continued)

Urban Place	County	Population 1900 (No.)	Population 1940 (No.)	Population 1960 (No.)	Population Growth 1940-60 (Percentage)	Employment Mfg. 1958 (No.)	Employment Tertiary 1960 (No.)	Tertiary Surplus 1960 (No.)	Median Family Income (1959) (In Dollars)
Boonville	Oneida	1,745	2,076	2,403	15.8	NA	NA	NA	NA
Brentwood[u]	Suffolk	NA	NA	15,387	NA	NA	NA	NA	$6,639
Brewster	Putnam	1,192	1,863	1,715	−7.9	NA	NA	NA	NA
Briarcliff Manor	Westchester	NA	1,830	5,105	179.0	186	521	−2,500	10,000+
Brightwaters	Suffolk	NA	1,562	3,193	104.4	25	644	27	6,896
Broadalbin	Fulton	NA	1,399	1,438	2.8	NA	NA	NA	NA
Brockport	Montgomery	3,398	3,590	5,256	46.4	995	2,225	5,869	NA
Brocton	Chautauqua	700	1,293	1,416	9.5	NA	NA	NA	NA
Bronxville	Westchester	579	6,888	6,744	2.1	49	6,267	24,591	10,000+
Brookville	Nassau	NA	204	1,468	619.6	NA	NA	NA	NA
Brownville	Jefferson	767	907	1,082	19.3	NA	NA	NA	NA
Brushton	Franklin	NA	487	553	13.6	NA	NA	NA	NA
Buchanan	Westchester	NA	1,600	2,019	26.2	NA	NA	NA	NA
BUFFALO URBAN SYSTEM (SMSA)		508,647	958,487	1,306,957	36.4	169,877	292,513	155,608	6,455
Urbanized Area		NA	NA	1,054,370	NA	NA	258,756	239,410	6,394
Buffalo	Erie	352,387	575,901	532,759	−7.5	70,827	186,336	398,921	5,713
Caledonia	Livingston	1,073	1,226	1,917	56.4	NA	NA	NA	NA
Cambridge	Washington	1,578	1,572	1,748	11.2	NA	NA	NA	NA
Camden	Oneida	2,370	2,021	2,694	33.3	NA	721	911	6,655
Camillus	Onondaga	567	1,133	1,416	25.0	NA	NA	NA	NA
Canajoharie	Montgomery	2,101	2,577	2,681	4.0	1,359	942	2,029	6,387
Canandaigua	Ontario	6,151	8,321	9,370	12.6	602	3,284	7,050	6,214
Canaseraga	Allegany	685	698	730	4.6	NA	NA	NA	NA
Canastota	Madison	3,030	4,150	4,896	18.0	835	1,060	404	5,532
Candor	Tioga	NA	66	956	44.6	NA	NA	NA	NA
Canisteo	Steuben	2,077	2,550	2,731	7.1	65	905	1,794	5,291
Canton	St. Lawrence	2,757	3,018	5,046	67.2	108	1,973	4,819	6,396
Cape Vincent	Jefferson	1,310	931	770	−17.3	NA	NA	NA	NA
Carmel[u]	Putnam	NA	NA	1,426	NA	NA	NA	NA	NA
Carthage	Jefferson	2,895	4,207	4,216	2	630	1,164	1,604	5,597
Cassadaga	Chautauqua	NA	514	820	59.5	NA	NA	NA	NA
Castile	Wyoming	1,088	902	1,146	27.1	NA	NA	NA	NA
Castleton-on-Hudson	Rensselaer	1,214	1,515	1,752	15.6	NA	NA	NA	NA
Catskill	Greene	5,484	5,429	4,825	7.3	NA	2,569	7,020	5,339
Cattaraugus	Cattaraugus	1,382	1,145	1,258	9.9	NA	NA	NA	NA
Cayuga	Cayuga	390	472	621	31.6	NA	NA	NA	NA
Cayuga Heights	Tompkins	NA	651	2,788	328.3	0	535	−113	10,000+
Cazenovia	Madison	1,819	1,689	2,584	53.0	NA	1,042	2,626	6,808
Cedarhurst	Nassau	NA	5,463	6,954	27.3	9	2,637	6,231	9,767
Celoron	Chautauqua	506	1,349	1,507	11.7	NA	NA	NA	NA
Centereach[u]	Suffolk	NA	NA	8,524	NA	NA	NA	NA	6,474
Center Moriches[u]	Suffolk	NA	NA	2,521	NA	NA	749	1,224	5,930
Centerport[u]	Suffolk	NA	NA	3,628	NA	NA	NA	NA	10,000+
Central Square	Oswego	364	568	935	64.6	NA	NA	NA	NA
Champlain	Clinton	1,311	1,354	1,549	14.4	NA	NA	NA	NA
Chateaugay	Franklin	973	1,183	1,097	−7.3	NA	NA	NA	NA
Chatham	Columbia	2,018	2,254	2,426	7.6	NA	NA	NA	NA
Chaumont	Jefferson	738	534	523	−2.1	NA	NA	NA	NA
Cheektowaga-Northwest[u]	Erie	NA	NA	52,362	NA	NA	NA	NA	6,806
Cheektowaga-Southwest[u]	Erie	NA	NA	12,766	NA	NA	NA	NA	6,233
Cherry Creek	Chautauqua	701	529	649	22.7	NA	NA	NA	NA
Cherry Valley	Otsego	722	704	668	−5.1	NA	NA	NA	NA
Chester	Orange	1,250	1,140	1,492	30.9	NA	NA	NA	NA
Chittenango	Madison	787	885	3,180	259.3	0	538	−490	6,132
Churchville	Monroe	505	601	1,003	66.9	NA	NA	NA	NA
Clarence[u]	Erie	NA	NA	1,456	NA	NA	NA	NA	NA

NA: Not Available.
[u] Indicates unincorporated urban places.

URBAN PLACE DATA (Continued)

Urban Place	County	Population 1900 (No.)	Population 1940 (No.)	Population 1960 (No.)	Population Growth 1940-60 (Percentage)	Employment Mfg. 1958 (No.)	Employment Tertiary 1960 (No.)	Tertiary Surplus 1960 (No.)	Median Family Income (1959) (In Dollars)
Clark Mills[u]	Oneida	NA	NA	1,148	NA	NA	NA	NA	NA
Clayton	Jefferson	1,913	1,999	1,996	−.2	NA	NA	NA	NA
Clayville	Oneida	568	711	686	−3.5	NA	NA	NA	NA
Cleveland	Oswego	689	440	732	66.4	NA	NA	NA	NA
Clifton Springs	Ontario	1,617	1,413	1,953	38.2	NA	NA	NA	NA
Clinton	Oneida	1,340	1,478	1,855	25.5	NA	NA	NA	NA
Clyde	Wayne	2,507	2,356	2,693	14.3	NA	877	1,692	$6,286
Cobleskill	Schoharie	2,327	2,617	3,471	32.6	335	1,398	3,519	5,454
Cohocton	Steuben	879	931	929	−.2	NA	NA	NA	NA
Cohoes	Albany	23,910	21,955	20,129	−8.3	2,737	2,864	−5,809	5,573
Cold Springs	Putnam	2,067	1,897	2,083	9.8	NA	NA	NA	NA
Cold Spring Harbor[u]	Suffolk	NA	NA	1,705	NA	NA	NA	NA	NA
Colonie	Albany	NA	1,407	6,992	396.9	783	1,249	−747	7,089
Commack[u]	Suffolk	NA	NA	9,613	NA	NA	NA	NA	7,782
Cooperstown	Otsego	2,368	2,599	2,553	−1.8	50	1,559	5,242	5,826
Copenhagen	Lewis	587	608	673	10.7	NA	NA	NA	NA
Copiague[u]	Suffolk	NA	NA	14,081	NA	NA	NA	NA	6,479
Corfu	Genesee	401	462	616	33.3	NA	NA	NA	NA
Corinth	Saratoga	2,039	3,054	3,193	4.6	1,589	482	−783	5,635
Corning	Steuben	11,061	16,212	17,085	5.4	6,079	4,409	4,960	6,719
Cornwall	Orange	1,966	1,978	2,785	40.8	174	795	1,190	7,244
Cornwall Southwest[u]	Orange	NA	NA	2,824	NA	NA	NA	NA	6,129
Cortland	Cortland	9,014	15,881	19,181	20.8	3,775	4,827	4,954	5,715
Coxsackie	Greene	2,735	2,352	2,849	21.1	431	647	386	5,492
Croghan	Lewis	NA	801	821	2.5	NA	NA	NA	NA
Croton-on-Hudson	Westchester	1,533	3,843	6,812	77.3	35	1,124	−1,192	8,113
Cuba	Allegany	1,502	1,699	1,949	14.7	NA	NA	NA	NA
Dannemora	Clinton	NA	4,830	4,835	.1	NA	837	−650	6,762
Dansville	Livingston	3,633	4,976	5,460	9.7	476	1,388	1,480	5,837
Deer Park[u]	Suffolk	NA	NA	16,726	NA	NA	NA	NA	6,901
Delevan	Cattaraugus	NA	554	777	40.3	NA	NA	NA	NA
Delhi	Delaware	2,078	1,841	2,307	25.3	NA	NA	NA	NA
Depew	Erie	3,379	6,084	13,580	123.2	1,257	1,729	−4,935	6,665
Deposit	Broome-Delaware	2,051	2,028	2,025	−.1	NA	NA	NA	NA
De Ruyter	Madison	623	526	627	19.2	NA	NA	NA	NA
Dexter	Jefferson	945	1,109	1,009	−9.0	NA	NA	NA	NA
Dobbs Ferry	Westchester	2,888	5,883	9,260	57.4	123	1,346	−2,530	8,313
Dolgeville	Fulton-Herkimer	2,685	3,195	3,058	−4.3	1,001	574	−188	5,192
Dryden	Tompkins	699	747	1,263	69.1	NA	NA	NA	NA
Dundee	Yates	1,291	1,168	1,468	25.7	NA	NA	NA	NA
Dunkirk	Chautauqua	11,616	17,713	18,205	2.8	4,424	3,800	795	5,973
Earlville	Chenango-Madison	711	864	1,004	16.2	NA	NA	NA	NA
East Aurora	Erie	2,366	5,253	6,791	29.3	494	2,339	4,904	7,506
East Elma[u]	Erie	NA	NA	52,362	NA	NA	NA	NA	NA
East Evans[u]	Erie	NA	NA	3,944	NA	NA	NA	NA	NA
East Greenbush[u]	Rensselaer	NA	NA	1,325	NA	NA	NA	NA	NA
East Hampton	Suffolk	NA	1,756	1,772	.9	NA	NA	NA	NA
East Herkimer[u]	Herkimer	NA	NA	1,068	NA	NA	NA	NA	NA
East Hills	Nassau	NA	343	7,184	1994.5	NA	252	−5,924	10,000+
East Massapequa[u]	Nassau	NA	NA	14,779	NA	NA	NA	NA	8,770
East Meadow[u]	Nassau	NA	NA	46,036	NA	NA	NA	NA	7,936
East Middletown[u]	Orange	NA	NA	8,752	NA	NA	NA	NA	NA
East Moriches[u]	Suffolk	NA	NA	1,210	NA	NA	NA	NA	NA
East Neck[u]	Suffolk	NA	NA	3,789	NA	NA	NA	NA	10,000+

NA: Not Available.
[u] Indicates unincorporated urban places.

URBAN PLACE DATA (Continued)

Urban Place	County	Population 1900 (No.)	Population 1940 (No.)	Population 1960 (No.)	Population Growth 1940-60 (Percentage)	Employment Mfg. 1958 (No.)	Employment Tertiary 1960 (No.)	Tertiary Surplus 1960 (No.)	Median Family Income (1959) (In Dollars)
East Northport[u]	Suffolk	NA	NA	8,381	NA	NA	NA	NA	$7,684
East Randolph	Cattaraugus	644	496	594	19.8	NA	NA	NA	NA
East Rochester	Monroe	NA	6,691	8,152	21.8	1,130	1,357	−1,367	6,625
East Rockaway	Nassau	739	5,610	10,721	91.1	240	938	−6,031	9,094
East Setauket[u]	Suffolk	NA	NA	1,127	NA	NA	NA	NA	NA
East Syracuse	Onondaga	2,509	4,520	4,708	4.2	1,563	1,688	3,732	6,208
East Williston	Nassau	NA	1,152	2,940	155.2	0	354	−1,170	10,000+
Eden[u]	Erie	NA	NA	2,366	NA	NA	NA	NA	NA
Edwards	St. Lawrence	373	624	658	5.4	NA	NA	NA	NA
Eggertsville[u]	Erie	NA	NA	44,807	NA	NA	NA	NA	8,895
Elba	Genesee	395	614	739	20.4	NA	NA	NA	NA
Elbridge	Onondaga	549	497	828	66.6	NA	NA	NA	NA
Elizabethtown	Essex	491	640	779	21.7	NA	NA	NA	NA
Ellenville	Ulster	2,879	4,000	5,003	25.1	680	1,796	3,977	5,509
Ellicottville	Cattaraugus	886	1,024	1,150	12.3	NA	NA	NA	NA
Elma[u]	Erie	NA	NA	12,766	NA	NA	NA	NA	NA
Elmira	Chemung	35,672	45,106	46,517	3.1	9,827	17,276	39,863	5,452
Elmira Heights	Chemung	1,763	4,829	5,157	6.8	2,597	913	−592	5,978
Elmira Southeast[u]	Chemung	NA	NA	6,698	NA	NA	NA	NA	5,953
Elmont	Nassau	NA	NA	30,138	NA	NA	NA	NA	7,494
Elmsford	Westchester	NA	3,078	3,795	23.3	99	1,589	4,150	7,084
Endicott	Broome	982[a]	17,702	18,775	6.1	13,998	6,938	15,915	6,304
Evans Center[u]	Erie	NA	NA	2,117	NA	NA	NA	NA	NA
Evans Mills	Jefferson	NA	523	618	18.2	NA	NA	NA	NA
Fair Haven	Cayuga	610	471	764	62.2	NA	NA	NA	NA
Fairport	Monroe	2,489	4,644	5,507	18.6	874	1,076	−127	7,151
Fairview[u]	Dutchess	NA	NA	8,626	NA	NA	NA	NA	6,894
Falconer	Chautauqua	1,136	3,222	3,343	3.8	1,221	595	−368	5,632
Farmingdale	Nassau	NA	3,524	6,128	73.9	2,088	4,606	16,902	7,650
Farmingville[u]	Suffolk	NA	NA	2,134	NA	NA	NA	NA	NA
Fayetteville	Onondaga	1,304	2,172	4,311	98.5	577	983	604	9,221
Fernwood[u]	Saratoga	NA	NA	2,108	NA	NA	NA	NA	NA
Fillmore	Allegany	NA	518	522	.8	NA	NA	NA	NA
Fishkill	Dutchess	589	720	1,033	43.5	NA	NA	NA	NA
Flanders	Suffolk	NA	NA	1,248	NA	NA	NA	NA	NA
Floral Park	Nassau	NA	12,950	17,499	35.1	205	4,339	4,196	8,463
Florida	Orange	NA	NA	1,550	NA	NA	NA	NA	NA
Flower Hill	Nassau	NA	666	4,594	589.8	0	741	−889	10,000+
Fonda	Montgomery	1,145	1,123	1,004	−10.6	NA	NA	NA	NA
Forestville	Chautauqua	623	692	905	30.8	NA	NA	NA	NA
Fort Covington	Franklin	822	813	976	20.0	NA	NA	NA	NA
Fort Edward	Washington	3,521	3,620	3,737	3.2	806	356	−1,957	6,409
Fort Johnson	Montgomery	NA	868	876	.9	NA	NA	NA	NA
Fort Plain	Montgomery	2,444	2,770	2,809	1.4	325	749	936	5,673
Frankfort	Herkimer	2,664	3,859	3,872	.3	239	301	−2,367	5,666
Franklin	Delaware	473	481	525	9.1	NA	NA	NA	NA
Franklin Square[u]	Nassau	NA	NA	32,483	NA	NA	NA	NA	7,973
Franklinville	Cattaraugus	1,360	1,884	2,124	12.7	NA	NA	NA	NA
Fredonia	Chautauqua	4,127	5,738	8,477	47.7	287	2,155	2,298	6,460
Freeport	Nassau	2,612	20,410	34,419	68.6	1,282	8,056	5,861	7,915
Freetown[u]	Suffolk	NA	NA	1,365	NA	NA	NA	NA	NA
Frewsburg[u]	Chautauqua	NA	NA	1,623	NA	NA	NA	NA	NA
Friendship	Allegany	1,214	1,148	1,231	7.2	NA	NA	NA	NA
Fulton	Oswego	8,206	12,462	14,261	6.7	4,617	3,223	1,854	5,954
Fultonville	Montgomery	977	806	815	1.1	NA	NA	NA	NA
Garden City	Nassau	NA	11,223	23,948	113.4	10,781	17,680	64,452	13,875
Garden City Park-Herricks[u]	Nassau	NA	NA	15,364	NA	NA	NA	NA	9,213

NA: Not Available. [u]Indicates unincorporated urban places.

URBAN PLACE DATA (Continued)

Urban Place	County	Population 1900 (No.)	Population 1940 (No.)	Population 1960 (No.)	Population Growth 1940-60 (Percentage)	Employment Mfg. 1958 (No.)	Employment Tertiary 1960 (No.)	Tertiary Surplus 1960 (No.)	Median Family Income (1959) (In Dollar)
Geneseo	Livingston	2,400	2,144	3,284	53.2	15	667	51	$6,620
Geneva	Ontario	10,433	15,555	17,286	11.1	2,115	5,204	8,734	5,627
Gilbertsville	Otsego	476	377	522	38.5	NA	NA	NA	NA
Glen Cove	Nassau	NA	12,415	23,817	91.8	2,518	5,370	3,033	7,549
Glen Park	Jefferson	494	523	561	7.3	NA	NA	NA	NA
Glenwood Park[u]	Orange	NA	NA	1,317	NA	NA	NA	NA	NA
Gloversville	Fulton	18,349	23,329	21,741	−6.8	4,972	5,035	3,434	5,432
Goshen	Orange	2,826	3,073	3,906	27.1	52	1,427	3,229	5,674
Gouverneur	St. Lawrence	3,689	4,478	4,946	10.5	NA	1,382	1,964	5,100
Gowanda	Cattaraugus-Erie	1,475	3,156	3,352	6.2	820	1,510	4,198	6,000
Granville	Washington	2,700	3,173	2,715	−14.4	61	850	1,535	5,054
Great Neck	Nassau	NA	6,167	10,171	64.9	373	3,687	8,264	10,622
Great Neck Estates	Nassau	NA	1,969	3,262	65.7	0	668	78	10,000+
Great Neck Plaza	Nassau	NA	2,031	4,948	143.6	NA	4,414	17,122	8,648
Greene	Chenango	1,236	1,431	2,051	43.3	NA	NA	NA	NA
Green Island	Albany	4,770	3,988	3,533	−11.4	1,488	392	−1,573	6,161
Greenlawn[u]	Suffolk	NA	NA	5,422	NA	NA	NA	NA	7,799
Greenport	Suffolk	2,366	3,259	2,606	−20.0	56	1,359	4,187	4,871
Greenwich	Washington	1,869	2,270	2,263	−.3	NA	NA	NA	NA
Greenwood Lake	Orange	NA	483	1,236	155.9	NA	NA	NA	NA
Groton	Tompkins	1,344	2,087	2,123	1.7	NA	NA	NA	NA
Hagaman	Montgomery	646	933	1,292	38.5	NA	NA	NA	NA
Halesite[u]	Suffolk	NA	NA	2,857	NA	NA	NA	NA	8,982
Hamburg	Erie	1,683	5,467	9,145	67.3	3,207	4,268	12,195	8,242
Hamburg-Lake Shore[u]	Erie	NA	NA	11,527	NA	NA	NA	NA	7,250
Hamilton	Madison	1,627	1,790	3,348	87.0	NA	1,120	2,252	6,709
Hammondsport	Steuben	1,169	1,112	1,176	5.8	NA	NA	NA	NA
Hampton Bays[u]	Suffolk	NA	NA	1,431	NA	NA	NA	NA	NA
Hancock	Delaware	1,283	1,581	1,830	15.7	NA	NA	NA	NA
Hannibal	Oswego	410	437	611	39.8	NA	NA	NA	NA
Harriman	Orange	NA	703	752	7.0	NA	NA	NA	NA
Harris Hill[u]	Erie	NA	NA	3,944	NA	NA	NA	NA	9,641
Harrisville	Lewis	639	832	842	1.2	NA	NA	NA	NA
Hastings-on-Hudson	Westchester	2,002	7,057	8,979	27.2	1,276	884	−4,559	9,030
Haverstraw	Rockland	5,935	5,909	5,771	−2.3	574	2,154	4,999	5,996
Head of the Harbor	Suffolk		255	524	105.5	NA	NA	NA	NA
Hempstead	Nassau	3,582	20,856	34,641	65.2	899	26,758	99,149	7,455
Henrietta-Northeast[u]	Monroe	NA	NA	6,403	NA	NA	NA	NA	7,873
Herkimer	Herkimer	5,555	9,617	9,396	−2.3	1,256	2,531	3,259	5,769
Hermon	St. Lawrence	503	487	612	25.7	NA	NA	NA	NA
Heuvelton	St. Lawrence	NA	620	810	30.6	NA	NA	NA	NA
Hewlett Bay Park	Nassau	NA	438	520	18.7	NA	NA	NA	NA
Hewlett Harbor	Nassau	NA	288	1,610	459.0	NA	NA	NA	NA
Hewlett Neck	Nassau	NA	252	507	101.2	NA	NA	NA	NA
Hicksville[u]	Nassau	NA	NA	50,405	NA	NA	NA	NA	7,908
Highland[u]	Ulster	NA	NA	2,931	NA	NA	NA	NA	6,188
Highland Falls	Orange	NA	3,711	4,469	20.4	NA	1,553	3,296	5,876
Hillburn	Rockland	824	1,161	1,114	−4.0	NA	NA	NA	NA
Hilton	Monroe	486	895	1,334	49.1	NA	NA	NA	NA
Hobart	Delaware	550	638	585	−8.3	NA	NA	NA	NA
Holbrook[u]	Suffolk	NA	NA	3,441	NA	NA	NA	NA	6,206
Holland Patent	Oneida	352	388	538	38.7	NA	NA	NA	NA

NA: Not Available.
[u] Indicates unincorporated urban places.

URBAN PLACE DATA (Continued)

Urban Place	County	Population 1900 (No.)	Population 1940 (No.)	Population 1960 (No.)	Population Growth 1940-60 (Percentage)	Employment Mfg. 1958 (No.)	Employment Tertiary 1960 (No.)	Tertiary Surplus 1960 (No.)	Median Family Income (1959) (In Dollars)
Holley	Orleans	1,380	1,230	1,788	45.4	NA	NA	NA	NA
Homer	Cortland	2,381	2,928	3,622	23.7	464	783	293	$6,018
Honeoye Falls	Monroe	1,175	1,274	2,143	68.4	NA	NA	NA	NA
Hoosick Falls	Rensselaer	5,671	4,279	4,023	−6.0	990	941	682	5,333
Hornell	Steuben	11,918	15,649	13,907	−11.1	998	5,913	15,658	5,541
Horseheads	Chemung	1,901	2,570	7,207	180.2	1,317	1,659	1,088	6,500
Hudson	Columbia	9,528	11,517	11,075	−3.8	1,475	3,365	5,750	5,263
Hudson Falls	Washington	4,473	6,654	7,752	16.5	2,198	1,054	−2,482	5,886
Huntington[u]	Suffolk	NA	NA	11,255	NA	NA	NA	NA	8,305
Huntington Bay	Suffolk	NA	408	1,267	210.5	NA	NA	NA	NA
Huntington Station[u]	Suffolk	NA	NA	23,438	NA	NA	NA	NA	7,192
Hyde Park[u]	Dutchess	NA	NA	1,976	NA	NA	NA	NA	NA
Ilion	Herkimer	5,138	8,927	10,199	14.2	3,395	1,708	−1,659	5,797
Interlaken	Seneca	NA	661	780	18.0	NA	NA	NA	NA
Inwood[u]	Nassau	NA	NA	10,362	NA	NA	NA	NA	5,922
Irvington	Westchester	2,231	3,272	5,494	67.9	253	891	−1,039	9,798
Island Park	Nassau	NA	1,531	3,846	151.2	75	951	909	6,712
Ithaca	Tompkins	13,136	19,730	28,799	46.0	1,898	17,168	57,041	6,125
Jamestown	Chautauqua	22,892	42,838	41,818	−1.9	10,627	12,085	18,607	5,607
Jericho[u]	Nassau	NA	NA	10,795	NA	NA	NA	NA	10,341
Johnson City	Broome	3,111	18,039	19,118	6.0	7,429	4,280	2,282	6,057
Johnstown	Fulton	10,130	10,666	10,390	−2.6	2,618	2,187	545	5,660
Jordan	Onondaga	1,118	1,115	1,390	24.7	NA	NA	NA	NA
Keeseville	Clinton-Essex	2,110	1,921	2,213	15.2	NA	NA	NA	NA
Kenmore	Erie	318	18,612	21,261	14.2	314	6,606	11,769	7,244
Kensington	Nassau	NA	933	1,166	25.0	NA	NA	NA	NA
Kinderhook	Columbia	913	745	1,078	44.7	NA	NA	NA	NA
Kings Park[u]	Suffolk	NA	NA	4,949	NA	NA	NA	NA	7,086
Kings Point	Nassau	NA	1,247	5,410	NA	0	852	−1,150	10,000+
Kingston	Ulster	24,535	28,589	29,260	2.3	3,027	9,128	16,380	5,875
Lackawanna	Erie	NA	24,058	29,564	22.9	14,215	6,227	1,571	6,058
Lacona	Oswego	388	413	556	34.6	NA	NA	NA	NA
Lake Carmel[u]	Putnam	NA	NA	2,735	NA	NA	644	485	5,813
Lake Erie Beach[u]	Erie	NA	NA	2,117	NA	NA	NA	NA	NA
Lake George	Warren	NA	803	1,026	28.8	NA	NA	NA	NA
Lake Katrine[u]	Ulster	NA	NA	1,149	NA	NA	NA	NA	NA
Lake Placid	Essex	NA	3,136	2,998	−4.4	29	1,599	4,997	4,900
Lake Ronkonkoma[u]	Suffolk	NA	NA	4,841	NA	NA	NA	NA	5,854
Lake Shenorock[u]	Westchester	NA	NA	1,402	NA	NA	NA	NA	NA
Lake Success	Nassau	NA	203	2,954	1355.2	NA	305	−1,429	10,000+
Lakewood	Chautauqua	574	2,314	3,933	70.0	106	552	−1,173	6,267
Lancaster	Erie	3,750	7,236	12,254	69.3	650	2,126	−1,624	6,999
Larchmont	Westchester	945	5,970	6,789	13.7	23	3,433	10,376	10,000+
Lattingtown	Nassau	NA	613	1,461	138.3	NA	NA	NA	NA
Laurel Hollow	Nassau	NA	110	834	658.2	NA	NA	NA	NA
Lawrence	Nassau	558	3,649	5,907	61.9	102	2,160	4,893	10,000+
Le Roy	Genesee	3,144	4,413	4,662	5.6	680	1,209	1,383	6,516
Levittown[u]	Nassau	NA	NA	65,276	NA	NA	NA	NA	7,467
Lewiston	Niagara	697	1,280	3,320	159.4	66	916	1,260	8,078
Liberty	Sullivan	1,760	3,788	4,704	24.2	65	1,892	4,756	5,758
Lima	Livingston	949	942	1,366	45.0	NA	NA	NA	NA
Limestone	Cattaraugus	732	558	539	−3.4	NA	NA	NA	NA
Lincoln Park[u]	Ulster	NA	NA	2,707	NA	NA	NA	NA	6,784
Lindenhurst	Suffolk	NA	4,756	20,905	339.6	1,144	3,073	−5,540	6,705
Little Falls	Herkimer	10,381	10,163	8,935	−12.1	2,340	1,925	690	5,518

NA: Not Available.
[u] Indicates unincorporated urban places.

URBAN PLACE DATA (Continued)

Urban Place	County	Population 1900 (No.)	Population 1940 (No.)	Population 1960 (No.)	Population Growth 1940-60 (Percentage)	Employment Mfg. 1958 (No.)	Employment Tertiary 1960 (No.)	Tertiary Surplus 1960 (No.)	Median Family Income (1959) (In Dollars)
Little Valley	Cattaraugus	1,085	1,234	1,244	.8	NA	NA	NA	NA
Liverpool	Onondaga	1,133	2,500	3,487	39.5	60	1,660	4,813	$7,442
Livingston Manor	Sullivan	NA	NA	2,080	NA	NA	NA	NA	NA
Livonia	Livingston	865	751	946	26.0	NA	NA	NA	NA
Lloyd Harbor	Suffolk	NA	603	2,521	318.1	NA	279	-1,126	10,000+
Lockport	Niagara	16,581	24,379	26,443	8.5	7,411	7,029	8,702	6,584
Locust Grove[u]	Nassau	NA	NA	11,558	NA	NA	NA	NA	9,963
Long Beach	Nassau	NA	9,036	26,473	193.0	260	3,702	-7,963	7,321
Lowville	Lewis	2,352	3,578	3,616	1.1	326	1,162	2,194	5,181
Lynbrook	Nassau	NA	14,557	19,881	36.6	1,054	7,157	15,904	8,090
Lyndonville	Orleans	NA	745	755	1.3	NA	NA	NA	NA
Lyons	Wayne	4,300	3,863	4,673	21.0	203	1,027	462	5,890
Lyons Falls	Lewis	470	818	887	8.4	NA	NA	NA	NA
Macedon	Wayne	592	557	645	15.8	NA	NA	NA	NA
Mahopac[u]	Putnam	NA	NA	1,337	NA	NA	NA	NA	NA
Malone	Franklin	5,935	8,743	8,737	-.1	537	3,072	6,623	5,109
Malverne	Nassau	NA	5,153	9,968	93.4	NA	889	-5,523	10,000+
Mamaroneck	Westchester	NA	13,034	17,673	35.6	882	4,864	6,647	7,642
Manchester	Ontario	711	1,330	1,344	1.1	NA	NA	NA	NA
Manlius	Onondaga	1,219	1,520	1,997	31.4	NA	NA	NA	NA
Manorhaven	Nassau	NA	484	3,566	636.8	127	128	-2,926	6,781
Marathon	Cortland	1,092	955	1,079	13.0	NA	NA	NA	NA
Marcellus	Onondaga	589	1,112	1,697	52.6	NA	NA	NA	NA
Margaretville	Delaware	640	812	833	2.6	NA	NA	NA	NA
Marilla	Erie	NA	NA	23,138	NA	NA	NA	NA	NA
Marlboro[u]	Ulster	NA	NA	1,733	NA	NA	NA	NA	NA
Massapequa[u]	Nassau	NA	NA	32,900	NA	NA	NA	NA	8,336
Massapequa Park	Nassau	NA	488	19,904	3978.7	NA	2,371	-8,049	8,082
Massena	St. Lawrence	2,032	11,328	15,478	36.6	185	4,304	6,042	6,856
Mastic Beach[u]	Suffolk	NA	NA	3,035	NA	NA	NA	NA	4,810
Mastic Shirley[u]	Suffolk	NA	NA	3,397	NA	NA	NA	NA	4,900
Matinecock	Nassau	NA	428	824	92.5	NA	NA	NA	NA
Mattituck[u]	Suffolk	NA	NA	1,274	NA	NA	NA	NA	NA
Maybrook	Orange	NA	1,189	1,348	13.4	NA	NA	NA	NA
Mayfield	Fulton	589	759	818	7.8	NA	NA	NA	NA
Mayville	Chautauqua	943	1,354	1,619	19.6	NA	NA	NA	NA
McGraw	Cortland	750	1,201	1,276	6.2	NA	NA	NA	NA
Mechanicville	Saratoga	4,694	7,449	6,831	-8.3	1,361	2,044	3,389	5,307
Medina	Orleans	4,719	5,871	6,681	13.8	842	2,260	4,619	6,202
Melrose Park[u]	Cayuga	NA	NA	2,058	NA	NA	NA	NA	NA
Menands	Albany	NA	1,764	2,314	31.2	NA	NA	NA	NA
Merrick[u]	Nassau	NA	NA	18,789	NA	NA	NA	NA	9,258
Mexico	Oswego	1,249	1,348	1,465	8.7	NA	NA	NA	NA
Middleburg	Schoharie	1,135	1,074	1,317	22.6	NA	NA	NA	NA
Middleport	Niagara	1,431	1,575	1,882	19.5	NA	NA	NA	NA
Middletown	Orange	14,522	21,908	23,475	7.2	2,754	8,114	17,095	5,622
Middleville	Herkimer	667	612	648	5.9	NA	NA	NA	NA
Milford	Otsego	532	460	548	19.1	NA	NA	NA	NA
Millbrook	Dutchess	1,027	1,340	1,717	28.1	NA	NA	NA	NA
Millerton	Dutchess	802	953	1,027	7.8	NA	NA	NA	NA
Mill Neck	Nassau	NA	101	701	594.1	NA	NA	NA	NA
Mineola	Nassau	NA	10,064	20,519	103.9	4,296	7,541	17,186	7,860
Mineville[u]	Essex	NA	NA	1,187	NA	NA	NA	NA	NA
Minoa	Onondaga	NA	902	1,838	103.8	NA	NA	NA	NA
Mohawk	Herkimer	2,028	2,882	3,533	22.6	213	398	-1,543	5,854
Monroe	Orange	796	1,616	3,323	105.6	39	1,334	3,347	6,533
Montgomery	Orange	973	844	1,312	55.5	NA	NA	NA	NA
Monticello	Sullivan	1,160	3,737	5,222	39.7	109	2,485	7,203	5,425

NA: Not Available.
[u] Indicates unincorporated urban places.

URBAN PLACE DATA (Continued)

Urban Place	County	Population 1900 (No.)	Population 1940 (No.)	Population 1960 (No.)	Population Growth 1940-60 (Percentage)	Employment Mfg. 1958 (No.)	Employment Tertiary 1960 (No.)	Tertiary Surplus 1960 (No.)	Median Family Income (1959) (In Dollars)
Montour Falls	Schuyler	1,193	1,345	1,533	14.0	NA	NA	NA	NA
Mooers	Clinton	527	477	543	13.8	NA	NA	NA	NA
Moravia	Cayuga	1,442	1,231	1,575	27.9	NA	NA	NA	NA
Morris	Otsego	553	599	677	NA	NA	NA	NA	NA
Morristown	St. Lawrence	466	540	541	.2	NA	NA	NA	NA
Morrisville	Madison	624	666	1,304	95.8	NA	NA	NA	NA
Mount Kisco	Westchester	1,346	5,941	6,805	14.5	402	2,959	7,990	$6,841
Mount Morris	Livingston	2,410	3,530	3,250	−7.9	121	1,287	3,185	5,633
Mount Vernon	Westchester	21,228	67,362	76,010	12.8	4,333	21,201	29,995	6,873
Munsey Park	Nassau	NA	1,456	2,847	95.5	0	383	−932	10,000+
Muttontown	Nassau	NA	335	1,265	277.6	NA	NA	NA	NA
Naples	Ontario	1,048	1,152	1,237	7.4	NA	NA	NA	NA
Nassau	Rensselaer	1,418	698	1,248	78.8	NA	NA	NA	NA
Nelliston	Montgomery	634	638	729	14.3	NA	NA	NA	NA
Nelsonville	Putnam	624	457	555	21.4	NA	NA	NA	NA
Nesconset[u]	Suffolk	NA	NA	1,964	NA	NA	NA	NA	NA
Newark	Wayne	4,578	9,646	12,868	33.4	942	9,288	33,572	6,487
Newark Valley	Tioga	818	949	1,234	30.0	NA	NA	NA	NA
New Berlin	Chenango	1,156	999	1,262	26.3	NA	NA	NA	NA
Newburgh	Orange	24,943	31,883	30,979	−2.8	6,185	11,882	28,431	5,363
Newfane	Niagara	NA	NA	1,423	NA	NA	NA	NA	NA
New Hartford	Oneida	1,007	1,914	2,468	28.9	NA	NA	NA	NA
New Hyde Park	Nassau	NA	4,691	10,808	130.4	2,620	7,728	27,832	8,155
New Paltz	Ulster	1,022	1,492	3,041	103.8	19	1,268	3,299	6,117
Newport	Herkimer	610	627	827	31.9	NA	NA	NA	NA
New Rochelle	Westchester	14,720	58,408	76,812	31.5	3,463	18,596	16,168	8,131
New Windsor[u]	Orange	NA	NA	4,041	NA	NA	NA	NA	6,673
NEW YORK URBAN SYSTEM[b] (SMSA)		3,992,103	8,984,127	11,086,097	23.4	1,112,836	3,495,490	6,391,353	NA
Urbanized Area		NA	NA	14,114,927	NA	NA	NA	NA	6,675
New York City		3,437,202	7,454,995	7,781,984	4.4	895,838	2,712,121	5,778,621	6,091
New York Mills	Oneida	NA	3,628	3,788	4.4	257	531	−1,133	6,133
Niagara Falls	Niagara	19,657	78,029	102,394	31.2	21,264	22,419	9,701	6,630
Nichols	Tioga	NA	541	663	22.6	NA	NA	NA	NA
Norfolk[u]	St. Lawrence	NA	NA	1,353	NA	NA	NA	NA	NA
North Bellmore[u]	Nassau	NA	NA	19,369	NA	NA	NA	NA	8,156
North Collins	Erie	NA	1,182	1,574	33.2	NA	NA	NA	NA
North Evans[u]	Erie	NA	NA	11,527	NA	NA	NA	NA	NA
North Hornell	Steuben	NA	589	917	55.7	NA	NA	NA	NA
North Merrick[u]	Nassau	NA	NA	12,976	NA	NA	NA	NA	8,835
North New Hyde Park[u]	Nassau	NA	NA	17,929	NA	NA	NA	NA	8,967
North Pelham	Westchester	684	5,052	5,326	5.4	347	557	−2,541	7,494
Northport	Suffolk	1,794	3,093	5,972	93.1	457	1,447	1,263	8,014
North Syracuse	Onondaga	NA	2,083	7,412	255.8	6	1,616	668	7,164
North Tarrytown	Westchester	4,241	8,804	8,818	.2	5,241	1,500	−1,318	7,277
North Tonawanda	Niagara	9,069	20,254	34,757	71.6	6,733	4,167	−13,922	6,554
North Valley Stream[u]	Nassau	NA	NA	17,239	NA	NA	NA	NA	8,794
Northville	Fulton	1,046	1,111	1,156	4.1	NA	NA	NA	NA
Norwich	Chenango	5,766	8,694	9,175	5.5	2,214	3,351	7,580	6,200
Norwood	St. Lawrence	1,714	1,905	2,200	15.5	NA	NA	NA	NA
Nunda	Livingston	1,018	1,077	1,224	13.6	NA	NA	NA	NA
Nyack	Rockland	4,275	5,206	6,062	16.4	248	3,272	10,298	6,221
Oakfield	Genesee	714	1,876	2,070	10.3	NA	NA	NA	NA
Oceanside[u]	Nassau	NA	NA	30,448	NA	NA	NA	NA	8,368
Odessa	Schuyler	NA	424	530	25.0	NA	NA	NA	NA
Ogdensburg	St. Lawrence	12,633	16,346	16,122	−1.4	1,036	3,506	1,408	5,606
Olcott[u]	Niagara	NA	NA	1,215	NA	NA	NA	NA	NA

NA: Not Available.
[u] Indicates unincorporated urban places.
[b] New York State portion only.

URBAN PLACE DATA (Continued)

Urban Place	County	Population 1900 (No.)	Population 1940 (No.)	Population 1960 (No.)	Population Growth 1940-60 (Percentage)	Employment Mfg. 1958 (No.)	Employment Tertiary 1960 (No.)	Tertiary Surplus 1960 (No.)	Median Family Income (1959) (In Dollars)
Old Brookville	Nassau	NA	356	1,126	216.3	NA	NA	NA	NA
Old Westbury	Nassau	NA	1,017	2,064	102.9	NA	NA	NA	NA
Olean	Cattaraugus	9,462	21,506	21,868	1.7	4,669	6,771	11,987	$5,643
Oneida	Madison	7,538	10,291	11,677	13.5	864	2,954	3,093	5,647
Oneida Castle	Oneida	291	556	754	35.6	NA	NA	NA	NA
Oneonta	Otsego	7,147	11,731	13,412	14.3	767	4,684	10,008	5,442
Orchard Park	Erie	NA	1,304	3,278	151.4	133	1,215	2,797	9,456
Oriskany	Oneida	NA	1,115	1,580	41.7	NA	NA	NA	NA
Oriskany Falls	Oneida	811	930	972	4.5	NA	NA	NA	NA
Ossining	Westchester	7,939	15,996	18,662	16.7	739	5,026	6,468	6,882
Oswego	Oswego	22,199	22,062	22,155	.4	2,078	4,971	2,700	5,689
Otego	Otsego	658	580	875	50.8	NA	NA	NA	NA
Otisville	Orange	NA	889	896	.8	NA	NA	NA	NA
Ovid	Seneca	624	578	789	36.5	NA	NA	NA	NA
Owego	Tioga	2,113	1,782	2,814	57.9	693	1,597	2,568	5,587
Oxford	Chenango	1,931	1,713	1,871	9.2	NA	NA	NA	NA
Oyster Bay Cove	Nassau	NA	466	988	112.0	NA	NA	NA	NA
Painted Post	Steuben	775	2,337	2,570	10.0	NA	256	−1,290	7,309
Palatine Bridge	Montgomery	360	585	578	−1.2	NA	NA	NA	NA
Palmyra	Wayne	1,937	2,709	3,476	28.3	120	776	404	6,596
Parish	Oswego	548	521	567	8.8	NA	NA	NA	NA
Patchogue	Suffolk	2,926	7,181	8,838	23.1	792	1,288	−2,398	6,115
Pawling	Dutchess	781	1,446	1,734	19.9	NA	NA	NA	NA
Peekskill	Westchester	10,358	17,311	18,737	8.2	2,050	8,152	22,023	6,427
Pelham	Westchester	303	1,918	1,964	2.4	NA	NA	NA	NA
Pelham Manor	Westchester	NA	5,302	6,114	15.3	1,547	1,337	571	10,000+
Penn Yan	Yates	4,650	5,308	5,770	8.7	823	1,699	2,725	5,368
Perry	Wyoming	2,763	4,468	4,629	3.6	1,124	961	176	5,103
Phelps	Ontario	1,306	1,499	1,887	25.9	NA	NA	NA	NA
Philadelphia	Jefferson	873	722	868	20.2	NA	NA	NA	NA
Philmont	Columbia	1,964	1,679	1,750	4.2	NA	NA	NA	NA
Phoenix	Oswego	1,532	1,757	2,408	37.1	NA	NA	NA	NA
Piermont	Rockland	1,153	1,876	1,906	1.6	NA	NA	NA	NA
Pine Bush[u]	Orange	NA	NA	1,016	NA	NA	NA	NA	NA
Pittsford	Monroe	1,000	1,544	1,749	13.3	NA	NA	NA	NA
Plainedge[u]	Nassau	NA	NA	21,973	NA	NA	NA	NA	7,972
Plainview[u]	Nassau	NA	NA	27,710	NA	NA	NA	NA	8,998
Plandome	Nassau	NA	897	1,379	53.7	NA	NA	NA	NA
Plandome Heights	Nassau	NA	317	1,025	223.3	NA	NA	NA	NA
Plandome Manor	Nassau	NA	262	705	169.1	NA	NA	NA	NA
Plattsburgh	Clinton	8,434	16,351	20,172	23.4	1,474	6,976	14,708	5,616
Pleasantville	Westchester	1,204	4,454	5,877	31.9	798	1,571	1,978	8,470
Poland	Herkimer	332	478	564	18.0	NA	NA	NA	NA
Pontiac[u]	Erie	NA	NA	1,078	NA	NA	NA	NA	NA
Port Byron	Cayuga	1,013	961	1,201	25.0	NA	NA	NA	NA
Port Chester	Westchester	7,440	23,073	24,960	8.2	4,254	6,127	5,675	6,644
Port Dickinson	Broome	379	2,436	2,295	−5.8	NA	NA	NA	NA
Port Ewen[u]	Ulster	NA	NA	2,622	NA	NA	NA	NA	5,979
Port Henry	Essex	1,751	1,935	1,767	−8.7	NA	NA	NA	NA
Port Jefferson[u]	Suffolk	NA	NA	2,336	NA	NA	NA	NA	NA
Port Jefferson Station[u]	Suffolk	NA	NA	1,041	NA	NA	NA	NA	NA
Port Jervis	Orange	9,385	9,749	9,268	−4.9	2,100	2,100	5,057	5,137
Port Leyden	Lewis	746	794	898	13.1	NA	NA	NA	NA
Portville	Cattaraugus	748	1,018	1,336	31.2	NA	NA	NA	NA
Port Washington[u]	Nassau	NA	NA	15,657	NA	NA	NA	NA	10,815

NA: Not Available.
[u] Indicates unincorporated urban places.

URBAN PLACE DATA (Continued)

Urban Place	County	Population 1900 (No.)	Population 1940 (No.)	Population 1960 (No.)	Population Growth 1940-60 (Percentage)	Employment Mfg. 1958 (No.)	Employment Tertiary 1960 (No.)	Tertiary Surplus 1960 (No.)	Median Family Income (1959) (In Dollars)
Port Washington North	Nassau	NA	628	722	15.0	NA	NA	NA	NA
Potsdam	St. Lawrence	3,843	4,821	7,765	61.1	149	2,386	4,165	$6,414
Poughkeepsie	Dutchess	24,029	40,478	38,330	-5.3	4,885	16,256	42,950	5,893
Plattsburgh	Steuben	713	635	690	8.7	NA	NA	NA	NA
Pulaski	Oswego	1,493	1,895	2,256	19.1	NA	NA	NA	NA
Quogue	Suffolk	NA	633	692	9.3	NA	NA	NA	NA
Randolph	Cattaraugus	1,209	1,321	1,414	7.0	NA	NA	NA	NA
Ravena	Albany	NA	1,810	2,410	33.1	NA	NA	NA	NA
Red Creek	Wayne	480	539	689	27.8	NA	NA	NA	NA
Red Hook	Dutchess	857	1,056	1,719	62.8	NA	NA	NA	NA
Remsen	Oneida	389	422	567	34.4	NA	NA	NA	NA
Rensselaer	Rensselaer	7,466	10,768	10,506	-2.4	2,760	3,103	5,009	5,590
Rhinebeck	Dutchess	1,494	1,697	2,093	23.3	NA	NA	NA	NA
Richfield Springs	Otsego	1,437	1,209	1,630	34.8	NA	NA	NA	NA
Richmondville	Schoharie	651	598	743	24.2	NA	NA	NA	NA
Ripley	Chautauqua	NA	NA	1,247	NA	NA	NA	NA	NA
Riverhead[u]	Suffolk	NA	NA	5,830	NA	NA	2,041	4,375	5,867
Riverside	Steuben	NA	643	1,030	NA	NA	NA	NA	NA
Rochdale	Dutchess	NA	NA	1,800	NA	NA	NA	NA	NA
ROCHESTER URBAN SYSTEM (SMSA)		217,854	438,230	586,387	33.8	100,876	131,607	71,728	7,147
Urbanized Area		NA	NA	493,402	NA	NA	126,465	138,923	7,098
Rochester	Monroe	162,608	324,975	318,611	-2.0	93,637	113,558	249,179	6,361
Rockville Center	Nassau	1,884	18,613	26,355	41.6	256	8,491	16,100	10,922
Rocky Point[u]	Suffolk	NA	NA	2,261	NA	NA	NA	NA	NA
Rome	Oneida	15,343	34,214	51,646	50.9	5,870	13,371	15,209	6,255
Ronkonkoma[u]	Suffolk	NA	NA	4,220	NA	NA	NA	NA	6,196
Ronkonkoma West[u]	Suffolk	NA	NA	1,446	NA	NA	NA	NA	NA
Roosevelt[u]	Nassau	NA	NA	12,883	NA	NA	NA	NA	7,111
Rosendale	Ulster	1,840	671	1,033	53.9	NA	NA	NA	NA
Roslyn	Nassau	NA	972	2,681	175.8	230	1,221	3,424	8,732
Roslyn Estates	Nassau	NA	464	1,289	177.8	NA	NA	NA	NA
Roslyn Harbor	Nassau	NA	303	925	205.3	NA	NA	NA	NA
Rotterdam[u]	Schenectady	NA	NA	16,871	NA	NA	NA	NA	6,537
Rouses Point	Clinton	1,675	1,846	2,160	17.0	NA	NA	NA	NA
Russell Gardens	Nassau	NA	556	1,156	107.9	NA	NA	NA	NA
Rye	Westchester	NA	9,865	14,225	44.2	134	2,999	770	11,205
Sackets Harbor	Jefferson	1,266	1,962	1,279	-34.8	NA	NA	NA	NA
Saddle Rock	Nassau	NA	69	1,109	1507.2	NA	NA	NA	NA
Sag Harbor	Suffolk	1,969	2,517	2,346	-6.8	NA	NA	NA	NA
St. James[u]	Suffolk	NA	NA	3,524	NA	NA	NA	NA	6,023
St. Johnsville	Montgomery	1,873	2,283	2,169	-5.0	NA	NA	NA	NA
Salamanca	Cattaraugus	4,251	9,011	8,480	-5.9	938	2,363	3,335	5,589
Salem	Washington	1,391	1,034	1,076	4.1	NA	NA	NA	NA
Sands Point	Nassau	NA	628	2,161	244.1	NA	NA	NA	NA
Sandy Creek	Oswego	692	646	697	7.9	NA	NA	NA	NA
San Remo[u]	Suffolk	NA	NA	11,996	NA	NA	NA	NA	7,300
Saranac Lake	Essex-Franklin	2,594	7,138	6,421	-10.0	125	2,370	5,429	4,753
Saratoga Springs	Saratoga	12,409	13,705	16,630	21.3	1,216	5,610	11,420	5,523
Saugerties	Ulster	3,697	3,916	4,286	9.4	614	1,416	2,794	5,964
Sauquoit[u]	Oneida	NA	NA	1,715	NA	NA	NA	NA	NA
Savannah	Wayne	573	601	602	.2	NA	NA	NA	NA
Savona	Steuben	611	653	904	38.4	NA	NA	NA	NA
Scarsdale	Westchester	NA	12,966	17,968	38.6	43	4,817	6,117	22,177
Schaghticoke	Rensselaer	1,061	603	720	19.4	NA	NA	NA	NA

NA: Not Available.
[u] Indicates unincorporated urban places.

URBAN PLACE DATA (Continued)

Urban Place	County	Population 1900 (No.)	Population 1940 (No.)	Population 1960 (No.)	Population Growth 1940-60 (Percentage)	Employment Mfg. 1958 (No.)	Employment Tertiary 1960 (No.)	Tertiary Surplus 1960 (No.)	Median Family Income (1959) (In Dollars)
Schenectady	Schenectady	31,682	87,549	81,682	-6.7	22,268	28,529	60,963	$5,925
Schoharie	Schoharie	1,006	941	1,168	24.1	NA	NA	NA	NA
Schuylerville	Saratoga	1,601	1,447	1,361	-5.9	NA	NA	NA	NA
Scotia	Schenectady	NA	7,960	7,625	-4.2	34	1,731	1,030	7,075
Scottsville	Monroe	NA	925	1,863	101.4	NA	NA	NA	NA
Scranton[u]	Erie	NA	NA	1,078	NA	NA	NA	NA	NA
Sea Cliff	Nassau	1,558	4,416	5,669	28.4	13	621	-2,564	8,321
Seaford[u]	Nassau	NA	NA	14,718	NA	NA	NA	NA	8,236
Selden[u]	Suffolk	NA	NA	1,604	NA	NA	NA	NA	NA
Seneca Falls	Seneca	6,519	6,452	7,439	15.3	2,721	1,633	726	6,385
Setauket[u]	Suffolk	NA	NA	1,207	NA	NA	NA	NA	NA
Sherburne	Chenango	899	1,192	1,647	38.2	NA	NA	NA	NA
Sherman	Chautauqua	760	675	873	29.3	NA	NA	NA	NA
Sherrill	Oneida	2,150	2,184	2,922	33.8	NA	261	-1,617	6,891
Shortsville	Ontario	922	1,316	1,382	5.0	NA	NA	NA	NA
Shrub Oak[u]	Westchester	NA	NA	1,874	NA	NA	NA	NA	NA
Sidney	Delaware	2,331	3,012	5,157	71.2	4,277	1,141	548	6,517
Silver Creek	Chautauqua	1,944	3,067	3,310	7.9	838	1,120	2,290	6,043
Silver Springs	Wyoming	667	766	726	-5.2	NA	NA	NA	NA
Sinclairville	Chautauqua	577	585	726	24.1	NA	NA	NA	NA
Skaneateles	Onondaga	1,495	1,949	2,921	49.9	59	1,124	2,699	7,669
Sloan	Erie	873	3,836	5,803	51.3	166	637	-2,618	6,378
Sloatsburg	Rockland	NA	1,771	2,565	44.8	NA	298	-1,075	6,956
Smithtown Branch[u]	Suffolk	NA	NA	1,986	NA	NA	NA	NA	NA
Sodus	Wayne	NA	1,513	1,645	8.7	NA	NA	NA	NA
Sodus Point	Wayne	NA	NA	868	NA	NA	NA	NA	NA
Solvay	Onondaga	3,493	8,201	8,732	6.5	3,511	1,421	-1,627	6,658
Sound Beach[u]	Suffolk	NA	NA	1,625	NA	NA	NA	NA	NA
Southampton	Suffolk	2,289	3,818	4,582	20.0	34	1,819	4,463	6,081
South Corning	Steuben	NA	681	1,448	112.6	NA	NA	NA	NA
South Dayton	Cattaraugus	NA	643	696	8.2	NA	NA	NA	NA
South Fallsburgh[u]	Sullivan	NA	NA	1,290	NA	NA	NA	NA	NA
South Farmingdale[u]	Nassau	NA	NA	16,318	NA	NA	NA	NA	7,864
South Floral Park	Nassau	NA	510	1,090	113.7	NA	NA	NA	NA
South Glens Falls	Saratoga	2,025	3,081	4,129	34.0	NA	755	-354	5,966
South Huntington[u]	Suffolk	NA	NA	7,084	NA	NA	NA	NA	7,710
South Nyack	Rockland	1,601	2,093	3,113	48.7	NA	604	-93	7,094
South Westbury[u]	Nassau	NA	NA	11,977	NA	NA	NA	NA	8,648
Spencer	Tioga	707	615	767	23.7	NA	NA	NA	NA
Spencerport	Monroe	715	1,340	2,461	83.7	NA	NA	NA	NA
Spring Valley	Rockland	NA	4,308	6,538	51.8	531	2,526	6,092	5,919
Springville	Erie	1,992	2,849	3,852	35.2	441	1,081	1,553	6,105
Stamford	Delaware	901	1,088	1,166	7.2	NA	NA	NA	NA
Stewart Manor	Nassau	NA	1,625	2,422	49.0	NA	NA	NA	NA
Stillwater	Saratoga	1,007	971	1,398	44.0	NA	NA	NA	NA
Stony Brook[u]	Suffolk	NA	NA	3,548	NA	NA	NA	NA	8,890
Stony Point[u]	Rockland	NA	NA	3,330	NA	NA	NA	NA	7,115
Stottville[u]	Columbia	NA	NA	1,040	NA	NA	NA	NA	NA
Suffern	Rockland	1,619	3,768	5,094	35.2	1,225	1,763	3,721	7,690
SYRACUSE URBAN SYSTEM (SMSA)		280,161	405,981	563,781	38.9	66,647	134,749	109,964	6,405
Urbanized Area		NA	NA	333,286	NA	NA	98,491	159,169	6,737
Syracuse	Onondaga	108,374	205,967	216,038	4.9	24,045	85,364	210,782	6,247
Tannersville	Greene	593	640	580	-9.4	NA	NA	NA	NA

NA: Not Available.
[u] Indicates unincorporated urban places.

URBAN PLACE DATA (Continued)

Urban Place	County	Population 1900 (No.)	Population 1940 (No.)	Population 1960 (No.)	Population Growth 1940-60 (Percentage)	Employment Mfg. 1958 (No.)	Employment Tertiary 1960 (No.)	Tertiary Surplus 1960 (No.)	Median Family Income (1959) (In Dollars)
Tarrytown	Westchester	4,770	6,874	11,109	61.6	1,007	3,539	6,586	$8,090
Theresa	Jefferson	917	908	956	5.3	NA	NA	NA	NA
Thomaston	Nassau	NA	1,159	2,767	138.7	NA	499	−272	10,000+
Ticonderoga	Essex	1,911	3,402	3,568	4.9	959	1,124	2,052	6,228
Tivoli	Dutchess	1,153	761	732	−3.8	NA	NA	NA	NA
Tonawanda	Erie	7,421	13,008	21,561	65.8	13,286	6,472	10,799	6,746
Tonawanda[u]	Erie	NA	NA	83,771	NA	NA	NA	NA	7,753
Troy	Rensselaer	60,551	70,304	67,492	−4.0	4,656	19,971	32,363	5,502
Trumansburg	Tompkins	1,225	1,130	1,768	56.5	NA	NA	NA	NA
Tuckahoe	Westchester	NA	6,563	6,426	−2.1	1,565	3,206	9,607	6,731
Tully	Onondaga	574	719	803	11.7	NA	NA	NA	NA
Tupper Lake	Franklin	NA	5,451	5,200	−4.6	412	1,250	1,050	4,955
Unadilla	Otsego	1,172	1,079	1,586	47.0	NA	NA	NA	NA
Uniondale[u]	Nassau	NA	NA	20,041	NA	NA	NA	NA	7,474
Union Springs	Cayuga	994	905	1,066	17.8	NA	NA	NA	NA
Unionville	Orange	454	387	511	32.0	NA	NA	NA	NA
Upper Brookville	Nassau	NA	456	1,045	129.2	NA	NA	NA	NA
Upper Nyack	Rockland	516	924	1,833	98.4	NA	NA	NA	NA
UTICA URBAN SYSTEM (SMSA)		183,849	263,163	330,771	25.7	40,251	72,783	33,144	6,022
Urbanized Area		NA	NA	187,779	N	NA	48,438	54,411	6,229
Utica	Oneida	56,383	100,518	100,410	−.1	16,570	31,816	58,670	5,873
Valatie	Columbia	1,300	1,208	1,237	2.4	NA	NA	NA	NA
Valley Falls	Rensselaer	NA	564	589	4.4	NA	NA	NA	NA
Valley Stream	Nassau	NA	16,679	38,629	131.6	1,069	11,741	20,076	8,021
Van Etten	Chemung	474	440	507	15.2	NA	NA	NA	NA
Vernon	Oneida	380	587	913	55.5	NA	NA	NA	NA
Vernon Valley[u]	Suffolk	NA	NA	5,998	NA	NA	NA	NA	7,591
Victor	Ontario	649	1,111	1,180	6.2	NA	NA	NA	NA
Victory Heights[u]	Chemung	NA	NA	2,528	NA	NA	NA	NA	5,530
Village of the Branch	Suffolk	NA	185	886	378.9	NA	NA	NA	NA
Voorheesville	Albany	554	717	1,228	71.3	NA	NA	NA	NA
Waddington	St. Lawrence	759	671	921	37.3	NA	NA	NA	NA
Walden	Orange	3,147	4,262	4,851	13.8	786	1,361	1,954	5,516
Wallkill[u]	Ulster	NA	NA	1,215	NA	NA	NA	NA	NA
Walton	Delaware	2,811	3,697	3,855	4.3	558	874	515	5,440
Wampsville	Madison	NA	282	564	100.0	NA	NA	NA	NA
Wantagh[u]	Nassau	NA	NA	34,172	NA	NA	NA	NA	8,671
Wappingers Falls	Dutchess	3,504	3,427	4,447	29.8	133	1,307	2,088	6,750
Warrensburg[u]	Warren	NA	NA	2,240	NA	NA	NA	NA	NA
Warsaw	Wyoming	3,048	3,554	3,653	2.8	NA	1,341	3,052	6,089
Warwick	Orange	1,735	2,534	3,218	27.0	109	1,083	2,197	6,577
Washington Heights[u]	Orange	NA	NA	1,231	NA	NA	NA	NA	NA
Washingtonville	Orange	667	801	1,178	47.1	NA	NA	NA	NA
Waterford	Saratoga	3,146	2,903	2,915	.4	730	762	895	5,810
Waterloo	Seneca	4,256	4,010	5,098	27.1	238	973	−233	5,974
Watertown	Jefferson	21,696	33,385	33,306	−.2	3,102	11,549	24,439	5,480
Waterville	Oneida	1,571	1,489	1,901	27.7	NA	NA	NA	NA
Watervliet	Albany	14,321	11,114	13,917	−13.6	3,307	2,205	−2,892	5,901
Watkins Glen	Schuyler	2,943	2,913	2,556	−12.3	171	1,035	2,362	5,573
Waverly	Tioga	4,465	5,450	5,950	9.2	NA	1,463	1,365	5,520
Wayland	Steuben	1,307	1,795	2,003	11.6	NA	NA	NA	NA
Webster	Monroe	NA	1,680	3,060	82.1	279	1,152	2,700	7,596
Weedsport	Cayuga	1,525	1,341	1,731	29.1	NA	NA	NA	NA
Wellsburg	Chemung	536	560	643	14.8	NA	NA	NA	NA
Wellsville	Allegany	3,556	5,942	5,967	.4	2,070	2,214	5,103	5,828

APPENDIX E URBAN PLACE DATA

URBAN PLACE DATA (Continued)

Urban Place	County	Population 1900 (No.)	Population 1940 (No.)	Population 1960 (No.)	Population Growth 1940-60 (Percentage)	Employment Mfg. 1958 (No.)	Employment Tertiary 1960 (No.)	Tertiary Surplus 1960 (No.)	Median Family Income (1959) (In Dollars)
Westbury	Nassau	NA	4,524	14,757	226.2	2,013	6,002	15,253	$9,466
West Carthage	Jefferson	1,135	1,767	2,167	22.6	NA	NA	NA	NA
West Elmira ᵁ	Chemung	NA	NA	5,763	NA	NA	NA	NA	8,503
West End ᵁ	Otsego	NA	NA	4,068	NA	NA	NA	NA	NA
Westfield	Chautauqua	3,882	4,638	5,498	18.5	473	1,073	1,487	5,750
West Glens Falls ᵁ	Warren	NA	NA	2,725	NA	NA	NA	NA	6,103
Westhampton Beach	Suffolk	NA	969	1,460	NA	NA	NA	NA	NA
West Haverstraw	Rockland	2,079	2,533	5,020	98.2	486	624	-1,900	6,832
West Hempstead-Lakeview ᵁ	Nassau	NA	NA	24,783	NA	NA	NA	NA	8,388
Westport	Essex	NA	654	723	10.6	NA	NA	NA	NA
West Seneca ᵁ	Erie	NA	NA	23,138	NA	NA	NA	NA	6,809
West Winfield	Herkimer	771	754	960	27.3	NA	NA	NA	NA
Whitehall	Washington	4,377	4,851	4,016	17.2	207	1,288	2,424	5,110
White Plains	Westchester	7,899	40,327	50,485	25.2	2,891	33,679	117,910	8,012
Whitesboro	Oneida	1,958	3,532	4,784	35.4	75	739	-1,089	6,522
Whitney Point	Broome	807	733	1,049	43.1	NA	NA	NA	NA
Williamson ᵁ	Wayne	NA	NA	1,690	NA	NA	NA	NA	NA
Williamsville	Erie	905	3,614	6,316	74.8	68	3,313	10,249	7,835
Williston Park	Nassau	NA	8,255	8,255	43.6	41	2,090	2,195	7,867
Wilson	Niagara	612	849	1,320	55.5	NA	NA	NA	NA
Windsor	Broome	739	766	1,026	33.9	NA	NA	NA	NA
Winona Lakes ᵁ	Orange	NA	NA	1,655	NA	NA	NA	NA	NA
Wolcott	Wayne	1,279	1,326	1,641	23.8	NA	NA	NA	NA
Woodmere ᵁ	Nassau	NA	NA	14,011	NA	NA	NA	NA	12,378
Woodridge	Sullivan	NA	854	1,034	21.1	NA	NA	NA	NA
Woodsburgh	Nassau	NA	702	907	29.2	NA	NA	NA	NA
Wurtsboro	Sullivan	450	487	655	34.5	NA	NA	NA	NA
Wyoming	Wyoming	NA	430	526	22.3	NA	NA	NA	NA
Yonkers	Westchester	47,931	142,598	190,634	33.7	33,760	39,484	6,786	7,471
Yorktown ᵁ	Westchester	NA	NA	3,576	NA	NA	NA	NA	8,224
Yorktown Heights ᵁ	Westchester	NA	NA	2,478	NA	NA	NA	NA	NA
Yorkville	Oneida	NA	3,311	3,749	13.2	1,733	460	-1,449	6,481
Youngstown	Niagara	547	799	1,848	131.3	NA	NA	NA	NA

NA: Not Available.
ᵁIndicates unincorporated urban places.
Source: U. S. Census; Richard T. Lewis, "The Measurement of Tertiary Activity," Unpublished Masters Thesis, Department of Geography, Syracuse University, 1964.

INDEX

Italic numbers refer to pages on which the entry is defined. "F," "P," "T," or "n" following a page number indicates that the item is found in a figure, photograph, table, or note respectively.

Accessibility, point of maximum, 334
Acetylene gas, 416
Acidity in soils, *105*, 106-10 *passim*
Adirondack Low Mountains (land form division), 25, 26F
Adirondack Mountain Peaks (land form division), 25, 26F, 20P
Adirondack North Country Forest Region, 228-29
Adirondack Region (recreation), 305
Adirondack State Park, 228, 229F
Adirondack Upland (land form region): 12P, 25, 26F, 115, 307P, 347, 368, 387, 478; prehistory, 19, 23; climate, 61, 67, 69, 71-72, 73-74; water, 86-87; forests, 96-98; lumbering, 98-100, 138, 167, 168, 178; manufacturing in, 100, 243; maple sugar production, 100; recreation in, 100-1, 302, 480; soils, 106, 108; power, 153, 289; minerals, 167, 168, 221, 223, 226; transportation, 173, 267, 269, 273, 469; agriculture, 217; economic health, 328
Adirondack Wilderness (uniform region), 371, 375, 376
Advertising, 436, 446
Agricultural areas. *See* Rural areas
Agricultural machinery, 179, 188, 246, 414, 482
Agricultural potential: 371; as understood by colonists, 138
Agricultural regions, 210-17, 212-13F, 371
Agriculture: 202-20; state's proportion of national, 6T; and soils, 106-10 *passim*; Indian, 117; early Dutch, 122-24; in 1785, 138; pioneer, 165; (1785-1855), 165-67; (1850-90), 177-178; (1890-1920), 183; (1920-60), 188; production, 202, 202F, 203, 209T; trends in, 202F, 203-4, 220F; types of production, 204-10; crop combinations, 210F; problems and prospects, 217-20; and city location, 347; Albany-Schenectady, 387; and New York City, 430; Rochester, 458; future of, 499. *See also* Farms, Rural landscape
Agriculture, subsistence, *203*
Air-conditioning equipment, 474
Aircraft and parts, 188, 239, 242, 251, 254, 416
Air passengers, 5F, 272, 274F
Air pollution in New York City, 455

Airports and suburbs, 475
Air transportation: 186-87, 188F, 193F, 271-73, 391; and manufacturing, 234; New York City, 434
Albany (Ft. Orange): 385-95 *passim*, 389P, 390P; climate, 63, 66, 68; pollution of Hudson, 88; Fort Orange, 122; and English rule, 128; c. 1700, 130; trade expansion, 131, 134-35, 152, 154-55, 156, 162; mid-18th century, 134; in 1771, 136; c. 1785, 138; and colonization movement, 143, 144, 145; and turnpikes, 157-58; impact of Erie Canal, 159, 162; manufacturing, 167-68, 178-79, 234, 242, 246; 1850, 169; and railroad development, 175; immigrants, 179; port, 234, 391, 395; and capital function, 262, 348; transportation, 263, 272, 274; services, 280; recreation, 304; form, 334, 338, 339; and urban spacing, 346, 347; and planning region, 373
Albany-Schenectady road, 136, 156
Albany-Schenectady Urban System: 385-95, 386F; manufacturing, 168, 178-79, 240; population (1850), 169; population (1890), 179; trade, 278, 282; and economic health, 328; urbanized area, 333, 386F; tertiary surplus, 375; urban frontier, 385, 386F; future, 493, 497
Albany-Schenectady-Troy SMSA. *See* Albany Urban System
Albany-Susquehanna Turnpike, 157
Albany and Susquehanna Railroad. *See* Delaware and Hudson Railroad
Albany Area Nodal Region, 375, 377
Albany County: population (1698), 130; population (1771), 135; cement, 224; manufacturing, 246; surplus, 256; SMSA, 385; airport, 391; tertiary surplus, 394
Albany, lake (glacial), 21, 22F, 47, 387
Alfalfa, 204
Algonkian language group, 114F
Algonquin Indians, 114-20, 114F
Algonquin Peak, 20P
All-American canal, 415
Allegany, 181
Allegany County: petroleum, 226
Allegany State Park: 45; forest use in recreation, 101; soils, 108
Allegheny Hills (land form division): 26F,

33; agriculture, 217; rural landscape, 368
Allegheny Indian Reservation, 141, 142F
Allegheny River, 152, 155
Aluminum, 242, 289
American Airlines, 391
American Locomotive Company, 387
Amherst, 407
Amsterdam: settlement, 131; population growth, 179, 189; manufacturing, 179, 234, 246
Amsterdam, Holland: compared with New Amsterdam, 126, 127
Anabaptists: in New Amsterdam, 127
Ancram, iron ore from, 136
Anglicans: in New Amsterdam, 127
Animals and recreation, 299
Ansco Corp., 401, 403
Anthony and Scoville, Inc., 401
Anticyclones, *64*
Antique shop districts, 446
Apartment districts, 342
Appalachian Country (uniform region), 371-72, 375, 376, 396
Appalachian Division (planning and development region), 377, 380F, 381
Appalachian Upland (land form region): 26F, 31-33, 32P, 70P, 271P; glaciation, 21-3; structure, 23, 221; climate, 62, 67, 69, 72, 74-77; water problems, 86, 87; vegetation, 95, 96; forest conditions and uses, 100, 229; soils, 106, 108; Indian settlement of, 115, 471; agriculture, 177, 178, 208, 210-11, 214, 500; salt, 225; manufacturing, 243; recreation, 305; and Finger Lakes, 309; and urban development, 347; rural landscape, 358, 363; and Binghamton, 397; and Buffalo, 407, 409; and Syracuse, 469; and Utica, 480, 483
Appalachian Upland Dairy Region, 210-11, 211P, 220
Appalachian Upland Low-Demand Region (power), 291
Apparel: functional interrelationships in, 442
Apparel Manufacturing: distribution of, 237, 239, 242; development of, 243, 246; and economic health, 248, 251, 319; Troy, 387, 392; New York City, 430, 431, 433, 436, 439, 440, 442, 446; Rochester, 458, 461, 463; Utica, 482
Apparel stores: location of, 334

INDEX 525

Apples, 125, 134, 205
Appliances, 248
Archaic Stage (Indian), 113
Architecture, colonial, 304
Area: *15*; New York's proportion of national, 6T
Area analysis tripod, 12, 12F
Areal differentiation: and transportation, 263
Argyle, 135
Arizona: New York compared with, 315–21 *passim*, 316–17T, 320–21T
Arkansas: New York compared with, 315–21 *passim*, 316–17T, 320–21T
Arks, 149, 155
Armonk, 226
Art gallerys, 446, 487
Arts: New York City as center of, 434
Ash trees, 229
Atlantic Coastal Lowland, 26F, 34–35
Atlantic Coast Axis: and New York City, 428, 430
Atlantic Ocean: and precipitation, 67; and vegetation, 93
Atomic energy: and power production, 288
Auburn: development of, 149; sugar beets, 208; manufacturing, 243; economic health, 328
Ausable Chasm, 303
Ausable River: lumbering, 28P, 168
Automation, 239, 251, 496
Automobile, effects of: 181–82, 184–87, 189; on agriculture, 188; urban, 338, 340; New York City, 438–39, 451; Syracuse, 473–75
Automobile and parts, manufacturing: 184, 239, 240, 254; in urban systems, 392, 416, 463, 473, 474
Automotive dealers: and retail sales, 278; location, 341–42, 446, 475
Available moisture, *70*
Avon (Hartford): development of, 150
Axes, of New Netherland, 122

Baldwinsville: canal to, 161
Ballston Spa: mineral waters, 52
Balmat: mining, 224
Baltic peoples: immigration, 184
Baltimore: trade competition, 149, 156, 429
Baltimore and Ohio Railroad, 174
Bank assets: external comparisons, 316–17T, 318, 320–21T
Banking: development of in New York City, 425, 430, 437; clustering of in New York City, 431, 441; and corporate offices in New York City, 432
Bank of the U. S. (Philadelphia): end of, and banking in New York City, 431
Banks: location of, 334, 340

Barge Canal: route, 182; and manufacturing, 232; and Niagara Frontier manufacturing, 240; freight on, 265; and recreation, 303–4, 420; and Albany, 391; and Rochester, 461; and Syracuse, 473; and Utica, 484; 271P, 272P, 409P
Barges: on Hudson at New York City, 433
Barley, 125, 134, 136, 166, 177, 204
Barnhart Dam (St. Lawrence), 285
Barriers: to transportation in New York City, 430
Basalt, 42, 225
Basic economic activities, *13*, *247*
Basic percentage, *256*
Batavia: development of, 152; impact of Erie Canal, 159; (1850), 169
Bath: development of, 150, 151; 1850, 169
Bausch, John, 461
Bayonne, New Jersey, 429P, 433
Bayside, 439
Beaches: on Long Island, 43
Beacon, 374
Beans, 125, 205, 207, 208, 208P
Beaverhill: flow, 80, 84F
Bedford-Stuyvesant area (New York City), 455
Bedrock, water-bearing, 81
Beech trees, 229
Beef, 136
Beer: New Netherland, 125
Beets, 208
Belleayre Mountain Ski-Center, 294, 309
Bendix Corp., 485
Benson Mines, 223
Bergen, New Jersey, 122F, 126
Bergen Co., New Jersey: 445; developing ring, 451; commuting, 452
Berkshire Hills (Massachusetts), 387, 388
Berkshire Thruway Extension, 391
Berries, 205, 207
Bethlehem Steel Co., 240P
Bingham, William, 399
Bingham Patent, 399
Binghamton (Chenango Point): 396–406 *passim*; climate, 63, 66, 68; Indian village, 115; development of, 148, 149–50, 169, 179, 184; transportation, 163, 269, 390, 391; clay, 225; manufacturing, 235, 242, 246; trade, 258, 278, 282–283; professional services, 262; recreation, 307; middle city, 334; and urban spacing, 347, 348; Appalachian Country, 371; Syracuse nodal region, 375; Syracuse trade area, 477–78
Binghamton SMSA. *See* Binghamton Urban System
Binghamton Urban System: 396–406, 396F, 397F, 398P, 400P; economic health, 328; urbanized area, 333, 396–97; future, 493, 497
Biochemical oxygen demand (BOD), *87*
Blackout of 1965, 291
Black River: power, 289

Black River Canal, 160–61, 177, 182
Black River Turnpike, 149
Black River Valley (land form division): 34, 26F; snowfall, 69; floods, 87; migration route, 145; settlement, 153; lumbering, 168
Black Rock: development of, 152, 410; rivalry with Buffalo, 159, 412, 413
Black Rock Channel, 414
Boating, 302, 420
Boonville: climate, 62, 69
Boston, Massachusetts: trade competition, 163, 425; and urban spacing, 346, 347
Boston and Albany Railroad, 389–90
Boston and Maine Railroad, 390
Boston Ten Towns, 141
Braddock Bay, 151
Brass, 242, 482
Brewing: development of, 136, 168, 246; New York City, 178, 431; Buffalo, 414
Brick-making, 243, 246, 387
Bridgehampton: precipitation, 67
Bridgeport, Connecticut, 424
Bridges: Chenango River, 399; Hudson River, 430
British: immigration, 179–81
British colonial policies: and New York manufacturing, 243
Broadalbin, 135
Broadway theatre district, 448
Bronx (borough and county): tertiary deficit, 256; bridges, 430; in Greater New York, 432; Jewish influx, 433; developed ring, 445, 450; commuting, 452; future population growth, 493
Brookhaven: atmospheric emissions, 66; water balance, 71, 72T; English village, 126
Brooklyn (Breuckelen): colonial village, 126, 127F; wharves, 154; bridges, 430; Greater New York, 432; piers, 433; service center, 433; historical remnants, 444; in developed ring, 450; 429P, 448P, 449P. *See also* Kings County
Brooklyn Bridge, 430
Broom corn, 166
Broome County: manufacturing, 242; SMSA, 397; airport, 404
Brownville, 148, 153
Bryant, William Cullen: and Central Park (New York City), 499
Buckwheat, 136–204
Buffalo (New Amsterdam): 407–22 *passim*, 240P; cloudiness, 58; temperature, 62, 63; winds, 65, 66; precipitation, 65, 67, 68, 70; development of, 148, 152, 169, 179; Erie Canal, 159; railroads, 162, 163, 174, 175; flour, 168, 178; immigrants, 169, 171, 181; clay, 225; power, 232, 285; harbor and port, 234, 407, 412, 413, 417P, 418P; manufacturing, 239, 246; trade and services, 256, 278, 279, 280,

282; transportation, 263, 274, 390; recreation, 304; urban form, 334, 339; and urban spacing, 346, 347; skyway, 417P, 421
Buffalo and State Line Railroad, 163–64
Buffalo Creek, 152, 410, 411, 412, 414
Buffalo nodal region, 375
Buffalo planning and development region, 380F
Buffalo River, 84, 415, 416, 417P, 418P
Buffalo Urban System: 407–22, 408F, 409P, 411F, 417P, 418P, 419P; Southern Ontario Plain, 34; manufacturing, 178, 188, 235, 265; vegetables, 207; industrial materials, 224; and Lockport, 243; power, 288; economic health, 328; urbanized area, 333, 407, 408F; Upstate Heartland, 373; tertiary surplus, 375; urban frontier, 407, 408F; future, 493, 497
Bundy Time Recorder Co., 401
Burlington, Vermont: and urban spacing, 346, 347
Buses, city, 182
Business equipment, 242, 251, 463
Business Fact Book, 385n
Business indicators: external comparisons, 318–20, 316–17T, 320–21T
Business services, 256, 280, 443
Butter: New Netherland, 125; Hudson Valley, 165; Orange County, 166; c. 1900, 177; decline in, 204
Buttermilk Falls, 303

Cabbage, 208
Caledonia: Highland Scots, 144
California: New York compared with, 6T, 315–21 *passim*, 316–17T, 320–21T
Calvinists, 127
Cambridge, 135
Campsites, 297–98T
Canada and Canadians: and glaciers, 19; refugees from, 141; lumber market, 153; immigration, 171, 189; and Niagara water, 415
Canadice Lake, 306
Canajoharie: Palatine Germans, 131, 133F
Canals: development of, 149, 158–62, 160F; and railroads, 162, 163, 176, 187; c. 1855, 165; and manufacturing, 167; 1860's–1890's, 176–77; abandonment, 177, 182; 1890's–1920's, 182–83; pattern, 192F; present status, 265; and recreation, 303–04; and Buffalo, 413; and New York City, 428–29; and Syracuse, 471–73
Canandaigua: site, 148; development of, 150, 151, 169, 179; and Erie Canal, 159
Canandaigua Lake, 43, 165, 306
Candles, 136, 179, 473
Cannonsville Reservoir, 86
Canton, 153, 178

Capital: for manufacturing, 246, 249, 250
Capital, state, 385, 386
Capital, national: effect on New York City of removal from Philadelphia, 428
Capital District, 186, 373, 385. *See also* Albany-Schenectady Urban System
Captree State Park, 309
Cargo: at Albany, 391
Carpets, 179, 188, 246, 248
Carriages, 400, 416
Carrier Corp., 476P, 477
Carthage: beginning, 153; lumber-building, 178
Cash crops, 208
Castorland, Constable's purchase, 153
Catholics: in New Amsterdam, 127; in central cities, 354
Catskill-Hudson Forest Region, 229
Catskill-Hudson Highlands (recreation region), 309
Catskill Mountains (land form division): 19, 23, 32–33, 26F, 34P, 373; climate, 61, 67, 69, 71–72; New York City water supply, 80; vegetation, 93, 96–98; economy, 100, 101, 217, 243, 328; soils, 108; Indians, 115; settlement, 128, 135, 138, 143, 145, 304, 347, 368; transportation, 165, 176, 267, 269; future, 493
Catskill State Park, 309
Cattaraugus County: petroleum and natural gas, 226; and glaciation, 373
Cattaraugus Hills (land form division): 33, 26F, 36P; rural landscape, 368
Cattaraugus Indian Reservation, 141, 142F
Cattle, 165, 166
Cayuga and Seneca Canal: development of, 160, 177, 183; and recreation, 303
Cayuga Indians, 114F, 115
Cayuga Lake: physical characteristics, 23, 43, 44; transportation, 156–57, 165; and recreation, 302, 306
Cazenovia: founding of, 149; in 1850, 169; and Syracuse, 475
Cement, 221, 226, 242
Cemetery area: New York City, 450
Census, U.S. Bureau of the, 491, 492, 496
Censuses, colonial, 130–31, 135–36
Central business district (CBD): *334*, 334–46 *passim*; engulfed, 346; and urban hierarchies, 348–51 *passim*; Albany, 394; Binghamton, 399, 402, 405; Buffalo, 407, 409P, 421; manufacturing, New York City, 433; changes in significance, New York City, 439–40; development, New York City, 444, 446–50; Rochester, 458, 462–63, 464–65P, 467; Syracuse, 474, 475, 478; future, 493
Central cities: population, 189; tertiary sector, 256–58; and suburbs, 354; problems, 354–56
Central Hudson Gas and Electric Corp., 285, 286F–87F

Centralization, economic; effect on New York City, 432, 434
Central Mixed Farming Region, 211, 214P, 215P, 220P
Central New York: settlement, 143, 148–50; north-south railroads, 175–76, 176F
Central Park (New York City): solar radiation, 55–58, 55F; value, 499
Central places: tertiary activities, 256–58, 259F, 260T–61T; hierarchy, 348–52; and city spacing, 346
Central Square: plank road, 165
Ceramics, 248
Cereals, 204
Chambers Street (New York City): wall, 425
Champlain, Lake: 30P; climate, 67, 74; settlement near, 143; and transportation, 165, 264, 391; and recreation, 303; and North Country, 374
Champlain, Samuel de, 115
Champlain Canal: interest in route, 136; development, 160, 177, 183, 265; and iron works, 153; and recreation, 303; and Albany area, 391
Champlain Lake Plain (land form division): 28, 26F, 30P; water balance, 71, 72; settlement, 135, 152–53; urban development, 347; transportation, 388, 390
Champlain Sea (glacial), 21, 22F
Changing role theory, 247
Charleston, South Carolina: rival to New York City, 425
Charlotte: port for Rochester, 151
Charlotte County, 135
Chasm Falls: temperatures, 62
Chateaugay ore (iron), 392
Chaumont Bay: fishing, 230
Chautauqua County: early development, 152; wine grapes, 177; natural gas, 226
Chautauqua Lake: pollution, 88; recreation, 302
Cheese: New Netherland, 125; Hudson Valley, 165; c. 1890, 177; Herkimer County, 177; decline, 204
Chelsea: tapping Hudson for New York City water, 88
Chemicals: Syracuse, 179, 239, 473, 477; Buffalo, 239, 240, 415; and economic health, 313; Albany area, 392; New York City, 440; to Rochester, 467
Chemung Canal: establishment, 161; abandoned, 177
Chemung County, 363P
Chemung Valley, 33P
Chenango Bridge, 396, 402
Chenango Canal: development, 161; abandoned, 177; and Binghamton, 399; and Utica, 484, 485
Chenango River, 398, 399, 402
Chenango Twenty Towns: establishment, 141, 142F; settlement, 149
Chenango Village, 398, 399

INDEX 527

Cherries, 125, 205, 207P
Cherry trees, 229
Cherry Valley: settlement, 134
Cherry Valley Turnpike, 156, 387
Chicago: and New York City, 430
Chicago Pneumatic Tool Co., 485
Chinatown (New York City), 446
Christmas tree production, 100, 227–28, 229
Cicero swamp, 477
Cigars: Binghamton, 400–1; New York City, 431
City: in urban hierarchy, 350–51
City-forming activities (basic), *441*, 441–44
City-serving activities (nonbasic), in New York City, *441*
Civil War: and agriculture, 177; and manufacturing, 178, 247
Clams, 229
Clarence (Clarence Hollow), 410
Clay: distribution, 221, 224, 225; and manufacturing, 242; Albany area, 387
Clay handpans, 405
Clayton: railroad, 176
Cleveland, Ohio: and urban spacing, 347
Climate: 54–78; *54*; processes, elements and controls, 54; and vegetation, 91, 93–96 *passim*; and soils, 104; Dutch understanding of, 124–25; and agriculture, 203, 207, 210–17 *passim*; and rural landscape, 358; and uniform regions, 371–75 *passim*
Climatic regions, 72–78, 75F, 76T, 371
Climatic stations, 58F
Clinton, 148
Clinton, Governor, 134
Clinton County: soils, 106; iron ore, 223; forest preserve, 292
Clinton Square (Syracuse), 475
Clinton Township, 141
Cloud-seeding, 81
Cloud cover, 58, 59T
Clover, 204
Coal: and canal development, 161; and Troy iron manufacture, 167; and railroad development, 175–76, 176F, 187; and power production, 288; and Binghamton, 399, 401; and Buffalo, 413, 415; Rochester, 461; Utica, 484
Coasts, 3, 264
Cochecton: settlement of, 134
Cohocton River: as migration route, 150
Cohoes: manufacturing, 168, 178–79, 392; immigrants, 179; population, 184, 385
Cohoes Falls, 136
Colden, Cadwallader: report as surveyor general, 128–29
Colgate-Rochester Divinity School, 467
Colonial period: 121–39; expansion, 130–31, 135; significance, 137–39
Colonie, 385
Colonization movement (post-Revolution): in Vermont, 140–41; magnitude, 143; sources, 143–44; patterns, 144–48; developments to 1825, 148–55
Columbia County: manufacturing, 167, 179; cement, 224
Columbus Circle (New York City), 446
Commerce. *See* Trade
Commerce, Department of, 385
Communication development: and New York City, 432
Communication equipment, 251
Community shopping district: *344*, 344–45; in urban hierarchy, 349, 350, 351; Syracuse, 475
Commuting: and electric railroad, 181; New York City patterns, 438, 451, 452–53. *See also* Railroads, commuter
Compages, 371
Competition in manufacturing, 239, 241, 242, 247, 248
Complementarity, *263*, 264
Composite region, concept of, 15, 370–75, *371*
Computers, 242, 248
Concentric patterns: in cities, 334–36, 335F, 340–42; New York City, 444–52, 445F
Conesus Lake, 306
Confederacy of the Five Nations, 114
Conklin, 402
Connecticut: and Long Island, 126; Dutch claims, 126; boundary problems, 129; source of colonists, 143–44
Conservation, 291–309
Conservation Department, 292–319 *passim*
Consolidated Edison Co. of New York, 285, 286F–87F
Constable, William, 153
Constantia (Rotterdam): establishment of, 153
Constant rate projections: and population projections, 492
Constitution (New York State): on Forest Preserve, 292
Construction: 250–51, 254; New York's proportion of national, 6T; Manhattan, 439, 450; Syracuse, 478
Consumer goods, *415*, 442
Consumption: origin of commodities, 4F
Continental Can Co., 473
Continental glaciers, 19
Convenience goods, *344*, 345, 349–51 *passim*
Cooper, Judge William: Landlord's axiom, 143; geographical predictions, 148; and ark, 155; and Hudson-Lake Erie connection, 159
Cooperstown, 148
Copper, 242, 482
"Copper City" (Rome), 480
Corn (maize): and Indians, 117; and colonists, 134, 136, 137; and development of agriculture, 166, 177, 183; in present agriculture, 204, 207, 208

Cornell University, College of Agriculture, 210, 220
Corning: and railroad, 163; population growth, 179, 184; manufacturing, 179, 235, 242; manufacturing belt, 404; and Elmira, 495
Corporate management function: in New York City, 431–50 *passim*
Corporation laws: and manufacturing development, 246
Cortland: manufacturing 239P, 242
Cortland County: future population, 493
Cost structure: in manufacturing, *249*
Cotton: early mills, 167; in Hudson-Mohawk Axis, 179; and Atlantic Coast Axis, 428; Little Falls, 482; Utica, 484
Counties: creation of ten, 129–30, 125F; mineral production in, 222T; tertiary activities in, 256–58, 257T, 258F; and urban problems, 356; future population, 493–95, 494T; economic health, 323–27, 322–23T, 324T, 325F, 326F, 327F. *See also* Appendix D
Cows, 183
Crime, New York City, 455
Crooked Lake Canal, 161, 171
"Cross channels," 21
Crystalline rocks, underlying, 23, 24F
Cropland: percentage of state in, 90F
Crops: colonial period, 125, 134
Culture areas and groups: creation of, 138–39; 1850, 169–71; patterns of distribution, 190–91
Cultural regions, 370
Cultural sites, 304
Cutlery, 242
Cyclonic storms, mid-latitude, *64*, 65F
Czechoslovakia: immigration from, 401

Dairy barns: and rural landscape, 359
Dairy belt, American, 204
Dairying: and forest vegetation, 98; development, 166, 177, 183; modern, 204–5, 204P, 205F, 211–17 *passim*, 215P; Rochester, 458
Dansville: canal, 161
Darby, William: on Canandaigua, 151; on Buffalo, 152; on Troy and Kingston, 154; on trade routes, 158, 158F
Decentralization, economic: effect on New York City, 434, 440, 454
Deficit percentage (tertiary), *256*
Delaware and Hudson Canal, 161
Delaware and Hudson (Albany and Susquehanna) Railroad, 175, 390, 399
Delaware County: forest preserve, 292
Delaware Hills (land form division), 33, 26F, 35P, 36P
Delaware, Lackawanna and Western Railroad, 174–75, 399
Delaware River: 35P, 36P; and New York City water supply, 86

Delaware River Basin Commission, 86
Delaware Valley: snowfall, 69; and New Netherland, 121; settlement in, 134, 135
Democrats: in central cities, 354
Department stores: location, 351; Manhattan, 434
Desalinization of sea water, 81
Developed ring (middle city): New York Urban System, *445*, 450–51
Developers, large scale: and suburbs, 440
Developing ring (urban fringe, suburbia): New York Urban System, *445*, 450, 451–52
Development theory, 14–15, 14F, 497
De Vries, David: and New Netherland, 125
Dewitt, 474, 475
Distance, friction of, 348
Distilling, 136, 168, 243
Disturbance, in vegetation development, 91, 93, 93–98 *passim*
Docks: New York City, 433
Dolomite, 221, 225
Downtown (New York City): shift of focus from, 433–34
Drainage, in soils: 104, *105*, 106–11 *passim*; and vegetation, 91, 93–96 *passim*; and sectors in cities, 339; suburban problems, 475
Droughts: and vegetation, 91
Drumlins, *20*
Dunkirk: 38P; establishment of, 152; Erie Railroad, 163, 175; immigrants, 179; electric railroad, 181; manufacturing, 242
Durable goods: *237*n; and Utica, 484, 487
Durand-Eastman Park (Rochester), 462
Dutch: beginning of colonization, 119, 121–24; basic features of colonization, 124; legacy, 127–28; under English rule, 128, 134, 135; areas at end of Revolution, 139; in post-Revolutionary colonization movement, 144, 152, 401; dairy stock, 166; oystering, 229; in Albany area, 387
Dutchess Country: growth, 131, 134, 135; agriculture, 166; cotton mills, 167; and New York Urban System, 424, 445, 452–53; future population, 493
Dwight, Timothy: on place names, 149
Dynamism: and New York City, 425, 440, 444, 453

Easements, scenic, 499
East Aurora, 407
Eastern Ontario Hills (land form division), 34, 26F
Eastman, George, 461, 462, 466
Eastman Kodak Corp., 348, 458–67 *passim*
Eastman School of Music, 462

East Meadowbrook Basin, 86
East River: and New Amsterdam, 122; docks, 154, 425, 433, 448P; bridges, 430
East Side, Lower (New York City), 431, 433, 449P
East Side, Upper (New York City), 433
East Syracuse, 473, 479
Eating and drinking places: and retail sales, 278; location, 334
Economic axis, principal, 232
Economic conditions: and population distribution, 6; development of patterns, 194–95; external comparison, 315–21; critical criteria, 321; internal spatial variations in, 321–29, 322–23T, 324T, 325F, 326F, 327F; mapping, 328–29; and urban systems, 354; in rural landscape, 364; and uniform regions, 371–75 *passim*; Utica, 480–82; and population projections, 491
Economies of scale: and New York port, 428
Edison, Thomas A., 184, 387
Education function: New York's proportion of national enrollment, 6T; growth, 256; Ithaca, 258; and social organization, 313; and urban spacing, 347; Albany area, 394–95; Rochester, 467
Education Department, 294, 304
Educational institutions: location in cities, 342; and population growth, 495
Edwards: mining, 224
Efficiency of labor: and manufacturing, 251; and future employment, 496
Eggs, 208
Electrical Machinery: development, 184, 188, 248, 251; Schenectady, 184, 242, 387, 392, 393; Rochester, 239, 458, 463; Niagara Frontier, 240; Binghamton, 403; Buffalo, 418; New York, 440; Syracuse, 477; Utica, 483, 484
Electric interurban railroads, 265
Electric power: impact, 184–84; in rural areas, 188; service regions, 285, 286–87F; transmission, 285, 416; generating stations, 286–87F, 288–89T; costs, 290T, 291
Electro chemicals, 239, 416
Electrometallurgical industry, 416
Electronics: development of, 188, 313; Rochester, 463; Syracuse, 474, 476P; Utica, 487
Elevation: and temperature, 61; and vegetation distribution, 91, 93–98 *passim*
Elevators, 432
Elizabeth, New Jersey, 424
Ellicott, Joseph, 151–52
Elm-Red Maple-Northern Hardwood Forest Zone, 95, 92F
Elmira (Newtown): development, 149–50, 169, 179; impact of railroad, 163; manufacturing, 242, 404; recreation, 307; and urban spacing, 347; and Corning, 495

Emery, 226
Empire State Building, 449P
Employee efficiency: external comparisons, 319–20, 316–17T, 320–21T
Employment: manufacturing, 167, 179, 236–42 *passim*, 246, 251–54, 252F, 253F; forest activities, 227; commercial fishing, 229; in tertiary activities, 255–56, 255T; retail trade, 278; and economic health, 318, 316–17T, 320–21T, 323, 322–23T, 324T; Albany area, 391–95; Binghamton, 403–4; Buffalo, 418, 420; New York City, 440, 442; Rochester, 463, 466; Syracuse, 477; Rome, 482; Utica, 484, 485, 487; forecasts, 496–98, 497T, 498T
Empty landscape, 360–61F, 368–69, 369P
Encroachment of incompatible land uses: and recreation, 299
Endicott, John B., 400
Endicott (Union): manufacturing, 242; and Binghamton Urban System, 396–405 *passim*
Endicott-Johnson Corp., 400, 401, 402, 403
Endwell, 396, 402
Energy, solar, 55–63
Engines, 246
English: and Indians, 119; in New Amsterdam, 127; west of Hudson, 134; immigration, 169, 189
English Puritans: in New Amsterdam, 127
Enrichment, in water, *87*
Entertainment services, 280, 434, 437
Entrepreneurs: and Rochester, 461, 462
Erie, Lake: basin characteristics, 37, 39; and climate, 65, 69, 70, 74; water characteristics and uses, 84, 86, 88; trade on, 151, 165; and agriculture, 205; fishing, 230; recreation, 305; and Buffalo, 375, 407–20 *passim*; as barrier, 469
Erie Canal: significance, 3; early interest in route, 136; development, 159–62, 165, 177; impact on agriculture, 166; and Barge Canal, 182; and manufacturing, 232, 240, 246; and complementarity, 263; and recreation, 303; and Albany area, 387, 389; and Binghamton, 399; and Buffalo, 411–12, 413; and New York City, 429; and Rochester, 461; and Syracuse, 471, 472, 473, 475; and Little Falls, 482; and Utica, 484
Erie County, settlement, 152; minerals, 221; natural gas, 226; and Buffalo Urban System, 407, 409, 414, 415, 420
Erie Indians, 114F, 115, 119
Erie-Lackawanna Railroad: merger, 187–88; Binghamton, 399; 404; significance, 267
Erie Lake Plain (land form division), 34, 26F, 100
Erie-Ontario Lowland (land form region): 33–34, 26F, 37P, 38P, 214P; climate, 67,

INDEX 529

Erie-Ontario Lowland—*Continued*
77; floods, 86; vegetation, 95, 98; soils, 106, 108; rural landscape, 358, 362; and Buffalo, 409; and Syracuse, 469

Erie (New York and Erie) Railroad: establishment of, 163; and Southern Tier agriculture, 166; competitors, 174–75; Binghamton, 399

Erratics, *19*

Essex County (New Jersey), 445, 451

Essex County (New York): minerals, 221, 223, 224, 225; manufacturing, 243; forest preserve, 292; recreation, 305

Established position: as asset, 453

Europeans: and Indians, 113, 119, 138

Evapotranspiration, 63, *66*, 70; actual, *70*; potential, *70*

Expansion, period of, 140–71

Expenditures, public: and urban problems, 354–56

Explosives, 242

Exports: from New Netherland, 124, 125; from New York colony, 136

Exurbia: *336*; problems, 356; New York City, 452; Rochester, 467; future growth, 493. *See also* Outer ring

Factories: size, 237, 238, 440; space for, 238; typical (New York City), 442; in CBD, 475

Fairfield County (Connecticut), 424, 446, 452, 453

Fairhaven, 176

Fairmount, 473, 474, 475

Farms: abandonment, 7, 74–77, 203, 211, 216, 217, 228, 362; land in, 178, 183, 202, 202F; number, 202, 202F, 363; non-commercial, 203; size, 203, 211–17 *passim*, 220, 362; range of status, 218P–19P; management, 220; landscape of, 358–62, 359P, 360–61F; relic features, 362–63; part-time, 368. *See also* Agriculture, Rural Landscape

Fayetteville, 473, 474

Federal arsenal, 178–79

Federal contracts: Niagara Frontier, 239, 240; and aircraft industry, 254; external comparisons, 319, 316–17T, 320–21T; and Utica, 487

Federal government: and urban problems, 356

Federalism: and urban problems, 356

Federal land survey system: similarities in New York, 143, 151

Federal Reserve System: and New York City, 434

Feeders, transportation, *264*

Felts, paper-making, 392

Fence posts: as a forest use, 100

Ferries, 430, 432–33

Fertility rates: and population projections, 491

Fifth Avenue (New York City), 446

Fifth-order trade centers, *284*, 349

Film, color, 467

Finance: and manufacturing, 232, 234; growth, 256; and New York City, 262, 434, 439, 443, as service, 280; and future, 497

Financial districts, 340, 446, 449P

Finger Lake Hills (land form division), 26F, 33, 373

Finger Lakes: formation, 21–23; physical characteristics, 43–44; canals, 265; recreation, 299, 302; and Rochester, 461

Finger Lakes area: climate, 67, 71, 72, 77; vegetation, 95, 98; settlement, 137, 145, 149; wine grapes, 177, 207P; growing degree months and agriculture, 205, 207; cash crops, 208; rural landscape, 358; and planning regions, 373, 375

Finger Lakes (recreation region), 306–9

First-order trade center, *281*, 282F, 375, 425

Fir trees, 227

Fiscal capacity: in urban systems, 354–56

Fish Creek (East Branch), 84F

Fishing: by Indians, 118; on Long Island, 138; Maine, 229; commercial, 229–30; fresh-water, 230; sport, 230, 299; licenses, 230, 294; hatcheries, 294

Fitchburg (Massachusetts) Railroad, 175

Flatbush, 122F, 126

Flatlands, 122F, 126

Flax, 134, 136, 166

Flemish: in New Amsterdam, 127

Floccules, *105*

Floodplain: manufacturing in Utica, 484

Floods, 86, 402

Florida: economic comparisons, 315–21 *passim*, 316–17T, 320–21T

Flour: development of milling, 136, 168, 178, 246; Genesee, 387; Binghamton, 400; Buffalo, 414, 416; Rochester, 458, 461; Little Falls, 482

"Flour City" (Rochester), 461

"Flower City" (Rochester), 461

Flower district (New York City), 446

Flushing: English village, 126

Fluviatile deposition, *21*

Fog: *66*; and highways, 391; and Broome County Airport, 404

Fonda: as Indian village site, 115

Food products: New York City, 237, 431, 440; Ontario Lake Plain, 239; Albany area, 392; Binghamton, 403; Buffalo, 415; Rochester, 458, 463, 467; Syracuse, 477; Utica, 484

Food stores: and retail sales, 278

Forage crops, 204

Ford Foundation, 436, 443

Forecasts, geographical, 491–500

Foreign-born population: (1855), 169–71, 171F; (1920), 184, 185F, 186F; (1960), 189, 190T

Foreign trade: and New York City, 428, 430, 431, 434–35

Forest condition and use regions, 228–29, 228F

Forest preserve: 98, 228, 229, 229F, 292, 294; recreation, 100–1, 305; Catskills, 229, 309; counties, 292; empty landscape, 368

Forests: distribution of, 90–93, 90F, 99F, 371; heavy, 96–98, *98*; ownership, 96–98, 100F; uses of, 98–101, 227–29; products, 136, 165; and economic health, 328; in exurbia, 336; future of, 499–500

Form, urban: *432*; modern reflections of old (New York City), 425; New York City (1815–1914), *432*; New York City (since 1914), 437; sources of (New York City), 444; problems related to (New York City), 454–55

Fort Edward, 134, 393

Forties, West (New York City), 448

Fort Niagara, 420

Fort Orange, 122

Fort Oswego, 131

Forts: English, 134; Niagara River, 410

Fort Stanwix: establishment of, 134; treaty of, and salt, 471

Fort Ticonderoga, 135

Forty-second Street (New York City): midtown, 433

Foundaries, 412

Foundations (philanthropic), 436, 443

Fourteenth Street (New York City), 425

Fourth-order trade centers, *283–284*, 283F, 350

Fragipan, *105*

Frame: in cities, *334*, 334–42 *passim*; in metropolis, 351; New York City, 444–45; Syracuse, 475

Franklin County: 208P; forest preserve, 292; and population growth, 495

Frankfort, 480, 482

Franklin Motor Co., 473

Fredonia: precipitation, 67; electric railroad, 181

Freight railroad: Albany, 390; yards, 390, 419P, 473; terminals, 430

French: trade with Indians, 119; colonial wars, 128, 130, 134; Protestants at New Paltz, 129; refugees, 144, 153

Front, weather, *64*

Frontier: in state, *143*

Frost, 62

Fruit, development as specialty, 166; modern, 205–7, 206F, 206P, 211–17 *passim*; Rochester, 458

Fruit belt, Niagara-Ontario, 177, 211

Fuel consumption: and heating degree-days, 63

Fuel oil, 264

Fuels, production of, 226

Fulton: Judge Cooper on, 148; flour, 168; and Syracuse Urban System, 243, 475

Fulton County: forest preserve, 292; economic health, 324; and population growth, 495
Functional region. *See* Nodal region
Fur district: New York City, 446
Furniture: Jamestown, 179, 242; New York City, 237; Albany area, 246, 387; and industrial aging, 248; Binghamton, 403; Utica, 484
Furniture districts: urban location of, 341
Fur trade: Indians, 119; early significance of, 124, 136; Albany, 130, 138, 387; Buffalo, 412

Galena (lead ore), 224
Galway, 135
Game farms, 294
Game management areas, 299
Garden City, 439, 444
Garnet, 225–26
Gasoline service stations: and retail sales, 278
Geddes, 149
General Electric Co.: Schenectady, 184, 387, 392, 393; and urban development, 348; Binghamton, 403; Syracuse, 476P, 477; Utica, 485
Genesee County: migration to, 143; advertising of, 145; settlement of, 150–51; development of, 152; barley, 177; in rural landscape, 358; flour and Erie Canal, 387; and Buffalo, 413
Genesee Flats, 150
Genesee Gorge, 303
Genesee River: and Valley Head Moraine, 44; and recreation, 302; at Rochester, 458, 462; port of Rochester, 461
Genesee Turnpike (Genesee Road): Indian trail, 119; and central New York, 149; extension to Lake Erie, 151; as turnpike, 156; and Buffalo, 410; and Syracuse, 471
Genesee Street (Syracuse): special districts, 475
Genesee Valley: landscapes, 45, 46–47P; snowfall, 69; Indian corn in, 118; settlement, 150; wheat, 166, 167
Genesee Valley Canal, 161, 177
Geneseo (Big Tree), 150
Geneva: temperatures, 60; migration through, 143; Judge Cooper on, 148; early development of, 150, 151; impact of Erie Canal, 159; (1850), 169; (1920), 184; and manufacturing region, 243
Geographical axes: *14*; in New York State, 14
Geographical conditions, 2000 A.D., 491–500
Geographical persistence: *15*; Indians and Europeans, 119; New York State culture areas, 139; and urban location, 348

Geographical perspective (1609–present), 190–95
Geographical prediction, *148*
Geographic regions, 371
Geographic tensions: development of, 171
Geography: *11*; analysis in, 3–7, 329; themes in, 7–16; interests and aims of, 11
Geologic features: and recreation, 299
Geologic history, 23, 25T
George, Lake, 41, 41F, 42P, 303
George Washington Bridge, 451
Georgian architecture, 338
Germans: immigration, 169, 179, 189; Lower East Side and Yorkville, 433
Germantown, 131, 133F
Glaciation: history, 19–23, 22F; and soils, 104, 106–10 *passim*
Glass products: Corning, 179, 242; Albany area, 246; Buffalo, 414; Utica, 483
Gleason, William, 461
Glens Falls: beginning, 138; lumber, 167, 178; growth, 179; ties to Albany area, 385, 387, 393
Gloversville, 179
Gloves, 242
Goshen, 181
Gouverneur: lumber-milling, 153, 178; talc, 225
Government: and recreation, 292–94; and economic health, 311–13; location of buildings, 341, 462; Broome County, 405; Monroe County, 467; Onondaga County, 479; local, in Utica area, 487
Government employment, 256, 262, 280, 497
Gowanda, 407
Grain: shipment of, 177, 390; farming, 183, 211; milling in Binghamton, 400; elevation in Buffalo, 407, 413, 416. *See also* specific crops
"Grand Canyon of the East," 45, 46–47P
Grand Central-Park Avenue area (New York City): office construction, 439; corporate management district, 446, 448
Grand Central Terminal: location of, 434
Grand Council Fire (or Iroquois), 471
Grand Island, 409
Granite, 225
Granville: slate, 226
Grapes, 177, 205, 207P
Grass, 204
Gravel: and manufacturing, 239, 242. *See also* Sand and gravel
Gravesend, 122F, 126
"Graybelt": development of, 439
Great Lakes: physical characteristics and development, 35–38, 79, 86; fishing, 38, 230; shipping, 38, 264, 265; climatic effects, 62, 65, 67; pollution, 85, 88; and manufacturing, 239–40; and Buffalo, 274. *See also* individual lakes
Great Lakes basin: Indian trade competi-

tion, 119; and Erie Canal, 263; New York City hinterland, division of, 429
Great Lakes-St. Lawrence Shore Line (recreation region), 305–6
Great South Bay: water pollution, 89
Greater New York: establishment of, 432–33
Greeks: immigration, 184
Green Island, 392
Greenport: ferries to New England, 164
Greenwich, 135
Greenwich Village (New York City): as suburb, 433; historical remnants, 444; as theater district, 448; residential development, 450
Green belts, 499
Greene County: cement, 224; forest preserve, 292
Grid, power, *285*, 291, 487
Grid pattern: in cities, 338, 402
Griffiss Air Force Base (Rome), 480, 482, 484, 487
Gross National Product (GNP): *198*; state's proportion of national, 3
Gross State Product (GSP), *198*, 199T
Ground moraine (till), *20*
Ground water, *79*, 81
Growing degree-days, *62*
Growing degree month: *63*; and vegetation distribution, 91, 93–96 *passim*; and agriculture, 205, 207
Growing season, *62*; and vegetation distribution, 91, 93–96 *passim*
"Growth" industries: 251–54, 253F; and Syracuse, 469
Growth rates, urban: hierarchy in, 352, 353F
Guns, 482
Gypsum, 221, 225, 226, 458

Hackensack, New Jersey: historical remnants, 122F, 444
Hackensack Meadows (New Jersey), 440, 451
Hackensack River (New Jersey), 429P
Hall, Basil: on Rochester (1827), 159
Hamburg, 407
Hamilton County: forest preserve, 292
Hamlet: in urban hierarchy, *349*
Hancock, 36P
Hanging valley waterfall, 44
Harbor Hill Moraine, 43
Hard goods, *237n*, 477, 480, 485
Hardwoods: uses of, 98, 229, 383
Harlem, 455
Harlem River: bridging of, 430
Harpersfield, 135
Hay, 166, 177, 183
Hay fields: and rural landscape, 359
Health resort and spa, public, 294
Heartland Division, Buffalo Region, 377

INDEX 531

Heating degree-days, *62*, 63T
Heavy manufacturing, 237n, 433
Helderberg Escarpment, 32, 47, 52, 387
Helderberg Hills (land form division), 26F, 33, 387
Hematite (iron mineral), 224
Hemlock bark: and tanning, 400, 414
Hemlock Lake, 306
Hemp, 134
Hempstead, 126, 439
Herkimer, General Nicholas, 482
Herkimer (German Flats): settlement, 131, 133F, 135; canal, 158; and Utica Urban System, 271P, 480; development of, 482
Herkimer County: cheese, 166, 177; airport, 273; forest preserve, 292; SMSA, 480; development of, 482
Hessian fly, 166
Hiawatha, 114
Hierarchies, urban, 348–52
Hierarchy: in manufacturing concentration, 236
High-demand regions (power), *285*
Highlands of the Hudson. *See* Hudson Hills
Highmarket: precipitation, 67
High-value-added manufacturing; *248*; and economic health, 313, 319; Rochester, 463; Utica, 487
Highways: development of, 182, 186, 187, 187F, 193F, 267–71; federal construction program, 182; and manufacturing, 188; mileage, 267; traffic flow, 269, 269F, 371; cities, 334, 340, 342; Albany area, 390–91; Binghamton, 405; Buffalo, 407, 421; New York City, 439, 452; Paramus, New Jersey, 439; Rochester, 462, 463, 464–65P; Syracuse, 473–75, 479; Utica, 487. *See also* Interchanges; Thruway; Interstate, U.S., and New York routes
Hillcrest, 396
Hills, 48F, 52
Historical remnants: and form of New York Urban System, 444
Historic sites, 294, 298T, 304
Hoboken, New Jersey: railroad terminal development, 430
Hoffmeister: precipitation, 73
Holland Land Company: Holland purchase, 142, 142F, 145, 151–52, 410; Lincklaen purchase, 142F, 149
Holland Purchase: acquired, 142, 142F; opened, 145; settled, 151–52, 410
Homer, 149
Honeoye Lake, 306
Hoosick Falls: manufacturing, 179
Hoosic River: and the Dutch, 135
Hoosic Tunnel (Massachusetts), 175
Hops, 166
Horizons: in soil, *105*
Hornell: transportation, 36P, 262, 274
Horseheads: canal, 161
Horticulture, 177, 183

"Hotbed" method: in land enterprises, *142*, 149
Hotels: location of, 334, 340; in urban hierarchy, 350, 351; New York City, 434, 439
Howe Caverns, 45–47
Hudson, Henry, 121, 122, 391
Hudson (Claverack Landing): growth, 154, 169, 179, 184; and turnpikes, 156; and railroads, 164; manufacturing region, 242
Hudson County (New Jersey), 445, 450
Hudson Falls, 393
Hudson Hills (Highlands of the Hudson) (land form division): 26F, 28, 31P, 373; heavy forests, 96–98; soils, 108
Hudson Mixed Farming Region (agricultural), 216
Hudson-Mohawk-Lake Plain axis: 7; and trade competition, 155; in canal era, 161–62; population development, 169–71, 179; manufacturing, 179; Thruway, 186; and transportation concentration, 232, 264–69 *passim*; transportation centers, 274; and urban development, 336, 347; New York City, 374; and Albany area, 388; and Buffalo, 407, 415
Hudson-Mohawk Lowland (land form region): 26F, 29; population (1850), 169; recreation, 305. *See also* Hudson-Mohawk-Lake Plain axis
Hudson River: 42; and Lake George, 41; flow, 84F; ship channel, 85; pollution, 85, 88; political division of estuary, 129; trade development, 136, 137; migration route, 144; as a barrier, 175, 430; oysters, 229; fishing, 230; and manufacturing, 234; transportation development, 264, 265; urbanization of shore, 270P, 374; recreation, 302; and Albany, 385–95 *passim*, 389P, 390P, estuary focus of New York Urban System, 423, 424; estuary shoreline residences, 433
Hudson Valley (land form division): 26F, 29; climate, 62, 67, 77–8; vegetation, 95, 229; soils, 106, 108; Indians, 114; in New Netherland, 121; land ownership, 125, 128; early settlement in, 126, 130, 131; flour, 134; mid-18th century, 134, 135–36; St. Lawrence rivalry, 134–35, 136, 159; in Revolution, 137; importance of political unity, 137–38; agricultural development, 138, 165–66, 216, 220; development (1785–1825), 154–55; opposition to Erie Canal, 159; and railroad development, 163, 175; fruit, 205–7; poultry, 208; cement, 224; clay, 225; manufacturing, 242, 243, 246; and trade complementarity, 263, 264; airport need, 273; power, 288; landscapes, 304, 338, 358; and Albany, 385–95 *passim*; and New York City, 425, 429

Hudson Valley (uniform region), 373–74, 375
Hudson Valley, upper: settlement, 134, 135; lumber, 167–68
Huguenots, 127
Human occupance: and regionalizing, 371
Hungarians: immigration, 184
Hunt's Point: "nuisance" industries, 433, 450
Hunting: by Indians, 118; licenses, 294; modern, 299
Huntington, 126
Huron Indians: location, 115, 114F; canoes, 118; monopoly on French trade, 119; dispersal by Iroquois, 119
Hurricanes, 65
Hydroelectric power, 284, 288–89, 414

Ice retreat, *20*
Ilion: manufacturing, 243, sales deficit, 279; and Utica Urban System, 480; development of, 482
Ilmenite (titanium-iron ore), 224
Immature soils, 105
Immigration: New Amsterdam, 127; Palatine Germans, 131; to 1855, 169–71; (1850's–90's), 179–81; (1890's–1920's), 184; (1920's–60's), 189–90; sources of, 189–90; Binghamton, 401; New York City, 431–36
Imports, colonial, 136
Income: and economic health, 311–13; external comparisons, 315–18, 316–17T, 320–21T, 322–23T, 323–24T; distribution, 316–17T, 317–18, 320–21T
Income, per capita: New York compared with nation, 6T; external comparisons, 315–17, 316–17T, 320–21, 322–23T, 324; Rochester, 458; Utica, 487
Income, personal: New York residents' proportion of national, 6T
Indian Lake: temperatures, 60, 62; airport, 273
Indian Point: atomic power plant, 288
Indian River, 303
Indians: historical geography, 113–20; distribution, 115–17, 116F; and Dutch, 122, 126; trade with, 126; wars with, 128–30; at end of Revolution, 138; territorial problems, 141; reservations, 141, 142F; and minerals, 221; oystering, 229; culture sites, 304; and empty landscape, 368; Binghamton, 398; Syracuse, 471. *See also* specific tribes
Industrial agglomeration theory, 249–50
Industrial cycle theory, 247–49, 247F, 248F
Industrial development: basis for, 232–34; and stock exchange development, 431; problems in Utica, 484, 487
Industrial districts: New York City, 433; Utica, 484–85

Industrial materials: and manufacturing development, 248, 249; Rochester, 458, 467; Utica, 484, 485
Industrial parks: New York City, 451; Utica, 485
Industrial Revolution, 167
Industry groups: relative significance of, 233F
Inner Loop (Rochester), 462, 463, 464, 465P
Inner zone of CBD, 334
Inns, 410
Institutional buildings: location of, 475
Instruments, scientific: Rochester, 179, 239, 458, 461, 463; New York City, 237, 440; growth in, 246; and industrial aging, 248; Binghamton, 403
Insurance: growth, 256; as service, 280; development in New York City, 425, 430, 437; location of firms in New York City, 434, 441; future, 497
Integrated store, *344*
Interaction, lines of, 265
Interchanges (super highway): impact, 186, 269–70; and manufacturing, 240; in cities, 340; Albany area, 391
Interchurch Center (New York City), 436–37
Interconnection, lateral, *264*
Interconnection, lines of, 499
Interconnection, patterns of: and planning and development regions, 376
International Business Machines Corp., 401, 402, 403, 241P
Interestate Highway system: establishment of, 186, 269; Route 81 (St. Lawrence–Pennsylvania), 186, 269, 493; Route 84 ("bridge route"), 186; Route 87 (Northway and Thruway), 186, 385, 390; Route 90 (Thruway and Berkshire extension), 186, 390, 391. *See also* Highways; Thruway
Interstate Sanitary Commission (New York City): establishment of, 437
Interurban electric railroad: and urban sector patterns, 340
Intervening opportunity, *263–64*
Investment in new plant and equipment: external comparisons, 319, 316–17T, 320–21T
Irish: immigration, 169, 179, 189; New York City, 433
Irondequoit, 458
Irondequoit Bay, 151
Iron manufacturing: development, 136, 168, 246; Northern New York, 153; Albany area, 167, 178–79, 387, 392; Buffalo, 414; Utica, 483, 484
Irone ore: mining, 136, 167, 221, 223–24, 392; and Seaway, 264; and Buffalo, 412, 413, 415
Iroquoian language group, 114F
Iroquois, Lake (glacial), 22F

Iroquois Indians: historical geography, 114–20, 114F; Binghamton, 398; Syracuse, 471; Upper Mohawk, 482. *See also* specific tribes
Irrigation, 71, 211
Italians: immigration, 181, 184; Binghamton, 401; New York City, 433
Ithaca: precipitation, 68; development of, 149; educational function of, 258, 280, 347; recreation, 307

Jackson, Andrew, and New York City banking, 428
Jamaica, 439
Jamestown: immigrants, 179, 181; manufacturing, 179, 242; population growth, 179, 184, 352; and urban spacing, 347; manufacturing belt, 404
Jefferson County, 106, 374
Jersey City, New Jersey, 424, 430, 433
Jersey Meadows (New Jersey), 429P, 444
Jersey shore, Hudson Estuary (New Jersey), 430, 433
Jewelry, 237
Jews: immigration, 127, 189, 431; residential areas, 433
Jogues, Father, 127
Johnson, Charles F., 401
Johnson, George F., 400
Johnson City (Lestershire), 396–405 *passim*
Jones Beach State Park, 43, 302P, 309
Journey-to-work survey (New York City), N.Y.U.S., 452–53

Kalm, Peter, 134
Kames, *21*
Katsbaan: Palatine German origin, 131, 133F
Kelsey-Hayes Co., 485
Kenmore, 407
Kennedy (John F.), International Airport, 272, 435P, 440
Kettles, *21*
Keuka Lake, 43, 165, 306, 309
Kieft's War, 119
Kill van Kull, 433, 451
Kilowatt hours produced, 284
Kings County (Brooklyn): market-gardening, 166; tertiary deficit, 256; and concentric pattern, 445; commuting, 452; future, 493. *See also* Brooklyn
Kingston (Esopus): early settlement (to 1700), 126, 128, 130; at end of colonial period, 136, 137, 138; status (1825), 154; and turnpikes, 156; growth, 169, 179, 184; and railroads, 175; manufacturing, 241P, 242; and urban spacing, 347; future, 374, 495
Knit goods: Cohoes, 392; Little Falls, 482; Utica, 483, 484

Knox, 135
Kodak Park, 238P, 466
Kortright, 135

Labor: agriculture, 214, 217–20; manufacturing, 246, 248–50 *passim*; Binghamton, 401; New York City, 436, 442; Rochester, 462
Labor force: future participation rate, 496–97
Lackawanna, 184, 235, 239, 240P, 407, 415
Lacustrine deposition, *21*
La Guardia Airport, 272, 440
Laissez-faire government: and New York City development, 428
Lake Ontario–Lake Erie Fruit and Vegetable Region, 211–14, 220
Lake Ontario Shore Railroad, 176
Lamprey, sea, 230
Land, speculation in: and geographical prediction, 148
Land-fill, 431, 440, 455
Land forms: 19–51, 49F, 50–51F; regions, 25–35, 26F, 371; categories, 48F
Land ownership: tracts, 142F; Northern New York, 153; forest lands, 228, 229, 292
Land policies: English, 128; during Revolution, 141; during colonization movement, 142–43; Rome, 149
Landscape: *332*; urban, 333–57; rural, 358–69; empty, 368–69
Land use: percentage distribution by type, 90F; and soils, 109F, 110; agricultural, 202; manufacturing, 236, 248; and railroads, 265; control and Catskills, 309; urban, 333–46 *passim*; sequential, 364–68, 365F, 366F, 367F; New York City, 447F, 450; future, 498–500. *See* Back-cover pocket (F)
Land values: and manufacturing, 238, 249; in cities, 335F, 336
Large-scale manufacturing, *237n*
Latham, 394
Latitude: and vegetation distribution, 91, 93–96 *passim*
L. C. Smith Typewriter Co., 473
Leaching, *105*
Lead, 224
Least cost point theory, 249
Leather products: New York City, 237; mid-Hudson region, 242; future, 251; Binghamton, 400, 403; Buffalo, 414
Lehigh Valley Railroad, 174–75
Le Roy, 243
Lester Shoe Co., 400
Letchworth State Park ("Grand Canyon of the East"), 45, 46–47P
Lewis County: forest preserve, 292
Lewiston, 148, 151, 407, 410, 416P
Life insurance: New York City, 431–32
Light manufacturing, *237n*

Limestone: formations, 23, 24F, 45, 52; exploitation, 221, 224, 225; and manufacturing, 239, 242; Albany area, 387; Buffalo, 415

Limited-access highways. *See* Highways

Lincklaen Purchase, 142, 142F

Linkages, urban, 340, 344, 345

Link Aviation, 403

Linseed oil, 414

Little Falls: Mohawk Gorge, 31; and transportation, 136, 158; and Utica Urban System, 243, 480; development, 482

Livelihood structure: *14*; Albany area, 391–94; Binghamton, 403–4; Buffalo, 415–20; New York City, 440–44; Rochester, 463–67; Syracuse, 477–78; Utica, 484–87

Liverpool, 149, 473, 474

Livestock and livestock products, 136, 204

Livingston County: settlement, 150; salt, 225; SMSA, 458

Local relief: *25*; and recreation, 299

Locational advantage and spatial interaction, 13–14

Location factors: manufacturing, 232, 236–42 *passim*, 248, *249*; urban, 346–48, 350–51, 482

Lockport: Erie Canal, 159; flour, 168; manufacturing, 239; and Buffalo Urban System, 243, 407; development, 412, 413

Locomotives: Schenectady, 168, 387, 392; Buffalo, 416

Loft buildings: manufacturing in, 440, 446

"Loftland" (New York City), 434

Logging: evidence of, 94P

Lomb, Henry, 461

London, England: and New York City, 425

Long Island (land form region): physical characteristics, 20, 23, 34, 43, 43F, 444; climate, 65, 67, 68, 77; agriculture, 85, 207, 208, 217; water problems, 86, 88–9; vegetation, 95, 98; soils, 108; Indians, 118; early settlement, 122F, 126; and New England, 126, 129; early economic activities, 127, 130, 138; and Erie Canal, 159; fishing, 229; manufacturing, 238; urbanization, 374, 440, 451; bridges to Manhattan, 430

Long Island Lighting Co., 285, 286F–87F

Long Island Railroad, 164, 175

Long Island recreation region, 309

Long Island Sound: Indians, 115; and New Netherland, 122; fishing, 230; shoreline residences, 433

Low-demand regions (power), *285*

Low-value-added manufacturing, *248*, 319, 487

Lowville: precipitation, 67

Lumber industry: New York's proportion of national, 65; as forest use, 98–100, 227–29; beginning, 136, 138; development, 153, 167–68, 178; and canals, 177; Albany area, 179, 246, 387; and urban location, 347; and empty landscape, 368; Binghamton, 400; Buffalo, 413; Little Falls, 482

Lutherans: in New Amsterdam, 127

Lyon Mountain, 392

Macomb, Alexander, 153

Macomb Purchase, 141, 142F

Machinery: Syracuse, 179, 473, 477; distribution, 237, 239, 242; Hudson-Mohawk Lowland, 246; Albany area, 393; Binghamton, 403; Buffalo, 415; New York City, 440; Rochester, 463; Little Falls, 482; Utica, 484

Machine tools, 242, 482

McKeever: temperatures, 62

Macropolis: in urban hierarchies, *351*, 352

Madison Avenue (New York City), 443, 446

Madison County, 469

Magnetite (iron ore), 223–24

Mail service: scheduled, 156

"Main Street": Manhattan as, 439, 448

Maintenance costs: in recreation, 294, 299

Malone: beginning, 153; railroad, 173

Malting, 136, 414

Mamaroneck, 129, 133F

Management: in agriculture, 203, 220, 362; in manufacturing, 247, 248; of forest resources, 368

Managerial Revolution: sign of, 443, 454

Manhattan (New York County): land form, 28, 374; wind speeds, 65, 66; Indians, settlement, 115; Dutch, 121, 122, 126; as urban core, 126, 429P, 444, 445, 446–50, 448P, 449P; population (1723), 130; (1770's), 135; urban expansion, 154; services, 256; wholesaling, 278; site qualities, 425; bridges, 430; land-fill, 431; and Greater New York, 432–33; historical remnants, 444; commuting, 452, 453; future, 493

Manhattan Hills (land form division), 26F, 28

Manlius, 149

Manufacturing: New York's proportion of national, 6T; (1774), 136; (to 1850's), 167–68; regional development, 168, 178–79, 236–42, 236F, 237F, 237T; (1850's–90's), 178–79; (1890's–1920's), 183–84; (1920's–60's), 188, 189; modern, 232–50; mapping, 234–35; production, 234–35T; hierarchy of concentrations, 236; descriptive terms in, *237n*; historical trends, 243–46, 244–45F; space, 248, 250; employment (external comparisons), 318, 219, 216–317T, 320–21T; urban location of, 334, 341, 342; and city location, 347, 348; and uniform regions, 371; Albany area, 392–93, 394; Binghamton, 399–405 *passim*; Buffalo Urban System, 412, 414, 421; New York City: development, 425–34 *passim*; suburbanization, 438; current, 440–41; city-serving, 441; city-forming, 442; in core, 446; and problems, 454; Rochester, 458, 461–62, 463–66; Syracuse, 473, 474, 477; Utica, 480, 483, 484–87, 486F; future, 251–54, 495, 497, 498

Manufacturing Belt, American, 167, 232, 236, 347, 430, 485

Manufacturing concerns: Albany area, 393T; Binghamton, 404T; Buffalo, 420T; Rochester, 466T; Syracuse, 478T; Utica, 485T

Manufacturing geography: theories, 247–50

Maple products: as a forest use, 100, 165

Maple trees, 229

Maps: use of, 11; and regionalization, 371

Marble, 225

Marine and casualty insurance, 431

Marine boilers and engines, 414, 416

Markets: for agriculture, 203, 207, 211–17 *passim*, 220; for dairy products, 204–5; for forest products, 227; for manufacturing, 232, 247, 248, 249; and rural landscape, 358, 362, 368; for electricity in Buffalo, 414; and Buffalo, 415, 416; and Rochester, 467; for salt, and road development in Syracuse, 471; and Little Falls, 482; and Utica, 484, 485

Market-gardening, 166

Marshes, 91

Massachusetts: territorial disputes with, 140–41; as source of colonists, 144

Massena: railroad, 176; power, 232; manufacturing, 242, 243, 374; and Seaway, 264

Mattydale, 473

Mature soils, *105*

Maude, John, 165

"Mauritius River" (Hudson), 122

Mayville, 152

Meat-packing, 178, 392

Meat-shipping: to Buffalo, 413

Mechanicville, 102P, 385, 387, 392, 393

Medical centers: urban location, 342

Medical district: Syracuse, 475

Medicine, 256

Medium-scale manufacturing, *237n*

Megalopolis, 455

Menhaden, 229

Merchant wholesalers, *278*

Mesophytic trees, *93*

Metals, fabricated: regional distribution, 237, 239, 242; development, 246; Albany area, 392, 393; Buffalo, 415; New York City, 440; Syracuse, 473, 474, 477; Utica, 482, 483, 484

Metals, mining of, 223–24

Metals, primary: regional distribution, 239, 240, 242; Buffalo, 415; Syracuse, 477; Utica, 484

Metropolis: in urban hierarchies, *351*, 352

Metropolitan areas: *336*, 340; and economic health, 328; problems in, 352–56, 479

Metropolitan New York, high-demand region (power), 288

Metropolitan New York (manufacturing region), 236–38, 251

Metropolitan New York (uniform region), 374–75

Metropolitan Regional Council (New York City), 423, 437, 438

Middleburgh, 131

Middle City: *334*, 334–42 *passim*; New York City, 452; Rochester, 467; Syracuse, 475

Middle Mohawk (uniform region): 373, 375; manufacturing, 241–42; economic health, 324, 328; urban growth, 352; and Albany area, 385; and Utica, 480

Middlesex County (New Jersey), 446, 452

Middletown: population, 179; electric railroad, 181

Midge, 166

Mid-Hudson manufacturing region, 242, 251

Midtown (New York City): new focus, 433–34, 440

Midtown Plaza (Rochester), 462–63

Migrant worker housing: and rural landscape, 362

Migration: into northern New York, 152; Wyoming Valley to Binghamton, 401; and population projections, 491; to fast-growing areas, 495

Military bases: and urban growth, 352

Military Tract: "New," 141, 142F, 149; "Old," 141, 142F

Milk: 177, 183; skim, 205

Milk shed, *204*

Mineral waters, 52, 221

Minerals and Rochester, 458

Mining: New York's proportion of national, 6T; production, 220–27; and urban location, 347; and rural landscape, 368. *See* back-cover pocket (F)

Minoa, 473

Mixed landscape of farms and abandoned land, 360–61F, 362–68, 363P

Mohawk (village), 271P, 480, 482

Mohawk Airlines, 391

Mohawk and Hudson Railway: establishment of, 162

Mohawk Escarpment Dairy Region (agriculture region), 214, 215P, 220

Mohawk Indians, 114F, 115, 138

Mohawk-Lake Plain axis: and Albany area, 391; and Rochester, 458, 461; and Syracuse, 469; and Utica, 480, 482

Mohawk River: and glaciation, 21, 22F, 37; land form, 31; pollution, 88; and Indian villages, 115; and trade, 136; bridge at Utica, 148; channel improvements, 158, 182; recreation, 302; and Albany area, 385–95 *passim*; early travel route, 410; and Utica, 480, 483, 484

Mohawk Valley (land form division): 31, 26F, 271P, 272P; forests, 98; soils, 106, 108; Indians, 114; settlement, 131, 135; impact of Revolution, 137; conditions after Revolution, 143; migration through, 144–45; settlement to 1825, 148–50; effect of turnpikes, 157; agriculture, 165–66, 214; manufacturing region, 234, 240–42; manufacturing development, 246; architecture, 304; Utica Urban System, 480, 482, 487

Mohawk Valley Community College, 487

Mohican (Mahican) Indians, 114F, 115, 119

Moisture, 66–72

Monadnock, *25*

Monazite, 467

Monmouth County (New Jersey), 424, 446

Monroe County, 458, 461, 467

Monroe County Airport, 461

Montgomery County: broom corn, 166

Montreal: and Northern New York, 153–54; trade compitition, 156; and Erie Canal debate, 159; and transportation routes, 388

Moraines, 444

Morris County (New Jersey), 446, 452

Morris Reserve, 142, 142F, 150–51

Mortality rates: and population projections, 491

Motor boats, 230, 294

Motor Boats, Division of, 302

Motor vehicles, registered, 267

Mountain glaciers, 19

Mountains, 48F, 52; selected summit elevations, Appendix A

Mount Colden, 20P

Mount Marcy, 19, 41, 305

Mount Van Hoevenberg bobsled run, 294, 305

Muck, *105*

Multiple nuclei, in cities, *342*, 342–46, 439–40

Municipal power companies, 285, 291

Munson-Williams-Proctor Institute (Utica), 487

Music: and New York City, 434

Nanticoke (town), 405

Napoleonic Wars: and manufacturing, 246

Nassau County, 250, 324, 445, 451, 493

National Park Service, 294, 298T, 304

Natural gas, 221, 226

Natural regions, 370

Natural Resources, Joint Legislative Committee on, 292

Natural-societal regions, 371

Negroes: in New Amsterdam, 127; population increase, 184, 190; in central cities, 354; in New York City, 443–44, 450, 455; in Rochester, 467

Neighborhood shopping district: *344*; in urban hierarchy, 349, 350; in Syracuse, 475

Net fishing, 230

Neversink Reservoir, 86

New Amsterdam, 126–27, 425. *See also* Dutch

Newark, New Jersey, 424, 433, 439, 444

Newark Airport (New Jersey), 272, 440

Newark Bay (New Jersey), 433

Newburgh: settlement, 131–34, 133F; development, 154; and turnpikes, 156; and railroads, 164, 175, 181; (1850), 169; manufacturing, 242; and urban spacing, 347; and urban growth, 374

New England: pressures on New York colony, 128, 129; as source of colonists, 143–44, 401; railroad interests, 163; railroad connections with 175, 175F

New Englanders: in New Netherland, 126; migration to New York, 129, 131; in frontier, 134, 135; in Upper Hudson and Mohawk valleys, 135; region of dominance (1785), 139; "Yankee invasion," 143–44; conflicts with Yorkers, 144, 169; and Hudson Valley towns, 154; region of dominance (1850's), 169

New England Upland (land form region): 26F; 28; climate, 74–77

New Hampshire: territorial dispute with, 140–41

New Harlem, 126, 122F

New Hartford, 148, 483

New Haven Railroad, 175

New Jersey: effect of political separation, 129; source of colonists, 144; economic comparisons, 315–21 *passim*, 316–17T, 320–21T; suburban development, 440. *See also* New York Urban System, counties, Jersey shore

New Netherland, 121–28, 123F. *See also* Dutch

New Paltz, 129, 133F

Newport, Virginia: rival to New York City, 425

New Rochelle, 129, 133F, 184

New Scotland, 135

Newtown Creek (New York City), 433, 450

New Utrecht, 126, 122F

"New Vermont," 144

New York (nodal region), 375

New York Central Railroad: establishment of, 163; competitors, 174–75, 176; and industrial development, 240, 271P, 272P; Albany area, 389; Rochester, 461; Syracuse, 473

New York City: 13P, 423–57 *passim*; impact on state, 3; climate, 62, 63, 68, 77; water supply, 80, 86; water management, 86; Dutch, 121–28; upstate tensions, 126, 171; effect of English, 128; in colonial period, 130–37 *passim*; in 1785, 138; effect of colonization, 144; interests in Northern New York, 153; growth to 1825, 154; and New York trade competition, 156; and Erie Canal, 159, 160; and railroad development, 163, 173–75; flour, 168; manufacturing development, 168, 178, 179, 188, 243, 246; Negroes, 184; migration to, 190; milk shed, 204; and New York State manufacturing, 232–34; manufacturing problems, 238; and Mid-Hudson manufacturing, 242; modern transportation development, 263, 264, 271, 272, 273–74; services, 280; power costs, 291; and recreation, 304; banking center, 318; apparel industry, 319; economic health, 323, 324; population, 333; middle city, *334*, 336; and urban spacing, 346, 347, 348; and uniform region, 375; and Binghamton, 404; population projections, 455

New York City Periphery Truck Crop Region (agriculture region), 216P, 217, 220

New York City Recreation Region, 309

New York Colony, 128–39

New York Metropolitan Region Study, 446n

New York-Northeastern New Jersey Standard Consolidated Area, 424

New York Ontario and Western Railroad, 174F, 176, 187

New York Route 5, 391, 410

New York Route 7, 157, 385, 391

New York Route 17, 307

New York SMSA, 424

New York State Electric and Gas Corp., 285, 286–87F, 291

New York Urban System: 423–57, 423F, 424T, 426–27F, 429P; population growth, 169, 179, 189, 191F, 333; agriculture, 207, 217; nonmetals, 224, 225; importance of harbor, 232; industrial center, 234; and peripheral manufacturing, 243; manufacturing importance in state, 246; tertiary surplus, 256, 375; parkways, 267; retailing, 278; wholesaling, 278, 320; trade order, 281; power companies, 285; and recreation, 209; economic health, 315, 328; urbanized area, 333, 424, 426–27F; urban frontier, 336, 424, 426–27F; urban growth, 352; metropolitan region, 423; SMSA, 424; standard consolidated area, 424; population projections, 455, 492; future, 493, 497

Niagara County, 407, 415, 420

Niagara Falls: general characteristics, 37, 38–39, 39P, 40F; regulation of flow and lake level, 84, 86; and Erie Canal debate, 159; power generation, 184, 285, 288; and manufacturing, 240; recreation, 299, 303, 305–6; and Buffalo Urban System, 414, 420; and Utica, 487

Niagara Falls (city): pulp and paper mills, 100; Canadians in, 171; and railroad development, 175, 176; (1920), 184; manufacturing, 235, 239; and urban location, 348; and Buffalo Urban System, 407, 409, 413, 415, 416, 421

Niagara Frontier, 410

Niagara Frontier Manufacturing Region, 239–40, 240P

Niagara Frontier Resource: high-demand region (power), 288–89

Niagara Gorge, 39, 306, 409, 414–15

Niagara limestone escarpment, 34, 410, 412

Niagara Mohawk Power Corp., 285, 286–87F, 291

Niagara River: erosion of, 37; in War of 1812, 152; fishing, 230; and manufacturing, 235; and recreation, 305; and Buffalo Urban System, 409, 410, 412, 414, 415, 420

Niagara River Power Project, 305

Nimmonsburg, 398

Nodal (functional) region: *15*; Dutch design of, 126; composite, 370–71, 375–76, 377F, 378F

Nonbasic economic activities, *13*, *247*

Nondurable goods, *237n*, 442, 487

Nonmetals, 224–26

North Atlantic Trade Route: and New York City, 428

North Country: climate, 74; Indians, 115; forests, 228–29; manufacturing, 243; services, 258; and Syracuse, 469, 478

North Country (uniform region), 374, 375

North Country Dairy Region (agricultural region), 214–16, 215P, 216P, 220

North Creek: garnet, 225

Northern Hardwood Forest Zone, 95–96, 92F, 94P

Northern New York: settlement, 143, 145, 152–54; railroads, 163, 173, 175, 176, 187; agriculture, 166; and Seaway, 264; lumber and Albany area, 387

Northern New York (planning and development region), 377, 380F

North Evans, 407

"North" River (Hudson River), 121, 122

North Syracuse, 473, 474, 475

North Tonawanda, 239, 407, 413

Northward migration: in Manhattan, 437

Northway (Interstate 87), 385, 390

Norwich, 149, 179

Norwood, 178

"Nuisance" industries: location in New York City, 433

Nunataks, *22F*

Nurseries: Rochester, 461

Oak Forest Zone, 93–95

Oak-Northern Hardwood Forest Zone, 92F, 95

Oak trees, 229

Oats, 125, 134, 136, 166, 177

Ocean transport: and New York City development, 428

Ochenang, 398

Off-Broadway theater districts, 448

Office for Regional Development, 376, 379F, 491, 492

Office: function in cities, 340, 350; space for, 439, 478

Ogdensburg: precipitation, 67; water balance, 71, 72T; and Judge Cooper, 148; development of, 153; and railroads, 163, 164, 175, 176; flour, 168; in 1850, 169; and Seaway, 264; and urban location, 348

Ohio Country, 157–58, 413

Ohio River system: drainage in New York State, 33

Oil-refining area (New York City), 450

Old-age industrial region, 319

"Old Americans" (New York City), 433

Olean: development of, 151, 152; electric railroad, 181; manufacturing, 242, 404

Oneida, 243

Oneida Country: as focus of settlement, 148

Oneida County: settlement, 143; hops, 166; cotton mills, 167; Welsh in, 181; hematite mining, 224; forest preserve, 292; airport, 487; SMSA, 480, 204P

Oneida Indian Reservation, 141, 142F, 149

Oneida Indians, 114F, 115, 118

Oneida Lake: pollution, 88; canal route, 161, 182; recreation, 303

Oneida Lake Plain (land form division), 26F, 34

Oneonta, 179, 181

Onions, 208, 208P

Onondaga County: agriculture, 166, 177; cement, 224; salt, 225; manufacturing, 243; SMSA, 262F, 469; government, 493

Onondaga Creek, 471

Onondaga Escarpment, 410

Onondaga Indian Reservation, 141, 142F, 471, 475, 477

Onondaga Indians, 114F, 471

Onondaga Lake, 182, 471, 472, 473

Onondaga Valley, 469

Ontario, Lake: basin characteristics, 20, 37; and climate, 65, 69, 70, 71, 73; pollution, 88; and agriculture, 205, 207; fishing, 230; and manufacturing, 239; and transportation, 153, 159, 461, 472; recreation, 305, 306; and North

Country, 374; and Buffalo, 410, 412, 420; and Rochester, 458, 461; as barrier, 469; and Syracuse, 472
Ontario County, attraction to settlement, 150
Ontario Drumlins (land form division), 34, 26F
Ontario Lake Parkway, 306
Ontario Lake Plain: and urban development, 347
Ontario Lake Plain (land form division): 26F, 34; agriculture, 206P, 208; Lockport, 412; Rochester, 461
Ontario Lake Plain (manufacturing region), 234, 238–39, 238P, 251
Ontario Ridge and Swampland (land form division), 34, 26F
Open space, urban, 451, 499
Optical goods: Rochester, 179, 184, 239, 458, 461, 463
Orange and Rockland Utilities, Inc., 285, 286F–87F
Orange County: colonization, 134, 135; agriculture, 166; resorts, 309; New York Urban System, 424, 445, 452; future, 493
Orchard Park, 407
Orchards, 362, 461
Orders of trade centers, 281–84
Ordinances, regulatory (New York City), 437
Ordnance equipment, 242, 392
Oriskany, 480
Oriskany, battle of, 482
Oriskany Creek, 149
Orleans County, 458
Oswego: climate, 60, 67, 69; and Erie Canal, 159; wheat and flour, 161, 168, 178; (1850), 169; Canadians in, 171; and railroad development, 176; (1890), 179; (1920), 184; and Syracuse, 243, 471, 472; and Seaway, 264; sales deficit, 279; atomic power plant, 288; economic health, 328; and urban location, 348
Oswego Canal, 160, 177, 183, 303
Oswego County: settlement, 143; dairying, 177; SMSA, 469
Oswego River: migration route, 145, 410; and Scriba's lands, 153; navigation on, 158; and Syracuse, 472, 475
Otisco Lake, 43, 306
Otsego County: hops, 166
Otsego Lake, 303
Outdoor recreation areas, 297–98T
Outer loop: Rochester, 463
Outer ring (exurbia), New York City, *445*, 452
Outer zone of CBD, 334
Outwash plains, *21*, 444
Overlay analysis, 11, 11F, 284
Owasco Lake: area, 43, 306
Owego: development, 149–50; impact of railroad, 163; tobacco area, 401; and urban expansion, 403

Owego Creek: flow, 80, 84F
Oxbow: and Judge Cooper, 148
Oystering, 138, 229
Ozalid Division, General Archive and Film Corp., 401

Palatine Germans, 131, 482
Paleo-Indian Hunters, 113
Palisades, 41–43, 299, 444, 451
Palisades Interstate Park, 309
Paper and paper products: development, 179, 246; Watertown, 242; Albany area, 387, 392; Binghamton, 403; Buffalo, 415; Syracuse, 477; Little Falls, 482
Paramus, New Jersey: shopping center development, 438, 439
Parent material: and soils, 104, 106–10 *passim*
Parish, David, 153
Parkways, 267, 269
Passaic, New Jersey, 438, 444, 452
Passaic County (New Jersey), 445, 451
Pasture land, 90F, 358, 359, 362
Paterson, New Jersey, 424, 438, 444, 452
Patroon system, 125
Pax Britannica: and New York City, 428
Peaches, 125, 134, 205
Pearlash, 165
Pears, 125, 134, 205
Peas, 125, 136, 208
Peat, 226
Peekskill, 226
Pelham, 126
Peneplain, *23*
Penetration, lines of, *264*
Pennsylvania: New York compared with, 6T; as source of colonists, 144; trade with, 149; economic comparisons, 315–21 *passim*, 316–17T, 320–21T. See also Coal, Wyoming Valley, Scranton–Wilkes-Barre area
Pennsylvania Railroad, 174
Pennsylvania Station, 434
Peripheral growth, urban, 338, 495
Personal services, 280, 334
Perth, 135
Petroleum: movement of, 187, 390; Rochester, 461
Pharmaceuticals, 237, 392
Phelps and Gorham Purchase, 141–42
Philadelphia, Pennsylvania: rivalry with New York City, 425, 428, 429
Photographic materials: Rochester, 179, 184, 239, 458, 461, 463; Binghamton, 242, 401, 403
Physical environment: interrelationship of man with, 12–13; and urban sectors, 339; and New York City development, 428, 440
Physical size, and form of New York Urban System, 444
Piermont, 163

Pine trees, 227
Pipelines, 187, 487
Pittsburgh: and trade competition, 157–58, 158F
Place consciousness, 7–11, *7*
Place names: of Dutch origin, 127; in Central New York, 149
Placid, Lake, 29P, 41, 62, 63, 369P
Plains, 48F, 52
Planes, 205
Plank roads, 165
Planning and development: in manufacturing, 247, 249, 250; and economic health map, 329; Buffalo, 421; in New York City, 437, 439, 451, 455; by corporations, 443; and future cities, 499
Planning and development regions: 370–81, 380F, 381T. See also Office for Regional Development
Plants: and recreation, 299
Plattsburgh, 153, 374, 390
Plows, 400
Poles: in Buffalo, 181; immigration, 184, 189; in Binghamton, 401
Political complexity: in urban systems, 354, 355F; Binghamton, 405; Syracuse, 479
Political reality: and planning and development regions, 376–77
Pollution: atmospheric, 66; of water, 85, *87*, 87–89, 405
Population: and geographic analysis, 3–7; New York's proportion of national, 6T; (1960), 8–9F; Indian, 115; New Netherland, 125; colonial period, 130–31, 130T; (1775), 135–36, 137; colonization movement and growth, 143; Genesee Country, 151; western New York, 152; Hudson Valley, 154; (1850), 168–71, 170F; (1850's–1890's), 179–81; (1880), 180F; (1890's–1920's), 184; (1920's–1960's), 189–90; (1780–1960); 189T; phases in evolution of patterns, 190; farm, 202, 203; and manufacturing, 235, 248; and tertiary sector, 256; and retail sales, 278–80; and economic health, 315, 316–17T, 320–21T, 322–23T, 323, 324T, 328; and urban hierarchies, 348, 349F; urban trends, 352; rural trends, 363–68; and uniform regions, 371–75 *passim*; and planning and development regions, 371, 377, 381T; future growth, 384, 491–96, 492T, 499; Albany area, 385, 395; Binghamton, 396, 399T, 402, 403T, 406; Buffalo, 407, 411, 413, 415, 421; New York Urban System, 424–25, 424T; and ubiquitous industries (New York City) 431; declining growth rate (New York City) 434; effect of shifts (New York City) 436; size and form of Urban System, 444; in concentric rings (New York City), 448–50, 451, 452; projections (New York City), 452, 455, 456T; related problems (New York

Population—*Continued*
  City), 455; Rochester, 458, 468; Syracuse, 469, 471, 473, 474, 474T, 475, 479; Utica, 480, 482, 482T, 484, 488
Population, threshold, *343*, 344–46 *passim*
Pork, 136
Portages, 136, 410
Port Jarvis (Minisink), 137
Portland, 152
Portland cement, 224–25, 387, 393
Ports: and manufacturing, 234; and urban development, 348; New York City, 425, 429, 430, 434, 437, 438; Buffalo, 412, 421
Port Watson, 149
Potash, 387
Potatoes, 134, 205, 208, 208P, 211, 458
Potsdam: and Judge Cooper, 148; origin 153; agriculture, 166; and railroad development, 176; lumber-milling, 178
Pottery, 179, 473, 484
Poughkeepsie: development of, 154; in 1850, 169; and railroad development, 175; manufacturing, 242; and urban spacing, 347; and urban growth, 374
Poultry, 165, 208–10, 216, 217
Power: 284–91; steam plants, 183; hydroelectric plants, 183–84, 240, 416P; and manufacturing, 249; Buffalo, 420, 421; New York City, 432
Power Authority of the State of New York (PASNY), 285, 289, 291
Power-generating regions, 285–91
Power transmission distance, 285
Precipitation: characteristics, 66–70, 68F, 69T; and vegetation distribution, 91
"Pre-emption line," 141
Prestige: and New York City, 431
Primary sector: 6T, *14*, *198*, 199T, 201–31, 201T; future, 230, 497, 498
Primate city, *425*, 434, 446
Principal economic axis (core area): *14*; New York in relation to the United States, 3, 14, 5F, 265, 374
Printing and publications: New York City, 237, 431, 436, 439, 440; growth in state, 251; Albany area, 392; Binghamton, 403; Buffalo, 415
Producer goods, *415*
Production: destination of New York State commodities, 4F
Professional services: significance, 255–56; in major cities, 258, 262; in services structure, 280; New York City, 443; future, 497
Protestants: in suburbs, 354; center in New York City, 437
Public housing: New York City, 455
Public relations: development in New York City, 436
Public Service Commission, 291
Publishing, 425, 434, 436, 439, 448. *See also* Printing and publications

Puerto Ricans: New York City, 190, 436, 443, 450, 455
Pulp, wood, 100, 102P, 229, 392
Pulteney Purchase, 142, 142F, 150–51
Putnam County: New England colonists in, 131; agriculture, 217; and New York urban system, 424, 445, 452–53; future population, 493
Pyrite, 224

Quakers, 127
Quartzite, 225
Quassaick Creek: Palatine Germans, 131
Quebracho bark, 400
Queens: swamps, 430; in Greater New York, 432; in developed ring, 445, 450; and commuting, 452; future growth, 493
Queens County (Queensborough): market-gardening in, 166
Queenston, Canada: and power blackout, 291

Racial problems, in central cities, 354, 455
Radial pattern: Buffalo streets, 407
Railroads: development of, 162–64, 164F, 265–67; track gauge, 163, 173; trunk line competition, 173–74, 174F; New England connections, 175, 175F; and canals, 177; and agriculture, 178, 183; electric, 181–82, 182F; and highways, 185; impact, 187–88; pattern, 192F; and manufacturing, 234, 246; (1960), 268F; problems, 272; and urban patterns, 340, 342; and planning regions, 371; Albany area, 387, 389–90; Binghamton, 399; Buffalo, 413, 415; and New York City port development, 428, 429–30; and stock exchange development, 431; connecting (New York City), 432, 433, 451, 455; New York City terminals, 433; in developed ring (New York City), 450; Rochester, 461; Syracuse, 473, 474–75; Utica area, 482
Ramapo Mountains, 29
Range of population projecting, 492, 492T
Rapid transit: and urban patterns, 340
Raquette River, 168, 289
Raspberries, 207
Raw materials, Albany area, 387
Real estate, 256, 280
Recessional moraine, *21*
Recreation: as a forest use, 100–1, 229; fishing, 230; and canals, 265; growth of, 256; in service structure, 280; and conservation, 291–309; attractions, 293F; data on facilities, 295–99T; and cities, 341, 347; and empty landscape, 368; Buffalo, 420; future, 495, 499–500
Recreation regions, 305–9, 306F

Redevelopment: in cities, 338–39; New York City, 450, 451; Rochester, 462–63
Redistribution theory, 250
Redistricting: in Utica, 487
Refining, sugar beet, 208
Refining industries: and New York City, 436
Reforestation, 227, 294
Region: concept of, *15*, *370*; process for delimiting, 15, 370–71; land form, 25–35, 26F; climatic, 72–78; 75F; forest (zones), 93–96, 92F; soil, 106–9, 107F; agricultural potential, 109, 110F; agricultural, *210*, 210–17, 211–12F; forest condition and use, 228–29, 228F; manufacturing, 236–42, 236F; electrical service, 285, 286–87F; power-generating, 285–91; recreation, 305–9, 306F; economic health, 321–29,·325F, 326F, 327F; landscape, 333–69, 360–61F; uniform, 371–75, 372F; nodal, 375, 378F; planning and development, 370–81, 380F
Regional consciousness: as problem in New York City, 455
Regional council: in planning and developing regions, 376
Regional Plan Association, 423, 437, 446, 452
Regional shopping districts: and urban hierarchies, *345*, 350, 351; Syracuse, 474, 475
Religion: of Indians, 117
Religious groups, 127
Religious institutions: location of, 341
Remsen: Welsh, 144; railroad, 173
Rensselaer, 385, 392
Rensselaer County: flax, 166; manufacturing, 246; SMSA, 385; tertiary sector, 394
Rensselaerswyck, 125, 125F
Repair services, 256, 280, 334
Representation, political: problems in Binghamton, 405
Republicans: in suburbs, 354
Research and development: in manufacturing, 249, 392, 421, 454
Residences: cities, 334–42 *passim*; in rural landscape, 359, 364; Binghamton, 399, 402, 405; Buffalo, 421; New York City, 433, 438, 449–50, 438; Rochester, 463, 467; Syracuse, 474, 475, 478, 479; Utica, 485
Resort area, 309
Resource base: and population distribution, 6
Resource regions (power), *285*
Resources, *201*
Restaurants: New York City, 441
Retail trade: 6T, 255, 278–80, 279F; employee efficiency (external comparisons), 316–17T, 319, 320–21T; areal variation in sales, 322–23T, 323, 324T; urban location of, 334; New York City, 436, 438, 439, 441, 451

Retirement: and rural nonfarmers, 364; and future employment, 496
Retsof, 225
Revolutionary War, 136–37
Rhinebeck, 131, 133F
Rhode Island: as source of colonists, 144; and Hudson (city), 154
Richards Atlas of New York State, 128
Richelieu Canal (Canada), 265, 388, 391
Richelieu River (Canada), 153
Richfield Springs, 52
Richmond County (Richmond borough, Staten Island), 445, 451, 452, 493. *See also* Staten Island
Ridge and Valley Province, 29, 31
Ridge Road, 410
Ritter, Frank, 461
Riverhead, 62, 81
Rivers. *See* Water, Appendix C, and specific rivers
Roads: Genesee Country, 150; Western New York, 151–52; Northern New York, 153; early 19th century, 157F; improved, 182, 183F, 185–86; rural, 188, 359; and forest preserve, 292; Binghamton, 398, 399; Buffalo, 410, 412; Syracuse, 471. *See also* Highways
Rochester: 458–68 *passim*, 459F, 464–65P; drumlins, 20; temperature, 63; cloudiness, 58; wind speed, 65; precipitation, 65, 67, 68, 70; snowfall, 70; and Judge Cooper, 148; development of, 151; and Erie Canal, 159; flour-milling, 166, 168, 178; (1850's), 169; and railroad development, 176; manufacturing, 184, 234, 235, 238P; (1890's), 179; tertiary surplus, 256; and transportation, 263, 270, 274; trade and services, 262, 278, 280; trade order, 282; recreation, 303, 304, 307; and urban form, 334, 339; and urban spacing, 347, 348; nodal region, 375
Rochester, University of, 467
Rochester Gas and Electric Corp., 285, 286–87F
Rochester Institute of Technology, 467
Rochester SMSA. *See* Rochester Urban System
Rochester Urban System: 458, 468, 459F; manufacturing, 238–39, 251; economic health, 328; urbanized area, 333, 375, 458, 459F; and Upstate Heart, 373; SMSA, 458; urban frontier, 458, 459F; port, 461; future, 492, 493, 497
Rockefeller Plaza, 448, 450
Rock formations, underlying, 24F
Rockland County: agriculture, 217; manufacturing, 238; economic health, 324; sandstones, 374; and concentric rings, 445, 450; commuting, 452; future population, 493
Rock salt, 23
Rome: and Judge Cooper, 148, 149; rivalry with Utica, 148–49, 176; canal to Wood Creek, 158; and railroad development rivalry of Utica, 176; manufacturing, 234, 242, 246; airbase and growth, 352; and Utica Urban System, 480, 482, 487
Rome, Watertown and Ogdensburg Railroad, 174F, 176
Ronkonkoma ridge, 43
Rotterdam, 385
Rubber goods, 251
Rugs, 242
Ruhr River (Germany): control of water pollution, 89
Runoff, average annual, 80, 80F
Rural Electrification Administration (REA), 285
Rural Landscape: 90F, 358–69, 360–61F, 371; tensions with cities, 171; population, 179, 493; manufacturing, 239P; and economic health, 328; nonfarm residences, 359; future employment, 497–98; and future cities, 499. *See also* Agriculture; Farms
Rural nonfarmers, 7, 217, 336, 359, 364
Russians: immigration, 184, 189
Rutland Railroad (Ogdensburg-Lake Champlain line): abandoned, 187
Rye: grain, 125, 134, 136, 166, 204; straw, in paper manufacturing, 179
Rye (village), 126

Sacandaga Camp Site, 307P
Sacandaga Reservoir, 303
Sackets Harbor: and Judge Cooper, 148; development of, 153; railroad, 176
St. John Fisher College, 467
St. Lawrence-Champlain Lowlands (land form region), 25–28, 26F
St. Lawrence County: soils, 106; minerals, 221, 223–24, 225, 226; forest preserve, 292; economic health, 324; forests, 374
St. Lawrence Hills (land form division), 25–28, 26F
St. Lawrence Marine Plain (land form division), 25–28, 26F, 30P
St. Lawrence Power Project, 85, 243, 284, 289, 305, 374, 487
St. Lawrence River: 30P, 40P; and glaciation, 37; and Thousand Islands, 39; and storm paths, 64; as trade route, 155, 264; power, 285, 289; recreation, 302, 305, 306; and North Country, 374; and Albany area, 388, 391; and western New York, 410
St. Lawrence Seaway: 37, 85; and manufacturing, 234, 243; impact, 264–65; Buffalo, 274, 415; and recreation, 305; and North Country, 374; and Albany area, 388, 391, 395; Rochester, 461
St. Lawrence Ten Towns, 141, 142F, 153
St. Lawrence Valley: 30P; ancient, 20; water deficit, 71; vegetation, 95, 98; soils, 106; dairying, 204; manufacturing, 243; and transportation, 265; urban development, 347
St. Lawrence Valley Resource Region (power), 289
St. Regis Indian Reservation, 141, 142F
Salamanca, 404
Salaries and wages, 251
Salem, 135
Sales activities, 256, 276–80
Salina, 149, 471, 472, 473
Salina and Chenango Turnpike, 471
Salmon, Atlantic, 230
Salt: deposition, 23; and development of Syracuse, 149, 471; importance in trade, 158, 159; and canals, 177, 387; decline, 179; and Indians, 221, 469; distribution, 225; extent of deposits, 226; and Rochester, 458; reservation, 471; manufacturing techniques, 472; peak, 473
"Salt City" (Syracuse), 469
Salt water: at Albany, 387
Sand, 387
Sand and gravel: as resource, 21P, 221, 225, 226; source of ground water, 81; Rochester, 458
Sand plains: in Albany area, 387, 393
Sandstone, 23, 24F, 225
Saranac Lake, 27P
Saratoga Battlefield, 304
Saratoga County: manufacturing, 246; forest preserve, 292; and Albany Urban System, 385, 394
Saratoga Springs: mineral waters, 52, 221; attacked by French, 134; public resort, 294; and Albany Urban System, 385
Satellite cities, 350
Sawmills, 100, 101F, 229, 243
Sawquoit Creek, 149, 480, 483
Sayville: solar radiation, 55–58, 55F
Schenectady: 385–95 *passim*; sacked and burned, 130; at end of colonial period, 136, 138; in colonization movement, 145; and Erie Canal, 159; locomotives, 168; in 1850, 169; manufacturing growth, 178–79, 246; beginning of General Electric Company, 184; population (1890–1920), 184; manufacturing, 234, 242, 392; and urban location, 339, 348; and uniform region, 373. *See also* Albany-Schenectady Urban System
Schenectady and Saratoga Railroad, 162
Schenectady County, 385, 392, 394
Schoharie, 131
Schoharie County: broom corn, 166; cement, 224; and Albany area, 385; and population growth, 493
Schoharie Valley: snowfall, 69; local colonization, 131; and revolution, 137
Schoolyears completed: state rank, 313
Schroon Lake, 62

Schuyler County: salt, 225
Scotch-Irish, 134, 135
Scotch pine, 102P
Scotia, 385
Scots: at Argyle, 135; at Caledonia, 144; distribution in 1855, 169
Scranton, Pennsylvania: and urban spacing, 346; and Binghamton, 399, 401, 404
Scriba, George, 153
Seaboard-Chicago Axis: and New York City, 428
Sea lamprey (eel), 38
Seasonality: and recreation, 294
Secondary sector, 6T, *14*, *198*, 232–54, 199T; and future, 493, 497–98
Second-hand store districts, 341
Second-order trade center: in urban hierarchy, *282*, 282F; and planning and development regions, 376; Albany area, 395; New York City, 425; Rochester, 467; Syracuse, 469
Secret Caverns, 45
Sector patterns: in cities, 181, 339–42
Sectors, economic: future employment, 497
Securities and exchange activities (New York City), 431, 432, 437
Seeds: Rochester, 461
Segregation, areal: in central cities, 354
Selkirk, 390
Seneca County: peat, 226
Seneca Falls, 243
Seneca Indians, 114F, 115, 118
Seneca Lake: 23, 43, 44P; transport on, 165; recreation, 302, 306, 307, 309
Seneca River: navigation on, 158
Seneca Turnpike, 149, 156, 471
Service activities: in state, 6T, 280–81, 280F, 281F; in New York City, 433, 436–37, 438, 441
Services, and taxes, 312
Setauket, 126
Settlement: problems in, 140; factors in, 145; spread of, 146–47F; and uniform regions, 371–75 *passim*
Sewage renovation, 81
Sewing machines, 482
Shale: formation, 23, 24F; as resource, 224, 225; Albany area, 387
Shawangunk Mountains (land form division), 26F, 31, 115
Sheep, 165, 166
Shellfish, 229
Sherburne, 149
Shipbuilding, 168, 243, 414, 416, 428
Shipping, 130, 134, 136, 155, 434
Shoes: colonial product, 136; in Binghamton Urban System, 242, 400–1, 403; and manufacturing development, 246, 248; Syracuse, 179; Utica, 484
Shopper's goods, *344*, 345, 349–51 *passim*
Shopping districts: and urban hierarchies, 342–46, 348–51 *passim*, 350F; planned, 346, 438, 439, 462–63

Shopping trips: and shopping districts, 344–46 *passim*
Sidney, 242
Silos: and rural landscape, 359
Silver: mining, 224
Site: and Indians, 115; New Amsterdam, 126; factors in manufacturing, 249; urban, 339, 348, *425*; and Albany area, 387, 388F, and Binghamton, 397F, 397–99; and Buffalo Urban System, 411F, 413; and New York City, 430–31, 440, 444, 454–55; and Rochester, 458, 460F; and Syracuse, 471, 472F; and Utica, 480, 483F, 485
Situation, urban, *425*, 428F, 425–30, 440, 453–54
Sixties, East (New York City), 443, 450
Skaneateles, 149
Skaneateles Lake, 43, 302, 306
"Skid Row," 341
Skiing, 294, 420
Skyscrapers, 432, 440
Slate, 221, 225
Slavs, immigration, 181, 184
Slide Mountain, 309
Slopes, excessive for use, 49F, 52
Slums: in cities, 334, 338, 433
Small-scale manufacturing, *237n*
Smith-Corona-Marchant Co., 239P, 477
Snow Belt mixed farming region (agricultural), 214
Snowfall: 69–70; snow shadow, *69*; snow belts, *70*; and highways, 267; and residences in Binghamton, 399
Soap, 136
Social geography: character of (New York City), 443–44
Social organization: and economic health, *311*, 311–15
Sodus Bay: and Judge Cooper, 148; as port, 150–51
Sodus Point, 176
Soft goods, *237n*
Soft goods manufacturing, 248, 250, 480, 484, 485
Softwoods, 98, 227, 229
Soil: *104*, 104–9; and vegetation, 91–93, 93–98 *passim*; profile, *105*; structure, *105*; texture, *105*; associations, *106*; types, *106*; regions, 106–10, 107F; and agriculture, 203, 207, 211–17 *passim*, 220; and rural landscape, 358, 362; and uniform regions, 371–75 *passim*; creep, 399
Solar radiation, 55–58, 472
Solvay, 473, 479
Solvay Process Co., 473
Somerset County (New Jersey), 446
"Soo" Canal: and Buffalo, 413
South: migration from, 184, 436
Southampton, 126, 129, 133F
Southern Ontario Plain (land form division): 26F, 34

Southern Tier: problems of development, 150; and railroad, 163, 265; cattle, 166; lumbering, 168; absence of immigrants, 171; poultry, 208; manufacturing, 235, 242; urban development, 347; and Binghamton, 396, 403
Southern Tier Manufacturing Region, 242
South Glens Falls, 393
Space: in suburbs, 354, 356
Spacing of urban systems, 346–48
Spafford, Horatio J., 149–55 *passim*
Spatial interaction: and locational advantage, 13–14
Specialization, areal: among Indians, 118–19; urban spacing, 347–48; in New York City, 433, 444, 446, 450
Speculation: in new land, 141, 142; in railroads, 162; in electric railroad, 181; in land around cities, 220, 336
Sperry Rand Corp., 485
Sphalerite (zinc ore), 224
"Spreadcity," 338
Springfield, Massachusetts: and railroad development, 175; and urban spacing, 347
Spruce-Fir Forest Zone, 92F, 96
Spruce trees, 227
Squalls, 64
Stage service, 156
Standard Metropolitan Statistical Area (SMSA), 384
State government: and commercial ventures, 163, 471; and recreation, 294; and urban development, 348; and urban problems, 356; and Albany, 394
Staten Island: land form, 20, 23, 34, 429P, 444; Dutch settlement, 122F, 126; ambiguity, 129; lack of bridges, 430; and Greater New York, 432–33
State Park System, 294, 295–96T
Statistical reality: and planning and development regions, 376–77
Steam: service on Lake Ontario, 153; service on Hudson River, 154; service on Cayuga Lake, 156–57; service on New York waterways, 165; tugs on Erie Canal, 177; in power production, 284, 288, 415, 484
Steel: Syracuse, 179; Niagara Frontier, 239, 240, 240P, 415; Dunkirk, 242; Southern Tier, 242; Albany area, 387, 392
Steuben County: and settlement, 150; potatoes, 208, 211, 216P; petroleum, 226
Stillwater: temperature, 60, 62
Stone, 221, 225, 226
Stone, clay, and glass products: Buffalo, 415
Storm mechanisms, 64–65
Strawberries, 207
Stream discharge, average annual, 80, 82–83F

Streetcars; beginning, 181; and urban patterns, 339–40; Buffalo, 414
Striations, *41*
Style center: New York City as, 431, 436
Submergence, 23
Suburbs (suburbia, urban fringe): *336*; and electric railroad, 181; growth, 189, 338, 340, 342; manufacturing in, 238, 248F, 249; in metropolis, 351; contrasts with central city, 354; problems, 356; and rural landscape, 364; Binghamton, 399; New York City, 433, 437–40 *passim*, 451–52; Rochester, 467; Syracuse, 473, 474T, 475–77, 479; future growth, 493
Subways, 440, 432, 433, 434, 455
Succession: in vegetation, 93
Suffolk County: construction, 250; parks, 309; and concentric rings, 445, 452; commuting, 452, 453; future, 493
Sugar beets, 208
Sugar refining, 136
Sullivan-Clinton Campaign (Revolutionary War), 115, 398
Sullivan County: agriculture, 216; forest preserve, 292; recreation, 309; future, 493
Sunken Meadow State Park, 309
Super highways: pattern, 193F; New York City, 438, 439–40. *See also* Highways
Surface configuration, 50, 51F, 52
Surface water, *79*, 79–80
Susquehanna Hills (land form division), 26F, 33
Susquehanna Indians, 114F, 115, 119
Susquehanna River: and glaciation, 37; migration route, 145; trade route, 155; recreation, 302; and Binghamton, 397–98, 399, 400P, 402
Susquehanna Valley and basin: penetration of settlement, 134, 135; and Revolutionary War, 137; settlement expansion, 149–50; tobacco, 401; and Binghamton, 397–405 *passim*; and Syracuse, 469
Swamps: New York City, 430, 444, 451; Syracuse, 471, 472
Swedes: in Jamestown, 181
Swimming, 302
Syracuse: 469–79, 470F, 472F; water supply, 44; cloudiness, 58; heating degree-days, 63; winds, 65; precipitation, 65, 67, 68; fogs, 66; snowfall, 70; establishment, 149; and railroads, 162, 163, 176; canals, 159, 182; salt manufacturing, 168; population characteristics (1850), 169; population (1890), 179; manufacturing development, 179, 234, 246; and super highways, 186, 269, 270; minerals, 221, 225; tertiary characteristics, 262, 262F, 278, 279, 280, 320; transportation development, 263, 272, 274; recreation, 303, 304, 307; urban patterns, 334, 339; and urban spacing, 346, 347
Syracuse (nodal region), 375, 377, 378F
Syracuse SMSA, 469
Syracuse University, 475
Syracuse Urban System, 37P, 469–79, 470F, 472F, 476P; land form, 20, 23, 34; manufacturing, 238, 239, 243, 251; construction, 250; tertiary sector, 256–58, 262F; trade order, 282; economic health, 328; urbanized area, 333, 469, 470F; urban frontier, 336, 469, 470F; political jurisdictions, 355F; and Upstate Heartland, 373; nodal region, 375; SMSA, 469; future, 492, 493, 497

Taconic Mountains (land form division): 26F, 28; formation of, 23, 25T; and Albany area, 387, 388
Tahawus, 223, 224
Talc, 221, 225
Tannin, 96, 400
Tanning, 136, 178, 243
Tappan, 129
Tariff, protective, 246
Tarrytown, 270P
Tax effort, *313*, 314T
Taxes: and manufacturing, 248, 249; and economic health, 311–13; corporate, 312–13, 312T; and urban problems, 354–56; Binghamton, 405
Taxicabs, 182
Taylor, George, 461
Taylor, Rev. John, 144, 153
Technology: and manufacturing, 247
Temperature, air, 59–63, 59T, 60F, 61F
Terminal moraine, *21*
Terrain: and vegetation distribution, 91, 93–98 *passim*; and soils, 104, 106–10 *passim*; and agriculture, 203, 211–17 *passim*, 220; and economic health, 328; and sectors in cities, 339, 342; and rural landscape, 358; and uniform regions, 371–75 *passim*; and Binghamton, 397–99, 402; and New York City, 440; and Syracuse, 474–75; and future agriculture, 499
Territorial disputes: after Revolution, 140–41
Tertiary sector: 6T, *14*, *198*, 199T, 255–310; and cities, 189; and changing role theory, 247, 250; components, 255, 255T, 263T, 276–77T; deficit, *256*; potential, *256*; surplus, *256*; external comparisons, 316–17T, 318, 320–21T; and planning and development regions, 371; Albany area, 393–94; Binghamton, 403–4; Buffalo, 418–20; New York City, 441, 443–44, 454; Rochester, 466–67; Syracuse, 477–78; Utica, 487; and future, 493, 497, 498
Tertiary surplus: *256*; and traffic flow, 269; and air travel, 272; and planning and development regions, 371
Teterboro Airport, 440
Texas, 153
Textiles: (1850's), 168; (1890's), 179; recent trends, 188, 251; distribution, 242; development, 246; and industrial old-age, 248; in upper Hudson, 393; Buffalo, 414; importing in New York City, 430, 431; manufactured in New York City, 440; Utica area, 482, 484
Theaters: urban location, 334, 340; New York City, 434, 439, 441, 446
Thermal inversion, *66*
Third Avenue (New York City), 446, 450
Third-order trade centers, *283*, 283F, 403, 495
Thirties, West (New York City), 448
Thompson's Lake: sink hole, 52
Thousand Islands, 39–40, 40P
Thousand Islands State Parks, 306
Threshold business size, *343*
Threshold population size, 348–52 *passim*; in manufacturing, 250; and Syracuse urban growth, 469
Thruway, Gov. Thomas E. Dewey: (Interstate Routes 87 and 90), 186, 270P, 271P, 272P
Thruway, New York State: and manufacturing, 240; and Mid-Hudson region, 242; and transportation network, 264, 267; geographic influence, 269; and Finger lakes, 307; Albany area, 390; and Buffalo, 415, 421; and urban expansion in New York City, 452; and Rochester, 458, 461, 463; and manufacturing in Syracuse, 474, 476P
Thunderstorms, 64, 65T
Tile, 242
Till. *See* Ground moraine
Tillable land, 211n, 211–17 *passim*
Timothy, 204
Tioga County: SMSA, 397
Tioga River: as migration route, 150
Titanium, 221, 223, 224
Tobacco: Indians, 118; in Susquehanna Valley, 401
Tomatoes, 208
Tompkins County: salt, 225
Tonawanda: lumber-milling, 178; development, 179, 184, 413; and Buffalo, 235, 407, 415; and Niagara Frontier region, 239
Tonawanda Indian Reservation, 141, 142F
Topographic features: and recreation, 299
Tornadoes, 64
Total regions, 371
Tourism: in Niagara Falls, 420
Town: in urban hierarchy, 350
Toys, 237
Tractors, 183, 188
Trade: among Indians, 118–19; between Indians and Europeans, 119; about

Trade—*Continued*
1770, 134, 136; state government in, 149; Buffalo, 413, 421

Trade, commodities of: in New Amsterdam, 126–27; in colonial New York City, 130; (1770), 136; on canals, 162

Trade and service centers: 281–84, 281F, 282F, 371; function, 349, 351; Buffalo, 410–12; Utica, 483, 484. *See also* central places

Trade area: *277*, 279, 281–84, 281F, 282F; and city spacing, 346; development in New York City, 425, 429, 431

Trade privileges: of patroons, 126

Trade routes: of Indians, 118–119; Dutch, 122; via natural waterway, 155; Albany area, 387

Traffic problems: New York Urban System, 455

Transferability, *264*

Transition zone: in cities, 334

Transit region, *265*

Transportation: 262–76, 275F, 270P, 271P, 272P, 273P; in Mohawk Valley, 31; and Indians, 115–17, 118; development of to 1855, 155–65; network (c. 1855), 164–65; (1850's–1890's), 172–77; (1890's–1920's), 181–84; (1920's–1960's), 185–88; development of patterns, 191–94; 192F–93F; and manufacturing, 232–50 *passim*; in tertiary structure, 262; development theory, 262–64, 266F; axis, *264*; in Adirondacks, 305; density and economic health, 328; and city growth, 338, 339–40; and urban sector patterns, 339–40; and city location, 347, 348, 350; in Albany area, 388–91, 390P, 395; in Binghamton, 398, 399, 404; in Buffalo, 410–12, 413–14, 415, 416; in New York City, 428–33 *passim*, 436, 438; in Rochester, 458, 461; in Syracuse, 469, 473, 475, 477; and Little Falls, 482; and Kingston, 495. *See also* individual media

Transportation equipment: Albany area, 178–79, 392; in Binghamton, 403; in Buffalo, 415; in New York City, 440; in Utica, 484; distribution, 237, 239, 242; trends, 246, 251; and industrial aging, 248

Trans World Airlines, 391

Trapping licenses, 294

Traprock, 225

Treaty of 1763, 134

Triangular trade: and New York City, 425

Triassic Lowland, 26F, 28–29

Triple Cities, 396

Trout Lake, 230

Troy: temperatures, 62; establishment of, 154; and Erie Canal, 162; lumber-milling, 167–68, 178; iron and steel, 167, 246; in 1850, 169; and immigration, 179; population trends, 184; 385–95 *passim*

Trucking: and agriculture, 188; and manufacturing, 234; Rochester, 461; Syracuse, 474

Trunk lines, *264*

Tryon, Governor, Report of 1774, 136, 138

Tryon County, 135

Tug Hill Upland (land form region): 26F, 33, sandstone, 23; influences on temperature, 61; precipitation, 67; water surplus, 71; snowfall, 69, 73; floods, 86–87; vegetation, 96–98; maple sugar production, 100; soils, 108; Indian rejection of, 115; and agriculture, 217; highways, 267, 269; rural landscape, 368; as barrier, 469; and Utica Urban System, 480

Tupper Lake: railroad through, 173

Turnpikes: promotion by Newburgh, 154; development of, 156–58, 157F; in relation to canal developments, 158–59; (c. 1855), 165; pattern of, 192F

Tuscarora Indian Reservation, 141, 142F

Tuscarora Indians, 115

Tully: salt, 225

Tuxedo Park: atmospheric emissions, 66

Typewriters, 179, 242, 482

Ulster County: settlement, 134, 135; agriculture, in, 166; cement, 224; forest preserve, 292; recreation, 309; future, 493

Unadilla River, 135

Unemployment: in construction, 251; external comparisons, 316–17T, 318, 320–21T, 322–23T, 323, 324T; in Pennsylvania coal fields and Binghamton, 401; in Rochester, 458; in Utica, 487; "unemployable," 496; technological, 496

Uniform regions, *15*, 370–71

Uniform regions, composite: 371–75, 372F; combined with composite nodal regions, 375–76, 377F, 378F

Union, 405

Union County (New Jersey), 445, 451

Union Square (New York City), 448

United Nations: as symbol of primacy, 434, 449P, 450

United States: economic comparison, 6T, 315–21 *passim*, 316–17T, 320–21T

United States Mint, 467

U.S. Route 1: on axis, 428

U.S. Route 9, 391

U.S. Route 9W, 391

U.S. Route 15, 307

U.S. Route 20, 156, 307, 391

U.S. Route 20 Alternate, 151–52

U.S. Route 104, 410

Unstandardized manufacturing: in New York City, 442, 454

Upper New York Bay, 433

Upstate Heartland (uniform region): 373, 375; and Buffalo, 407; and Rochester, 469; and Utica, 480

Upstate New York: tensions with New York City: 126, 171; as "colony" of New England, 143; and Syracuse, 256, 258; power companies, 285

Upstate New York Medical Center (Syracuse), 475

Urban Landscape, 6, 90F, 333–57; urbanized areas, *333*; urban population, *333*; urban frontier, *336*, 337F; urban fringe, *336*; in New Netherland, 126, 127; at end of colonial period, 135–36, 137–38; and manufacturing, 167–68, 178–79, 183–84, 188, 247; developments by 1850's, 169; tensions with rural areas, 171; (1850's–1890's), 179–81; and agricultural development, 183, 220; (1890's–1920's), 184; (1920's–1960's), 189–90; renewal and construction, 250; recreation attractions, 304–5; "urban sprawl," 338; urban renewal, 339; frontier and rural landscape, 364; and the future, 492–99 *passim*; general data, Appendix E. *See also* specific urban places

Urban systems: *384*, 337F; and manufacturing, 232, 248F, 249, 250; construction in, 250; and interchanges, 270; and economic health, 322; growth in, 352, 353F, 471T; and Upstate Heartland, 373; and planning and development regions, 376; population projections, 492–93, 493T; future employment, 497, 498T; and green belts, 499. *See also* Urban Landscape and individual urban systems

Utica: 480, 481F, 483–88, 483F, 486F; pollution of Mohawk, 88; beginning of Genesee Road, 119; in colonization movement, 143; "metropolis" of the west, 144; and Judge Cooper, 148; rivalry with Rome, 148–49; and Chenango Canal, 161; arrival of railroad, 162; manufacturing development, 168, 179, 234, 241P, 246; population characteristics (1850's), 169, 171; and railroad development, 176; in 1890's, 179; wholesaling, 278; services, 280; central place status, 258; and transportation development, 263; trade order, 282–83; canal, 303; and urban patterns, 334, 339; and urban spacing, 347, 348; and urban growth, 352; and Syracuse nodal region, 375, 477–78

Utica and Black River Railroad, 176

Utica College of Syracuse University, 487

Utica-Rome SMSA, 480

Utica Urban System: 480–88, 481F, 483F; manufacturing, 240, 242; construction, 250; urbanized area, 333, 480, 481F; and Upstate Heartland, 373; SMSA, 480; urbanized frontier, 480, 481F; future, 492, 493; "Valley" (New York City), 434, 446

Valley Head Moraine, 44

Valley trains, *21*

Value added by manufacture: *232*, 236–42

passim; trends in, 251T; external comparisons, 316–17T, 319, 320–21T; internal variations, 323, 322–23T, 324T; Albany area, 392; Utica, 484
Vegetable production: 205, 207–8, 211–17 *passim*; and rural landscape, 362; Rochester area, 458
Vegetation: 90–103; freezing conditions for, 62; associations, 91F; zones (regions), 92F, 93–96; and soils, 104; and uniform regions, 371–75 *passim*
Vermont: Yankee colonization in, 135; dispute over, 140–41; as source of colonists, 144; economic comparisons with, 315–21 *passim*, 316–17T, 320–21T
"Vermont sufferers," 141, 142F
Vestal, 396, 398, 400P, 402
Victorian period: urban growth in, 338
Villages: in urban hierarchy, *349*–50; Dutch development of, 126; in Holland Land Co. lands, 152; of Onondaga Indians, 471
Vineyards, 38P, 362

Wages: internal variations, 322–23T, 323, 324T; New York City, 442–44; Utica, 485–87
Wagons: Buffalo, 416
Walden, 181
Wall Street (New York City), 425, 433, 439
Wallkill River, 29
Wallkill Valley (land form division): 29–31; settlement in, 130, 133F; and Appalachian Country, 373
Walloons, 127
Wappinger Creek: flow, 80, 84F
Warehouses, 334, 341, 407, 475
War Memorial (Rochester), 462
War of 1812: effect in Buffalo area, 152, 411; effect on ironworks, 153; and interest in Erie Canal, 159
Warren County: cement, 224; garnet, 225; forest preserve, 292; Capital District, 385
Washington, D.C.: as Capital, and New York City, 428, 454
Washington County: flax, 166; slate, 226; manufacturing, 246; forest preserve, 292; Capital District, 385
Waste disposal: and water, 85
Water: 79–89; balance, 70–72, 72T; hard water, *84*; bodies and vegetation distribution, 91, 93–96 *passim*; power, 285; bodies and sectors in cities, 339; Binghamton, 405; and recreation, 299–304, 305–10 *passim*; power in Lockport, 412; power in Rochester, 458; power in Little Falls area, 482; power in Utica area, 483, 484; lines in Utica, 487; bodies of, general data on, Appendix B. *See also* specific bodies
Waterford, 391

Waterfowl: and Long Island, 309
Waterfront Commission (New York City), 437
Watertown: establishment of, 153; agriculture, 166; (1850), 169; and railroad development, 176; dairying, 177; population development (1920), 184; manufacturing, 242; trade order, 283; and Seaway, 264; and urban spacing, 347; and North Country, 374
Water transportation: 155–58, 264–65; early importance of, 148; in 1850's, 164–65, 192F; pattern of, 192F; and Albany area, 391; and New York City, 430, 433, 450; and Rochester, 461
Watervliet: impact of Erie Canal, 159; manufacturing, 178–79, 392; and Albany Urban System, 385
Watkins Glen, 44P, 300P, 307
Watson, Elkanah: in Utica, 143
Watson, Thomas J., 401
Waverly: and railroad development, 175
Wayne County: SMSA, 458
Weather, *54*
Welfare: and rural nonfarmers, 364; Binghamton, 401; New York City, 455
Welland Ship Canal, 161, 415, 461
Welsh: at Remsen, 144; in 1855, 169; Oneida County, 181
Wertenbaker, Thomas: on New Amsterdam, 127
Westchester County: New England colonists in, 131; agriculture, 217; resources, 226; manufacturing, 238; construction, 250; attractiveness, 374, 444; in developing ring, 445, 451; future, 493; parks, 499
Western Adirondack Hills (land form division), 25, 26F
Western Inland Lock Navigation Co., 158, 161
Western New York: colonization in, 143; north-south railroads in, 175–76, 176F; and trade complementarity, 263
Western New York Forest Region, 229
West India Company (Dutch): first report of, 122; policies, 124
West Indies: trade with, 136, 243
Westmoreland, 148
West Shore Railroad, 174, 389
West side, lower (New York City): wholesaling in, 433
Westvale, 474
West Virginia: economic comparisons with, 315–21 *passim*, 316–17T, 320T–21T
Wetlands, 306
Whaling, 138
Wheat: in colonial period, 125, 136; Hudson Valley, 165; in central and western New York, 166; decline in, 177; in current agriculture, 204, 208; and Seaway, 264; and Buffalo, 413, 416; and Rochester, 458

Whiteface Mountain, 27P, 28P, 29P, 305
Whiteface Ski Center, 294, 308P
Whitefish, 230
Whitehall (Skenesborough): settlement, 135; canal, 391
White pine, 97P
White Plains: and suburbanization, 438, 452; as subsidiary focus, 439, 444
Whitesboro: beginning, 148; and Chenango Canal, 161; and Utica Urban System, 480, 483
Whitney, Joshua, III, 399
Wholesale trade: 6T, 255, 277–78, 278F; employee efficiency (external comparisons), 316–17T, 319–20, 320–21T; urban location of, 334, 341; and New York City, 430–41 *passim*
Wilderness zoning, 292
Wildlife refuges, 294, 299T
Williamsburg, 150
Williamson, Charles: Pulteney agent, 150–51
Williamsport, Pennsylvania: road to, 150
Willsboro, 135
Wind, 63–66, 64T, 66T
Wine, 177
Wollastonite, 221, 226
Wood Creek, 136, 158
Woodland Stage: of Indian occupance, 113
Woodlawn, 407
Wood products: and empty landscape, 368
Woolen mills, 482
World economic organization: and problems of New York City, 453–54
World War I: and Negro migration, 184; and transportation, 185
World War II: and air travel, 185; and titanium, 223
Wyoming County: development of, 152; potatoes, 208, 211; salt, 225
Wyoming Valley (Pennsylvania): as source of colonists, 145

Xerophytic trees, 91
Xerox Corporation: Rochester, 466
Xerox Square (Rochester), 463

Yarns, 242
Yates Company: retail sales, 279
Yonkers: population growth, 179, 184; population (1960), 424; subsidiary focus of urban system, 439, 444; and suburbia, 452; region of dominance, 169
Yorkville, 480, 483
Yorkville section (New York City), 433

Zone of trade competition: via natural waterways, 155; railroads in, 163
Zoning; in cities, 340, 342
Zinc, 221, 224

SUPPLEMENT

Geography of New York State

PAPERBACK EDITION

John H. Thompson, Editor

Professor of Geography
Syracuse University

THIS SUPPLEMENT to *Geography of New York State* cites some of the important trends of the last decade and briefly speculates upon their conceivable effect during the years ahead. At least five trends deserve special emphasis: (1) slowing of population growth, (2) development of the fiscal dilemmas of New York City and the state, (3) intensification of a previously strong tendency toward more and more dominance of the state's economy by the sales and service (tertiary) sector, (4) emergence of a national energy crisis and its impact, and (5) environmental concerns—the rise and growth of a statewide interest in ecological issues.

Seven tables from the hardcover edition have been updated and included in this Supplement—numbers 16, 21, 22, 23, 24, 25, and 26. Four tables have been added at the end of this Supplement from the New York State Department of Commerce *Business Fact Book,* 1976 supplement.

Copyright © 1977 by Syracuse University Press
Syracuse, New York 13210
Second printing 1980

Manufactured in the United States of America

SLOWING OF POPULATION GROWTH

Both the number of people and their distribution will greatly affect the nature of the state's future geography. It now seems likely that New York can expect a substantially smaller future population than was indicated in the mid-sixties and that the great urban systems will become less powerful as attractors of people than they have been in the past.

Birth and fertility rates have fallen nationally to their lowest levels on record, and New York is feeling the impact. In the past these rates have fluctuated widely, reflecting economic conditions, general attitudes toward family size, and many other factors. It is difficult to be certain they will not rise substantially again, but most demographic experts now expect them to stay lower than could have been predicted only a few years ago. Also, net in-migrations from the South and elsewhere are down and may not make the additions to New York's population that they did in past decades.

Table 56, page 492, contains past population figures and forecasts to the year 2000. A footnote to that table warns of the slowing of growth being discussed here. The census count for 1970 turned out to be 18,241,000 lower than the *low* series on Table 56. Using this figure as a base year, the New York Office of Planning Services is now projecting the population for 1980 to 19.6 million, for 1990 to 21.2 million, and for the year 2000 to 22.7 million. United States Bureau of the Census estimates, made in 1972, place the 1980 figure at 20.3 million and that for 1990 at 22.9 million. These figures vary somewhat but are below and near, respectively, the low series on Table 56. They may even turn out to be high, for recent estimates of the United States Census Bureau and the New York State Department of Commerce vary, respectively, from slight decline to slight gain in the 1970-75 period. As Tables 57 through 61 (pages 493-98) were based on the *medium low* series of Table 56, downward adjustments in each are clearly in order.

There are those who despair in the slowing of population growth in New York State. They see, related to it, a slowing of business growth, a decline in the relative significance of the state in the nation, and a less bright employment picture. On the other side of the coin many are arguing that "bigger is not necessarily better," that "quality not quantity could well be the essence of the state's future." They see in a slower population growth less congestion, less pollution, less crime, less of the general set of problems recently associated with life in cities.

To argue whether "bigger is better" is not the intent here. But that the State of New York strive for a reputation of a good business climate and a good living environment is important regardless of population or other growth trends. These things are important to the well-being of the state's citizens and will influence those from afar who aspire to life and livelihood here. Future net migrations, in amount and in kind, will be influenced by the state's comparative reputation with other parts of the nation and the world.

In-migrations of productive people who participate in the job market probably may be judged as good for the state; by contrast, in-migrations which swell welfare rolls become a burden which is difficult to bear. Growing, healthy corporations foster the former; uncommonly generous social service programs may contribute to the latter.

It is not easy to do much about out-migrations which reflect New York's reputation for rainy springs and cold, snowy winters, but great care should be exercised by state and local governments to avoid the development of a reputation which tends to drive people and corporations out of the state because of comparatively unfavorable tax conditions or repetitive severe fiscal crises. A recent sampling by this writer of thirty people who were moving from New York to Florida showed climate to be, as expected, the primary reason for moving, followed by state income taxes, employment opportunities, estate taxes (important among retirees), recreational opportunities, and real estate taxes. If intensive investigations into reasons for migration of both individuals and businesses were to become important bases for governmental policy, the state's reputation should profit.

Much was said in *Geography of New York State* about the overwhelming strength of the great urban systems in dominating the job market and thus the geography of population in the state (page 492 and elsewhere). Although the job market is dispersing somewhat in the state, as witness the out-migration of employers from New York City to Long Island and Westchester County, still the great urban systems, because of their locational attributes and massive infrastructures, probably will continue to dominate job generation even though more and more of that generation may take place in their peripheries. What is particularly interesting is the apparent growing antipathy by people toward life in the great urban systems. This is not to say that people will ignore the geography of jobs; it is to say they will exert great energy to get away from urban living unless urban problems can be worked out.

For years I have asked Syracuse University students about locational preferences for living. The question has been informally posed in the classroom as follows: Excluding differences in employment opportunities, which one of the following three locations would you prefer for your residence? Here are the results:

Locational Preferences for Residences

	Older City (Middle City and Frame of page 334)	Urban Fringes (Urban fringe and exurbia of page 336)	Rural or Small Town (beyond urban frontier, page 336)
Student Response Mid-1960s (%)	12	70	18
Student Response Mid-1970s	5	40	55

There is no certainty that Syracuse University students are representative of New York State's population as a whole or even that students in my classes are representative of college students generally. But the expressed rising affinity for things rural seems to parallel recent trends for larger older cities to lose populations, with frequently expressed displeasure with the commonplace uniformity of suburbs and with the perceived better situation in the uncongested, closer to nature, more individualist opportunities outside large urban areas. Census data already reflect this kind of human response and signal an end to rural population decline. Sounder futures for smaller cities and towns in the state should be expected. This will be even more likely if increasing numbers of employers find it feasible to locate outside the great urban systems. As a final note regarding the growing interest in rural New York, almost none of the 55 percent of the students now preferring rural or small-town living intended to be farmers. Essentially all plan to earn a living as professional or business people.

THE FISCAL DILEMMA

Both New York City and New York State have received widespread publicity about existing and impending fiscal crises. Both have spent widely and borrowed widely, with presumed anticipation that larger-sized future economies would take care of rising debts. Economic growth has slowed, but earlier established expenditure patterns, particularly in social services, education, and public payrolls, have relentlessly expanded. This was not an apparent major problem in the mid-1960s, although many of the processes were already underway at that time. How well the fiscal dilemma is dealt with in the future will have a powerful effect on the state's image as a place in which to work and live. In the 1970s fiscal difficulties are hurting that image. A geography written in the 1980s will be strongly influenced by what happens.

New York City is the nation's primate city, its principal business and economic center, the world's greatest port, an outstanding cultural focus, a unit statistically dominant within New York State. Yet despite these superlatives, plus many others which might be justified, it stands at the time of this writing as one of the most financially distressed political units anywhere. It is skirting the edge of bankruptcy; it is weakening general confidence in the municipal bond market; it has lost the confidence of bankers, and businesses are departing from it in alarming numbers. In order to survive it is calling for help from Albany and Washington.

New York City of course is not the only urban unit having fiscal problems, but because it is the nation's largest city and has problems that are unique in size and complexity, and undoubtedly broadest in national and state impact too, it is receiving maximum attention.

Simply stated, New York's expenditures have far exceeded revenue for so long that the city's debt and the cost of carrying it have become unmanageable. The administrations of at least three mayors—Wagner, Lindsay, and Beame—were engaged in a spend-now-pay-later philosophy. And, while many cities for good reasons are short of funds, the accusation being directed at New York is that it has been especially irresponsible fiscally. Admonitions aimed at New York City include the following: (1) New York has more municipal workers per 1,000 residents and at higher pay scales than other major cities; (2) it spends three times more per capita on health and hospitals than most other municipalities; (3) its per capita outlays for welfare are incredibly high when compared to other urban places with over a million population; (4) it has generously supported a huge, free-tuition, open-enrollment university.

The only way these kinds of expenditures could have taken place at all over the years was for the city to receive increasing assistance from the state and federal governments and to engage in budgetary subterfuge. The subterfuge broadly involved not paying off debts in past years except by incurring more and larger debts. The problem now is that the money supply for those larger debts is threatening to dry up despite the state-created-and-guaranteed Municipal Assistance Corporation ("Big MAC") designed to borrow money for the city.

Late in 1975 New York City was snatched from the brink of default by President Gerald Ford's pledge of federal loans, plus a large financing package backed by New York State. The federal government agreed to loan up to $2.3 billion a year to New York State so that it could meet the city's short-term cash needs until mid-1978. The loans must be paid by the close of each fiscal year. As a requirement for receipt of this loan the federal government stipulated that New York City must raise taxes, cut expenditures, and insist upon increased contributions to pension plans by city workers.

Also, city pension funds are to buy additional city securities, New York City banks are to retain maturing notes at reduced interest rates, and investors are to swap soon-to-mature notes for new ten-year ("Big MAC") bonds. At the time of this writing the city appears to be failing in its efforts to meet financial commitments. At least some experts are doubtful that New York City can extricate itself. In any case its difficulties placed fiscal burdens on New York State, which has massive burdens of its own.

New York State's fiscal dilemma emerges, too, from policies of the last decade or so which promoted debt increase at rates far greater than economic conditions have proven to warrant. In recent years personal income in New York State has grown less rapidly than in many other parts of the country and population growth has been small. At the same time state debts have grown more rapidly than in most other states. New York State debt expansion, including state agencies and authorities, more than tripled between 1964 and 1974, while personal income only doubled and population increased by but a few percentage points.

State debts fall into three categories: long-term guaranteed debt, short-term debt, and long-term nonguaranteed debt. Long-term guaranteed debts pay for assets such as schools or highways as those assets are "consumed." They are secured by the full faith and credit of the state's taxing power and require approval of voters in a referendum. Long-term guaranteed bonds are the traditional borrowing vehicle of most state governments, and because of voter control, seldom lead to serious excesses. Between 1964 and 1974 this widely approved type of borrowing increased in New York State by a smaller percentage than the average for all states, while less-desirable techniques involving short-term and long-term nonguaranteed loans proceeded at a rate more rapid than for the average of all other states.

Short-term debt covers periodic gaps between, say, quarterly inflow of sales tax revenues and outflow of salaries paid weekly. In New York State short-term borrowing has increased 700 percent in the last decade, more than twice the rate of increase of all other states. Principal payments and interest payments on this massive $2 billion liability have become a major burden.

Long-term nonguaranteed debt includes that of the moral-obligation authorities such as the Urban Development Corporation (UDC), Housing Finance Agency (HFA), and the State Dormitory Authority (SDA). In theory, revenues from projects of these authorities are supposed to take care of debt retirement in the long run, but in order to induce investment confidence, moral-obligation provisions expressing the state's intention to make good on bonds in case of default are written into the legislation creating the agencies. Not requiring voter approval, this type of "backdoor" indebtedness has come to dominate state finances in the past decade. New York has seen its long-term nonguaranteed debt size grow much more rapidly than that of most other states.

It is unexplainable that New York State should have embarked on financial policies during the past decade which are turning out to be so troublesome. In any case a sound fiscal policy for the future will require shouldering responsibility for existing debts and following wiser procedures in taking on new ones. Involved, no doubt, will be both tax increases and expenditure cuts. The former, however, could seriously endanger an already precarious economic climate in the state; the latter, to be large enough to be effective, must include rational handling of the ever-rising demands for social service and education spending.

Social services and education are so overwhelming in their demands on the state budget and their costs are increasing so rapidly that they offer perhaps the only effective places for saving the large amount of money needed to make the state fiscally solvent. If new legislation is required to put the state in a rational ranking with other states in terms of social services and education expenditures, then that legislation should be forthcoming immediately. Especially in the case of welfare, it would seem wise for the state to institute reduction of payments over a period of years to a point at least no higher than the average of all states. This would probably meet with approval at the local level, at least outside of New York City. And, in New York City, the need for such policy may be greater even than elsewhere.

Not all of New York State's fiscal problems are self-induced. Much comment in the last year or two has been directed toward the important fact that Northern states in general, and New York in particular, are suffering from unfavorable balances of federal spending. When the South was poor and the Midwest and Northeast relatively well-off, a net flow of federal funds southward may have been justified. Now, however, the South and Southwest, profiting from this flow for years, have broadened and deepened their economic base, and they exhibit few of the money, unemployment and welfare problems which plague New York.

The sixteen states of the South enjoyed a total "balance-of-payments" *surplus* of over $11 billion in fiscal 1975, while New York alone suffered a *deficit* of $3.4 billion. Pressure for Congressional correction of this kind of imbalance would surely seem to be in order.

So, a two-pronged internal and external attack on New York's fiscal dilemma should be pursued in the interest of achieving a better image in terms of business climate and general economic health. The very significant tertiary sector especially, but manufacturing and agriculture too, to say nothing of governments at all levels, should be able to exist in an economic climate free from serious recurring fiscal crises.

NEW YORK STATE'S ECONOMY
AND THE DOMINANCE OF THE TERTIARY SECTOR

Dominance of New York's economy by the tertiary sector is clearly documented in Part III, and it was predicted on page 497 that this dominance would increase through the rest of the twentieth century. Census data show the tertiary sector to be making even faster than expected relative gains in recent years. In 1960 it accounted for 62.5 percent of the employed labor force; in 1970 it had risen to 69.7 percent and added about one million workers. During this decade the secondary sector declined from 35.4 percent to 28.9 percent and lost 309,000 workers. The primary sector, small in New York State for many years, became less significant still, dropping from 2.1 percent in 1960 to 1.4 percent in 1970 and losing 34,000 workers. These trends of the last decade come into perspective when compared to the long-term conditions shown in Figure 6, page 14. It follows from the employment figures that the tertiary sector is continuing to dominate the production structure of the state. Estimates indicate that it contributed 71.7 percent of the gross state product in 1970 as compared to 65.9 percent in 1960, while the secondary sector accounted for only 27.1 percent in 1970 as compared to 32.6 percent in 1960, and the primary sector 1.2 percent in 1970 and 1.5 percent in 1960. Some additional details of how the three economic sectors performed in the last decade follow.

TABLE 16 (p. 199)

Estimated New York Gross State Product, 1970

	1970 GNP (In Billions $)	NYS Employment as a % of U.S. Employment, 1970	GSP* Col. 1 x Col. 2 (In Millions $)	% of GSP
Agriculture, Forestry, Fishing	$ 31.6	2.79	880.7	.9
Mining	16.9	1.84	311.1	.3
Primary Sector Total	48.5		1191.8	1.2
Manufacturing	252.3	8.68	21,906.6	23.4
Construction	46.6	7.42	3,456.9	3.7
Secondary Sector Total	298.9		25,363.5	27.1
Wholesale & Retail Trade	166.4	9.08	15,109.1	16.1
Finance, Ins, Real Estate	137.8	13.85	19,091.8	20.4
Transportation	38.5	11.33	4,361.4	4.7
Communication	22.7	12.10	2,746.7	2.9
Pub. Utilities	22.6	8.83	1,996.3	2.1
All Services	114.0	10.28	11,723.5	12.6
Government & Gov't. Enterprise	129.4	9.31	12,051.5	12.9
Rest of World	4.6	—	—	—
Tertiary Sector Total	636.0		67,080.3	71.7
Total GNP 1970	983.4			
Total GSP 1970			93,635.6	100.0
N.Y.S.'s Per Cent of GNP	$\frac{93.6}{983.4} = 9.52$			

Sources: GNP—Statistical Abstract of the United States, 1974. Employment—U.S. Census of Population, 1970

The Primary Sector

Made up of agriculture, forestry, fishing, and mining, the primary sector has long taken a distinct back seat in employment and total production to the secondary and tertiary sectors. In spite of this it does utilize a good deal of the state's land area. Agriculture, and particularly dairying, is commonly viewed as a major New York State industry. The picture of agriculture during recent years, though, is one of continued decline. Farms employed only 98,000 workers in 1971 as compared to 123,000 in 1959 and occupied roughly 33 percent of the state's land area as compared to 44 percent a dozen years earlier. Approximately 12 percent of the state was planted and harvested in 1971 as compared to 16 percent in 1959, and the number of commercial farms declined by 22,000 during the same period. The amount of land in all farms has declined by about 1/3 million acres per year for the last ten years, a rate of decline even greater than during previous decades. In spite of such trends production in constant dollars has held relatively steady and even increased in some years. This has been accomplished on fewer, larger farms, employing constantly higher levels of technology and greater capitalization.

The issue of greater needed capitalization has become particularly apparent for the dairy farm. While the average one in 1960 needed perhaps a total of $1,500 investment in land, buildings, animals, and equipment for each milker maintained, the figure in the mid-1970s is close to $3,000. And, fifty milkers is probably now the minimum-sized herd for economic viability. A farm today that returns a good net profit may require seventy-five milkers and a total capital investment of over $200,000. Unless a farm is inherited, acquiring such a large amount of capital or handling current interest and principal payments on a substantial mortgage could be very difficult. It seems reasonable to predict that the recent trend for increasing percentages of dairy products to come from ever-larger privately operated farms and corporate units will continue.

Mining, lumbering, and fishing remain as very minor contributors to the state's economy, but unlike agriculture, they do employ slightly larger (together, approximately 2,200 more people) work forces than a decade ago.

TABLE 21 (p. 223)

Major Items in New York State Mineral Production

Item	Per Cent of State's Mineral Production as Measured by Value		
	1956	1962	1972
Cement (est.)	21	25	26
Stone	15	19	24
Salt	11	13	14
Sand and Gravel	12	13	12
Iron Ore	17	10	8 (est.)
Zinc	7	5	7
Petroleum	5	3	2

The Secondary Sector

Manufacturing continues to account for approximately 5/6 of the secondary sector, and New York ranked first in manufacturing among the states until California passed it by in the mid 1970s. In 1971 New York had 1,701,900 factory workers, 9.77 percent of the United States. This is a drop from 11.2 percent of the nation's total in 1961. Actual employment in the state fell by about 65,000 during this period, but value added by manufacture, measured in constant dollars, increased substantially.

Among major industry groups, apparel and related products continue to rank first as an employer (see Figure 72, page 233) but accounted for less than 15 percent of the state's factory workers in 1971 as compared to 19 percent in 1961. Printing and publishing, machinery, and food remain about at their status of a decade ago. Only one of the major industry groups—rubber and plastic products—employed more workers in 1971 than in 1961 (compare with Figure 79, page 252). The major industry groups, at best, are exhibiting a state of stagnation when measured by employment.

Stagnation does not describe the situation, however, when value added in constant dollars is used as a measure. According to this measure the state increased at an average annual rate of 3.9 percent per year, producing nearly $6 billion more in 1971 than in 1961. All but leather and leather products and food and kindred products showed gains, and two groups, instruments and related products and printing and publishing, together accounted for $2.3 billion in gains. Other substantial gainers have been machinery, except electrical, transportation equipment, and electrical machinery in that order. It should be noted that value added figures on Table 22 are not in constant dollars. The trend for factories with the same number of workers, or even a smaller number, to turn out products of greater value means manufacturing in New York can

TABLE 22 (pp. 234-35)

Manufacturing Production

Code	State and Industry	1971			
		All Employees (1000)	Adjusted* Value Added ($1,000,000)	Value Added as % of U.S. Industry Value Added	Industry Value Added as % of State Total Value Added
	New York Total	1701.9	28,862.2	9.19	100.00
20	Food & Kindred Products	97.7	2286.6	6.70	7.92
22	Textile Mill Products	46.5	611.1	6.11	2.12
23	Apparel & Related Products	250.8	3132.5	25.16	10.85
24	Lumber & Wood Products	13.3	147.9	2.19	.51
25	Furniture & Fixtures	30.0	370.8	7.09	1.28
26	Paper & Allied Products	55.7	809.4	6.93	2.80
27	Printing & Publishing	173.9	4109.9	22.72	14.24
28	Chemical & Allied Products	57.6	1982.4	6.74	6.87
30	Rubber & Plastic Products (not elsewhere classified)	30.9	471.0	4.95	1.63
31	Leather & Leather Products	36.4	345.8	12.53	1.20
32	Stone, Clay & Glass Products	36.6	722.7	7.23	2.50
33	Primary Metal Industries	59.6	982.9	4.65	3.41
34	Fabricated Metal Products	80.7	1348.4	6.14	4.67
35	Machinery, Except Electrical	136.1	2477.8	8.01	8.58
36	Electrical Machinery	167.8	2620.9	9.40	9.08
37	Transportation Equipment	88.5	2066.9	5.93	7.16
38	Instruments & Related Products	91.5	3122.7	37.24	10.82
39	Miscellaneous Manufacturing	78.7	1065.8	18.67	3.69

*Adjusted value added figures are not in constant dollars, so one year can not be compared with another.

Source: *Annual Survey of Manufactures,* 1971.

TABLE 25 (p. 251)

Trends in Value Added by Manufacture for New York State and Selected Areas, 1961-1971

(In Constant Dollars*)

	1961 ($1,000,000)	% of State Total	% Average Annual Growth 1961-71	1971 ($1,000,000)	% of State Total
New York State	14,408.7	100.0	3.9	20,113.0	100.0
New York City	8,627.7	59.9	2.3	10,623.9	52.8
Niagara Frontier	1,442.3	10.0	4.8	2,136.5	10.6
Rochester & Syracuse SMSAs	1,623.2	11.3	11.2	3,435.5	17.1

*The wholesale price index (1947-9=100) for all manufactured goods published by the U.S. Bureau of Labor Statistics is used to establish a constant dollar value.

Source: *Annual Survey of Manufactures,* 1961, 1971.

increase contribution to the economy without employment growth. In one sense, at least, this is evidence in support of the viability of a slow- or no-growth situation.

Current data show interesting changes in employment trends for specific types of manufacturing. If the major industry groups are dissected as they were in Figure 80, page 253, some of these changes become apparent. For example the big gainer in employment in the 1947-61 period on Figure 80, aircraft and parts, lost 10,000 workers between 1961 and 1971. Communication equipment, the second fastest grower in Figure 80, lost 13,000 workers. The largest gainer in employment in the 1961-71 period was electronic components, which picked up 19,000 workers, and books ranked second with a 7,000 gain. Neither were among leaders in the 1947-61 period. Women's and misses outerwear, the major loser in Figure 80, continued to be the largest loser between 1961 and 1971, dropping in employment by 25,000. It is not the intent here to attempt a detailed explanation for these changes in trend, but such things as military markets, national changes in consumer demand for certain products, net interstate migration of producers, individual corporate success or failure with competitors in

TABLE 23 (p. 236)

Manufacturing in Metropolitan Areas 1971*

SMSA	Employees (1000)	Value Added ($ Millions)	Employees as % of NYS Employees	Value Added as % of NYS Value Added
New York State	1701.9	28,862.2		
Albany-Schenectady-Troy	60.2	1008.4	3.5	3.5
Binghamton	40.1	655.4	2.4	2.3
Buffalo	155.7	3065.9	9.2	10.6
New York	992.9	15,245.3	58.3	52.8
Rochester	140.7	3905.2	8.3	13.5
Syracuse	58.8	1024.8	3.5	3.6
Utica-Rome	39.1	737.5	2.3	2.6

*Value-added figures are not in constant dollars, so 1971 cannot be compared with earlier years. Employee and value-added figures are in a different form from earlier years due to changes in reporting format.
Source: *Annual Survey of Manufactures,* 1971.

TABLE 24 (p. 237)

The Manufacturing Regions

	Employment 1971 (1000)	Employment 1961 (1000)	Percentage Growth 1961-71	Absolute Growth or Decline 1961-71 (1000)	Percentage State Manufacturing Employment 1971
Metropolitan New York	992.9	1,081.7	− 8.2	−88.8	58.34
Ontario Lake Plain	208.1	192.2	8.3	15.9	12.22
Niagara Frontier	155.7	158.7	− 1.9	− 3	9.14
Mohawk Valley	105.6	111.1	− 4.9	− 5.5	6.20
Eastern	54.0	55.1	− 2.0	− 1.1	3.17
Middle	27.5	26.3	4.6	1.2	1.61
Western	24.1	29.8	−19.1	− 5.7	1.42
Southern Tier	64.6	71.4	− 9.5	− 6.8	3.79
Mid-Hudson	60.0	58.2	3.1	1.8	3.52
Total for State	1701.9	1,767.4	− 3.7	−65.5	100

other parts of the country, and the degree of automation development in a particular industry all effect employment levels.

The manufacturing regions in the state also experienced different employment trends in the 1961-71 period. Referring to Figure 74, page 236, the Ontario Lake Plain exhibited an 8.3 percent gain (16,000 employees) largely as a result of growth in the Rochester area. The Mid-Hudson region showed a modest gain of 1800 employees. All of the other regions declined. Metropolitan New York's loss of 89,000 manufacturing jobs, and the fact that losses within New York City's borders were still greater, contribute to the fiscal crisis in the city.

Construction employment in the state has risen over the last decade, but this is a following of the national trend. Actually New York lagged a bit behind that national trend.

The Tertiary Sector

The addition of over one million workers to the tertiary labor force and the sector's attendant rise to a point in 1970 when it accounted for nearly 70 percent of the total labor force may be considered indicative of the direction of New York's economy. A strong rise was forecast in the 1960s, but the size of the rise which occurred in this short period was not anticipated. Instead of the tertiary sector providing 76 percent of the new jobs for the rest of this century as was predicted on page 497, the decade of the 1960s suggests that not only will all of future net gains be generated by the tertiary sector but that losses in the secondary and primary sectors will have to be made up by gains in the tertiary sector if the employment level is to rise.

So great was the 1960 tertiary employment increase that all of the sector's major parts, as exhibited on Table 26, page 255, participated in the gains except personal services which continued its earlier decline. Fifty-six percent of the sector's net growth between 1960 and 1970 is accounted for by professional and related services. This was also the largest gainer between 1950 and 1960, and its increase of 565,000 employees during the 60s was impressive indeed. In 1970 it became the largest tertiary employer, displacing the earlier leadership of wholesale and retail trade. Education, medicine,

TABLE 26 (p. 255)

The Tertiary Sector New York State

	1970 Employment	% of Sector Total	Absolute Gains or Losses 1960-70
Transportation, Communication, and Public Utilities	578,131	11.7	+ 69,559
Wholesale and Retail Trades	1,395,166	28.1	+ 192,523
Finance, Insurance and Real Estate	531,798	10.7	+ 122,555
Business and Repair Services	292,946	5.9	+ 82,147
Personal Services	287,925	5.8	− 70,162
Entertainment and Recreation Services	75,555	1.5	+ 12,016
Professional and Related Services	1,407,925	28.4	+ 565,525
Public Administration	391,314	7.9	+ 77,541
TOTAL	4,960,760	100.0	+1,051,704

Source: U.S. Census of Population

law, and similar fields grew rapidly as society became more complex and sophisticated. And, as enrollments in the state's colleges and universities rose, more and more individuals were prepared to enter, and did enter, the professional fields.

Wholesale and retail trade which, because of mechanization and self-service orientations, had lost employment in the 1950-60 decade, was the second-largest gainer between 1960 and 1970, adding 192,000 new jobs. This increase probably reflects the growing amount of spendable income and the widespread proliferation of shopping districts in suburban areas. Finance, insurance, and real estate ranked third as a new job generator, adding 123,000 new employment opportunities in the 1960-70 decade. Business and repair services, public administration, transportation, communication and public utilities, and entertainment and recreation services followed, in that order.

The broad span of growth through the various parts of the tertiary sector as compared to the 1950s interestingly suggests some improvement in opportunities for the moderately skilled and unskilled job hunter. Substantial growth in wholesale and retail trade and business and repair services, particularly (275,000 jobs in all), indicate this. Yet it would surely be incorrect to assume that all, or even most, of the jobs in these types of employment have only low skill requirements. Very significant, in contrast, is the amount of growth in professional and related services and finance, insurance, and real estate (688,000 in all), where high skill requirements parallel the growing complexity of American culture.

The potential impact of New York's trends in employment amount and structure is difficult to predict. Worry can be expressed that the total job market is not growing more rapidly in the state, that unemployment and welfare rolls are higher than desirable, that a declining labor force in manufacturing is troublesome, that the payroll for public employees is excessive, that fewer and fewer people are being employed in agriculture, or that the general demand for low-skill jobs is poor. Some of these problems are national in character, with New York being no worse off than many other states; some, too, are related to the fact that New York is located in the older Northeast where economic expansion is, as should be expected, lagging behind certain other sections of the nation; still others may be related to state laws and regulations or environmental conditions.

Good signs can be seen too. The vigorous trends for employment in the tertiary sector is encouraging. The tertiary sector in general, and the fast-growing professional and related services portion in particular, generate better than average incomes. In 1975 only five states exceeded New York in personal income per capita, with New York's figure standing at just over $6,500. By contrast, the United States average is $5,658 for that year. There is evidence, too, that New York's per-capita income situation weathered the recession from the last quarter of 1973 to the first quarter of 1975 better than most states and rose during this period more than the United States average. None of these figures conclusively show where New York is going economically in the years just ahead, nor do they say anything about the likelihood of certain sections of the state doing much better than other sections. What they do suggest is that New York's economy has its strong as well as weak points.

THE ENERGY CRISIS

Chapter 12 details the distribution of power production, defines power regions, and identifies New York State as both a major producer and consumer of electrical energy. Except for a brief discussion of the Great Blackout of 1965 no suggestion of possible energy shortages is made and the problem of the rising costs has not been identified. In 1966 many steam-generating plants were converting from coal to oil because of both cost and air pollution considerations, and almost no one was concerned about the cost of home heating oil and natural gas or gasoline for motor vehicles.

Although New York State is a comparatively large producer of hydroelectric energy, it has little in the way of fossil fuels and so must bring these across its borders. Even if it had coal and petroleum within its borders, however, the cost of energy would be little improved, for the prices of these products tend to be determined largely by federal policies and national marketing conditions. Consumption or demand issues, on the other hand, do have a regional dimension, at least to the extent they are influenced by heating costs dictated by climate. The record cold winter of 1976-77 dramatically drove this point home as natural gas supplies for heating and industrial power in New York State became inadequate to the point of closing schools and laying off industrial and commercial work forces.

The national energy problem has been just beyond the horizon for many years, but it was not until late 1973 that manipulation of international supplies and prices of petroleum by producing nations in the Middle East began to be strongly felt. Scarcities appeared and prices rose; an energy shock was experienced. "Energy shock" is a new term used to describe the unusual effects of increase in petroleum prices on employment, the gross national product, and the economy at large. The shock occurs because a substantial part of the price rise in oil, unlike that for many commodities, goes to foreigners or into profits of domestic oil producers. In neither case is the money plowed *back* immediately into the domestic spending stream. The result is that an oil price increase not only raises the price indexes like any other price increase, but it also reduces total purchasing power and the gross national product.

A continuing energy crisis can reduce New York State's and the nation's normal course of economic development and produce debilitating inflation. New York, because it uses relatively more energy due to being in a cold region, may be more adversely affected than states favorably located as to winter temperatures. Whether this could become sufficient to tip the balance in favor of business, industry, and individuals locating in the South or the Pacific Coast is difficult to predict. In any case, the greater the energy price rise the greater will be the relative disadvantage of northern states like New York.

Short-range solutions to the energy crisis no doubt will include changeovers from oil and gas to coal, with perhaps attendant air pollution aggravation. Long-range solutions should involve moving away from the use of fossil fuels and toward nuclear plants (fusion types would be best) and possibly installations utilizing wind and solar sources. Certainly, New York's relatively cloudy climate is not particularly conducive to the development of direct solar energy on a large scale. Conserving energy use will be a major thrust of the coming generation.

Decisions about the location of energy installations of the distant future will have impacts on the state's landscapes which cannot be clearly seen at this time. For example, will numerous nuclear plants replace fossil fuel plants in the vicinity of New York City, and where exactly will they be situated? Will the windy east shore of Lake Ontario or the Tug Hill spawn massive installations of wind generators, or will these be more common on promontories throughout the Appalachian uplands or the Adirondacks? Will ordinary structures have self-contained energy systems combining small-wind generators and roof solar-energy cells; and if so what would this do to architecture and the appearance of settlements generally? Will gasoline service stations give way to battery replacement centers serving electrical automobiles?

ENVIRONMENTAL CONCERNS

Environmental conditions are now generally viewed as important by most people. Involved are economic, social-political, ecologic, and esthetic issues, the last often exhibiting an overlap with the other three. Since the mid-60s ecological concerns have come relatively more to the forefront. Worry about pollution of lakes and rivers, about the quality of the air we breathe, about the noise generated around us, even about visual pollution and land preservation has intensified. The scientific community, the press, legislative bodies, businesses, and a substantial portion of the populace have been swept up in this ecological movement. The impact, much of it beneficial, should be extensive and lasting.

In 1965 the people of New York State overwhelmingly approved the Pure Waters Bond Proposition for more than $1 billion. Again in 1972 they approved a similar amount which extended state expenditures beyond water treatment into air pollution problems, solid waste disposal, and acquisition of public forests and wetlands. Federal assistance funds are increased because of such a large effort at the state level. The State of New York's Department of Environmental Conservation is probably as active as any department of its kind in the country. It publishes *The Conservationist* bimonthly and *Environment* monthly. In 1975 alone a number of new state environmental bills were signed into law, including the Environmental Quality Bill, Freshwater Wetlands Bill, Wild, Scenic and Recreational Rivers Bill, and the Cumulative Impact Bill.

Water

A report, *Pure Waters Progress,* issued by the Department of Environmental Conservation categorically states that as of 1974 New York's rivers, streams, and lakes are the cleanest in modern times and getting cleaner. The Hudson River, Lake Erie, and other waters have improved during the last decade. Several hundred municipal wastewater treatment projects have been or are being constructed.

Industries in the state must now provide satisfactory wastewater treatment. Several hundred installations have been expedited by the State Pure Waters program, and the state's industries have generally been responsible in their cooperation.

A network of numerous monitoring stations as well as water-watcher volunteers keep tabs on potential polluters. Legal machinery is available to force compliance with regulations. It is estimated that more than 2/3 of the major polluters in the state outside New York City are abated. Problems related to sludge disposal, storm water treatment, and non-point sources such as agricultural field drainage remain challenges.

One of the difficulties of managing water quality is that new problems from new pollutants seem to continually arise. Examples of these in New York include Mercury, PCBs (polyclorinate biphenyl), and Mirex. Each has come to the front from a position of apparent obscurity; each has been identified as significant; and each has enlarged the scope and complexity of water-pollution control.

Air

The environmental bond approved in 1972 gave particular attention to public facilities such as refuse incinerators, schools, and hospitals. It provided 50 percent of the cost of upgrading or improving these facilities. Air pollution from industrial sources, automobiles, and homes so far has been largely the responsibility of private enterprise operating under federal standards and state law. The Department of Environmental Conservation has engaged in a public information effort to give motor vehicle owners a more informed and critical attitude toward the automobile. It also engages in inspection and enforcement efforts using mobile units but operating under federal standards and aimed largely at the gross emitter.

Federal air quality standards for certain pollutants are not being universally met in the state's cities now, and the national energy crisis will be inclined to aggravate the air pollution problem. Fortunately for New York State prolonged stationary high pressure systems and temperature inversions—the atmospheric conditions most conducive to serious air contamination—are not especially common.

Noise

The ecology movement only fairly recently has included an emphasis on noise, and up to now it is viewed by many as far less a problem than water and air pollution. Yet, in this mechanical age, noise is one of the most widespread forms of environmental deterioration. Slight noise may interrupt sleep or just be annoying while one is awake; moderate noise levels can interfere with speech; high-level noise can cause permanent hearing loss and perhaps nervous system damage.

Recently I was enjoying a beautiful view across an Adirondack lake and suddenly realized I could hear neither birds in the trees nor voices from a camp a short distance away. These sounds were being completely masked by the whine of high-speed motor boats on the lake, aircraft overhead, and a procession of cars driving at high speeds on a road not far away. All this mechanically generated and at that moment unwanted noise, as far as I was concerned, was striking my ears at about a 60-65 decibel [dB(A)] level. By comparison a whisper is 20 dB(A), a vacuum cleaner about 80, and a jet plane at 200 feet about 120. To give meaning to these figures it should be pointed out that on the dB(A) scale an increase of 10 represents a doubling of the noise level.

Some noise problems arise from stationary sources such as factories, construction sites, and mines, but perhaps the greatest noise problem emerging across the entire nation and in New York State is from traffic on the Interstate Highway network and other major routeways. These highways carry large volumes of traffic at high speeds, including many trucks, and generate a *noise belt* ranging from a mile or so to many miles in width depending on location and conditions near the highway. The nature and impact of highway noise belts were not seriously considered when the Interstate system was engineered. Noise adjacent to these highways is particularly apparent in country locations where the normal noise level is fairly low. They are less noticeable in cities where other noises mask or drown out the effect of the highway.

Airports, too, produce noise anomalies so intense that large areas are rendered undesirable for residential and many other uses. Maps of average or peak noise levels adjacent to airports are not readily available, so the average person needs to take particular cognizance of noise before investing in a residence or selecting a "quiet" place for a vacation.

In 1971 New York State added noise to the list of air contaminants. The Department of Environmental Conservation now has a Bureau of Noise Control which is fostering programs to control noise pollution. The federal government has preempted the state in regulation of aircraft noise, but in 1965 New York became the first state to specify a maximum allowable noise level for motor vehicles. It has established an 88 dB(A) maximum at 50 feet when motor vehicle speed is less than 35 miles per hour. Unfortunately this is too high a figure for that speed because vehicles routinely travel faster than 35 miles per hour, and so can legally generate very high levels of noise. Furthermore, there has been very little control of violators. Trucks are by far the greatest contributors to the highway noise problem because they commonly produce two, three, or even four times the amount of noise of the properly operated passenger car. Only strict legislation controlling engine, muffler, and tire design as well as maximum speed can significantly abate noise in the Interstate Highway noise belts. Even then, state and local enforcement agencies will have to be substantially more effective in the control of violators. Until proper control and enforcement exists people in thousands of square miles of the Interstate Highway belts in New York State will suffer from noise pollution.

Snowmobiles manufactured and sold after June 1, 1974, must have mufflers that limit noise to 73 dB(A) at 50 feet. The same restriction does not apply, however, to older snowmobiles and other off-the-road vehicles. In many sections of the state substantial animosity exists between off-the-road-vehicle operators on one hand and non-operators and landowners on the other. Some of the problem has simply to do with trespassing, but in most instances noise is a central issue. Stricter local codes and enforcement may be required to provide reasonable environmental protection against this kind of noise pollution.

Appearance of the Environment

Individuals vary widely in what they perceive to be tolerable levels of visual pollution, but almost everyone has begun in recent years to object to such things as open dumps, large amounts of dirt and debris in the streets, garish signs along highways, graffiti here and there, junk automobiles strewn over the landscape, run-down residences, and filthy industrial districts.

State legislation has attacked some of these problems, often under the aegis of health control. Local zoning and building codes, which are now widely in existence, commonly contain controls also. Enforcement against abuses of visual pollution is not always easy and is often construed to be encroachment on the private rights of the individual, but generally great strides seem to have been made in convincing people that visual pollution indeed detracts from their surroundings and thereby makes living less pleasant. The total effect of this on the State of New York has steadily become more apparent over the last decade. The exception seems to be in some locations in inner cities where renters greatly outnumber owner-occupants. Establishment of more local improvement associations or similar organizations which would take interest in visual pollution, among other things, seems to be a good idea.

Land Preservation

New York State, more than any other state in the nation, has been active for many years in acquiring and preserving open space and wilderness. The great Adirondack and Catskill preserves plus many smaller public land parcels provide a priceless but fragile resource for present and future generations to use and enjoy. The Environmental Bond Issue of 1972 carries on this tradition by enabling the state and municipalities to acquire and preserve additional forest wetlands, urban parks, and natural areas because of their ecological importance or because of their value for outdoor recreation. This is an important step during a period when wetlands are being lost to drainage and development. Private lands in the Adirondacks, Catskills, and elsewhere are being bought up by real estate developers, and open lands near cities are being bulldozed and blacktopped.

The ecological movement was needed and already has made significant contributions to the improvement of New York State's environment. However, there is some uncertainty and disagreement about tolerable levels of environmental control, about the benefits of pollution abatement, and land-use restrictions, on the one hand, and resulting curtailment of economic endeavor and encroachment on individual freedoms to use land, on the other. Any state which has "excessive" controls, as defined in some quarters, may have to expect a certain amount of emigration away from those controls. Weighing the economic costs of this emigration against the ecological benefits of a better environment is not simple and requires serious inputs from all segments of the state's society.

A PERSPECTIVE

Generally, the above trends seem to present a somewhat pessimistic picture of the great State of New York. Serious fiscal difficulties need correction; the means of adjusting to a slower-growth economy, particularly one away from manufacturing and agriculture, must be found; cities need to regain attractiveness as places in which to live; and the state generally must put up with what probably will be regional disadvantages accruing from the national energy crisis.

Not all is dark, however. Substantial improvement is taking place in the maintenance of the environment, and a slower growth in population and the economy may be viewed as an aid to that maintenance. While the primary and secondary sectors are exhibiting declines in employment, the tertiary sector is expanding. A note of optimism emerges from the fact that the tertiary sector is the most productive and promotes the highest salaries and wages.

It may be safely hypothesized that political decisions and legislative action will play increasingly crucial roles in New York State's future. If shaky fiscal policies continue, if a situation persists which makes New York more attractive to those who do not compete in a productive sense than to those who do, if an economic climate is maintained which repels rather than attracts business, then the state will be in trouble. On the other hand, if political action is oriented toward establishing wise economic, social, and ecological conditions, toward support of good business climates and attractive living environments, then New York's inherent advantages of location, large sophisticated population, and diverse physical base should enable it to remain an impressive giant among states.

Syracuse, New York
January 1977

John H. Thompson

POPULATION ESTIMATES, 1975 and 1974

Area	July 1, 1975	July 1, 1974	Percent Change from 1974
NEW YORK STATE	18,409,200	18,327,800	+ 0.4
DOWNSTATE	11,605,200	11,549,700	+ 0.5
REST OF STATE	6,804,000	6,778,100	+ 0.4

DOWNSTATE

Area or Community	July 1, 1975	July 1, 1974	Percent Change from 1974	Area or Community	July 1, 1975	July 1, 1974	Percent Change from 1974
NEW YORK CITY	7,567,100	7,567,100	0.0	NASSAU–SUFFOLK DISTRICT	2,770,200	2,730,200	+ 1.5
BRONX COUNTY	1,393,200	1,393,200	0.0	NASSAU COUNTY	1,428,500	1,429,200	—
KINGS COUNTY	2,448,200	2,448,100	0.0	EAST ROCKAWAY	11,650	11,650	0.0
NEW YORK COUNTY	1,440,200	1,440,200	0.0	FLORAL PARK	19,050	18,950	+ 0.5
QUEENS COUNTY	1,962,700	1,962,700	0.0	FREEPORT	41,800	41,550	+ 0.6
RICHMOND COUNTY	322,700	322,700	0.0	GARDEN CITY	25,850	25,850	0.0
				GLEN COVE	25,900	25,900	0.0
WESTCHESTER–ROCKLAND– PUTNAM DISTRICT	1,267,900	1,252,400	+ 1.2	GREAT NECK	10,850	10,850	0.0
				HEMPSTEAD	35,300	35,550	− 0.7
PUTNAM COUNTY	69,400	66,900	+ 3.7	LONG BEACH	36,000	35,550	+ 1.3
ROCKLAND COUNTY	274,500	266,200	+ 3.1	LYNBROOK	25,450	25,000	+ 1.8
SPRING VALLEY	24,550	23,300	+ 5.4	MALVERNE	10,150	10,150	0.0
WESTCHESTER COUNTY	924,000	919,300	+ 0.5	MASSAPEQUA PARK	21,400	21,600	− 0.7
DOBBS FERRY	10,450	10,500	− 0.5	MINEOLA	21,400	21,550	− 0.7
MAMARONECK	18,450	18,500	− 0.5	NEW HYDE PARK	9,000	9,250	− 2.7
MOUNT VERNON	71,400	71,800	− 0.6	ROCKVILLE CENTRE	27,600	27,600	0.0
NEW ROCHELLE	74,650	74,950	− 0.4	VALLEY STREAM	41,550	41,300	+ 0.6
OSSINING	21,900	21,850	+ 0.2	WESTBURY	15,400	15,450	− 0.3
PEEKSKILL	19,650	19,650	0.0				
PORT CHESTER	26,500	26,400	+ 0.4	SUFFOLK COUNTY	1,341,700	1,301,000	+ 3.1
RYE	16,300	16,200	+ 0.6	BABYLON	12,700	12,800	− 0.8
SCARSDALE	19,950	19,850	+ 0.5	LINDENHURST	32,450	31,750	+ 2.2
TARRYTOWN	10,100	10,300	− 1.9	PATCHOGUE	14,850	14,250	+ 4.2
WHITE PLAINS	49,550	49,700	− 0.3				
YONKERS	202,600	203,150	− 0.3				

REST OF STATE

Area or Community	July 1, 1975	July 1, 1974	Percent Change from 1974	Area or Community	July 1, 1975	July 1, 1974	Percent Change from 1974
BINGHAMTON AREA	412,300	413,800	− 0.4	CAPITAL DISTRICT	950,400	943,200	+ 0.8
BINGHAMTON METRO AREA (N.Y. PORTION)	264,800	266,100	− 0.5	ALBANY–SCHENECTADY–TROY METRO. AREA	818,600	812,400	+ 0.8
BROOME COUNTY	216,200	217,900	− 0.8	ALBANY COUNTY	288,900	289,500	− 0.2
BINGHAMTON	55,500	57,500	− 2.8	ALBANY	104,300	107,300	− 2.8
ENDICOTT	15,100	15,350	− 1.6	COHOES	17,550	17,350	+ 1.2
JOHNSON CITY	14,500	15,200	− 4.6	WATERVLIET	11,250	11,450	− 1.7

(continued on next page)

POPULATION ESTIMATES, 1975 and 1974 *(Continued)*

Area or Community	July 1, 1975	July 1, 1974	Percent Change from 1974	Area or Community	July 1, 1975	July 1, 1974	Percent Change from 1974
				REST OF STATE (continued)			
CHENANGO COUNTY	47,000	46,900	+ 0.2	MONTGOMERY COUNTY	59,500	58,800	+ 1.2
DELAWARE COUNTY	44,600	44,700	- 0.2	AMSTERDAM	25,800	25,750	+ 0.2
OTSEGO COUNTY	55,900	56,100	- 0.4	RENSSELAER COUNTY	156,500	155,900	+ 0.4
ONEONTA	15,400	15,650	- 1.6	TROY	9,750	9,900	- 1.5
					58,450	59,400	- 1.6
TIOGA COUNTY	48,600	48,200	+ 0.8	SARATOGA COUNTY	148,900	143,900	+ 3.5
BUFFALO AREA	1,579,900	1,581,800	- 0.1	SARATOGA SPRINGS	22,500	21,950	+ 2.5
BUFFALO METRO. AREA	1,354,600	1,355,300	- 0.1	SCHENECTADY COUNTY	164,800	164,300	+ 0.3
				SCHENECTADY	75,600	76,250	- 0.9
CATTARAUGUS COUNTY	80,600	81,100	- 0.6	SCHOHARIE COUNTY	26,000	25,800	+ 0.8
OLEAN	17,250	17,700	- 2.5				
CHAUTAUQUA COUNTY	144,700	145,400	- 0.5	WARREN COUNTY	51,600	51,300	+ 0.6
DUNKIRK	15,750	15,950	- 1.3	GLENS FALLS	16,400	16,600	- 1.2
FREDONIA	10,450	10,600	- 1.4	WASHINGTON COUNTY	54,200	53,700	+ 0.9
JAMESTOWN	38,200	38,550	- 0.9				
ERIE COUNTY	1,120,100	1,120,300	*	ELMIRA AREA	345,300	345,200	*
BUFFALO	421,900	431,100	- 2.1	ELMIRA METRO. AREA	100,400	100,500	- 0.1
DEPEW	26,650	25,850	+ 3.1				
HAMBURG	10,450	10,950	+ 1.4	ALLEGANY COUNTY	47,600	47,500	+ 0.2
KENMORE	20,450	20,550	- 0.4				
LACKAWANNA	27,700	27,950	- 0.9	CHEMUNG COUNTY	100,400	100,500	- 0.1
LANCASTER	13,100	13,200	- 0.8	ELMIRA	35,600	36,350	- 2.1
TONAWANDA	21,400	21,550	- 0.7				
				SCHUYLER COUNTY	17,400	17,200	+ 1.2
NIAGARA COUNTY	234,500	235,000	- 0.2	STEUBEN COUNTY	99,000	99,200	- 0.2
LOCKPORT	24,800	24,950	- 0.6	CORNING	14,450	14,750	- 2.0
NIAGARA FALLS	79,750	81,600	- 2.5	HORNELL	10,600	10,950	- 3.2
NORTH TONAWANDA	35,500	35,600	- 0.3				
				TOMPKINS COUNTY	80,900	80,800	+ 0.1
				ITHACA	24,600	25,400	- 3.1
MID-HUDSON AREA	781,300	770,800	+ 1.4	ROCHESTER AREA	1,184,700	1,172,900	+ 1.0
POUGHKEEPSIE METRO. AREA	236,200	234,500	+ 0.7	ROCHESTER METRO. AREA	1,030,600	1,019,600	+ 1.1
COLUMBIA COUNTY	53,100	52,800	+ 0.6	GENESEE COUNTY	60,600	60,400	+ 0.3
				BATAVIA	16,600	16,750	- 0.9
DUTCHESS COUNTY	236,200	234,500	+ 0.7				
BEACON	12,950	13,300	- 2.6	LIVINGSTON COUNTY	57,900	57,600	+ 0.5
POUGHKEEPSIE	27,150	28,200	- 3.7				
				MONROE COUNTY	767,500	758,700	+ 1.2
GREENE COUNTY	34,200	33,900	+ 0.9	ROCHESTER	279,050	281,650	- 0.9
ORANGE COUNTY	251,600	245,700	+ 2.4	ONTARIO COUNTY	84,100	83,200	+ 1.1
MIDDLETOWN	20,250	20,400	- 0.7	CANANDAIGUA	11,300	11,200	+ 0.9
NEWBURGH	24,200	24,550	- 1.4	GENEVA	16,000	16,200	- 1.2
SULLIVAN COUNTY	55,400	54,900	+ 0.9	ORLEANS COUNTY	38,700	38,400	+ 0.8
ULSTER COUNTY	150,800	149,000	+ 1.2	SENECA COUNTY	35,200	35,100	+ 0.3
KINGSTON	22,850	23,350	- 2.1				

Area	1970	1975	% change
MOHAWK VALLEY AREA	400,500	400,900	− 0.1
UTICA-ROME METRO. AREA	342,900	343,300	− 0.1
FULTON COUNTY	52,800	52,800	0.0
GLOVERSVILLE	18,400	18,650	− 1.3
JOHNSTOWN	9,750	9,850	− 1.0
HAMILTON COUNTY	4,800	4,800	0.0
HERKIMER COUNTY	67,200	67,400	− 0.3
ONEIDA COUNTY	275,700	275,900	− 0.1
ROME	55,250	54,200	+ 1.2
UTICA	83,750	85,350	− 1.9
NORTHERN AREA	371,100	373,200	− 0.6
CLINTON COUNTY	75,100	75,100	0.0
PLATTSBURGH	18,400	18,650	− 1.3
ESSEX COUNTY	33,500	33,800	− 0.9
FRANKLIN COUNTY	42,100	42,600	− 1.2
JEFFERSON COUNTY	86,800	87,200	− 0.5
WATERTOWN	28,550	29,000	− 1.6
LEWIS COUNTY	23,300	23,400	− 0.4
ST. LAWRENCE COUNTY	110,300	111,100	− 0.7
MASSENA	12,250	13,000	− 2.3
OGDENSBURG	13,400	13,600	− 1.5
POTSDAM	10,800	10,800	0.0
WAYNE COUNTY	82,400	81,700	+ 0.9
WYOMING COUNTY	38,600	38,100	+ 1.3
YATES COUNTY	19,700	19,700	0.0
SYRACUSE AREA	778,500	776,300	+ 0.3
SYRACUSE METRO. AREA	649,000	647,700	+ 0.2
CAYUGA COUNTY	81,600	80,800	+ 1.0
AUBURN	35,950	35,650	+ 0.8
CORTLAND COUNTY	47,900	47,800	+ 0.2
CORTLAND	19,650	19,900	− 1.3
MADISON COUNTY	66,700	66,100	+ 0.9
ONEIDA	11,400	11,450	− 0.4
ONONDAGA COUNTY	474,700	474,700	*
SYRACUSE	178,150	182,050	− 2.1
OSWEGO COUNTY	107,600	106,700	+ 0.8
FULTON	13,550	13,650	− 0.7
OSWEGO	19,550	20,100	− 2.7

*Less than 0.05 percent.
1. New York City and Nassau, Putnam, Rockland, Suffolk and Westchester counties.

POPULATION *Source:* New York State Department of Health.

Data are estimates as of July 1, based on the Census count in 1970, recorded births and deaths, and estimated net migration. The latter is based on observed net migration between 1960 and 1970 in most counties outside New York City and, in some counties where special censuses were taken, between the date of the latest special census and 1970. The estimate for Canandaigua City is based on a special census conducted on April 30, 1971. The estimates for Saratoga Springs and Hamburg have been adjusted to include the population of the areas annexed by those places on February 1, 1971 and October 21, 1971, respectively. The 1975 estimate of population for New York City is the same as the provisional 1974 figure, prepared by the U.S. Census Bureau. All data have been adjusted to include institutional population, and all estimates were corrected for a small underenumeration of children under five, then rounded.

Source: *New York State Business Fact Book,* 1976 Supplement.

EMPLOYMENT, 1975 and 1974 (Nonagricultural Establishments)

NEW YORK STATE

Industry	1975 (000)	1974 (000)	Industry	1975 (000)	1974 (000)
TOTAL	6,791.1	7,070.2	CONTRACT CONSTRUCTION	202.7	257.6
			GEN'L. CONTRACTORS-BUILDING	44.9	62.2
MANUFACTURING	1,407.1	1,573.9	GEN'L. CONTRACTORS-NONBLDG.	29.9	37.9
			SPECIAL TRADES CONTRACTORS	127.6	157.5
DURABLE GOODS	700.6	787.8			
ORDNANCE, ACCESSORIES	3.5	4.0	TRANSPORTATION, PUBLIC UTILITIES	432.5	456.9
LUMBER, WOOD PROD. EXC. FURNITURE	11.7	13.9	RAILROAD TRANSPORTATION	23.0	25.1
FURNITURE, FIXTURES	23.9	29.3	AIR TRANSPORTATION	53.6	55.7
STONE, CLAY, GLASS PRODUCTS	37.5	43.8	OTHER TRANSPORTATION	166.9	181.7
PRIMARY METAL INDUSTRIES	59.9	72.1	COMMUNICATION	131.3	135.9
FABRICATED METAL PRODUCTS	71.9	82.1	ELECT., GAS, SANIT. SERVICE	57.7	58.5
MACHINERY, EXC. ELECTRICAL	155.9	171.1			
ELECTRICAL MACHINERY	145.0	162.7	WHOLESALE, RETAIL TRADE	1,395.8	1,441.7
TRANSPORTATION EQUIPMENT	73.6	82.8	WHOLESALE TRADE	416.0	437.0
INSTRUMENTS, RELATED PRODUCTS	117.8	125.5	RETAIL TRADE	979.8	1,004.7
			BLDG. MATERIALS, HARDWARE,		
NONDURABLE GOODS	706.5	786.1	FARM EQUIPMENT	30.8	33.2
FOOD, KINDRED PRODUCTS	96.6	102.2	GEN'L. MERCHANDISE	206.8	216.1
TOBACCO MANUFACTURES	2.3	2.4	FOOD STORES	171.2	174.8
TEXTILE MILL PRODUCTS	43.0	52.9	AUTO DEALERS, SERV. STATIONS	88.1	89.8
APPAREL, OTHER FINISHED PRODUCTS	173.7	198.6	APPAREL, ACCESSORIES	83.2	88.9
PAPER, ALLIED PRODUCTS	47.2	53.0	FURNITURE, HOME FURN., EQUIP-		
PRINTING, PUBLISHING, ALLIED INDUS.	147.5	154.7	MENT STORES	40.6	43.9
CHEMICALS, ALLIED PRODUCTS	73.6	77.8	EATING, DRINKING PLACES	246.4	244.6
PETROLEUM REFINING, RELATED INDUS.	8.5	8.7	MISC. RETAIL STORES	112.7	113.5
RUBBER, MISC. PLASTICS PRODUCTS	23.4	29.1			
LEATHER, LEATHER PRODUCTS	27.9	33.0	FINANCE, INSURANCE, REAL ESTATE	580.0	584.5
MISCELLANEOUS MFG. INDUSTRIES	62.8	73.6	BANKING	194.0	189.6
			SECURITY, COMMODITY BROKERS	69.3	72.2
MINING	7.4	7.5	INSURANCE CARRIERS	119.7	124.4
			INSURANCE AGENTS, BROKERS	41.6	41.6
GOVERNMENT	1,326.2	1,298.6	REAL ESTATE INS. COMB., ETC.	122.2	124.1
			OTHER FINANCE AGENCIES	33.3	32.7
			SERVICES AND MISCELLANEOUS	1,439.6	1,449.6

NEW YORK CITY

Industry	1975 (000)	1974 (000)	Industry	1975 (000)	1974 (000)
TOTAL	3,275.9	3,444.6	CONTRACT CONSTRUCTION	77.9	99.9
MANUFACTURING	527.8	602.1	GEN'L. CONTRACTORS-BUILDING	16.1	19.9
			GEN'L. CONTRACTORS-NONBLDG.	8.6	11.7
			SPECIAL TRADES CONTRACTORS	53.3	68.4
DURABLE GOODS	116.7	140.0			
ORDNANCE, ACCESSORIES	—	0.2	TRANSPORTATION, PUBLIC UTILITIES	268.4	282.7
LUMBER, WOOD PROD. EXC. FURNITURE	3.4	4.0	RAILROAD TRANSPORTATION	7.0	7.8
FURNITURE, FIXTURES	11.0	13.7	AIR TRANSPORTATION	48.6	50.6
STONE, CLAY, GLASS PRODUCTS	5.2	6.1	OTHER TRANSPORTATION	105.4	114.8
PRIMARY METAL INDUSTRIES	8.8	9.9	COMMUNICATION	81.1	83.6
FABRICATED METAL PRODUCTS	23.6	27.6	ELECT., GAS, SANIT. SERVICE	26.5	25.9
MACHINERY, EXC. ELECTRICAL	16.6	19.8			
ELECTRICAL MACHINERY	29.9	35.6	WHOLESALE, RETAIL TRADE	635.3	665.1
TRANSPORTATION EQUIPMENT	5.2	8.4	WHOLESALE TRADE	252.7	266.2
INSTRUMENTS, RELATED PRODUCTS	13.0	14.9	RETAIL TRADE	382.6	398.8
			BLDG. MATERIALS, HARDWARE,		
NONDURABLE GOODS	411.1	462.1	FARM EQUIPMENT	6.9	7.5
FOOD, KINDRED PRODUCTS	37.3	39.6	GEN'L MERCHANDISE	85.5	88.0
TOBACCO MANUFACTURES	2.3	3.2	FOOD STORES	63.4	66.8
TEXTILE MILL PRODUCTS	25.1	31.5	AUTO DEALERS, SERV STATIONS	17.8	18.6
APPAREL, OTHER FINISHED PRODUCTS	142.1	162.7	APPAREL, ACCESSORIES	45.9	49.3
PAPER, ALLIED PRODUCTS	16.0	18.3	FURNITURE HOME FURN., EQUIP-		
PRINTING, PUBLISHING, ALLIED INDUS.	91.8	97.2	MENT STORES	17.1	18.3
CHEMICALS, ALLIED PRODUCTS	27.8	29.4	EATING, DRINKING PLACES	102.3	105.3
PETROLEUM REFINING RELATED INDUS.	5.6	5.3	MISC. RETAIL STORES	43.6	45.2
RUBBER, MISC. PLASTICS PRODUCTS	6.8	8.2			
LEATHER, LEATHER PRODUCTS	14.2	17.9	FINANCE, INSURANCE, REAL ESTATE	422.1	425.8
MISCELLANEOUS MFG. INDUSTRIES	42.1	49.2	BANKING	138.6	134.4
			SECURITY, COMMODITY BROKERS	66.5	69.4
MINING	1.5	1.4	INSURANCE CARRIERS	81.1	84.6
			INSURANCE AGENTS, BROKERS	23.6	23.4
GOVERNMENT	572.1	582.7	REAL ESTATE, INS. COMB., ETC.	91.3	92.9
			OTHER FINANCE AGENCIES	21.0	21.0
			SERVICES AND MISCELLANEOUS	770.7	785.0

(continued on next page)

EMPLOYMENT, 1975 and 1974 (Nonagricultural Establishments) - continued

NASSAU-SUFFOLK LABOR AREA
Nassau and Suffolk counties

Industry	1975 (000)	1974 (000)
TOTAL	785.0	803.6
CONTRACT CONSTRUCTION	32.4	41.5
TRANSPORTATION, PUBLIC UTILITIES	34.5	35.0
WHOLESALE, RETAIL TRADE	206.0	210.8
FINANCE, INSURANCE, REAL ESTATE	42.3	43.2
SERVICES, MINING, MISCELLANEOUS	164.0	160.1
GOVERNMENT	166.0	161.0
MANUFACTURING INDUSTRIES-TOTAL	139.0	151.8
DURABLE GOODS	92.4	100.4
FABRICATED METALS, INC. ORDNANCE	10.2	12.0
INSTRUMENTS, ELECTRICAL MACHINERY	37.3	40.9
TRANSPORTATION EQUIPMENT	28.7	28.5
OTHER DURABLE GOODS	16.2	19.1
NONDURABLE GOODS	46.6	51.4
FOOD, KINDRED PRODUCTS	3.4	3.5
TEXTILE MILL PRODUCTS	4.7	4.7
APPAREL, OTHER FINISHED PRODUCTS	7.9	9.1
PRINTING, PUBLISHING, ALLIED INDUS.	12.7	12.9
OTHER NONDURABLE GOODS	18.8	21.3

WESTCHESTER LABOR AREA
Westchester County

Industry	1975 (000)	1974 (000)
TOTAL	307.4	312.5
CONTRACT CONSTRUCTION	13.6	17.2
TRANSPORTATION, PUBLIC UTILITIES	17.9	18.2
WHOLESALE, RETAIL TRADE	71.2	72.2
FINANCE, INSURANCE, REAL ESTATE	17.0	16.5
SERVICES, MINING, MISCELLANEOUS	73.0	71.0
GOVERNMENT	51.7	50.6
MANUFACTURING INDUSTRIES-TOTAL	63.0	66.9
DURABLE GOODS	31.6	34.0
FABRICATED METALS	2.4	2.8
MACHINERY, EXCEPT ELECTRICAL	12.7	12.6
ELECTRICAL MACHINERY	6.1	6.6
TRANSPORTATION EQUIPMENT	2.4	2.9
INSTRUMENTS, RELATED PRODUCTS	4.4	4.3
OTHER DURABLE GOODS	3.8	4.7
NONDURABLE GOODS	31.4	32.9
FOOD KINDRED PRODUCTS	9.2	9.7
APPAREL, OTHER FINISHED PRODUCTS	4.3	4.6
PRINTING, PUBLISHING, ALLIED INDUS.	6.9	7.0
CHEMICALS, ALLIED PRODUCTS	6.0	5.7
OTHER NONDURABLE GOODS	4.9	6.0

ROCHESTER LABOR AREA
Livingston, Monroe, Ontario, Orleans and Wayne counties

Industry	1975 (000)	1974 (000)
TOTAL	383.5	396.0
CONTRACT CONSTRUCTION	11.8	15.2
TRANSPORTATION, PUBLIC UTILITIES	12.9	13.7
WHOLESALE, RETAIL TRADE	73.0	72.8
FINANCE, INSURANCE, REAL ESTATE	15.3	14.8
SERVICES, MINING, MISCELLANEOUS	68.8	68.9
GOVERNMENT	58.3	55.9

SYRACUSE LABOR AREA
Madison, Onondaga and Oswego counties

Industry	1975 (000)	1974 (000)
TOTAL	234.0	244.7
CONTRACT CONSTRUCTION	10.0	12.3
TRANSPORTATION, PUBLIC UTILITIES	13.7	14.3
WHOLESALE, RETAIL TRADE	52.4	53.3
FINANCE, INSURANCE, REAL ESTATE	14.3	14.3
SERVICES, MINING, MISCELLANEOUS	43.6	44.0
GOVERNMENT	46.8	44.2

	1975 (000)	1974 (000)
MANUFACTURING INDUSTRIES-TOTAL	143.4	154.7
DURABLE GOODS	113.1	121.5
PRIMARY, FAB. METALS, INC. ORDNANCE	7.1	8.0
MACHINERY, EXCEPT ELECTRICAL	14.0	15.8
ELECTRICAL MACHINERY	11.4	13.3
INSTRUMENTS, RELATED PRODUCTS	76.0	78.8
OTHER DURABLE GOODS	4.5	5.5
NONDURABLE GOODS	30.3	33.2
FOOD, KINDRED PRODUCTS	8.4	9.4
APPAREL, OTHER FINISHED PRODUCTS	3.4	4.1
PRINTING, PUBLISHING, ALLIED INDUS.	7.6	7.9
OTHER NONDURABLE GOODS	10.9	11.7

	1975 (000)	1974 (000)
MANUFACTURING INDUSTRIES-TOTAL	53.2	62.3
DURABLE GOODS	35.0	41.9
STONE, CLAY, GLASS PRODUCTS	1.7	1.7
PRIMARY METAL INDUSTRIES	4.1	5.5
FABRICATED METALS	1.9	2.2
MACHINERY, INC. ELECTRICAL	21.0	25.1
TRANSPORTATION EQUIPMENT	4.5	5.3
OTHER DURABLE GOODS	2.1	2.2
NONDURABLE GOODS	18.2	20.5
FOOD, KINDRED PRODUCTS	4.9	5.3
PAPER, ALLIED PRODUCTS	2.4	2.2
PRINTING, PUBLISHING, ALLIED INDUS.	2.7	2.8
OTHER NONDURABLE GOODS	8.1	9.6

BINGHAMTON LABOR AREA
Broome and Tioga (N.Y.) and Susquehanna (Pa.) counties

Industry	1975 (000)	1974 (000)
TOTAL	106.5	108.0
CONTRACT CONSTRUCTION	4.3	4.4
TRANSPORTATION, PUBLIC UTILITIES	4.6	4.7
WHOLESALE, RETAIL TRADE	19.3	19.4
FINANCE, INSURANCE, REAL ESTATE	3.6	3.6
SERVICES, MINING, MISCELLANEOUS	13.8	13.6
GOVERNMENT	22.4	21.3
MANUFACTURING INDUSTRIES-TOTAL	38.4	41.1
DURABLE GOODS	29.2	31.5
PRIMARY, FAB. METALS, ORDNANCE AND MACHINERY, INC. ELECTRICAL	22.8	24.7
OTHER DURABLE GOODS	6.3	6.8
NONDURABLE GOODS	9.2	9.6

ELMIRA LABOR AREA
Chemung County

Industry	1975 (000)	1974 (000)
TOTAL	37.0	38.8
CONTRACT CONSTRUCTION	1.3	2.1
TRANSPORTATION, PUBLIC UTILITIES	1.5	1.6
WHOLESALE, RETAIL TRADE	7.9	8.3
FINANCE, INSURANCE, REAL ESTATE	1.0	1.0
SERVICES, MINING, MISCELLANEOUS	6.8	6.2
GOVERNMENT	6.8	6.1
MANUFACTURING INDUSTRIES-TOTAL	12.5	13.5
DURABLE GOODS	9.3	9.9
PRIMARY, FABRICATED METALS AND MACHINERY, INC. ELECTRICAL	5.4	5.9
OTHER DURABLE GOODS	3.9	4.0
NONDURABLE GOODS	3.1	3.6

(continued on next page)

EMPLOYMENT, 1975 and 1974 (Nonagricultural Establishments) - continued

UTICA-ROME LABOR AREA
Herkimer and Oneida counties

Industry	1975 (000)	1974 (000)
TOTAL	109.5	113.3
CONTRACT CONSTRUCTION	2.8	3.5
TRANSPORTATION, PUBLIC UTILITIES	4.0	4.5
WHOLESALE, RETAIL TRADE	19.9	20.3
FINANCE, INSURANCE, REAL ESTATE	4.9	5.1
SERVICES, MINING, MISCELLANEOUS	17.9	17.3
GOVERNMENT	29.1	27.5
MANUFACTURING INDUSTRIES-TOTAL	30.9	35.2
DURABLE GOODS	21.3	24.6
PRIMARY METAL INDUSTRIES	4.1	5.2
FABRICATED METALS, INC. ORDNANCE		4.0
MACHINERY, INC. ELECTRICAL AND TRANSPORTATION EQUIPMENT	11.3	13.0
OTHER DURABLE GOODS	2.0	2.3
NONDURABLE GOODS	9.5	10.6
FOOD, KINDRED PRODUCTS	1.8	2.0
TEXTILE MILL PRODUCTS	1.0	1.0
APPAREL, OTHER FINISHED PRODUCTS		1.0
LEATHER, LEATHER GOODS	1.5	1.5
OTHER NONDURABLE GOODS	4.3	5.0

ALBANY-SCHENECTADY-TROY LABOR AREA
Albany, Montgomery, Rensselaer, Saratoga and Schenectady counties

Industry	1975 (000)	1974 (000)
TOTAL	305.3	313.2
CONTRACT CONSTRUCTION	11.1	13.6
TRANSPORTATION, PUBLIC UTILITIES	15.1	16.1
WHOLESALE, RETAIL TRADE	60.8	62.0
FINANCE, INSURANCE, REAL ESTATE	13.6	13.6
SERVICES, MINING, MISCELLANEOUS	57.9	56.7
GOVERNMENT	86.7	84.1
MANUFACTURING INDUSTRIES-TOTAL	60.1	67.1
DURABLE GOODS	33.4	36.4
STONE, CLAY, GLASS PRODUCTS	3.3	4.7
PRIMARY, FAB. METALS AND ORDNANCE	3.8	4.3
MACHINERY, INC. ELECTRICAL AND TRANSPORTATION EQUIPMENT	25.2	26.2
OTHER DURABLE GOODS	1.0	1.2
NONDURABLE GOODS	26.8	30.6
FOOD, KINDRED PRODUCTS	5.3	5.4
TEXTILE MILL PRODUCTS	3.9	4.6
APPAREL, OTHER FINISHED PRODUCTS	3.5	4.5
PAPER, ALLIED PRODUCTS	3.2	4.1
PRINTING, PUBLISHING, ALLIED INDUS.	4.6	4.9
CHEMICALS, ALLIED PRODUCTS		
OTHER NONDURABLE GOODS	2.3	2.8

POUGHKEEPSIE LABOR AREA
Dutchess County

Industry	1975 (000)	1974 (000)
TOTAL	87.1	87.6
MANUFACTURING	28.4	30.2
CONTRACT CONSTRUCTION	3.0	3.3
TRANSPORTATION, PUBLIC UTILITIES	2.8	2.8
WHOLESALE, RETAIL TRADE	14.9	14.5
FINANCE, INSURANCE, REAL ESTATE	2.5	2.6
SERVICES, MINING, MISCELLANEOUS	14.9	14.2
GOVERNMENT	20.5	19.9

BUFFALO LABOR AREA
Erie and Niagara counties

Industry	1975 (000)	1974 (000)
TOTAL	481.2	502.6
CONTRACT CONSTRUCTION	15.4	18.7
TRANSPORTATION, PUBLIC UTILITIES	27.0	29.7
WHOLESALE, RETAIL TRADE	105.7	109.0
FINANCE, INSURANCE, REAL ESTATE	19.8	20.2
SERVICES, MINING, MISCELLANEOUS	86.4	85.8
GOVERNMENT	87.5	83.0
MANUFACTURING INDUSTRIES-TOTAL	139.4	156.3

ROCKLAND COUNTY

Industry	1975 (000)	1974 (000)
TOTAL	72.2	72.3
CONTRACT CONSTRUCTION	1.9	3.0
TRANSPORTATION, PUBLIC UTILITIES	3.7	3.8
WHOLESALE, RETAIL TRADE	15.8	15.0
FINANCE, INSURANCE, REAL ESTATE	2.5	2.4
SERVICES, MINING, MISCELLANEOUS	15.1	14.9
GOVERNMENT	19.6	18.7
MANUFACTURING INDUSTRIES	13.5	14.4
DURABLE GOODS	94.7	106.0
STONE, CLAY AND GLASS PRODUCTS	6.2	7.0
PRIMARY METAL INDUSTRIES	25.6	30.1
FABRICATED METALS, INC ORDNANCE	11.9	13.5
MACHINERY EXCEPT ELECTRICAL	12.8	13.2
ELECTRICAL MACHINERY	12.1	12.2
TRANSPORTATION EQUIPMENT	21.5	24.5
OTHER DURABLE GOODS	4.7	5.4
NONDURABLE GOODS	44.6	50.3
FOOD, KINDRED PRODUCTS	9.6	10.5
TEXTILES, APPAREL	3.4	3.7
PAPER ALLIED PRODUCTS	4.1	4.6
PRINTING, PUBLISHING, ALLIED INDUS.	7.9	8.3
CHEMICALS, ALLIED PRODUCTS	10.7	12.1
RUBBER, MISC. PLASTICS PRODUCTS	4.5	5.7
OTHER NONDURABLE GOODS	4.4	5.4

EMPLOYMENT *Source:* New York State Department of Labor.

Estimates of employment in nonagricultural establishments are made for the State and the larger labor areas. The industries for which estimates are provided individually are the principal ones in each area. The totals do not include the self-employed, the military, farm workers and the employees of international organizations or private households.

Employment covered by unemployment insurance is less inclusive. It excludes not only the groups omitted from nonagricultural employment, but also workers for employers with less than $300 and those for household workers with less than $500 in payroll for a calendar quarter, and for interstate railroads and government. Data from this source ordinarily become available six to eight months later than the estimates of nonagricultural employment. However, they provide greater detail by industry than is available from the estimates, and give separate figures for each county — those combined in the multicounty industrial areas of the nonagricultural employment estimates as well as the smaller counties outside these areas. For any industry group, data are withheld in any county, metropolitan area or economic area for which the total might disclose information about a single establishment. In any county or area with only one nondisclosable industry, information has been withheld from a second to prevent subtraction of detail from the total to derive a withheld figure. Data in the table *Business Establishments, March 1975*, are the reporting units covered by unemployment insurance distributed according to the number of employees per unit. A reporting unit is normally a single place of business, but it may be only part of one when functions in more than one industry, carried on by a firm at one location, are reported as separate units, and it may be several places of business when the branches of a firm, within one county and in the same industry, are reported as one unit.

The classification by industry in both series is that of the 1972 Standard Industrial Classification; employment in central administrative offices, research laboratories and other auxiliary facilities of manufacturing companies is included in the appropriate manufacturing industries; the totals include some duplication because dual job holders are counted twice; and all the figures are as of the place of employment, not the place of residence.

Source: *New York State Business Fact Book, 1976 Supplement.*

MANUFACTURING EMPLOYMENT, 1974 (Covered by Unemployment Insurance)

Area or County	Ordnance and Accessories	Food and Food Products	Textile Mill Products	Apparel, Cloth Products	Lumber and Products	Furniture and Fixtures	Paper and Paper Products	Printing and Publishing	Chemical Products	Petroleum and Coal Products
NEW YORK STATE	4,158	102,714	53,231	198,678	13,956	29,454	53,002	154,663	78,180	10,331

ECONOMIC AREAS (map on inside back cover)

Area or County	Ordnance	Food	Textile	Apparel	Lumber	Furniture	Paper	Printing	Chemical	Petroleum
DOWNSTATE	896	53,199	37,949	176,985	5,396	17,844	25,413	118,670	46,394	6,196
NEW YORK CITY	241	39,550	31,630	162,773	4,003	13,663	18,315	97,227	29,479	*
NASSAU–SUFFOLK DISTRICT	*	3,469	4,694	9,060	1,068	3,162	4,860	12,879	5,822	*
WESTCHESTER–ROCKLAND–PUTNAM DISTRICT	*	10,180	1,625	5,152	325	1,019	2,238	8,564	11,093	*
REST OF STATE	3,262	49,271	15,238	21,664	8,519	11,600	27,486	35,778	31,710	2,477
BINGHAMTON AREA	0	2,383	580	747	916	1,227	352	2,652	2,000	78
BUFFALO AREA	0	13,839	1,304	2,840	1,857	4,525	*	9,172	12,864	947
CAPITAL DISTRICT	0	5,627	5,069	5,553	1,217	*	6,894	4,552	15,926	15
ELMIRA AREA	*	3,765	5,200	244	496	933	*	1,426	469	109
MID–HUDSON AREA	*	2,737	2,976	4,316	357	1,283	1,520	4,128	2,324	1,148
MOHAWK VALLEY AREA	*	2,109	2,108	*	712	*	790	1,241	92	16
NORTHERN AREA	*	1,809	230	639	821	366	5,936	1,139	865	*
ROCHESTER AREA	*	10,991	871	*	928	826	3,924	8,207	2,254	*
SYRACUSE AREA	*	6,011	1,900	4,926	1,215	479	2,683	3,261	4,916	53

STANDARD METROPOLITAN STATISTICAL AREAS[1]

Area or County	Ordnance	Food	Textile	Apparel	Lumber	Furniture	Paper	Printing	Chemical	Petroleum
ALBANY–SCHENECTADY–TROY	0	5,376	*	486	473	229	4,186	4,302	4,894	*
BINGHAMTON (NY PORTION)	*	1,271	*	2,566	239	1,027	352	1,820	154	32
BUFFALO	0	10,549	1,183	*	800	1,234	4,565	8,269	12,274	937
ELMIRA	*	2,286	*	*	171	*	428	714	53	*
NASSAU–SUFFOLK	0	3,469	4,694	9,060	1,068	3,162	4,860	12,879	5,822	*
NEW YORK, NY–NJ (NY PORTION)	241	49,730	33,255	167,925	4,328	14,682	20,553	105,791	40,572	6,028
POUGHKEEPSIE	*	9,448	507	791	100	301	443	2,204	37	1,047
ROCHESTER	*	9,644	516	*	*	612	3,708	7,917	*	*
SYRACUSE	*	5,437	1,040	487	783	340	2,660	2,824	4,778	26
UTICA–ROME	*	1,975	1,190	988	517	1,110	765	1,062	37	*

COUNTIES

Area or County	Ordnance	Food	Textile	Apparel	Lumber	Furniture	Paper	Printing	Chemical	Petroleum
ALBANY	0	2,376	1,178	1,207	154	138	1,204	2,751	523	0
ALLEGANY	0	207	*	*	*	0	*	22	*	0
BRONX	0	3,598	1,170	7,620	434	2,187	1,291	2,314	718	*
BROOME	0	1,108	*	377	50	775	352	1,802	*	*
CATTARAUGUS	0	420	*	*	555	578	152	259	538	*
CAYUGA	*	463	*	105	89	*	*	337	65	*
CHAUTAUQUA	0	2,871	4,694	273	502	2,713	*	644	*	*
CHEMUNG	*	2,286	33,255	19	171	*	*	714	53	0
CHENANGO	0	202	507	*	111	*	428	154	1,786	*
CLINTON	*	297	*	116	17	0	*	128	*	0
COLUMBIA	0	133	498	187	*	*	1,275	117	87	0
CORTLAND	0	111	*	*	11	*	285	101	73	*
DELAWARE	*	739	0	639	343	*	*	546	*	*
DUTCHESS	0	448	507	791	100	200	443	2,204	37	1,047
ERIE	*	8,940	889	2,479	472	1,127	2,798	6,916	5,028	760

County									
ESSEX	0	41	0	0	181	*	*	52	*	0
FRANKLIN	0	183	0	178	263	0	*	70	0	0
FULTON	0	134	919	70	144	63	25	179	56	*
GENESEE	0	620	*	*	8	0	205	205	*	*
GREENE	0	122	*	149	21	*	0	106	0	0
HAMILTON	*	0	412	*	51	476	*	*	0	0
HERKIMER	*	369	412	281	238	476	195	83	0	0
JEFFERSON	0	524	*	187	101	*	1,507	289	43	*
KINGS (BROOKLYN)	*	13,460	14,256	34,185	1,481	6,335	6,025	4,568	5,953	0
LEWIS	0	309	*	*	107	*	973	276	*	0
LIVINGSTON	0	726	0	0	79	0	*	500	212	*
MADISON	*	286	0	0	*	58	*	165	*	*
MONROE	0	5,456	516	3,602	443	385	1,262	6,729	1,009	0
MONTGOMERY	*	1,831	1,643	1,088	*	*	1,233	377	144	*
NASSAU	0	1,078	2,172	4,306	444	2,036	3,178	10,356	3,613	0
NEW YORK (MANHATTAN)	*	9,690	9,397	107,797	1,251	2,137	5,606	84,970	18,829	5,358
NIAGARA	*	1,609	294	87	329	107	1,767	1,353	7,246	177
ONEIDA	0	1,606	778	707	278	634	979	979	*	*
ONONDAGA	*	3,032	228	447	328	282	1,028	2,379	4,752	24
ONTARIO	0	854	*	*	*	*	374	374	506	20
ORANGE	0	1,229	1,708	2,036	67	578	302	946	1,550	*
ORLEANS	0	777	*	*	33	*	697	124	*	0
OSWEGO	*	2,119	811	*	376	*	1,470	279	*	*
OTSEGO	0	171	0	261	128	0	*	133	34	*
PUTNAM	*	*	*	18	6	*	*	481	0	*
QUEENS	*	12,476	6,807	11,806	814	2,955	5,315	5,017	2,617	72
RENSSELAER	0	322	1,036	1,770	183	38	375	589	2,595	*
RICHMOND	0	326	*	1,366	24	50	379	358	1,363	36
ROCKLAND	0	445	*	557	16	*	1,102	1,165	5,476	*
ST. LAWRENCE	0	455	0	*	153	353	1,155	324	*	*
SARATOGA	0	216	674	*	110	*	2,123	183	1,228	0
SCHENECTADY	0	632	*	218	158	0	251	403	405	0
SCHOHARIE	0	40	0	*	81	0	*	55	*	0
SCHUYLER	0	60	0	0	*	*	0	22	*	*
SENECA	0	199	*	0	235	899	0	36	*	*
STEUBEN	*	1,170	2,522	4,754	624	1,126	1,682	286	0	*
SUFFOLK	0	2,391	*	59	12	145	*	2,523	2,209	*
SULLIVAN	0	418	*	109	189	252	*	105	*	*
TIOGA	*	163	*	*	*	*	0	18	*	*
TOMPKINS	*	43	217	*	147	140	95	381	*	0
ULSTER	0	387	*	1,095	418	*	*	651	462	0
WARREN	0	162	*	697	167	*	1,127	161	*	*
WASHINGTON	0	*	*	258	148	163	1,582	34	8	*
WAYNE	*	50	*	*	303	774	2,059	190	80	0
WESTCHESTER	0	1,832	854	4,577	303	*	1,136	6,918	5,617	191
WYOMING	0	243	*	524	107	*	*	39	*	0
YATES	0	283	0	*	*	*	0	10	*	0

(continued on next page)

MANUFACTURING EMPLOYMENT, 1974 (Covered by Unemployment Insurance)
(Continued)

Area or County	Rubber and Plastics Products	Leather Products	Stone, Clay and Glass Products	Primary Metals	Fabricated Metal Products	Machinery (except electrical)	Electrical Machinery	Transportation Equipment	Instruments, Photo Goods	Miscellaneous	Other
NEW YORK STATE	28,966	32,999	43,626	72,265	82,170	171,403	160,341	84,993	125,681	73,902	2,428

ECONOMIC AREAS (map on inside back cover)

DOWNSTATE	13,938	19,270	9,398	16,046	43,135	43,508	71,005	39,679	32,417	57,610	2,361
NEW YORK CITY	8,230	17,906	6,086	*	27,719	19,956	36,009	8,272	14,860	49,344	17,906
NASSAU-SUFFOLK DISTRICT	4,033	786	2,036	2,363	11,289	10,644	27,342	28,424	13,295	5,591	823
WESTCHESTER-ROCKLAND- PUTNAM DISTRICT	1,675	578	1,276	*	4,127	12,908	7,565	2,983	4,262	2,675	4,166
REST OF STATE	15,030	13,704	34,212	56,211	38,974	127,835	89,271	45,288	93,257	16,276	67
BINGHAMTON AREA	1,125	4,216	300	745	1,560	11,585	11,585	3,040	5,840	94	0
BUFFALO AREA	*	614	8,278	*	17,849	18,891	14,310	25,724	*	4,258	46,832
CAPITAL DISTRICT	1,744	370	5,361	*	1,425	18,684	7,789	2,576	1,525	845	3,593
ELMIRA AREA	*	*	9,445	747	*	*	3,958	2,033	111	219	15,229
MID-HUDSON AREA	1,705	1,959	2,599	1,364	*	12,869	12,869	2,786	1,008	*	24,571
MOHAWK VALLEY AREA	528	4,525	441	5,376	1,779	8,007	3,632	2,073	*	3,105	4,962
NORTHERN AREA	260	637	486	4,987	373	*	980	1,797	398	233	1,303
ROCHESTER AREA	1,659	322	5,587	2,483	7,381	18,689	19,001	989	79,082	4,192	410
SYRACUSE AREA	1,785	*	1,715	5,675	*	18,098	14,947	6,270	1,187	*	6,939

STANDARD METROPOLITAN STATISTICAL AREAS[1]

ALBANY-SCHENECTADY-TROY	*	*	4,700	2,833	1,311	18,036	5,842	2,562	458	800	12,521
BINGHAMTON (NY PORTION)	824	3,734	221	683	*	11,591	7,560	2,339	5,125	46	2,573
BUFFALO	5,724	303	7,055	29,974	13,408	13,215	12,075	*	3,452	4,177	24,539
ELMIRA	*	*	1,841	568	1,342	1,354	2,736	1,812	*	98	144
NASSAU-SUFFOLK	4,033	786	2,036	2,363	11,289	10,644	27,342	28,424	13,295	5,591	823
NY, NY-NJ (NY PORTION)	9,905	18,484	7,362	13,683	31,846	32,864	43,663	11,255	19,122	52,019	2,361
POUGHKEEPSIE	*	85	349	37	363	12,061	10,271	373	79,066	3,867	7,069
ROCHESTER	1,548	96	3,863	1,087	2,196	12,628	13,273	*	1,135	12,693	
SYRACUSE	1,453	*	1,591	*	1,596	7,760	12,504	*	305	3,028	13,407
UTICA-ROME	134	1,488	*	*	*	*	*	*	*	*	*

COUNTIES

ALBANY	737	*	3,927	1,350	756	658	224	1,469	128	282	10
ALLEGANY	*	0	*	179	1,200	1,174	*	*	0	103	764
BRONX	1,517	1,507	652	219	4,550	1,499	*	*	*	2,875	3,899
BROOME	*	3,734	*	*	1,311	11,554	1,920	*	5,125	46	3,922
CATTARAUGUS	*	*	509	*	546	3,198	1,944	*	*	46	949
CAYUGA	*	495	111	186	132	1,623	1,930	0	340	*	971
CHAUTAUQUA	206	*	714	1,979	3,896	2,478	491	1,012	*	36	107
CHEMUNG	*	0	1,841	568	1,342	1,354	2,736	1,812	*	98	144
CHENANGO	*	*	*	*	1,231	1,263	602	*	0	*	860
CLINTON	199	0	67	*	130	*	*	0	*	12	931
COLUMBIA	*	0	393	*	*	357	*	*	*	470	766
CORTLAND	*	0	13	*	*	365	81	*	*	731	*
DELAWARE	*	*	*	*	*	3,847	*	1,033	553	*	85
DUTCHESS	910	85	349	37	363	12,061	3,150	131	127	145	1,164
ERIE	5,129	*	3,209	27,760	12,082	11,427	10,271	103	*	*	*
									14,905	9,239	3,363	3,462	343

30

County									
ESSEX	*	.	.	.	0	.	0	.	1,056
FRANKLIN	637	.	.	0	0	.	54	.	354
FULTON	394	3,037	.	183	248	.	78	.	415
GENESEE	103	226	1,100	521	1,049	.	.	.	2,181
GREENE	.	.	63	217	.	175	.	.	648
HAMILTON	0	0	.	488	0	0	.	.	32
HERKIMER	.	1,029	.	.	4,125	.	210	.	2,876
JEFFERSON	0	.	2,011	.	713	.	.	381	2,020
KINGS (BROOKLYN)	3,400	4,589	2,226	11,000	5,422	6,211	1,702	.	2,360
LEWIS	0	.	.	.	899	.	.	.	357
LIVINGSTON	424	.	291	1,241	9,745	.	10,196	.	573
MADISON	.	.	39	53	275	.	.	.	374
MONROE	1,040	76	1,383	5,115	13,562	101	1,734	78,839	401
MONTGOMERY	451	322	36	.	224	21	466	.	323
NASSAU	2,230	328	887	6,267	14,133	20,810	3,967	10,473	748
NEW YORK (MANHATTAN)	1,006	10,844	1,191	4,524	12,448	.	26,200	5,015	818
NIAGARA	595	459	3,846	1,326	2,836	.	715	89	9,594
ONEIDA	1,029	.	430	1,108	3,178	1,568	2,817	305	1,168
ONONDAGA	335	.	1,469	1,800	12,499	5,136	1,203	1,219	568
ONTARIO	588	1,591	382	75	226	.	.	220	198
ORANGE	.	.	849	1,674	876	268	391	459	1,499
ORLEANS	.	.	11	164	.	.	.	.	52
OSWEGO	129	.	83	82	806	.	.	.	26
OTSEGO	0	.	75	343	273	570	29	.	97
PUTNAM	.	.	39	.	81	.	47	.	963
QUEENS	2,263	0	1,562	36	65	807	83	7,347	312
RENSSELAER	98	.	94	7,480	6,963	.	10,044	185	739
RICHMOND	44	73	455	341	724	493	33	.	1,239
ROCKLAND	.	.	693	166	143	122	29	.	128
ST. LAWRENCE	.	.	346	1,282	652	.	577	.	.
SARATOGA	99	.	.	192	246	.	.	.	1,113
SCHENECTADY	.	.	90	137	51	.	.	.	370
SCHOHARIE	.	.	554	86	437	518	38	.	210
SCHUYLER	0	.	177	.	4,844	.	.	.	1,046
SENECA	.	.	941	.	30	0	0	.	1,928
STEUBEN	0	.	.	.	16,278	.	.	.	652
SUFFOLK	1,803	457	7,516	554	1,562	0	.	.	75
SULLIVAN	.	.	1,149	5,023	3,166	.	0	.	250
TIOGA	.	.	42	54	4,274	7,614	0	2,823	405
TOMPKINS	.	.	1,042	.	31	.	.	.	756
ULSTER	0	.	63	94	37	.	14	.	480
WARREN	.	.	489	555	4,628	.	1,623	.	2,014
WASHINGTON	0	.	191	.	7,102	83	0	.	3,088
WAYNE	0	.	293	.	1,508	.	184	.	480
WESTCHESTER	67	.	.	145	186	.	63	.	1,438
WYOMING	1,478	505	1,796	.	432	.	16	.	2,858
	.	.	545	2,809	887	.	.	.	1,397
YATES	0	0	36	.	6,604	.	2,015	4,218	574
WYOMING	.	.	0	3,125	12,597	.	.	.	.
YATES	.	.	0	0	2,752	0	.	.	.

*Data withheld to avoid disclosure.

NOTE: The New York State totals include employment which was not assigned to any county; totals and subtotals for specific industries include employment withheld for counties or areas, to avoid disclosure. For counties or areas where employment for some industries is withheld, the amount is combined with employment in establishments not classified in a specific industry in the "Other" column.

Source: *New York State Business Fact Book, 1976 Supplement.*

PERSONAL INCOME, 1974

Area or County	Total	Farms	Contract Construction	Manu-facturing	Wholesale and Retail Trade	Finance, Insurance & Real Estate	Trans-portation	Communica-tion & Public Utilities	Services	Govern-ment	Other
NEW YORK STATE	76,633.0	130.0	3,530.0	18,165.0	12,777.0	6,794.0	3,496.0	2,805.0	13,802.0	14,828.0	306.0

ECONOMIC AREAS (map on inside back cover)

Area or County	Total	Farms	Contract Construction	Manu-facturing	Wholesale and Retail Trade	Finance, Insurance & Real Estate	Trans-portation	Communica-tion & Public Utilities	Services	Govern-ment	Other
DOWNSTATE	52,570.0	15.7	2,338.4	9,575.5	9,485.5	5,956.9	2,724.3	2,095.4	10,874.3	9,335.3	168.8
NEW YORK CITY	40,283.4	0.4	1,481.*	6,774.1	7,050.2	5,315.1	2,448.2	1,656.0	8,699.0	6,726.4	132.4
NASSAU-SUFFOLK DIST.	8,076.1	13.1	560.2	1,730.8	1,696.6	448.3	169.9	239.0	1,381.8	1,817.0	19.4
WESTCHESTER-ROCKLAND-PUTNAM DISTRICT	4,210.5	2.3	296.7	1,070.6	738.6	193.5	106.2	200.4	793.4	791.9	16.9
REST OF STATE	24,063.0	114.3	1,191.6	8,589.5	3,291.5	837.1	771.7	709.6	2,927.7	5,492.7	137.2
BINGHAMTON AREA	1,399.6	7.3	68.0	562.4	173.3	38.4	38.0	36.1	137.5	333.9	4.7
BUFFALO AREA	5,756.8	15.0	279.0	2,217.7	827.5	196.9	271.6	159.4	657.5	1,109.7	22.5
CAPITAL DISTRICT	3,552.3	9.0	195.6	891.0	522.7	135.5	126.0	118.7	519.0	1,020.7	14.0
ELMIRA AREA	1,176.1	7.7	62.3	431.1	125.2	28.2	39.5	35.2	166.2	272.6	8.2
MID-HUDSON AREA	2,454.8	21.6	106.8	801.1	338.9	81.3	61.0	75.4	322.8	634.4	11.4
MOHAWK VALLEY AREA	1,225.3	4.0	42.6	391.4	151.2	44.6	24.5	36.8	123.3	400.0	7.0
NORTHERN AREA	1,032.5	10.3	41.4	244.0	124.2	25.1	26.1	30.2	125.0	376.4	29.9
ROCHESTER AREA	4,687.4	29.3	216.9	2,202.3	581.6	151.2	83.5	111.4	531.5	752.7	27.0
SYRACUSE AREA	2,778.2	10.2	179.0	848.5	446.9	135.8	101.5	106.5	344.9	592.3	12.6

STANDARD METROPOLITAN STATISTICAL AREAS[1]

Area or County	Total	Farms	Contract Construction	Manu-facturing	Wholesale and Retail Trade	Finance, Insurance & Real Estate	Trans-portation	Communica-tion & Public Utilities	Services	Govern-ment	Other
ALBANY-SCHENECTADY-TROY	3,181.2	5.2	178.2	774.2	467.6	121.5	114.0	108.7	475.1	925.2	10.6
BINGHAMTON (NY PORTION)	1,010.5	1.6	52.0	444.6	122.2	26.8	27.0	27.1	89.0	217.0	3.1
BUFFALO	5,054.9	8.5	255.6	1,939.4	738.6	178.0	248.1	143.2	590.0	938.7	15.0
ELMIRA	349.5	0.9	25.6	135.6	55.1	8.8	9.7	8.2	37.0	67.9	0.7
NASSAU-SUFFOLK	8,076.1	13.1	560.2	1,730.8	1,696.6	448.3	169.9	239.0	1,381.8	1,817.0	19.4
N.Y., NY-NJ (NY PORTION)	44,493.9	2.6	1,778.2	7,844.7	7,788.9	5,508.5	2,554.4	1,856.4	9,492.5	7,518.2	149.4
POUGHKEEPSIE	961.1	4.0	39.2	426.0	108.2	22.8	11.5	22.2	104.6	216.5	6.0
ROCHESTER	4,247.5	21.1	203.3	2,033.5	528.7	142.1	73.3	103.5	498.9	621.1	22.0
SYRACUSE	2,447.7	6.5	166.5	727.0	407.6	126.9	92.7	96.1	307.8	505.3	11.3
UTICA-ROME	1,089.7	3.7	38.6	347.1	132.2	41.2	24.1	28.4	-109.4	358.3	6.8

COUNTIES

Area or County	Total	Farms	Contract Construction	Manu-facturing	Wholesale and Retail Trade	Finance, Insurance & Real Estate	Trans-portation	Communica-tion & Public Utilities	Services	Govern-ment	Other
ALBANY	1,665.0	1.3	114.7	215.0	255.7	80.9	80.5	76.1	205.9	629.5	5.3
ALLEGANY	118.5	1.0	8.2	37.4	9.9	3.2	1.6	1.4	13.4	39.7	2.8
BRONX	Separate county data not available. See Economic Areas above for New York City total.										
BROOME	865.3	0.7	49.4	347.4	113.5	25.4	25.9	26.3	81.9	192.0	2.7
CATTARAUGUS	242.7	1.6	5.9	80.3	29.8	5.5	12.2	5.4	23.0	65.8	4.8
CAYUGA	183.8	2.2	7.1	58.3	22.6	6.3	7.8	7.3	15.2	50.2	0.8
CHAUTAUQUA	459.7	4.9	17.5	190.0	59.1	13.4	11.3	10.9	44.5	105.3	2.7
CHEMUNG	349	0.9	25.6	135.6	55.1	8.8	9.7	8.2	37.0	67.9	0.7
CHENANGO	129.	1.6	3.5	50.2	18.2	4.1	2.8	2.7	18.1	35.1	0.4
CLINTON	232.9	2.6	9.5	33.2	26.3	3.9	8.6	5.0	23.3	120.1	0.3
COLUMBIA	125.8	3.7	5.3	30.6	18.9	2.7	6.6	4.4	18.0	34.4	1.2
CORTLAND	146.7	1.5	5.4	63.2	16.7	2.7	1.0	3.1	15.9	36.8	0.4
DELAWARE	130.6	2.1	5.6	54.2	12.6	3.1	1.2	2.3	22.6	10.7	0.5
DUTCHESS	961.1	4.	39.2	426.0	108.2	22.8	11.5	22.2	104.6	216.5	6.0
ERIE	4,207.7	5.*	213.6	1,488.4	658.7	164.4	231.0	130.2	517.9	785.6	12.9
ESSEX	85.3	0.6	3.6	16.4	10.0	1.9	1.1	2.2	15.9	29.0	4.5
FRANKLIN	83.6	1.3	2.9	9.4	13.1	1.9	1.1	3.2	12.3	38.2	0.1
FULTON	123.3	0.3	3.5	43.7	18.0	3.3	0.4	8.3	12.5	33.1	0.3
GENESEE	190.0	2.8	4.9	82.0	25.8	4.1	3.6	3.8	17.0	43.7	2.1
GREENE	85.5	0.7	3.3	17.6	12.3	2.9	5.2	3.8	10.0	29.4	0.3
HAMILTON	12.3	0.0	0.5	0.6	1.0	0.2	*	0.1	1.4	8.5	*
HERKIMER	178.5	1.2	2.3	100.0	15.5	3.3	1.2	1.7	9.6	42.5	1.2
JEFFERSON	253.4	2.5	10.0	62.3	36.3	5.1	1.1	11.0	35.6	72.7	1.3
KINGS (BROOKLYN)	Separate county data not available. See Economic Areas above for New York City total.										
LEWIS	50.7	1.2	1.2	19.2	4.8	0.8	0.5	0.1	3.2	19.4	0.3
LIVINGSTON	138.8	3.0	2.4	35.9	15.0	2.3	5.8	3.4	8.9	56.9	5.2
MADISON	116.7	2.3	6.3	15.2	18.7	2.8	1.3	1.8	19.5	48.1	0.7
MONROE	3,607.1	3.5	176.4	1,856.2	433.0	129.0	55.4	88.7	442.0	409.7	13.3
MONTGOMERY	141.5	1.3	3.1	60.5	16.8	4.3	5.9	3.8	12.9	32.4	0.3
NASSAU	5,065.8	2.4	374.0	1,128.9	1,146.0	317.5	109.3	149.2	919.1	907.8	11.6
NEW YORK (MANHATTAN)	Separate county data not available. See Economic Areas above for New York City total.										
NIAGARA	847.2	3.4	42.0	451.0	79.9	13.7	17.1	13.0	72.1	153.1	2.0
ONEIDA	911.2	2.5	36.3	247.1	116.7	37.8	22.9	26.7	99.7	315.8	5.6
ONONDAGA	2,084.5	2.5	131.7	634.0	364.4	120.2	88.6	82.6	269.9	381.4	9.3
ONTARIO	208.4	3.0	15.5	42.2	40.7	5.4	5.2	6.3	29.8	58.9	1.4
ORANGE	693.6	6.5	35.6	152.6	111.0	30.5	26.4	26.1	87.5	216.2	1.2
ORLEANS	78.9	4.4	3.3	21.0	10.3	1.2	3.0	1.5	5.8	27.0	1.4
OSWEGO	246.4	1.8	28.4	77.9	24.4	3.8	2.8	11.6	18.4	75.8	1.4
OTSEGO	128.6	2.0	6.9	13.4	20.3	4.4	5.8	4.2	26.0	45.3	0.4
PUTNAM	94.8	0.2	12.1	10.2	18.0	2.6	2.8	3.9	15.5	28.8	0.7
QUEENS	Separate county data not available. See Economic Areas above for New York City total.										
RENSSELAER	318.9	1.2	22.2	86.2	49.1	11.2	1.1.2	10.2	59.7	67.0	0.8
RICHMOND	Separate county data not available. See Economic Areas above for New York City total.										
ROCKLAND	688.3	0.5	38.3	158.4	104.6	23.4	12.6	32.7	109.6	199.9	8.3
ST. LAWRENCE	326.8	2.1	14.2	103.4	33.7	5.5	3.8	9.0	34.7	97.1	23.3
SARATOGA	277.0	1.1	11.7	75.1	44.6	5.3	8.5	3.8	31.0	95.0	0.9
SCHENECTADY	778.8	0.4	26.4	337.3	101.4	19.8	8.6	14.8	165.6	101.3	3.3
SCHOHARIE	49.4	1.5	2.4	5.8	7.0	1.1	1.2	1.4	6.7	21.8	0.4
SCHUYLER	33.8	0.7	2.4	12.4	3.0	0.7	0.9	0.1	3.3	10.3	0.1
SENECA	109.6	0.8	4.3	39.2	11.1	1.5	0.8	0.9	5.6	45.2	0.2
STEUBEN	373.0	3.6	14.8	182.9	34.2	4.1	2.3	6.2	27.8	77.2	1.1
SUFFOLK	3,010.4	10.7	186.2	601.8	550.6	130.8	60.6	89.8	462.7	909.2	7.9
SULLIVAN	151.3	2.7	7.9	8.9	28.2	5.9	2.3	6.0	51.3	37.9	0.3
TIOGA	145.3	0.9	2.6	97.2	8.7	1.5	1.1	0.8	7.5	25.1	0.4
TOMPKINS	301.3	1.5	13.7	62.9	26.8	9.3	4.1	19.2	85.3	77.6	3.5
ULSTER	437.5	4.0	15.5	165.3	60.4	16.5	9.1	13.0	51.4	100.0	2.3
WARREN	185.3	0.1	10.4	49.0	36.9	11.5	2.9	7.7	29.9	34.6	2.5
WASHINGTON	136.4	2.2	4.7	62.0	11.3	1.4	7.1	0.8	7.4	39.2	0.5
WAYNE	214.2	7.2	5.7	78.2	29.6	4.3	3.9	3.6	12.4	68.7	0.6
WESTCHESTER	3,427.4	1.6	246.3	901.9	616.1	167.5	90.7	163.8	668.4	563.2	7.9
WYOMING	97.0	2.6	2.8	40.3	9.8	2.7	2.3	1.2	4.6	30.3	0.4
YATES	43.3	2.0	1.5	7.3	6.2	0.8	3.4	2.0	5.4	12.5	2.3

*Less than $50,000.

NOTE: Details may not add to totals, due to rounding.

Source: *New York State Business Fact Book*, 1976 Supplement.

SYRACUSE UNIVERSITY PRESS • Syracuse, New York 13210

Supplement and 3 color maps   ISBN 0-8156-2185-X $6.95